12th Joint Conference on Chemistry 2017

IOP Conference Series: Materials Science and Engineering
Volume 349

Semarang, Indonesia
19 – 20 September 2017

ISBN: 978-1-5108-6265-4
ISSN: 1757-8981

Printed from e-media with permission by:

Curran Associates, Inc.
57 Morehouse Lane
Red Hook, NY 12571

Some format issues inherent in the e-media version may also appear in this print version.

Printed by Curran Associates, Inc. (2018)

For permission requests, please contact the Institute of Physics
at the address below.

Institute of Physics
Dirac House, Temple Back
Bristol BS1 6BE UK

Phone: 44 1 17 929 7481
Fax: 44 1 17 920 0979

techtracking@iop.org

Additional copies of this publication are available from:

Curran Associates, Inc.
57 Morehouse Lane
Red Hook, NY 12571 USA
Phone: 845-758-0400
Fax: 845-758-2633
Email: curran@proceedings.com
Web: www.proceedings.com

Table of contents

Volume 349

The 12th Joint Conference on Chemistry

19–20 September 2017, Indonesia

Accepted papers received: 9 April 2018
Published online: 2 May 2018

Preface

Papers

IOP Conf. Series: Materials Science and Engineering **349** (2018) 011001 doi:10.1088/1757-899X/349/1/011001

12th Joint Conference on Chemistry

JCC12

2017

Crystall Ballroom, Aston Hotel and Convention Centre, Semarang, Indonesia

19-20 September 2017
Semarang, Indonesia

Organized by:

UNNES HKI

IOP Conf. Series: Materials Science and Engineering **349** (2018) 011001 doi:10.1088/1757-899X/349/1/011001

PREFACE

The Joint Conference on Chemistry (JCC) is annual society meeting of the Chemistry department consortium of Universitas Negeri Semarang (UNNES), Universitas Diponegoro (UNDIP), Universitas Sebelas Maret Surakarta (UNS), Universitas Jenderal Soedirman (UNSOED), and Universitas Kristen Satya Wacana (UKSW) in colaboratio with Indonesian Chemical Society or Himpunan Kimiawan Indonesia (HKI). The 12th JCC 2017 meeting was held in Semarang from 19th to 20th September 2017. It was co-organized by Chemistry Department, Faculty of Mathematics and Natural Sciences of UNNES.

The Conference adopts a timely theme Contributions of chemical sciences, engineering, education, and technology in solving global challenges Chemistry is one of the keys to create a sustainable society, and for that purpose we need to reinforce the chemistry science, technology and education and to extend our finding to innovation

Approximately 150 participants have attended the conference. Among the participants, they were 6 plenary speakers, 11 invited speakers, and 115 contributed speakers. Based on the country where the participants came from, they were from 10 (six) countries namely Singapore, Japan, Indonesia, Hong Kong, Malaysia, and Thailand

The speakers of the meeting have submitted their full papers for possible publications in a proceeding of IOP Publishing (Journal of Physics: Conference Series). After peer reviewed by experts in related fields, finally, the scientific boards decide to accept 77 papers.

The JCC acknowledges the enthusiasms of the participants, all members of the committees, the international advisory board, Chemistry Department Faculty of Mathematics and Natural Sciences of UNNES, and all those who have contributed to the success of the 12th JCC 2017.

The next JCC event will be held in 2018. We are looking forward to welcoming you to the 13th JCC 2018!

Cepi Kurniawan, PhD

Chairman of the 12th Joint Conference on Chemistry 2017

IOP Conf. Series: Materials Science and Engineering **349** (2018) 011001 doi:10.1088/1757-899X/349/1/011001

COMMITTEE

Chairman	: Cepi Kurniawan, PhD (Universitas Negeri Semarang, Indonesia)
Secretary	: Ella Kusumastuti, M.Si (Universitas Negeri Semarang, Indonesia)
Treasurer	: Dr. Triastuti S, M.Si (Universitas Negeri Semarang, Indonesia)
Program	: Willy Tirza E, M.Sc, Apt (Universitas Negeri Semarang, Indonesia)
	Dante Alighiri, M.Sc (Universitas Negeri Semarang, Indonesia)
Scientific	: Sri Kadarwati, PhD (Universitas Negeri Semarang, Indonesia)
	Uyi Sulaiman, Ph.D. (Universitas Jenderal Soedirman, Indonesia)
	M. Alauhdin, PhD (Universitas Negeri Semarang, Indonesia)
	Adi Darmawan, PhD (Universitas Diponegoro, Indonesia)
	Dr. Dwi Hudayanti (Universitas Diponegoro, Indonesia)
	Dr. Eddy Heraldy (Sebelas Maret University, Indonesia)
	Dr. Jumaeri (Universitas Negeri Semarang, Indonesia)
	Dr. Veinardi Suwendo (Institut Teknologi Bandung)
	Dr. rer. nat. A. Heru Wibowo (Universitas Negeri Sebelas Maret, Indonesia)
	Dr. Endang Susilaningsih (Universitas Negeri Semarang)

IOP Conf. Series: Materials Science and Engineering **349** (2018) 011001 doi:10.1088/1757-899X/349/1/011001

INTERNATIONAL ADVISORY BOARD

IOP Conf. Series: Materials Science and Engineering **349** (2018) 011001 doi:10.1088/1757-899X/349/1/011001

DOCUMENTATION

Registration of participants

Photo session of Plenary Speakers with the Dean of Mathematics and Natural Sciences Faculty of UNNES

IOP Conf. Series: Materials Science and Engineering **349** (2018) 011001 doi:10.1088/1757-899X/349/1/011001

Photo session with all participants

Plenary Session: Assoc. Dr. David Harding (Walailak University, Thailand)

IOP Conf. Series: Materials Science and Engineering **349** (2018) 011001 doi:10.1088/1757-899X/349/1/011001

Question and answer session

Parallel session

IOP Conf. Series: Materials Science and Engineering **349** (2018) 011001 doi:10.1088/1757-899X/349/1/011001

Pictures of Participants and Poster Display

City tour excursion of some participants (Goa Kreo, Semarang, Indonesia)

IOP Conf. Series: Materials Science and Engineering **349** (2018) 011002 doi:10.1088/1757-899X/349/1/011002

Peer review statement

All papers published in this volume of *IOP Conference Series: Materials Science and Engineering* have been peer reviewed through processes administered by the proceedings Editors. Reviews were conducted by expert referees to the professional and scientific standards expected of a proceedings journal published by IOP Publishing.

IOP Conf. Series: Materials Science and Engineering **349** (2018) 012001 doi:10.1088/1757-899X/349/1/012001

Adsorption kinetics of surfactants on activated carbon

Arnelli, Aditama WP, Z Fikriani, Y Astuti*

Department of Chemistry, Faculty of Sciences and Mathematics, Diponegoro University

*Corresponding author: yayuk.astuti@live.undip.ac.id

Abstract. A study on the adsorption of both cationic and anionic surfactants using activated carbon as well as the investigation of the adsorption isotherms and adsorption kinetics has been conducted. The results showed that the adsorption of sodium lauryl sulfate (SLS) by activated carbon was Langmuir's adsorption isotherm while its adsorption kinetics showed pseudo-second order with an adsorption rate constant of 2.23 x 10^3 g mg^{-1} $hour^{-1}$. Meanwhile, the adsorption of HDTMA-Br by activated carbon showed that the isotherm adsorption tended to follow Freundlich's isotherm and was pseudo-second order with an adsorption rate constant of 89.39 g mg^{-1} $hour^{-1}$.

1 Introduction

Activated carbon is a porous material composed of carbon with a surface area and pores growing well in the presence of an activator. The well-developed surface and pores on carbon may be used as adsorbent [1, 2]. One of the raw materials in the production of activated carbon is rice husk due to its high content of organic compounds [3, 4]. Research on carbon activation using H_3PO_4 has been undertaken by Huang et al [5] , which indicated that the carbon pores were more open. Consequently, the adsorption ability of the activated carbon would increase.

The adsorption of surfactants is necessary to be undertaken since the pollution generated by surfactants as a result of their use now shows an increase in proportion to the increase in the human's population. The pollution is caused by surfactants in the environment such as the foam emergence and their limited biodegradation by bacteria or microorganisms resulting in the disturbance of the water ecosystems.

Schouten et al [6] have studied the kinetics of adsorption of surfactant linier alkylbenzene sulfonate (LAS) by granular activated carbon derived from norit (Netherlands). The result showed that the kinetics order of this adsorption process was first order while Yakout et al [7] reported that the order reaction of the CTAB surfactant adsorption on carbon (from corn cob and char) activated by KOH was a pseudo-second order.

This research studied the adsorption isotherm and adsorption kinetics of sodium lauryl sulfate (SLS) and hexadecyltrimetilammonium bromide (HDTMA-Br) surfactant adsorbed by activated carbon made from rice husks. By determination of adsorption isotherm and adsorption kinetics, the adsorption order, adsorption rate constant and adsorption isotherm models can be specified. Order adsorption and adsorption rate constant associated with the rate of adsorption of surfactant by activated carbon. It is known that the factors affecting the rate of adsorption, such as the concentration of adsorbate and adsorbent, temperature, the surface area of adsorbent and catalyst, can be used to accelerate the rate of adsorption. Adsorption isotherm was studied to estimate that the adsorbate was trapped onto the adsorbent in monolayer or multilayer pattern.

2 Materials and Method

2.1 Materials. Materials used in this research were rice husks, H_3PO_4, H_2SO_4, NaOH, phenolphthalein, phosphate-washing solution, SLS and HDTMA-Br surfactants, distilled water, bromophenol blue, phosphate buffer pH 8 and chloroform. All the reagents are analytical grade and were purchased from Merck, Indonesia.

2.2 Adsorption of surfactants by activated carbon. This research experiment consisted of several stages. The first stage was carbonization of rice husk, the second step was activation of carbon, and the last stage was adsorption of surfactants onto the activated carbon. Carbonization of rice husk was conducted at 300°C by pyrolysis (decomposition process of organic substances resulting in tar, charcoal (carbon) and volatile compounds by heating with or without oxygen). The carbon generated had been then activated using H_3PO_4 60% at 420°C for 1 hour. This step aimed to remove impuritities covering the pores of the adsorbent so that pores become bigger. The adsorption process was conducted by contacting 1 g of activated carbon with 25 ml of SLS surfactant (60 ppm) in different times 3, 4, 5, 6 and 7 hours at room temperature. The mixture was then separated by simple filtration using filter paper (Whatmann no. 1, diameter 15 mm, pore zise 11 μm). The filtrate was then analyzed using methylen blue active substance (MBAS) method to determine the concentration of SLS that was not adsorbed [3]. This method is applied for determination of anionic surfactant unadsorbed by adsorbent. This procedure was also applied to the adsorption of 300 ppm HDTMABr by 1 gram of activated carbon. However, the concentration of unabsorbed HDTMABr surfactant was determined using the bromphenol blue method [8] (method applied for determiantion of cationic surfactant concentration).

3 Results and Discussion

Surfactant adsorption on activated carbon was conducted by contacting the activated carbon with both SLS and HDTMABr surfactants in the range of time 3, 4, 5, 6 and 7 hours. The data obtained were in the form of correlation between adsorption contact time and adsorbed-surfactant concentration and also adsorption capacity. The data were then used to determine the adsorption types and adsorption kinetics of both surfactants trapped onto activated carbon. Table 1 indicates that the adsorption SLS and HDTMA-Br by activated carbon occurred at the optimum time of 4 hours. The adsorption of both surfactants increased insignificantly after 4 hours. It shows that at the time of 4 hours, the activated carbon optimally absorbed surfactants.

Table 1. The concentration of surfactant adsorbed by activated carbon

No.	Time (Hours)	Adsorbed surfactant (ppm)		Adsorption capacity (mg / g)	
		HDTMABr	SLS	HDTMABr	SLS
1.	3	298.79	59.95	7.47	1.4987
2.	4	299.11	59.98	7.48	1.4995
3.	5	299.11	59.98	7.48	1.4996
4.	6	299.22	59.97	7.48	1.4992
5.	7	299.28	59.96	7.48	1.4991

3.1 Adsorption isotherms

Determination of the types of adsorption occurring in the adsorption of surfactants by activated carbon used Langmuir (1) and Freundlich (2) equations.

$$\frac{C}{N} = \frac{C}{N_m} + \frac{1}{K}N \qquad \text{...} (1)$$

N = moles of surfactant adsorbed per gram of activated carbon; C = final concentration of surfactant in moles/liter; K = Langmuir's constant and Nm = the number of moles required to make a single layer of surfactants on the activated carbon.

$$\frac{X}{m} = kC^n \longrightarrow \quad log\frac{X}{m} = logk + nlogC \qquad \text{................} (2)$$

X = weight of substances (solute) adsorbed (g); m = weight of adsorbent (g); C = concentration of the solution after adsorption (after equilibrium); k = Freundlich's constant; n = another constant in which k and n are related with mass of adsorbent and temperature of adsorbate, respectively.
Tables 2 and 3 show the data that will be used to determine the type of adsorption occurred in the SLS adsorption process by activated carbon. This data was obtained from data calculation in Table 1. Plotting data between N vs. C / N (Figure 1) and log C vs. log X/m (Figure 2) is then applied to determine the type of adsorption (Langmuir or Freundlich) by looking at the linearity of the curve (R^2).

Table 2. The component data of SLS surfactant adsorbed on the activated carbon for the determination of Langmuir's kinetics.

C (mol / liter)	C / N	N (mol / g)
$1.73x10^{-7}$	$3.33\ x10^{-2}$	$5.1970x10^{-6}$
$6.93x10^{-8}$	$1.33\ x10^{-2}$	$5.1990x10^{-6}$
$6.93x10^{-8}$	$1.33\ x10^{-2}$	$5.1990x10^{-6}$
$1.04x10^{-7}$	$2.00\ x10^{-2}$	$5.1988x10^{-6}$
$1.39x10^{-7}$	$2.67\ x10^{-2}$	$5.1980x10^{-6}$

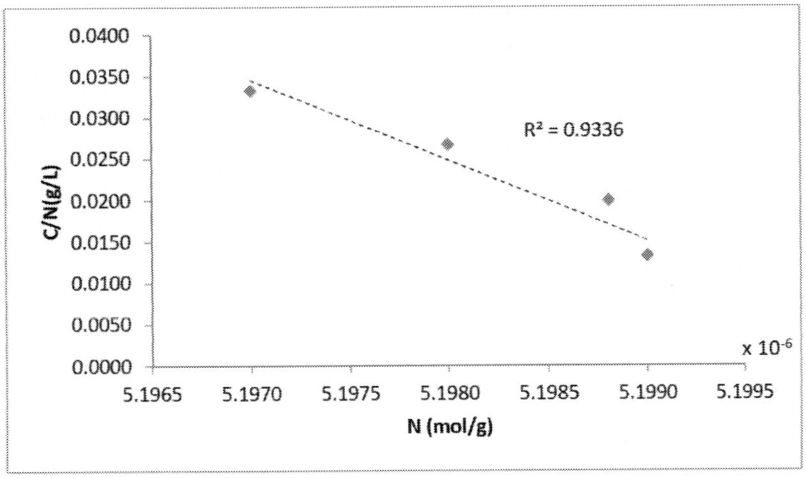

Figure 1. The Langmuir adsorption isotherms model of SLS adsorption on the activated carbon.

IOP Conf. Series: Materials Science and Engineering **349** (2018) 012001 doi:10.1088/1757-899X/349/1/012001

Table 3. The component data of SLS surfactant adsorbed on the activated carbon for the determination of Freundlich's kinetics.

C (mol)	X (g)	m (g)	X / m (g / g)	log C	log X / m
$4.33x10^{-9}$	$1.49870x10^{-3}$	1	$1.50x10^{-3}$	-8.36351	-2.82429
$1.73x10^{-9}$	$1.49930x10^{-3}$	1	$1.50x10^{-3}$	-8.76195	-2.82412
$1.73x10^{-9}$	$4.99220x10^{-3}$	1	$4.99x10^{-3}$	-8.76195	-2.30171
$2.60x10^{-9}$	$1.49890x10^{-3}$	1	$1.50x10^{-3}$	-8.58503	-2.82423
$3.46x10^{-9}$	$1.49813x10^{-3}$	1	$1.50x10^{-3}$	-8.46092	-2.82445

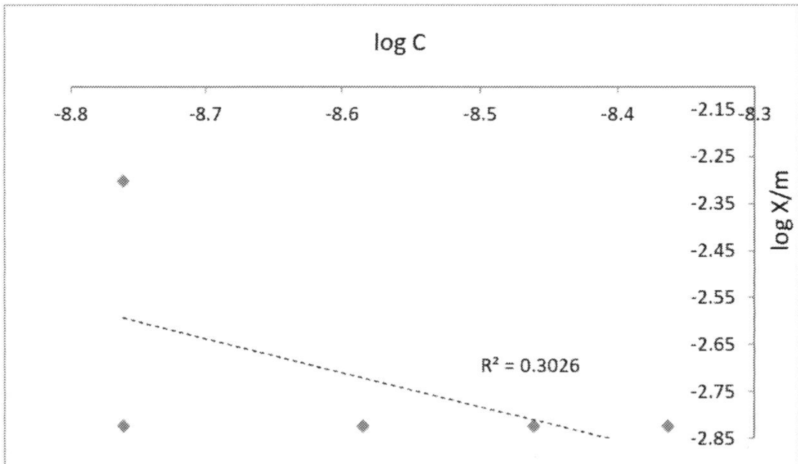

Figure 2. The Freundlich adsorption isotherms model of SLS adsorption on activated carbon.

Figures 1 and 2 show the model of the adsorption isotherms. Langmuir's adsorption isotherm model in Figure 1 provides a linearity (R^2) of 0.9336 greater than the linearity of Freundlich adsorption isotherm (R^2 = 0.3026) (Figure 2). **The Figures 1 and 2** indicate that the SLS adsorption by activated carbon followed the Langmuir's isotherm rather than Freundlich's isotherm. Based on the hypothesis of the Langmuir's adsorption, SLS surfactant was adsorbed by activated carbon by forming a monolayer.

Tables 4 and 5 are the data used to determine the type of adsorption occurred in the adsorption process of HDTMABr by activated carbon. In the same way as the determination of adsorption SLS, Figures 3 and 4 show the type of Langmuir's and Freundlich's adsorption isotherms in the adsorption HDTMABr.

Table 4. The component data of HDTMABr surfactant adsorbed on the activated carbon for the determination of Langmuir's kinetics.

C (mol / liter)	C / N	N (mol / g)
$3,20x10^{-6}$	0,156	$2,050x10^{-5}$
$2,44x10^{-6}$	0,119	$2,051x10^{-5}$
$2,44x10^{-6}$	0,119	$2,051x10^{-5}$
$2,14x10^{-6}$	0,104	$2,052x10^{-5}$
$1,97x10^{-6}$	0,096	$2,053x10^{-5}$

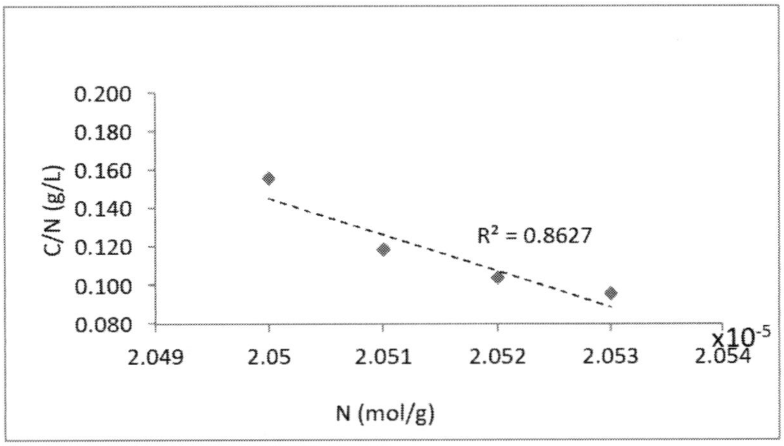

Figure 3. The Langmuir's adsorption isotherms model of the adsorption of HDTMABr on the activated carbon

Table 5. The component data of HDTMABr surfactant adsorbed on the activated carbon for the determination of Freundlich's kinetics.

C (mol)	**X (g)**	**m (g)**	X / m (g / g)	log Csisa	log X / m
8,3x10^{-8}	1,2x10^{-3}	1	1.20x10^{-3}	-7.08092	-2.92082
6.1x10^{-8}	8,9x10^{-4}	1	8.9x10^{-4}	-7.21467	-3.05061
6,1x10^{-8}	8,9x10^{-4}	1	8.90x10^{-4}	-7.21467	-3.05061
5,35x10^{-8}	7,8x10^{-4}	1	7.80x10^{-4}	-7.27165	-3.10791
4,93x10^{-8}	7,1x10^{-4}	1	7.10x10^{-4}	-7.30715	-3.14874

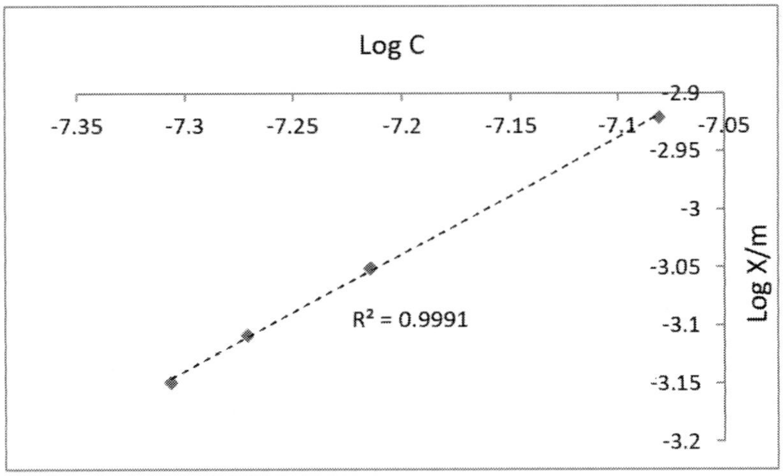

Figure 4. The Freundlich's adsorption isotherms model of the adsorption of HDTMABr on the activated carbon

It can be seen in Figures 3 and 4 that HDTMA-Br adsorption on the activated carbon followed Freundlich's isotherm with R^2 of 0.999. Compared to SLS adsorption process, the HDTMABr adsorption was in the form of multilayer, which means that after HDTMABr molecules were adsorbed and formed a monolayer, they attracted their neighboring molecules.

3.2 Adsorption Kinetics

The data in Table 6 and 7 were then used to determine the adsorption kinetics of SLS and HDTMABr adsorption process by activated carbon. By using the kinetics equation of Lagergren's pseudo second-order $\frac{t}{q_t}=\frac{1}{kq_{e^2}}+\frac{1}{q_e}t$ the linear graphs between t/qt vs. t as presented in Figures 5 and 6 were generated.

Table 6. The component data of SLS surfactant adsorbed on the activated carbon for the determination of Lagergren's kinetics.

Kinetics of the adsorption of SLS by activated carbon			
Time (hours)	[SLS] adsorbed (mg / L)	t/qt	qt
3	55.95	2:00	1:50
4	55.98	2.67	1:50
5	55.98	3:33	1:50
6	55.97	4:00	1:50
7	55.93	4.67	1:50

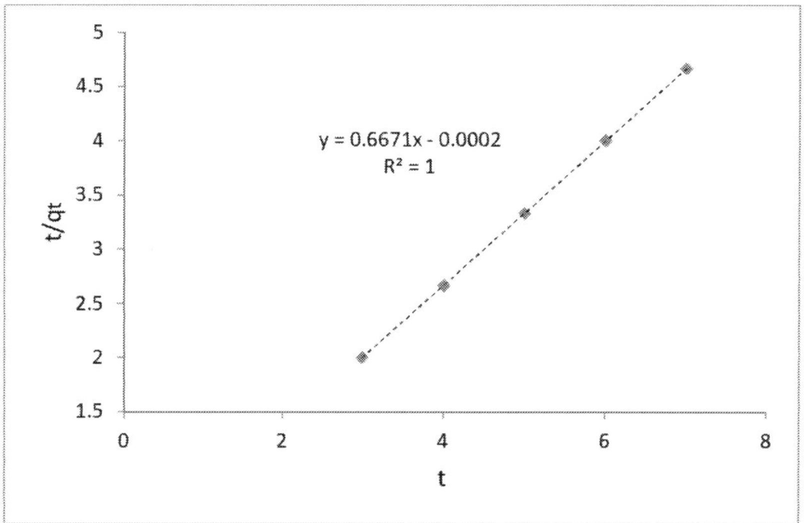

Figure 5. The pseudo second-order kinetics of the adsorption of SLS on activated carbon

Table 7. The component data of HDTMABr surfactant adsorbed on the activated carbon for the determination of Lagergren's kinetics.

Adsorption kinetics HDTMA-Br activated carbon			
Time (hours)	[HDTMA-Br] adsorbed (mg/L)	t/qt	qt
3	298.79	0.400534045	7.4900
4	299.11	0.535475234	7.4700
5	299.11	0.669344043	7.4700
6	299.22	0.802139037	7.4800
7	299.28	0.935578722	7.4820

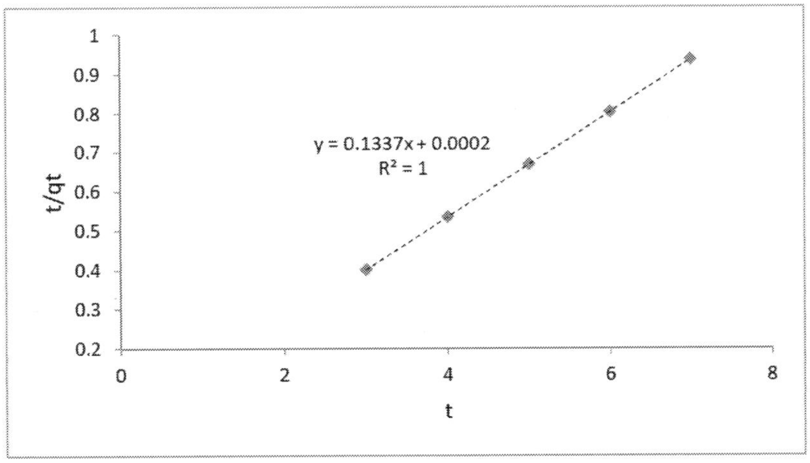

Figure 6. The pseudo second-order kinetics of HDTMA-Br adsorption on activated carbon

Based on the Lagergren's kinetics equation, the adsorption kinetics of both SLS and HDTMA-Br adsorption by the activated carbon were pseudo-second order because both adsorptions showed a linearity (R^2) of 1. By equation of a simple second-order reaction rate, $v = k [A]^2$, the concentration of the reactant A is magnified two times if the rate of reaction is four times of the original rate. Also, based on the Lagergren's formula, the rate constant of adsorption for SLS and HDTMABr adsorption can be calculated. The result showed that the adsorption rate constant of the adsorption of SLS and HDTMABr by activated carbon was 2.23×10^3 $gmg^{-1}hour^{-1}$ and 89.39 $gmg^{-1}hour^{-1}$, respectively. This result suggests that the adsorption rate of SLS on the activated carbon is more effective than HDTMABr.

4 Conclusion

Surfactant adsorption process both SLS and HDTMABr by activated carbon shows that the process followed the model of Langmuir and Freundlich adsorption isotherms, respectively. The adsorption kinetics of both surfactants showed pseudo second order with a rate constant for adsorption of SLS and HDTMABr is consecutively 2.23×10^3 $gmg^{-1}h^{-1}$ and 89.39 $gmg^{-1}h^{-1}$.

Acknowledgment

This research was supported by the Ministry of Research, Technology and Higher Education of the Republic of Indonesia with the grant no. 022 / SP2H / LT / DRPM / II / 2016.

References

1. AL-Doury, M. and S.S. Ali, *Removal of Phenol and Parachlorophenol from Synthetic Wastewater Using Prepared Activated Carbon from Agricultural Wastes.* International Journal of Sustainable and Green Energy, 2015. **4**(3): p. 92-101.
2. Ali, I.O., et al., *Synthesis and characterization of ZSM-5 zeolite from rice husk ash and their adsorption of Pb 2+ onto unmodified and surfactant-modified zeolite.* Separation and Purification Technology, 2011. **83**: p. 38-44.
3. Al-Latief, D., Arnelli, and Y. Astuti. *Synthesis of sodium lauryl sulphate (SLS)-modified activated carbon from risk husk for waste lead (Pb) removal.* in *AIP Conference Proceedings*. 2015. AIP Publishing.

IOP Conf. Series: Materials Science and Engineering **349** (2018) 012001 doi:10.1088/1757-899X/349/1/012001

4. Zeng, D., et al., *Synthesis porous carbon-based solid acid from rice husk for esterification of fatty acids.* Microporous and Mesoporous Materials, 2016. **219**: p. 54-58.
5. Huang, Y., et al., *Adsorption of Pb (II) on mesoporous activated carbons fabricated from water hyacinth using H_3PO_4 activation: adsorption capacity, kinetic and isotherm studies.* Applied Surface Science, 2014. **293**: p. 160-168.
6. Schouten, N., et al., *Kinetic analysis of anionic surfactant adsorption from aqueous solution onto activated carbon and layered double hydroxide with the zero length column method.* Separation and Purification Technology, 2009. **68**(2): p. 199-207.
7. Yakout, S. and A. Nayl, *Removal of cationic surfactant (CTAB) from aqueous solution on to activated carbon obtained from corncob.* Carbon: Science and Technology, 2009. **2**(2): p. 107-116.
8. Wibowo, W., T. Utari, and R.T. Yunarti, *Anion exchange capacity of chromate on modified zeolite clinoptilolite with HDTMA-Br and its regeneration.* Makara Journal of Science, 2011.

IOP Conf. Series: Materials Science and Engineering **349** (2018) 012002 doi:10.1088/1757-899X/349/1/012002

Aluminium - Cobalt-Pillared Clay for Dye Filtration Membrane

A Darmawan[1,*], Widiarsih[1]

[1]Chemistry Department, Faculty of Sciences and Mathematics, Diponegoro University Jl. Prof. Soedarto, SH., Tembalang, Semarang, Indonesia

* adidarmawan@live.undip.ac.id

Abstract. The manufacture of membrane support from cobalt aluminium pillared clay has been conducted. This research was conducted by mixing a clay suspension with pillared solution prepared from the mixture of $Co(NO_3)_2.6H_2O$ and $AlCl_3.6H_2O$. The molar ratio between Al and Co was 75:25 and the ratio of $[OH^-]$/[metal] was 2. The clay suspension was stirred for 24 hours at room temperature, filtered and dried. The dried clay was then calcined at 200°C, 300°C and 400°C with a ramp rate of 2°C/min. Aluminium-cobalt-pillared clay was then characterized by XRD and GSA and moulded become a membrane support for subsequent tests on dye filtration. The XRD analysis showed that basal spacing (d_{001}) value of aluminium cobalt was 19.49 Å, which was higher than the natural clay of 15.08Å however, the basal spacing decreased with increasing calcination temperature. The result of the GSA analysis showed that the pore diameter of the aluminium cobalt pillared clay membrane was almost the same as that of natural clay that were 34.5Å and 34.2Å, respectively. Nevertheless, the pillared clay has a more uniform pore size distribution. The results of methylene blue filtration measurements demonstrated that the membrane filter support could well which shown by a clear filtrate at all concentrations tested. The value of rejection and flux decreased with the increasing concentration of methylene blue. The values of dye rejection and water flux reached 99.89% and 5. 80 x 10-6 kg min^{-1}, respectively but they decreased with increasing concentration of methylene blue. The results of this study indicates that the aluminium-pillared clay cobalt could be used as membrane materials especially for ultrafiltration.

1. Introduction

The separation technology using membrane is a component separation technique with a very specific way by passing one of the components more quickly than the others. The use of membrane technology as separation or purification process can diminish energy consumption and can gain better results [1], The membrane is a semi-permeable layer that can withstand and skip certain material movement of a fluid flow which being passed to the membrane itself [2], Membranes are classified into three which are organic membranes, inorganic (ceramic membrane) and a mixture of both. Inorganic membranes can be described as asymmetric porous ceramic[3] which generally consists of a macroporous support and a few thin layers of porous with pore sizes getting smaller which is the end serves as a separator [4], The advantages of inorganic membranes include high thermal stability, higher separation efficiency, durable, easy to clean and environmentally friendly[5].

The quality of the membrane support determines the quality of the membrane itself. Membrane support provides the mechanical strength for the top-layer membrane that can withstand stress and must simultaneously have a low resistance to the filtrate flow [6]. Membrane support should have a small pore size, mechanical robustness and low damage to the feed flow as well as lower production costs [5, 7]. All this time, the manufacture of the membrane support predominantly by using alumina (γ-alumina and α-alumina). However, the prices of γ-alumina and α-alumina are very expensive hence the production costs will be higher [8]. The α-alumina membranes have fragile properties as generally ceramic material which limit them in applications and shorten the lifetime [9] and they take a high sintering temperature up to 1700°C even more to get a good mechanical strength [4]. Therefore, a material modification in manufacturing membrane support is necessary. The modification of the main material in the membrane support manufacture using pillared clay is expected to produce a membrane support with good character and economically valuable.

Instead, the abundant occurrence of clay in nature can be used as an innovation in the membrane support manufacture with low production costs and the resulting good quality membrane support [1, 5]. The natural clay has several drawbacks, among others are the unstable pore, low thermal resistance, low adsorption capacity and the resulting small basal spacing [10]. The clay performance can be maximized if it pillarized [1]. The excellences of pillared clays are able to improve thermal stability, increasing the pore volume, pore diameter and specific surface area, stabilising the basal spacing and enhancing the acidity of the clay surface [11]. The pillared clays were synthesized by adding a metal-hydroxy polycation into the clay interlayer. Afterwards, by calcination, the poly cation will transform into rigid metal oxides clusters thereby inhibiting the damage of the clay interlayer gallery [8].

Al/Co pillared clays were widely used as catalysts and adsorbents [12]. Bertella and Pergher [13] stated that clay pillarization using aluminium and different cations could cause increasing and stabilizing of clay basal spacing. A mixture of aluminium cobalt had greater interlayer spacing and thermal resistance than pillared clay with one metal only. The basal spacing of Al/Co pillared clay were 16.0Å-19.3Å. Even though the Co pillared clay's thermal resistance was very low which was less than 200°C however the Al/Co pillared clay could withstand a temperature of 450°C [13]. Based on these excellence properties of Al/Co pillared clay, this research tried to develop Al/Co pillared clay as a membrane support for dye filtration.

2. Methodology

Natural clay was cleaned from impurities, dried, pulverized and sieved with a 170 mesh sieve. Al/Co pillared clays were prepared by mixing the Al/Co pillaring solution with a clay suspension. The molar ratio of $[Al^{3+}]/[Co^{2+}]$ was 75:25 [13]. A specific amount of $Co(NO_3)_2 \cdot 6H_2O$ and $AlCl_3 \cdot 6H_2O$ were dissolved in the water and stirred thus the concentration of the metal was 1 M. Then NaOH was added dropwise and stirred to achieve a molar ratio $[OH^-]/[metal] = 2$. The solution was stirred at 60°C for 24 hours.

2.1. Pillarization Process

Pillaring solution was added to the 10% clay suspension while stirring at room temperature for 24 hours and then aged for 24 hours. Then, the solid fraction was separated and washed with water to remove Cl^- and dried. The dried intercalated clay then calcined to temperature of 100°C, 200°C, 300°C and 400°C with calcination rate of 2°C/min. The natural and pillared clays were analysed by X-ray diffraction to determine the optimum calcination temperature.

2.2. Membrane Fabrication

The Al/Co pillared clay at optimum calcination temperature was added PEG-400 and water with a weight ratio of 8:3:1. The mixture was stirred until homogeneous and placed on a membrane mould, then pressed with a pressure of 60 kg/cm^2 for 10 minutes. Pillared clay membrane was then heated at 70°C for 12 hours, and then calcined at 600°C for 10 hours with a ramp rate 1°C/min. Further, Al/Co pillared clay membranes were analysed by Gas Sorption Area Analyser (GSA).

2.3. Membrane Performance Testing

The membrane performance was tested on a filtration process of methylene blue dye. The methylene blue concentrations were varied of 10 ppm, 20 ppm, 40 ppm and 80 ppm. The methylene blue solution was passed to the filtration equipment as shown at Figure 1 in crossflow stream for 2 hours. Subsequently, the rejection and flux were measured using the formula below.

$$\text{Rejection} = \frac{(\text{Initial adsorbance - final adsorbance})}{\text{Initial adsorbance}} \times 100\%$$

$$\text{Flux} = \frac{\text{flowrate}}{\text{surface x pressure}}$$

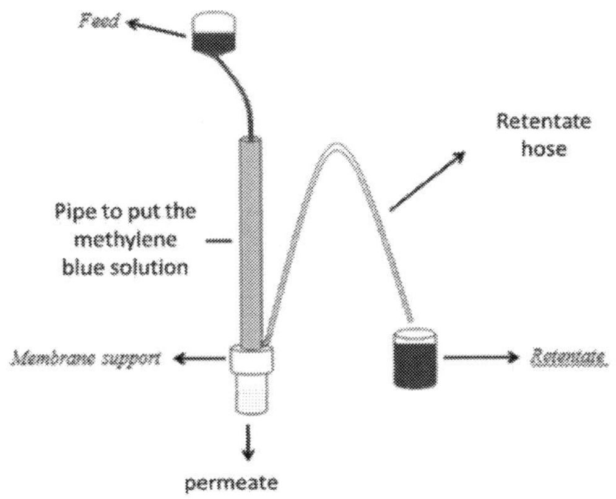

Figure 1. Schematic picture of membrane filtration

3. Results and Discussion

3.1. X-ray Diffraction (XRD)

Figure 1 and Table 1 are presenting the results of the XRD characterization of Al/Co pillared clays in the 2θ region of 2° -20°, which typically indicate clay basal spacing of d_{001}.

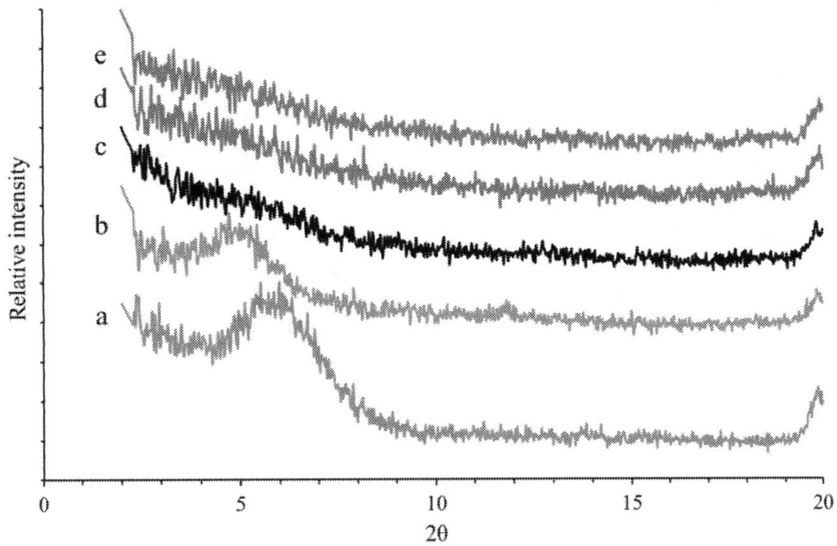

Figure 2. XRD diffractogram of Al/Co pillared clays (a) natural clay, Al/Co pillared clay calcined at (b) 100°C (c) 200°C (c) 300°C and (d) 400°C

Table 1. Basal Spacing on nature and pillared clays

Sample	2θ	Basal spacing (d_{001})
Natural clay	5.86°	15.08Å
Al/Co pillared clay (100°C)	5.9°	18.3Å
Al-pillared clay Co. (200°C)	4.81°	19.3Å
Al-pillared clay Co. (300°C)	4.6°	18.34Å
Al-pillared clay Co. (400°C)	4.82°	15.09Å

The Al/Co pillarization phenomenon in the clay can be observed with the existing of the d_{001} peak (2θ <10°) which shows the basal spacing of clay. The existence of basal spacing needs to be assessed to determine whether the pillarization process occurs and to determine the stability of the interlayer region when the clay calcined as well as to determine the clay interlayer spacing changes during the formation of pillar. The diffractograms in Figure 2 show the basal spacing formed of the Al/Co pillared clays.

Figure 1 and Table 1 show an increase in basal spacing values on the clay heated on 100°C and 200°C which are 18.3Å and 19Å respectively. This is consistent with what was reported [1] that the basal spacing value rises after pillarization process in natural clay where most clay pillared by alumina produced basal spacing values between 17Å and 18Å. The increase in basal spacing value due to Al/Co molecules that are relatively larger size were capable to intercalate in the clay interlayer region. In previous research [13], basal spacing values generated on Al/Co pillared clays were 17.6Å-19.3Å. In addition, there was resemblance of basal spacing decrease at 300°C and 400°C. The decreased basal spacing after pillarization most likely caused by the demolition of the clay interlayer structure. According to Bertella and Pergher [13], cobalt as a pillaring agent did not stand at temperatures over 250°C. In addition, the Co^{2+} ions acidity is lower than Al^{3+} ions henceforth when Al^{3+} was mixed with Co^{2+}, the pH increased. The increase in pH inhibited the formation of Al_{13} Keggin ions and deterred the pillarization process. The optimum pH range for the formation of Al polynuclear species is 3.5-4.5 [14]. The clay basal spacing after calcination increased compared with the natural clay. The basal spacing obtained at calcination temperature of 200°C was 19.3Å which was in line with previous research [13]. The calcination transformed Co^{2+} and Al^{3+} ions into stable metal oxides that subsequently became pillars between the layers.

XRD diffractogram also pointed out that by the rising of the heating temperature, there was a decrease in the crystallinity of the resulting pillared clays which characterized by the lower intensity and peak widening. The decline in intensity was predicted to occur due to the reducing of pillars formed therefore the number of unidirectional beam reflected from the crystal planes were also reduced. The sample, which has a high crystallinity, reflects directional beam thus the XRD peak has high intensity. Consequently, the sample, which has a low peak intensity, could be concluded to have low crystallinity.

3.2. Gas sorption Analyser

The characterization of the Al/Co pillared clay membrane pore was conducted by a gas sorption analyser in which the surface area was determined using BET method, while the pore diameter, pore volume and pore distribution were examined using BJH (Barret Joyner Hallendra). Adsorption isotherms of natural clays and Al/Co pillared clay membranes are presented in Figure 3.

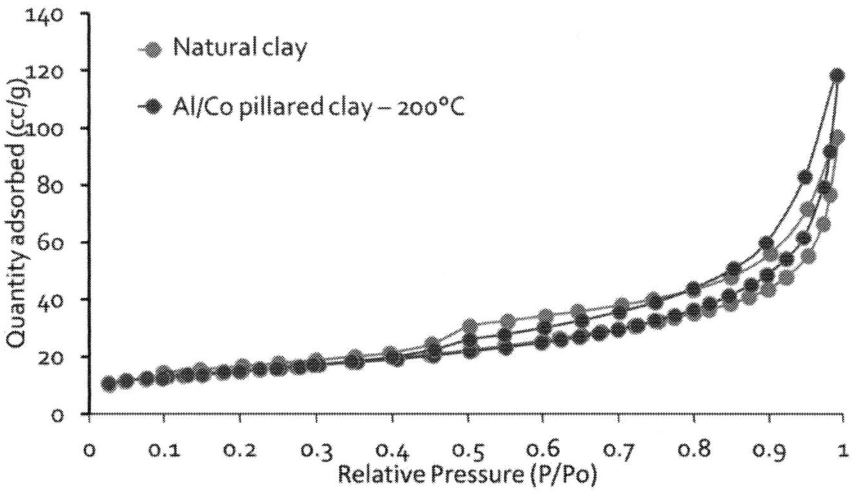

Figure 3. Adsorption-desorption isotherms curve of natural clay and Al/Co pillared clay membrane

Figure 3 shows that the adsorption-desorption isotherm curve is a type IV in which this type occurs in mesoporous materials and commonly happens in pore size between 1.5 and 100 nm. Both natural clays and Al/Co pillared clays show the hysteresis loop wherein the amount of desorption gas was not equal to the amount of adsorbed gas at the beginning. Figure 3 shows at the pressure $P/P_o<0.3$, the adsorbed gas is scarce. When the pressure increased ($P/P_o> 0.4$), N_2 gas adsorption that saturated the monolayer occurred. The amount of gas adsorbed on the Al/Co pillared clay membranes was more than natural clay. This marks that the Al/Co pillared clay membrane pore volume is greater than the natural clay. This result is consistent with the data presented in Table 2. Table 2 shows data for surface area, pore diameter and pore volume of natural clay and Al/Co pillared clay membranes

Table 2. Surface Area, Pore Diameter and Pore Volume

Sample	Surface area (m^2g^{-1})	Pore Diameter (Å)	Pore Volume (mL g^{-1})
Natural clay	52.45	34.214	0,137
Al/Co pillared clay membrane	52.04	34.457	0.173

The surface area and pore diameter of natural clays and Al/Co pillared clay membranes are almost identical, however the total pore volume of the Al/Co pillared clay membranes is greater (approximately 3.6%) compared with natural clay. This indicates that the clay pillarization increased the total pore volume generated that implies the formation of new spaces in the area between the clay layers.

The pore distributions of natural clays and Al/Co pillared clay membranes are presented in Figure 4. Based on the IUPAC classification, the pore sizes are divided into three groups, i.e. (1) micropores (<2 nm) (2) mesoporous (2-50 nm) and (3) the macropores (>50 nm). An average pore diameter of natural clay and Al/Co pillared clay membranes Al/Co were about 34 Å that were in category of mesoporous. The pore distribution of both natural clay and Al/Co pillared clay membranes were not uniform, however the Al/Co pillared clay membrane slightly increased in the number of pores in the area of 10-14Å.

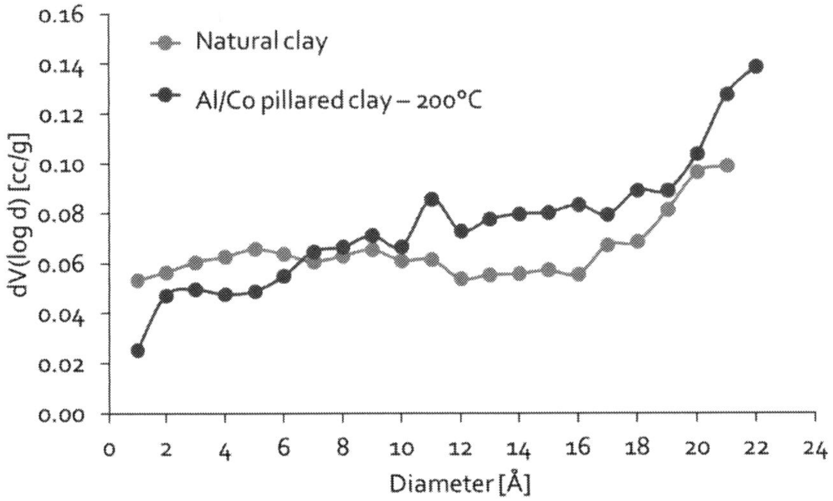

Figure 4. Pore Size Distribution of natural clay and Al/Co pillared clay membrane

3.3. Al/Co pillared clay membrane Performance on methylene blue filtration

Al/Co pillared clay membrane performance tests were carried out on the methylene blue dye filtration process. The filtration process was conducted by varying the methylene blue concentration of 10, 20, 40, and 80 ppm. The filtration time was fixed for 2 hours. The Al/Co pillared clay membranes filtration was through a sieving mechanism in which particles or molecules that are smaller than the filter media pore size will pass through the membrane, whilst the larger than the pore size are retained on the feed side [9].

The performance of membrane filtration was measured by flux (J) and rejection (R). The fluxes were calculated from the flow rate per membrane surface area unit. Rejection is a ratio permeate concentration that divided by feed concentration. Figure 5 presents the relationship between the feed concentration with the flux and rejection values generated from the methylene blue filtration.

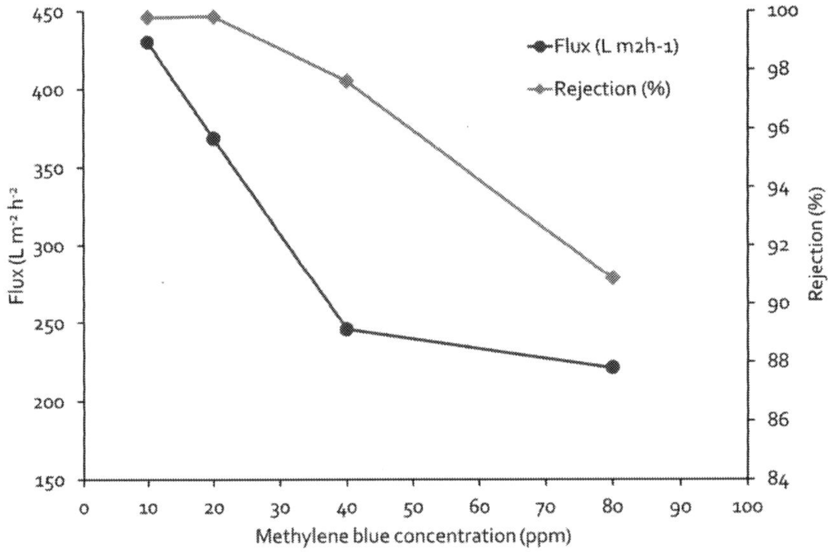

Figure 5. The relationship between the feed concentration with the flux and rejection of methylene blue filtration

Figure 5 shows the concentration is inversely proportional to the flux, where the greater the concentration, the smaller the flux. Concentration of 80 ppm had the lowest flux whilst it was dur to the solution concentration. Increasing the feed concentration decreased the filtering process. High turbidity caused the pore spaces in membrane grains rapidly clogged. The presence of flocculants in the membrane affected the membrane lifetime and degraded membrane performance. The decline in flux is the result of an escalation in membrane resistance by the presence of particles which blockage the pore and the formation of a cake layer on the membrane surface. A layer cake formation creates additional resistance to permeate flow. Pore blockage depends on the shape and the relative size of the particle. Blockage occurs faster when the shape and size of the particle and pore are similar [15]. Usually, the pore blockage occurs faster than the layer cake formation [16]. As a result, the permeate flux decreases with the passage of time. The highest flux was at a concentration of 10 ppm where the concentration was the smallest and solution density was the lowest.

Figure 6. Methylene Blue solution before and after filtration

Figure 6 shows significant differences between the methylene blue before and after filtration. All solution obtained from the filtration process was pellucid by visual observation. The rejection was correlated with the membrane pore average diameter of 34.4 Å (mesoporous). The filtration process generally can occur due to the membrane pore size smaller than that of the particles. However, the pore size had a greater value than the size of methylene blue diameter that was 15-25 Å. The reason why Al/Co pillared clay membranes could sieve methylene blue as the existing pores on the Al/Co pillared clay membrane piled or stratified one another therefore the large pore size covered each other. As a result, the membrane can filter methylene blue that was smaller than the average pore diameter of pillared clay membrane.

Methylene blue solution with the highest concentration had the least rejection. The same filtration time and the pore diameter greater than the diameter of methylene blue molecules caused the rejection capability decreased with increasing of methylene blue concentration. The rejection resulted at all methylene blue feed concentrations were above 90% and the optimum rejection occurred at a concentration of 10 ppm that was 99.89%. Thus, it was proved that the Al/Co pillared clay membranes could filter methylene blue dye well

4. Conclusion

Al/Co pillared clay membranes can be fabricate from the clay pillared by aluminium-cobalt solution. The XRD results showed the optimum calcination temperature Al/Co pillared clays was 200°C with a basal spacing of 19.3Å. The Al/Co pillared clays surface area and pore diameter were similar to the natural clay however they had larger pore volumes. The flux and rejection on the filtration process were inversely proportional to the concentration of methylene blue dye. The greater the concentration of methylene blue solution, the flux and rejection were getting smaller and the rejection for all concentrations of more than 90%.

Acknowledgments

Adi Darmawan gratefully acknowledge financial support from Diponegoro University via the Research for International Scientific Publications Award (Number: 315-04/UN7.5.1/PG/2015).

References

[1] S. Khemakhem, R.B. Amar, Grafting of fluoroalkylsilanes on microfiltration Tunisian clay membrane, Ceramics International, 37 (2011) 3323-3328.
[2] S. Notodarmojo, A. Deniva, Penurunan Zat Organik dan Kekeruhan Menggunakan Teknologi Membran Ultrafiltrasi dengan Sistem Aliran Dead-End (Studi Kasus: Waduk Saguling, Padalarang), Journal of Mathematical and Fundamental Sciences, 36 (2004) 63-82.
[3] A. Harabi, F. Bouzerar, Fabrication of Tubular Membrane Supports from Low Price Raw Materials, Using Both Centrifugal Casting and/or Extrusion Methods, Expanding Issues in Desalination, (2011) 253-274.
[4] Q. Chang, Y. Wang, S. Cerneaux, J.-e. Zhou, X. Zhang, X. Wang, Y. Dong, Preparation of microfiltration membrane supports using coarse alumina grains coated by nano TiO2 as raw materials, Journal of the European Ceramic Society, 34 (2014) 4355-4361.
[5] S. Vercauteren, M. Vayer, H. Van Damme, J. Luyten, R. Leysen, E.F. Vansant, The preparation and characterization of ceramic membranes with a pillared clay top layer, Colloids and Surfaces A: Physicochemical and Engineering Aspects, 138 (1998) 367-376.
[6] F. Bouzerara, A. Harabi, B. Ghouil, N. Medjemem, B. Boudaira, S. Condom, Elaboration and Properties of Zirconia Microfiltration Membranes, Procedia Engineering, 33 (2012) 278-284.
[7] M. Elma, C. Yacou, D.K. Wang, S. Smart, J.C. Diniz da Costa, Microporous silica based membranes for desalination, Water, 4 (2012) 629-649.
[8] P. Monash, G. Pugazhenthi, Effect of TiO2 addition on the fabrication of ceramic membrane supports: A study on the separation of oil droplets and bovine serum albumin (BSA) from its solution, Desalination, 279 (2011) 104-114.
[9] G. Li, H. Qi, Y. Fan, N. Xu, Toughening macroporous alumina membrane supports with YSZ powders, Ceramics International, 35 (2009) 1641-1646.
[10] C.M. Urruchurto, J.G. Carriazo, C. Osorio, S. Moreno, R.A. Molina, Spray-drying for the preparation of Al–Co–Cu pillared clays: A comparison with the conventional hot-drying method, Powder Technology, 239 (2013) 451-457.
[11] A. Gil, S.A. Korili, R. Trujillano, M.A. Vicente, A review on characterization of pillared clays by specific techniques, Applied Clay Science, 53 (2011) 97-105.
[12] G. Centi, S. Perathoner, Catalysis by layered materials: A review, Microporous and Mesoporous Materials, 107 (2008) 3-15.
[13] F. Bertella, S.B.C. Pergher, Pillaring of bentonite clay with Al and Co, Microporous and Mesoporous Materials, 201 (2015) 116-123.
[14] J.B. Fein, C.J. Daughney, N. Yee, T.A. Davis, A chemical equilibrium model for metal adsorption onto bacterial surfaces, Geochimica et Cosmochimica Acta, 61 (1997) 3319-3328.
[15] F. Shebl, M.A. El-Khalik, F. Helmy, Theoretical studies on the standard but misleading adsorption isotherm I: True adsorption isotherm, Surface technology, 19 (1983) 321-334.

IOP Conf. Series: Materials Science and Engineering **349** (2018) 012002 doi:10.1088/1757-899X/349/1/012002

[16] A. Abdelrasoul, H. Doan, A. Lohi, Fouling in membrane filtration and remediation methods, Mass Transfer-Advances in Sustainable Energy and Environment Oriented Numerical Modeling, (2013) 195-218.

Synthesis of Eugenol–Lauryl Methacrylate Copolymers via Cationic Polymerization

A Fajrin[1], SD Marliana[1], D S Handayani[1*]

[1]Department of Chemistry, Faculty of Mathematics and Natural Sciences, Sebelas Maret University, Jl. Ir. Sutami 36A Surakarta, Central Java 57126 Indonesia

*Corresponding author, E-mail: dsuci72@gmail.com/desisuci@staff.uns.ac.id

Abstract. Eugenol is one of the most abundant natural resources in Indonesia. The recently bio-based polymer resin is created based on eugenol because eugenol is functionalized with the polymerizable group. In order to improve the functional properties of eugenol, in this research Eugenol–Lauryl Methacrylate copolymers (co-poly(Eg-LMA)) were synthesized by cationic polymerization using H_2SO_4 as an initiator under the nitrogen atmosphere. Structure identification of the copolymer showed the absorption of the vinyl group from the monomers disappear at the analysis through FTIR at the wave number 1637-1639 and 985-995 cm^{-1} and also ^{1}H-NMR on the chemical shift 5,97 and 5,08 ppm. The resulting copolymers obtained brown powder in 32.03 % yieldsand melting point at 96 – 97 °C. Solubility test of the co-poly(Eg-LMA) showed that the polymer couldnot soluble in water but soluble in chloroform, diethyl ether, and benzene. Average molecular weight of co-poly(Eg-LMA) Led Ostwald viscometry was obtained 42020 with the degree of polymerization by 200.

1 Introduction

Eugenol (4-allyl-2-methoxyphenol) is a major component of clove oil. It can be extracted from clove (*Syzygium aromaticum syn. Eugenia aromaticum*) [1]. These compounds are particularly widely used in various fields such as pharmaceutical industry, food, and beverage industry, as well in material synthesis [2]. The development of eugenol-based materials is now widely practiced because eugenol is a very useful compound with functional groups that can be polymerized. The eugenol structure has three functional groups, namely hydroxy groups, methoxy groups and allyl groups [3].So that, eugenol can be used as starting material for the synthesis of new compounds [4].

Eugenol-based polymer modification has been widely used to improve the function, properties, and application of the material. Some modifications have been made by combining several other monomers such as Eugen oxy acetate acid [5] and methacryloyl chloride [1] or by the addition of cross-linking agents such as DVB [6]. However, to extend the utilization of eugenol, it is necessary to modify it with other monomers having specific functions, such as alkyl acrylates which are widely used for an oil absorbent [7].

The alkyl acrylate derivative compounds have many types with various functional properties. Recently, Jang & Kim [7] has modified various alkyl acrylates such as ethyl hexyl acrylate (EHA), lauryl acrylate (LA), lauryl methacrylate (LMA) and stearyl acrylate (SA) by copolymerization with styrene. Copolymer styrene with various alkyl acrylates is applied as an absorbent oil.Based on Jang & Kim [7] the best absorption activity is a copolymer styrene with LMA or SA. LMA is now widely developed by copolymerization with other monomers. The nature of poly-lauryl methacrylate has a good absorption activity, but the nature of long chain acrylic acid polymers such as LMA tends to form crystalline phases. The high crystallinity of the polymer material causes its low swelling power [7]. Styrene is a petroleum derivative that structurally resembles eugenol. By copolymerization of eugenol with lauryl methacrylate compound is expected to form polymer stability with an amorphous phase.

The polymerization process can carry out in three ways: condensation polymerization, addition or radical polymerization. The most reactive polymerization process is radical polymerization [8]. However, in the polymerization process of phenolic compounds, these compounds have antioxidant activity which can inhibit the reaction of radical polymerization [9]. Several previous studies which carried out polymerization of eugenol with cationic addition polymerization with concentrated sulfuric acid catalyst [10], concentrated nitric acid [11], or with $BF_3O(C_2H_5)_2$[12].

This study was conducted to produce a new material that can extend the utilization of eugenol.Eugenol which structurally similar to styrene is expected to be copolymerized with LMA using cationic polymerization method with the concentrated sulfuric acid catalyst. It is expected that with this method can be produced a new material that can be utilized more widely and can improve the characteristic of the resulting polymer.

2 Materials and Methods

2.1. *Materials*

Eugenol(Eg) was obtained from PT. Indesso Aroma, Indonesia was used without further purification. Lauryl methacrylate (LMA, Aldrich) was shaken before polymerization with a 5% of sodiumhydroxide and then with distilled water, subsequent drying over Na_2SO_4. Sulfuric acid (H_2SO_4, Aldrich). Natrium Chloride, Chloroform, and Na_2SO_4were obtained from Merck. Nitrogen gas also used.

2.2. *Synthesis of co-poly(Eg-LMA)*

Polymerization was carried out by a cationic mechanism, in a 100 mL three-neck flask equipped with magnetic stirrer and saturated with nitrogen gas, 1.970 g (0.012 mol) of the Eugenol and 2.035 g (0.008 mol) of the LMA were included. The mixture is stirred until the two solutions are mixed and then added 500 μL of sulfuric acid catalyst twice with an interval of 1 hour. Two hours after the last addition the copolymerization process was discontinued by the addition of 500 μL of methanol.

The resulting copolymer was then dissolved into water and extracted with chloroform to neutral pH. The organic phase extracted was added Na_2SO_4 then filtered and evaporated.The product was identified with FTIR and ^{1}H-NMR spectrophotometers.

2.3. *Calculated average molecular weight of co-poly(Eg-LMA) using Ostwald viscometry*

Determination of average molecular weight of co-poly(Eg-LMA) was performed by measuring the flow time using an Otswald viscometer with a concentration of 0.01 g / mL; 0.005g / mL; 0.0025 g / mL and 0.00125 g/mL and the pure solvent time of chloroform. Flow time measurements are carried out at room temperature. The result of the measurement was obtained by the time flow data of co-poly(Eg-LMA) solution and pure chloroform solvent, then calculated relative viscosity (ηr) and specific viscosity (ηsp). The data obtained are then made in the graph $\eta sp/C$ versus C for obtaining an intrinsic viscosity $[\eta]i$. The relative molecular period is calculated by the Mark-Houwink equation, with $k = 11 \times 10^{-3}$ and $a = 0.725$.

2.4. *FTIR analysis*

The spectra of the co-poly(Eg-LMA)were recorded at room temperature as KBr pellets on Shimadzu type FT-IR 8201 PC spectrophotometer. The light source of transmittance was in the range of 4000-500 cm^{-1}. The films were mounted directly in the sample holder.

2.5. *^{1}H-NMR analysis*
^{1}H-NMR spectra were taken in$CDCl_3$at room temperature, on Anglient instrument 400 MHz. Chemical shift was reported in δ (ppm) using TMS as aninternalstandard, and coupling constants were expressed in hertz.

3 Result and discussion

3.1. *Synthesis of co-poly(Eg-LMA)*
The polymerization process of eugenol and LMA is carried out via a non-media cationic polymerization using H_2SO_4 as an initiator under the nitrogen atmosphere. The polymerization process appeared with a change of color from a clear solution to a thick purple gel then solidifies. In general, the cationic polymerization takes place through three stages of initiation, propagation, and termination [13], this process is a series of addition reaction of the vinyl group by the opening of the C=C double bond to a single bond. The success of this process can be identified via FTIR and ^{1}H-NMR analysis.
The identification of co-poly(Eg-LMA) with both FTIR and ^{1}H-NMR spectrophotometers showed that the polymerization process has successfully combined the two monomers. Based on the results of the identification it was seen that the reacting process only via an opening of the vinyl group because the other groups such as –OH group, methoxy (–OCH_3), or –C=O carbonyl are not changed. Based on the results of the identification and analysis carried out the cationic polymerization reaction step of eugenol and LMA proposed is shown in Figure 1.

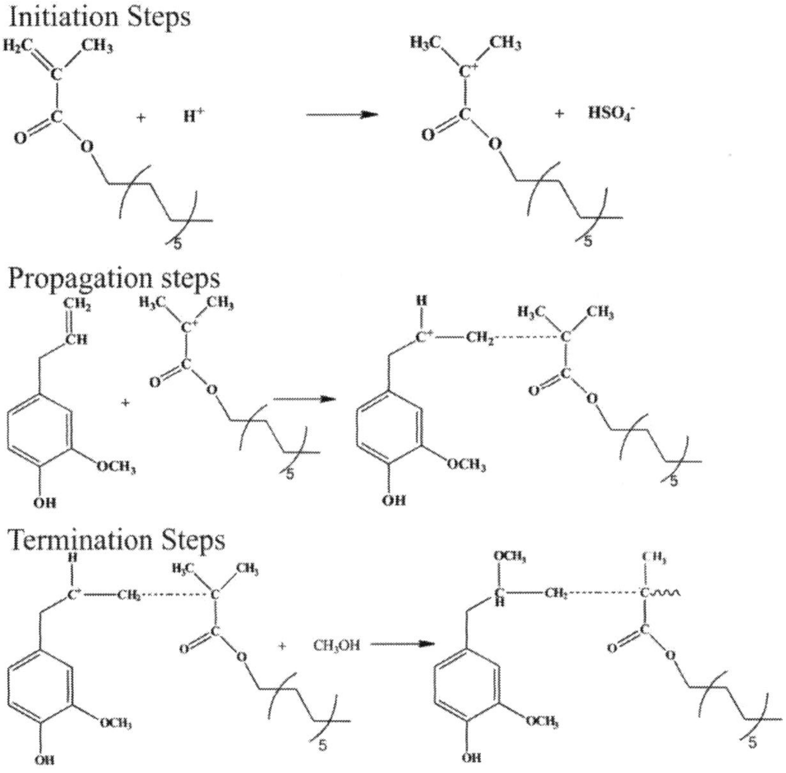

Figure 1. Scheme of the synthesis of cationic polymerization of eugenol and LMA

The initiation stage is carried out by the formation of carbocation on a tertiary C vinyl group from LMA due to initiation of H_2SO_4. This initiation stage can occur either through the eugenol monomer or LMA because both have a vinyl group that can form a carbocation. However, the propagation stage will be greatly influenced by the stability of the formed carbocation; the more stable ion will increase the reaction rate [10]. In both monomers can form a carbocation, but according to McMurry [14], the carbocation on the tertiary C atoms is more stable than at the C atom of the secondary and the primary. Therefore initiation of LMA is more proposed to form an active carbocation. The second stage is the propagation stage; the elongation takes place through the end of the reactive carbocation that initiates the other monomer continuously until the termination or chain runs out. At the propagation stage, the eugenol monomer and LMA opened the double bond and form a covalent bond with reactive carbocation at polymer end [10]. This propagation process will affect the length of the polymer chain and its relative molecular weight. The last stage of the polymerization process is the termination stage, at this stage,the chain elongation process is stopped by the addition of methanol. The $-OCH_3$ group of methanol will interact with the carbocation at the end of the polymer and stopped the formation of covalent bonds of the vinyl group by carbocation [6].Physical identification of the co-poly(Eg-LMA) from polymerizationwere obtained brown powder in 32.03 % yields and melting point at 96 – 97 °C. Thesolubility test of the co-poly(Eg-LMA) showed that the polymer could not soluble in water but soluble in nonpolar solvent (chloroform, diethyl ether, and benzene).

3.2. *FTIR analysis*

The FTIR spectra of eugenol, LMA, and co-poly(Eg-LMA) monomers are shown in figure 2. IR spectra of the eugenol monomer had characteristics of hydroxyl group uptake at 3448 cm^{-1}, double bond benzene ring at 1612 cm^{-1} and methoxy group ($-O-CH_3$) at 1034 cm^{-1}. The peak for absorption of C=C vinyl groups was present in wave numbers 1637, 995 and 914 cm^{-1}. While in LMA monomer spectra the identified characteristic waveforms were C=O carbonyl acryl absorption at 1722 cm^{-1}, –C–O–C–methacrylate at 1296 and 1165 cm^{-1}. The LMA also had a vinyl group uptake at 985 cm^{-1}.

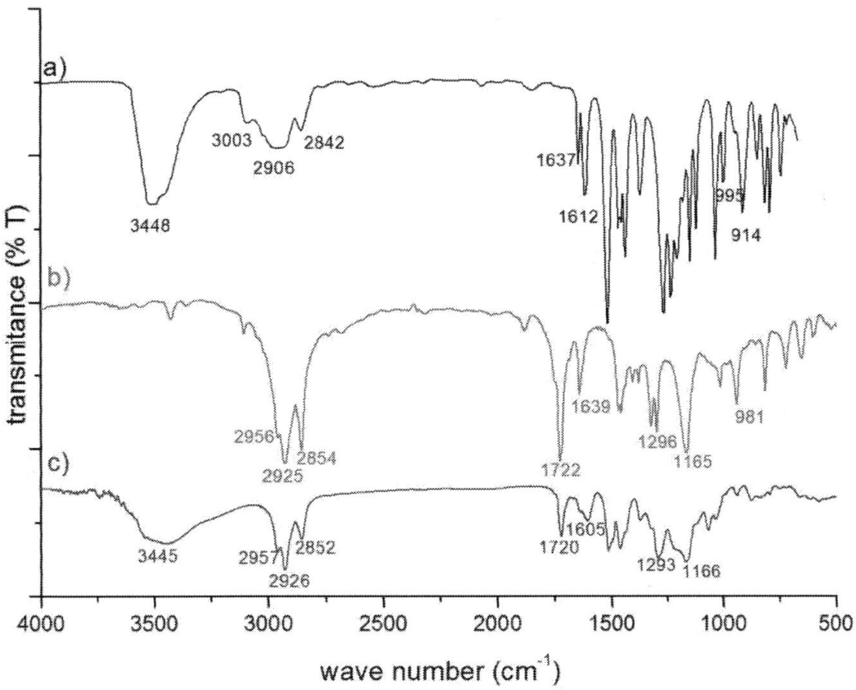

Figure 2. Infra-red spectrum of (a) eugenol, (b) LMA and (c) co-poly(Eg-LMA)

The spectra of co-poly(Eg-LMA) showed a hydroxyl group at 3445 cm^{-1} which are estimated to be –OH of eugenol because the LMA has no hydroxyl group. The sharp peak at the 1720 cm^{-1} wave number indicated that the presence of a C=O carbonyl group that was initially absent in the eugenol monomer. The peaks of the vinyl double bonds seen in the eugenol and LMA monomer spectra of wave numbers 1637, 1639, 995 and 985 cm^{-1} are no longer present in the co-poly(Eg-LMA) spectra. While the uptake of the vinyl monomer group at 914 cm^{-1} shifts to 936 cm^{-1} and is not completely lost. Aran [15], indicates that the group vinyl is not completely lost and shifted because the end of the chain is more or less disproportionately. Based on the existing FTIR spectra analysis, it can be said that the eugenol polymerization process with LMA is successfully done by opening the double bond C=C.

3.3. ^{1}H-NMR analysis

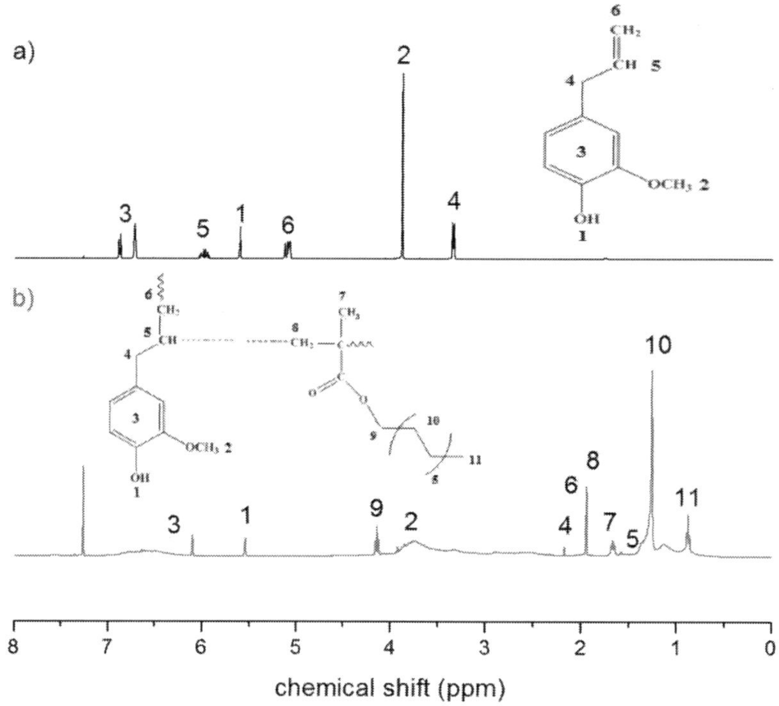

Figure 3. ^{1}H-NMR spectrum of (a) eugenol and (b) co-poly(Eg-LMA)

Eugenol and co-poly(Eg-LMA) monomers identified with ^{1}H-NMR can be seen in Figure 3. The proton NMR showed chemical shift signals at 0.88 and 1.26 ppm from the methyl of acrylic chain and overlapping of amethylene ($-CH_2-$) chain. Also, the chemical shift at 3.92 ppm is a proton signal of the methoxy($-O-CH_3$) group, the signal at 5.54 ppm for the hydroxyl (−OH) group, and the proton signal at 6.10 ppm from abenzenering. These proton signals werecharacteristic functional groups of eugenol in the identified polymer. The identification of successful polymerization processes in co-poly(Eg-LMA) was demonstrated by the loss proton signal of vinyl eugenol group at 5.97, and 5.08 ppm and the appearance of chemical shifts at 2.50 - 3.00 ppm which are proton signals for double bonding changes from vinyl group becomes a single bond [16]. Identification Based on the identification made known that co-poly(Eg-LMA) has been successfully established.

3.4. *Average Molecular Weight of co-poly(Eg-LMA)*

The molecular weight of a polymer is a very important thing because it will affect the physical properties and its application in life. The determination of average molecular weight ($\overline{M}$) of co-poly(Eg-LMA) was performed based on the measurement of its viscosity with an Ostwald viscometer. Measurements were made on 4 solutions with different concentrations: 0.01; 0.005; 0.0025; and 0.00125 g/mL. The results obtained from the measurement of the flow time using an Ostwald viscometer at the five concentration variations are shown in table 1. Measurements are made at room temperature.

Tabel 1. Flow time of copoly(Eg-LMA) using an Ostwald viscometer

Concentration (g/mL)		0	0.00125	0.0025	0.005	0.01
Copoly (Eg-LMA)	t (s)	15.77	16.16	16.58	16.97	17.14
		15.80	16.28	16.60	17.04	17.25
		15.81	16.31	16.66	17.20	17.33
	$t_{average}$(s)	15.793	16.250	16.613	17.070	17.240

The calculation using the Mark-Houwink equation based on the data from the measurement of the flow time using an Ostwald viscometer. Average molecular weight ($\overline{M}$) of co-poly(Eg-LMA) is 42020 with the degree of polymerization by 200.

4 Conclusion

Copoly (Eg-LMA) have been successfully synthesized via cationic polymerization using H_2SO_4 as an initiator under anitrogen atmosphere without any reaction medium. FTIR analysis revealed that the absorption of the vinyl group from the monomers disappear at the wave number 1637-1639 and 985-995 cm-1 and also 1H-NMR on the chemical shift 5.97 and 5.08 ppm. Based on the results analysis it is seen that the polymerization process reacted only via an opening of the vinyl group. Copoly(Eg-LMA) were obtained brown powder in 32.03 % yields and melting point at 96 – 97 °C, which can not soluble in water but soluble in chloroform, diethyl ether, and benzene. The average molecular weight of co-poly(Eg-LMA) also investigated using Ostwald viscometry obtained theaverage molecular weight of co-poly(Eg-LMA) of 42020 with the degree of polymerization by 200 Variation amount of the each component and the physical properties still need to be evaluated toproduce material with better functional properties.

Acknowledgments
Author(s) would like to acknowledge their thanks tothe supervisor Dr. Desi Suci Handayani., M.Si for the financial support and organic material research group.

References

[1] Towaha J 2012 *Prespektif.* **279** – 90
[2] Rojo L, Blanca V, Juan P, Antonio L B, Sanjukta D and Julio S R 2006 *Biomacromolecules.* **7** 2751 – 2761
[3] Chowdhry B Z, Ryall J P, Dines T J and Mendham A P 2015 *The Journal of Physical Chemistry A***119** 11280 – 11292
[4] Djunaidi M C, Lusiana R A, Wibawa P J, Siswanta D and Jumina 2010 *Reactor.***13** 16 – 23
[5] Mahawati D 2005 *Skripsi S-1.* Universitas Sebelas Maret
[6] Handayani D S, Kusumaningsih T and Yuli M 2004 *Biofarmasi.* **2** 53 – 57
[7] Jang J and Kim, B S 2000 *J. App. Polym. Sci.***77** 903 – 9013
[8] Odian G 2004*Principle of polymerization 4th ed.* New York: John Wiley & Sons
[9] Satoh K, Saitoh S and Kamigaito, M 2007 *Journal ACS.* **129** 9586 – 9587
[10] Ngadiwiyana 2005 *JSKA.* **8** 1-10
[11] Suirta I W, Rustini N L, and Prakosa T I 2012 *Jurnal Kimia.* **6** 37 – 46
[12] Kiswandono A A, Siswanta D, Aprilita N H, Santosa S J and Hayasitha T 2014 *Journal of Environmentally Friendly Processes.* **2** 57 – 68
[13] Sudarlin and Haryadi W 2017 *Kimia Valensi.* **3** 50 – 58
[14] McMurry J 2010 *Fundamentals of Organic Chemistry 7th ed.* Belmont: Cengage Learning
[15] Aran B 2004*Thesis* Middle East Technical University
[16] Iswanto P and Susilawati S S 2003 *Prosiding Simposium Nasional Polimer.* **4** 56 – 60

IOP Conf. Series: Materials Science and Engineering **349** (2018) 012004 doi:10.1088/1757-899X/349/1/012004

Preliminary Study of *Hyptis pectinata* (L.) Poit Extract Biotransformation by *Aspergillus niger*

D S Rejeki[1], A L N Aminin [1], M Suzery [1]

1Chemistry Department, Faculty of Sciences and Mathematics, Diponegoro University, Indonesia.

E-mail: meiny.suzery@live.undip.ac.id

Abstract. One alternative approach to increase the content of bioactive compounds is fermentation. *Hyptis pectinata* (L.) Poit is a plant that can be found in tropical area and potentially as anticancer, anti-inflammatory, insect repellant, antiviral and antioxidant. In this research, efforts have been made to increase bioactive plant capacity of *Hyptis pectinata* (L.) Poit through submerged fermentation using *Aspergillus niger*. The study was performed by adding methanol extract of *Hyptis pectinata* (L.) Poit on two conditions, that was added at the beginning of fermentation and while entering a phase of death. *Aspergillus niger* growth rate in both conditions was observed by determining the dry weight of cells every 24 hours. The transformation profil of extract was observed after 24 hours of extract addition in early death phase by the TLC method. The results show that the addition of *Hyptis pectinata* (L.) Poit extract at log phase triggers the cells to growth faster, whereas the addition at the early death phase precisely accelerates cell death. TLC profile shows the emergence of new compounds suspected as the products of transformation of Hyptis pectinata (L.) Poit extract on day 8 after addition of extract.

1. INTRODUCTION

Nowadays, the discovery of bioactive compounds from the herbs is very developed and be the best choice for many people. The people have various alternatives to treat some diseases. Because of the phenomena, it is needed a certain effort to explore the natural materials well; not only for inhibiting, but also for providing effective cures without causing excessive side effects.

One effort that can be used to increase the content of bioactive compounds in plants is by fermentation [1]. This process can cause the decomposition and or biotransformation of complex substrates into compatible components by changing the properties of the product or altering the quantity of certain bioactive compounds [2]. Biotransformation is considered an economically competitive technology in synthetic organic chemistry because it provides new production routes for fine chemical, pharmaceutical, and agrochemical compounds [3].

Microorganisms are known to perform a variety of reactions, even if they are not chemically manageable [4]. The efficiency of using microorganisms as one of the biocatalyst agents is known to be more effective and productive, due to its ability to metabolize substrates. Biotransformation by using microbes is more specific than ordinary chemical compounds, because it allows the addition, deletion or modification of certain functional groups. Catalyzed reactions include oxidation, dehydrogenation, hydroxylation, dehydration, condensation, decarboxylation, deamination, amination and isomerization [5].

Aspergillus niger considered as a safe microorganism and the FDA (Food and Drug Administration) in the United States classifies it as a GRAS (*Generally Recognized As Safe*) microbe for not producing mycotoxins and in its metabolism *Aspergillus niger* is able to produce citric acid [6]. In some studies, *Aspergillus niger* is used in the biotransformation process [7]. Among them, it is a biotransformation in isoflavone compounds, namely 7.4'- *dimethoxyisoflavon* and 7.4' *diacethoxyflavon* which can produce a new metabolite that is daidzein. Potentially as antihemolytic, antioxidative, and antitumor [8]. Biotransformation research was developed against Budlein A lactone compound, which can produce 3 new metabolites that can reduce tumor cell growth and cell viability in low doses [9]. The next 13 metabolites derived from 20 (S)- Protopanaxatriol compounds that potentially inhibit the growth of 7 cancer cells with an average IC_{50} below 100 (μmol/Liter) [10] catechin biotransformation [11] and flavonoid compounds that have higher antioxidant biotransformation results [12].

Hyptis pectinata Poit is a plant that thrives in wild land and has not been widely used in Indonesia [7]. In Brazil, the farming of this plant has developed, and is known to treat inflammation, pain, injury, cancer, bacterial infections, and for healing wounds [13] anti-inflammatory and the decoction of leaves used as tea drinks [14]. While in India and Mexico, this plant is known as an herbal plant that has a versatile drug quality in the treatment of fever, skin diseases and stomach disorders. In the development of potency of *Hyptis pectinata* (L) Poit, research has been done on isolation, transformation, activity test of the pure compound and essential oil isolation from *Hyptis pectinata* Poit extract [15]. Until now no research has reported on the effect of fermentation on the methanol extract of *Hyptis pectinata* Poit using *Aspergillus niger*. This research is very important to do, considering it can be used as an alternative to increase the content of bioactive compounds.

This research was conducted at the Laboratory of Organic Chemistry and Biochemistry, Faculty of Science and Mathematics, Chemistry Department of Diponegoro University. This research starts with the preparation of *Hyptis pectinata* Poit leaf extract, *Aspergillus niger* rejuvenation, *Aspergillus niger* growth profile measurement, *Hyptis* leaf extract fermentation with *Aspergillus niger* as preliminary test and and transformation profile analysis using TLC method.

2. METHOD

2.1. Extraction

The 108 grams of *Hyptis pectinata* Poit leaves were collected from Kanayakan Village, East Dago-Bandung. The first step was covering the leaves with black cloth and drying under the sunlight for three days. Then, the dried leaves have been crushed and made into powder. The extraction process was performed by maceration. *Hyptis pectinata* Poit powder was macerated with 2 L methanol for 3 days and filtered. The filtrate was concentrated with *a rotary vacuum evaporator* up to one-third of volume, then added warm aquadest (1:1) and left for 1 x 24 hours until it formed into two layers. The top layer was a yellow methanol-water, while the bottom layer was a green chlorophyll deposit. The top layer was concentrated again with *a rotary vacuum evaporator* until the liquid extract was obtained. To remove the residual solvent, evaporation was performed over the *water bath*, and measured rendement to the extract [16].

2.2. Preparation of *Aspergillus niger* Culture

The black colonized *Aspergillus niger* stock was preserved on PDA *(Potato Dextrosa Agar)* medium which is made with tilted agar and incubated for 2-3 days. For inoculum preparation, culture is rejuvenated in PDB (Potato Dextrosa Broth) media as *a starter*, and observations are made until the spores looked solid (1 x 10^7) spores/mL [9].

2.3. Measurement of *Aspergillus niger* Growth Profile

The mold isolates from the culture stock of PDB medium can be rejuvenated by inoculating each 10 mL of *starter* solution into 10 flask containing 100 mL of PDB medium. Then they were incubated in a *shaker incubator* at 30° Celsius and 125 rpm speed. The measurement of *Aspergillus niger* mold growth profile was performed on the basis of dry cell weight and determined once every 24 hours on screening results. Because of the increasing acid content in the media, thus the mycelium was added with 1% Sodium hydroxide and washed back with distilled water until neutral pH was obtained. The mycelium was dried in an oven at 80 °C for 24 hours, the weighs the mycelium to obtain a constant weight. The relationship between the weight of the dry cell and the incubation time was used to make the growth curve.

2.4. Fermentation of *Hyptis Pectinata* Poit leaf methanol extract with *Aspergillus niger*

An amount of 10^7 spores / mL of *Aspergillus niger* in 100 mL PDB of 1% was added by 1 gram of the *Hyptis pectinata* Poit leaf extract. Then it was incubated at 30°C and at 125 rpm speed. Measurement of *Aspergillus niger* mold growth profile in PDB medium with the addition of *Hyptis* leaf extract, was performed by measuring the weight of dry cells determined every 24 hours. Furthermore, the same steps were done as in the *Aspergillus niger* growth profile measurements.

2.5. Identification of Biotransformation Compounds

To observe the biotransformation process on the fermented production, identified using TLC. The filtrate was separated by centrifugation at 2500 rpm for 25 minutes. To remove the remaining solvent, *a rotary vacuum evaporator* was applied to get concentrated solvent. Crude extract was analyzed further at every 24 hours once using the Gel 60 F_{254} gel plate on 4 various eluents. The measurement of stain spots was performed under UV detector lamps with λ_{366} nm. The Rf svalue generated from the incubation time variation would be compared. The presence of changes in Rf values indicated the occurrence of transformation of compounds in fermentation products.

3. RESULTS AND DISCUSSION

3.1. *Aspergillus niger* Growth Profile

The *Aspergillus niger* growth profile shows information about growth phases that occurred in *Aspergillus niger*, which can be used to determine the exact time for fermentation (Figure 1).

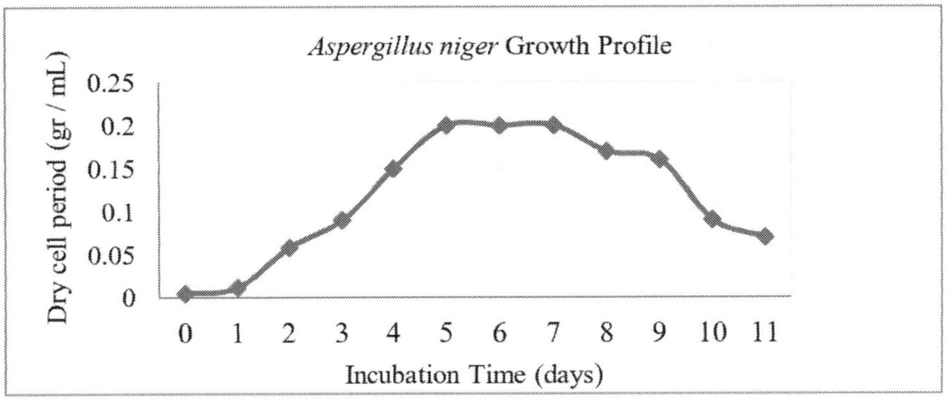

Figure1. *Aspergillus niger* Growth Profile

From the resulting data, it is seen that the initial phase of adaptation (lag) occurs on days 0-1, in this phase there is an adjustment of cells with the enzyme environment to break down the substrate. The next phase of acceleration (logarithmic) occurs on days 1-3, because in this phase the cells begin to divide and become more active. Then the exponential phase occurs on days 3-5. During this phase, the cell can divide maximally due to the availability of many nutrients, so that cell activity is greatly increasing, and this phase is an important phase for fungi life. At the beginning of this phase many enzymes can be produced. Meanwhile, on day 5-7, there is a stationary phase seen from the dry cell period remains. In that phase the number of cells that grow is relatively the same as the number of dead cells. The profile in this phase is a horizontal straight line and there are many secondary metabolites. Furthermore, during days 8-11, it is the death phase of *Aspergillus niger,* which can be seen from the decline in the dry cell mass. In this phase, the nutrients have begun to decrease and the presence of other metabolites that are toxic and can inhibit cell growth, so that the growth rate of *Aspergillus niger* begins to decline.

3.2. Fermentation Profile Methanol Extract of *Hyptis pectinata* Poit Leaves with *Aspergillus niger*

From the pattern formed between the addition of *Hyptis* extract and without the addition of *Hyptis* extract, there is a prominent difference at the beginning of the growth phase. It is characterized by an increasing number of cells in the logarithmic phase. While at the end of the stationary phase, the addition of *Hyptis* extract actually decreases the number of cells sharply. As shown in Figure 2:

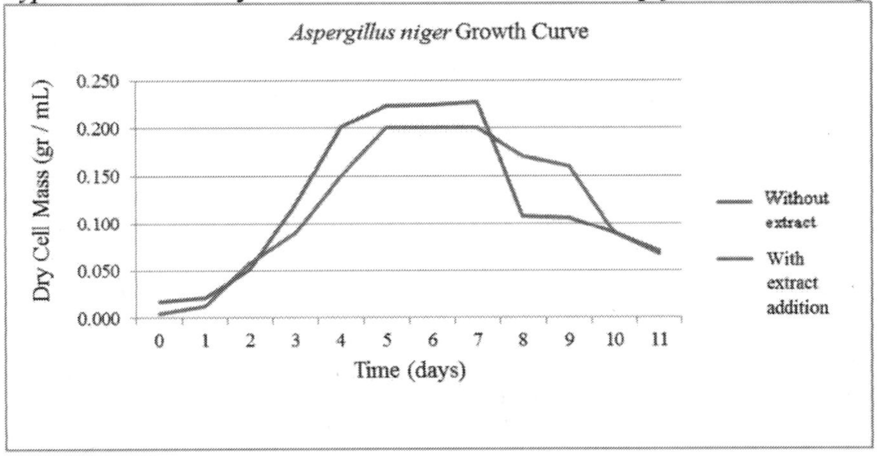

Figure2. *Aspergillus niger* Growth Profile without extract and with the addition of extract.

The results show that the addition of *Hyptis* extract at the beginning of the growth phase provides a significant increase in the number of cells in the logarithmic phase compared to the number of cells without the addition of *Hyptis* extract. The increasing number of cells at the beginning of the growth phase is assumed because of the molecules secreted by *Aspergillus niger*. In general, in the logarithmic phase of enzyme production is abundant. The ability of hydrolytic extracellular enzymes in *Aspergillus niger* can decompose the polymer into its monomers, thus in this phase facilitating the absorption of nutrients well.

In addition, when viewed from its nature, *Aspergillus* lives as a decomposer and a *heterotrophic symbiont* that gets its nutrients or nutrients by absorption. In general, the surface area in liquid culture also supports very rapid growth and adapts *Aspergillus niger* to obtain the perfect absorptive nutrients. The addition of *Hyptis* leaf extract to the liquid inoculum of *Aspergillus niger*, suppresses the hormone synthesis that promotes the addition of cells. According to another report that *Aspergillus utus* in a liquid culture treatment can synthesize the presence of auxin and gibberellin hormones that can promote the growth and development of host plants [17].

According to its phytochemical content, the compounds of *Hyptis pectinata* Poit plants are composed of α,β-unsaturated lactone skeletons [8].Whereas if it is seen from the structure of the major compound lactone, there is an atom C in the cyclical position which is easily replaced by another atom (e.g. atom O). The existence of lactones resembles the structure of kinetin hormone which is thought to be synthesized by *Aspergillus niger*. Kinetin hormone is a derivative of cytokinin hormone that serves to spur cell division. Thus, in the presence of similar structures, it can induce the occurrence of faster cell division and can increase the number of cell densities.

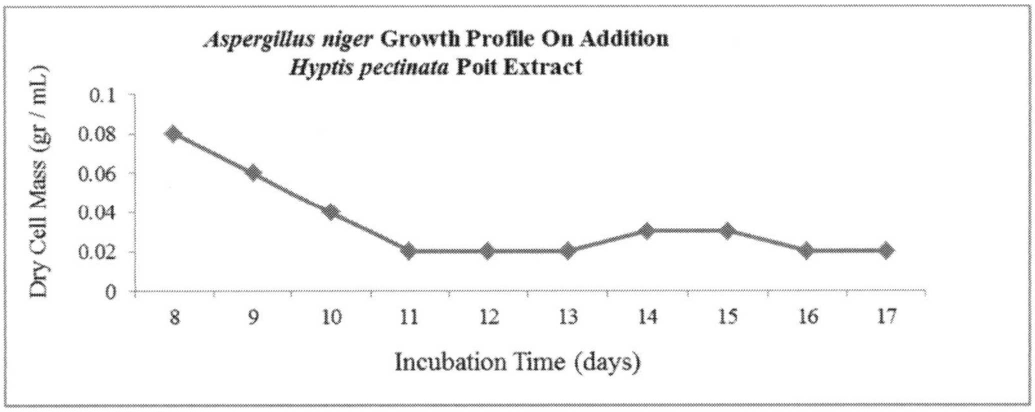

Figure 3. *Aspergillus niger* Growth Profile on the early addition of extract at 8 hours

Unlike the case at the beginning of the death phase, the addition of *Hyptis* extract can decrease the number of *Aspergillus niger* cells. The decline in the number of cells at the beginning of the dead phase is because the supply of nutrients that have been exhausted and also because the formed transformant can increase the toxicity for the growth of *Aspergillus niger*. According to another report, one component of *Hyptis pectinata* (L.) Poit is an essential oil (Pectinolida), which consists of Pectinolida A-G. The bioactive compound of pectinolide in *Hyptis* extract is potential as an antimicrobial [18].

3.3. Identification of Biotransformation Compounds

Table 1. Rf values of the biotransformation of the compound

Observation Day to	Eluen			
	n-heksan	Chloroform	Ethyl acetate	Methanol
	Rf			
8	-	0.80	-	0.87
9	-	0.82	-	0.86
10	-	0.85	-	0.87
11	-	0.84	-	0.84
12	-	0.82	-	0.86
13	-	0.82	-	0.84
14	-	0.78	-	0.83
15	-	0.31; 0.34	0.73; 0.77; 0.82	0.87
16	-	0.66	0.97	0.83
17	-	0	-	0.87

On the 15th day observation, it is seen a change of stain from some eluent. Of the chloroform eluents there are two different stains. While on the ethyl acetate eluent, which initially cannot elute at all, but on 15th day, it has formed three different stains. Similarly, on the 16th day, chloroform and ethyl acetate eluents also still develop a stain with different Rf. While on the next day, in the chloroform eluent, there is no change of stain, the distance of the stain that appears is proportional to the distance in the eluent. Based on the qualitative TLC test it can be concluded that the start formation of the transformational result is at the 15th day, so the fermentation process can be stopped at that time. The process of fermentation in liquid culture generally has a shorter time than fermentation in solid medium, but the influence of adding a certain substrate, can increase the length of fermentation time. This is due to the more complex substrate degradation process because of the addition of substance to the environment where microbe grows [17].

4. CONCLUSION

In Addition of *Hyptis pectinata* (L.) Poit extract at early growth increase the cell growth of *Aspergillus niger* cells that are affected by the molecular secretion in the form of hormone cells growth. While the significant decline of cells in early death phase is influenced by secondary metabolites that are pathogenic to *Aspergillus niger* growth. It is assumed that the transformation on the 8th day after the addition of *Hyptis pectinata* (L.) Poit extract was marked by a change in the number of stains formed.

ACKNOWLEDGMENTS

Research of Development and Application (RPP) of PNBP UNDIP.

BIBLIOGRAPHY

[1]Martins, S., Mussato, S.I ., Mart nez avula, G., Monta ez-Saenz, J., Aguilar, C.N., & Teixeira, J.A (2011). Bioactive phenolic compounds: Production and extraction by solid–state fermentation. Areview. *Biotechnology Advances*, 29 (3), 365-373.

[2]Hussain, A., Bose, S., Wang, J. H., Yadav, M. K., Mahajan, G. B., & Kim, H. (2016). Fermentation, a feasible strategy for enhancing bioactivity of herbal medicines. *Food Research International*, 81, 1–16.

[3]Adelin.E., Martin, M.T., Bricot, M.Francoise., Cortial, S., Retailleav, P., Ouazzani, J. (2012). Unexpect thio conjugation of Sch-642305 with 3-mercaptolactate catalyzed by Aspergillus niger ATCC 16404, *Biotransformation of natural compounds*, 84, 135-140.

[4]Falcao, R.A., L.A, Patricia., Nascimento, do., De Souza, S.A., Da Silva, T.M.G., De Queiroz, A.C., Da Matta, C.B.B., Moreira, M.S.A., Camara, C.A., Silvan, T.M.S, 2013, Antilesishmanial Phenylpropanoids from the Leaves of Hyptis Pectinata (L) Poit, (1-7).

[5]Walker, J.M. and Rapley, R. 2002. Molekular Biologi and Biotechnology Britain: Athenacum Pr.

[6]Sasmitaloka, K. S. (2017). Produksi asam sitrat oleh *Aspergillus niger* pada kultivasi media jamur, *Integration Journal Process*,3(6), 116-122.

[7]Parshikov, i.A., & Sutherland, J.B (2014). *The use of Aspergillus niger cultures for biotransformation of terpenoid.* Proces Biochemistry, *49 (12), 2086-2100.*

[8]Miyazawa, M., Ando, H., Okuno,.Y., Araki, H., (2004), Biotransformation of isoflavones by Aspergillus niger as biocalalyst, *Journal of Molecular catalysis,* B : Enzymatic, 27, 91-95.

[9]Arakawa,N.S., Neto,L.G., Ambrosio, S.R., Antonvcci, G.A., Sampaio, S.V., Pupo, M.T., Sard.S., Schmidt, T.J., Da Costa, F.B, (2013) Unnusual biotransformation Products of the sesquiterpene lacton Budlein A by Aspergillus species, *Journal of Phytochemistry,* 96, 92-100.

[10]Cao, H., Chen, X., Reza, A., & Xiao, J. (2015). Microbial biotransformation of bioactive flavonoids. *Biotechnology Advances*, *33*(1), 214–223

[11]Ni, H., Chen, F., Dong, Z., Ying, M., & Fan, Y. (2015). LWT - Food Science and Technology Biotransformation of tea catechins using Aspergillus niger tannase prepared by solid state fermentation on tea by product. *LWT - Food Science and Technology*, *60*(2), 1206–1213.

[12]Chen, G., Song, Y., Ge, H., Ren,J., Yang, X., and Li, J., (2015), Microbial Biotransformation of (20) S-protopanaxatriol by Aspergillus niger and the cytotoxicity of the resulting metabolites. *Phytocemistry Letters,* 11 (19), 111-115.

[13]Bispo M., Moura R., Franzotti E., Bomfim K., Arrigoni-Blank M., Moreno M., Marchioro M., Antoniolli A., (2001), Antinociceptiveand antiedematogenic effects of the aqueous extract of experimental animals, *Journal of Ethnopharmacology* 76, 81-86.

[14]Arigoni,B.F., Antoniolli A., Caetano, L.C., Campos, D.A., A.F.Blank., 2008, Antinociceptive activity of the volatile oils of *Hyptis pectinata* Poit. (Lamiaceae) genotypes, *Phytomedicin*, 334-339

[15]Suzery, M., & Cahyono, B. (2014). Jurnal Sains dan Matematika Evaluation of Cytotoxicity Effect of Hyptis pectinata Poit (Lamiaceae) extracts using BSLT and MTT methods *Jurnal Sains dan Matematika*, 22(3), 84–88.

[16]Suzery M., ,Gultom M., and Cahyono B. (2013). Senyawa Hiptolida dan Pektinolida dalam fraksi Diklorometana dari daun Hyptis pectinata Poit. *Jurnal Sains & Matematika*, 21(2), 31–34.

[17]Marina,S., Angel.M., Silva.M-Flores., Cervanter.M.G-Badillo, Rosales.M,T-Saavedra, Islas.M.A-Osura., Casas.S-Flores (2011), The Plant Growth Promoting Fungus Aspergillus ustus Promotes Growth and Induces Resistance Againts Different Lifestyle Pathogen In Arabidopis thaliana. *J.Microbiol. Biotechnol*, 21 (7), 686-696

[18]Pereda-Miranda, R., Hernandez, L., Villavicencio, M.J., Novelo, M., Ibarra, P., Chai, H., Pezzuto, J.M., (1993). Structure and stereochemistry of pectinolides A–C, novel antimicrobial and cytotoxic 5,6-dihydro-a-pyrones from *Hyptis pectinata. J. Nat. Prod.* 56, 583-593.

IOP Conf. Series: Materials Science and Engineering **349** (2018) 012005 doi:10.1088/1757-899X/349/1/012005

The Initial Comparison Study of Sodium Lignosulfonate, Sodium Dodecyl Benzene Sulfonate, and Sodium p-Toluene Sulfonate Surfactant for Enhanced Oil Recovery

Argo Khoirul Anas[1], Nurcahyo Iman Prakoso[1], Dilla Sasvita[1]

[1]Chemistry Department, Universitas Islam Indonesia, Yogyakarta, Indonesia

E-mail: argokhoirulanas@uii.ac.id

Abstract. Surfactant (surface active agent) exhibit numerous interesting properties that enable their use as additional component in mobilising of residual oil from capillary pore after secondary recovery process using gas injection and water flooding. In this study, Sodium Lignosulfonate (SLS) surfactant was successfully synthesized by applying batch method using lignin from oil palm empty fruit bunches as precursor. Furthermore, its performance in reducing interfacial tension of crude oil and formation water colloidal system was compared with commercial available surfactant including Sodium Dodecyl Benzene Sulfonate (SDBS) and Sodium p-Toluene Sulfonate (SpTS). The synthesized SLS surfactant was characterized by using Fourier Transform Infrared (FTIR) spectroscopy. Meanwhile, its performance in reducing interfacial tension of crude oil and formation water colloidal system was analyzed by using compatibility test, phase behaviour analysis, and interfacial tension (IFT) measurement. The compatibility test shows that SLS, SDBS, and SpTS surfactants were compatible with formation water. In addition, the phase behaviour analysis shows that SLS surfactant was better than SpTS surfactant, while SDBS surfactant generates the highest performance proved by the best microemulsion formation resulted by SDBS. Furthermore, the optimum concentration of SLS, SDBS, and SpTS surfactants in reducing the interfacial tension of crude oil and formation water was 1.0%. The IFT measurement indicates that the performance of SLS with the value of 1.67 mN/m was also better than SpTS surfactant with the value of 3.59 mN/m. Meanwhile, SDBS surfactant shows the best performance with the IFT value of 0.47 mN/m.

1 Introduction

Enhanced oil recovery (EOR) is a well-known technology that usually applied to improve oil production in the petroleum industry. EOR based on chemical compound called as surfactant (surface-active agent) exhibit numerous interesting properties that enable their use as additional component in mobilising of residual oil from capillary pore [1-2] Surfactants are used in various field due to their interesting physicochemical characteristics such as foaming, detergency, dispersion, and emulsification ability. It is an amphiphilic compound consists of non-polar lyophobic part called as tails group and polar lyophilic part called as head group. This amphiphilic structure is responsible for their role as a third component that adsorbs onto oil and water interface to reduce their interfacial tension. [3]

One type of surfactant that potential to develop is Sodium Lignosulfonate (SLS) which can be synthesized using lignin. Lignin is a compound that can be isolated from oil palm empty fruit bunches which is a waste or main byproduct of palm oil industries. This waste is not optimally used, and then it is possible to use it as a raw material in the synthesizing of SLS surfactant. This isolation process can be conducted using batch method with the highest yield of lignin in the amount of 14%.[4] Furthermore, the SLS surfactant can be resulted from the reaction between lignin as a precursor and sodium hydrogen bisulfite (NaHSO3) as a sulfonating agent. The optimum condition on the synthesizing method of SLS surfactant has been previously reported, [5] hence is important to compare synthesized SLS surfactant with another surfactant, in order to observe their ability in reducing the surface tension of crude oil and formation water.

In this case, the SLS surfactant is can be compared with the commercial surfactant such as Sodium Dodecyl Benzene Sulfonate (SDBS) and Sodium p-Toluene Sulfonate (SpTS). SDBS is a famous surfactant that has been used in various field such as carbon nanotube dispersing agent [6] and heavy metal adsorption. [7] On the other hand, SpTS is a commercial surfactant that has similar structure with SDBS, but SpTS has shorter hydrophobic chains than SDBS. When compared to SLS surfactant, SDBS and SpTS have smaller hydrophobic chains. These structural differences are interesting to study, especially in their application for EOR technology.

Here, we report the initial comparison study of SLS surfactant with two commercially available surfactants including SDBS and SpTS. In the researcher knowledge, there is no reported study on the comparison of these three surfactants in EOR technology. This study is important to provide additional information about the suitable surfactant structure in EOR technology. The compatibility test, phase behaviour analysis, and interfacial tension (IFT) measurement were used to observe the surfactant ability in reducing interfacial tension of crude oil and formation water colloidal system.

2 Experimental

2.1 Material

The chemical used in this study were $NaHSO_3$ (Sigma Aldrich), NaOH (Sigma Aldrich), H_2SO_4 (Merck), and methanol. SDBS and SpTS were purchased from Sigma Aldrich and directly used without purification.

2.2 Synthesis of SLS Surfactant

Oil palm empty bunches were put into autoclave, which then added by the mixture of NaOH solution and ethanol with the ratio of 1:1. The system was heated at temperature of 170 °C (15 atm) for 5 hours. After filtration process, the resulted black liquor was acidified using H_2SO_4 10% until pH 1. The mixture was then allowed to produce lignin precipitation for 8 hours. In addition, 0.1 M of dried lignin was sulfonated by $NaHSO_3$ solution (0.3 M) using reflux method for 4 hours at the temperature of 97 °C. The sulfonation product was then dried in the oven to produce brown powder of SLS surfactant.

2.3 Characterization

The SLS surfactant was characterized by Fourier Transform Infrared spectroscopy using KBr pellet technique (FTIR, Shimadzu Prestige-21). Furthermore, the performance of all surfactant in reducing interfacial tension of crude oil and formation water were analyzed by using compatibility test, phase behaviour analysis, and interfacial tension measurement (IFT meter TX500D).

IOP Conf. Series: Materials Science and Engineering **349** (2018) 012005 doi:10.1088/1757-899X/349/1/012005

3 Result and Discussion

3.1 FTIR Analysis

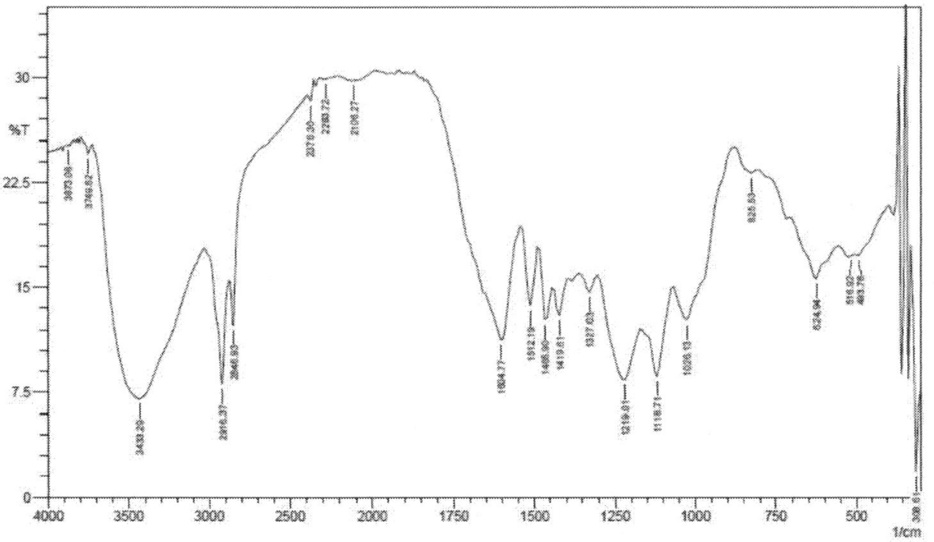

Figure 1. The FTIR spectra of SLS surfactant.

FTIR spectroscopy was used to understand the substitution of hydroxyl group (OH) by sulphonate group (-SO_3) in the benzylic carbon of lignin structure. As can be seen in Figure 1, this substitution can be observed by the presence of FTIR absorption peak at 1465 cm^{-1} corresponds to SO_3 bending vibration. In addition, it is also can be identified by the band at 1118, 1026, and 826 cm^{-1} which are resulted from S=O asymmetric stretch, S-O stretch (1), and S-O stretch (2), respectively, indicates that this compound is sodium lignosulfonate.

3.2 Compatibility Test

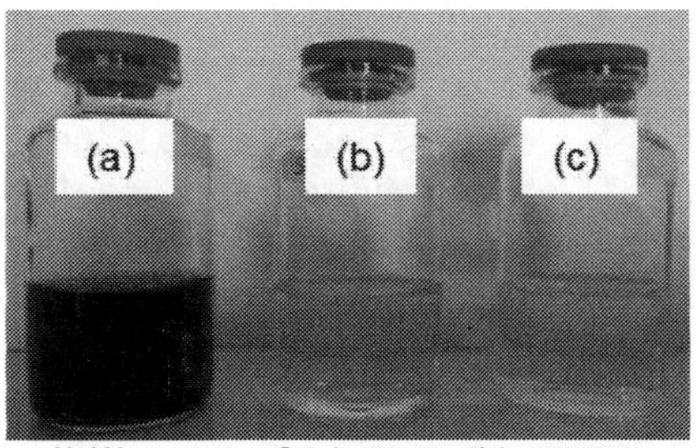

Figure 2. The compatibility test of (a) SLS, (b) SDBS, and (c) SpTS surfactant.

This test was performed to observe the compatibility of surfactant and formation water in a reservoir. It is expected that the formed mixture is a perfect solution or colloid. The suspension is not allowed due to it can be a barrier when the surfactant solution injected into the rock.[8] As can be seen in Figure 2, the SDBS and SLS surfactants produced a perfect solution, while SpTS surfactant generates a fine solution with small amount of un-soluble surfactant residue. Hence, it can be identified that SLS, SDBS, and SpTS surfactants were compatible with formation water.

3.3 Phase Behaviour Analysis

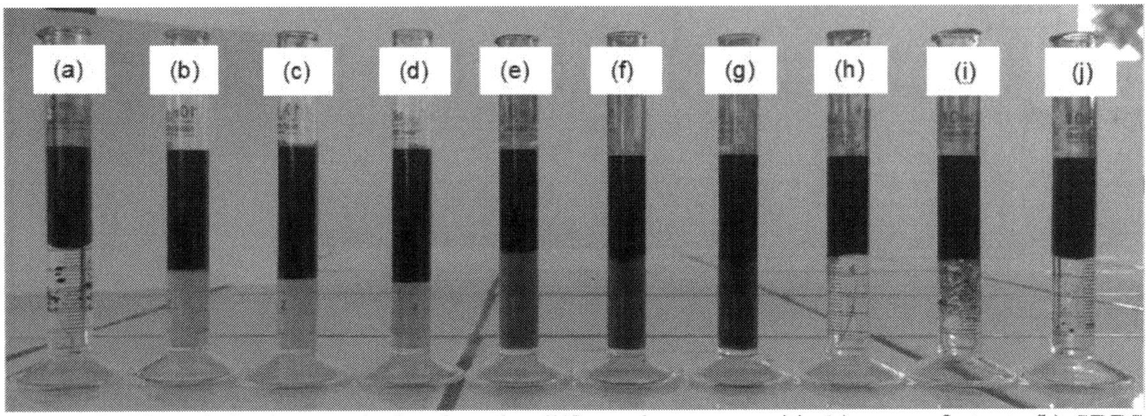

Figure 3. Phase behaviour analysis of crude oil/formation water with (a) no surfactant, (b) SDBS 0.5%, (c) SBDS 0.75%, (d) SDBS 1.0 %, (e) SLS 0.5%, (f) SLS 0.75%, (g) SLS 1.0%, (h) SpTS 0.5%, (i) SpTS 0.75%, and (j) SpTS 1.0%.

The phase behaviour analysis of SLS, SDBS, and SpTS in the crude oil and formation water colloidal system were shown in Figure 3. The concentration of SLS, SDBS, and SpTS surfactants were varied in 0.5; 0.75; and 1.0%. It is known that SLS surfactant was better than SpTS surfactant, while SDBS surfactant generates the highest performance proved by the best microemulsion formation. Furthermore, the optimum concentration of SLS, SDBS, and SpTS surfactants is 1.0%.

3.4 Interfacial Tension (IFT) Measurement

The interfacial tension (IFT) measurement is an important method to identify the effect of surfactant in the reducing interfacial tension of crude oil and formation water colloidal system. In this respect, the presences of surfactants have a vital role to reduce the interfacial tension between crude oil and formation water. The surfactant concentration used in this measurement was 1%. As can be seen in Table 1, the result shows that the performance of SLS surfactant was better than SpTS surfactant with the value of 1.67 and 3.59 mN/m, respectively. Meanwhile, SDBS surfactant shows the highest performance with the IFT value of 0.47 mN/m.

Table 1. Interfacial Tension Measurement

No.	Surfactant Name	Surfactant Concentration (%)	Surfactant Density (g/cm^3)	IFT (mN/m)
1	SLS	1	1.0076	1.67
2	SDBS	1	1.0075	0.47
3	SpTS	1	1.0005	3.59

4 Conclusion

In conclusion, SLS surfactant was successfully synthesized by applying batch method using lignin from oil palm empty fruit bunches as a precursor. Its performance in reducing interfacial tension of crude oil and formation water colloidal system was analyzed by using compatibility test, phase behaviour analysis, and interfacial tension measurement, which then compared with two commercial surfactants called as Sodium Dodecyl Benzene Sulfonate (SDBS) and Sodium p-Toluene Sulfonate (SpTS). It is observed that the ability of that surfactants in reducing interfacial tension of crude oil and formation water follows the order of SDBS > SLS > SpTS.

5 Acknowledgments

The authors thank the Direktorat Penelitian dan Pengabdian Masyarakat (DPPM), Universitas Islam Indonesia, for the financial support in this research project.

6 References

[1] Sheng, J.J., (2015). Status of surfactant EOR technology. *Petroleum*, *1*(2), pp.97-105.

[2] Kamal, M.S., Sultan, A.S. and Hussein, I.A., (2015). Screening of amphoteric and anionic surfactants for cEOR applications using a novel approach.*Colloids and Surfaces A: Physicochemical and Engineering Aspects*, *476*, pp.17-23.

[3] Tadros, T. F. (2006). *Applied Surfactants: Principles and Applications*. Berkshire: John Wiley & Sons.

[4] Prakoso, N.I., Purwono, S., and Rochmadi, Study on Lignin Isolation from Oil Palm Empty Fruit Bunches, *Eksakta: Jurnal-jurnal Ilmu MIPA*, Yogyakarta, 2016, e. ISSN: 2503-2364 Vol. 16, pp. 45-54

[5] Prakoso, N. I., Purwono, S., & Rochmadi. (2017). Synthesis of sodium lignosulphonate from oil palm empty fruit bunches' lignin. In *AIP Conference Proceedings* (Vol. 1823, No. 1, p. 020037). AIP Publishing.

[6] Mohamed, A., Anas, A.K., Bakar, S.A., Ardyani, T., Zin, W.M.W., Ibrahim, S., Sagisaka, M., Brown, P. and Eastoe, J., (2015). Enhanced dispersion of multiwall carbon nanotubes in natural rubber latex nanocomposites by surfactants bearing phenyl groups. *Journal of colloid and interface science*, *455*, pp.179-187.

[7] Mi-Na, Zhang., Xue-Pin, L. and Bi, S., (2006). Adsorption of surfactants on chromium leather waste. *Journal-Society of Leather Technologists and Chemists*, *90*(1), p.1.

[8] Eni, H., Syahrial, E., dan Sugihardjo. (2010). Screening Test dan Karakterisasi Surfaktan yang Efektif untuk Injeksi Kimia. *Lembaran Publikasi Lemigas*, Vol 44, No 22, page108 – 116

Radical Scavenging Activity from Ethanolic Extract of Malvaceae Family's Flowers

A N Artanti, N Rahmadanny, F Prihapsara

Department of Pharmacy, Faculty of Mathematics and Natural Sciences, Universitas Sebelas Maret, Indonesia

*E-mail : anif.apt@staff.uns.ac.id

Abstract Sea hibiscus flower (*Hibiscus tiliaceus* L.), shoe flower (*Hibiscus rosa-sinensis* L.), and turk's cap flower (*Malvaviscus arboreus* Cav.) are a plant that belongs to the same family, Malvaceae. There are expected contain of anthocyanins as active compound. Several studied shows that some flowers could protect human body from free radical danger exposure. This study has been done to examine ethanolic extract from malvaceae family's which has potency as radical scavenger. Antiradical activity assay was determined by DPPH method with IC_{50} value as parameter. Based on the study the malvaceae family's flower was contain of tannins, polyphenols, saponin, and anthocyanine. The radical scavenging activity respectively from the lowest to the higest activity are vitamin c (4,05 ppm ± 0,094), Turk's cap flower (6,80 ppm ± 0,22), shoe flower (14,62 ppm ± 0,104) and sea hibiscus flower (38,8 ppm ± 0,086). The three of the extract was having strong antioxidant activity.

1. Introduction

Free radicals in the body basically play a role in health maintenance because of its reactive nature to bind or react with foreign molecules that enter the body, for example to kill bacteria that enter the body. But if free radicals in large amounts, can cause disruption of the metabolic system, this is due to the nature of free radicals that can attack lipids, DNA, protein components of cells and tissues. Therefore, its existence must be controlled by the antioxidant system [1]. *Hibiscus rosa-sinensis* L. is widely grown as an ornamental plant in tropical and subtropical area. This plant has large flowers and is red. The leaves, flower and root contain flavonoids. The leaves contain saponins and polyphenols, as well as taraksyl acetate. Flowers contain polyphenols, cyanidin diglucosides, hibisetin, bitter and mucilage substances, vitamins, thiamin, riboflavin and ascorbic acid as well as alkaloids and saponins. While the root contains tannins and saponins [2]. The sea hibiscus flower (*Hibiscus tiliaceus* L.) is known to have many uses as a reliever of fever, hair grower, cough medicine and bloody or slimy diarrhea drug [3]. *Malvaviscus arboreus Cav* is a plant belonging to the family of Malvaceae. It is thought to have potential as an antioxidant agent, cytotoxic, and thrombolytic, with a total fenolic compound on leaves value of 20.06 ± 0.87, and an LC_{50} value of 3.82 ± 0.08 [4]. The success of this study is expected to provide an overview of the selection of plants that have antioxidant activity correlated with anticancer activity. Untill now, the result for cancer treatment is still unsatisfactory. Therefore, the study of cancer drug discoverynis still intensively conducted. One of the medicinal plants that become the object of study is the flower of Malvaceae family.

2. Method

2.1 Chemicals and reagents
The sample of *Hibiscus rosa-sinensis* L., *Hibiscus tiliaceus* L., and *Malvaviscus arboreus* Cav. was collected from Balai Besar Penelitian dan Pengembangan Tanaman Obat dan Obat Tradisional (B2P2TOOT), Tawangmangu, Karanganyar, Central Java, Indonesia DPPH (2,2-diphenyl-1-picrylhydrazyl) (Sigma), etanol p.a (Emsure), technical etanol, aqua bidestilata (Merck), Thin Layer Chromatography (TLC) plates Silica GF_{254} (Sigma).

2.2 Sample Preparation
The sample was extracted using 70% ethanol solvent. The maseration is carried out for 3 days to obtain the filtrate which extracts the active ingredient effectively. The obtained filtrate was then evaporated with a rotary evaporator at a temperature between 55-60 °C.

2.3 Identify phytochemicals
Chemical identification test in this research was done by using thin layer chromatography (TLC). The purpose of this identification test is to determine whether there are secondary metabolite compounds including tannins, polyphenols, saponins, and anthocyanins. For the stationary phase used is silica gel GF_{254}, while for the mobile phase, adjusted to the degree of polarity of the extract. The tannin test used the mobile phase of n-butanol: glacial acetic acid: water (14: 1: 5 v/v/v) [5], polyphenol test used glacial acetic acid: acetone: water (40: 20: 4 v/v/v) [6], while anthocyanin test used a mobile phase of n-butanol: glacial acetic acid: water (4: 5:15 v/v/v) [7].

2.4 Antioxidant Activity Assay (DPPH method)
The free radical scavenging method was performed based on Kikuzaki *et al* (2002) [8]. In this assay, sample from the stock solution of ethanolic extract from *Hibiscus rosa-sinensis L., Hibiscus tiliaceus L.,* and *Malvaviscus arboreus Cav.* were prepared in five different concentration (15,30,60,90 and 120 μM). Each of sample added by 0.7 mL DPPH 0.4 mM and ethanol added up to 5.0 mL. This mixture was homogenized by mixing for 30 second and incubated for 30 minutes. The sample absorbance were measured by UV-Vis spectrofotometer (Shimadzu) with λ_{max} 515 nm. The sample absorbance also compared with the control solution containing 0.7 mL DPPH 0.4 mM diluted in ethanol. The radical scavenging activity was calculated using the following formula :

$$\textbf{Radical scavenging activity (\%)} = \frac{Ac-As}{Ac} \times 100\ \%$$

Where Ac is absorbance of control (DPPH free radical without the addition of the sample solution), As is sample absorbance (absorbance of DPPH free radical after the addition of sample solution). The percentage (%) of antiradical activity were measured by linier regretion between concentration curve versus antiradical activity precentage were obtained. Then, the linier regretion formula and sample concentration at 50% activity were determined.

2.5 Cytotoxic Activity
$1x10^4$ T47D cells/well were grown in 96-well plate before being exposured to drug treatment. For cell viability assay, cells were treated for 24 hours with increasing concentration of extract. At 100 μg/ml of tetrazolium salt MTT [3-(4,5-dimetiltiazol-2-yl)-2.5-diphenyl terazolium bromide] solution (0.5 mg/ml in Phosphat Buffer Saline) was added to each well continued with incubation for 3 hours at 37°C. The reaction was stopped by dilution with 10% (w/v) Sodium Dodecyl Sulphate in 0.01 N HCl, and cells were incubated overnight. The absorbance was determined by using ELISA reader at λ 595 nm.

3. Result and Discussion

3.1 TLC Analysis

The extraction with 800 grams of drying flowers of *Hibiscus rosa-sinensis* L., *Hibiscus tiliaceus* L., and *Malvaviscus arboreus* Cav.were gave viscous extract of 35.23 gram (7.046), 18.07 gram (3,614%) and 14, 06 gram (2,812%)respectively. TLC analysis showed result was obtained based on Table 1.

Table I. The comparison of the Rf value of tannin, plyphenol and antocyanin from the extract

Extract	Tannin (Rf)	Polyphenol (Rf)	Antocyanin (Rf)
Hibiscus rosa-sinensis	0.5	0.74	0.25
Hibiscus tiliaceus	0.66	0.83	0.16
Malvaviscus arboreus Cav.	0.63	0.82	0.47
Standard	0.53-0.8	0.82	0.1-0.4

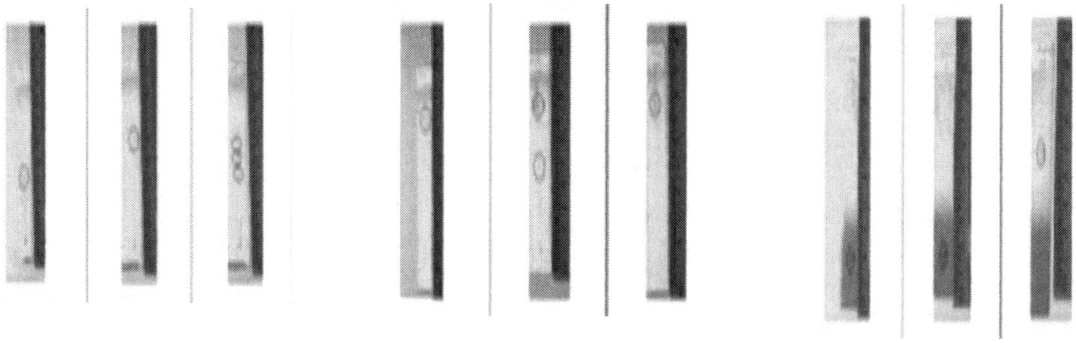

(A) (B) (C)

Figure 1. The result of Qualitative Phytochemistry Assay by TLC showed that *Hibiscus rosa-sinensis* L., *Hibiscus tiliaceus* L., and *Malvaviscus arboreus* Cav. contain tannin (A) with Rf value are 0.5 ; 0.66; 0.63 respectively after eluated with mobile phase of n-butanol: glacial acetic acid: water (14: 1: 5 v/v/v), polyfenol (B) with Rf value are 0.74; 0.83 and 0.82 respectively after eluated with mobile phase glacial acetic acid: acetone: water (40: 20: 4 v/v/v) and antosianin (C) with Rf value are 0.25; 0.16; and 0.47 respectively after eluated with mobile phase n-butanol: glacial acetic acid: water (4: 5:15 v/v/v)

3.2 Radical scavenging activity

Antioxidant Activity Test of *Hibiscus rosa-sinensis* L., *Hibiscus tiliaceus* L., and *Malvaviscus arboreus* Cav. determined by the ability of a compound contained in extract to reduce the purple color intensity of DPPH radical in its maximum wavelenght. The reduction of purple colour intensity of DPPH radical is caused by the decrease of chromophore or conjugated double bond in DPPH compound. It caused by potency of extract compound which scavenging the radical by donating hydrogen atom to DPPH structure so that become reducted DPPH-H. DPPH-H is a compound which have yellow colour [9]. The principle reaction as bellow (Figure 2.) Antioxidant activity divide into 4 categories i.e active antioxidant if its IC_{50} value is lower than 50 µg/ml, medium antioxidant if its IC_{50} value between 50-100 µg/ml, less active in antioxidant activity if its IC_{50} value between 100-200 µg/ml and if its IC_{50} value is more than 200 µg/ml showed that the compound is not active as antioxidant [10].

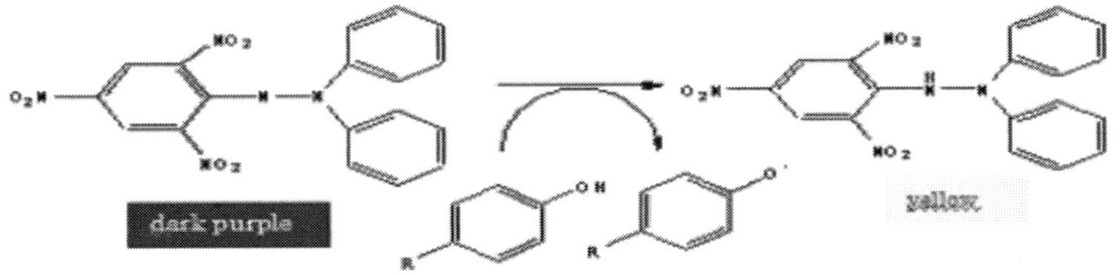

Figure 2. The principle reaction of DPPH (Molyneux, 2004) [11]

Table II. The comparison of the antiradical activity of extract

Extract	Concentration (µg/mL)	Antiradical activity (%)	IC_{50} (µg/mL)
Hibiscus rosa-sinensis	15	50,72	
	30	53,77	
	60	64,98	14,62
	90	73,84	
	120	83,25	
Hibiscus tiliaceus	15	35,27	
	30	46,67	
	60	62,85	38,89
	90	78,01	
	120	83,71	
Malvaviscus arboreus Cav.	15	47,54	
	30	61,39	10,45
	60	69,31	
	90	80,65	
	120	83,68	

The parameter used in the interpretation of radical activity is the value of inhibitory concentration 50% (IC_{50}) were obtained by linier regretion between the concentration versus the precentage of antiradical activity [12]. The radical scavenging activity respectively from the lowest to the higest activity are *Malvaviscus arboreus* Cav (10,45 µg/mL) ; *Hibiscus rosa-sinensis* L.(14,62 µg/mL) and *Hibiscus tiliaceus* L. (38,89 µg/mL) (Fig.2). With this IC50 value, all of extracts can be classified into active antioxidant [10] which states that crude extract with IC_{50} less than 50 µg/mL

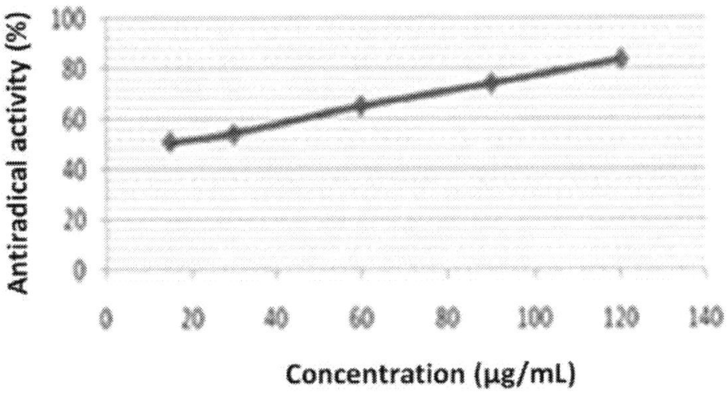

(A)

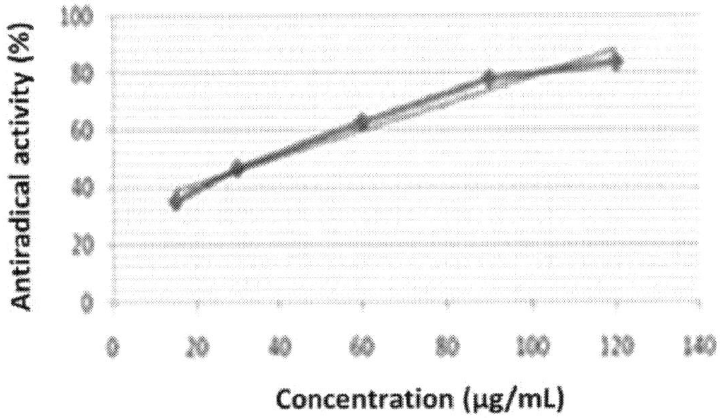

(B)

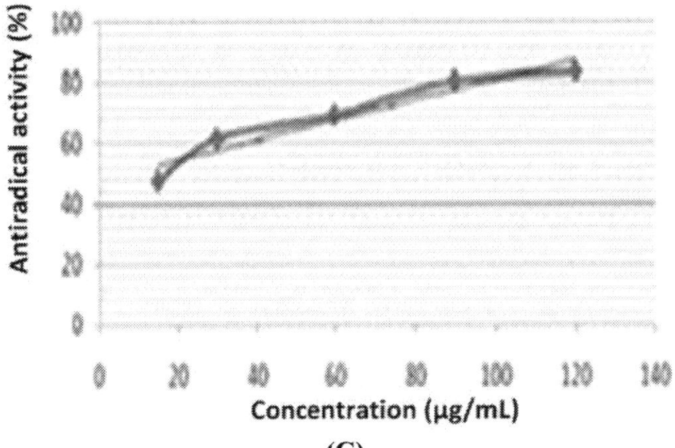

(C)

Figure 3. The IC_{50} curve of ethanolic extract of *Hibiscus rosa-sinensis* L.(A), *Hibiscus tiliaceus* L. (B), and *Malvaviscus arboreus* Cav. (C). The sample absorbance were measured by UV-Vis spectrofotometer with μ_{max} 515 nm.

3.3 Cytotoxic activity

Cytotoxic activity was tested by in vitro to determine the potential of a compound such as cytotoxic anticancer drugs. The parameter used in the interpretation of expressed in IC_{50} value that can inhibit cell proliferation by 50% of the population. This study using MTT assay to determine the cytotoxic potential of *Hibiscus rosa-sinensis L., Hibiscus tiliaceus L.,* and *Malvaviscus arboreus Cav.*The method is used to measuring the color intensity that occur as a result of reaction a substrate by living cells into colored product. The reaction of tetrazolium salt MTT [3-(4,5-dimetiltiazol-2-yl)-2,5-diphenyl tetrazolium bromide] reduced to formazan by the succinate tetrazolium reductase system, which is included in the mithocondria of living cells. Formazan intensity was measured using ELISA reader at a wavelength of 595 nm. Color intensity is proportional to the number of living cells. Sample extract was supplied with various concentration from 10 µg/ml to 150 µg/ml. Cell morphology was observed after treatment with extract for 24 hours, thus it being treated with MTT reagent. Treated cells showed cytotoxic effect with change of cell morphology and decreased cell viability. Viable cells had epithelial shape, but after being treated with certain concentration of samples they gave spherical-shape and shrunk cell wall. By using ELISA reader, absorbance value were obtained. IC_{50} was calculated using the linier regression between extract concentration and precentage of living cell. The IC_{50} value respectively are *Malvaviscus arboreus Cav* (152,45 µg/mL) ; *Hibiscus rosa-sinensis L.*(173,65 µg/mL) and *Hibiscus tiliaceus L.* (144,76 µg/mL) (Fig 3). With this IC_{50} value, the extract can be classified into a potent chemotherapeutic agent which states that crude extract with IC_{50} less than 100 µg/mL of highly potent as an anticancer agent [13].

Table III. The comparison of the cytotoxic activity of extract

Extract	**Concentration (µg/mL)**	**Cell viability (%)**	**IC_{50} (µg/mL)**
Hibiscus rosa-sinensis	10	91,753	
	50	88,781	
	110	64,706	173,65
	130	63,008	
	150	56,034	
Hibiscus tiliaceus	10	97,5	
	50	75,2	
	110	65,2	144,76
	130	50,4	
	150	46,2	
Malvaviscus arboreus Cav.	10	101,58	
	30	95,88	
	50	84,78	
	70	84,72	152,45
	90	90,05	
	110	86,11	
	130	75,99	
	150	53,92	

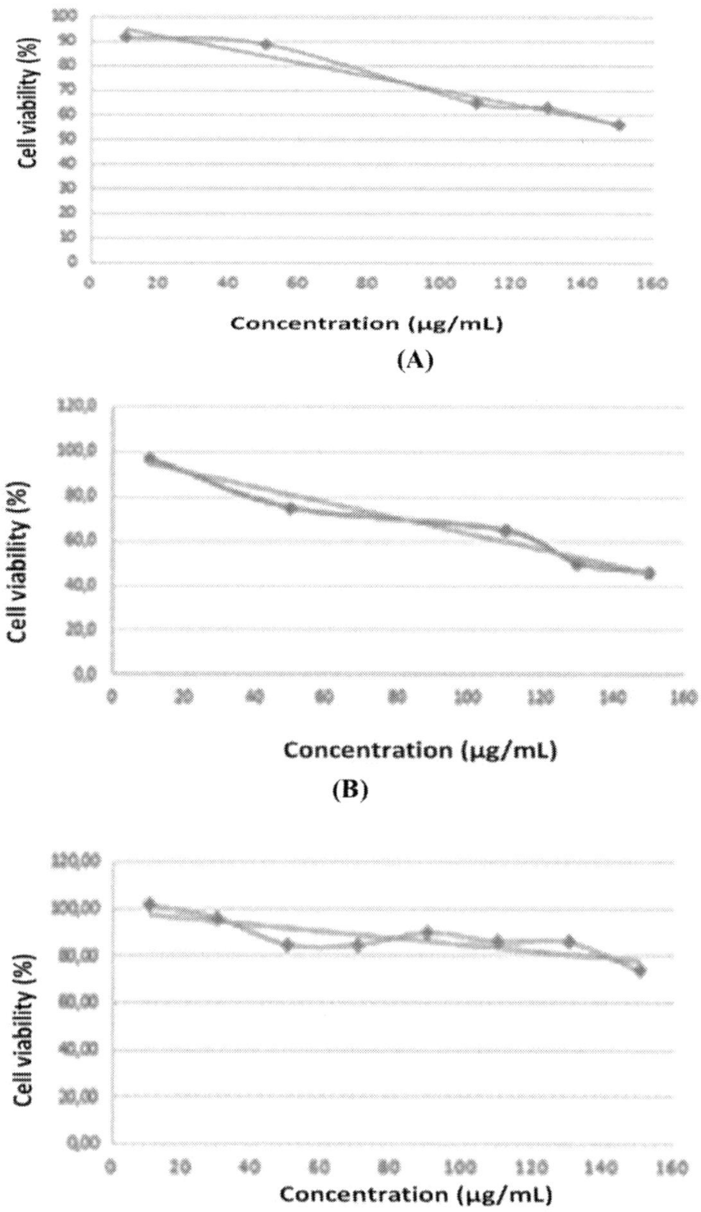

(C)

Figure 4. The effect of ethanolic extract of *Hibiscus rosa-sinensis* L.(A), *Hibiscus tiliaceus* L. (B), and *Malvaviscus arboreus* Cav. (C) on the viability cells. Graph showed the treatment effect of various concentration of extract on T47d breast cancer cell viability.

Based on IC_{50} value from antioxidant activity and IC_{50} value from cytotoxic activity it indicated that the antioxidant might somehow contribute to growth inhibition in T47d cells. But further studies on the role of antioxidants growth inhibitory effect on T47d cell line is needed to be done. It also important to explore the mechanistic function between antioxidant and anticancer activity of flowers.

Conclusion

Antiradical activity of *Malvaviscus arboreus* Cav.is known to have highest activity than the other malvaceae family's flowers. Its IC_{50} value of *Malvaviscus arboreus* Cav. is 10,45 μg/mL. Also the cytotoxic activity of *Malvaviscus arboreus* Cav. is 152,45 μg/mL. Based on Husna *et al* (2013) [14] reported that antosianin compound in *Malvaviscus arboreus* Cav. is the compound responsible to the antiradical and cytotoxic activity, meanwhile the main compound fron the extracts used in this study that is responsible in the antiradical and cytotoxic activity are not been known yet.

References

[1] Limbono, S. 2013 Antioxidant activity of Ethanolic Extract from *Canarium indicum* L. with *1,1-Diphenyl-2- picrylhidrazyl (DPPH)* method, *Calyptra*, Universitas Surabaya, **2**(2)

[2] Syamsuhidayat, S.S and J.R Hutapea, 1991, *Indonesia Medicinal Plant Inventory. 1st Edition.* Jakarta : Agency for Research and Development of Health Department

[3] Heyne K, 1987, Useful Plants from Indonesia 3rd Edition. Jakarta : Agency for Forestry Research

[4] Yeasmin, Zokhroof, Sharif Tanvir, Tasnuva Sharmin, Ridwan Bin Rashid, Md. Al Amin Sikder., 2014, Bioactivities of *Malvaviscus arboreus* var *drumomondii* and *Phyllanthus reticulatus* Poir. *Dhaka Univ J. Pharm. Sci.* **13**(2): 143-147

[5] Kamilah H., Fasyah A., Sa'adah L. 2010. Fractination and identification of tannin compound from Avverhoa bilimbi L. leaves. J.chem 4(2) :193-200

[6] Sulistyani. 2011. Antifungi activity of ethanolic extract from Anredera cordifolia (Tenore) Steen and chemical identification compound. Jurnal Ilmiah Kefarmasian, 1 (2): 51 – 62.

[7] Windriyati, Y.N, Budiarti A, Syahida I.A. (2011). Mucolitic activity in vitro study of Piper crocotum and identification of chemical compound. *Procceding*. National Conference of Herbal Contribution on Degenertive Disease, Universitas Wahid Hasyim.

[8] Kikuzaki, H., Hisamoto, M., Hirose, K., Akiyama, K. and Taniguchi, H. 2002. Antioxidant Isolated from Leaf Wax of Eucalyptus Leaves. *Journal of Agricultural and Biology Chemistry* .45: 735-739

[9] Huang, D., Ou,B., and Prior, R.L, 2005, The chemistry behin Antioxidant Capacity Assays, J. Agr.Food Chem., 53, 1841-1856

[10] Reynertson, K.A., 2007, Phytochemical Analysis of Bioactive Constituens from Edible Myrtaceae Fruit, *Dissertation*, The City University of New York, New York.

[11] Molyneux, P., 2004, The Use of the stable free radical diphenylpicrylhydrazyl (DPPH) for estimating antioxidant activity, Songklanakarin *J. Sci. Technol.*, 26 (2) : 211-219

[12] Rohman, A., and Riyanto, S., 2006, Aktivitas Antiradikal Bebas Ekstrak Kloroform Buah Mengkudu dan Fraksi-fraksinya, Artocarpus, 6 (1), 39

[13] Boik, J, 2001, Natural compound in cancer therapy : Oregon Medical Press.

[14] Husna E N, Novita M, Rohaya S, 2013, Antosianin Content and Activity Sweet Potatoes Purple Fresh and Processed Products, Agritech, Vol 33 (3).

IOP Conf. Series: Materials Science and Engineering **349** (2018) 012007 doi:10.1088/1757-899X/349/1/012007

Isolation and Identification of Active Compounds from Papaya Plants and Activities as Antimicrobial

A T Prasetya[1], S Mursiti[2], S Maryan[3], N K Jati[3]

[1]Doctoral Degree of Postgraduate, Semarang State University, Indonesia, Kampus Kelud-Petompon-Gajah Mungkur Semarang Indonesia 50237
[2]Chemistry Department, Faculty of Mathematics and Natural Science, Semarang State University, Indonesia, Sekaran Gunungpati Semarang Indonesia 50229
[3]Chemistry Student, Chemistry Department, Faculty of Mathematics and Natural Science, Semarang State University, Indonesia, Sekaran Gunungpati Semarang Indonesia 50229

agungchem@mail.unnes.ac.id

Abstract. Extraction and isolation of papaya seeds and leaves (*Carica papaya L*) has been performed using *n*-hexane and ethanol solvents. Further isolation of the extract obtained using ethyl acetate and diethyl ether solvents. The result of the phytochemical test of papaya extract obtained by mixture of an active compound of flavonoids, alkaloids, tannins, steroids, and saponins. Ethyl acetate isolates containing only flavonoids and diethyl ether isolates contain only alkaloids. Extracts and isolates from papaya plants had gram-positive antibacterial activity greater than the gram-negative bacteria, but both did not have antifungal activity. Papaya extracts have greater antibacterial activity than flavonoid isolates and alkaloid isolates. Strong antibacterial inhibitory sequences are extracts of papaya plants, flavonoid isolates, and alkaloid isolates.

1. Introduction

Indonesia is one of the countries that have biodiversity, but few existing plants have been utilized in the field of medicine. Most of these plants are utilized as traditional medicinal ingredients that have proven efficacy derived from generation to generation [1]. There is very little scientific literature that examines the chemical components and composition of the medicinal plants. One of the most widely used plants in the field of traditional medicine namely papaya (*Carica papaya L*), especially on the seeds and leaves [2]. Traditionally papaya plants have been used as an anthelmintic, digestive disorders, diarrhea, skin diseases, male contraceptives, and raw materials for colds [3].

Artanti *et al*., [4] state that a number of flavonoid-containing medicinal plants have been reported to have antioxidant, antibacterial, antiviral, anti-inflammatory, antiallergic and anticancer activity. According to Sudibyo [5], part of the papaya plant has bioactive and pharmaco-active properties such as secondary metabolite compounds which are often used in the pharmaceutical field. In papaya leaf extract contained *papain* enzyme that has proteolytic and antimicrobial activity, and alkaloids as antibacterial, flavonoids as an anti-inflammatory [6], [7]. In addition, papaya plants also contain several active antioxidant components [8]. Based on research conducted Delphin *et al*., [9] papaya seed extract contains various kinds of secondary metabolite compounds such as alkaloids, flavonoids, saponins, and tannins.

Infectious diseases are the most common types of diseases suffered by the population of developing countries, including Indonesia [10]. Infections are caused by pathogenic microbes such as bacteria, viruses, fungi, protozoans, or some other minority group (microplasma, rickets, and chlamydia) [11]. Pathogenic microbes, such as *Escherichia coli*, can cause acute diarrhea, as well as a major cause of urinary tract infections. Diseases caused by these infections are usually treated with antibiotics [12]. The content of active compounds in plants is usually in small quantities, so it needs to be isolated and concentrated with various methods, one of the extraction. In order for the desired active compound to be extracted, the choice of solvent used is necessary. Selection of suitable solvents is an important factor in the extraction process. The solvent used is a solvent which can filter out most of the secondary metabolites present in the simplified. The most commonly used solvents are alcohols such as methanol and ethanol [13]–[18]. This study aims to determine the active compounds contained in seeds and leaves of papaya and to test its effectiveness as an antimicrobial.

2. Method

Instruments used in this study are rotary vacuum evaporator (*Heidolph*), *Fourier Transform Infrared Spectrometer* (FT-IR from *Perkin Elmer Frontier Series*), and chemicals with grade pro-analyst made by *Merck*. The bacteria cultures *Escherichia coli*, *Staphylococcus aureus*, *Bacillus subtilis* and *Candida albicans* fungi obtained from RSUD Dr. Kariadi Semarang city.

Making Simplicia is done by drying the plant part (seeds and leaves) and then mashed. Mixed Simplicia in maceration using *n*-hexane solvent for 3x24 hours. The obtained residue is dried and macerated back with ethanol solvent. The obtained filtrate was concentrated using a rotary vacuum evaporator. Some of the extracts obtained were isolated using ethyl acetate and diethyl ether solvents to concentrate and isolate further active compounds [3], [6]. Extracts and isolates obtained by phytochemical screening included alkaloid, flavonoid, saponin, tannin, steroid, and terpenoid test, and further identification using FT-IR spectrometer. The extracts and the resulting isolates were then tested for antimicrobials using paper disc method with a long incubation of 1x24 hours.

3. Result and Discussion

The research begins with the collection of seeds and leaves of papaya from Sekaran, Gunungpati, Semarang. Once identified, the samples obtained are then processed into a powder of Simplicia. The next step is extraction using *n*-hexane solvent to remove nonpolar compound components such as oil and fat. The extraction followed by ethanol solvent to take polar compounds such as flavonoids, alkaloids, and others. Some of the extracts obtained are then isolated using ethyl acetate and diethyl ether solvents to extract and concentrate desired polar compounds.

The extract of ethanol (seeds and leaves of papaya) obtained tested its chemical content qualitatively. This test is treated to compare the results obtained with the literature search results, as well as to compare with the results of isolation. Based on the literature, papaya leaf ethanol extract contains compounds such as flavonoids, alkaloids, tannins, saponins [9]. The results of phytochemical tests on extracts and isolates as listed in Table 1.

Table 1. Phytochemical test results

No.	Class test	Papaya seeds		Papaya leaf	
		Extract	Isolate	Extract	Isolate
1.	Flavonoid	+	+	+	-
2.	Alkaloid	+	-	+	+
3.	Tanin	+	-	+	-
4.	Saponin	-	-	+	-
5.	Steroids	+	-	+	-
6.	Terpenoid	+	-	-	-

Table 1 shows that the active compounds of both papaya seed extract and papaya leaf extract are almost the same, which contains flavonoids, alkaloids, tannins, and steroids. The difference is, papaya seed extract does not contain saponins and papaya leaf extract does not contain terpenoid. It differs precisely in the further isolation results, isolation using ethyl acetate solvent will separate the active compound flavonoids (papaya seed isolates), while isolation with diethyl ether solvent will separate the alkaloid active compound (papaya leaf isolate). In this stage, the choice of solvent type becomes very important depending on what active compounds will be taken. After phytochemical testing, the next step is to identify functional groups using FT-IR instruments to strengthen the results obtained from the phytochemical test. More results as shown in Figure 1.

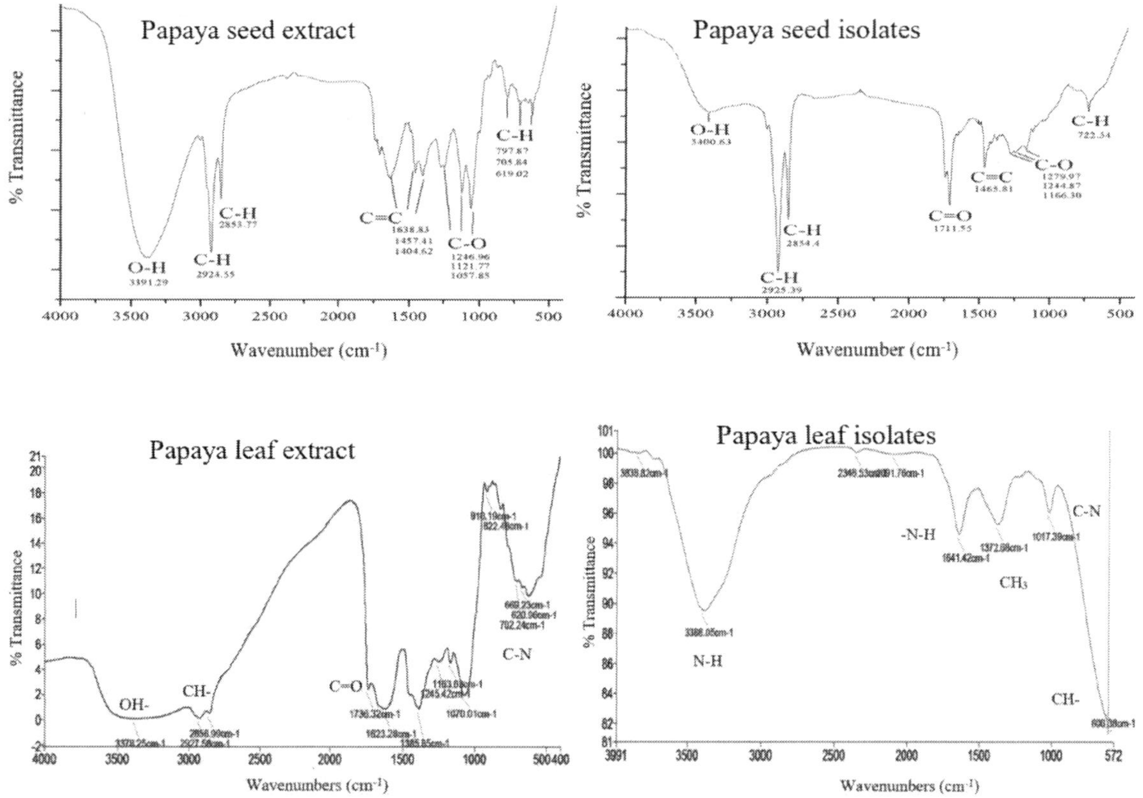

Figure 1. Infrared spectra of extract and isolate from seed and papaya leaf

The result of FT-IR analysis of papaya seed extract showed the presence of -OH group, ie the absorption of widening at wavenumber 3991.29 cm^{-1}. The presence of an aromatic C=C group is indicated by absorption at wavenumbers 1457.41 and 1404.62 cm^{-1}. The presence of vibration C-O is indicated by absorption in the area 1246.96, 1121.77 and 1057.85 cm^{-1}. The presence of an aromatic C-H group out of the field is indicated by the absorption of the fingerprint region in the wavenumber 797.87, 705.84 and 619.02 cm^{-1}. The results of FT-IR analysis of papaya seed isolate showed widespread uptake at 3400.63 cm^{-1} with weak intensity, presumably an uptake of O-H. The sharp aliphatic C-H stretch absorption appears at 2925.39 and 2854.4 cm^{-1} waves. The presence of carbonyl or ketone group (C=O) as one of the common characteristics of flavonoid group compounds is indicated by the absorption of wavenumbers 1711.55 cm^{-1}. The aromatic C=C uptake occurs at the wavenumber 1465.81 cm^{-1}, the vibration of C-O in the wavenumbers 1279.97, 1244.87 and 1166.30 cm^{-1}. Absorption at the wavenumber 722.34 cm^{-1} the presence of aromatic C-H groups out the plane. So it can be concluded that the functional groups contained in papaya seed isolates are the O-H, C=O, C-O, C=C aromatic, C-H aliphatic and C-H aromatic groups.

The results of leaf extract showed absorption at wavenumber 3378.25 cm^{-1} from -OH. The absorption at wavenumbers 2927.58 and 2856.99 cm^{-1} are -CH aliphatic. The absorption at wavenumber 1736.32 cm^{-1} shows C=O. The C=C bond is indicated by absorption at the wavenumber 1623.28 cm^{-1}. The absorption at the waveform region 1385.85 cm^{-1} shows $-CH_3$. Cluster -CO- is indicated by absorption at wavenumber 1245.42 cm^{-1}. The absorption at wavenumber 1070.01 cm^{-1} shows C-N. So from the infrared spectra data, it shows that papaya leaf extract has functional groups -OH, -CH aliphatic, C-O, C=C, $-CH_3$, -CO-, C-N. Infrared spectra data of papaya leaf isolates may contain several functional groups such as N-O recombination at wavenumber of 3388.05 cm^{-1}, this uptake is supported by the presence of absorption at the 1641.42 cm^{-1} wavenumbers indicating the presence of a bending and at 1017.39 cm^{-1} wavenumbers indicating the presence of CN groups supported also by uptake at wavenumber 1372.68 cm^{-1} showing $-CH_3$ and at wave numbers 606.38 cm^{-1} showing -CH.

After extracts and isolates from papaya seeds and leaves were tested for phytochemicals and their functional groups, the next step was tested for antimicrobial ability, the results are summarized in Table 2.

Table 2. Results of observation of antimicrobial inhibition zone

Antimicrobial source		Type of microbes	Inhibitory zone (mm)		
			Sample	Control (-)	Control (+)
Papaya seeds	Extract	*B. subtilis*	15.9	2.3	10.5
		E. coli	14.3	0.8	9.0
		C. albicans	Resistant	2.4	14.3
	Isolate (Flavonoids)	*B. subtilis*	14.2	2.3	10.5
		E. coli	13.8	0.8	9.0
		C. albicans	Resistant	2.4	14.3
Papaya leaf	Extract	*S. aureus*	14.3	2.1	12.1
		E. coli	11.9	0.8	9.0
	Isolate (Alkaloid)	*S. aureus*	12.6	2.1	12.1
		E. coli	9.7	0.8	9.0

The results of the antibacterial test against *B. subtilis*, *E. coli*, and antifungal against *C. albicans* showed that papaya seed extract was more effective to inhibit the growth of *E. coli* and *B. subtilis* bacteria than the flavonoid isolate. Papaya leaf extract is also more effective in inhibiting the growth of *E.coli* and *S.aureus* bacteria than its alkaloid isolates. The results showed that papaya extract and isolate were more effective in inhibiting the growth of gram-positive bacteria (*B. subtilis* and *S. aureus*) compared to gram-negative bacteria (*E. coli*). The existence of different types of samples in the form of extracts and isolates and types of bacteria suspected to affect antimicrobial activity. In different bacteria produce different inhibitory power. According to Edziri, *et al.*, [17] gram-positive bacteria have a single-layer cell wall structure and are composed of peptidoglycan and low-grade lipids (1-4%) so that they are easily penetrated by the flavonoid or alkaloid compounds present in the extract/isolate. While the gram-negative bacteria, has a cell wall structure is layered three and the composition of the cell wall has a high lipid content (11-22%) so difficult to be penetrated by flavonoids or alkaloids contained in polar extract/isolate. This is in accordance with the results of research Wardhani and Supartono [19], which has reported rambutan fruit skin extract more effectively inhibit the growth of *B. subtilis* compared to *E. coli*.

Papaya extract (from seeds and leaves) has a greater inhibitory effect than flavonoids and alkaloids. This is due to the synergistic between the components of the compounds in the extract so as to provide greater antimicrobial activity when compared with isolated flavonoid and alkaloids. Papaya seed extract and isolated flavonoid were inactive to inhibit the growth of *C. albicans* fungus. This is because *C. albicans* has a more complex composition of cell membranes so that the chemicals in extracts and isolates are unable to penetrate the sterol membrane of the cell wall and inhibit chitin synthesis in rigid fungal walls.

4. Conclusion

The active compounds contained in seed extract and papaya leaf are flavonoids, alkaloids, tannins, steroids, and saponins. Extracts and isolated from papaya plants had gram-positive antibacterial activity greater than the gram-negative bacteria, but both did not have antifungal activity. Papaya extracts have greater antibacterial activity than isolated flavonoid and isolated alkaloid. Strong antibacterial inhibitory sequences are extracts of papaya plants, isolated flavonoid, and isolated alkaloid.

References

[1] A. A. S. K. Darmawati, I. G. A. G. Bawa, and I. W. Suirta, "Isolasi dan Identifikasi Senyawa Golongan Flavonoid pada Daun Nangka (Artocarpus heterophyllus Lmk) dan Aktivitas Antibakteri terhadap Bakteri Staphylococcus aureus," *J. Kim.*, vol. 9, no. 2, pp. 203–210, 2015.
[2] P. Malathi and S. R. Vasugi, "Evaluation of mosquito larvicidal effect of Carica Papaya against Aedes Aegypti," vol. 2, no. 3, pp. 21–24, 2015.
[3] J. A. O. Okeniyi, T. A. Ogunlesi, O. A. Oyelami, and L. A. Adeyemi, "Effectiveness of Dried *Carica papaya* Seeds Against Human Intestinal Parasitosis: A Pilot Study," *J. Med. Food*, vol. 10, no. 1, pp. 194–196, 2007.
[4] N. Artanti, Y. Ma'arifa, and M. Hanafi, "Isolation and Identification of Active Antioxidant Compound from Star Fruit (Averrhoa carambola) Mistletoe (Dendrophthoe pentandra (L.) Miq.) Ethanol Extract," *J. Appl. Sci.*, vol. 6, no. 8, pp. 1659–1663, 2006.
[5] R. S. Sudibyo, "Metabolit Sekunder: Manfaat dan Perkembangannya dalam Dunia Farmasi. Pidato Pengukuhan Jabatan Guru Besar pada Fakultas Farmasi." Fakultas Farmasi Universitas Gadjah Mada, Yogyakarta, 2002.
[6] B. V. Owoyele, O. M. Adebukola, A. A. Funmilayo, and A. O. Soladoye, "Anti-Inflammatory Activities of Ethanolic Extract of Carica Papaya Leaves," *Inflammopharmacology*, vol. 16, no. 4, pp. 168–173, 2008.
[7] J. F. Rehena, "Uji Aktivitas Ekstrak Daun Pepaya (Carica papaya . LINN) sebagai Antimalaria in vitro," *J. Ilmu Dasar*, vol. 11, no. 1, pp. 96–100, 2010.
[8] D. B. S. Utama, Y. M. D. Arina, and M. N. Amin, "Pengaruh Ekstrak Daun Pepaya terhadap Jumlah Sel Limfosit pada Gingiva Tikus Wistar Jantan yang Mengalami Periodontitis," *e-Jurnal Pustaka Kesehat.*, vol. 2, no. 1, pp. 50–57, 2014.
[9] D. V Delphin, R. Haripriya, S. Subi, D. Jothi, and P. T. Vasan, "Phytochemical Screening of Various Ethanolic Seed Extracts," *World J. Pharm. Pharm. Sci.*, vol. 3, no. 7, pp. 1041–1048, 2014.
[10] M. Radji, *Mikrobiologi Panduan Mahasiswa Farmasi & Kedokteran*. Jakarta: Penerbit Buku Kedokteran EGC, 2011.
[11] D. Gould and C. Brooker, *Mikrobiologi Terapan untuk Perawat*. Jakarta: Penerbit Buku Kedokteran EGC, 2003.
[12] Jawetz, J. L. Melnick, and E. A. Adelberg, *Mikrobiologi Kedokteran*. Jakarta: Salemba Medika, 2005.
[13] R. Aksara, W. J. A. Musa, and L. Alio, "Identifikasi Senyawa Alkaloid dari Ekstrak Metanol Kulit Batang Mangga (Mangifera indica L)," *J. Entropi*, vol. 8, no. 1, pp. 514–519, 2013.
[14] J. Alka, K. Padma, and J. Chitra, "Antifungal Activity of Flavonoids of Sida acuta Bum f. Against Candida albicans," *Int. J. Drug Dev. Res.*, vol. 4, no. 3, pp. 92–96, 2012.
[15] M. M. Cowan, "Plant Products as Antimicrobial Agents," *Clin. Microbiol. Rev.*, vol. 12, no. 4, pp. 564–582, 1999.
[16] J. H. Doughari and S. Manzara, "In Vitro Antibacterial Activity of Crude Leaf Extracts of Mangifera indica Linn," *African J. Microbiol. Res.*, vol. 2, no. 2, pp. 67–72, 2008.
[17] H. Edziri, M. Mastouri, M. A. Mahjoub, Z. Mighri, A. Mahjoub, and L. Verschaeve, "Antibacterial, Antifungal and Cytotoxic Activities of Two Flavonoids from Retama raetam Flowers," *Molecules*, vol. 17, no. 6, pp. 7284–7293, 2012.
[18] M. Masibo and Q. He, "In vitro Antimicrobial Activity and the Major Polyphenol in Leaf Extract of Mangifera indica L.," *Malays. J. Microbiol.*, vol. 5, no. 2, pp. 73–80, 2009.
[19] R. A. P. Wardhani and Supartono, "Uji Aktivitas Antibakteri Ekstrak Kulit Buah Rambutan (Nephelium lappaceum L.) pada Bakteri," *Indones. J. Chem. Sci.*, vol. 4, no. 1, pp. 46–51, 2015.

Generic Science Skills Enhancement of Students through Implementation of IDEAL Problem Solving Model on Genetic Information Course

A Zirconia[1*], F M T Supriyanti[2], A Supriatna[3]**

[1] Chemistry Education Study Program, Graduate School of Indonesia University of Education
[2,3] Chemistry Education Department Indonesia University of Education

*e-mail: zirconia.ai@gmail.com

Abstract: This study aims to determine generic science skills enhancement of students through implementation of IDEAL problem-solving model on genetic information course. Method of this research was mixed method, with pretest-posttest nonequivalent control group design. Subjects of this study were chemistry students enrolled in biochemistry course, consisted of 22 students in the experimental class and 19 students in control class. The instrument in this study was essayed involves 6 indicators generic science skills such as indirect observation, causality thinking, logical frame, self- consistent thinking, symbolic language, and developing concept. The results showed that genetic information course using IDEAL problem-solving model have been enhancing generic science skills in low category with <g> of 20,93%. Based on <g> result for each indicator, showed that there are indicators of generic science skills classified in the high category.

Keywords: genetic information course, IDEAL problem-solving model, generic science skills.

1 INTRODUCTION

One of the problems faced by students in the Biochemistry 2 course is the concept of genetic information (protein biosynthesis). One of problem that faced by students in biochemistry 2 class is the concept of genetic information flow (protein biosynthesis). The genetic information is the change of DNA into proteins where the genetic code changes the DNA sequences into amino acid sequences. Proteins are arranged of hundreds of amino acids bound by peptide bonds [1].

The DNA molecule is a double helix-shaped polynucleotide chain that has a nitrogen purine and pyrimidine bases. DNA molecules can replicate, transcribe and translate, thereby determining the sequence of amino acids in protein formation. It determines the role of the DNA molecule as a material carrier of genetic information or the nature of heredity in living organisms, such as hereditary diseases such as type I diabetes that can be passed from parent to child.

Genetic information includes a few unobservable and interdependent mechanisms, it is a subject that is poor to be understood by students [2]. In survey previously, genetic information subject has been defined as the sixth the most difficult subject to learn among thirty subjects [3]. It has been determined that students lack information on understanding the relationships between every stage of genetic information process although students can connect between gene and protein synthesis. In another study mentioned in addition to the biochemical process, students are also difficulties in determining the order of DNA, RNA and amino acid genetic information flow process results [4].

The lack authority of concept causes the students' cognitive learning outcomes to be low. Low cognitive learning outcomes of students also led to the basic capabilities of students to be low anyway. This basic ability is known as generic science skills. According to Liliasari, the generic science skills are constantly evolving along with the increase of high order thinking skills, generic science and creative skills including being in it, so that generic science skills can be seen to relate to students' cognitive learning outcomes [5].

According to Brotosiswoyo generic science skills gained from the learning process, beginning with the observation of natural phenomena (1) observation (direct or indirect), (2) awareness of scale, (3) symbolic language, (4) logical frame (5) logical consistency, (6) causal law, (7) mathematical modeling, and (8) developing concepts [6]. Generic skills consist of fundamental skills, people related skills, personal skills, thinking skills, business related skills, and community-related skills [7]. These skills are ultimately less able by the students if the lecturer uses conventional methods.

One of lecture models that can be used to improve students' generic science skills is the problem solving model. In this study, problem solving model used is the IDEAL problem solving model. The IDEAL problem solving model is a problem solving model introduced by Bransfort and Stein to improve thinking ability and problem solving. This problem-solving model has steps such as (1) identify problem (2) define the goal, (3) explore the solution, (4) act the strategy, (5) Look back and Evaluate the effect [8].

This article discussion about research results related to improving students' generic science skills after implementation IDEAL problem solving model in the genetic information course.

2 METHOD

Method of this research was mixed method, with pretest-posttest nonequivalent control group design. Subjects of this study were chemistry students enrolled in biochemistry course, consisted of 22 students in the experimental class and 19 students in control class. Instrument in this study was essay involves 6 indicators generic science skills such as indirect observation, causality thinking, logical frame, self- consistent thinking, symbolic language, and developing concept.

3 RESULTS AND DISCUSSION

The purpose of this study is determine generic science skills enhancement of students after the implementation of IDEAL problem solving on genetic information lecture. The findings obtained from test result in the form of pretes and postes, observation sheets, and student worksheets.

3.1. Finding based on score test result

The purpose of this subtopic is to get information about enhancement generic science skills enhancement of students based on test result. The test scores included pretest and postest obtained from a essay test consisting of 10 questions which developed from 6 generic science skills indicators. Results of average scores pretest and posttest generic science skills students in the experimental class and control obtained can be seen in Figure 1.

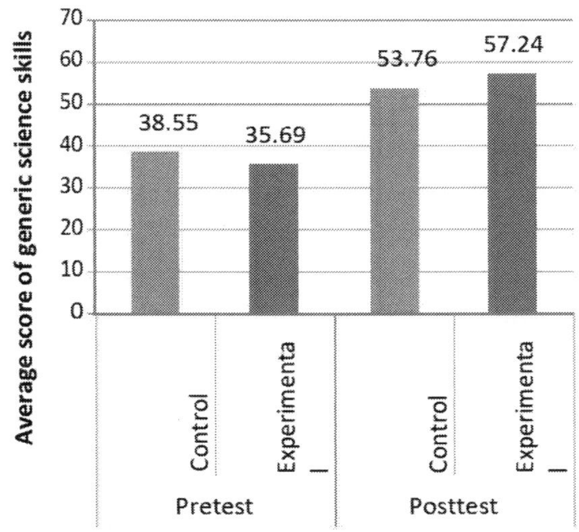

Figure 1. Results of average scores pretest and posttest generic science skills students in the experimental class and control

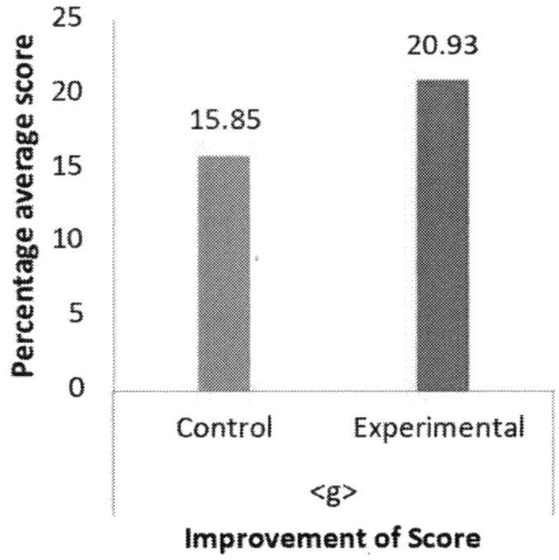

Figure 2. Percentage average scores <g> generic science skills student experimental and control class

Based on Figure 1 can be seen that experimental class pretest score less than the control class pretest, whereas experimental class posttest score higher than the control class pretest. It means IDEAL problem solving model can increase generic science skills students better than expository class. If the <g> experimental class score and <g> control class score is compared and both are obtained <g> score according to Figure 2.

Based on Figure 2 can be seen that <g> experimental class score with <g> of 20,93% in the low category better than <g> control class score with <g> of 15,85% in the low category. This proves that the result of the improvement of generic science skills in experimental class is better than control class. In the IDEAL problem solving model, the generic science skills students can developed using stages according to problem solving model developed by Bransford [8] appropriate Table 1.

Table. 1 Relationship IDEAL Problem Solving Stages and Generic Science Skills

IDEAL Problem Solving Stages	Generic Science Skill Indicators
Identify problem	1. Indirect Observation 2. Causal Law
Define the goal	1. Logical Frame 2. Self- Consistent Thinking
Explore solution	1. Self- Consistent Thinking 2. Symbolic Language
Act the strategy	1. Causal Law 2. Developing Concept
Look back and evaluate the strategy	1. Developing Concept

In Table 1, it can be seen that IDEAL problem-solving model genetic information course can be used to develop 6 generic science skill indicators according Brotosiswoyo[6] and Ramlawati et al[11] while 3 other indicators cannot be developed that is direct observation, awareness of scale and mathematical modeling.

In Figure 1, the average Postes and Pretes scores on generic science skills of experimental class and control class students have increased, but each class needs to be further analyzed on generic science skills improvement in both classes and whether there is a significant difference.

Table. 2 Test Results Statistics of Generic Science Skills on Pretest

Class	Normality Test	Homogeneity Test	Sig	Ref
Experimental	0,715	0,141	0,355	No diff.
Control	0,194			

In Table 2 can be seen that t test result obtained of 0,705 at significance level of 5%, because 0,705 > 0.05, there were no significant differences between the experimental class and control class. It means the ability of Generic Science Skills student in experimental class and control student at the beginning of the term did not differ significantly. Furthermore, the significance test on <g> total for Generic Science Skills can be seen in Table 3.

Table 3 Test Results Statistics of Generic Science Skills on <g>

Class	Normality Test	Homogeneity Test	Sig	Ref
Experimental	0,003	0,388	0,266	No diff.
Control	0,000			

Based on Table 3, it can be seen that the average score of pretest and postes is increasing, but not a significant improvement. This is evidenced from the U test results obtained for 0,266 at significance level of 5%, because 0.266 > 0.05 there were no significant differences between the experimental class and control class. After testing the significance of differences <g> total between the experimental class and control class is done, then the pretest-postest score of each indicator in the experimental class was tested for improvement through <g>.

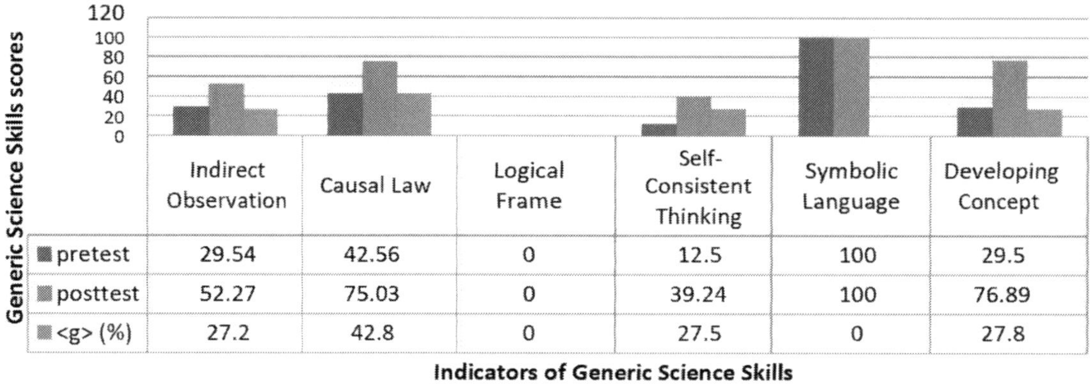

Figure 3. The score of generic science skills students experimental class at pretest and postest for each indicator.

Based on Figure 3 it can be seen that there are 6 indicators of Generic Science Skills which improved. Indicators of Generic Science Skills are indirect observation, causal law, logical frame, self-consistent thinking, symbolic language, and developing concept. The score of <g> showed just indicator causal law classified in middle category while score of <g> for 5 other indicators developed are low. Causal law indicators are observed at the identify problem and act the strategy stage problem-solving. Causal law is the ability to express the relationship between two or more variables in the phenomenon of heredity and estimate the cause, where there is a series of relationships between the various factors of the phenomena observed.[10] The causal law can be assessed from students' ability to state the relationship between two or more variables in a natural phenomenon and be able to estimate the cause of the natural phenomenon.[6] In these findings, causal law is assessed from students' ability to analyse the cause of heredity and determine the complementary DNA sequences, RNA sequences and amino acid sequences of the translation. This condition is suspected because students are less able to estimate the causes of heredity. Different results obtained by Agustin where in her research on improving the generic skills of science on the concept of interaction between molecules found that the causal law indicators classified into the low category of 27.62%. This is because Agustin's research emphasizes the modelling of interactions between molecules using interactive media.[9]

Indicators of generic science skills that have increased the most low is logical frame and symbolic language. Logical frame indicator can be assessed based on students' ability to connect between phenomena with a concept.[6] The logical frame indicator in this finding is the ability of students to be able to reveal the logical connection between genes and the process of genetic information through the central dogma concept. Based on the assessment, <g> the lowest of the generic science skill indicators developed, one of which lies in the logical frame indicator. Based on the results, score for logical frame indicator of 0% on low criteria. This is because students cannot express the physical meaning of the concept of central dogma and cause students to be less able to express the logical connection between genes and the process of genetic information through concept of central dogma. Students are only able to explain the concept of the central dogma verbally, but difficult to apply it to everyday problems. Unlike Anwar's research on chemistry kinetic lectures where in his research found that the indicators of obedience framework is classified in the middle category. This is due to the subject of Anwar's research emphasizing the chemical kinetics count so that students easily find a logical relationship between two variables.[13]

Indicators of generic science skills that lowest increase secondly is an indicator of symbolic language that observed in the exploration strategy stage. In this finding, the ability of the symbolic language in question is that students can determine the complementary DNA sequences. Based on the research, IDEAL problem-solving model cannot improve student symbolic language with <g> 0%. This is because students have been able to estimate the sequence of complementary DNA on pretest. This is evidenced from the score of student pretes with a maximum score of 100. Different results obtained from research Rafiuddin where the research on generic science skills on the subject of protein metabolism indicators showed relatively symbolic language middle category of 68.44% for the the dehydrogenation reaction concept and 50.02% for the concept of amino acid catabolism.[14]

4 CONCLUSION

The results showed that genetic information course using IDEAL problem solving model have been enhance generic science skills in low category with <g> of 20,93%. Based on <g> result for each indicator, showed that there is indicators of generic science skills classified in the high category.

REFERENCES

1. Merrick W C 2010 Eukaryotic Protein Synthesis: Still a Mystery. *The Journal Of Biological Chemistry* **285** 28 21197–21201
2. Lewis J & Robinson C W 2000 Genes, Chromosomes, Cell Division and Inheritance- Do Students See Any Relationship? *International Journal of Science Education* **22** 2 177-195
3. Venville G J, & Treagust D F 1998 Exploring Conceptual Change in Genetics Using Multidimensional Interpretive Framework *Journal of Research in Science Teaching* **35** 9 1031-1055
4. Concannon J, & Buzzetta M 2010.Students conceptualizing transcription and translation from a celluar perspective. *Science Activities* **47** 83-88.
5. Liliasari 2011 *Membangun Masyarakat Melek Sains Berkarakter Bangsa Melalui Pembelajaran* Makalah Seminar Nasional Unnes tahun 2011.
6. Brotosiswoyo, B.S. 2001. *Hakikat Pembelajaran Fisika di Perguruan Tinggi,* Dalam hakikat pembelajaran MIPA dan Kiat Pembelajaran Fisika di Perguruan Tinggi Jakarta : Proyek Pengembangan Universitas Terbuka, DEPDIKNAS
7. Singh, Harshvardhan dan Gera, Manju. 2015. Developing Generic Skills in Higher Education. *Indian Journal Of Applied Research.* **5** 6 824-826
8. Bransford *et al* 1998 *The IDEAL Workplace: Strategies for Improving Learning, Problem Solving, and Creativity* (Woshington DC: Nashville)
9. Agustin, R R 2013 Pengembangan Keterampilan Generik Sains melalui Penggunaan Multimedia Interaktif. Jurnal Universitas Pendidikan Indonesia. **18** 2 253-257
10. Muh Tawil dan Liliasari, 2014. *Keterampilan- keterampilan sains dan implementasinya dalam pembelajaran IPA (* Badan penerbit UNM : Makassar)
11. Ramlawati, Liliasari, Martoprawiro, Muhamad A, and Wulan A R 2014 The Effect of Electronic Portofolio Assessment Model to Increase of Students' Generic Science Skill in Practical Inorganic Chemistry *Journal of Education and Learning.* **8** 3 179-186.
12. Anwar M 2014. The Effect of Active-Cooperative Learning on Science Generic Skills of Students in Chemical Kinetics Course for Prospective Teacher *Journal of Education and Practice* **5** 13 149-154
13. Moerwani P C, Radiman S, Akhmad E, Ratnaningsih 2001 Hakikat Pembelajaran MIPA di Perguruan Tinggi. (Jakarta: Pusat Antar Universitas Untuk Peningkatan dan Pengembangan aktivitas Instruksional Universitas Terbuka).
14. Rafiuddin 2012 Pengembangan Program Pembekalan Kecakapan Generik Biokimia pada Materi Pokok Metabolisme protein bagi mahasiswa Calon Guru Disertasi SPS UPI: Tidak dipublikasikan

IOP Conf. Series: Materials Science and Engineering **349** (2018) 012009 doi:10.1088/1757-899X/349/1/012009

Development of ultrasonic-assisted extraction of antioxidant compounds from Petai (*Parkia speciosa Hassk.*) leaves

Buanasari[1,*], P D Palupi [1], Y Serang [1], B Pramudono [2], S Sumardiono[2]

[1]Akademi Farmasi Nusaputera, Semarang, Indonesia
[2]Chemical Engineering Department, Faculty of Engineering, Diponegoro University, Semarang, Indonesia

*Corresponding author: buanasari@akfarnusaputera.ac.id.

Abstract. Research on Petai (*Parkia speciosa Hassk.*) suggests it has an antihypertension, antidiabetic, analgesic, and antiulcer effects. In the present study, an ultrasonic-assisted extraction method was developed for the effective extraction of active compound from petai leaves. Some parameters such as ethanol concentration (0, 20, 40, 60, 70, 80, 100 %v), solid-to-liquid ratio (1:5; 1:10; 1:15; 1:20; 1:25; 1:30; 1:35; 1:40; 1:50 g/mL), extraction time (15, 20, 25, 30, 35, 40, 50 minutes) and extraction temperature (40, 45, 50, 55, 60, 65, 70°C) were studied and evaluated base on extract yield and 1,1-diphenyl-2-picry hydrazyl (DPPH) scavenging activity. The result showed that the highest extract yield was obtained at 40% ethanol concentration, 1:30 (%w/v) of solid-to-liquid ratio, 30 minutes and 65°C of temperature with DPPH scavenging activity 92.53 ± 0.87% and extract yield 21.25 ± 2.38%. The result obtained is helpful for the utilization of Petai leaves, and also indicated that ultrasonic-assisted extraction is a very recommended method for the extraction of active compounds from plant material.

1. Introduction

Parkia speciosa Hassk., known as petai leaves, are a Mimosaceae family that contains amino acids such as lysine, aspartine, glucine, alanine, valine and leucine [1] and carbohydrates content of 13g/100g [2]. Petai leaves are also rich in mineral content such as calcium, phosphorus, magnesium, manganese [2] and sulfur [3]. Petai leaves (*Parkia speciosa Hassk.*) can also be used for the treatment of hypertension, diabetes, and headaches [4,5,6]. Petai leaves can function as antiulcer [7], while petai seed leaves contain vitamin C which is quite a lot and phenolic compound which can perform an antioxidant. Petai leaves can be utilized as a natural source of antioxidant [8].

Extracts of petai plants have been widely performed both for the skin of the fruit [9,10,11], the leaf part [7,12], and seed [13]. Common extractions to take antioxidants from petai plants are soaking extraction at room temperature or maceration. The maceration extraction is straightforward and economical, but the disadvantage is that it requires a lot of solvents and a long time. Previous studies have compared between microwave assisted and ultrasonic-assisted extractions to extract antioxidant compounds in petai leaves, and ultrasonic-assisted extraction provides an alternative method that can be developed to take the active compound of petai leaves [12].

Another advantage of ultrasonic-assisted extraction is surface contact between the solids and the wider liquid, due to the direct contact between particles and ultrasonic waves [14]. Extraction of the antioxidant compound from *Eucommiae folium* with ultrasound-assisted extraction resulted the flavor content of flavonoids for 72 minutes reached 17.6% while the maceration with heating for 2.5 hours only reached 12% [15]. An antioxidant study of red grapes showed extrapolation using ultrasonic method had a DPPH radical binding ability of 70% higher than that of maceration [16].

Research on the extraction of ultrasonic assisted petai plants (*Parkia speciosa Hassk.*) by finding process conditions regarding solvent concentration, solids: solvents ratio, time, and temperature for optimal yield and scavenging activity have never been reviewed before.

2. Methods

Materials

The main ingredient of this research is petai leaves (*Parkia spesiosa* Hassk.) obtained from Semarang area, Central Java. Other materials which used: methanol reagents (Merck), DPPH (Sigma-Aldrich, 90% purity), ethanol (Merck, 96% purity), and aquadest.

Equipment

The tools which used in this study are electric scales (Sartorius), oven (Memmert), moisture analyzer (Radwag MAC50), digital ultrasonic cleaner (Branson 5210), UV-Vis spectrophotometer (Shimadzu 2480), rotary evaporator (Scilogex), Whatman filter paper, 100 mesh filter, vacuum pump and aluminum foil.

Raw Material Preparation

Petai leaves are uniformed with the size of 100 mesh and water content less than 10%w/w. Petai leaves (*Parkia speciosa Hassk.*) are cleaned and aerated without sun for 5-10 days. The dried leaves are smoothed with 100 mesh size followed by drying until the moisture content is less than 10%w/w.

Ultrasonic-assisted Extraction (UAE)

The extraction method is used ultrasonic-assisted extraction with ethanol solvent. The ultrasonic instrument used in digital ultrasonic cleaner (Branson 5210) with 200 W ultrasonic power and 40 kHz frequency, equipped with digital temperature and time control. Petai leaves powder (5.0 g) was weighed carefully in a glass tube, then added with solvent and homogenized. The tube was sealed and inserted into the water in digital ultrasonic cleaner that has been regulated process conditions. Extraction was performed to determine the yield of extract and scavenging activity produced by ultrasonic-assisted extraction with solvent concentration variables (0, 20, 40, 60, 70, 80, 100 percent of volume), solid (g): solvent (mL) ratio(1:5; 1:10; 1:15; 1:20; 1:25; 1:30; 1:35; 1:40; 1:50), time (15, 20, 25, 30, 35, 40, 50 minutes), and temperature (40, 45, 50, 55, 60, 65, 70°C). After extraction, sample was centrifuged and filtered using vacuum pump with Whatman. The solvent was separated from extract by rotary evaporator.

1,1 Diphenyl 2-Picryl Hydrazyl Scavenging Activity (DPPH-SA)

DPPH scavenging activity test was carried out by the method of Banerjee *et al.*, 2005 [17]. The 1.0 mL sample extract solution was added 3.0 mL 0.1 mM DPPH solution. The absorbance was measured by spectrophotometer (Shimadzu Japan) at 514 nm. The extract activity on DPPH is expressed as:

$$\text{DPPH radical scavenging sctivity (\%)} = \frac{\text{A Blank - A sample}}{\text{A Blank}} \text{ X } 100\%$$

3. Result and Discussion

Effect of Ethanol Concentration on Extracts and DPPH-SA

Solvents that are often used for extraction include methanol, ethanol, and water. In this study used ethanol because it was safer for medicine and food than methanol. Water is also a safe and cheap solvent, but its effectiveness in extracting is lower than ethanol-water. The effect of ethanol concentration (0, 20, 40, 60, 70, 80, 100 percent volume) on yield extract and DPPH scavenging activity was studied with

fixed parameters: 1:10 solid-liquid ratio; time 20 minutes, temperature 40°C with frequency 40 Hz. The yield of extract is presented in Figure 1A. When ethanol concentration was increased from 0 to 40% extract yield increased to 10.04 ± 1.09%w. After concentration above 40%, yield tends to decrease even for 100% concentration was obtained yield 5.50 ± 0.53%w.
The results in this variable also tested its DPPH-SA and presented in Figure 2A. The graph profile shows the same characteristics as extract yield calculation. DPPH-SA increased from 0-40% ethanol concentration of 71.90±0.03 to 88.25±0.09% and decreased with increasing ethanol concentration above 40%. Zou *et al.* 2014 use mixture of water-ethanol to take the active compound on mango leaves and the best results with 40% ethanol. The results of this extraction variable also found an excellent ethanol-water mixture to take active substance content in petai leaves, and 40% ethanol gave the largest extract yield and DPPH-SA which will be used for further research [18].

Effect of Solid: Liquid Ratio against Extract Yield and DPPH-SA
The influence of solid-liquid ratio on extract yield and DPPH-SA was studied by variation (1:5; 1:10; 1:15; 1:20; 1:25; 1:30; 1:35; 1:40; 1:50 g/mL) and fixed parameters: ethanol concentration 40%, extraction time 20 minutes, temperature 40°C with frequency 40 Hz. The results are presented in Figure 1B (effect on extract yield). When the solid: liquid ratio is raised from 1:5 to 1:30,extract yield also increases, and after that, it tends to remain even down at 1:50 ratio. In general, the greater use of solvents increases extracted active substance [19]. But the use of too high solvents will be much wasted and ineffective, the lack of solvent leads to less optimum extraction performed [20]. So the selection of solvent amount is crucial.

The results of DPPH-SA measurements are shown in Figure 2B. The 1,1-diphenyl-2-picrylhydrazyl (DPPH) compound is a stable free radical having nitrogen free radicals commonly used to measure antioxidant activity of extract. Free radicals are unpaired unstable electrons, and their molecules are highly reactive. DPPH radical scavenging activity by antioxidants occurs through provision of protons to radicals. Compounds capable of donating protons show strong radical scavenging activity. Some phenolic compounds in the extract release their hydrogen atom to bind 1,1-diphenyl-2-picrylhydrazyl and then convert to 1,1-diphenyl-2-picryl hydrazine (nonradical). Once stable, these compounds do not attack other molecules to form new radicals. An increase of 1,1-diphenyl-2-picryl hydrazine nonradical amount will be characterized by color change from purple to yellow. Figure 2B showed an increase of the solid: liquid ratio from 1:5 to 1:30 will increase DPPH-SA even though not significantly, and increasing of further solvent gave DPPH-SA value fixed and tended to decrease. The results of this study are in line with other UAE studies. Oancea *et al.* 2013 reported a not-so-great difference in the anthocyanin yield of a gold black sweet cherry at 10:1 - 20:1 (v/w) ratio for 20 minutes using UAE [21]. Petigny *et al.* 2013 reported an insignificant effect of solvent ratios at 4-10% on Boldo leaves yield [22]. Wardhani *et al.* 2013 reported the effect of *Eucheuma cottonii* seaweed on solid solvent ratio at 3:1-10:1 (v/w) ratio is also insignificant [23]. The highest DPPH-SA in solid: liquid ratio 1:30 equal to 88.19±0.92%. Solid: solventratio 1:30 gives the largest extract yield and DPPH-SA and will be used for further research. The results of this ratio are in line with research of Zou *et al.* 2014 to take active compounds through UAE [18].

Effect of Extraction Time on Extract Yield and DPPH-SA
The effect of extraction time was done with time variation (15, 20, 25, 30, 35, 40, 50 minutes), and as fixed parameter of 40% ethanol concentration, solid: liquid ratio 1:30 g/mL, 40°C with frequency 40 Hz. The results are presented in Figure 1C. Yield continues to increase from 15 to 30 minutes. Longer extraction times allow more contact time for cavitation bubbles broke sample cells, increasing extracted active substances [24]. Solid-fluid extraction is a mass transfer phenomenon which compounds in a solid matrix migrate into the solvent through diffusion and osmotic ultrasound-induced mechanisms. The ultrasound application in the form of solid-liquid extraction forms a cavitation bubble which disconnects the sample cell to solvent facilitate penetration into the cell. Maximum yield result is 92.85±0.61% at

30 minutes. Yield decreasedmore than 30 minutes. The results show that ultrasound can accelerate formation of equilibrium for active substance dissolution between sample wall and extraction solvent in short time. This effect is an advantage of UAE compared to conventional extraction methods [25]. Prolonging the extraction time no longer affects active compound amount of *Orthosiphon* which is significantly removed when the diffusion process reaches equilibrium [26]. In this study, the maximum is 30 minutes and will be used for further research.

DPPH-SA for time variation is shown in Figure 2C and showed an upsurge of activity, although not too large from 15-30 minutes after which the DPPH-SA decline. Extraction time extension after the equilibrium time causes phenolic active ingredient, exposed to light and oxygen for longer time. This exposure allows the phenolic compounds to become oxidized which in turn decreases the obtained DPPH-SA. Also, prolonging the extraction time may lead to the accumulation of compounds that may increase oxidation [27].

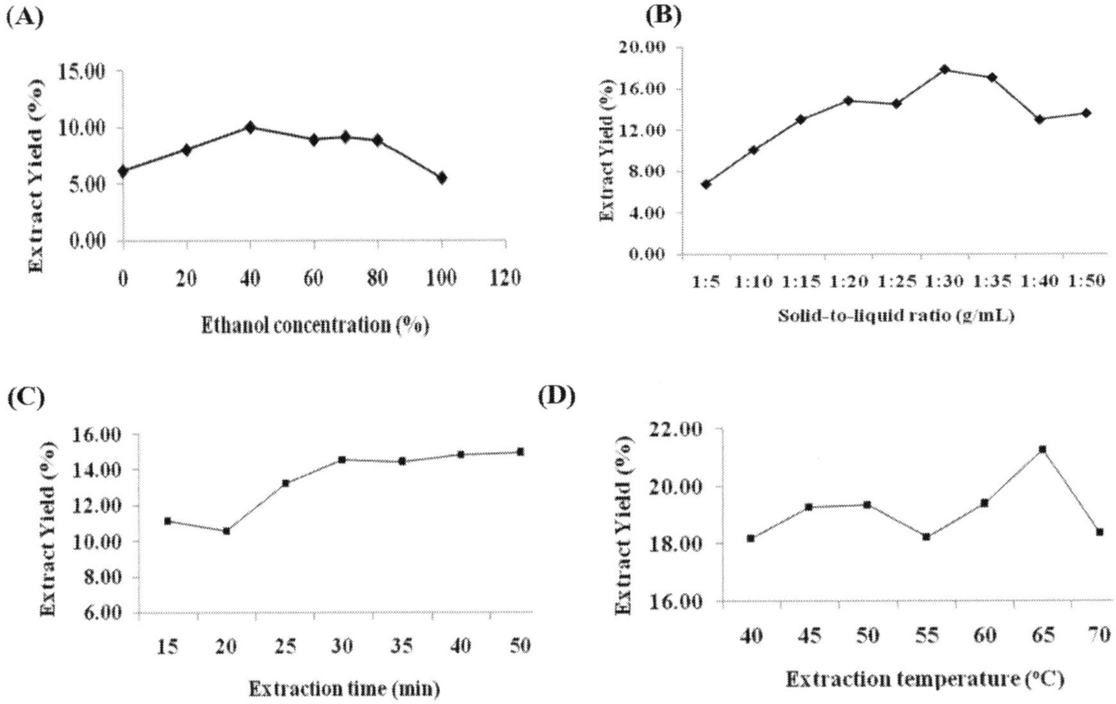

Figure 1. Effect of extraction parameters on extract yield. **(A)** Effect of ethanol concentration on extract yield; **(B)** Effect of solid-liquid ratio on extract yield; **(C)** Effect of time on extract yield; **(D)** Effect of temperature on extract yield.

Effect of Extraction Temperature on Extract Yield and DPPH-SA

The influence of extraction time was carried out with temperature variations (40, 45, 50, 55, 60, 65, 70°C), and as fixed parameter of 40% ethanol concentration, solid: liquid ratio of 1:30 g/mL, 30 minutes with frequency 40 Hz. The results are shown in Figure 1D for extract yield and 2D for DPPH-SA results. The best temperature in this study was found 65°C with yield of 21.25 ± 2.38% and DPPH-SA of 92.53 ± 0.87%.Heo et al. 2005 found 70°C as the best extraction temperature of various seaweed antioxidant compounds [28]. Wardhani *et al.* 2013 found the best temperature for extraction of phenolic compounds at 55°C with methanol solvent [23]. Zou *et al.* 2014 found an optimum temperature of 60°C to extract mango leaves with 40% ethanol solvent [18].

The best extraction temperature difference shows that ultrasound application reduces phenolic extraction optimum temperature compared with conventional extraction. In this process, the heat gradually propagates from solvent to solid. Meanwhile, direct contact between solid and ultrasonic waves in the UAE allows reaching optimum temperatures that do not damage extracts compared to conventional extractions. Ultrasound assists mechanical effects with compression cycles and continuous evacuation that allow greater solvent absorption into sample matrix, increasing contact surface area between solid and liquid phases, so that the solute diffuses fast from solid phase to solvent [29].Phenolic compounds are unstable at higher temperatures. Thus high extraction temperatures may increase extract impurity due to joint extraction of other undesirable compounds [30]. However, several previous studies have shown that ultrasound induction can maintain extracts with a high-temperature range between 60-80°C.

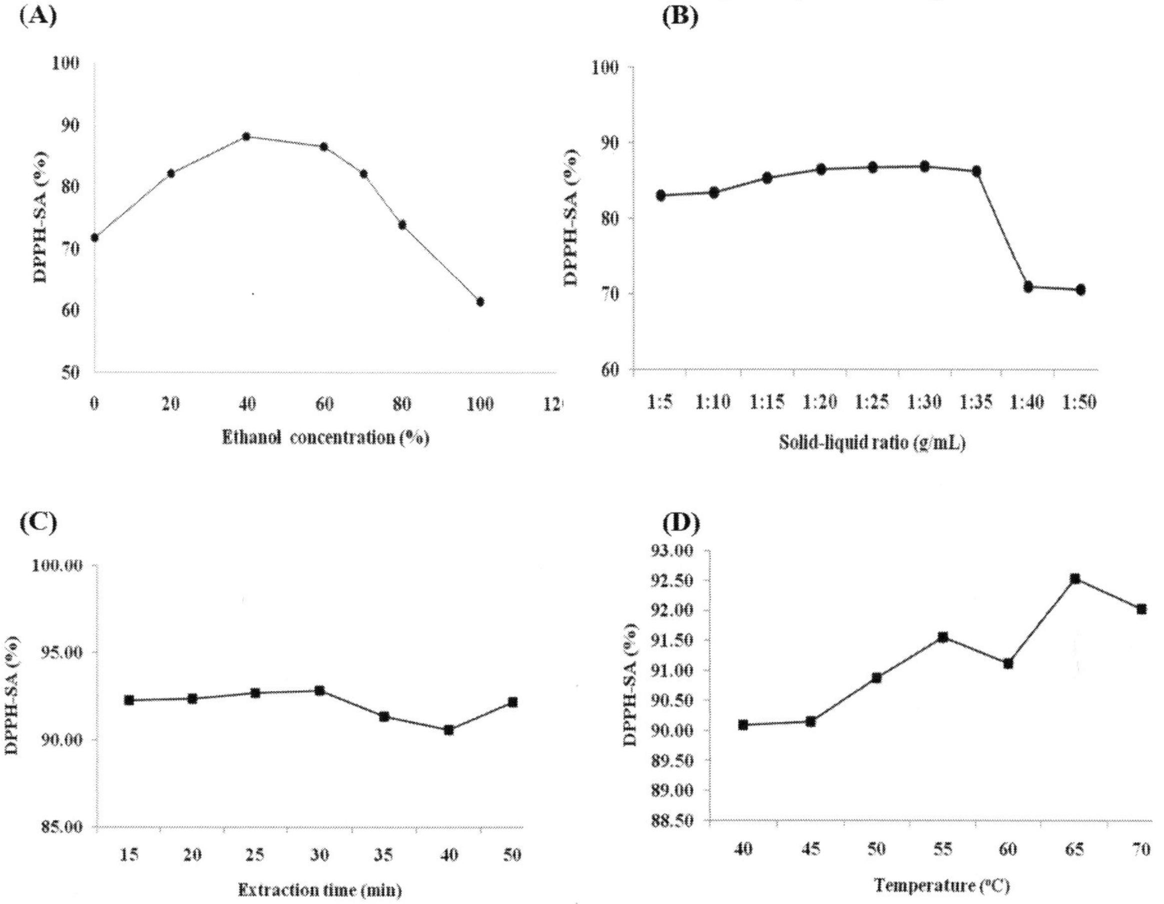

Figure 2.Effects of extraction parameters on DPPH-SA. (A) Effect of ethanol concentration on DPPH-SA; (B) Effect of solid: liquid ration on DPPH-SA; (C) Effect of time on DPPH-SA; (D) Effect of temperature on DPPH-SA.

4. Conclusion

In this study, an ultrasonic-assisted extraction method has been developed for extraction of Petai leaves active compound (*Parkia speciosa Hassk.*). Ultrasonic waves are tools that can efficiently improve extraction results. The results showed that ethanol concentration, solid-liquid ratio, and extraction time affected extraction result in extract yield and DPPH-SA value. The best-operating conditions are 40% ethanol, 1:30 solid-liquid ratio, and 30 minute extraction time at 65°C with 40Hz ultrasound frequency. The highest yield value reached 21.25±2.38% and DPPH-SA of 92.53±0.87%.

Acknowledgment

The authors acknowledge PEKERTI Fund from Kemenristekdikti for financial support that makes possible the development of our research. The authors also would like to acknowledge Phytochemistry Laboratory of Akademi Farmasi Nusaputera Semarang, UPT UNDIP Semarang, and Chemical Engineering Department UNDIP Semarang for providing the equipment and supporting this research.

Reference

[1] Suvachittanont, W. and Pothirakit, P. 1988, Protein in *Parkia speciosa* seeds. *Sangkia Med.J.*, 6, 23-30.
[2] Mohamed, S., Shamsuddin, A.B.D., Rehman, Sulaiman, S., and Abdullah, F.1987, Some nutritional and anti nutritional components in jering (*Pithecellobium jeringa*), kerdas (*Pithecellobium microcarpum*) and petai (*Parkia speciosa*). *Pertanika*. 10(1), 1-68.
[3] Suvachittanont, W., Yukiko, K., Hiroyasu, E. and Mitsuhiro, T. 1995, Formation of thiazolidine-4-carboxylic acid (thioproline), an effective nitrate-trapping agent in human body, in *Parkia speciosa* seeds and other edible leguminous seeds in Thailand. *Food Chemistry*. 55(4), 359-363.
[4] Jamaluddin, F., Mohamed, S., and Lajis, M. N. 1994, Hypoglycaemic effect of *Parkia speciosa* seeds due to the synergistic action of β-sitosterol and stigmasterol. *Food Chemistry*. 49, 339–345.
[5] Jamaluddin, F., Mohamed, S., and Lajis, M. N, 1995, Hypoglycaemic effect of stigmast- 4-en-3-one, from *Parkia speciosa* empty pods. *Food Chemistry*. 54, 9–13.
[6] Suvachittanont, W. and Peutpaiboon, A. 1992, Lectin from *Parkia speciosa* seeds. Phytochemistry. 31,4065-70.
[7] Al-Batran, R., Al-Bayaty, F., Jamil, Al-Obaidi, M.M., Abdualkader, A.M., Hadi, H.A., Ali, H.M., Abdulla, R., A., 2013, In Vivo Antioxidant and Antiulcer Activity of *Parkia speciosa* Ethanolic Leaf Extract against Ethanol-Induced Gastric Ulcer in Rats. *Pub-Med*. USA.
[8] Amarnath, B., 2004, A Study on Antioxidant Nature of Petai (*Parkia speciosa*). *Thesis*. National University of Singapore.
[9] Gan, C.Y., Manaf, N.A., and Latiff, A.A., 2010, Physico-chemical properties of alcohol precipitate pectin-like polysaccharides from *Parkia speciosa* pod. *Food Hydrocolloids*. 24, 471–478.
[10] Gan, C.Y., Manaf, N.A., and Latiff, A.A., 2010, Optimization of alcohol insoluble polysaccharides (AIPS) extraction from the *Parkia speciosa* pod using response surface methodology (RSM). *Carbohydrate Polymers*. 79, 825–831.
[11] Gan, C.Y., and Latiff, A.A., 2011, Antioxidant *Parkia speciosa* pod powder as potential functional flour in food application: Physicochemical properties characterization. *Food Hydrocolloids*. 25, 1174e1180.
[12] Buanasari, Eden, W.T., Solichah, A.I., 2017, Extraction of Phenolic Compounds from Petai Leaves (*Parkia speciosa Hassk.*) using Microwave and Ultrasound Assisted Methods. *Jurnal Bahan Alam Terbarukan*.6(1), 35-37.
[13] Gan, C.Y., and Siow, H.L., 2013, Extraction of antioxidative and antihypertensive bioactive peptides from *Parkia speciosa* seeds. *Food Chemistry*. 141, 3435–3442.
[14] Wiyarno, B., Rosli, M.Y., and Maizzirwan, M., 2010, Ultrasound Extraction Assisted (UEA) of Oil From Microalgae (Nannochoropsis).*International Journal of Science Engineering and Technology*. 3 (1), 55-59.
[15] Huang, W., Xue, A., Niu, H., Jia, Z., and Wang, J.W., 2009, Optimized Ultrasonic assisted Extraction of Flavonoids from Folium Eucommiae and Evaluation of Antioxidant Activity in Multi-Test System In Vitro. *Food Chem*. 114, 1147-1154.
[16] Morelli, L.L.L., and Prado, 2012, Extraction Optimization For Antioxidant Phenolic Compounds in Red Grape Jam using Ultrasound With a Respons Surface Methodology.*Ultrasonic Sonochemistry*. 19, 1144

[17] Banerjee, A., Dasgupta, N. and Dee, B., 2005, In Vitro Study off Antioxidant Activity of *Syzigium cumini* Fruits.*Journal Food Chemistry*.

[18] Zou, T.B., Xia, E.Q., He, T.P., Huang, M.Y., Jia, Q., and Li, H.W., 2014, Ultrasound-Assisted Extraction of Mangiferin from Mango (*Mangifera indica* L.) Leaves Using Response Surface Methodology, *Molecules Journal*, 19, 1411-1421.

[19] Li, H., Chen, B., Yao, S., 2005, Application of ultrasonic technique for extracting chlorogenic acid from *Eucommia ulmodies* Oliv. (*E. ulmodies*). *Ultrason. Sonochem.* 12, 295–300.

[20] Valachovic, P., Pechova, A., Mason, T.J., 2011,Towards the industrial production of medicinal tincture by ultrasound assisted extraction. *Ultrason. Sonochem.* 8, 111–117.

[21] Oancea, S., Grosu, C., Ketney, O., and Stoia, M., 2013, Conventional and Ultrasound-Assisted Extraction of Anthocyanins from Blackberry and Sweet Cherry Cultivars.*Acta Chim. Slov.* 60(2), pp. 383-389.

[22] Petigny, L., Périno-Issartier, S., Wajsman, J., and Chemat, F., 2013, Batch and Continuous Ultrasound Assisted Extraction of Boldo Leaves (*Peumus boldus* Mol.).*International Journal of Molecular Sciences*. 14(3), pp. 5750-5764.

[23] Wardhani, D. H., Sari, D.K., and Prasetyaningrum, A., 2013, Ekstraksi Berbantu Ultrasonik Senyawa Fenolik Antioksida dari *Eucheuma cottoni*.*Reaktor*.14(4).

[24] Wang, X., Wu, Q., Wu, Y., Chen, G., Yue, W., and Liang, Q., 2012, Response Surface Optimized Ultrasonic-Assisted Extraction of Flavonoids from Sparganii Rhizoma and Evaluation of Their in Vitro Antioxidant Activities. *Molecules*. 17(6), pp. 6769- 6783.

[25] Tabaraki, R. and Nateghi, A., 2011, Optimization of ultrasonic-assisted extraction of natural antioxidants from rice bran using response surface methodology. *Ultrason. Sonochem.* 18, 1279–1286.

[26] Chew, K.K., Ng, S.Y., Thoo, Y.Y., Khoo, M.Z., Wan, A.W.M., and Ho, C.W., 2011, Effect of Ethanol Concentration Extraction Time and Extraction Temperature on The Recovery of Phenolic Compounds and Antioxidant Capacity of *Centella asiatica* Extracts.*International Food Research Journal*.18, pp. 571-578.

[27] Bazykina, N.I, Nikolaevskii, A.N., and Fillipenko, T.A., 2002, Optimization of conditions for the extraction of natural antioxidants from raw plant materials.*Pharmaceutical Chemistry Journal*. 36(2), pp. 46-49.

[28] Heo, S.J., Park, E.J., Lee, K.W., and Jeon, Y.J., 2005, Antioxidant activities of enzymatic extracts from brown seaweeds.*Bioresource Technology*. 96(14), pp. 1613-1623.

[29] Rostagno, M., Palma, M., and Barroso, C.G., 2007, Ultrasound-assisted extraction of isoflavones from soy beverages blended with fruit juices.*Analytica Chimica Acta*. 597(2), pp. 265-272.

[30] Liu, Q.M., Yang, X.M., Zhang, L., and Majetich, G., 2010, Optimization of ultrasonic-assisted extraction of chlorogenic acid from Folium eucommiae and evaluation of its antioxidant activity.*Journal of Medicinal Plants Research*. 4(23), pp. 2503-2511.

Characteristics of eugenol loaded chitosan-tripolyphosphate particles as affected by initial content of eugenol and their in-vitro release characteristic

B Cahyono[1], Qurrotu A'yun[1], M Suzery[1], Hadiyanto[2]

[1]Departement of Chemistry, Faculty of Science and Mathematics, Diponegoro University, Prof Sudharto SH street, Tembalang, Semarang 50275 (phone/fax 024-76480824)

[2]Departement of Chemical Engeneering, Diponegoro University, Prof Sudharto SH street, Tembalang, Semarang 50275

Email: cahyono@live.undip.ac.id

Abstract: The aim of this research was to determine encapsulation efficiency, loading capacity and controlled release of eugenol loaded chitosan-tpp products which prepared by coaservation method. The characteristic of eugenol-loaded chitosan showed that %EE and % LC increased by increasing the initial eugenol content. The optimum of %EE (72.63%) and %LC (43.96%) were obtained at the ratio of chitosan to eugenol of 1:1.5. The FTIR spectrum showed the characteristic peaks of eugenol appearing on spectrum of eugenol encapsulated and blue-shift in the hydroxyl band from 3425.58 cm^{-1} in chitosan-tpp to 3417.86 cm^{-1} and 3394.72 cm^{-1} in eugenol loaded chitosan-tpp indicating that eugenol was successfully encapsulated. The surface morphologies of freeze-dried particles with the optimum %EE showed that more surface roughness and porosity than plain particles. Furthermore, the *in vitro* release of particles with minimum and optimum %EE were also investigated in acid (Simulated Gastric Fluid) and base (Simulated Intestinal Fluid) medium at ambient temperature.

1. Introduction

Eugenol ($C_{10}H_{12}O_2$), is phenolic compound with IUPAC name 4-allyl-2-methoxyphenol. Eugenol is the major component (76.23-85.3%) of clove oil (*Syzygium aromaticum*) [1],[2]. The eugenol has been reported with many advantages including antioxidant, antifungi, antiseptic and anticancer [3]. However, as other most plant-derived bioactive compounds, eugenol is volatile and less in bioavailability [4]. Therefore, a technique to overcome these problems is required. The encapsulation technique has been reported as an efficient method to reduce these limitations [4].

Encapsulation is a process by which a sensitive material is coated or entrapped within another material to protect the sensitive substance from the influences of external environment. Another advantage of encapsulation is, to mask the organoleptic properties like colour, taste, odour of the substance and to obtain controlled release of the drug susbtance [5]. Encapsulation of eugenol have been reported in previous studies; there were eugenol loaded-β-cyclodextrin and 2-hydroxypropyl-β-cyclodextrin by inclusion method [4], eugenol loaded β-cyclodextrin grafted chitosan by inclusion complex [6]. Eugenol loaded chitosan by coacervation method has been repoted by Woranuch and Yoksan [7], but the encapsullation efficiency was not more than 25%.

In this study, the eugenol encapsulation has been deeply studied. The encapsulation was done by using chitosan-NaTPP as coating and crosslinker by coacervation method. The in-vitro release characteristic of this encapsulation process is one of the significant findings of this research.

2. Material and method

2.1 Materials

Chitosan (CV. M&H Farm, Indonesia), molecular weight was 624.739 kDa, determined by intrinsic viscosity (Kasaai, 2007), degree of deacetylation (DD) was 64.2 % calculated using baseline method (Khan, 2002). Eugenol 99% (CV. Happy Green, Jakarta), sodium tripolyphosphate (Bratachem,Indonesia), buffer solution pH 7.4 (Flexylab) and buffer pH 1.2; acetic acid (Merck).

2.2 Preparation of eugenol-loaded chitosan particles by coacervation method.

Eugenol-loaded chitosan particles were prepared according to a modified method [8]. Chitosan solution (1.2% (w/v)) was prepared by agitating chitosan in an aqueous acetic acid solution (1% (v/v)) at ambient temperature until it reached homogenous solution. Tween 20 (0.3 g) was added to the chitosan solution (40 ml), and the mixture was stirreed for 30 min to obtain a homogeneous solution. Eugenol was gradually dropped into the stirred mixture, and agitation was carried out for 20 min. The concentration of eugenol was varied (25%, 50%, 75%, 100%, 125%, 150% and 175% (w/w of chitosan)) to obtain different weight ratios of chitosan to eugenol (1:0.25. 1:0.5. 1:0.75. 1:1. 1:1.25. 1:1.5 and 1:1.75. respectively). Tripolyphosphate (TPP) solution 0.5% (40 ml) was slowly dropped into an o/w emulsion while stirring at ambient temperature for 30 min. The formed particles were collected by centrifugation at 7000 rpm for 10 min and subsequently washed several times with deionized water. The susupensions were immediately freeze-dried at -100 °C for 24 h.

2.3 Determination of encapsulation efficiency and loading capacity.

The content of eugenol-loaded in chitosan particles was determined by UV-vis spectrophotometry [9]. Eugenol-loaded particles (±20 mg) sample were taken into methanol (20 ml) and stirrer for 1 h. The supernatant was collected and the eugenol concentration was measured using UV-vis spectrophotometer at wavelenght of 282.5 nm. The amount of eugenol was calculated by appropiate calibration curve of free eugenol in methanol. The encapsulation efficiency (EE) and loading capacity (LC) of eugenol were calculated from Eqs. (1) and (2) respectively :

$$\text{EE (\%)} = \frac{Total\ amount\ of\ loaded\ eugenol}{initial\ amount\ of\ eugenol} x\ 100 \quad (1)$$

$$\text{LC (\%)} = \frac{Total\ amount\ of\ loaded\ eugenol}{weight\ of\ particles\ after\ freeze\ drying} x\ 100 \quad (2)$$

2.4 Instrumental analyses.

FTIR ananlyses for pure chitosan, chitosan-TPP, pure eugenol and eugenol loaded particles were recorded from wavenumber 400-4000 cm^{-1}. Samples were prepared by grinding the dried nanoparticles with KBr and pressing them to form disks. The morphology of the freeze dried particles was studied by scanning electron microscopy (SEM).

2.5 In vitro release studies.

In vitro release was conducted by dissolving 20 mg freeze-dried particles in 20 ml solution containing 60% buffer solution and 40% methanol under gentle agitation. The buffer solution was prepared at pH 1.2 (Simulated Gastric Fluid) and pH 7.4 (Simulated Intestinal Fluid). At specific time intervals, a specific volume of supernatant was sucked out for analysis, and was replaced with an equivalent volume of fresh media. To calculate the total cumulative amount of eugenol release loaded chitosan particles, eugenol concentration (ppm) in the release medium was measured at sampling time intervals by a UV-vis spectrophotometer at 281 nm [10]. Cumulative percentage of eugenol released was obtained by dividing the cumulative amount of eugenol release at each sampling time point to the initial weight of the eugenol-loaded in the sample:

$$\%\text{Release} = \frac{Eugenol\ released\ at\ that\ time}{intial\ weight\ of\ eugenol - loaded\ in\ sample} x\ 100\%$$

3. Results and discussion

3.1 Encapsulation efficiency and loading capacity

The percentages of EE and LC of different formulation were demonstrated in Figure 1. The amount of loaded eugenol was determined using UV-vis spectrophotometry from its absorbance at 282.5 nm. The EE (%) tends to increase by increasing initial eugenol content. The maximum of EE (72.63%) and maximum of LC (43.96%) were obtained for the sample prepared using a weight ratio of chitosan to eugenol of 1:1.5. This result was higher than a study conducted by Woranuch and Yoksan [7]. A decrease of EE at ratio of 1.75 was found due to the loading limitation. These findings are in accordance with previous studies [8], the EE and LC maximum do not necessarily occur at the same point.

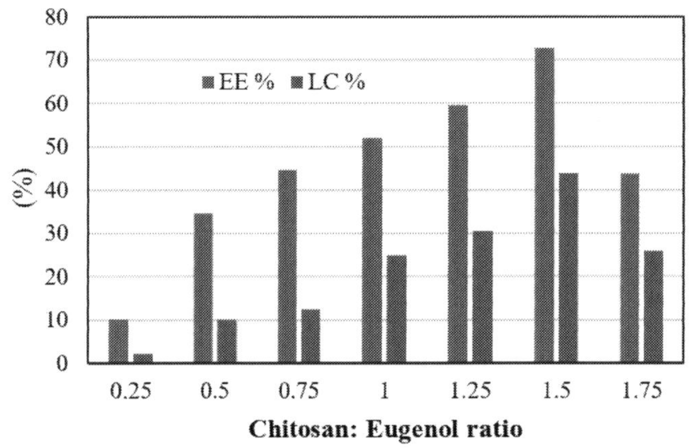

Figure 1. Encapsulation efficiency and loading capacity of eugenol encapsulate

3.2 FTIR characterization

Figure 2 shows FTIR spectra of chitosan powder, chitosan-TPP, eugenol and eugenol-loaded chitosan particles (1:0.25; 1:0.75 and 1:1.5). In general, chitosan powder show characteristic peaks at 3425.58 cm^{-1} (-OH and $-NH_2$ stretching), 2877.79 cm^{-1} (-CH stretching), 1651.07 cm^{-1} (amide I/C=O stretching), 1597.06 cm^{-1} (amide II/-NH bending), 1072.42 cm^{-1} (C-O-C stretching) and 894.97 cm^{-1} (pyranoside ring stretching vibration) [8],[9]. For chitosan-TPP the peak of hydoxyl group is already broader after crosslinked with NaTPP. The peak of amide I (C=O) 1651.07 cm^{-1} and amide II (-NH) 1597.06 cm^{-1} shifted to 1635.64 cm^{-1} and 1543.05 cm^{-1} respectively,and new peaks appeared at 1219.01 cm^{-1} (P=O), implying the complex formation via electrostatic interaction between NH^{3+} groups of chitosan and phosphoric groups of TPP within the particles [8] (N *et al.*, 2012; Yoksan *et al.*,2010). Pure eugenol spectra shows sharp characteristic peaks at 3518.16 cm^{-1} (-OH stretching), 3070.68 cm^{-1} (-CH stretching/sp^2), 2738.92-2908. cm^{-1} (-CH stretching/ sp^3), 1464. cm^{-1} (-CH bending) and 1512.19-1604.77 cm^{-1} (C=C aromatic ring) [7].

The FTIR spectrum of eugenol loaded chitosan (1:0.75 and 1:1.5) show characteristic peaks of eugenol at 1435.08-1458.18 cm^{-1} (-CH_3 bending). The spectrum of product showsed the shifted wavenumber in the –OH stretching to 3417. 86 $cm^{-1.}$ 3417.86 cm^{-1} and 3394.72 $cm^{-1.}$ respectively, which showed the possible interaction of eugenol with chitosan-TPP.

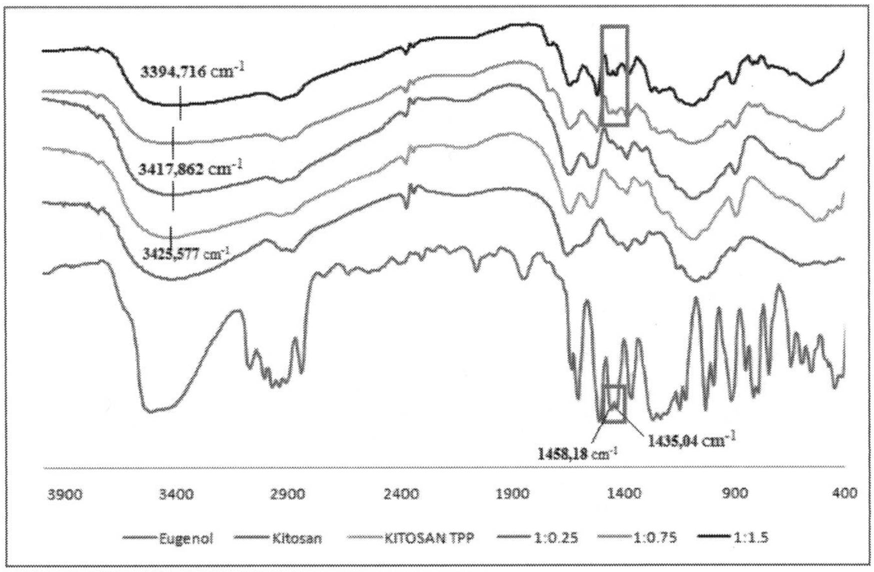

Figure 2. FTIR spectra of eugenol, chitosan , chitosan-TPP and eugenol-loaded chitosan particles 1:0.25; 1:0.75 and 1:1.5

3.3 Scanning Electrom Microscopy (SEM)

The morphology of the particles was observed by using SEM. The SEM images of the chitosan particles and eugenol-loaded chitosan particles (1:1.5). At 100 x magnification, it appears that chitosan-TPP and encapsulated eugenol was flaky and not uniform, this result is similiar to previous study by Kaasgaard and Keller [10] on freeze dried particles. The clumping of particles occured due to effect of centrifugation and freeze drying which cause the agglomerated particles. The freeze drying results of the colloidal suspension produce a porous surface, the pores being generated from water crystals formed during freezing, which are subsequently removed by sublimation [11]. The surface morphology of the eugenol-loaded particles, at 1000 x magnification was rough and more pores and hollow form than the chitosan-TPP.

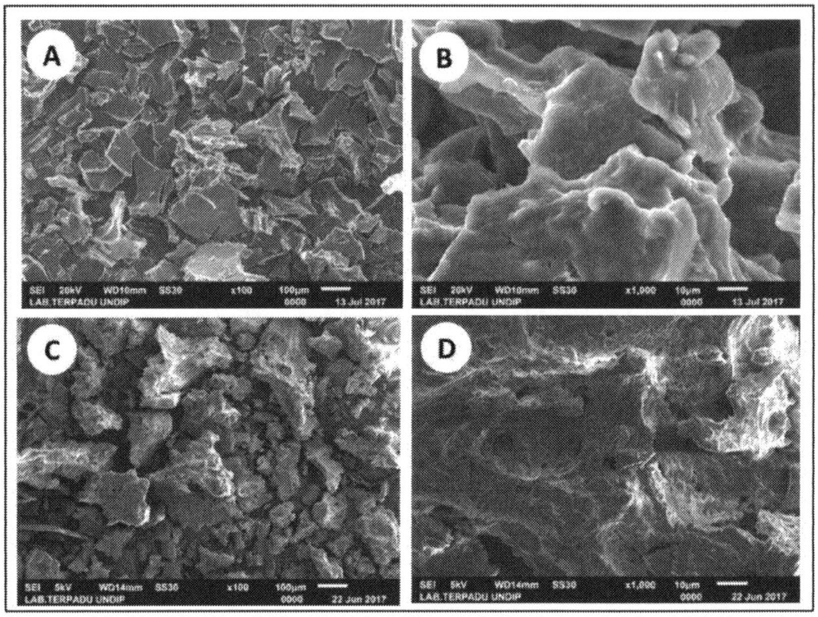

Figure 3. Surface morphology of chitosan-TPP (A) 100x, (B) 1000x and eugenol loaded chitosan particles (C) 100 x and (D) 1000 x

3.4 In vitro release study

The in vitro release profiles of eugenol from particles, prepared using different weight ratio of chitosan to eugenol (1:0.25 and 1:1.5) were shown in Fig 4. The amount of eugenol released at different times was measured at 281 nm. At low concentration of eugenol (25% w/w chitosan) in Simulated Gastric Fluid (SGF) brust effect was 99.3 %, occured within 15 min while at concentration of eugenol 150% was about 86.34%. While at the same concentrations, Eugenol released in Simulated Intestinal Fluid (SIF) wasabout 92.65% and 85.04%, respectively. This initial burst release was attributed to the eugenol molecules adsorbed on the surface of the particles and oil entrapped near the surface [12]. In addition, the very early release of eugenol was due to the presence of methanol in the release medium, wherein eugenol has good solubility in organic solvent. This profil show that the initial burst effect of low concentration of eugenol (25%) was lower than higher concentration of eugenol (150%) in particles. These findings is in accordance with previous studies [13].

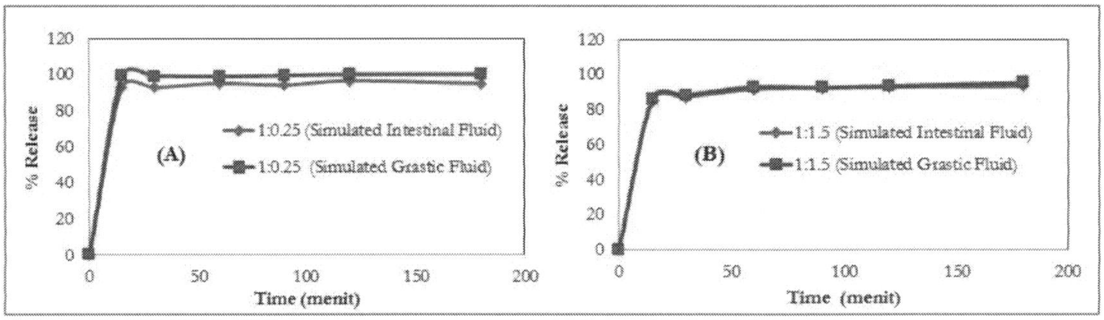

Figure 4. In vitro release profiles of eugenol from chitosan particles prepared using different weight ratio of chitosan to eugenol 1:0.25 (A) and 1:1.5 (B). Values were expressed as mean (n = 2).

In the second stage, the release rate was relatively slow, this might be due to the diffusion of the eugenol dispersed into the polymer matrix as the dominant mechanism. The diffusion occurs in three stages, the first step is penetration of the buffer solution into the particles, causing the polymer swell/ the second step is the change of glassy polymer to rubber matrix then the third stage is the diffusion of the compound from the swollen rubber matrix [14]. The release of eugenol in SGF was faster than SIF due to in acid state (pH <4) almost all free amine groups in chitosan will be protonated, so there will be repulsions between the groups with another amino group on neighboring chain. This causes swelling of the chitosan and the elution of the compound more quickly [15].

4. Conclusion

We found that %EE and %LD of eugenol on citosan-NaTPP encapsulation were higher than previous studied conducted by other researchers. Howeever, in vitro release of eugenol in this encapsulation product has been nearly complete both in the stomach and in the intestinein

Acknowledgments:

The authors would like to acknowledge the financial support from Diponegoro University through PNBPScheme research grant (RPP UNDIP 2017)

References

[1] Wang H F, Wang Y K. and Yih K H. 2008 *Journal of cosmetic science* **6** 509–22.
[2] Jirovetz L, Buchbauer G, Stoilova I, Stoyanova A, Krastanov A and Schmidt E 2006 *Journal of Agricultural and Food Chemistry* **54** 6303–6307
[3] Yoo C, Han K, Cho K, Ha J and Park H. 2005 *Cancer Letters* **225** 41–52.
[4] Choi M, Soottitantawat A, Nuchuchua O, Min S and Ruktanonchai U 2009 *Food Research International* **42** 148–156
[5] Jyothi N, Venkata N, Prasanna P M, Sakarkar S N, Prabha K S, Ramaiah P S and Srawan G Y. 2010 *Journal of Microencapsulation* **27** 187–197
[6] Phunpee S, Saesoo S, Sramala I, Jarussophon S, Sajomsang W, Puttipipatkhajorn S, Ruktanonchai U R 2016 *International Journal of Biological Macromolecules* **84** 472-480.
[7] Woranuch S and Yoksan R 2013 *Carbohydrate Polymers* **86** 578–585
[8] Yoksan R, Jirawutthiwongchai J and Arpo K 2010. *Colloids and Surfaces B: Biointerfaces*, **79** 292–297.
[9] Akyüz L, Duman F and Kaya M 2017 *The Turkish Journal of Pharmaceutical Sciences*, **14** 34–39.
[10] Kaasgaard T and Keller D 2010 *J Agric Food Chem* **58** 2446-54
[11] Sowasod N, Nakagawa K, Tanthapanichakoon W and Charinpanitkul T 2012 *Materials Science and Engineering* **32.** 790–798
[12] Anitha A, Deepagan V G, Rani V V D, Menon D, Nair S V and Jayakumar R 2011 *Carbohydrate Polymers*. **84** 1158–1164.
[13] Fakhreddin S, Zandi M, Rezaei M and Farahmandghavi F 2013 *Carbohydrate Polymers*. **86** 50–56
[14] Agnihotri S A, Mallikarjuna N N and Aminabhavi T M 2004 *Journal of Controlled Release*, **100** 5–28
[15] Magdalena G and Ostrowska-Czubenk J 2016 *Progress on Chemistry and Application of Chitin and its Derivatives* **21** 55–62.

The Role of Pectin in Pb Binding by Carrot Peel Biosorbents: Isoterm Adsorption Study

B Hastuti, F Totiana, R Winiasih

Department of Chemistry Education, Faculty of Faculty of Teacher Training and Education, Universitas Sebelas Maret, Jl. Ir. Sutami 36A Surakarta, 57126, Indonesia

E-mail:budihastuti@staff.uns.ac.id Telp. +628121504044

Abstract. Cheaply and abundantly biosorption available materials such as carrot peels can be a cost-efficient method for removing heavy metals from wastewater. To investigate the role pectin plays in metal binding by carrot peels, commerce pectin was compared. FTIR spectra confirmed the presence of carboxyl and hydroxyl groups in commerce pectin and carrot pectin. Isoterm experiments showed that all materials could remove Pb (II) ion. All of materials binding Pb (II) follow Freundlich models adsorption. The commerce pectin bindsPb (II) by involving energy 16.6 KJ/mole whereas pectin from carrot peel involves energy 21.09 KJ/mole. It indicates that commerce pectin binds the Pb (II) by physics adsorption whereas pectin from carrot peel by physics and chemical adsorption.

1 Introduction

The availability of heavy metals in soil ecosystems and aquatic ecosystems due to anthropogenic activities such as mining, smelting and agriculture operations [1]. Metal species whose its existence in the aquatic environment exceeds the threshold will be a contaminant in the ecosystem [2]. Heavy metals are categorized as environmental pollutants such as Cd, Co, Cr, Pb, As and Ni. Heavy metals contamination in aquatic ecosystems is an important issue, as the impact of heavy metal toxicity and its accumulation in aquatic habitats [3]. One of the dangerous heavy metals is lead (Pb). Lead can cause damage to the central nervous system. Lead also can damage the kidneys, liver and reproductive system, basic cellular processes and brain function. Symptoms of poisoning are characterized by anemia, insomnia, headache, dizziness, irritability, muscle weakness, hallucinations and kidney damage [4].

Waste treatment technologies commonly used and are developing in waste treatment are filtration, precipitation,ion exchange and adsorption. Adsorption especially used when the products from other industries are used as biosorbents [5,6].Compared with other methods, adsorption is one method of an efficient,low-cost alternative waste treatment. Adsorption is process of absorption of a substance (molecule or ion) on the surface of an adsorbent. One of potential biosorbent used as a heavy metal binder is pectin.

Pectin is an anionic plant cell wall polysaccharide based on a-(1–4) linked D-galacturonic acid[7].Pectin widely available in the middle lamella of plant cell walls. Pectin has many active groups. The main functional groups contained in the pectin is hydroxyl, carboxyl, amide, and methoxy [8].Lately, the pectin was widely used in pharmaceuticals and cosmetics, in the food industry and as adsorbent to remove heavy metal ions[9,10]. Pectin contains a lot of active groups, that can be used as a source biosorbent [8]. The main functional groups of pectin are hydroxyl, carboxyl, amide, and methoxyl. These functional groups can be used to bind heavy metals, especially hydroxyl groups [11].

2 Methods

2.1 Materials and Method

Materials used Carrots peel, universal indicator, hydrochloric acid, alcohol96% and alcohol 70%, Pb (II), nitric acid, distilled water.

2.2. Research Method

2.2.1 Isolation of Pectin from Carrot by Extraction Method.

This dry carrot peel is crushed then and mixed, and the product is called carrot flour, and it is used as the raw material for all the assays made concerning pectin extraction and characterization. 1000 grams of carrot flour washed, dried initially at room temperature and then at 50-60°C for about 24 hours. Pectin was extracted under reflux in a condensation system at 90°C for 60 min (solute/solvent 1:20), using water acidified with hydrochloric acid to pH 2.0, using carrot flour as raw material. Next, the flour was classified with metallic sieves (600 μm) in a shaker device.

2.2.2 Precipitation and Washing of Pectin.

The acid pectin extract is precipitated by alcohol acidic solution (1 liter of 96% alcohol solution added with 2 ml of concentrated HCl) at a ratio 1:1(v/v) for 24 hours. The floating pectin precipitate is filtered and be separated from the filtrate. Then the pectin precipitate is washed with alcohol up to several times. The pressed pectin is dried to constant weight at 40-50°C for approximately 6 hours. Cooled it in a desiccator and the yield calculated on a dry weight basis (initial weight of sample). The hard pectin cake was broken up, ground and sieved to obtain powdered pectin.

2.2.3 Isoterm Adsorption Determination.

Pectin20 mg was immersed in a 20 ml Pb (II) metal solution. Concentrations of each solution were variations of 5, 10, 15 and 20 ppm. After being interrupted for 120 minutes, then filtered, the contacting filtrate was tested with AAS.

3 Results and Discussion

3.1 Characterization of the Adsorbent

The isolated pectin is tested with FT-IR spectroscopy to determine the esterification degree of carrot pectin and compare with commerce pectin spectra. According to Manrique&Lajolo[12], theesterificationdegreeis determined by counting the number of active group of carboxylates (located in about 1730 cm^{-1} area) divided by the total number of carboxylate groups (the areas ware between 1730 and 1600 cm^{-1}) then multiplied by 100.

From FT-IR spectra in Fig.1can be shown that the spectra changes of carrots powder before extracting, pectin carrots and spectra from Commercial pectin. Pectin spectra has typical absorption of a strong band between 1760-1730 and 1630-1600 cm^{-1} which is a carbonyl ester group and a stretching band of carboxylic ions.

In the carrot spectra before extracting there is no absorption in 1700 cm^{-1}indicate the presence of an esterified carboxylic group. Whereas after extracting the absorption appeared in 1751cm^{-1}indicate the presence of an esterified carboxylic group and the spectra in the 1636 cm^{-1}indicate the presence of a carboxylic group.

Pectin has a fingerprint region of 1300-800 cm^{-1} (there is a COO-stretching group). Other important absorption bands on pectin are C-H bending at 1380 cm^{-1} and C=O stretching at 1300-1000 cm^{-1}.From

FT-IR spectra can be calculated the esterification degree of carrot is 56,82%, and commerce pectin is 59,32%. The esterification degree (DE) also called the methoxylation degree is the proportion of galactic acid in the form of a methyl ester which is usually expressed by the percentage or percentage of the carboxyl group that was methylated with methanol. DE affects of a pectinproperties. Otheridentification of pectin properties,i.e., moisture content andthe moisture content of carrot pectin was 14%.

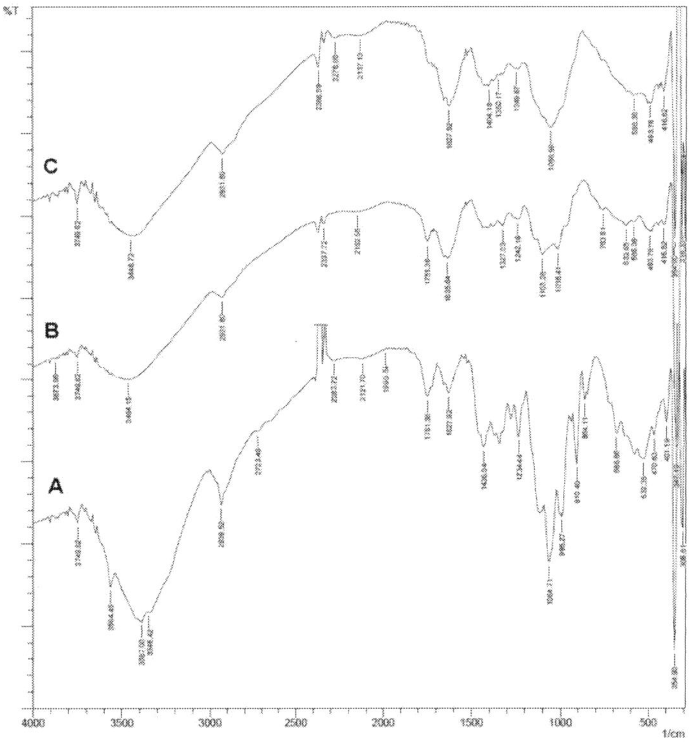

Figure 1. FTIR Spectra of A. carrot powder, B. carrot pectin, and C. commerce pectin

3.2 Isoterm Adsorption Determination

The adsorption isotherm saw the relationship between the amounts of adsorbed ions per weight of the adsorbent with the ion equilibrium concentration in the solution. By determination of the adsorption isotherm model, an adsorption mechanism of Pb (II) canbe predicted on the adsorbent. The mechanism adsorption of pectin was studied using Langmuir and Freundlich isotherm adsorption models. The Langmuir isotherm adsorption model describes monolayer adsorption on a homogeneous surface with the same energy at all the active sites involved. The Freundlich isotherm adsorption model describes multilayer adsorption on heterogeneous surfaces. The determination of the adsorption isotherm model is shown in Table 1.

From Table 1 it appears that the Freundlich correlation coefficient is greater than the Langmuir correlation coefficient, so it can be assumed that both pectin adsorbents in adsorbing Pb (II) ions follow the Freundlich isotherms models in multilayer surface. It indicates that the reaction between the adsorbatein adsorption of Pb (II) metal ions occursby physical mechanism, adsorption happensin a multilayer. The energy (ΔG) of commerce pectin and carrot pectin was 16.60 and 21.09kJ /mole respectively. The adsorption energy value under 20 kJ/mole indicates that adsorption process occurs by physical adsorption. The adsorption energyvaluebetween 20-80 kJ/mole indicates that adsorption process occurs by physical and chemical adsorption [13]. In this research, commerce pectin, to bind the Pb (II) occurring by physical adsorption, The carrot peel pectin, adsorp to Pb (II) by physical and chemical adsorption Process. The Pectin and Chitosan adsorbent generate energy for 16,6 and 19,54 kJ/mol respectively.

Adsorbent	**Parameter**						
	Langmuir				Freundlich		
	b (mol/g)	K (L/mol)	R^2	ΔG (kJ/mol)	n	K_f	R^2
Carrot pectin	$1.08x10^{-4}$	4.698	0.3921	21.09	1.46	0.912	0.962
Commerce pectin	$4.167x10^{-3}$	0.187	0.15	16.60	1.04	0.936	0.866

4 Conclusion

From the research results can be concluded that: Pectin can be isolated from carrot peel that proved by FT-IR test results and test of water content on the adsorbent. The esterification degree of pectin from carrots is 56.82%, and commerce pectin is 59.32%. Carrots pectin can be used to adsorb lead metals. The adsorption isotherm pattern occurring in pectin commerce and carrot pectin to lead metal tends to follow Freundlich isotherm equation.

References

1 Gowdhaman, P., Antonyraj, K., and Annamalai, V2015 *Inter. J.of Advances in Scientific Research* **8** 322–28.
2 Hameed, A., Obaidy, M. J. Al, Talib, A. H., and Zaki, S. R. 2014 *Inter. J. of Advances in Scientific Research* **8** 947–52
3 Muley, D. V. 2013 *Universal Journal of Environmental Research and Technology* **6**690–94
4 Fu, F., & Wang, Q. 2011 *Journal of Environmental Management*, ***92***(3) 407–18.
5 Niragu, J.O. and Sprague J.B., 1987Cadmium in the Aquatic Environment, Wiley-Interscience, New York.
6 Bailey, S.E. Olin, T.J. Bricka, R.M. Adrian, D.D. 1999 *Water Research* **33** (11) 2469–79.
7. Renard, C.M.G.C Crepeau,M.J. andThibault, J.F. 1995*Carbohydrate Research*, **275**, 155–165,
8. Mata, Y.N. Blázquez, M.L. Ballester, A.González ,F. and Mu˜noz, J.A. 2009*Chemical Engineering Journa l***150**, 289–301
9. Fares , M.M., Yahya R.,Tahboub, Khatatbeh,S.T., Yousef M, AbulH., 2011 *Journalof Polymers and Environment***19**431–39
10. Schiewer, S.andIqbal, M.2010, *Journal of Hazardous Materials* **177**899–907
11. Kupchik, LA., Kartel, N.T, Bogdanov, E.S, Bogdanov,O.V.,and Kupchik, M.P, 2006, *Journal of Macromolecul Chemical and Polymer Material* **79** 457-60
12. Wong, W.W., Phuah, E.T., Al-Kharkhi, A., Liong, M.T., Nadiah, W.A., Rosma, A., and Easa, A.M.2008.Proceedings International Conference on Environmental Research and Technology: Penang (Malaysia), 92-4
13. Solener, M., Tunali, S., Ozcan, A. and Gedikbey, T. 2008 *Desalination* **223** 308-22

New Silica Magnetite Sorbent: The Influence of Variations of Sodium Silicate Concentrations on Silica Magnetite Character

C Azmiyawati[1], P I Pratiwi[1], and A Darmawan[1]

[1]University of Diponegoro, Indonesia

Email: choiril.azmiyawati@live.undip.ac.id

Abstract. The adsorption capacity of an adsorbent is determined by the adsorbent and the adsorbate properties. The character of the adsorbent will play a major role in its ability to adsorb the corresponding adsorbate. Therefore, in this study we looked at the effects of variations of sodium silicate concentrations on the resulting magnetite silica adsorbent properties. The application of silica coating on the magnetite was carried out through a sol-gel process with sodium silicate and HCl precursors. Based on the characterization data obtained, it was found that the silica coating on magnetite can increase the resistance to acid leaching, increase the particle size, but decrease the magnetic properties of the magnetite. Based on Gas Sorption Analyzer (GSA) and X-ray Difraction (XRD) data it can successively be determined that increase in concentration of sodium silicate will increase the surface area and amorphous structure of the Silica Magnetie.

1 Introduction

Magnetite is a particle that has super paramagnetic properties often used in a variety of applications; one of which is as an adsorbent support material [1]. Magnetite can be pulled strongly by a magnetic field [2], thereby facilitating the separation process between adsorbent and adsorbate by utilizing the help of external magnetic field. Unfortunately, the main composition of magnetite is iron which is highly vulnerable to oxidation. The silica coating on the magnetite in addition to preventing corrosion also generates the active side of the silanol and siloxane groups in the resulting silica magnetite. It is known that silica containing silanol and siloxane groups can bind metal ions. In addition, the silica coating on magnetite provides inert, non-toxic, and biocompatible properties to the human body thus its application is often used in biotech and biomedical practices.

Studies of modified silica magnetics have previously been conducted. A study previously conducted by Mao et al. [3] analysed the synthesis of Fe_xO_y @ silica-pillared clay (SPC) by sol-gel method to control drug release and release. Another study was also conducted by Merdekani [4] concerning the synthesization of Fe_3O_4 / SiO_2 nanocomposite particles using coprecipitation methods. According to Merdekani, the nature of Fe_3O_4 / SiO_2 nanocomposite particles will decrease as heating and increase time increases. Other studies were also performed by incorporating triethylenetetramine on the surface of nanomagnetic iron oxide by using nanosilicon oxide and sorbent yields with covalent triethylenetetramine bonds.

One of the most popular methods used to coat magnetite with silica is the sol-gel method [6]. The sol-gel method is chosen because it has several advantages: better homogeneity, high purity, relatively low temperature, no reaction with residual compound, and minimal evaporation loss. [7] Silica gel can be synthesized via a sol-gel process through condensation using a solution of sodium silicate in an acidic atmosphere. The thickness of the coating and the number of functional groups present on the magnetite surface will affect the adsorption ability of an adsorbent. Therefore, this study will discuss the effect of sodium silicate on the silica layer of the magnetite. Subsequently, with the information gained each character of the silica magnetite formed and the optimum sodium silica concentration can be determined and used as a reference for further researches.

2 Experimental

2.1 Chemicals and reagent.

Magnetite particles used as the core material in the synthesis of magnetic silica were obtained from the shores of Marina beaches. The magnetite particle size used is 100-170 mesh. Other ingredients used werere HCl and Na_2SiO_3 from Merck.

2.2 Instrumentation

The tools used in the extraction of magnetite from natural sand were permanent rod magnets and Vibrational Ball Mill (VBM), Fourier transform infrared spectroscopy (FT-IR) Prestige-21 Shimadzu, powder X-Ray Difractometer (XRD) Shimadzu XRD-6000, and Gas Sorption Analyzer Quantachrome Instruments versions 2.2 .

2.3 Synthesis of adsorbent

In order to create a solid shell on the surface of the magnetite particles (Fe_3O_4), 5 g Fe_3O_4 were mixed with 400.0 mL of sodium silicate 5% (v/v). Next, HCl 1 M was added dropwise as the mixture was stirred until pH = 7 at room temperature. The gel was aged for further 24 h, and dried in an oven for 2 h at 80°C. The gel was thoroughly washed with distilled water. The particles were separated with a magnet and was air dried. The same steps were applied for each sodium silicate concentrations of 6.7%, 10%, and 20% (v/v). The varying sodium silicate concentrated particles formed and the blank were then coded according to the concentration of sodium silicate, i.e., MS 5%, MS 6.7%, MS 10%, MS 20%, and blank as S. Determination of the characteristics of each products was carried out using FTIR, XRD and GSA to determine the functional group, crystallinity, surface area and pore size contained in the product.

3 Results and discussions

Not only does the silica layer on the magnetite increases its magnetic stability, but it also increases the absorption and desorption ability of the magnetic silica against metal ions. The silica layer on the surface serves as a shield against detrimental environmental effects. Thus, the magnetite becomes more stable and the conversion to FeOOH is curbed. Silica can block magnetic polarization attraction between particles, produce particles dispersed in a liquid medium and protect from damage due to acidic atmosphere. In addition, the presence of silanol groups on the silica coating helps in the adsorption process, especially for metal cations. Aside from being a cation-bonding medium, the silanol group is also a binding site for other functional groups such as sulphonates, thiazols, and surfactants, for instance cetyltrimethylammonium bromide, to facilitate the application of magnetite as an adsorbent.

Visually, the magnetite-silica sorbent produced takes is in a form of grey powder. Figure 1 shows samples of magnetite powder (Figure 1a), silica gel (Figure 1b) and magnetic silica (Fe_3O_4 / $mSiO_2$) (Fig. 1c). The colour change that occur indicate the presence of magnetite in silica. The black magnetite and the white silica gel subsequently produce magnetite-silica that is grey. The weight of the obtained magnetic silica produced is shown in Table 1.

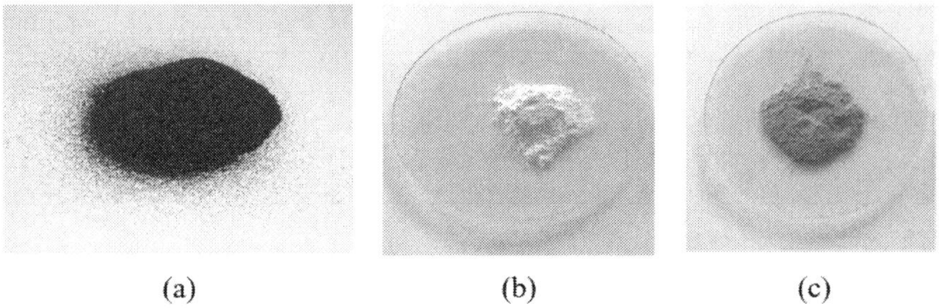

Figure 1 Powder image of magnetite (a), silica gel (b), and magnetite-silica (c)

Based on the data acquired, the greater the concentration of sodium silicate used, the greater is the weight of the resulting silica magnetic particles. Due to the increased concentration of Sodium silicate, silicic anion is more present in the solution. The silicate anion will condense and form a silicate polymer so that more gel is formed. Figure 2 shows the overall reaction of the silica incorporated magnetite layer.

Table 1. Product yield

Sample	Yield **(gram)**
S	10.153
MS 20%	20.0303
MS 10%	11.285
MS 6.67**%**	8.978
MS 5%	7.524

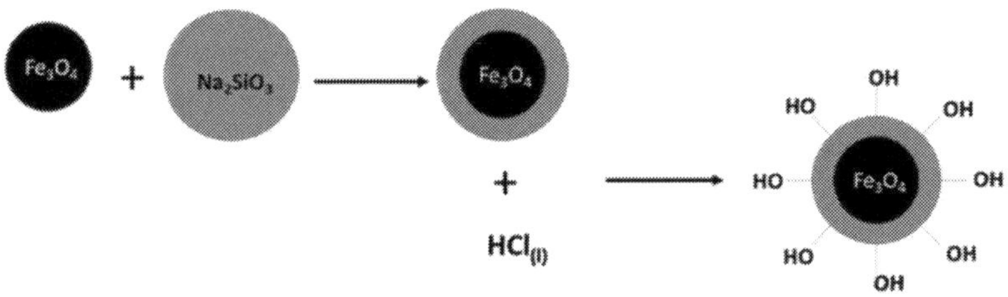

Figure 2 Overall reaction of the silica incorporated magnetite layer

3.1 Functional Groups of the Silica Magnetite

The presence of functional groups on the silica magnetite was determined by FTIR spectra analysis shown in Figure 3. The absorption capacity for silica without magnetite appears at 3400-3450 cm^{-1} wave number, indicating the presence of vibrations of the free -OH group. The wide and sharp absorption of the wave number 1072-1126 cm^{-1} shows the asymmetric Si-O group of Si-O-Si. The 1620-1640 cm^{-1} wave number denotes the vibrations of the -OH group of water molecules, while the wave number 950-960 cm^{-1} shows the movements of the Si-O group of Si-OH. The absorption at 787-802 cm^{-1} shows the symmetric Si-O group of Si-O-Si and the uptake at 450-470 cm^{-1} indicates the bending vibration of the Si-O-Si group.

Compared to silica without magnetite, magnetic silica has a pattern similarity, followed by a decrease in the intensity of multiple bands, multiple band shifts and the appearance of new bands in absorption around 300 cm-1 wavenumbers indicating the presence of Fe metals. The presence of magnetite particles is shown by the peak of the Fe-O group vibration absorption derived from the bonds in Fe_3O_4 molecules at 1000-1200 cm^{-1} wavelengths and Si-O groups at 970 cm^{-1}. [8] In addition, the silica magnetite coating is shown by the emergence of vibrations corresponding to the Si-O-Si bonds in the 1000-1200 wave numbers and the Si-O groups at 970 cm^{-1} [8].

Clearly observed from the peak level of absorption in the infrared spectrum, a sample of magnetic silica with 20% Na-silicate is the best sample of magnetic silica used as a heavy metal ion adsorbent. This is because the absorption peak of -OH of Si-OH at wave number 3448 cm^{-1} has the largest area. This shows that there are a number of -OH terminal groups in the magnetic silica structure. The -OH terminal group will then bind heavy metal chemically (chemisorption).

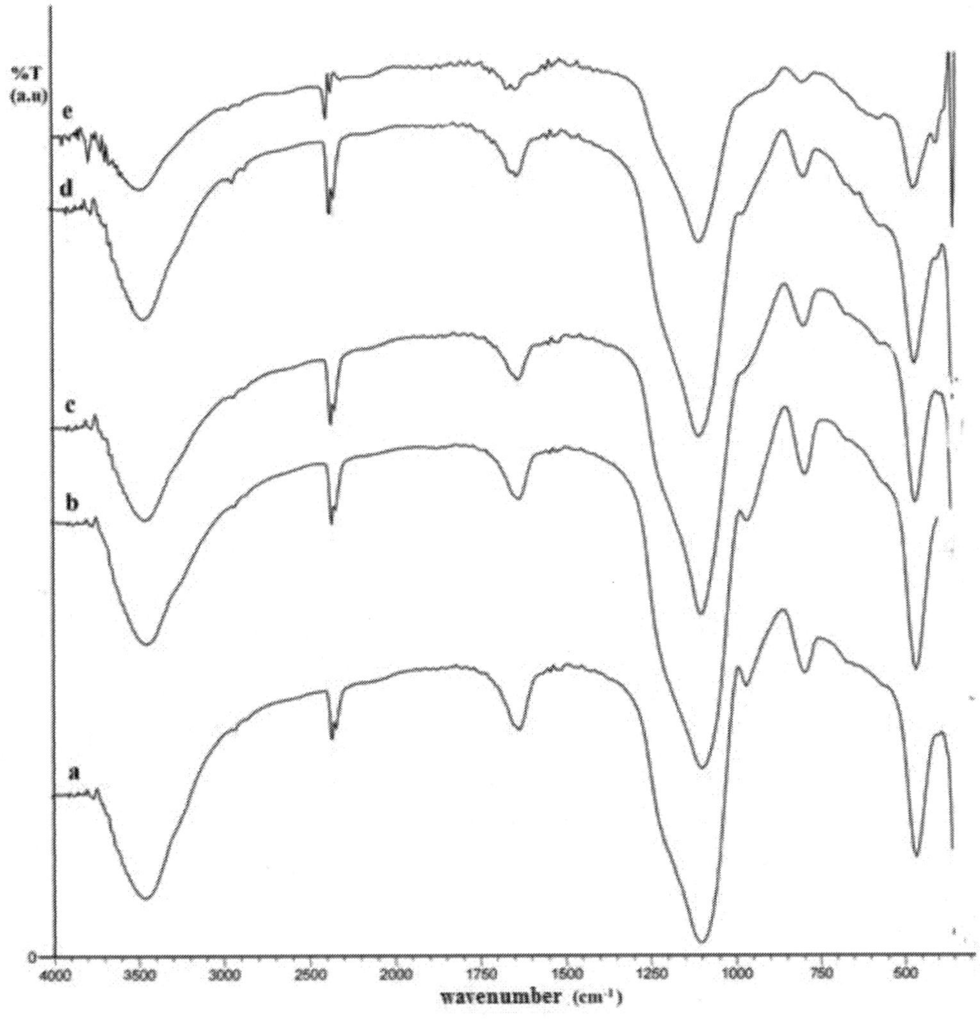

Figure 3 IR spectra of silica gel (S) a), b) MS 20%, c) MS 10%, d) MS 6.67%, and e) MS 5%

3.2 Character of Silica Magnetite

The success of the silica layer on the magnetite can be seen from the seven diffractograms in Figure 4. It can be seen that the presence of silica in magnetite is indicated by the larger peak of amorphous magnetic silica having a greater silica concentration. Therefore, it can be said that the addition of sodium silicate concentration increases the amorphous properties of the adsorbent. That is, the formed silica layer gets bigger with the addition of sodium silicate.

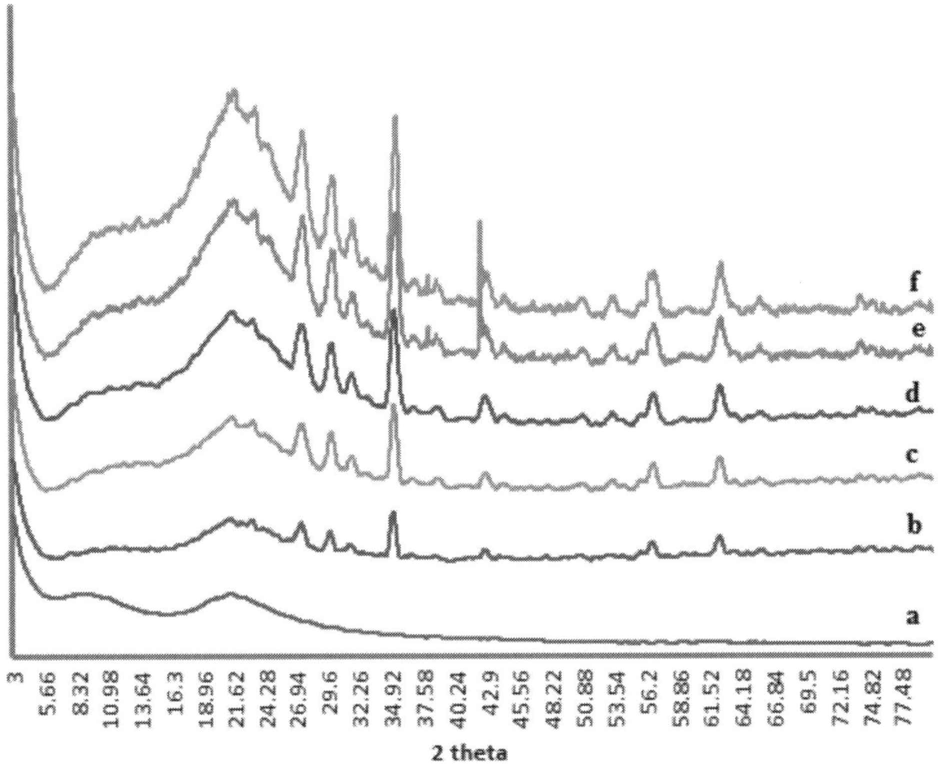

Figure 4 Diffractogram of a) Silica, b) Natural magnetite, c) MS 5%, d) MS 6.67%, e) MS 10%, and f) MS 20%

This silica-coated magnetite is expected to be used in the process of desorption of metal ions by retracting the adsorbent using an external magnetic field so that the adsorbent can be separated from the adsorbed and reusable metal ions. Increasing the amount of silica present in the magnetite layer will forge the active side of the adsorbent in the form of a silanol group thus the adsorbent is expected to bind more metal ions. The addition of the silanol active group can be seen from the results of the FTIR spectra for each of the adsorbents. As can be seen, MS 20% has the sharpest and widest band of about 960 cm^{-1} which shows more silanol groups than the same absorption bands on other spectrums.

Based on the adsorption isotherm curve in Fig. 5, the higher the concentration of sodium silicate, the bigger is the surface area of the magnetite-silica. Overall, the magnitude of surface area, pore size and pore volume are shown in Table 2.

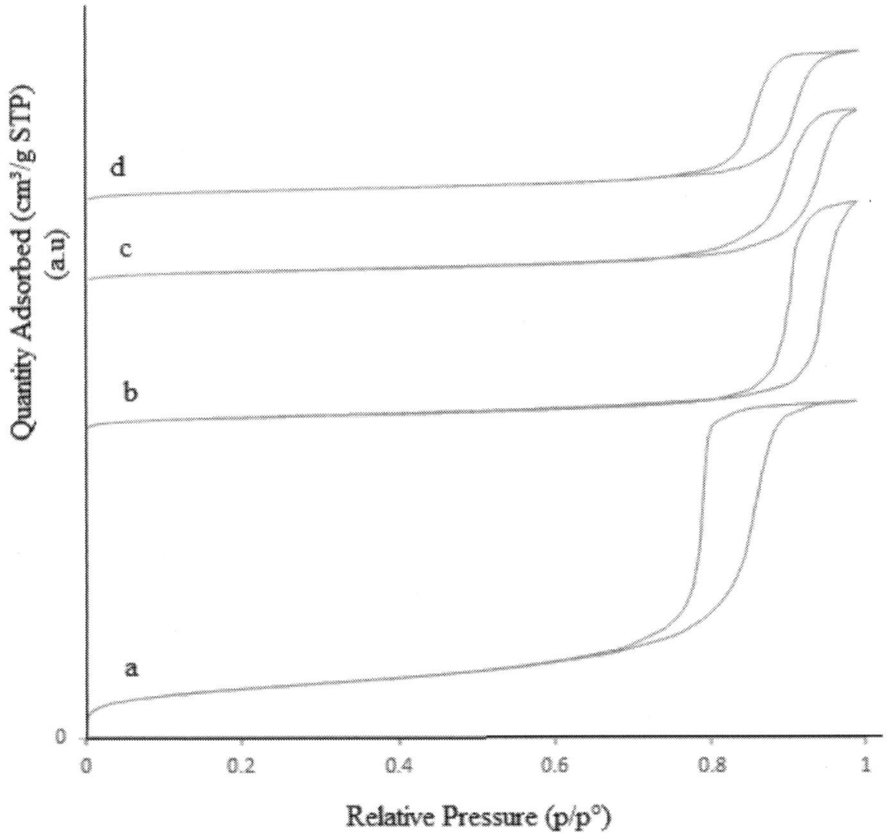

Figure 5 Isotherm adsorption of a) silica, b) MS 20%, c) MS 10%, d) MS 5%

Table 2: Surface area, pore volume and pore size of adsorbent

Adsorbent	Surface Area (m^2/g)	Pore Volume (cm^3/g)	Pore size (nm)
S	266.5461	0.822	12.092
MS 20%	102.3459	0.509	19.511
MS 10%	88.3644	0.414	18.410
MS 5%	86.6157	0.380	17.265

Without magnetite, silica has a larger surface area and the volume than magnetite-silica. On the other hand, the pore size of silica is smaller than silica magnetite. The addition of silica to magnetite will create a product with substantial surface area of the as well as the pore volume. Increase in the surface area of the magnetite signifies the increasing amount of silica gel that lines the magnetite. With the addition of sodium silicate concentration, the amount of silica gel coating will increase.

4 Conclusion

Magnetite modified silica was successfully prepared as proven by diffractogram results. The magnetite already exists in magnetic silica compounds. Increased concentrations of sodium silicate affect the number of Si-OH groups and the surface area of magnetite-silica produced. The results explain that the increase in the number of Si-OH group and the surface area will affect the adsorption capacity of the product.

Reference

[1] Eivari, H. A., Rahdar, A., 2013, Some Properties of Iron Oxide Nanoparticles Synthesized in Different Conditions, *World Appl. Programming*, 3, 52-55

[2] Del Campo, A., Sen, T., Lellouche, J.P., Bruce, I. J., 2005, Multifunctional Magnetite and Silica-Magnetite Nanoparticles: Synthesis, Surface Activation and Applications in Life Scince, *J. Magn. Mater*, 293, 33-40

[3] Mao, H., Liu, X., Yang, J., Li, B., Yao, C., & kong, Y., 2014, Synthesis of Magnetic Fe_xO_y@silica-Pillared Clay (SPC) Composites Via a Novel Sol-Gel Route for Controlled Drug Release and Targeting, *Materials Science and Engineering: C,40*, 102-180

[4] Merdekani, sera., 2013, Sintesis Partikel Nanokomposit Fe_3O_4/SiO_2 dengan Metode Popresipitasi, *Prosiding Seminar Nasional Sains dan teknologi nuklir PTNBR – BATAN*, 472-477.

[5] Mahmoud, M.E., Abdelwahab, M.S., & Fathallah, E. M, 2013, Design of Novel Nano-Sorbents Based on Nano-Magnetic Iron Oxide-Bound-Nano-Silicon Oxide-Immobilized Triethylenetetramine for Implementation in Water Treatment of Heavy Metals, *Chemical Engineering Journal*, 223, 318-327

[6] Duhan, S., Devi, S., 2010, Synthesis and Structural Characterization of Iron Oxide-silica Nanocomposites Prepared by the sol-gel Method, *International Journal of Electronic Engineerin*, 2, 89-92.

[7] Kazemzadeh, H., Ataie, A., Rashci, F., 2011, In situ Synthesis of Silica coated Magnetite Nanoparticles by Reverse Coprecipitation Method, J. *Supercon. Nov. Magn*, 25, 2803-2808

[8] Nuryono, N. M. Rosiati., S.C.W. Sakti dan S. Tanaka., 2012, Coating of Magnetite with Mercapto Modified Silica in a One-pot Process. *Indonesian Ministry of Education and Culture through the International Collaboration Research Grant No. 180/SP2H/PL/Dit.litabmas/IV/2012,* 1-19

IOP Conf. Series: Materials Science and Engineering **349** (2018) 012013 doi:10.1088/1757-899X/349/1/012013

Surface and Groundwater Interactions: Cikapundung Bandung, Kanal Banjir Timur Semarang and Cisadane Tangerang

D E Irawan[1*], E Sulistyawati[1], A A Midori[1], B Faisal[1], A Darul[2], A Agustin[3]

[1]Institut Teknologi Bandung, [2]Institut Teknologi dan Sains Bandung, [3]Independent researcher

E-mail: dasaptaerwin3@gmail.com

Abstract. In most Asia countries, the riverbank area is mostly inhabited by the low-income population, due to the shortage of formal housing. Most of the settlement areas are not equipped with proper sanitation system. Hence, the water quality gets lower over time with the increasing number of inhabitants around the riverbank. Th water quality gets worse with the close hydrological connection between surface water and the shallow groundwater. We compare the state of water quality based on our three case studies: Cikapundung Bandung, Kanal Banjir Timur Semarang, and Cisadane Tangerang. In each location, we gathered the following data: water level measurements, water flow mapping, and water quality samples. Then we make maps to evaluate existing status. The comparison will be made based on the physical and chemical properties that we get from the field. On all locations, we find very close interactions between surface water and groundwater. The hydrological connections are different in direction from upstream to downstream: gaining stream, combined stream or perched stream, and losing stream. However different river gradient gives a slightly different length of hydrological zonations. All samples show a high bicarbonate from rain water, the dissolution of carbonate minerals from the rocks and soils, and also organic species from microbial activities, which induced by domestic wastes. However, we need to make a carbonate balance calculation to break down the components. All samples also have high nitrate and nitrite concentration which come from domestic waste along the river and fertilizer from the rice fields upstream (only in Cikapundung river). For further research, we suggest chemical modeling to break up the contamination components and possible sources.

1 Introduction

In most Asia countries, the riverbank area is mostly inhabited by the low-income population, due to the shortage of formal housing. Most of the settlement areas are not equipped with proper sanitation system. Hence, the water quality gets lower over time with the increasing number of inhabitants around the riverbank. The water quality gets worse with the close hydrological connection between surface water and the shallow groundwater.

Several studies on surface and groundwater interactions have been done in several Metropolitans in Asia (references) and SE Asia (references). Aside to that we also found several existing studies from Indonesia: Cikapundung, Ciliwung, Cisadane, and Bengawan Solo (references) (Figure 1). All studies identify that surface water and groundwater have close connections as most of the metropolitans are located in lowlands, close to the coastal area, on top of a fluvial system. They also point out that water quality is closely controlled by anthropogenic contaminants, households waste, and agricultural waste.

In this paper, we compare the state of water quality based on our three case studies: Cikapundung Bandung, Kanal Banjir Timur Semarang, and Cisadane Tangerang. We need to get typical characteristics of river bank based on geological and water resources properties.

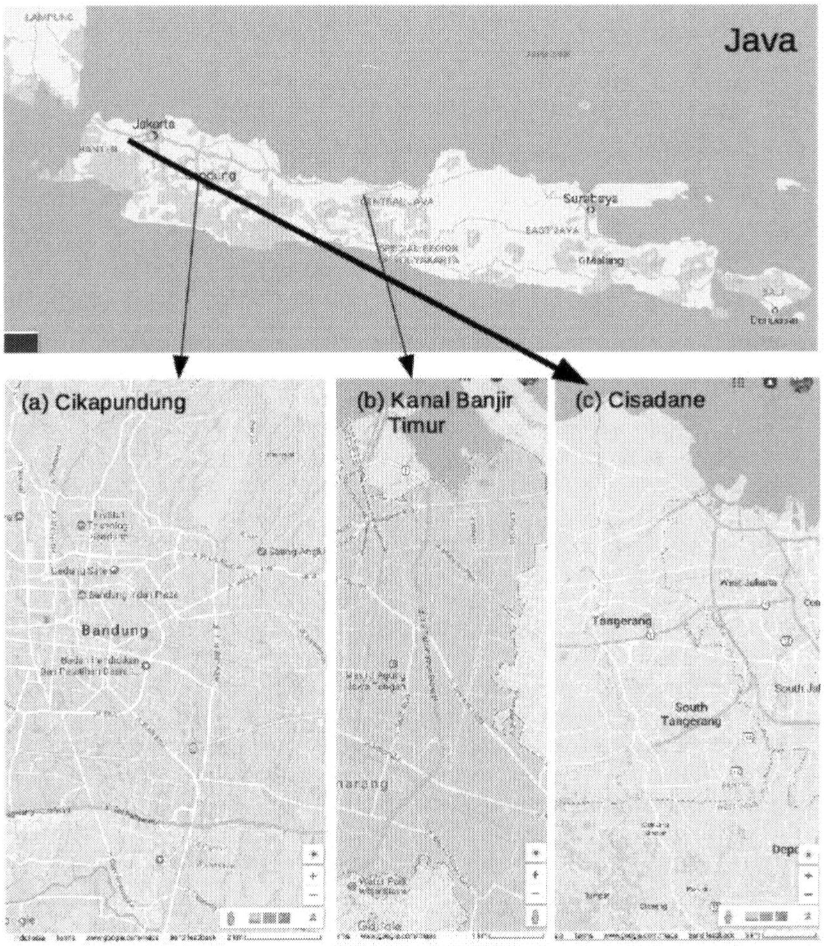

Figure 1. Insert location maps: (a) Cikapundung-Bandung, (b) Kanal Banjir Timur-Semarang, (c) Cisadane-Tangerang.

2 Materials and methods

In each location, we gathered the following data: water level measurements, water flow mapping, and water quality samples (Irawan, 2017). Then we make maps to evaluate existing status. The comparison will be made based on the physical and chemical properties that we get from the field (Table 1). All data will be made public after the submission of the full paper on OSF repository. For water level measurements, we used Solinst water level meter, and for TDS, temperature, and pH measurements we used Hanna Instruments portable equipment. All equipment was calibrated before field measurements.

Table 1. Data

Locations	Parameters	Sampling/ observation points
Cikapundung	Water level	50
	Water samples	20
Kanal Banjir Timur	Water level	10
	Water samples	11
Cisadane	Water level	50
	Water level	10

3 Results and discussions

The hydrological connections are different in direction from upstream to downstream: gaining stream, combined stream or perched stream, and losing stream. However different river gradient gives a slightly different length of hydrological zonations. All samples show a high bicarbonate from rain water, the dissolution of carbonate minerals from the rocks and soils, and also organic species from microbial activities, which induced by domestic wastes. However, we need to make a carbonate balance calculation to break down the components. All samples also have high nitrate, nitrite, and sulfate concentration which come from domestic waste along the river and fertilizer from the rice fields upstream (only in Cikapundung river and some part in Kanal Banjir Timur river). Another similarity is all three rivers serve as one of the sources for public water supply managed by local government water company (PDAM). The condition should give more reason for the government to endorse a more strict water conservation law in the areas. The comparison summary of the three rivers/channels is presented in Table 2.

3.1 Cikapundung

Cikapundung is a good example showing the evolution of water quality from highlands to lowlands in a short distance, less than 100 km. The morphological features quickly change from sharply eroded mountainous terrain to a flat urban city landscape. The high gradient gives the advantage of self-natural flushing during the wet season. Looking at this feature, Cikapundung has a more favorable condition compared to the other two sites. However, it holds another risk. With poor drainage management, this rapid natural flushing system will quickly turn into flash floods in seconds. Several authors observed the connection between groundwater and river water in the river bank. It shows that the change of connection from upstream to downstream: no connection (zero flux), gaining stream, mixed stream, and then losing stream [1-3].

3.2 Cisadane

Cisadane riverbank has the longest distance from the upstream compared to Cikapundung dan KBT. The water flows through several cities, Bogor, Depok, and Jakarta with heavy urban housing in the riverbank. This setting gives a severe contamination potential to this river. A set of 60 observations along the riverbank which shows that the TDS values are three or four times higher in the downstream than upstream [4], which is due to the active large and home-scale industries along its riverbank, including electro-plating home industries.

3.3 Kanal Banjir Timur (KBT) Channel

Kanal Banjir Timur has a length of 27.82 km which flows from the south of Semarang City to the Java Sea on the north of Semarang City. In the upper reaches, there are plantation areas and residential areas along the river, while on the downstream, there are residents and rice fields along the river. To know the relationship between river water and groundwater, then test water quality. There are 11 points of water sampling along KBT. 7 points are water from citizen well, and 4 points are water from river KBT. Water sampling is conducted in the downstream river area (data set is available at [5]). Here we also see a close similarity of groundwater with river water. Of which, we see an increase of TDS, nitrite, nitrate, and sulfate values three times from upstream and downstream. The KBT channel has a more reserved riverbank as no settlements are allowed to be developed in its river bank.

3.4 Regulations, references, data loggers

The water quality flowing on the three rivers and the groundwater flowing in the aquifer in the riverbank area doesn't getting better unless we make efforts to manage the water. Recently the only law concerning water management (UU No. 7/2004) was evaluated and withdrawn with no revised version up to now. The no revised version would a drawback for water regulators as no other national law (UU) that regulates water resources as such details as the withdrawn law [6]. However, the references written in this here are limited, due to the limited online availability of published scholarly works. More detail literature study should be done using text mining and reference manager to visualize all researches that have been done in all three areas [7]

Technically, we see that all water level observations were not done rapidly to make a good time series data. We propose more researchers in these areas should use data logger tools to record the water level and water temperature in the riverbank intensively to understand the interactions between groundwater and surface water.

Table 2. A summary of river comparison.

No	Rivers	Length (km)	Basins (km2)	Min elevation (mdpl)	Max elevation (mdpl)	Gradient
1	Cikapundung	28.5	0.4	600	1500	31.6
2	Cisadane	137.6	1367	0	3002	21.8
3	KBT	27.82	74.07	0	1300	46.7

No	Rivers	Water similarity*	TDS increase	Season when observed	Morphological setting	
1	Cikapundung	Similar	2 times	Combined	Highlands	
2	Cisadane	Similar	3 times	Combined	Coastal	
3	KBT	Similar	3 times	Combined	Coastal	

4 Conclusions

Water quality is highly controlled by land use and anthropogenic activities. Therefore the authorities must have a strong control on those aspects to conserve water quality, surface water as well as groundwater. The results of this paper show a strong control of anthropogenic activities to steer water quality. However, due to limited time series data, we still cannot determine the exact component in the process. Hence for further research, we suggest chemical modeling to break up the contamination components and possible sources.

Moreover, we need to gradually, restore the role of river bank zone as a buffer zone, using more engineering efforts. The efforts are due to the vast development of unregistered settlements along the riverbank. We also suggest for the authorities to compose more integrated rules on the subject, to replace the withdrawn UU 7/2004.

Acknowledgments
Authors appreciate the students who helped with data collecting in each area. We also thank ITB as grant funder for this research.

References

[1] Satrio, Paston S., Sum, LC., and Syafalni, S., Groundwater Dynamic and Its Interrelationship with River Water of Bandung Basin Using Environmental Isotopes (18O, 2H, 14C), Modern Applied Science, Vol. 6, No. 11, 2012, link.

[2] Darul, A., Irawan, DE., and Trilaksono, NJ., Groundwater and river water interaction on Cikapundung River: Revisited, AIP Conference Proceedings, 1677, 110007, 2015, link.

[3] Darul, A., Irawan, DE., Trilaksono, NJ., Pratama, A., Fitria, UR., Conceptual model of groundwater and river water interactions in Cikapundung riverbank, Bandung, West Java IOP Conference Series: Earth and Environmental Science 29 (1), 2016, link.DE Irawan, DJ Puradimaja, D Yeni, AA Kuntoro, MM Julian, Decreasing groundwater quality at Cisadane river banks: groundwater-surface water approach, ArXiv preprint, 2016, link.

[4] Irawan, DE., A Priyambodho, Rachmi, CN., D Wibowo, A Fahmi, Bibliometric study to assist research topic selection: a case from research design on Jakarta's groundwater (part 1), Journal of Research Ideas and Outcomes Journal, 2016, link.

[5] Midori, AA. and Irawan, DE., Dataset for JCC paper -- Surface and Groundwater Interactions: Cikapundung Bandung, Kanal Banjir Timur Semarang, and Cisadane Tangerang, 2017, link.

[6] Faisal, B., Midori, AA., Irawan, DE., Optimalisasi Fungsi Sistem Kanal Banjir Timur Semarang sebagai Ruang Terbuka Hijau dan Pengendali Banjir, Jurnal Ilmu Pengetahuan, Teknologi dan Seni Institut Teknologi Bandung, 2017, link.

[7] Irawan, DE., A Darul, H Sumadi, AA Kuntoro, TA Argo, Y Nurhayati, Beban ganda pengelolaan air tanah di kab/kota pasca pembatalan UU No 7/2004 tentang Sumber Daya Air, Jurnal Ilmu Sosial dan Humaniora 6 (1), 83-96, 2017, link.

Simultant encapsulation of vitamin C and beta-carotene in sesame (*Sesamum indicum l.*) liposomes

D Hudiyanti[1], H Fawrin[2], and P Siahaan[1]

[1]Chemistry Department, Diponegoro University, Semarang, Indonesia
[2]Undergraduate Program, Chemistry Department, Diponegoro University, Semarang, Indonesia

E-mail: dwi.hudiyanti@live.undip.ac.id

Abstract. In this study sesame liposomes were used to encapsulate both vitamin C and beta-carotene simultaneously. Liposomes were prepared with addition of cholesterol. The encapsulation efficiency (EE) of sesame liposomes for vitamin C in the present of beta-carotene was 77%. The addition of cholesterol increased the encapsulation efficiency. The highest encapsulation efficiency was 89% obtained in liposomes with 10% and 20% cholesterol. Contrary to that, the highest beta-carotene encapsulation efficiency of 78%, was found in the sesame liposomes prepared without the added cholesterol. Results showed that sesame liposomes can be used to encapsulate beta-carotene and vitamin C simultaneously. When beta-carotene and vitamin C were encapsulated concurrently, cholesterol intensified the efficiency of vitamin C encapsulation on the contrary it diminished the efficiency of beta-carotene encapsulation.

1. Introduction

Ascorbic acid or vitamin C is an antioxidant needed in human bodies [1-5]. Vitamin C serves to reduce the risk of cancer (such as prostate and gastric cancer) and cardiovascular diseases [6-8]. Therefore, vitamin C is an important nutrient that is widely consumed [9]. Not only vitamin C, beta-carotene also acts as an antioxidant which is needed by human bodies [4, 5]. Beta-carotene is a pro vitamin A that prevents vitamin A deficiency [10-12]. Beta-carotene serves to reduce the risk of lung cancer and cardiovascular diseases, and eye deseases [5, 13, 14]. However, vitamin C and beta-carotene have disadvantages which may reduce their benefits. The role of vitamin C may be disrupted due to its high reactivity and its poor stability in solutions. Vitamin C may also easily be oxidized by the presence of oxygen, which is catalyzed by transition metal ions such as iron and copper [15]. Beta-carotene is a highly hydrophobic compound with high melting point, low chemical stability, and low bioavailability [16, 17]. These disadvantages can be overcome by the encapsulating vitamin C and beta-carotene in liposomes. Liposome is able to protect vitamin C from being oxidized and increases its stability [18, 19]. Liposome may increase the bioavailability and the stability of beta-carotene [20].

Liposome is a lipid vesicle which contains hydrophilic core and hydrophobic bilayer membrane [21, 22]. Its biocompatible structure and characteristics make it possible to be utilized as a drug delivery system [23-25]. Liposomes can encapsulate hydrophobic compounds inside its bilayer membrane and hydrophilic compound inside its core [21]. Some evidences have demonstrated that liposome is able to encapsulate beta-carotene and vitamin C singly [18, 19, 26]. Liposome is mostly built up by amphiphilic molecules such as phospholipids [27]. Phospholipid can be isolated from natural materials, like egg, soybean, coconut, and sesame seeds [28-30]. Phospholipid will create double layers similar to biologic membrane [31, 32]. The stability of liposomes can be optimized by adding cholesterol [33, 34]. Cholesterol is able to reduce the permeability of the membrane and rigidify the fluidity of the intravesical interaction [35, 36].

This research was conducted to investigate the capability of the sesame liposomes in encapsulating both beta-carotene and vitamin C simultaneously. The liposomes analysis was done by calculating the value of encapsulation efficiency (EE) and analyzing the influence of cholesterol towards the encapsulation efficiency value. Our new finding was that sesame liposomes can be used to encapsulate both beta-carotene and vitamin C simultaneously. When beta-carotene and vitamin C were encapsulated concurrently, cholesterol amplified the efficiency of vitamin C encapsulation but cut down the efficiency of beta-carotene encapsulation.

2. Materials and method

2.1. Materials

The materials used in this research were isolated sesame phospholid (Sesamum Indicum L.), chloroform p.a (Merck No.Batch 1.02445.2500), ethanol 96% (Brataco), cholesterol (Sigma Aldrich No. 11145-50G), vitamin C uncoated (Brataco No. Batch 1140870083), beta-carotene (Sigma No. C4582-25MG), soy phospholipid (Sigma Aldrich No. P3644-100G), $Na_2HPO_4.2H_2O$ (Merck No. Batch 2121439), $NaH_2PO_4.2H_2O$ (Merck No.Batch 1.06342.0250), demineralized water (Brataco).

2.2. Encapsulation beta-caroten and vitamin C in liposomes

The preparation of liposomes was done according to the method by Hudiyanti et al. (2015) with the addition of cholesterol 0%, 10%, 20%, 30%, 40% (w/w). Before creating the liposomes, the partition coefficient of vitamin C and beta-carotene was determined using MarvinSketch program. It was done by making thin layer, hydration, and ultrasonication. Beta-carotene was added to phospholipid with ratio 20% (w/w). Cholesterol was added to the mixture with ratio as above. The mixture was dissolved in chloroform to prepare thin layer. Vitamin C in phosphate buffer solution was added in the hydration process. Liposome's capability in encapsulating beta-carotene and vitamin C, and the influence of cholesterol towards its efficiency can be calculated by encapsulation efficiency value. The solution was centrifuged in order to separate the encapsulated molecules from those which was not encapsulated inside the liposomes. The supernatant which contained unencapsulated molecules (C_t) was analyzed by using spectrometer UV-Vis at the wavelength of vitamin C (265 nm) and beta-carotene (453 nm). The encapsulation efficiency of the vitamin C and beta-carotene was determined by equation 1.

$$EE = [1 - (C_t / C_0)] \times 100\% \qquad (1)$$

3. Result and discussion

3.1. The encapsulation efficiency of vitamin c

The encapsulation efficiency value of vitamin C in the sesame liposomes was 88%. Cholesterol reduced the encapsulation efficiency of vitamin C as shown in the Figure 1. This reduction maybe caused by the sludge resulted from centrifugation. During the centrifugation process, the particles inside the tube were separated and spread according to the specific weight of each particle by centrifugal force [37]. This sludge was unstable and could not be dispersed back to the supernatant. It affected the absorbance of vitamin C at 265 nm wavelength. The measured absorbance becomes higher hence the unencapsulated vitamin C concentration increased. As a result the encapsulation efficiency values were reduced.

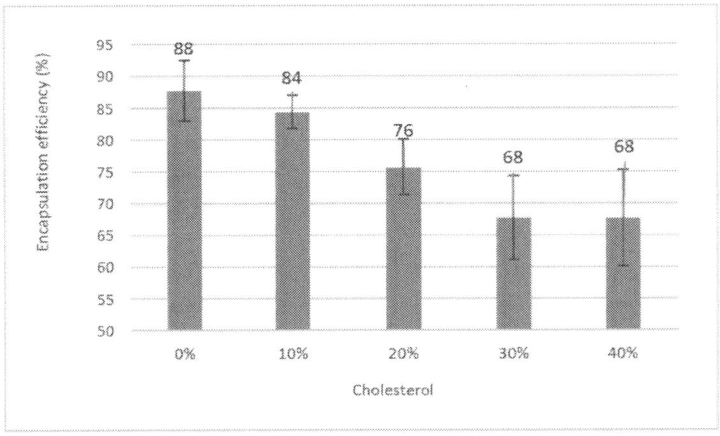

Figure 1. The encapsulation efficiency value of vitamin C in sesame liposomes.

The addition of 20% beta-carotene lowered the encapsulation efficiency value of the vitamin C inside the sesame liposomes to 77%. It was due to the big structure of beta-carotene with several double bonds. Beta-carotene interacted with sesame phospholipid which also contains many double bonds and created liposomes in the big shapes. As a result, there were less liposomes particles and encapsulated vitamin C. Cholesterol also unexpectedly increased the encapsulation efficiency of the vitamin C inside the sesame liposomes which contains beta-carotene as shown in Figure 2. Phospholipid which has been mixed by cholesterol has more well-ordered membrane. This regularity reduced the permeability of the membrane and rigidified the fluidity of the intravesical interaction, made it more effective in encapsulating a molecule [33].

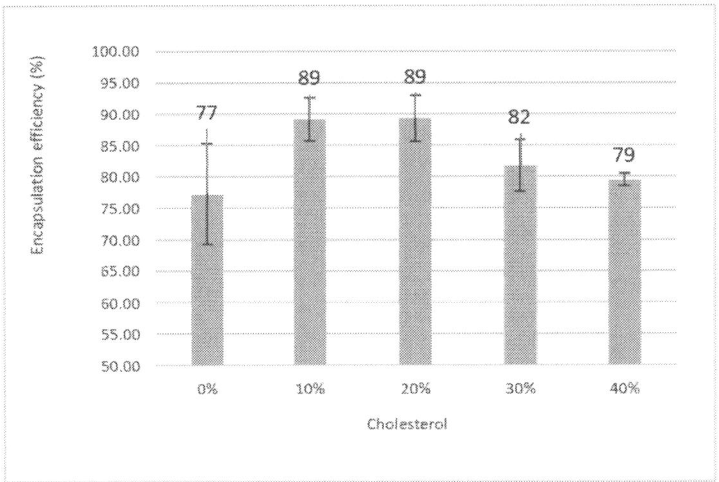

Figure 2. The encapsulation efficiency of vitamin C in sesame liposomes which contains beta-carotene.

Figure 3. Sesame liposomes solution after centrifugation. The yellow solution represents sesame liposomes containing beta-carotene and the white one was sesame liposomes solution which do not containing beta-carotene.

The effect of cholesterol towards the encapsulation efficiency of vitamin C in the sesame liposomes which contains beta-carotene was contradictory with the sesame liposomes with out beta-carotene. It could happen because beta-carotene could increase the specific weight of the sediment after centrifugation process, so the sediment became more stable (Figure 3).

The encapsulation efficiency value of vitamin C inside non cholesterol soy liposomes without and with beta-carotene was 78% and 84% respectively. Cholesterol increased the encapsulation efficiency value of the soy liposomes with or without beta-carotene as seen in Figure 4.

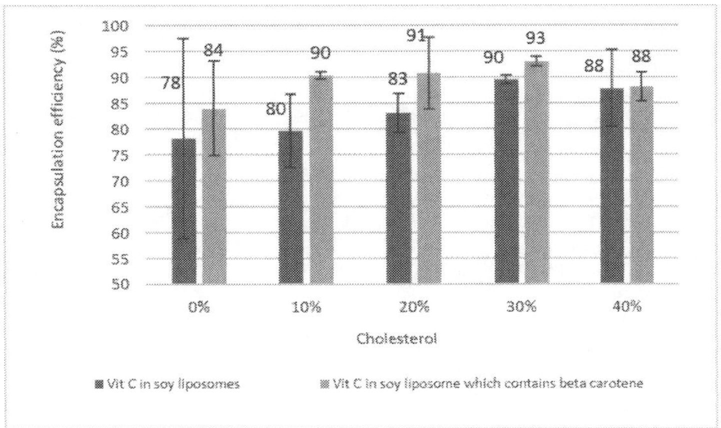

Figure 4. The encapsulation efficiency of vitamin C in soy liposomes.

The encapsulation efficiency of vitamin C inside sesame and soy liposomes with beta-carotene has increased compared to those without beta-carotene. The partition coefficient of beta-carotene which was 11,12 suggested that this compound would resided inside liposomes membrane. Meanwhile vitamin C with partition coefficient -1,91 would be in the hydrophilic part or liposomes core. Beta-carotene can alter membrane microviscosity, hydrophobicity, membrane permeability, and also protect phospholipid from being oxidized [38].

3.2. The encapsulation efficiency of beta-carotene

The encapsulation efficiency of beta-carotene inside sesame liposomes was 79%. The addition of cholesterol reduced the encapsulation efficiency of beta-carotene inside sesame liposomes as shown in Figure 5. This reduction was caused by the high competition between cholesterol and beta-carotene to be in the liposomes membrane [38]. It would decreased the concentration of beta-carotene inside the sesame liposomes membrane. The position of beta-carotene and cholesterol in the sesame liposomes membrane was predicted as shown as in Figure 6. Although both cholesterol and beta-carotene competed to be in the bilayer membrane, cholesterol could easily adapt compared to beta-carotene, due to its tiny shape and amphiphilic characteristic like phospholipid [38, 39].

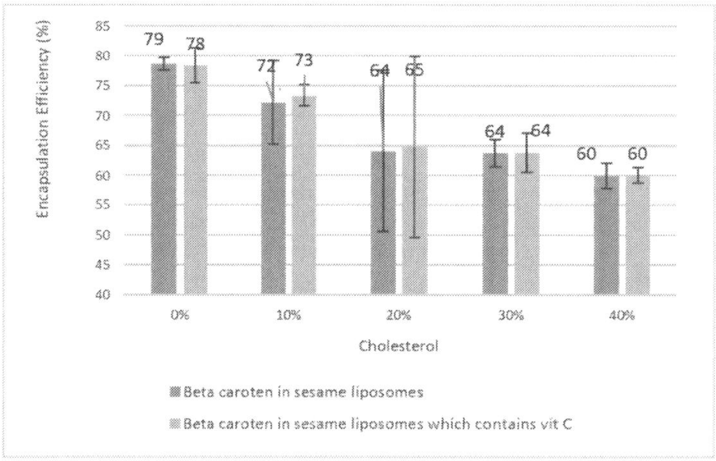

Figure 5. The encapsulation efficiency of beta-carotene in sesame liposomes.

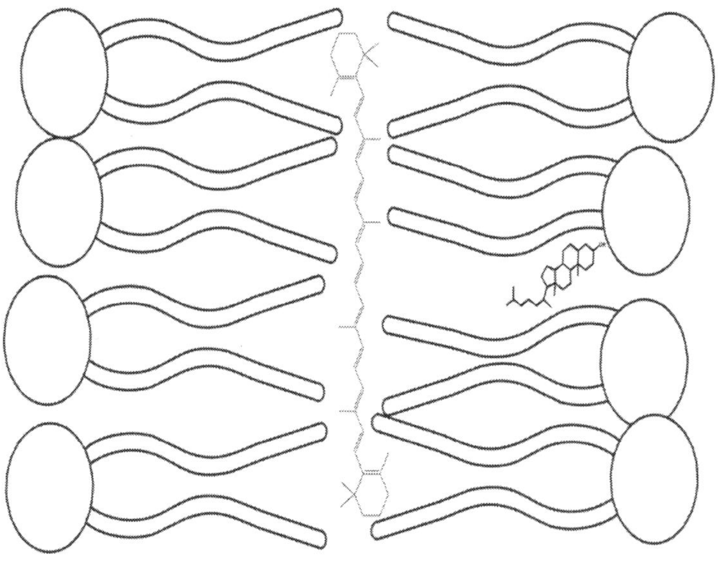

Figure 6. Prediction of position of beta-carotene and cholesterol in sesame liposomes membrane.

The results also showed that the encapsulation efficiency value of beta-carotene in sesame liposomes with or without loaded vitamin C was similar (Figure 5). It can be concluded that vitamin C did not affect the encapsulation efficiency value of beta-carotene.

Cholesterol showed the same effect on soy liposomes and sesame liposomes encapsulation efficiency of beta-carotene. Cholesterol reduced the encapsulation efficiency value of beta-carotene as depicted in Figure 7. These results can be explained by the similar lipophilic part of the phospholipids [30]. The decrease in beta-carotene encapsulation efficiency is possibly due to competition between cholesterol and beta-carotene to reside in the liposomal bilayer membrane.

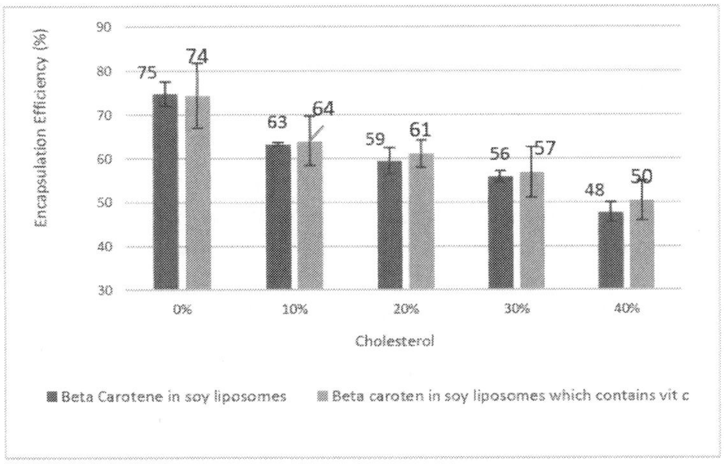

Figure 7. The encapsulation efficiency of beta-carotene in soy liposomes

4. Conclusion

The encapsulation efficiency of sesame liposomes for vitamin C in the present of beta-carotene was 77%. Addition of cholesterol increased the encapsulation efficiency. The highest was 89% obtained in liposomes with 10% and 20% cholesterol. On the contrary, the highest beta-carotene encapsulation efficiency, i.e. 78%, was found in the sesame liposomes prepared without cholesterol. Sesame liposomes can be used to encapsulate both beta-carotene and vitamin C simultaneously. When beta-carotene and vitamin C were encapsulated concurrently, cholesterol increased the efficiency of vitamin C encapsulation but lowered the efficiency of beta-carotene encapsulation.

Acknowledgments

This work was supported by The PTUPT Research Grant 2017 From The Ministry of Research, Technology and Higher Education of the Republic of Indonesia.

References

[1] Proteggente A R, Pannala A S, Paganga G, Van Buren L, Wagner E, Wiseman S, Van De Put F, Dacombe C and Rice-Evans C A 2002 The antioxidant activity of regularly consumed fruit and vegetables reflects their phenolic and vitamin C composition *Free Radic. Res.* **36** pp 217–233

[2] Padayatty S J, Katz A, Wang Y, Eck P, Kwon O, Lee J, Chen S, Corpe C, Dutta A, Dutta S K and Levine M 2003 Vitamin C as an Antioxidant: Evaluation of Its Role in Disease Prevention *J. Am. Coll. Nutr.* **22** pp 18-35

[3] Halliwell B 1996 Vitamin C : Antioxidant or Pro-Oxidant In Vivo? **25** pp 439–454

[4] Sies W and Stahl H 1995 Vitamins E and C, β-carotene, and other carotenoids as antioxidants *Am. J. Clin. Nutr.* **62** 1315S

[5] Prakash D and Gupta K R 2009 The Antioxidant Phytochemicals of Nutraceutical Importance **109** pp 20-35

[6] Moser M A and Chun O K 2016 Vitamin C and heart health: A review based on findings from epidemiologic studies *Int. J. Mol. Sci.* **17** p 1328

[7] Bai X Y, Qu X, Jiang X, Xu Z, Yang Y, Su Q, Wang M and Wu H 2015 Association between Dietary Vitamin C Intake and Risk of Prostate Cancer: A Meta-analysis Involving 103,658 Subjects *J Cancer.* **6** pp 913-21

[8] Jenab M. *et al.* 2006 Plasma and dietary vitamin C levels and risk of gastric cancer in the European Prospective Investigation into Cancer and Nutrition (EPIC-EURGAST) *Carcinogenesis* **27** pp 2250–2257

[9] Foote J A, Murphy S P, Wilkens L R, Hankin J H, Henderson B E and Kolonel L N 2003 Factors associated with dietary supplement use among healthy adults of five ethnicities: The multiethnic cohort study *Am. J. Epidemiol.* **57** pp 888–897

[10] van Jaarsveld P J, Faber M, Tanumihardjo S, Nestel P, Lombard C J and Benadé A J S 2005 Beta-carotene-rich orange-fleshed sweet potato improves the vitamin A status of primary school children assessed with the modified-relative-dose-response test *Am. J. Clin. Nutr* **81** pp 1080–1087

[11] M J Haskell 2012 The challenge to reach nutritional adequacy for vitamin A: β-carotene bioavailability and conversion—evidence in humans 1–4 *Am J Clin Nutr* **96** pp 1193–203

[12] Grune T, Lietz G, Palou A, Ross A C, Stahl W, Tang G, Thurnham D, Yin S and Biesalski H K 2010 β -Carotene Is an Important Vitamin A Source *J. Nutr.* **140** pp 2268S-2285S

[13] Krinsky N I and Johnson E J 2005 Carotenoid actions and their relation to health and disease *Mol. Aspects Med.* **26** pp 459–516

[14] Omenn G S, Goodman G E, Thornquist M D, Balmes J, Cullen M R, Glass A, Keogh J P, Meyskens F L, Valanis B, Williams J H, Barnhart S, Hammar S 1996 Effects of a combination of beta-carotene and vitamin A on lung cancer and cardiovascular disease *N Engl J Med* **334** pp 1150–1155

[15] Liao M L and Seib P A 1988 Chemistry of L-ascorbic acid related to foods *Food Chem* **30** pp 289–312

[16] Qian C, Decker E A, Xiao H, and McClements D J 2012 Physical and chemical stability of β-carotene-enriched nanoemulsions: Influence of pH, ionic strength, temperature, and emulsifier type *Food Chem.* **132** pp1221–1229

[17] Boon C S, McClements D J, Weiss J and Decker E A 2010 Factors influencing the chemical stability of carotenoids in foods *Crit Rev Food Sci Nutr* **50** pp 515–532

[18] Wechtersbach L, Ulrih N P, and Cigic B 2012 Liposomal stabilization of ascorbic acid in model systems and in food matrices *LWT - Food Sci. Technol.* **45** pp 43–49

[19] Kirby C J, Whittle C J, Rigby N, Coxon D T and Law B A 1991 Stabilization of ascorbic acid by microencapsulation in liposomes *Int. J. Food Sci. Technol.* **26** pp 437–449

[20] Jeong S H, Park J H and Park K 2007 *Role of Lipid Excipients in Modifying Oral and Parenteral Drug Delivery* ed K M Wasan (John Wiley & Sons, Inc) chapter 2

[21] A Akbarzadeh, Rezaei-sadabady R, Davaran S, Joo S W and Zarghami N 2013 Liposome : classification , preparation , and applications *Nanoscale Res. Lett.* **8** pp 1–9

[22] Allen T M and Cullis P R 2013 Liposomal drug delivery systems: From concept to clinical applications *Adv. Drug Deliv. Rev.* **65** pp 36–48

[23] Wang F, Chen L, Zhang R, Chen Z, and Zhu L 2014 RGD peptide conjugated liposomal drug delivery system for enhance therapeutic efficacy in treating bone metastasis from prostate cancer *J. Control. Release*, **196** pp 222–233

[24] Kaminskas L M, McLeod V M, Kelly B D, Sberna G, Boyd B J, Williamson M, Owen D J, Porter C J H 2012 A comparison of changes to doxorubicin pharmacokinetics, antitumor activity,

and toxicity mediated by PEGylated dendrimer and PEGylated liposome drug delivery systems *Nanomedicine Nanotechnology, Biol. Med.* **8** pp 103–111

[25] Malam Y, Loizidou M, and Seifalian A M 2009 Liposomes and nanoparticles: nanosized vehicles for drug delivery in cancer *Trends Pharmacol. Sci.* **30** pp 592–599

[26] Moraes M, Carvalho J M P, Silva C R, Cho S, Sola M R, and Pinho S C 2013 Liposomes encapsulating beta-carotene produced by the proliposomes method: Characterisation and shelf life of powders and phospholipid vesicles *Int. J. Food Sci. Technol.* **48** pp 274–282

[27] Farhang B, Kakuda Y, and Corredig M 2012 Encapsulation of ascorbic acid in liposomes prepared with milk fat globule membrane-derived phospholipids *Dairy Sci. Technol.* **92** pp 353–366

[28] Lang J, Vigo-Pelfrey C, and Martin F 1990 Liposomes composed of partially hydrogenated egg phosphatidylcholines: fatty acid composition, thermal phase behavior and oxidative stability *Chem. Phys. Lipids* **53** pp 91–101

[29] Thomas A H, Catalá Á and Vignoni M 2016 Soybean phosphatidylcholine liposomes as model membranes to study lipid peroxidation photoinduced by pterin *Biochim. Biophys. Acta - Biomembr.* **1858** pp 139–145

[30] Hudiyanti D, Raharjo T J, Narsito N and Noegrohati S 2012 Isolasi dan karakterisasi lesitin kelapa dan wijen **32** pp 23–26

[31] Begu S, Aubert-Pouessel A, Polexe R, Leitmanova E, Lerner D A, Devoisselle J and Tichit D 2009 New layered double hydroxides/phospholipid bilayer hybrid material with strong potential for sustained drug delivery system *Chem. Mater.* **21** pp 2679–2687

[32] Hudiyanti D, Raharjo T J, Narsito N and Noegrohati S 2015 Study on leakage of sesame (Sesamum indicum L.) and coconut (Cocos nucifera L.) liposomes *Orient. J. Chem.* **31** pp 435-439

[33] Sułkowski W W, Pentak D, Nowak K, and Sułkowska A 2005 The influence of temperature, cholesterol content and pH on liposome stability *J. Mol. Struct.* **744–747** pp737–747

[34] Abea K, Higashi K, Watabe K, Kobayashi A, Limwikrant W, Yamamoto K, Moribe K 2015 Effects of the PEG molecular weight of a PEG-lipid and cholesterol on PEG chain flexibility on liposome surfaces *Colloids Surfaces A Physicochem. Eng. Asp.* **474** pp 63–70

[35] Briuglia M L, Rotella C, McFarlane A and D A Lamprou 2015 Influence of cholesterol on liposome stability and on in vitro drug release," *Drug Deliv. Transl. Res.* **5** pp 231–242

[36] Liang X, Mao G, and Ng K Y 2004 Mechanical properties and stability measurement of cholesterol-containing liposome on mica by atomic force microscopy *J. Colloid Interface Sci.* **278** pp 53–62

[37] Faatih M 2009 Isolasi dan digesti DNA kromosom *J Penelit. Sains dan Teknol.* **20** pp 61–67

[38] Socaciu C, Jessel R, and Diehl H A 2000 Competitive carotenoid and cholesterol incorporation into liposomes: Effects on membrane phase transition, fluidity, polarity and anisotropy *Chem. Phys. Lipids* **106** 79–88

[39] Matsuoka S and Murata M 2002 Cholesterol markedly reduces ion permeability induced by membrane-bound amphotericin B *Biochim. Biophys. Acta - Biomembr.* **1564** pp 429–434

IOP Conf. Series: Materials Science and Engineering **349** (2018) 012015 doi:10.1088/1757-899X/349/1/012015

Quantitative analysis of Curcuminoid collected from different location in Indonesia by TLC-Densitometry and its antioxidant capacity

D S C Wahyuni[1], A N Artanti[1], Y Rinanto[2]

[1] Pharmacy Department, Faculty of Mathematics and Natural Sciences, Universitas Sebelas Maret, Jl. Ir. Sutami 36 A, Surakarta, Central Java, Indonesia 57126
[2] Biology Education Program, Faculty of Teacher Training and Education, Universitas Sebelas Maret, Jl. Ir. Sutami 36 A, Surakarta, Central Java, Indonesia 57126

*E-mail: dinarsari_cw@staff.uns.ac.id

Abstract. Curcuminoid, consisting of curcumin, demethoxycurcumin and bis demethoxycurcumin, is the active compound in *Curcuma longa* L. rhizome. It yielded vary depend on the environment and varieties. Curcumin has been known to have a potent antioxidants activity. The present work was conducted to assess the curcuminoid content and antioxidant capacity in the crude extracts of *C. longa* L. collected from eight different locations in Indonesia. Samples were collected from eight locations including Java island, Sumatera island and Nusa Tenggara island. All samples were extracted using 96% ethanol and analyzed quantitatively using TLC-Densitometry. Antioxidant activity was assessed using diphenylpicrylhydrazyl (DPPH) radical scavenging assay and analyzed at 517 nm using a spectrophotometer. Curcuminoid content in *C. longa* varied among eight different locations (0.53±0.05 - 5.33±0.12 % w/w). The highest curcumin, demethoxycurcumin and bisdemethoxycurcumin yield were found in the samples from Magetan (5.33±0.12%, 1.54±0.05%, 0.46±0.02% w/w, respectively). In contrast, the lowest curcumin, demethoxycurcumin and bisdemethoxycurcumin yield were found in the sample from Demak (0.53±0.05%, 0.17±0.05%, 0.17±0.05% w/w, respectively). Antioxidant capacity showed similar for all places unless sample from Magetan which exhibited two times lower than other locations. It is apparently curcuminoid content and antioxidant activity varied among places. These results are useful information for curcuminoid standardization method in pharmaceutical products.

1. Introduction

Curcuma longa L. is a rhizomatous small perennial plant belongs to Zingiberaceae family. It is a popular herb used as a household medicine [1], dye [2] and seasoning in East Asian countries for many years. Curcuminoid, i.e. curcumin, demethoxycurcumin and bisdemethoxycurcumin are active constituents in *C. longa* L. These have been studied as antioxidant [3], antiinflamatory, antimicrobial effects [4].

C. longa L. has been widely cultivated in Indonesia. Total curcuminoids should have more than 5% (w/w) of dried *C. longa* L. as recommended by Standard of ASEAN Herbal Medicine (1994). Curcuminoid content varied in *C.longa* grown in India [6] and in Thailand [7]. The difference of curcuminoid yield among the various lines of *C. longa* was due to variation of genotype-phenotype [8] and geography [6]. Thus, *C.longa* rhizome which grow in different geography need to be further investigated to identify curcuminoid yield.

Curcumin were previously reported as antioxidant activity [9], [10]. Antioxidant can decrease oxidative stress due to excessive free radicals and reactive oxygen species (ROS). These free radicals could lead to detrimental diseases such as cancer [11]. Antioxidant activity of curcumin has been shown due to either the β–diketone group in its structure (Wright, 2002 ; Priyadarsini et al., 2003). Curcumin could enhance the activity of many antioxidant enzymes such as heme oxygenase [14], catalase, superoxide dismutase (SOD) and glutathione peroxidase [15]. Curcumin as the major compound present in *C. longa* L. and its analogs showed strong relationships to the structure and antioxidant activity [16]. However, antioxidant activity of curcuminoids in the extract could have either enhance or decrease due to synergistic, additive or antagonistic effect of mixtures. Thus, this study aimed (1) to quantify curcuminoids in the extract of *C. longa* L. and (2) to measure the antioxidant activity of *C. longa* L. extracts. Curcuminoid quantitative analysis used thin layer chromatography (TLC)-Densitometric due to its rapid, simple and reliable method as shown by Pothitirat and Gritsanapan (2005).

2. Materials and methods

2.1. Chemicals and reagents
Curcuminoid standard including curcumin, demethoxycurcumin and bisdemetoxycurcumin obtained from CV. Chemonix Pratama as commercial standard. All chemicals and reagents used in this research were analytical grade.

2.2. Plant material
The rizhomes of *C. longa* were collected from eight different locations in Indonesia i.e. Wonogiri, Sukoharjo, Magetan, Demak, Banten, Nusa Tenggara Timur, Nusa Tenggara Barat and Jambi.

2.3. Preparation of sample solutions, standard solutions and calibration curves
A sample of 50 mg of dry material was transferred to a microtube to which 500 ul of ethanol 96% were added. The mixture was vortexed at room temperature for 3 min and centrifuged at 30.000 rpm for 20 min. The supernatant (5 ul) was taken for TLC-densitometry. Standard curves of curcuminoid were prepared in ethanol at 2000, 1000, 500, 250 and 125ppm. Five microliters of each curcuminoid solutions were spotted into the precoated silica gel aluminium plate 60GF$_{254}$ (20 x 10 cm; E. Merck, Germany). Plate was developed using saturated mobile phase of dichloromethane : methanol (97:3). The length of each chromatogram was 8 cm. densitometric scanning was performed on Camag TLC Scanner 4 in the reflectance-absorbance mode at 420 nm, perated by CATS software. The amount of each curcuminoid presented in the sample was calculated using peak area with linier regression. Each curcuminoid content was calculated using calibration curve. The content of curcumin was expressed as 100 grams of the dried sample.

2.4. *Antioxidant activity, DPPH (diphenylpicrylhdrazyl) radical scavenging assay*
Antioxidants react with DPPH, a stable fee radical which is reduced to DPPH-H and as a consequence the absorbance is decreased from the DPPH radical to the DPPH-H form. The degree of discoloration indicates the scavenging potential of the antioxidant compounds or extracts in terms of hydrogen donating ability.

The free radical scavenging capacity of the *C. longa* was determined using DPPH according to the method of Blois. DPPH (0.004% w/v was prepared freshly in 97% ethanol and was added to sample solution (10 ppm) in ethanol. The mixture was allowed to stand at room temperature in dark for 20 mins. Then the mixture was vortexed and the absorbance was read at 517 nm using a spectrophotometer. Lower absorbance of the reaction mixture indicated higher free radical scavenging activity. All tests were performed in duplicates. Antioxidant activity (%) was calculated at 100 μg/mL.

2.5. *Data analysis*

Differences between curcuminoid, curcumin, demethoxycurcumin, bisdemethoxycurcumin related to antioxidant capacity were analyzed by one-way ANOVA. Relationship between each metabolites and antioxidant capacity were conducted using Pearson correlation.

3. Result and discussion

3.1. *Quantification of curcuminoid yield*

Thin layer chromatography analysis showed the separation of curcuminoid content, curcumin, demethoxycurcumin and bisdemethoxycurcumin, nicely depicted on Figure 1. Rf value of these curcumin, demethoxycurcumin and bisdemetoxycurcumin were 0.81, 0.5, 0.3, respectively. Curcuminoid content varied among eight different locations (0.53±0.05% to 5.33±0.12% w/w) (Table 1). Curcumin, demethoxycurcumin and bisdemethoxycurcumin were highly significant among eight different location ($F = 21.325$, $df = 7$, $P = 0.000$; $F = 4.733$, $df = 7$, $P = 0.008$; $F = 20.010$, $df = 7$, $P = 0.000$; respectively). The highest curcumin, demethoxycurcumin and bisdemethoxycurcumin yield were found in the samples from Magetan (5.33±0.12%, 1.54±0.05%, 0.46±0.02% w/w, respectively). In contrast, the lowest curcumin, demethoxycurcumin and bisdemethoxycurcumin yield were found in the sample from Demak (0.53±0.05%, 0.17±0.05%, 0.17±0.05% w/w, respectively).

Curcuminoid content in *C. longa* varied as reported by Ashraf et al.(2012) and Pawar et al. (2014) collected in India, Pothitirat and Gritsanapan (2005) collected in Thailand. *C. longa* from Magetan showed the total curcuminoid above 5% which fulfilled the ASEAN standard. Besides, it was the highest curcuminoid among others. Meanwhile, the curcuminoid content of other samples varied below the standards. Anandaraj et al. (2014) reported that 70% and 16% variation of curcuminoid yield in *C. longa* was attributed to environments and genotype effect, respectively. Altitude, temperature, rainfall and soil type were taken into account for curcuminoid yield. Moreover, genotypic factor of variation and heritability were highly parameter for yield and curcumin content [17]. Hayakawa et al. (2011) suggested the hybridization and introgression with other Curcuma species could effect curcuminoid content.

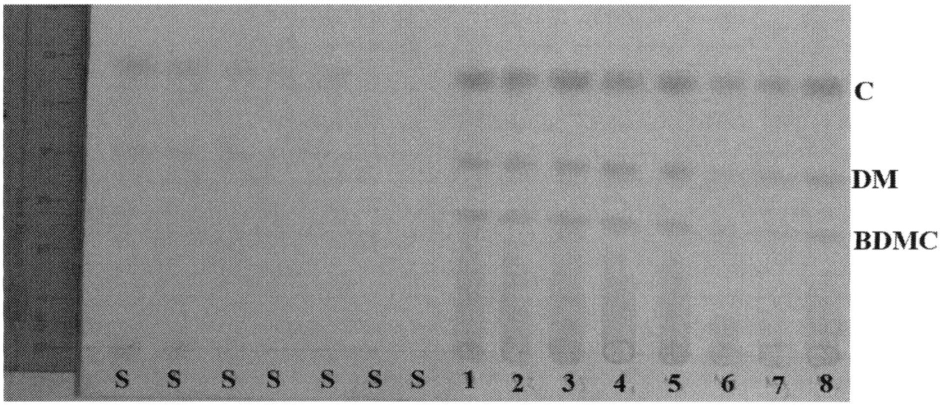

Figure 1. TLC profile of curcuminoids (Curcumin (C), demethoxycurcumin (DMC), bisdemethoxycurcumin (BDMC)) in the extracts of *Curcuma longa* L. collected from various locations. Stationary phase: $60GF_{254}$; solvent system: dichloromethane-methanol (97:3). S = standard curcuminoid, 1 = Jambi, Sumatera, 2 = Sukoharjo, Middle Java, 3 = Magetan, Middle Java, 4 = Banten, West Java, 5 = Wonogiri, Middle Java, 6 = Demak, West Java, 7 = Nusa Tenggara Timur and 8 = Nusa Tenggara Barat

Table 1. The content of curcuminoid in ethanolic extracts of *Curcuma longa* collected from different location in Indonesia by TLC-densitometry. Analysis was done in triplicate as mean ± SE

Location	Each curcuminoid content (% w/w in dried powder)			Total curcuminoid content (% w/w in dried powder)
	C	DMC	BDMC	
Wonogiri	0.68±0.28	0.59±0.31	0.16±0.05	1.66±0.21
Sukoharjo	1.32±0.02	1.05±0.02	0.34±0.02	3.14±0.09
Magetan	3.26±0.01	1.54±0.05	0.46±0.02	5.33±0.12
Demak	0.27±0.08	0.17±0.05	0.06±0.03	0.53±0.05
Banten	1.78±0.02	1.52±0.02	0.43±0.01	3.83±0.11
NTT	0.81±0.13	0.74±0.14	0.24±0.02	1.89±0.13
NTB	1.10±0.32	0.84±0.44	0.26±0.02	2.47±0.28
Jambi	1.41±0.06	1.24±0.03	0.32±0.01	3.09±0.11

3.2. Antioxidant activity

Antioxidant capacity was determined by the ability of substances to donate a hydrogen atom to a free radical DPPH into a nonradical DPPH [19]. It reduced the violet colour of a free radical DPPH into a residual pale yellow colour from the picryl group .

Antioxidant capacities of *C. longa* ethanolic extracts are presented in Table 2. Antioxidant capacity was significantly different ($F = 36.440$, $df = 7$, $P = 0.000$). It showed similar activity for all places unless sample from Magetan which exhibited two times lower than other locations. The more potent antioxidant shows the higher antioxidant capacity value. Reynertson et al. (2007) classified antioxidant into strong, medium and weak activity with IC_{50} value of <50 µg/mL, 50-100 µg/mL and 100-200 µg/mL, respectively. Present study showed the antioxidant capacity of curcuminoid was 54.76 µg/mL. These exhibited medium antioxidant capacity. In contrast, curcumin showed to have IC_{50} of 12.88 µg/mL [21].

Curcumin, a major phenolic compounds in *C.longa* L., has been known as a strong antioxidant agent [22]. Antioxidant capacity crude extract *C.longa* L. has been shown by Singh (2010) to have 86-92% inhibition against DPPH radical. Besides, major constituents of essential oil such as aromatic-turmerone, alpha-turmerone, beta-turmerone, oleoresin played a role in antioxidant activity in *C.longa* L. [23].

Table 2. Antioxidant activity of ethanolic extracts of *Curcuma longa* using DPPH method.

Location	Concentration (µg/mL)	Antioxidant capacity (%)
Wonogiri	0.1	56.27 ± 0.80
Sukoharjo	0.1	66.82 ± 0.86
Magetan	0.1	35.98 ± 0.96
Demak	0.1	63.30 ± 1.13
Banten	0.1	61.71 ± 0.85
NTT	0.1	56.03 ± 1.53
NTB	0.1	59.57 ± 0.20
Jambi	0.1	64.45 ± 0.70
Curcuminoids	0.1	54.76

To understand the correlation between curcuminoid yield and antioxidant activity, we do the correlation statistical analysis. It resulted no correlation between demethoxycurcumin and bisdemethoxycurcumin to antioxidant capacity ($R= -0.262$, $df= 8$, $P= 0.531$ and $R= -0.294$, $df= 8$, $P= 0.479$, respectively). However, curcumin showed little correlation with antioxidant activity ($R= -0.715$, $df= 8$, $P= 0.046$) (Fig. 2). In general, the total curcuminoid content and antioxidant showed no significant correlation ($R= -0.562$, $df= 8$, $P= 0.147$). Curcumin was reported as a strong antioxidant agent [3, 24]. However, *C. longa* L. extract as whole constituent system, the antioxidant activity is influenced by other component as synergistic or antagonistic actions. Thus, instead of curcumin, other constituents i.e essential oils and oleoresin as reported by Zaeoung et al. [23].

4. Conclusion

These results are useful information for curcuminoid standardization method in pharmaceutical products. It has to fulfill the requirement of total curcuminoid above 5% based on the ASEAN standard. Thus, the standardized cultivation for related herbal medicines needs to be done to produce an acceptable herbal material for pharmaceutical industry.

Acknowledgement

The authors are grateful to our students at Natural Medicine Research Study for the collaboration to finish the work. First author thankful to PT. Javaplant for a valuable contribution and sample collection. The work is fully funded by MRG project, 2016, Universitas Sebelas Maret. The authors declared no potential conflicts of interest in this study.

References

[1] M. J. Bhandary and K. R. Chandrashekar, "Diversity and use of ethnomedicinal plants in coastal Karnataka, India," *ISSN/E-ISSN*, p. 89, 2014.

[2] B. Sutradhar, D. Deb, K. Majumdar, and B. K. Datta, "Short communication: traditional dye yielding plants of Tripura, Northeast India," *Biodiversitas*, vol. 16, pp. 121–127, 2015.

[3] T. Masuda, T. Maekawa, K. Hidaka, H. Bando, Y. Takeda, and H. Yamaguchi, "Chemical studies on antioxidant mechanism of curcumin: analysis of oxidative coupling products from curcumin and linoleate," *J. Agric. Food Chem.*, vol. 49, no. 5, pp. 2539–2547, 2001.

[4] S. Julie and M. T. Jurenka, "Anti-inflammatory properties of curcumin, a major constituent," *Altern. Med. Rev.*, vol. 14, no. 2, 2009.

[5] A. Countries, "Standard of ASEAN herbal medicine. Vol. II," 1994.

[6] K. Ashraf, M. Mujeeb, A. Ahmad, M. Amir, M. N. Mallick, and D. Sharma, "Validated HPTLC analysis method for quantification of variability in content of curcumin in Curcuma longa L (turmeric) collected from different geographical region of India," *Asian Pac. J. Trop. Biomed.*, vol. 2, no. 2, pp. S584–S588, 2012.

[7] W. Pothitirat and W. Gritsanapan, "Quantitative analysis of curcumin, demethoxycurcumin and bisdemethoxycurcumin in the crude curcuminoid extract from Curcuma longa in Thailand by TLC-densitometry," *Mahidol Univ. J. Pharm. Sci.*, vol. 32, no. 1–2, pp. 23–30, 2005.

[8] M. Anandaraj *et al.*, "Genotype by environment interaction effects on yield and curcumin in turmeric (Curcuma longa L.)," *Ind. Crops Prod.*, vol. 53, pp. 358–364, 2014.

[9] N. Aftab and A. Vieira, "Antioxidant activities of curcumin and combinations of this curcuminoid with other phytochemicals," *Phyther. Res.*, vol. 24, no. 4, pp. 500–502, 2010.

[10] O. Tokusoglu, A. Simsek, M. Parvizi, and D. Eymen, "Turmeric curcuminoid polyphenolics as antioxidant and anticarcinogenic agents," *Nat. Sci. Discov.*, vol. 1, no. 3, pp. 56–61, 2015.

[11] V. Lobo, A. Patil, A. Phatak, and N. Chandra, "Free radicals, antioxidants and functional foods: Impact on human health," *Pharmacogn. Rev.*, vol. 4, no. 8, p. 118, 2010.

[12] J. S. Wright, "Predicting the antioxidant activity of curcumin and curcuminoids," *J. Mol. Struct. THEOCHEM*, vol. 591, no. 1, pp. 207–217, 2002.

[13] K. I. Priyadarsini *et al.*, "Role of phenolic OH and methylene hydrogen on the free radical reactions and antioxidant activity of curcumin," *Free Radic. Biol. Med.*, vol. 35, no. 5, pp. 475–484, 2003.

[14] G.-S. Jeong *et al.*, "Comparative effects of curcuminoids on endothelial heme oxygenase-1 expression: ortho-methoxy groups are essential to enhance heme oxygenase activity and protection," *Exp. Mol. Med.*, vol. 38, no. 4, pp. 393–400, 2006.

[15] A. C. P. Reddy and B. R. Lokesh, "Effect of dietary turmeric (Curcuma longa) on iron-induced

lipid peroxidation in the rat liver," *Food Chem. Toxicol.*, vol. 32, no. 3, pp. 279–283, 1994.
[16] K. S. Bhullar, A. Jha, D. Youssef, and H. P. Rupasinghe, "Curcumin and its carbocyclic analogs: structure-activity in relation to antioxidant and selected biological properties," *Molecules*, vol. 18, no. 5, pp. 5389–5404, 2013.
[17] H. Pawar, M. Karde, N. Mundle, P. Jadhav, and K. Mehra, "Phytochemical evaluation and curcumin content determination of turmeric rhizomes collected from Bhandara District of Maharashtra (India)," *Med. Chem. (Los. Angeles).*, vol. 2014, 2014.
[18] H. Hayakawa, Y. Minaniya, K. Ito, Y. Yamamoto, and T. Fukuda, "Difference of curcumin content in Curcuma longa L.(Zingiberaceae) caused by hybridization with other Curcuma species," *Am. J. Plant Sci.*, vol. 2, no. 2, p. 111, 2011.
[19] P. Molyneux, "The use of the stable free radical diphenylpicrylhydrazyl (DPPH) for estimating antioxidant activity," *Songklanakarin J Sci Technol*, vol. 26, no. 2, pp. 211–219, 2004.
[20] K. A. Reynertson, "Phytochemical analysis of bioactive constituents from edible Myrtaceae fruits." The City University of New York, 2007.
[21] P. Somparn, C. Phisalaphong, S. Nakornchai, S. Unchern, and N. P. Morales, "Comparative antioxidant activities of curcumin and its demethoxy and hydrogenated derivatives," *Biol. Pharm. Bull.*, vol. 30, no. 1, pp. 74–78, 2007.
[22] J. Miquel, A. Bernd, J. M. Sempere, J. Dıaz-Alperi, and A. Ramırez, "The curcuma antioxidants: pharmacological effects and prospects for future clinical use. A review," *Arch. Gerontol. Geriatr.*, vol. 34, no. 1, pp. 37–46, 2002.
[23] G. Singh, I. P. S. Kapoor, P. Singh, C. S. De Heluani, M. P. De Lampasona, and C. A. N. Catalan, "Comparative study of chemical composition and antioxidant activity of fresh and dry rhizomes of turmeric (Curcuma longa Linn.)," *Food Chem. Toxicol.*, vol. 48, no. 4, pp. 1026–1031, 2010.
[24] K. M. Youssef, M. A. El-Sherbeny, F. S. El-Shafie, H. A. Farag, O. A. Al-Deeb, and S. A. A. Awadalla, "Synthesis of curcumin analogues as potential antioxidant, cancer chemopreventive agents," *Arch. Pharm. (Weinheim).*, vol. 337, no. 1, pp. 42–54, 2004.

IOP Conf. Series: Materials Science and Engineering **349** (2018) 012016 doi:10.1088/1757-899X/349/1/012016

Simple approach in understanding interzeolite transformations using ring building units

D Suhendar[1,2], Buchari[1], R R Mukti[1], Ismunandar[1]

[1] Program Studi Kimia, Fakultas Matematika dan Ilmu Pengetahuan Alam, Institut Teknologi Bandung, Jl. Ganesha no. 10 Bandung 40132, Indonesia,
[2] Program Studi Kimia, Fakultas Sains dan Teknologi, Universitas Islam Negeri Sunan Gunung Djati, Jl. A. H. Nasution no. 105 Bandung 40614, Indonesia

dede.suhendar@uinsgd.ac.id

Abstract. Recently, there are two general approaches used in understanding interzeolite transformations, thermodynamically represented by framework density (FD) and kinetically by structural building units. Two types of structural building units are composite building units (CBU's) and secondary building units (SBU's). This study aims to examine the approaches by using interzeolite transformation data available in literature and propose a possible alternative approach. From a number of cases of zeolite transformation, the FD and CBU approach are not suitable for use. The FD approach fails in cases involving zeolite parents that have moderate or high FD's, while CBU approach fails because of CBU's unavailability in parent zeolites compared with CBU's in their transformation products. The SBU approach is most likely to fit because SBU's are units that have basic form of ring structures and closer to the state and shape of oligomeric fragments present in zeolite synthesis or dissolution cases. Thus, a new approach can be considered in understanding the interzeolite transformation, namely the ring building unit (RBU) approach. The advantage of RBU approach is RBU's can be easily derived from all framework types, but in SBU approach there are several types of frameworks that cannot be expressed in SBU forms.

1. Introduction

The interzeolite transformation is an interesting phenomenon to be studied because it is potentially developed for more efficient synthesis of zeolites, can be performed faster and/or lower temperatures than conventional synthesis [1] [2]. This phenomenon can be utilized to add variety of values to natural sources of abundant natural aluminosilicates. Although the phenomenon of inter-zeolite transformation has long been emerging and studied, to date this phenomenon is still not fully understood. Due to the metastability of the zeolite framework, the constraint in understanding the transformation comes from understanding the structural relationship between the parent zeolite and its product transformation. The difficulty to fully understand the phenomenon of interzeolite transformation involves many variables that affect it, such as reaction conditions (pH, temperature, and time) and other chemical involvements such as the presence of templates, seeds, organic directing agents, and the presence of ionic types and concentrations.

To solve the problem of understanding interzeolit transformation, several approaches can be done. To date there are two common approaches considered for understanding the inter-zeolite transformation, i.e. the framework density (FD) and the building unit relationships involved in the transformation. The two most commonly used framework building units are composite building units (CBUs) [1] [3] and secondary building units (SBU's) [2] [4]. Since each of these approaches is often shown as a separate case, it is interesting to examine its application for all transformation cases in the literature. In this study will be tested and discussed both approaches and possible approaches that may apply to zeolites in general, namely the ring-building unit (RBU) approach of the zeolite couples involved in transformation, as the parent and transformation product.

2. Evaluation of framework density approach in interzeolite transformations

Although zeolite has been recognized as a metastable material, the background of framework density has often been a major consideration in the zeolite transformation efforts, according to the majority of transformation cases using FAU parent zeolite as shown in Table 1, and the fact is FAU is one type of zeolite among the zeolites having the lowest FD's. The consideration of framework density was proposed by Goel et al., although the zeolite-to-zeolite transformation is not a one-stage process because in equal system with time and temperature differences can produce a variety of zeolites and do not always produce zeolite types with higher FD's [3]. Based on the data in Table 1 it is clear that by using zeolite types with high FD's, the FD approach fails to understand the transformation from one to other zeolites.

Table 1 Framework density data in selected interzeolite transformation cases

Parent zeolite		Transformation product		Ref.[b]	suitability[c]
Type	FD[a] / T·1000A^{-3}	Type	FD[a] / T·1000 Å^{-3}		
Using parent zeolite with low FD (< 15 T)					
FAU	12.7	CHA	14.5	[5]	+
FAU	12.7	GME	14.6	[6]	+
FAU	12.7	GIS	15.3	[7]	+
FAU	12.7	PHI	15.8	[6]	+
FAU	12.7	MER	16.0	[8]	+
FAU	12.7	LTL	16.3	[7]	+
FAU	12.7	HEU	17.1	[6]	+
FAU	12.7	MOR	17.2	[7]	+
FAU	12.7	BRE	17.3	[6]	+
FAU	12.7	STF	17.3	[1]	+
FAU	12.7	MFI	17.9	[1]	+
FAU	12.7	ANA	18.5	[4] [8] [6]	+
FAU	12.7	ABW	19.0	[8]	+
FAU	12.7	MTW	19.4	[1]	+
FAU+FER	12.7+(17.8)	GIS	15.3	[7]	+
FAU+BEA	12.7+(15.1)	MFI	17.9	[1]	+
LTA	12.9	GIS	15.3	[9]	+
LTA	12.9	MWW	16.5	[10]	+
LTA	12.9	SOD	17.2	[11] [12] [13]	+
GME	14.6	ANA	18.5	[14]	+
Using parent zeolite with high FD (>15 T)					
BEA	15.1	MFI	17.9	[7]	+
LEV	15.2	CHA	14.5	[2]	–
GIS	15.3	ANA	18.5	[15]	+
GIS	15.3	CHA	14.5	[16]	–
MWW	16.5	FER	17.8	[17]	–
SOD	17.2	CAN	16.6	[18]	–
MTW	19.4	ITW	18.1	[19]	–
TON	19.7	ITW	18.1	[20]	–

[a] FD obtained from reference [21], expressed by T, number of heteroatoms per 1000 Å^3.
[b] Source data of the parent zeolite type and its transformation product.
[c] the suitability of framework density approach for the transformation, "+" and "–" are suitable and not suitable, respectively.

According to Maldonado et al. (2013) [3], the FD approach can only be used in a series of framework transformations in the reaction system that pass through the formation of some zeolites. Despite having the same framework type, the difference in Si and Al atomic composition of the parent zeolite can produce zeolite with increase or decrease in FD [3]. The failure of the FD approach is clear evidence that zeolite transformation involves not only thermochemical controls such as temperature, enthalpy, and entropy, but also the framework chemistry, dissolved fragments, basic concentrations, structure directing agents, and solvents.

3. Evaluation of composite and secondary building unit approaches in interzeolite transformations

Some research groups mentioned explicitly the phenomenon of transformation is the relationship of framework building units, both in general and in particular as composite building units (CBU's) or secondary building units (SBU's), between the parent zeolite and its transformation product [1] [2] [4] [7] [8] [13] [20]. The claims of the relationship of these building units are in fact still limited in their proofs, especially in the involvement of d4r and d6r CBU's, or 4-4 and 6-6 SBU's [1] [2] [7] [14], which consecutively contain 8 and 12 units of $TO_{4/2}$. The CBU or SBU number correlated still limited in the number of existing CBU's and SBU's, of which at least 47 and 23 respectively [21]. The majority of CBU's are fairly large blocks of the structure with a range of 5 - 48 units of $TO_{4/2}$, whereas SBU's in 3 - 16 $TO_{4/2}$ [21].

If the parent zeolite has CBU/s or SBU/s are correlated to or correlated with its transformation product/s, the most feasible relationship can be expressed by "one of CBU/SBU constructing a transformed zeolite framework must be available in one of the CBU/SBU constructing its parent zeolite framework". By using the constraints of this building unit relationship, the most (at least 19 of 35 cases) of transformations have no CBU relationships as shown in Table 2, and little (at least 7 of 35 cases) of transformations still have no SBU relationships as shown in Table 3.

Based on Tables 2 and 3, SBU has more suitability than CBU in connecting the parent zeolite framework with its transformation product. We concluded that CBU approach couldnot be used in understanding transformation, the CBU fact is only a structural concept that connects large blocks between framework types [21], rarely match with dissolved fragments in zeolite or silica dissolution cases. Otherwise, the SBU has more compatibility to understand the transformation because the number of atoms of T involved in zeolite or silica dissolution cases is relatively small and has more resemblance to the fragments. In cases of zeolite or silica dissolutions, the dissolved fragments are more similar to SBUs, such as ring 4, 5 and 6-membered T [22] [23] [24] [25] [26] [27]. Both CBU and SBU are basically composed of ring structure units, but the difference is that the rings in SBU's are simpler and more likely to resemble the zeolite and silica dissolved fragments in basic solution.

4. Ring building unit (RBU) as an alternative approach

Basically, each SBU consists of one or a combination of several rings. Table 4 presents some cases of zeolite transformations that have no SBU relationship, especially in cases of transformation with support of seed or SDA. To explain transformation cases that do not have SBU relationship in transformations using seed or organic directing agent (SDA) can be explained by ring building approach (RBU). There is the fact that the interzeolite transformation always yields parent zeolite dissolution which not all dissolved fragments produce new zeolite type crystals [2] [7] [28], with dissolution of Si tends greater than Al [29] [30] [31]. Based on the data in Table 4 it can be seen the role of seeds in the case of FAU transformation into HEU, MFI, MTW and STF, because the existence of seeds can facilitate the crystal nucleation of dissolved silicate fragments. FAU does not contain a 5-ring but still produces a framework that has a 5-ring like MFI, MTW, and STF. Ring 5T is characteristic of high silica zeolites, because the most stable 5-ring contains only Si heteroatoms (pentasil). This characteristic can be proven by RBU's belong to high silica zeolite as shown in Table 5 and 6. The always larger dissolution of Si than Al has provided enough Si to form these 5-rings.

Table 2 Composite building unit relationships between parent zeolite and its transformation product

Parent zeolite		Transformation product		Ref.[b]	Availability[c]
Framework type	CBU's[a]	Framework type	CBU's[a]		
Without template, structure directing agent, or seed					
*BEA	*mor, bea*	MFI	*mor, cas, mel, mfi*	[7]	+
FAU	*d6r, sod*	ANA	?	[4]	?
FAU	*d6r, sod*	GIS	*dcc, gis*	[7]	–
FAU	*d6r, sod*	MOR	*mor*	[7]	–
FAU	*d6r, sod*	LTL	*dsc, d6r, can, ltl*	[7]	+
FAU+*BEA	*(d6r, sod) + (mor, bea, mtw)*	MFI	*mor, cas, mel, mfi.*	[1]	+
FAU+FER	*(d6r, sod) + (fer)*	GIS	*dcc, gis*	[7]	–
GIS	*dcc, gis*	ANA	?	[15]	?
LEV	*d6r*	CHA	*d6r, cha*	[2]	+
LTA	*d4r, sod, lta*	FAU	*d6r, sod*	[3]	+
LTA	*d4r, sod, lta*	GIS	*dcc, gis*	[9]	–
LTA	*d4r, sod, lta*	SOD	*sod*	[11] [12] [13] [3]	+
SOD	*sod*	CAN	*dzc, can*	[18]	–
supported with template, structure directing agent, or seed					
*BEA	*mor, bea, mtw*	MOR	*mor*	[32]	+
FAU	*d6r, sod*	ABW	*dzc, abw*	[8]	–
FAU	*d6r, sod*	ANA	?	[8] [6]	?
FAU	*d6r, sod*	BRE	*bre*	[6]	–
FAU	*d6r, sod*	CAN	*dzc, can*	[33]	–
FAU	*d6r, sod*	CHA	*d6r, cha*	[1] [8] [28] [6] [5] [34]	+
FAU	*d6r, sod*	GME	*dcc, d6r, gme*	[6]	+
FAU	*d6r, sod*	HEU	*bre*	[6]	–
FAU	*d6r, sod*	MER	*dcc, d8r, pau*	[8]	–
FAU	*d6r, sod*	MFI	*mor, cas, mel, mfi*	[1]	–
FAU	*d6r, sod*	MTW	*jbw, cas, bik, mtw*	[1]	–
FAU	*d6r, sod*	MWW	*d6r, mel*	[35]	+
FAU	*d6r, sod*	PHI	*dcc double crankshaft chain*	[6]	–
FAU	*d6r, sod*	SOD	*sod*	[33]	+
FAU	*d6r, sod*	STF	*cas, stf*	[1]	–
LTA	*d4r, sod, lta*	MWW	*d6r, mel*	[10]	–
GME	*dcc, d6r, gme*	ANA	?	[14]	?
GIS	*dcc, gis*	CHA	*d6r, cha*	[36] [16]	–
MTW	*jbw, cas, bik, mtw*	ITW	*d4r*	[19]	–
MWW	*d6r, mel*	FER	*fer*	[17]	–
TON	*jbw, mtt, bik, ton*	ITW	*d4r*	[20]	–

[a] CBU's obtained from reference [21], ANA have no specific CBU's; [b] source data of the parent zeolite type and its transformation product; [c] Availability of CBU/s in parent zeolite, signs of "+" and "–" are available and not available, respectively, in one, some, or all of CBU's, whereas "?" is not specified yet in reference [21].

The case of FAU can be transformed into HEU because the HEU zeolite type contains 4 and 6 rings available in the FAU framework. The first number on the SBU codes can represent the ownership of the RBU from a zeolite framework type, such as FAU with SBU 6-6, 6-2, 6, 4-2, 1-4-1, and 4 can be simplified into RBU 6 and 4, while in HEU 4-4=1 can be simplified into SPC 4. However, SBU codes does not always represent all existing RBU, as in HEU, besides containing RBU 4 also contains RBU 6. RBU can be easily determined directly from the structure framework image available in literature, while specifying the SBU requires sufficient time, unless it is stated in the framework type data. Thus, beside the transformation of FAU into HEU is supported by seed, it is also supported by the relationship of both RBU's.

Table 3 Secondary building unit relationships between parent zeolit and its transformation product

Parent zeolite		Transformation product		Ref.[b]	availability[c]
Framework type	SBU's[a]	Framework type	SBU's[a]		
Without template, structure directing agent, or seed					
*BEA	5-1	MFI	5-1	[7]	+
FAU	6-6 / 6-2 / 6 / 4-2 / 1-4-1 / 4	ANA	6-2 / 6 / 4[1,1] / 1-4-1 / 4	[4]	+
FAU	6-6 / 6-2 / 6 / 4-2 / 1-4-1 / 4	GIS	8 / 4	[7]	+
FAU	6-6 / 6-2 / 6 / 4-2 / 1-4-1 / 4	LTL	6 / 4-2	[7]	+
FAU	6-6 / 6-2 / 6 / 4-2 / 1-4-1 / 4	MOR	5-1	[7]	–
FAU+*BEA	(6-6 / 6-2 / 6 / 4-2 / 1-4-1 / 4)+(5-1)	MFI	5-1	[1]	+
FAU+FER	(6-6 / 6-2 / 6 / 4-2 / 1-4-1 / 4)+(5-1)	GIS	8 / 4	[7]	+
GIS	8 / 4	ANA	6-2 / 6 / 4[1,1] / 1-4-1 / 4	[15]	+
LEV	6	CHA	6-6 / 6 / 4-2 / 4	[2]	+
LTA	8 / 4-4 / 6-2 / 6 / 1-4-1 / 4	FAU	6-6 / 6-2 / 6 / 4-2 / 1-4-1 / 4	[3]	+
LTA	8 / 4-4 / 6-2 / 6 / 1-4-1 / 4	GIS	8 / 4	[9]	+
LTA	8 / 4-4 / 6-2 / 6 / 1-4-1 / 4	SOD	6	[11] [12] [3]	+
SOD	6	CAN	12 / 6 / 4	[18]	+
Supported with template, structure directing agent, or seed					
*BEA	5-1	MOR	5-1	[32]	+
FAU	6-6 / 6-2 / 6 / 4-2 / 1-4-1 / 4	ABW	8 / 4	[8]	+
FAU	6-6 / 6-2 / 6 / 4-2 / 1-4-1 / 4	ANA	6-2 / 6 / 4[1,1] / 1-4-1 / 4	[8] [6]	+
FAU	6-6 / 6-2 / 6 / 4-2 / 1-4-1 / 4	BRE	4	[6]	+
FAU	6-6 / 6-2 / 6 / 4-2 / 1-4-1 / 4	CAN	12 / 6 / 4	[33]	+
FAU	6-6 / 6-2 / 6 / 4-2 / 1-4-1 / 4	CHA	6-6 / 6 / 4-2 / 4	[1] [8] [28] [6] [5] [34]	+
FAU	6-6 / 6-2 / 6 / 4-2 / 1-4-1 / 4	GME	12 / 6-6 / 8 / 6 / 4-2 / 4	[6]	+
FAU	6-6 / 6-2 / 6 / 4-2 / 1-4-1 / 4	HEU	4-4=1	[6]	–
FAU	6-6 / 6-2 / 6 / 4-2 / 1-4-1 / 4	MER	8-8 / 8 / 4	[8]	+
FAU	6-6 / 6-2 / 6 / 4-2 / 1-4-1 / 4	MFI	5-1	[1]	–
FAU	6-6 / 6-2 / 6 / 4-2 / 1-4-1 / 4	MTW	5-[1,1]	[1]	–
FAU	6-6 / 6-2 / 6 / 4-2 / 1-4-1 / 4	MWW	?	[35]	?
FAU	6-6 / 6-2 / 6 / 4-2 / 1-4-1 / 4	PHI	8 / 4	[6]	+
FAU	6-6 / 6-2 / 6 / 4-2 / 1-4-1 / 4	SOD	6	[33]	+
FAU	6-6 / 6-2 / 6 / 4-2 / 1-4-1 / 4	STF	5-3	[1]	–
GIS	8 / 4	CHA	6-6 / 6 / 4-2 / 4	[36] [16]	+
GME	12 / 6-6 / 8 / 6 / 4-2 / 4	ANA	6-2 / 6 / 4[1,1] / 1-4-1 / 4	[14]	+
LTA	8 / 4-4 / 6-2 / 6 / 1-4-1 / 4	MWW	?	[10]	?
MTW	5-[1 /1]	ITW	1-4-1 / 4-[1,1]	[19]	–
MWW	?	FER	5-1	[17]	?
TON	5-1	ITW	1-4-1 / 4-[1,1]	[20]	–

[a] SBU's obtained from reference [21], The "/" sign separating SBU's states the alternative use of SBU, SBU's constructing MWW type is not specified yet in the reference [37].
[b] source data of the parent zeolite type and its transformation product,
[c]Availability of SBU/s alternatives in parent zeolite, the "+" and "–" marks state that one of them is available and not available in the parent zeolite, respectively.

In Table 4 there is also a case of FAU transformation into a MOR that can be explained by RBU relationship. The MOR framework not only has RBU 5 (representation of SBU 5-1), but also has RBU 4 and 6. The RBU 5 which is not available in the FAU can be explained from the characteristics of RBU 5 that can be identified with high-silica zeolites. Honda et al. (2013) succeeded in transforming FAU into MOR without the aid of template or SDA due to using high-silica FAU (Si/Al ratio = 76), while using lower silica FAU, with Si/Al = 25 only produced GIS (containing only RBU 4) and LTL (containing RBU 4 and 6) [7], but RBUs are both available in FAU. Sufficient silicate fragments of FAU dissolution allow formed MOR transformation products containing RBU 5 as well.

Table 4 Some cases of transformation that have no SBU relationship

Parent zeolite		Transformation product		Ref.[b]	Support[c]
Type	SBU[a]	Type	SBU[a]		
FAU	6-6, 6-2, 6, 4-2, 1-4-1, 4	MOR	5-1	[7]	–
FAU	6-6, 6-2, 6, 4-2, 1-4-1, 4	HEU	4-4=1	[6]	seed
FAU	6-6, 6-2, 6, 4-2, 1-4-1, 4	MFI	5-1	[1]	seed
FAU	6-6, 6-2, 6, 4-2, 1-4-1, 4	MTW	5-[1,1]	[1]	seed
FAU	6-6, 6-2, 6, 4-2, 1-4-1, 4	STF	5-3	[1]	seed
MTW	5-[1,1]	ITW	1-4-1, 4-[1,1]	[19]	SDA
TON	5-1	ITW	1-4-1, 4-[1,1]	[20]	SDA

[a] source data SBU of zeolite type from reference [21].
[b] source data of the parent zeolite type and its transformation products.
[c] support in the transformation.

Table 5 Types of aluminosilicate and silica zeolites framework with SBU or RBU having 5-membered ring.

Type	Si/Al ratio[a]	SBU[b]	RBU				
BIK	2.0	5-1	5 and 6	TON	11.0 - infinite	5-1	5 and 6
MAZ	2.6	5-1 / 4-2	4 and 5	CDO	silica	5-1	5 and 6
STI	2.6	4-4=1	4 and 5	IMF	silica	5-1	4, 5 and 6
DAC	3.8	5-1	4, 5 and 6	NSI	silica	5-1	5 and 6
BOG	4.3	6 / 5-1 / 4	4, 5 and 6	DON	silica	5-3	4, 5 and 6
CAS	5.0	5-1	5 and 6	GON	silica	5-3	4, 5 and 6
FER	5.0	5-1	5 and 6	SFF	silica	5-3	4, 5 and 6
MOR	5.0	5-1	4, 5 and 6	SGT	silica	5-3	5 and 6
ESV	8.37	5-1	4 and 5	STF	silica	5-3	4, 5 and 6
*BEA	8.14	?	4, 5 and 6	STT	silica	5-3	4, 5 and 6
EON	4.26 - 15.67	5-1	4 and 5	RTE	silica	6 / 5-1	4, 5 and 6
MFI	2.5 - infinite	5-1	4, 5 and 6	ISV	silica	6-2	4, 5 and 6
EUO	4.89 - infinite	1-5-1	4, 5 and 6	IFR	silica	6-2	4, 5 and 6
MEL	5.0 - infinite	5-1	4, 5 and 6	RWR	silica	6-2	5 and 6
MTW	10.2 - infinite	5-[1,1]	4, 5 and 6	RRO	silica	4-1=1	4, 5 and 6
MTT	11.0 - infinite	5-1	5 and 6	DOH	silica	?	4, 5 and 6

a and b source data taken from reference [21], *BEA and DOH are not specified yet in the reference

The transformation of MTW or TON into ITW can be explained also by the RBU relationship. MTW contains RBU 4, 5 and 6, while TON contains RBU 5 and 6. ITW itself consists only of RBU 4 and 6. TON which is a high-silica zeolite will have a considerable dissolution of silicates allowing the RBU 4 available to construct ITW framework. ITW is a bit of an exceptional silica zeolite that does not have RBU 5.

5. Relationship evidence of ring building units in zeolite classification based on silicon to aluminum molar ratio

Until now there has been no approach to understand framework types based on RBU relationships. Although SBU has basic ring forms, but not all rings within the zeolite framework can be represented in SBUs. Although there are 3 to 20 T-rings [21], only 4, 5, and 6 rings are most commonly found in all zeolites, as well as those found in the zeolite and silica dissolved fragments [22] [23] [24] [25] [26] [27].

The existence of 5-ring becomes the characteristic of high-silica aluminosilicate or silica zeolites, it is not surprising that the use of the term "pentasil" is often associated with rings 5 of high-silica zeolites, which means that the 5-ring tend to have only Si atoms, as in MFI, MEL, and *BEA [37]. In addition to having a single SBU 5-1, the three framework types are also dominated by RBU 5. From Table 5 and 6 we can compare the RBU approach with and without RBU 5.

Tabel 6 Types of aluminosilicate and silica zeolites without 5-membered ring

Tipe	Si/Al[a]	SBU[b]	RBU	Tipe	Si/Al[a]	SBU[b]	RBU
GIS	1.0	4	4	GME	2.0	12 / 6-6 / 8 / 6 / 4-2 / 4	4 and 6
LIO	1.0	6 / 4	4 and 6	LAU	2.0	6 / 1-4-1	4 and 6
GIU	1.0	6 / 4	4 and 6	KFI	2.2	6-6 / 6-2 / 8 / 6 / 4-2 / 4	4 and 6
FRA	1.0	6 / 4	4 and 6	FAU	2.3	6-6 / 6-2 / 6 / 4-2 / 1-4-1 / 4	4 and 6
FAR	1.0	6	4 and 6	OFF	2.6	6 / 4-2	4 and 6
CAN	1.0	12 / 6 / 4	4 and 6	LTL	3.0	6 / 4-2	4 and 6
TSC	1.0	8-8 / 8 / 6-6 / 6 / 4-2 / 4	4 and 6	ERI	3.0	6 / 4	4 and 6
SOD	1.0	6	4 and 6	EAB	3.0	6 / 4	4 and 6
MAR	1.0	8 / 4	4 and 6	MOZ	3.5	6 / 4-2	4 and 6
LOS	1.0	6-2 / 6 / 4	4 and 6	HEU	3.5	4-4=1	4 and 6
AFG	1.0	6 / 4	4 and 6	EMT	3.57	6-6 / 6-2 / 6 / 4-2 / 1-4-1 / 4	4 and 6
LTA	1.0	8 / 4-4 / 6-2 / 6 / 1-4-1 / 4	4 and 6	TER	5.5	2-6-2 / 4-1	4 and 6
LTN	1.0	6 / 4-2	4 and 6	MSO	13.3	2-6-2 / 4-1	4 and 6
ANA	2.0	6-2 / 6 / 4-[1,1] / 1-4-1 / 4	4 and 6	ITW	silica	1-4-1 / 4-[1,1]	4 and 6
CHA	2.0	6-6 / 6 / 4-2 / 4	4 and 6				

a and b obtained from reference [21]

From Table 5 it can be seen that most SBU's that have 5-rings are high-silica zeolites. However, there are low Si/Al framework types although they contain SBU 5-1 or RBU 5 as presented in Table 7. In Table 7, there are framework types that have RBU 5 but are classified as having low to moderate Si/Al, i.e. BIK, MAZ, DAC, BOG, CAS, FER, MOR, and STI. For FER and MOR, the presence of 5-rings assumed to be possessed only by zeolites that have a moderate to high Si/Al can be proven because both of these have recently been shown to be synthesized with high Si/Al [38]. Furthermore, in Table 7 it is seen that STI have only SPS 4-4=1, but the zeolite can be synthesized with high-silica [39] because STI has RBU 5. There is a possibility of BIK, MAZ, DAC, BOG, and CAS can be found in medium or high Si/Al as it does on FER, MOR, and STI.

As shown in Table 6, MSO and ITW are two types of high-silica zeolites among zeolite types that do not have RBU 5. However, siliceous zeolites or high-silica zeolites without 5-ring can be produced from synthesis using HF mineralizer and SDA, such as ITW [40] and LTA A [41], even in the lower H_2O/SiO_2 ratio indicate more 4-ring zeolites [40]. The tendency of fragmentation in the fluoride solution media is analogous to fragmentation in the base medium, a concentrated base concentration causes silicate fragmentation tends to be in ring forms 4 rather than 5 and 6, or more containing smaller rings [2] [42].

The cases of aluminosilicate and silica zeolite transformation generally employed in an alkaline solution medium. There are few cases of transformation studies using fluoride solution media so there is still difficulty in understanding the role of CBU, SBU, and RBU in these cases. To date, there have been at least two cases of transformation using fluoride solution media, i.e., MTW to ITW [19], and TON to ITW [20], both using SDA (see Table 4). The use fluoride media tends to be able synthesized siliceous zeolites easier than hydroxide media [40] [41], but always using SDA [19] [40] [41] [43] [44]. These facts suggested that zeolite transformation in fluoride media occur by parent zeolite dissolution completely. Thus the RBU approach in understanding the transformation in fluoride media cannot be used. Transformations in the fluoride solution media tend to be determined by the SDA. Because of the total dissolution, the process of forming a new zeolite in the fluoride solution medium takes a relatively long time and the temperature is relatively high [19] [20] [40] [41] [43] [44].

Table 7 The relationship between SBU and RBU with low-silica zeolites but has 5-ring

Type	Si/Al ratio			SBU ownership		RBU ownership	
	Previous findings[a]	Recent findings	Ref.[b]	SBU	suitability[c]	RBU	suitability[c]
BIK	2.0	n.a.		5-1	?	5 and 6	?
MAZ	2.6	n.a.		5-1 / 4-2	?	4 and 5	?
DAC	3.8	n.a.		5-1	?	4, 5 and 6	?
BOG	4.3	n.a.		6 / 5-1 / 4	?	4, 5 and 6	?
CAS	5.0	n.a.		5-1	?	5 and 6	?
FER	5.0	46	[39]	5-1	+	5 and 6	+
MOR	5.0	60	[38]	5-1	+	4, 5 and 6	+
STI	2.6	45	[39]	4-4=1	+	4 and 5	+

[a] Si/Al value based on reference [21]; [b] reference of recent findings in Si/Al ratio; [c] a "+" sign matching the Si/Al mole ratio, "?" states that the match is not certain yet.

In Table 8, it can be seen that the RBU is better able to distinguish zeolite classification than SBU approach based on Si/Al, i.e. on zeolite types with non-SBU 5-ring (STI, DOH, IFR, ISV, RRO, and RWR). However, there are still some cases of RBU incompatibilities that also occur in their SBU, especially in MSO and ITW, both of which are high-silica but do not have 5-ring. The opposite of the case is on BIK, MAZ, DAC, BOG, and CAS, are low-silica but have 5-ring.

Table 8 The comparison between the SBU and RBU approaches in less common Si/Al ratio zeolites

Tipe	Si/Al[a]	SBU approach		RBU approach	
		SBU[b]	suitability[c]	RBU	suitability[c]
MSO	13.3	2-6-2 / 4-1	–	4 and 6	–
STI	2.6 or 45	4-4=1	+	4 and 5	+
ITW	silica	1-4-1 / 4-[1,1]	–	4 and 6	–
DOH	silica	?	?	4, 5 and 6	+
IFR	silica	6-2	–	4, 5 and 6	+
ISV	silica	6-2	–	4, 5 and 6	+
RRO	silica	4-1=1	–	4, 5 and 6	+
RWR	silica	6-2	–	5 and 6	+

[a] Si/Al value based on reference [21]; [b] SBU data obtained from reference [21], "?" states it is not be ascertained yet; [c] "+" and "-" marks express suitable or not suitable with the Si/Al mole ratio, "?"states can not be ascertained suitability.

Based on the ratio of Si/Al mole ratios, zeolite classification is divided into three categories, namely low (Si/Al = 1.0 ~ 1.5), medium (Si/Al = 2 ~ 5), and high (Si/Al = 10 ~ 100) siliceous zeolites [37]. This categorization is actually still rough but quite useful in tracing the relationship of the chemical composition of zeolite and its framework with its ring structure units. Using the RBU 5 limits, there is a fairly wide slice of the Si/Al ratio between medium- and low- or high-silica zeolites. Based on the facts in RBU 4, 5, and 6, the following provisions can be made:

1. Under Lowenstein's rule, the lowest Si/Al ratio in RBU 4 and 6 only allows the ratio of Si/Al = 1, this corresponds to the fact that all zeolites with Si/Al = 1 can be found in zeolite frameworks having 4 and 6-ring only,
2. RBU 5 is the most common feature present in high-silica aluminosilicate zeolites, while the ownership of RBU 5 allows Si/Al> 1,
3. The characteristic of zeolites having RBU 5 extends to a range of 1 < Si/Al < infinite, so that the majority of siliceous zeolites will have the distinctive features of RBU 5 ownership as well,
4. Lowenstein's rule allows the formation of silica zeolites without having SPC 5,
5. The combined repeat of RBU 4 and 6 will have T = 6n (n = repetition), as shown in the following scheme,

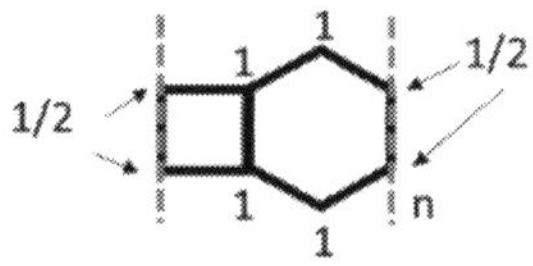

MCM-61 (MSO), having the chemical formula $|K_{2.1}C_{12}H_{24}O_6|$ $[Al_{2.1}Si_{27.9}O_{60}]$ [21], with T = 30 per unit of formula (with Si/Al = 13.29), has a population ratio of RBU 4 relatively equal to RBU 6 (directly calculated from the framework description available in literature). The amount of T = 30 per unit of the formula represents the repetition of RBU 4 and 6 combinations by 5 times, so that some possible Si/Al mole ratios are 29/1 = 29, 28/2 = 14, ..., 15/15 = 1. If MCM-61 is an aluminosilicate zeolite having the highest Si/Al ratio among zeolites without RBU 5, the Si/Al = 14 ratios can be considered as the highest Si/Al limit of the medium-silica zeolite category. Based on the assumptions above, zeolites can be categorized based on Si/Al ratio as follows:

Criteria	Zeolite category			
	Low-silica zeolites	Medium-silica zeolites	High-silica zeolites	Siliceous zeolites
RBU	4 or 6	4, 5, or 6	Always contains RBU 5	4, 5, or 6
Chemical framework	Si/Al = 1	2 ≤ Si/Al ≤ 14 or 2 ≤ Si/Al ≤ 29	Si/Al > 14 or Si/Al > 29	Infinite Si/Al, the majority always contain RBU 5, except ITW only contains RBU 4 and 6

Based on the above categorization, it can be explained why FAU (RBU 4 and 6) can be transformed to MOR (RBU 4, 5, and 6) without the aid of SDA, template, or seed (see Table 3). According to Honda et al. (2013), the dealuminated FAU reaches Si/Al = 76 transformed to MOR, whereas with FAU on Si / Al = 25 produces a mixture of MOR and GIS (containing only SPC 4), only GIS or LTL (RBU 6 and 4) [7]. This proves that RBU 5 is the characteristics of high-silica zeolite, RBU 5 can obtain from dealuminated-FAU dissolution and can build a MOR framework zeolite.

The presence of 5-ring-SBU and RBU 5 are generally identical with high-silica zeolites and can correlate the structure between the parent zeolite and its transformation product. However, the RBU has more suitability with the Si/Al ratio than the SBU approaches. In addition, the RBU is more suitable than SBU approaches for early consideration in choosing parent zeolites and predicting several transformation products, based on the facts as follows:

1. In general, the transformation product may be attributed to the availability of its RBUs in the parent zeolite, either do not change or reduced in RBU numbers, of both its type and its dominance,
2. The attributable RBUs are mainly 4, 5, and 6,

3. The higher the Si/Al ratio in parent zeolites the higher chance of producing transformation products that have RBU 5 without the aid of templates, SDAs, or seeds,
4. The process of transformation by using an alkaline solution will always decrease the Si/Al ratio, the higher alkaline concentration the less chances of obtaining RBU 5 in its transformation product if there is no template or SDA aid, and
5. An increase in concentration of base will increasingly provide an opportunity to produce a product with RBU 4 more dominance than RBU 6 on the framework because an increase in base concentration tends to produce smaller fragments.

6. Conclusion

Understanding the transformation between zeolites can be performed by using the ring building units (RBUs) approach as an alternative to using existing approaches, such as framework density (FD), composite building unit (CBU) and secondary building unit (SBU). The FD approach has many discrepancies because the zeolite is metastable material so that the transformation is more affected by the type and population of soluble oligomers partially or completely. The SBU approach has more suitability than CBU approaches because the basic forms of SBU have more resemblance to the basic ring forms in zeolite dissolution cases and in conventional zeolite synthesis. The RBU approach can overcome the shortcomings of the SBU approach in transformation cases involving 5-ring contained zeolites. The RBU approach has been shown to distinguish zeolite classification based on its Si/Al molar ratio, namely low-silica, medium-silica, high-silica, and siliceous zeolites. The RBU approach can be a simple and easy preliminary consideration in selecting zeolite parent and predicting zeolite types based on RBU's ownership of the parent's zeolite, with or without the aid of a template, structure directing agent, or seed. Thus, the success of transformation can be considered as the success of keeping the type and population of RBU's available in the parent zeolite required to construct transformation product framework.

Acknowledgement

Financial support by Ministry of Religious Affairs (Bantuan Biaya Penyelesaian Studi Pascasarjana 2016) is gratefully acknowledged.

References

[1] S. Goel, S. I. Zones dan E. Iglesia, "Synthesis of zeolites via interzeolite transformations without organic structure-directing agents," *Chemistry of Materials,* vol. 27, no. 6, pp. 2056-2066, 2015.

[2] I. Goto, M. Itakura, S. Shibata, K. Honda, Y. Ide, M. Sadakane and T. Sano, "Transformation of LEV-type zeolite into less dense CHA-type zeolite," *Microporous and Mesoporous Materials,* vol. 158, p. 117–122, 2012.

[3] M. Maldonado, M. D. Oleksiak, S. Chinta dan J. D. Rimer, "Controlling crystal polymorphism in organic-free synthesis of Na-zeolites," *Journal of the American Chemical Society,* vol. 135, pp. 2641-2652, 2013.

[4] Y. Wang, X. Li, Z. Xue, L. Dai, S. Xie dan Q. Li, "Preparation of zeolite ANA crystal from zeolite Y by in situ solid phase iso-structure transformation," *Journal of Physical Chemistry B,* vol. 114, p. 5747–5754, 2010.

[5] T. Takata, N. Tsunoji, Y. Takamitsu, M. Sadakane dan T. Sano, "Nanosized CHA zeolites with high thermal and hydrothermal stability derived from the hydrothermal conversion of FAU zeolite," *Microporous and Mesoporous Materials,* vol. 225, pp. 524-533, 2016.

[6] O. Chiyoda dan M. E. Davis, "Hydrothermal conversion of Y-zeolite using alkaline-earth cations," *Microporous and Mesoporous Materials,* vol. 32, pp. 257-264, 1999.

[7] K. Honda, M. Itakura, Y. Matsuura, A. Onda, Y. Ide, M. Sadakane dan T. Sano, "Role of structural similarity between starting zeolite and product zeolite in the interzeolite conversion process," *Journal of Nanoscience and Nanotechnology,* vol. 13, pp. 3020-3026, 2013.

[8] L. V. Tendeloo, E. Gobechiya, E. Breynaert, J. A. Martens and C. E. Kirschhock, "Alkaline cations directing the transformation of FAU zeolite into five different framework types," *Chemical Communications,* vol. 49, pp. 11737-11739, 2013.

[9] B. Subotic, I. Smit, O. Madzija dan L. Sekovanic, "Kinetic study of the transformation of zeolite A into zeolite P," *Zeolites,* vol. 2, pp. 135-142, 1982.

[10] E. Xing, Y. Shi, A. Zheng, J. Zhang, X. Gao, D. Liu, M. Xin, W. Xie, F. Zhang, X. Mu dan X. Shu, "Transformation from NaA to MCM-49 zeolite and its catalytic alkylation performance," *Industrial and Engineering Chemistry Research,* vol. 54, no. 12, pp. 3123-3135, 2015.

[11] B. Subotic, D. Skrtic, I. Smit and L. Sekovanic, "Transformation of zeolite A into hydroxysodalite: I. An approach to the mechanism of transformationa nd its experimental evaluation," *Journal of Crystal Growth,* vol. 50, pp. 498-508, 1980.

[12] R. I. Walton, F. Millange, D. O'Hare, A. T. Davies, G. Sankar dan C. R. A. Catlow, "An in situ energy-dispersive X-ray diffraction study of the hydrothermal crystallization of zeolite A. 1. Influence of reaction conditions and transformation into sodalite," *Journal of Physical Chemistry B,* vol. 105, no. 1, pp. 83-90, 2001.

[13] L. Ding, H. Yang, P. Rahimi, O. Omotoso, W. Friesen, C. Fairbridge, Y. Shi dan S. Ng, "Solid transformation of zeolite NaA to sodalite," *Microporous and Mesoporous Materials,* vol. 130, pp. 303-308, 2010.

[14] P. N. Joshi, A. Thangaraj and P. V. Shiralkar, "Studies on zeolite transformation of high-silica gmelinite into analcime," *Zeolite,* vol. 11, pp. 164-168, 1991.

[15] S. N. Azizi, A. A. Daghigh dan M. Abrishamkar, "Phase Transformation of Zeolite P to Y and Analcime Zeolites due to changing the time and temperature," *Journal of Spectroscopy,* vol. 2013, 2013.

[16] S. I. Zones, "Direct Hydrothermal Conversion of Cubic P Zeolite to Organozeolite SSZ-13," *Journal of the Chemical Society, Faraday Transactions,* vol. 86, no. 20, pp. 3467-3472, 1990.

[17] S. Xie, S. Liu, N. Gao, X. Li, Y. Gao, K. Liu and L. Xu, "Interzeolite conversion of zeolite MCM-49 into zeolite ZSM-35 in cyclohexylamine–hexamethyleneimine–Na2O–H2O containing systems," *New Journal of Chemistry,* vol. 38, pp. 2514-2521, 2014.

[18] M. C. Barnes, J. Addai-Mensah and A. R. Gerson, "The mechanism of the sodalite-to-cancrinite phase transformation in synthetic spent Bayer liquor," *Microporous and Mesoporous Materials,* vol. 31, p. 287–302, 1999.

[19] A. Rojas, M. L. San-Roman, C. M. Zicovich-Wilson dan M. A. Camblor, "Host-guest stabilization of a zeolite strained framework: In situ transformation of zeolite MTW into the less dense and more strained ITW," *Chemistry of Materials,* vol. 25, pp. 729-738, 2013.

[20] C. M. Zicovich-Wilson, F. Gandara, A. Monge dan M. A. Camblor, "In Situ Transformation of TON Silica Zeolite into the Less Dense ITW: Structure-Direction Overcoming Framework Instability in the Synthesis of SiO2 Zeolites," *Journal of American Chemical Society,* vol. 132, pp. 3461-3471, 2010.

[21] C. Baerlocher, L. B. McCusker dan D. H. Olson, Atlas of zeolite framework types, 6th penyunt., Amsterdam: Elsevier, 2007.

[22] I. H. Lim, W. Schrader dan F. Schüth, "The formation of zeolites from solution – Analysis by mass spectrometry," *Microporous and Mesoporous Materials,* vol. 166, pp. 20-36, 2013.

[23] M. Castro, M. Haouas, I. Lim, H. J. Bongard, F. Schuth, F. Taulelle, G. Karlsson, V. Alfredsson, E. Breyneart, C. E. Kirschhock dan W. Schmidt, "Zeolite beta formation from clear sols: silicate

speciation, particle formation and crystallization monitored by complementary analysis methods," *European Journal of Chemistry,* vol. 22, pp. 1-14, 2016.

[24] L. I. Meza, M. W. Anderson, B. Slater dan J. R. Agger, "In situ atomic force microscopy of zeolite A dissolution," *Physical Chemistry Chemical Physics,* vol. 10, pp. 5066-5076, 2008.

[25] B. B. Schaack, W. Schrader dan F. Schuth, "Detection of structural elements of different zeolites in nucleating solutions by electrospray ionization mass spectrometry," *Angewandte Chemie International Edition,* vol. 47, pp. 9092-9095, 2008.

[26] L. Ren, C. Li, F. Fan, Q. Guo, D. Liang, Z. Feng, C. Li, S. Li dan F.-S. Xiao, "UV–raman and NMR spectroscopic studies on the crystallization of zeolite A and a new synthetic route," *Chemistry: A European Journal,* vol. 17, pp. 6162-6169, 2011.

[27] Y. Xiao, N. Sheng, Y. Chu, Y. Wang, Q. Wu, X. Liu, F. Deng, X. Meng dan Z. Feng, "Mechanism on solvent-free crystallization of NaA zeolite," *Microporous and Mesoporous Materials,* vol. 237, pp. 201-209, 2017.

[28] M. Itakura, I. Goto, A. Takahashi, T. Fujitani, Y. Ide, M. Sadakane and T. Sano, "Synthesis of high-silica CHA type zeolite by interzeolite conversion of FAU type zeolite in the presence of seed crystals," *Microporous and Mesoporous Materials,* no. 144, pp. 91-96, 2011.

[29] A. Cizmek, B. Subotic, R. Aiello, F. Crea, A. Nastro dan C. Tuoto, "Dissolution of high-silica zeolites in alkaline solutions I. Dissolution of silicalite-1 and ZSM-5 with different aluminum content," *Microporous Materials,* vol. 4, pp. 159-168, 1995.

[30] A. Cizmek, B. Subotic, I. Smit, A. Tonejc, R. Aiello, F. Crea dan A. Nastro, "Dissolution of high-silica zeolites in alkali solutions II. Dissolution of 'activated' silicalite-1 and zsm-5 with different aluminum content," *Microporous Materials,* vol. 8, pp. 159-169, 1997.

[31] A. Petushkov, J. Freeman dan S. C. Larsen, "Framework stability of nanocrystalline NaY in aqueous solution at varying pH," *Langmuir,* vol. 26, no. 9, pp. 6695-6701, 2010.

[32] H. Zhang, H. Zhang, P. Wang, Y. Zhao, Z. Shi, Y. Zhang dan Y. Tang, "Organic template-free synthesis of zeolite mordenite nanocrystals through exotic seed assisted conversion," *RSC Advances,* vol. 6, pp. 47623-47631, 2016.

[33] J.-C. Buhl, "Enhanced methods of crystallization: The crossover synthesis from gel to melt flow - A case study on sodalites," *Microporous and Mesoporous Materials,* vol. 236, pp. 13-20, 2016.

[34] X. Xiong, D. Yuan, Q. Wu, F. Chen, X. Meng, R. Lv, D. Dai, S. Maurer, R. McGuire, M. Feyen, U. Müller, W. Zhang, T. Yokoi, X. Bao, H. Gies, B. Marler, D. E. De Vos, U. Kolb, A. Moini dan F.-S. Xiao, "Efficient and rapid transformation of high silica CHA zeolite from FAU zeolite in the absence of water solvent," *Journal of Materials Chemistry A,* vol. 5, no. 19, pp. 9076-9080 , 2017.

[35] E. Xing, Y. Shi, W. Xie, F. Zhang, X. Mu dan X. Shu, "Size-controlled synthesis of MCM-49 zeolite from NaX for liquid phase alkylation of benzene with ethylene," *Microporous and Mesoporous Materials,* vol. 236, pp. 54-62, 2016.

[36] S. I. Zones dan R. v. Nordstrand, "Novel zeolite transformations: The template-mediated conversion of cubic P zeolite to SSZ-13," *Zeolites,* vol. 8, pp. 166-174, 1988.

[37] R. Xu, W. Pang, J. Yu, Q. Huo dan J. Chen, Chemistry of Zeolites and Related Porous Materials: Synthesis and Structure, Singapore: John Wiley & Sons (Asia) Pte Ltd, 2007.

[38] A. Lv, H. Xu, H. Wu, Y. Liu and P. Wu, "Hydrothermal synthesis of high-silica mordenite by dual-templating method," *Microporous and Mesoporous Materials,* vol. 145, pp. 80-86, 2011.

[39] T. M. Davis, C.-Y. Chen, N. Zilková, D. Vitvarová-Procházková, J. Cejka and S. I. Zones, "The importance of channel intersections in the catalytic performance of high silica stilbite," *Journal of Catalysis,* vol. 298, pp. 84-93, 2013.

[40] S. I. Zones, S.-J. Hwang, S. Elomari, I. Ogino, M. E. Davis and A. W. Burton, "The fluoride-based route to all-silica molecular sieves; a strategy for synthesis of new materials based upon close-packing of guest–host products," *C. R. Chimie,* vol. 8, pp. 267-282, 2005.

[41] T. Moteki and R. F. Lobo, "A General Method for Aluminum Incorporation into High-Silica Zeolites Prepared in Fluoride Media," *Chemistry of Materials,* vol. 28, no. 2, pp. 638-649, 2016.

[42] M. D. Oleksiak, J. A. Soltis, M. T. Conato, R. L. Penn dan J. D. Rimer, "Nucleation of FAU and LTA zeolites from heterogeneous aluminosilicate precursors," *Chemistry of Materials,* vol. 28, no. 14, pp. 4906-4916, 2016.

[43] D. F. Shantz, A. Burton and R. F. Lobo, "Synthesis, structure solution, and characterization of the aluminosilicate MCM-61: the first aluminosilicate clathrate with 18-membered rings," *Microporous and Mesoporous Materials,* vol. 31, pp. 61-73, 1999.

[44] C. Jo, W. Park and R. Ryoo, "Synthesis of mesoporous zeolites in fluoride media with structure-directing multiammonium surfactants," *Microporous and Mesoporous Materials,* vol. 239, pp. 19-27, 2017.

IOP Conf. Series: Materials Science and Engineering **349** (2018) 012017 doi:10.1088/1757-899X/349/1/012017

Isolation, Identification, and Xanthine Oxidase Inhibition Activity of Alkaloid Compound from *Peperomia pellucida*

E Fachriyah, M A Ghifari, K Anam

Chemistry Department, Faculty of Sciences and Mathematics
Diponegoro University

Corresponding author : enny.fachriyah@live.undip.ac.id

Abstract The research of the isolation and xanthine oxidation inhibition activity of alkaloid compound from *Peperomia pellucida* has been carried out. Alkaloid extract is isolated by column chromatography and preparative TLC. Alkaloid isolate is identified spectroscopically by UV-Vis spectrophotometer, FT-IR, and LC-MS/MS. Xanthine oxidase inhibition activity is carried out by in vitro assay. The result showed that the alkaloid isolated probably has piperidine basic structure. The alkaloid isolate has N-H, C-H, C = C, C = O, C-N, C-O-C groups and the aromatic ring. The IC_{50} values of ethanol and alkaloid extract are 71.6658 ppm and 76.3318 ppm, respectively. Alkaloid extract of *Peperomia pellucida* showed higher activity than ethanol extract.

1. Introduction

Indonesia is a developing country with the highest number of uric acid sufferers in Asia. Approximately 1.7% of Indonesia's population suffers gout [1][2]. Uric acid is caused by serum uric acid (SUA) levels that exceed the normal conditions. When uric acid levels in the blood exceed 6.8 mg / dL, monosodium urate crystals (MSU) will form and accumulate in joints, tendons, and other tissues. The pile of that crystals causes gout [3][4]. In addition, abnormal levels of SUA are correlated with cardiovascular disease, hypertension and renal disease [3]. Several therapies can be attempted to treat gout. The most effective therapy is to lowering uric acid levels, one of which can be used is a xanthine oxidase inhibitor [5][6].

Peperomia pellucida is a plant of the Piperaceae family and Peperomia genus. Traditionally, it is used by Indonesian and Filipino to treat kidney and gout problems [7][8]. So far has not been reported its bioactive compounds that have a reducing activity of uric acid. Phytochemical screening of *Peperomia pellucida* shows secondary metabolite content such as alkaloids, saponins, tannins, flavonoids, steroids, and triterpenoids [9]. Alkaloids have been known potential as an anti-gout by inhibiting the activity of xanthine oxidase enzyme [6]. Alkaloid compounds such as berberine, costinone A and costinone B have been shown inhibition activity of xanthine oxidase enzymes to reduce uric acid levels [10][11]. Based on that description, this research aims to isolate alkaloid compounds from *Peperomia pellucida* to know its inhibitory activity against xanthine oxidase enzyme.

2. Materials and Method

2.1 Materials

The materials in this research are *Peperomia pellucida*, methanol, ethanol, ethyl acetate n-hexane, chloroform, ammonia 25%, concentrated sulfuric 98%, distilled water, hydrochloric acid 2 N, Dragendorff reagent, Mayer reagent, silica gel TLC plate 60 GF_{254}, silica 60, DMSO, xanthine substrate X0626-5G, xanthine oxidase from bovine milk lyophilized powder X4736-5UN, allopurinol, aqua DM, potassium dihydrogen phosphate and potassium hydrogen phosphate.

2.2. Method

2.2.1 Alkaloid Isolation

Dried plants powder which has 2.5 kg of its weight was macerated with 96% ethanol. The extract was concentrated by a rotary evaporator (Buchi Rotavapor R-3) to obtain ethanol extract. Then chlorophyll was removed by adding hot distilled water (1:1). The presence of alkaloids in the ethanol-water extract was tested by Mayer and Dragenddorff reagents. Furthermore, the extract was partitioned with n-hexane. The n-hexane layer and ethanol-water layer were separated. Ethanol-water layer is added HCl 1M till pH 2 and partitioned with ethyl acetate to form two layers [12][13]. The alkaloid salt layer is added NH_4OH 25% till reach pH 9-10 and partitioned with ethyl acetate. The ethyl acetate layer was concentrated using a rotary evaporator to obtain alkaloid extract. Then, alkaloid extract is weighed and analyzed by TLC under the UV light 254 and 365 nm (CAMAG UV Cabinet 4) and Dragendorff reagent.

Alkaloid extract which has 2 g for its weight was subjected to column chromatography eluting with n-hexane, ethyl acetate, and methanol in increasing order of polarity. Eluate that produces the same pattern is combined into one fraction. Alkaloid fraction was isolated by preparative TLC using ethyl acetate and methanol as eluent. The alkaloid bands are scraped and dissolved in methanol. The alkaloid isolates were identified using UV-Vis spectroscopy (Shimadzu UV-1280), FTIR (Perkin Elmer 10.4.00), and LC-MS / MS.

2.2.2 Xanthine oxidase inhibition assay

Sample (ethanol extract, alkaloid extract, and allopurinol) which has 10 mg for its weight were dissolved in a phosphate buffer pH 7.5 to obtain 100, 50, 20, 10 and 5 ppm solutions. Allopurinol is used as a positive control. The assay solution consisted of 1 mL of sample solution (5-100 ppm), 2.9 mL of 0.05 M potassium phosphate buffer (pH 7.5) and 0.1 mL of 0.1 unit/mL xanthine oxidase enzyme is pre-incubated 30 minutes at 25 ° C. Then the mixture solution was added 1 mL xanthine 0.1 5 mM and incubated for 30 min at 25 ° C. After incubation, the mixture solution was added 1 mL of HCl 1 N to stop the reaction and measured its absorbance at a wavelength of 290 nm using a UV-Vis spectrophotometer [14]. The inhibition activity of xanthine oxidase enzyme is calculated using the equation:

$$\%\,inhibition = \left(\frac{(A-B)-(C-D)}{(A-B)}\right) \times 100\%$$

Where A is an enzyme activity without the addition of a sample, B is the control of A without samples and enzymes, C and D are the sample enzyme activity with and without enzymes respectively. The value of IC_{50} is calculated from the percentage of inhibition.

3. Result and Discussion

3.1 Alkaloid Isolation

Ethanol extract was obtained dark green color as much as 539.22 g. Ethanol-water extract gave a positive reaction of alkaloids to Mayer and Dragendorff reagents which formed a white mist and red deposits [15].

The ethanol-water extract is partitioned to obtain an alkaloid extract. Alkaloid extract was obtained in brown solids form as much as 7.24 g.

The TLC analysis of alkaloid extract using ethyl acetate: n-hexane (4: 1) obtained 10 spots shown in ***Figure 1***. The Dragendorff spotting reagent showed that the spots with Rf 0.253; 0.172 and 0.103 are alkaloid compounds [15]. TLC analysis of eluate using ethyl acetate: n-hexane (4: 1) obtained 8 fractions. Based on the alkaloid test, the 8th fraction gives positive alkaloid spot. Furthermore, isolation of alkaloid compounds in fraction 8 was done using preparative TLC method. The first isolation using ethyl acetate: methanol (7: 2) and continued using methanol eluent. Alkaloid isolate was obtained as much as 5.1 mg.

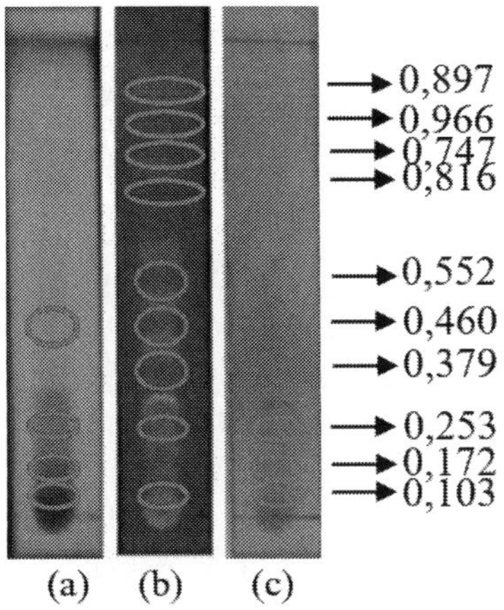

Figure 1. TLC profile of alkaloid extract in silica gel 60 GF_{254} and ethyl acetate: n-hexane (4: 1) under (a) UV lamp $\lambda = 254$ nm (b) UV lamp $\lambda = 366$ nm and (c) Dragendorff reagent

UV-Vis spectroscopy analysis showed that the isolate had a maximum absorbance at 206 nm as shown in ***Figure 2***. Absorbance at 206 nm indicates the possibility of $n \rightarrow \sigma^*$ transition. This transition occurs in saturated compounds which have heteroatoms such as nitrogen (C-N) [16]. Most of the alkaloids found in the Piperaceae family are piperidine [17]. Nigramide L (λ_{max} 209 nm), Nigramide N (λ_{max} 208 nm) and Nigramide N (λ_{max} 206 nm) are piperidine-type alkaloid compounds that have been isolated from Piperaceae family [18]. Based on the chemotaxonomy and maximum absorbance comparison of known piperidine alkaloid, it is suspected that the alkaloid isolate has a piperidine base structure.

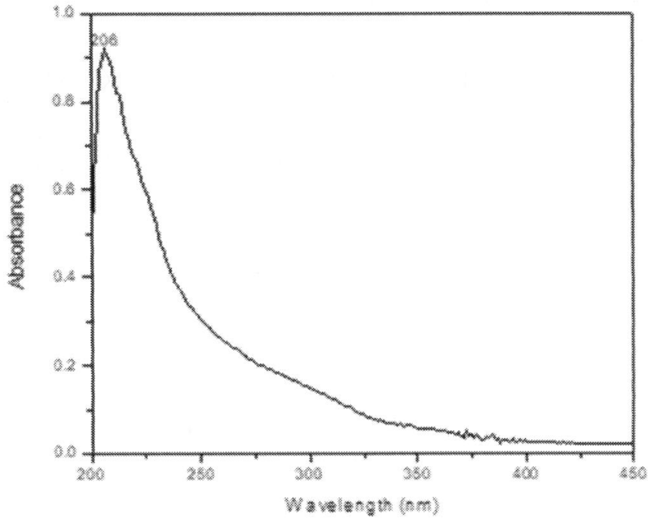

Figure 2. UV-Vis spectra of alkaloid isolate

The FT-IR spectra of alkaloid isolate shown in Fig. 3. Based on the results, the alkaloid isolate has the N-H, C-H, C = C, C = O, C-N, C-O-C group and aromatic rings, it showed in Table 1. The presence of C-N group on the FT-IR spectra strengthen the possibility that the alkaloid isolate has a piperidine base structure.

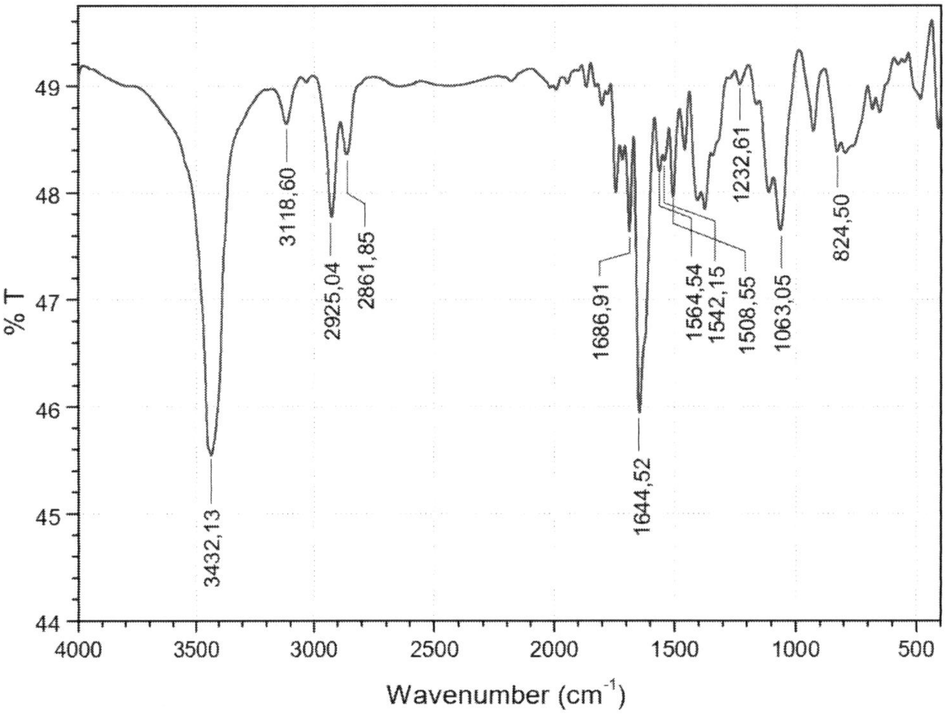

Figure 3. FT-IR spectra of alkaloid isolate

Table 1. FT-IR data of alkaloid isolate

Bilangan Gelombang (cm^{-1})	Jenis Vibrasi
3432.13	N-H stretching vibrations
3118.60	C-H aromatic stretching vibrations
2925.04	C-H aliphatic asymmetry stretching vibrations
2861.85	C-H aliphatic symmetry stretching vibrations
1686.91	C=C alkene stretching vibrations
1644.52	C=O amide stretching vibrations
1564.54	C=C aromatic stretching vibrations
1542.15	C=C aromatic stretching vibrations
1508.55	N-H bending vibrations
1232.61	C-N stretching vibrations
1063.05	C-O ether stretching vibrations
829.50	C-H aromatic out of plane bending vibrations

3.2 Xanthine oxidase inhibition assay

Ethanol and alkaloid extract were assayed for their xanthine oxidase inhibition activity using allopurinol as a positive control. Based on the inhibition, it is known that ethanol and alkaloid extract have the potential to inhibit the activity of xanthine oxidase enzyme. The percentage of inhibitory activity of xanthine oxidase enzyme from ethanol extract, alkaloid extract, and allopurinol is shown in the graph in ***Figure 4***.

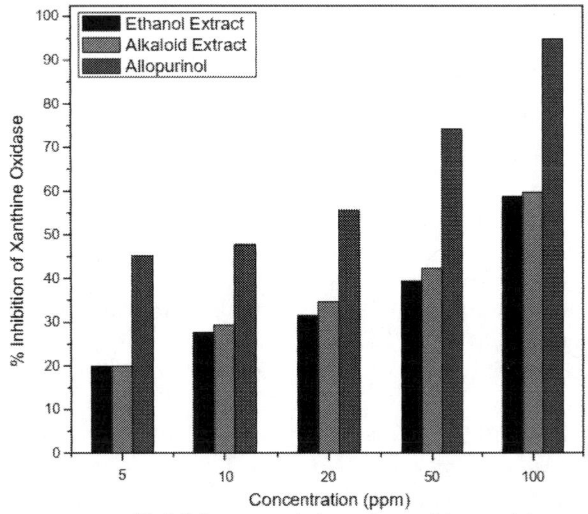

Figure 4. Graphs Percentage of inhibitory activity of xanthine oxidase enzyme from ethanol extract, alkaloid extract *Peperomia pellucida* and allopurinol

Based on the graphs, allopurinol has the greatest inhibition activity. At concentrations of 100 ppm allopurinol can inhibit up to 94.8598%. While ethanol extract and alkaloid extract *Peperomia pellucida* able to inhibit 58.6995% and 59.6782%.

The inhibitory activity of allopurinol, ethanol extract, and alkaloid extract obtained IC_{50} value are 11.4954, 76.3318 and 71.6658 ppm respectively. The IC_{50} of alkaloid extract was lower than ethanol extract which showed that the alkaloid extract was more active than ethanol extract. This is suitable for the statement that alkaloid compounds are active inhibit the xanthine oxidase enzyme [6, 10, 11].

4. Conclusion

The obtained alkaloid isolate suspected to has a piperidine base framework. The alkaloid isolate has N-H, C-H, C = C, C = O, C-N, C-O-C groups and there are aromatic rings. Based on the inhibition activity assay of enzyme xanthine oxidase, alkaloids extract (IC_{50} 71.6658 ppm) has higher activity than the ethanol extract (IC_{50} = 76.3318).

References

[1] Darmawan J, Valkenburg H, Muirden K, Wigley R. The Epidemiology of Gout and Hyperuricemia in a Rural Population of Java. *Journal of Rheumatology*. 1992;19(10):1595-9.
[2] Kuo CF, Grainge MJ, Zhang W, Doherty M. Global epidemiology of gout: prevalence, incidence and risk factors. *Nature Reviews Rheumatology*. 2015;11(11):649-62.
[3] Chen C, Lu J-M, Yao Q. Hyperuricemia-Related Diseases and Xanthine Oxidoreductase (XOR) Inhibitors: An Overview. Medical Science Monitor: *International Medical Journal of Experimental and Clinical Research*. 2016;22:2501.
[4] Maiuolo J, Oppedisano F, Gratteri S, Muscoli C, Mollace V. Regulation of uric acid metabolism and excretion. *International journal of cardiology*. 2016;213:8-14.

[5] Schlesinger N. Management of acute and chronic gouty arthritis. Drugs. 2004;64(21):2399-416.
[6] Ling X, Bochu W. A review of phytotherapy of gout: the perspective of new pharmacological treatments. Die Pharmazie-An *International Journal of Pharmaceutical Sciences*. 2014;69(4):243-56.
[7] Majumder P, Abraham P, V S. Ethno-medicinal, Phytochemical and Pharmacological review of an amazing medicinal herb *Peperomia pellucida* (L.) HBK. *Research Journal of Pharmaceutical, Biological, and Chemical Sciences*. 2011;2(4):358-64.
[8] Wijayakusuma HMH. *Atasi Rematik dan Asam Urat Ala Hembing*. Jakarta: Puspa Swara; 2006. 78 p.
[9] Majumder P. Evaluation of Taxo-Chemical Standardization and Quality Control Parameters of *Peperomia pellucida* (Family : Piperaceae): A Multi Valuable Medicinal Herb. *Journal of Pharmaceutical and Scientific Innovation*. 2012;1(6):7-12.
[10] Ahmad I, Ijaz F, Fatima I, Ahmad N, Chen S, Afza N, et al. Xanthine oxidase/tyrosinase inhibiting, antioxidant, and antifungal oxindole alkaloids from *Isatis crosta*. *Pharmaceutical biology*. 2010;48(6):716-21.
[11] Kong LD, Yang C, Ge F, Wang HD, Guo YS. A Chinese herbal medicine Ermiao wan reduces serum uric acid level and inhibits liver xanthine dehydrogenase and xanthine oxidase in mice. *Journal of Ethnopharmacology*. 2004;93(2):325-30.
[12] Harborne JB. *Phytochemical methods*. 2nd ed. London Chapman and Hall publications; 1984. 288 p.
[13] Lenny S, Barus T, Sitompul EY. Isolasi Senyawa Alkaloid dari Daun Sidaguri (*Sida rhombifolia* L.). *Jurnal Kimia Mulawarman*. 2010;8(1):40-3.
[14] Umamaheswari M, Asokkumar K, Sivashanmugam A, Remyaraju A, Subhadradevi V, Ravi T. In vitro xanthine oxidase inhibitory activity of the fractions of *Erythrina stricta* Roxb. Journal of Ethnopharmacology. 2009;124(3):646-8.
[15] Farnsworth NR. Biological and Phytochemical Screening of Plants. *Journal of Pharmaceutical Sciences*. 1966;55(3).
[16] Sastrohamidjojo H. *Dasar-Dasar Spektroskopi*. Yogyakarta: Liberty; 2001.
[17] Gutierrez YV, Yamaguchi LF, de Moraes MM, Jeffrey CS, Kato MJ. Natural products from Peperomia: occurrence, biogenesis, and bioactivity. *Phytochemistry Reviews*. 2016:1-25.
[18] Wei K, Li W, Koike K, Chen Y, Nikaido T. Nigramides A–S, Dimeric Amide Alkaloids from the Roots of *Piper nigrum*. *The Journal of Organic Chemistry*. 2005;70(4):1164-76.

IOP Conf. Series: Materials Science and Engineering **349** (2018) 012018 doi:10.1088/1757-899X/349/1/012018

Development of performance assessment instrument based contextual learning for measuring students laboratory skills

E Susilaningsih *, K Khotimah, S Nurhayati

Chemistry Department, Universitas Negeri Semarang

E-mail: endang.arkan@gmail.com

Abstract. The assessment of laboratory skill in general hasn't specific guideline in assessment, while the individual assessment of students during a performance and skill in performing laboratory is still not been observed and measured properly. Alternative assessment that can be used to measure student laboratory skill is use performance assessment. The purpose of this study was to determine whether the performance assessment instrument that the result of research can be used to assess basic skills student laboratory. This research was conducted by the Research and Development. The result of the data analysis performance assessment instruments developed feasible to implement and validation result 62.5 with very good categories for observation sheets laboratory skills and all of the components with the very good category. The procedure is the preliminary stages of research and development stages. Preliminary stages are divided in two, namely the field studies and literature studies. The development stages are divided into several parts, namely 1) development of the type instrument, 2) validation by an expert, 3) a limited scale trial, 4) large-scale trials and 5) implementation of the product. The instrument included in the category of effective because 26 from 29 students have very high laboratory skill and high laboratory skill. The research of performance assessment instrument is standard and can be used to assess basic skill student laboratory.

1. Introduction

Assessment is a process document through a measurements, knowledge, skills, and attitude students. Assessment, in other words, is the activities systematic to obtain information about what is known, done, and worked by students. Assessment is seen as the particular characteristic and determination of value or the value of an object [1]. Performance assessment can be regarded as a test action, in this assessment, students are expected to practice and to implement of activities and assessed students based on a guided assessment. Performance assessment lab work pharmacy is assessment who asked students did some performance in lab work pharmacy. Performance assessment is reviewed some aspects of assessment, the result and lab work. The assessment had a role as program assessment process, the advancement of learning and result student learning [2].

Assessment is very important to use in the learning process, as written by Adiguzel [3] "The increased use of approaches used for assessment along the importance of knowing and boosting the academic progress of student". Definition of performance assessment has been defined by several figures. One of them by Atac [4] "performance assessments call upon the examinee to demonstrate specific skills and competencies, that is, to apply the skills and knowledge they have mastered".

The tendency of current learning back to the idea that children would be better if the environment is created naturally. Learn to be more meaningful if the students are experiencing what they are learned. Learning that oriented mastery of material proved successful in short-term given the competition but failed to provide children solve problems in the long term life. Contextual learning is the concept of learning which helps the teacher link the matter between the real world situation students. Contextual learning encourages students to make the connection between their knowledge and application in their life as members of the family and society. The concept of contextual learning expected more meaningful for students.

The problem that has been happened students are not able to do lab work independently and didn't understand lab work matter. One of the causes of the problem that students have not mastered the skills of performance for pharmacy lab work, so the teacher only regard based on the result and cognitive value only. Performance skills are very important in lab work pharmaceutical to grow confidence in learning pharmaceutical critically and creatively. The success of laboratory skills depends on the quality of training and assessment program. Instrument performance assessment is needed to resolve the problem.

Performance assessment provides some benefits on the assessment process objective, measurable, and comprehensive on the ability of the end student learning [5]. Other research from Izza [6] showed that performance assessment had been analyzed can be used to assess basic skills laboratory. Research conducted by Ardli [2] regarding the implementation of performance assessment, obtained the conclusion that the development of performance assessment improve students interest in practical activities, motivate students in learning and effectively help teachers measure skills and attitude students. Research conducted by Puspitasari [7] concluded that lab work with performance assessment rubric could achieve mastery learning, students character can be built for practical activities include discipline, honesty, independence, responsible and cooperate. Another research conducted by Oktriawan [8] about developing of performance assessment instruments. Results of the development have characteristics consist of instrument performance assessment, assessment rubric is simple, and has level legibility and excellent construction.

The result of observations in several pharmacy schools in Rembang obtained that during the lab work held, the teacher already has practical assessment guidelines that apply nationally, but the assessment has not been developed and only apply for some materials as science lab recipes. One material lab work has not assessment guide is Pharmacognosy. Assessment of the student skills while still carried with an estimated, it makes difficulty in assessing because there is no assessment guide in this aspect.

Respons to the problems and research has been conducted by experts, it is necessary to research and development that related assessment instrument with the title "Development of performance assessment instrument based contextual learning for measuring students laboratory skills" has purpose to facilitate teachers to assess students ability so that advances the ability of students can be seen clearly and accurately. Problem formulation in this research is: Is performance assessment feasible and effective in measuring students laboratory skills. Purpose of the research: obtain an instrument performance assessment appropriate and effective to measure students laboratory skills.

2. Methods

This research was conducted at SMK Kesehatan Bina Mandiri dan SMK Avicenna Lasem. This research uses Research and Development Methode (R & D). Development research not using general population, but only using limited subject. The subject is students of pharmacy grade X academic year 2015/2016. Schools are intended to test small class, large scale to the implementation of developed performance assessment instruments. The subject research at small scale test amounted ten students.The subject research at large scale test amounted 26 students and the subject research at implementation amounted 29 students.

This research uses Research and Development (R&D) method. R & D research method has three stages of development that includes defining, design, and develop (**Figure 1**). Define phase is to identify potential problems that cover analytical instruments used in schools and pharmaceutical process analysis lab performed. After identifying the problem and formulate issues and gather information. The design phase is compiling the design of instrument performance assessment to measure students' laboratory skills current pharmacy lab work followed with expert validation and performance assessment instrument design revisions. Develop phase is testing instruments, namely the first stage of small-scale testing and revision of assessment instruments. Then followed the second phase of testing instruments with large-scale test and revision of assessment instruments. After the third stage carried out an assessment instrument is ready to implement. Research and Development (R&D) method used to produce a product, and testing the effectiveness the products [9]. This research needs analysis and tes the effectiveness of product development required to produce a particular product.

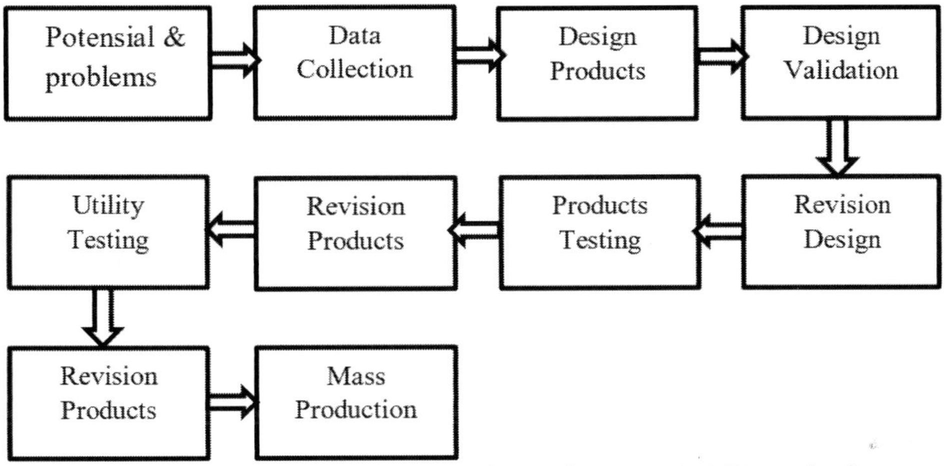

Figure 1. Step of Research and Development (R&D) method

Instrument reliability assessment process on observation sheet student lab skills assessments was analyzed using inter-rater reliability. Inter-rater reliability formula to analyze the reliability of laboratory skills assessment observation sheet on this below:

$$r_{11} = \frac{V_p - Ve}{V_p + (k+1)V_e}$$

Reliability item questionnaire can be measured by using Alpha Cronbach. Instrument reliability calculation of questionnaire sheet conducted on the small-scale test, large-scale test, and implementation. Instrument said reliable if score $\propto \geq 0.70$. Alpha Cronbach formula uses to analyze the reliability of the questionnaire sheet on this below:

$$\propto = \frac{k}{k-1}\left(1 - \frac{\sum \sigma^2{}_b}{\sigma_t{}^2}\right)$$

The effectiveness of assessment instrument determines achievement of developing performance assessment instrument. Determining the effectiveness instruments carried out by calculating the proportion of students who have laboratory skill of high and very high from a total number of all students, can also percentage achievement of all students for each item, then inserted using the formula below:

IOP Conf. Series: Materials Science and Engineering **349** (2018) 012018 doi:10.1088/1757-899X/349/1/012018

$$K = \frac{\sum ni}{N} \times 100\%$$

Methods of data collection were conducted by interview and observation method. The interview used to take document or data that supports research. Observation methods used to see raters understanding. The validity of instrument calculated from expert judgement. Assessment instrument is said to be effective if 75% of the students are in the criteria high laboratory skills and very high laboratory skills.

3. Results and discussion

Assessment instrument that used in this research is performance assessments instrument for measuring laboratory skills students. Instrument considered valid if the result means score by expert judgements achieve good and very good categories. During process validation instrument repairs for obtaining performance assessment instrument better. Detail of the analysis expert for validation performance assessment instrument can be seen in Table 1.

Table 1. Detailed Results Analysis Instrument Validation for performance Assessment Instrument

No	Validator	Score	Category
1	Syllabus	28	Very Good
2	Interview Sheet	29,5	Very Good
3	Questionnaire Responses and Users Sheet	29	Very Good
4	Observation Students Lab Skills Assessment Sheet	62.5	Very Good

Table 1 shows the average score obtained by each instrument achieve very good category. Performance assessment instrument and the other instrument that supports valid and feasible to use [6]. The result of the validation instruments appropriates with the opinions [10] the validity with good categories if coefficients categories ranged between good and very good.

Validation results by four validators concluded that instrument performance assessment dan and supporting instruments be valid. Instrument performance assessment and developed supporting instruments eligible for use in the research process. Improvement performance assessment instrument conducted during the validation process by an expert to get a good instrument, valid and feasible for use. Summary of revision developing performance assessment instrument can be seen in Table 2.

Table 2. Summary of revision developing performance assessment instrument

No	Instrument	Improvement Notes
1	Interview Sheet	Using standard sentence for each item
2	Observation sheet, rubric and procedure laboratory skills instrument	Observation sheet made more detail discussion section Rubric aspect on 15. 15. 17 replaced with indicator of laboratory skills Scoring column in merge cell Improvement the rubric and made in context
3	Questionnaire responses of teacher, observer, and student sheet	Using standard sentence for each item

Small-scale test conducted to determine deficiencies and legibility or the instruments. The subject of the small-scale test is ten students with nine female students and one male student class X F 1 SMK Kesehatan Bina Mandiri Rembang. The process of small-scale test carried out on active learning hours. The calculation result of the data analysis of laboratory skills assessment obtained reliability 0.833. Observation sheet laboratory skills of the students said to be reliable because of more than 0.70.

Large-scale test with some subject 26 students. The subject from class X F 2 SMK Avicenna Lasem. The process of large-scale test performed on active learning hours. The calculation result of the data analysis of laboratory skills assessment obtained reliability 0.887. Observation sheet laboratory skills of the students said to be reliable because of more than 0.70.

Implementation phase with some subject 29 students. The subject from class X F 1 SMK Avicenna Lasem. The process of large-scale test performed on active learning hours. The calculation result of the data analysis of laboratory skills assessment obtained reliability 0.903. Observation sheet laboratory skills of the students said to be reliable because of more than 0.70.

Improvement performance assessment instrument was conducted during the validation process by an expert to get a good instrument, valid and feasible for use. Improvement after small-scale test, the base for instruments to be tested again on a large scale. Data obtained and analyzed for reliability. Detailed result reliability observation sheet laboratory skill students can be seen in Table 3.

Table 3. Detailed result reliability observation sheet laboratory skill students

No	Stage of Development	Reliability	Description
1	Small Scale Test	0.833	Reliable
2	Large Scale Test	0.877	Reliable
3	Implementation	0.903	Reliable

Table 3 show that observation sheet laboratory skills of students stated reliable because of the value of the reliability more than 0.70. Instruments that valid and reliable can be used to assist and motivate students more excited to improve their skill in practical competence [1]. Assessment instrument is valid and reliable highly effective for measuring result student learning [11].

Questionnaire responses teacher, observer and students are also required in development this research to determine the criteria assessment instruments developed. Analysis results show that performance assessment instrument developed has very good criteria. The result of the analysis questionnaire of teacher, observer, and student about developing of performance assessment instrument can be seen in Table 4.

Table 4. Data Result Questionnaire Response Develop of Performance Assessment instrument

No	Instrument	Reliability		Description
		Students Questionnaire	Teacher Questionnaire	
1	Small Scale Test	0.758	0.803	Reliable
2	Large Scale Test	0.807	0.828	Reliable
3	Implementations	0.815	0.835	Reliable

The effectiveness of an assessment sheet determining achievement of performance assessment instrument development. How to determine the effectiveness of assessment instruments to the assessment done by calculating the proportion of the number of students who have the skills of the high and very high laboratory of the total number of students there, and can by the percentage achievement of all students on each item. Assessment instrument is said to be effective if 75% of the students are on the criteria of laboratory skills of high and very high. The effectiveness of developing performance assessment instrument can be seen in Figure 1.

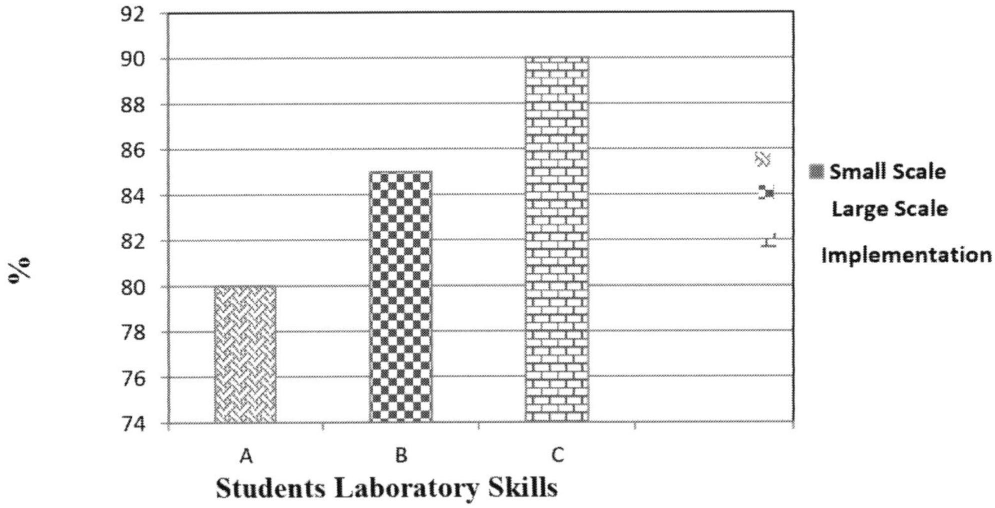

Figure 2 Percentage of Students Laboratory Skills Assessment in Development Phase

Figure 2 shows that in the small scale test around 80% of students or 8 from 10 students have high minimum laboratory skills. The large-scale test showed that 84.61% or 22 of 26 students have high minimum laboratory skills. Implementation phase showed that 89.65% or 26 of 29 students have high minimum laboratory skills.

Practicality instruments are obtained through the spreadsheet questionnaire on teachers, and the student observer at the trial stage until the implementation. The results of the analysis of students' responses to the instrument performance assessment showed that students responded positively to the instruments developed. The response of students to the instruments during the test to meet the criteria of the practical implementation phase. Students are interested and agreed to carry out practical Pharmacognosy using performance assessment instruments.

The result of the teacher and observer feedback on performance assessment developed indicates that the user has to provide an assessment with practical until very practical category with score range between 42-53. The questionnaire results showed that instruments developed in general have practical qualification level. The result consistent with research Haksani [10] devices developed get a positive response by the user with questionnaire responses, so the device developed stated practically.

The assessment activities in the lab can help the teacher to determine laboratory skills of the students. Authentic assessment intended to measure laboratory skills of the students with actual. Authentic assessment can be held in various ways either by using a performance assessment. Implementation using performance assessment can be held during a lab work in pharmaceutical laboratories to measure students laboratory skills. Performance assessment instrument directed during lab work proses that are currently observing, analyzing, interpreting data and accumulated during the lab work process. The material had been obtained students during learning pharmacognosy materials can be used in social life.

The result of developing has increased teacher task when the lab work held. The teacher should explain the matter and observe all of the students to assess their skills laboratory. This obstacle can be overcome by companion assessment techniques, self-assessment by the student's peer assessment can also be used to overcome this problem. Performance assessment instrument is also still to be developed by the creativity and needs, so the teacher can easily record the student's ability to achieve educational goals.

4. Conclusion

Based on this research performance assessment instrument feasible to implement and validation result is 62.5 with very good categories for observation sheet, and all of the support instrument has very good categories. The instrument is included in the category of effective because 26 from 29 students have very high laboratory skill and high laboratory skill. The research of performance assessment instrument is standard, and it can be used to assess basic skill student laboratory.

References

[1] Ardli, I., Gafar, A., & Mudjalipah, S. 2012. Perangkat Penilaian Kinerja Untuk Pembelajaran Teknik Pemeliharaan Ikan. *Jurnal INVOTEC*, 8(2): 147-166.

[2] Adiguzel, T. 2011. Use of Audio Modification in Science Vocabulary Assessment. *Eurasia Journal of Mathematics, Science, Technology Education*, 7(4): 215-225).

[3] Doctor, J. dan Heller, K. 2009. Robust Assessment Instrument for Student Problem Solving, *Prosiding the NARST 2009 Annual Meeting*, Minnesota University.

[4] Amalia, N.F. 2014. Pengembangan Instrumen Penilaian Keterampilan Berpikir Kritis Siswa SMA pada Materi Asam Basa. *Jurnal Inovasi Pendidikan Kimia*, 8(2): 1380-1389.

[5] Atac, B. 2012. Foreign Language Teachers Attitude toward Authentic Assessment in Language Teaching. *The Journal of Language and Linguistic Studies*, 8(2): 7-19.

[6] Haksani. 2013. Pengembangan Perangkat Assessment Berbasis Keterampilan Generik Sains pada Mata Kuliah Praktikum Kimia Dasar Lanjut. *Jurnal Chemica,* 14 (1): 27-37.

[7] Oktriawan, T. 2015. Pengembangan Instrumen Asesmen Kinerja Pada Praktikum Pengaruh Luas Permukaan Terhadap Laju Reaksi. *Jurnal Pendidikan dan Pembelajaran Kimia,* 4(2): 593-604.

[8] Susila, I K. 2012. Pengembangan Instrumen Penilaian Unjuk kerja (Performance Assessment) Laboratorium pada Mata Pelajaran Fisika sesuai Kurikulum Tingkat Satuan Pendidikan SMA Kelas X di kabupaten Gianyar. *Tesis.* Denpasar: Pascasarjana Universitas Pendidikan Ganesha.

[9] Yustika, A., Eko, B., & Murbangun, N. 2014. Uji Kriteria Instrumen Peniaian Hasil Belajar Kimia. *Jurnal Inovasi Pendidikan Kimia*, 8(2): 1380-1389.

IOP Conf. Series: Materials Science and Engineering **349** (2018) 012018 doi:10.1088/1757-899X/349/1/012018

[10] Puspitasari, N., Widiarti, N., & Haryani, S. 2014. Pengembangan Rubrik Performance Assessment Pada Praktikum Hidrolisis Garam. *Skripsi*. Semarang: Universitas Negeri Semarang.
[11] Izza, L.N. 2014. Analisis Instrumen Performance Assessment denga Metode Generalizability Coefficient Pada Ketermpilan Dasar Laboratorium. *Jurnal Chemistry in Education*, 3(1): 30-36.
[12] Sudrajat, A. 2011. Pengembangan Rubrik Asesmen Kinerja untuk Mengukur Kompetensi Mahasiswa Melakukan Praktikum Kimia Analisis Volumetri. *Jurnal Chemica*, 2(1): 1-8.
[13] Sugiyono, 2013. *Metode Penelitian Kuantitatif, Kualitatif dan R&D*. Bandung: Alfabeta.

IOP Conf. Series: Materials Science and Engineering **349** (2018) 012019 doi:10.1088/1757-899X/349/1/012019

Sunlight-assisted synthesis of colloidal silver nanoparticles using chitosan as reducing agent

E Susilowati[1*], Maryani[2], Ashadi[1]

[1]Chemistry Education Department, Faculty of Teacher Training and Education, Universitas Sebelas Maret Surakarta, Indonesia
[2]Medical Doctor Program, Faculty of Medicine, Universitas Sebelas Maret Surakarta, Indonesia
Jl. Ir Sutami 36 A Surakarta Indonesia 53126

[*]Email: endwati@yahoo.co.id

Abstract. The present study we explore an environmentally friendly colloidal silver nanoparticles preparation using chitosan as reducing agent and stabilizer. It develops a new strategy on preparation of silver nanoparticles through the gel phase using sodium hydroxide (NaOH) as accelerator reagent. Sunlight irradiation was employed to assisted reducing process of silver ions to silver nanoparticles. Localized surface plasmon resonance (LSPR) phenomenon of silver nanoparticles was investigated using UV-Vis spectrophotometer. The shape and size of silver particles were analyzed using TEM. The formation of silver nanoparticles was confirmed by the appearance of LSPR absorption peak at 396.0–412.0 nm. The absorption peak of LSPR were affected by NaOH amount, time of sulight irradiation and concentration of $AgNO_3$. The produced silver nanoparticles were spherical with dominant size range of 5 to 8 nm as shown by TEM images. All colloidals were stable without any aggregation for 30 days after preparation.

1. Introduction

Metallic nanoparticles have possible applications in diverse areas such as electronics, cosmetics, coatings, packaging, and biotechnology [1]. Among nanostructured noble metals the most important commercialized nanoparticles are silver nanoparticles. Silver nanoparticles have attracted considerable attention for their application in antimicrobial materials [2], medical therapy [3], biomolecular detection [4] and catalysts [5]. Silver nanoparticles can be synthesized through 2 approaches with bottom up and top down approaches. The bottom up approach is more developed because it is easier to do with simple tools. Chemical reduction is one of the methods in bottom-up approach that have carried out by many previous researchers

The silver synthesis of nanoparticles is effective when reduction using strong reducing agents eg $NaBH_4$ but not environmentally friendly [6]. If using a weak reducing agent requires a high temperature and a long time [7, 8]. Therefore it is necessary to do alternative methods that are environmentally friendly and economical. Several eco-friendly methods that can be developed include the use of photon-assisted reduction methods of gamma ray irradiation [9], UV rays [10] and sunlight [11]. Sunlight-assisted reduction is very interesting to study because it is more efficient and does not require complicated equipment. Samples containing precursor salts are irradiated by direct sunlight in the presence of a stabilizer. Some polymers can be used as stabilizers in the synthesis of silver nanoparticles such as chitosan [12], Poly vinyl alchohol [13] and polyethylene glycol [14]. Chitosan is one biopolymer that is potential to be used as stabilizing agent because it has antibacterial activity and also environmentally friendly [12, 15].

In this study, silver nanoparticles were synthesized using chitosan as stabilizers and sodium hydroxide as accelerators and sunlight as a trigger for chemical reduction processes. The use of NaOH accelerators has been done by previous studies on weak reducing agents [15, 16]. Actually chitosan itself can act as a stabilizer as well as a weak reductor, but require high temperatures to be able to act as a reducer [7]. The use of sodium hudroxide accelerates the formation of silver nanoparticles at low temperatures. The combination of chitosan, sodium hydroxide and sunlight irradiation on the formation of silver nanoparticles is expected to be more efficient and effective. The formation of silver nanoparticles in the chitosan solution system is here in after referred to as colloidal silver nanoparticles. In this study also examined the effect of accelerator count, reaction time and the amount of salt per cursor. In addition, stability of the colloidal silver nanoparticles produced during 4 weeks storage was observed.

2. Experimental

Chitosan with molecular degree of deacetylation (DD) of 73.44 % and weight (MW) of 1,077,919.28 Da was purchased from Biotech Surindo Cirebon Indonesia. Silver nitrate ($AgNO_3$), acetic acid (CH_3COOH), sodium hydroxide (NaOH) were purchased from Merck.

The preparation of colloidal silver nanoparticles was carried out by chemical reduction methods with $AgNO_3$ as salt precursors, and chitosan as stabilizers. A solution of chitosan (1 % w/v) in acetic acid solution (1 % v/v) was firstly prepared. Due to the poor solubility of chitosan, the mixture was stirred to achieve complete dissolution, and then kept overnight at room temperature. Aqueous solution of $AgNO_3$ (0.012 g/mL) of 0.5 mL was added to 12.5 mL of chitosan (1 % w/v) in a flask. These solution was stirred at room temperature for 10 min. Furthermore, various mole of aqueous solution of NaOH were added to prepare various samples i.e., 0.0 (S0); 2.0 (S1); 2.5 (S2); 3.0 (S3); 3.5 (S4) and 4.0 mL (S5). Solution is exposed to sunlight for 2 hours. In this process will form a brown gel and then dissolved in 1% chitosan of 47.5 mL and stirred until all the gel is soluble and homogeneous solution is formed. The resulting colloidal silver nanoparticles solution is stored in glass bottles at room temperature. The same method was performed using 0.5 mL $AgNO_3$ and 1.75 mL NaOH without sun exposure. To determine the effect of solar radiation time, the variation of exposure time was 30 (T1), 60 (T2), 120 (T3) and 180 (T4) minutes using 0.5 mL $AgNO_3$ and 1.75 mL NaOH. Then to know the influence of concentration of salt precursor, similar procedur was carried out at vaious of concentration of $AgNO_3$ namely 0.25 (A1); 0.50 (A2); 0.75 (A3) and 1.00 mL (A4). Furthermore, observation of nanoparticle silver stability based on LSPR on samples with variation of $AgNO_3$ amount.

Characterization of colloidal silver nanoparticles was performed with UV3150 Shimtzu-UV3150 Ultraviolet-visible (UV-Vis) spectroscopy at 300 - 600 nm wavelengths with 2 times dilution with aquadest. Study on the size, shape and distribution of Ag nanoparticles was performed by using TEM (Transmission electron microscopy) with JEM-1400 EX microscope at 120.0 kV voltage.

3. Result and Discussion

According to Pestov *et.al.* chitosan may act as a reducing agent in the manufacture of gold nanoparticles under alkaline conditions through hydrolysis and degradation mechanisms to produce glucosamine molecules [17]. Furthermore, this glucosamine will reduce the metal ions as in glucose reducers under alkaline conditions [12, 15, 16]. The presence of OH^-ions from NaOH, chitosan will be able to act as a reducing agent that converts Ag + ions to Ag metal. The crystal growth of Ag will be confined by chitosan polymer so that Ag can be obtained by nano size (<100 nm). The formation of this nanoparticle Ag is characterized by the change of gel color to brown. The addition of 1% chitosan solution to gel containing Ag nanoparticles yields a solution hereinafter called colloidal colloidal silver nanoparticles. The silver position of these nanoparticles will be covered by hydroxyl and amino groups of chitosan.

The effect of the number of moles of NaOH used as an accelerator on the preparation of colloidal silver nanoparticles against absorbance is shown in Figure 1 with UV-Vis spectra showing localized surface plasmone resonance (LSPR) in Figure 2. In general, absorbance increases as the amount of NaOH is greater. This increased intensity of absorbance indicates that the amount of silver nanoparticles in colloidal increases [18]. Figure 5.2 (b) shows a 5 mmol NaOH addition of a considerable increase in absorbance which indicates a significant increase in the number of silver nanoparticles. The addition of NaOH amounts greater than 5 mmol did not significantly increase the amount of silver nanoparticles. Even absorbance decreased at the addition of 8 mmol NaOH. This is probably due to the formation of a more dense gel in the addition of more NaOH, thus inhibiting the process of reducing Ag^{+} ions to Ag nanoparticles.

Meanwhile, for the absorption peak there is a shift towards a longer wavelength (red-shifted) on the use of NaOH to 3.5 mmol. The red-shifted indicates that there is an increase in particle size that is likely due to the increasing number of OH^{-} ions that impose the increasing number of silver particles produced. However, further addition of NaOH of 4.0 mmol causes a shift towards a shorter wavelength (blue shift) which indicates a decrease in particle size. This may be due to an increase in the amount of NaOH until saturation leads to changes in chitosan structure so that it becomes more compressed and more effective as a stabilizer.

Figure 1. Photograph of colloidal silver nanoparticles sythesized ot different moles of NaOH

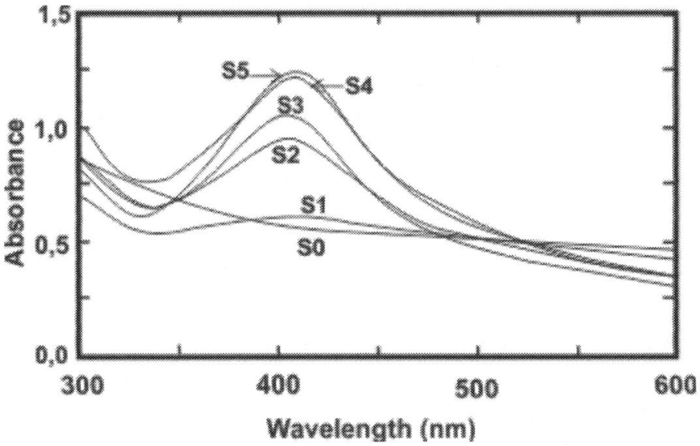

code	λ_{max} (nm)	Abs
S0	-	-
S1	404.00	0.597
S2	404.50	0.929
S3	405.50	1.055
S4	407.00	1.200
S5	406.00	1.220

Figure 2. The UV-Vis spectra of colloidal silver nanoparticles prepared at different mole of NaOH

The effect of sunlight irradiation to chemical reduction on the formation of silver nanoparticles was observed by comparing with the chemical reduction process without sunlight. The result of observation on LSPR can be seen in Figure 2. In Figure 2 shows that in the absence of sunlight irradiation, the chemical reduction process can take place, but goes slowly

From the color of the solution, it is seen that without the irradiation of colloidal sun the colloidal silver nanoparticles produced is yellow which indicates the amount of silver produced is less than if using sunlight irradiation. In the use of sunlight irradiation shows the formation of brown colloids that indicate more silver nanoparticles produced. This is supported by the observation of LSPR phenomenon, in which the colloidal produced without sun irradiation has a much lower absorbance value than colloidal with sunlight irradiation. This proves that the irradiation of sunlight trigger the reduction process so that more silver nanoparticles produced

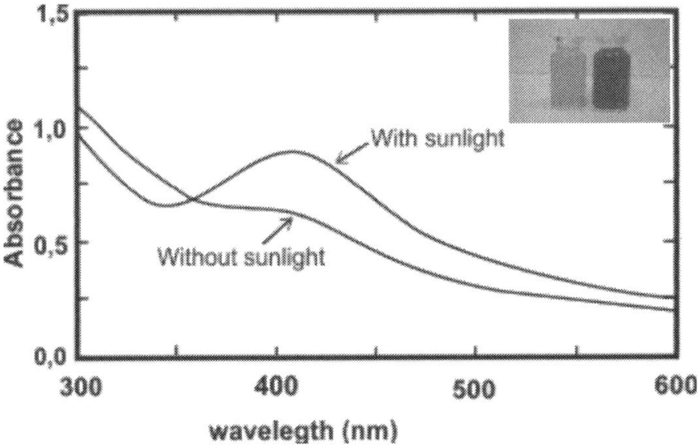

Figure 3. The UV-Vis spectra of colloidal silver nanoparticles prepared with sunlight irradiation and without sunlight irradiation

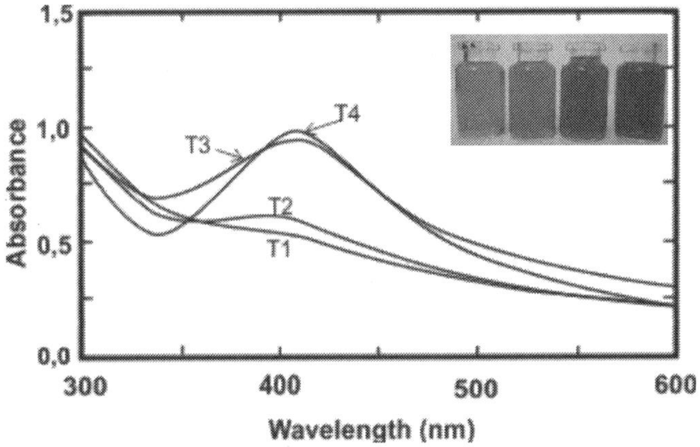

code	λ_{max} (nm)	Abs
T1	396.00	0.502
T2	396.50	0.605
T3	407.00	0.940
T4	408.00	0.978

Figure 4. The UV-Vis spectra of colloidal silver nanoparticles prepared at different time (h) of: 0.5 (T1); 1.0 (T2); 2.0 (T3) dan 3.0 (T4)

The sunlight-assisted reduction process is affected by irradiation time as shown in Figure 4. Figure 4 shows that in the duration of 15 minutes to 180 minutes indicate that the longer the irradiation of sunlight, the reduction process is more effective. The absorbance at 120 to 160 minutes it does not appear to be significantly different so for the reduction process by using sunlight irradiation can occur quite optimally within 120 minutes.

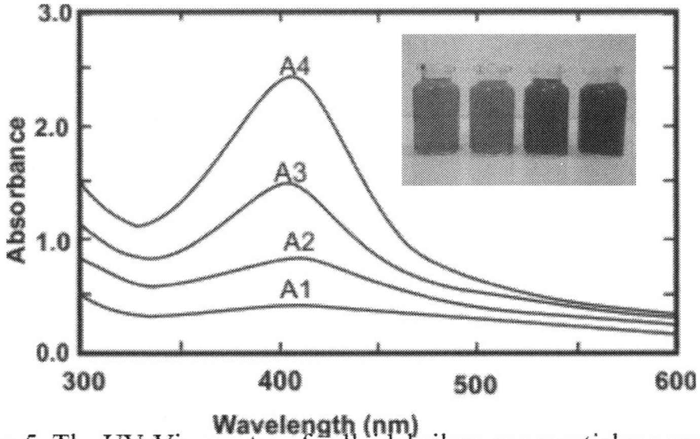

Code	λ_{max} (nm)	Abs
A1	412.50	0.419
A2	405.50	0.820
A3	403.50	1.465
A4	407.00	2.408

Figure 5. The UV-Vis spectra of colloidal silver nanoparticles prepared at different $AgNO_3$ concentration

This study also observed the effect of $AgNO_3$ concentration on silver LSPR nanoparticles. Variation of $AgNO_3$ concentration represented by variation of AgNO3 volume with concentration of 0.012 g / mL ie 0.25 mL (A1); 0.50 mL (A2); 0.75mL (A3); 1.00 mL (A4). Visually colloidal and UV-Vis columns of the Ag / Kit nanocomposite spectra are presented in Figure 5. Color gradations of light brown gradually dark brown show that the amount of silver nanoparticles produced increases with the concentration of $AgNO_3$ used. Figure 5 shows that absorbance increases with increasing concentrations of $AgNO_3$ which indicates that the more AgNO3 is added the more nanoparticle silver is formed [18]. Viewed from the position of LSPR, initially (A1-A3) occurs blue-shifted, then followed by red-shifted on A4.

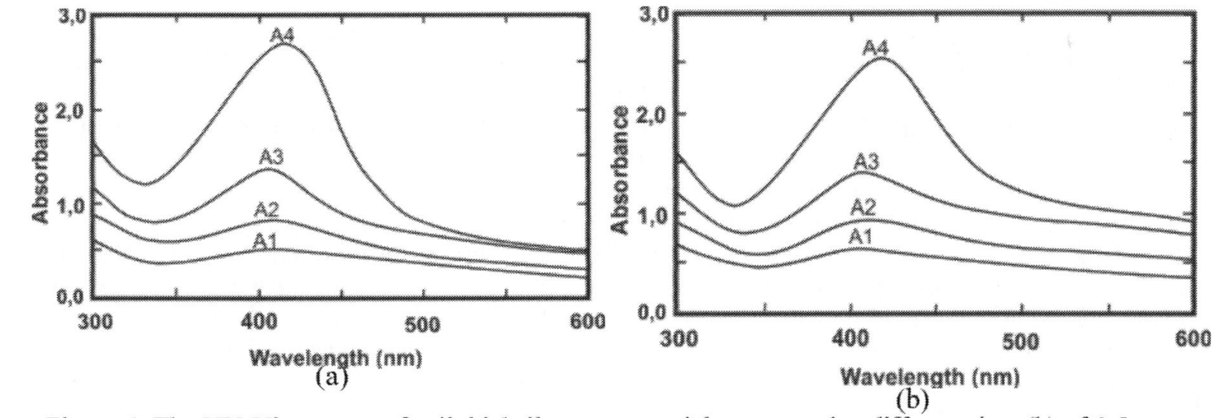

Figure 6. The UV-Vis spectra of colloidal silver nanoparticles prepared at different time (h) of 0,5 (T1); 1,0 (T2); 2,0 (T3) dan 3,0 (T4) at storage at (a) 2 weeks (b) 4 weeks

Stability of colloid solutions is the result of a potential barrier due to competition between van der Waals attraction forces and Coulomb repulsive forces. The metal soles are aggregated in two ways: particle-cluster aggregation and cluster aggregation [19]. Chitosan can act as a stabilizer that controls crystal growth and prevents the aggregation of already formed particles. In this study, stability observation of colloidal silver nanoparticles during storage on glass bottles in closed condition by monitoring LSPR based on UV-Vis spectra for 4 weeks. The UV-Vis spectra of the colloidal silver nanoparticles before and after storage for the duration of 2 weeks and 4 weeks can be seen in Figure 6, with absorbance values and peaks position (λ_{max}) shown in Table 1.

Table 1. Changes in absorbance and peak position of UV-Vis spectra of colloidal silver nanoparticles at different storage time

Code	λ_{max} (nm)			Absorbansi		
	initialy	2 weeks	4 weeks	initialy	2 weeks	4 weeks
A1	412.50	410.50	409.50	0.419	0.525	0.493
A2	405.50	407.00	408.00	0.820	0.820	0.776
A3	403.50	406.00	405.50	1.465	1.345	1.229
A4	407.00	414.50	416.00	2.408	2.696	2.370

From Table 1, the average absorbance intensity decreased and the average λmax value increased. This decrease in absorbance intensity indicates that during storage the number of particles decreases slightly with increasing time. This decrease is probably due to the aggregation of silver particles of nanoparticles during storage and will precipitate due to the instability of the stabilizer agent. This phenomenon is likely due to the chitosan that has experienced this degradation decreases its ability as a stabilizer agent. When compared between the spectra in Figures 5, 6 (a) and 6 (b), there is a baseline increase of the spectra when the colloidal sample is stored. This is probably due to the decrease in the amount of silver nanoparticles is followed by the formation of larger particles

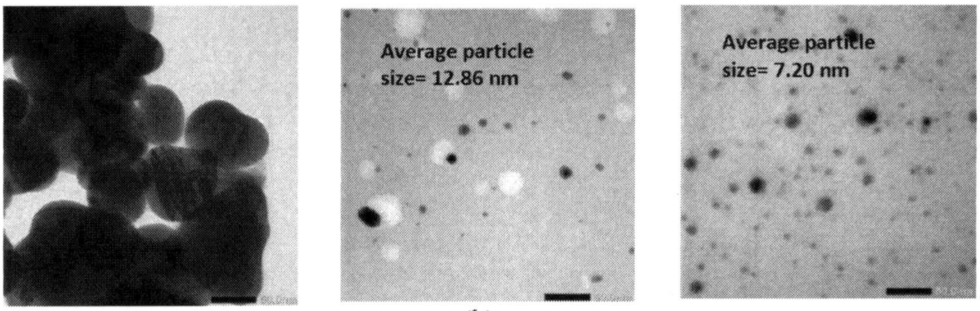

(a) (b) (c)

Figure 7. TEM image of silver nanoparticles and particles size dsirtibution (a) S0, (b) A2 dan (c) A4

Figure 7 shows the particle size distribution based on the TEM image for S0, A2 and A4 samples. In S0 sample it is seen that nano-size silver particles are not formed. This means, without the use of NaOH accelerators, silver nanoparticles are not formed. This corresponds to the color of the solution in Figure 1 which shows the gray color of the solution. This gray color is probably related to the formation of silver oxide. When correalated with the UV-Vis spectra in Fig. 5, then based on the absorbance data, there is conformity with the TEM image in Fig. 7. The appearance of a single absorption of LSPR in this 400 nm range indicates that the particles are spherical [20]. In the same area there was an increase in the number of detected particles of 20 particles to 100 particles with an increase in absorbance from 0.820 to 2.408.

4. Conclusion

The results show that the colloidal silver nanoparticles can be prepared using a reduction method assisted sunlight irradiation with NaOH as accelerator and chitosan as stabilizer. The formation of silver nanoparticles is characterized by the appearance of LSPR phenomenon in an area of approximately 400 nm. The particles is spherical in shape with dominant size at 5-8 nm. The amount of silver nanoparticles produced is influenced by the number of accelerators, reaction time and concentration of silver precursors. The colloidal silver nanoparticles are quite stable at 30 days of storage. Colloidal silver nanoparticles may have many potential applications as antibacterial agents in the field of medicine

Acknowledgements

The author would like to thank to DRPM Ristek Dikti who gave financial support through PUPT Grant 2017 of Universitas Sebelas Maret.

References

[1] Thakkar K N, Mhatre, S S, Parikh R Y, 2010. Nanomed.-Nanotechnol. **6** 257–262.
[2] Liong M, France B, Bradley K A, Zink J I 2009. Adv. Mater. **21** 1684–1689.
[3] Shrivastava S, Bera T, Singh S K, Singh G, Ramachandrarao P, Dash D 2009. ACS Nano 3 1357–1364.
[4] Wei H, Chen C, Han B, Wang E 2008. Anal. Chem. **80** 7051–7055
[5] Christopher P, Xin H, Linic S 2011. Nat. Chem. **3** 467–472.
[6] Shameli K, Ahmad M B, Zargar M, Yunus W M Z W, Ibrahim N A, Shabanzadeh P, Ghaffari M, Moghaddam 2011 *Int. J. Nanomed.* 271-284.
[7] Wei D, Sun W, Qian W, Ye Y and Ma X 2009 *Carbohydr. Res.* **344** 2375-2382.
[8] Ratyakshi and Chauhan R P 2009 *Asian J. Chem.* **21(10)** 113-116.
[9] Shameli K, Ahmad M B, Yunus W M Z W, Ibrahim N A, Gharazebi Y, Sedagha S 2010 *Int. J. Nanomed.* **5** 1067-1077.
[10] Ahmad M B, Shameli K, Darroudi M, Yunus W M Z W, Ibrahim N A 2009 *Am. J. Appl Sci.* **6(12)** 2030-2035
[11] Vimala K, Mohan Y M, Varaprasad K, Redd, N N, Ravindra S, Naidu N S, Raju K M, 2011 *J.Biomater. Nanobiotechnol.* **2** 55-64
[12] Susilowati E, Maryani and Ashadi 2015a *Indonesian J. Pharm.* **26(1)** 37 – 44
[13] Raneesh B, Pragatheeswaranp P, Dhanasekaran, Chandrasekaran P, Kalarikkal N. 2010 *J. Ovonic. Res.* **6(4)** 187-191
[14] Ahmad M B, Tay M Y, Shameli K, Hussein M Z and Lim J J 2011 *Int. J. Mol. Sci.* 2011, **12(8)** 4872–4884
[15] Susilowati E, Triyono, Santosa S J and Kartini I 2015b *Indones. J. Chem.* **15(1)** *29 - 35*
[16] Darroudi M, Ahmad M B, Abdullah A H, Ibrahim N A and Shameli K 2010 *Int. J. Mol. Sci.* **11** 3898-3905
[17] Pestova A, Nazirov A, Modinc E, Mironenko A dan Bratskaya B 2015 *Carbohydr. Polym.* **117** 70–77
[18] Šileikaitė A, Prosyčevas I, Puišo J, Juraitis A, Guobienė A 2006 *Mater. Sci. (Medžiagotyra)* **12(4)** 287-291
[19] Šileikaitė A, Puišo J, Prosyčevas I and Tamulevičius S 2009 *Mater. Sci.(Medžiagotyra)* **15(1)** 21-27
[20] Guzmán M G, Dille J and Godet S 2009 *Int. J. Chem. Biol. Eng.* **2(3)** 104-111

The Effect of Cellulose Acetate Concentration from Coconut Nira on Ultrafiltration Membrane Characters

E Vaulina[1], S Widyaningsih[1], D Kartika[1], M P Romdoni[1]

[1]Chemistry Department, University of Jenderal Soedirman, Purwokerto, Indonesia

E-mail: vaulinaeva@gmail.com

Abstract. Cellulose acetate is one of material in produce ultrafiltration membrane. Many efforts have been done to produce cellulose acetate from natural product to replace commercial one. In this research, ultrafiltration membrane has been produced from coconut flower water (nira). Ultrafiltration membrane is widely used in separation processes. This research aims to determine the characteristics of ultrafiltration membrane at a various concentration of cellulose acetate. The ultrafiltration membrane is conducted by phase inversion method at various concentration of cellulose acetate. The cellulose acetate concentration was 20%, 23% and 25% (w/w) with formamide as additives. The results showed that the greater the concentration of cellulose acetate, the smaller the flux value. The highest flux was a membrane with 20% cellulose acetate concentration with water flux value 55.34 L/(m^2. h). But the greater the concentration of cellulose acetate the greater the rejection. The highest rejection value was on a membrane with 25% cellulose acetate concentration of 82.82%. While from the tensile strength test and the pore size analysis, the greater the cellulose acetate concentration the greater the tensile strength and the smaller the pore size

1. Introduction

Membrane technology is widely used in the separation process. Membrane technology is chosen because the process is very simple, low energy consumption, does not damage the material, not use additional chemicals and not produce new waste hence it is classified as clean technology [1]. In the previous research [2], a membrane was made from nira, but the resulting membrane performance was low. The performance of the separation process depends on the characteristics of the membrane structure. Several factors which influencing membrane structure characteristics are the polymer type and polymer concentration, solvent type, membrane-making method, and the addition of additive [3].

In order to produce membrane with good characteristic, this research was used various polymer concentration. The polymer was cellulose acetate (CA). CA was obtained as a result of cellulose acetylation from a variety of sources, including bacterial cellulose [4]. Bacterial cellulose is a cellulose that is produced by microorganisms, especially the genus Acetobacter. Bacterial cellulose has superior purity properties compared to wood cellulose, including highly hydrophilic, high mechanically physical properties in wet and dry, uniquely smooth and distinct wicker form and can be produced from a variety of relatively inexpensive substrates [5]. A wide variety of substrates that have high glucose levels can be used for the manufacture of bacterial cellulose. One of them is nira. Nira is a water from coconut flowers that has a high enough glucose level. The content of glucose in nira was around 15.40% [6]. Nira is also a fertile medium for microorganisms to growth so that the process of glucose conversion into cellulose by *Acetobacter xylinum* bacteria is higher. This research used coconut nira as a source of bacterial cellulose. Acetylation of bacterial cellulose produce cellulose acetate which was used as the basic material of ultrafiltration membranes. Membrane characterization included the measurement of flux, rejection, tensile strength, and pore size. This study is expected providing information on the effect of cellulose acetate concentration made from coconut palm to the membrane characteristics produced.

2. Experimental

2.1 Materials
The materials used were coconut nira, *Acetobater xylinum*, sugar, ammonium sulfate, glacial acetic acid, sodium hydroxide, concentrated sulfuric acid, acetic anhydride, formamide where obtained from Merck while dextran T-500 from Sigma.

2.2 Preparation of cellulose acetate
Nira microbial cellulose was carried out by boiling 5 L of nira. Then 500 g of sugar and 25 g of ammonium sulfate were added until dissolved. The solution had an acidity in pH of 4. The solution was poured into plastic trays each containing 400 mL, closed with sterile newspapers, and aged 24 hours at room temperature. Then 10% of starter bacteria *Acetobacter xylinum* was added and incubated at room temperature for 7 days to obtain microbial cellulose.

The microbial cellulose was purified by boiling for 20 minutes, then soaked in a solution 1% of NaOH for 24 hours at room temperature. It was soaked again with 1% of acetic acid for 24 hours at room temperature and washed with running water. Then the microbial cellulose was pressed and dried.

In the acetylation experiment, there were three stages i.e. activation, acetylation, and hydrolysis. The activation stage was carried out by mixing and stirring 5 g of microbial cellulose and 12 mL of glacial acetic acid for 60 min. The acetylation stage was conducted by adding 20 mL of glacial acetic acid and 0.09 mL of concentrated sulfuric acid as catalyst and stirred again for 45 min. The mixture was cooled until its temperature reached 16°C and 13.5 mL of acetic anhydride 98% was added. Another mixture containing 20 mL of glacial acetic acid and 0.6 mL of concentrated sulfuric acid was added into the first mixture and stirred for 20h.. Then the hydrolysis stage was carried out by adding 30 mL of acetic acid 67% drop by drop within 2 h at 38°C. The hydrolysis reaction was allowed to continue for 20 h. The product was poured into water with strong agitation and the precipitated was washed with water until the pH became neutral and finally dried at 50°C.

2.3 Preparation of ultrafiltration membranes

The membrane was carried out by phase inversion method. Casting solution composed of cellulose acetate, acetone, and formamide as additive. The composition of cellulose acetate was made in 3 variations i.e. 20, 23, and 25%. The mixture was stirred for 24 h at room temperature until homogeneous. The dope was aged for 24 h to get rid of air bubbles, then casted on a glass plate. After evaporation for 10 s, the glass plate was gently immersed into cold water at 4°C. The membrane was formed and then washed with deionized water for several hours until all the solvent and additive have been removed.

2.4 Characterization of ultrafiltration membrane
The flux and rejection of the membrane were measured in dead-end test cell under a constant applied pressure 3 kgf/cm^2. Water and dextran T-500 were used as feed. The flux water and dextran was measured every 10 m until the volume of permeate was constant. For rejection measurement, the feed and permeate concentration of dextran T-500 were determined by mixing the dextran with 5% of phenol and concentrated sulfuric acid in ratio of 1:1:5, then the absorbance was determined by spectrophotometric at the wave length of 490 nm. The flux was determined using equation: $J_v = \frac{V}{A.t}$ with Jv = flux (L/(m^2.h)); V = permeate volume; A = surface dimension, and t = time. The rejection was determined using equation: % R = (1- C_p/C_f) x 100 % with R = rejection; Cp = permeate concentration; and Cf = feed concentration.

The membrane morphology were observed by scanning electron microscope JSM-6510 and the mechanical properties were measured by using Autograph Shimadzu AGS-500D.

3. Results and Discussion

3.1 Characteristics of cellulose acetate

Microbial cellulose was produced from coconut nira through the fermentation process by *Acetobacter xylinum* [7]. The fermentation process produced yellowish-white coloured nata with an average thickness of 1 cm. Nata that had been dried had 4% water content. According to Syamsu & Kuryani (2014) the water content required for the acetylation process is 3-7%. Acetylation of cellulose was carried out to replace part or all of the hydroxyl (OH) groups in the cellulose molecule with the acetyl group (CH-CO) of the acetic anhydride to form cellulose acetate [8].

The cellulose acetate is a white solid with a 38% acetyl content and a relative molecular weight of 9.29×10^4 g/mol. Measurement of acetyl content was performed to determine the type of cellulose acetate and the corresponding solvent on membrane preparation. The acetyl content of 38% was cellulose diacetate which dissolves in acetone [1].

3.2 Characteristics of Ultrafiltration Membranes

Flux is one of the important parameters used in the performance of membrane filtration. The flux value is expressed as the volume of the feed solution which can pass through the membrane per unit area of membrane (L/m^2.hours). Flux is directly related to the amount of pore and pore size of the membrane surface. The increase in cellulose acetate concentration the increase the viscosity of the mixture, and as a consequence is the decrease the diffusivity between the components in the system during the solidification process of the dope solution, then the precipitation became slower resulting in smaller pore membranes. At lower cellulose acetate concentrations, more voids formed inside the membrane and the increase of the speed of molecules passing through the membrane leaded to higher flux. Figures 1 and 2 show the relationship between the flux values and various concentrations of cellulose acetate.

Figures 1 and 2 show that the value of flux decreases as the cellulose acetate concentration increases. Water fluxes of cellulose acetate concentration of 20%, 23% and 25% are 55.03; 45.86; and 30.57 L/m^2.h respectively whereas the dextran fluxes for the same cellulose acetate concentration are 33.02; 32.10 and 20.79 L/m^2. hour. The water fluxes are higher than the dextran flux as dextran has higher molecular weight hence it took longer time to pass through the membrane pores. The highest flux value was obtained by a membrane with a concentration of 20% cellulose acetate. The greater the concentration of cellulose acetate added, the smaller the flux.

The rejection was determined by passing dextran on the three variations of the ultrafiltration membrane. Feed and permeate were measured by spectrophotometer at wavelength of 490 nm. The value of rejection for each cellulose acetate concentration is displayed at Table 1.

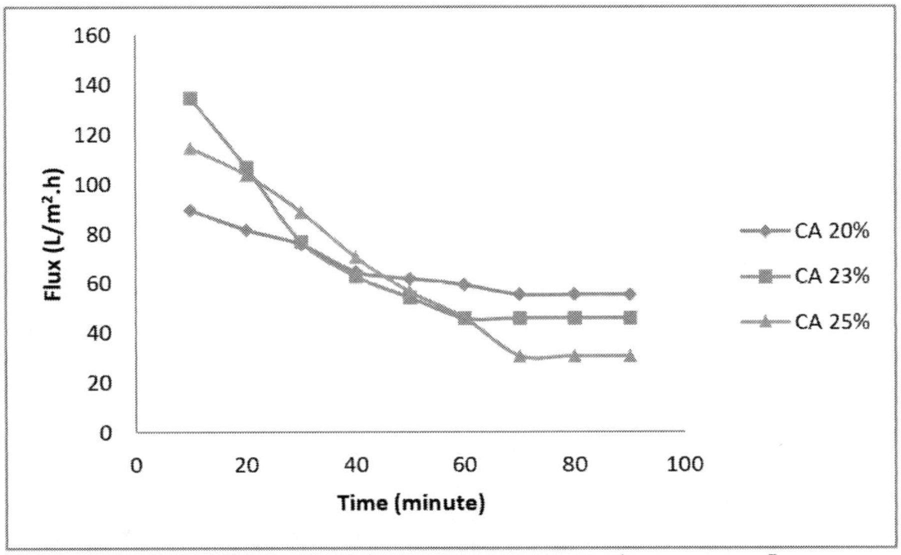

Figure 1. Effect of cellulose acetate concentration on water flux

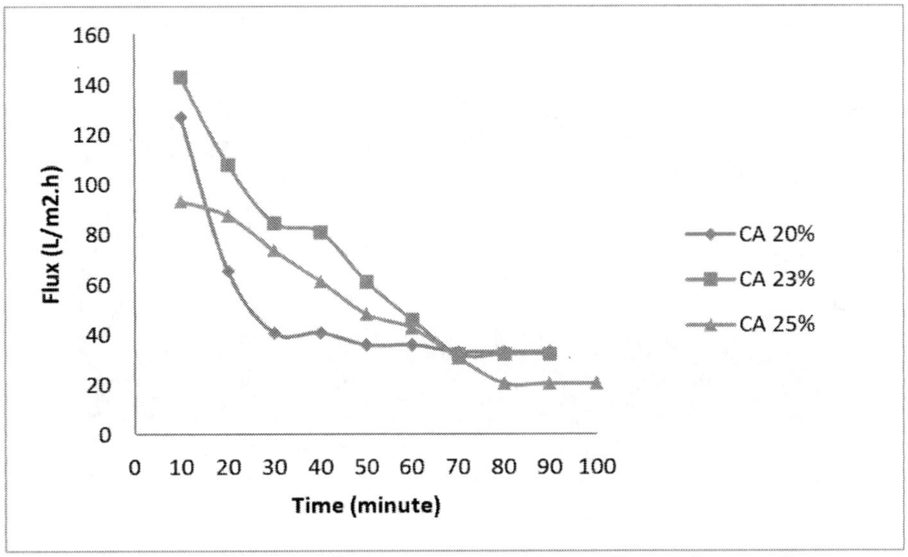

Figure 2. Effect of cellulose acetate concentration on dextran flux

Table 1. The value of dextran rejection at various concentrations of cellulose acetate

Membrane with cellulose acetate concentration (%)	Rejection (%)
20	79,15
23	79,47
25	82,82

Increasing concentrations of cellulose acetate increased the percent of rejection. The highest rejection value was obtained on the membrane with 25% of cellulose acetate. Membrane rejection was closely related to membrane selectivity [9]. Selectivity can be used to determine the ability of the membrane to retain or pass a particle. The selectivity depended on the interaction of the membrane with the dissolved particles, the pore size of the membrane, and the size of the particles passed through the membrane pores [10]. The more concentrated cellulose acetate, the smaller the pore size and the more particles were retained on the membrane hence the percentage of rejection increases.

3.3 Mechanical properties of ultrafiltration membranes

The tensile test results supported the flux and rejection measurements. The membrane with a 25% cellulose acetate concentration provided the best tensile strength value. The value of the tensile strength of the membranes for the compositions 20, 23, and 25% were 0.672, 0.973, and 1.233 MPa, respectively. The membrane morphology determined with SEM shown at Figure 3, 4, and 5. It showed that increasing concentration of cellulose acetate produce membrane with smaller pores. The membranes pore sizes are in range 0.05 until 0.1μm. Ultrafiltration membrane has size pores ranging from 0.01 until 2 μm [11].

Figure 3. SEM photo of membrane with 20% of cellulose acetate surface

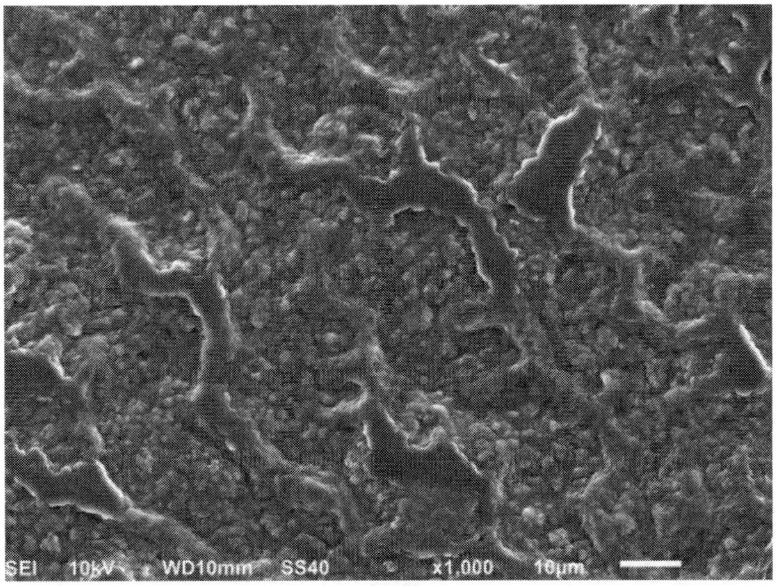

Figure 4. SEM photo of membrane with 23% of cellulose acetate surface

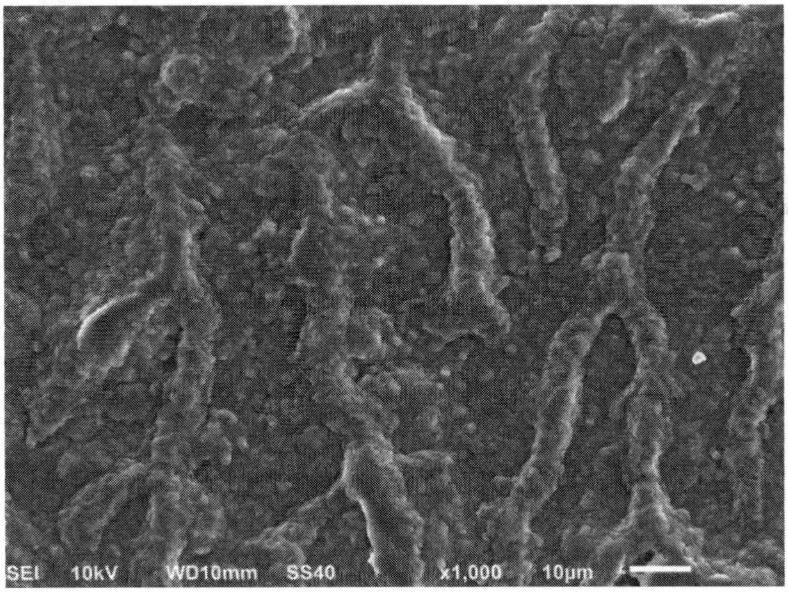

Figure 5. SEM photo of membrane with 23% of cellulose acetate surface

The authors should give deep analysis regarding the SEM images and correlate these images with the membrane performance

4. Conclusion

The higher the concentration of cellulose acetate in the preparation of ultrafiltration membranes, the higher the value of rejection and tensile strength, but the lower the flux value and pore size.

Acknowledgements
Part of this work has been funded by Jenderal Soedirman University, for which the authors are very grateful.

References
[1] Lindu M, PuspitasariT, and Ismi E, 2010, *Jurnal Sains Materi Indonesia* ***12*** (1) : 17-23.
[2] WidyaningsihS, Chasani M, dan Zusfahair, 2016, Unsoed Purwokerto
[3] Ciptaraharja I and Praptowidodo V S, 2006, *Jurnal Teknik Kimia Indonesia* *5*(3): 478-489.
[4] Widyaningsih S, Rastuti U, dan Arifah N I, 2012, *Prosiding Seminar Nasional Kimia dan Pendidikan Kimia Tahun 2012* Purwokerto
[5] GustianI, Sutanto T D, and Adfa M, 2006, *Jurnal Gradien 2* (1): 126-129
[6] RukmanaRand Yuniarsih Y, 2001, *Membuat Kecap: Tempe Busuk, Nira, dan Air Kelapa* (Yogyakarta: Kanisius)
[7] Piluharto B, 2003 *Jurnal Ilmu Dasar* **4** (5): 52-57
[8] Radiman C L, 2004, *Kimia Polimer* (Bandung: Penerbit ITB) p 5-20
[9] Radiman C L, Widyaningsih S, and Sugesty S 2008 *Journal ofMembrane Science* **315**: 141-146
[10] SilviaV, Pinem J A, and Irianty R A 2016 *Jom FTEKNIK* **3** (1): 1-9
[11] Meijuan Z, Nemade P R, Lu X, Zeng X, Hatakeyama E S, Noble R D, and Gin D L 2007 *Journal of The American Chemical Society* **129**: 9574-9575.

IOP Conf. Series: Materials Science and Engineering **349** (2018) 012021 doi:10.1088/1757-899X/349/1/012021

Assessment of Drug Binding Potential of Pockets in the NS2B/NS3 Dengue Virus Protein

F Amelia[1], Iryani[1], P Y Sari[1], A A Parikesit[2], R Bakri[2], E P Toepak[2], U S F Tambunan[2]*

[1]Department of Chemistry, Universitas Negeri Padang, Air Tawar, Padang, West Sumatera, Indonesia

[2]Bioinformatics Research Group, Departement of Chemistry, Faculty of Mathematics and Natural Science, Universitas Indonesia, Kampus UI Depok, West Java 16424, Indonesia

* Email:usman@ui.ac.id

Abstract. Every year an endemic dengue fever estimated to affect over 390 million cases in over 128 countries occurs. However, the antigen types which stimulate the human immune response are variable, as a result, neither effective vaccines nor antiviral treatments have been successfully developed for this disease. The NS2B/NS3 protease of the dengue virus (DENV) responsible for viral replication is a potential drug target. The ligand-enzyme binding site determination is a key role in the success of virtual screening of new inhibitors. The NS2B/NS3 protease of DENV (PDB ID: 2FOM) has two pockets consisting of 37 (Pocket 1) and 27 (Pocket 2) amino acid residues in each pocket. In this research, we characterized the amino acid residues for binding sites in NS3/NS2B based on the hydrophobicity, the percentage of charged residues, volume, depth, $\Delta G_{binding}$, hydrogen bonding and bond length. The hydrophobic percentages of both pockets are high, 59 % (Pocket 1) and 41% (Pocket 2) and the percentage of charged residues in Pocket 1 and 2 are 22% and 48%, and the pocket volume is less than 700 Å^3. An interaction analysis using molecular docking showed that interaction between the ligand complex and protein in Pocket 1 is more negative than Pocket 2. As a result, Pocket 1 is the better potential target for a ligand to inhibit the action of NS2B/NS3 DENV.

1. Introduction

Dengue virus (DENV) infection has become a global concern. There were over 50 million infection cases per year.Not only that, approximately 2.5% people from 500.000 severe dengue cases were reported to die[1]. The developing of an effective DENV treatment is needed to repress the mortality rate caused by DENV infection. In efforts to find a treatment for DENV, characterization of the enzyme binding site is an important step during the virtual screening of new inhibitors. One of the DENV essentials enzymes that essential to be characterized is NS2B/NS3 protease. This enzyme is known to has an essentials role in the maturation of DENV virion and viral replication process [2].

The previous research on NS2BCF-Gy-NS3PRO protease using metapocket analysis and molecular dynamics detected three amino acid residues in NS2B DENV protein and 19 amino acid residues in NS3 DENV protein as binding sites [3]. Other researchers working on NS2B-NS3 2FOM found 12 amino acid residues as binding sites using *CastP* [4]. Altering the structural conformation of DENV NS2B/NS3 has been proposed as a potential means of deactivating the protease, but some uncertainty remains regarding the exact structural arrangement of NS2B/NS3 [5]. However, it is estimated that 60% of proposed or potential blocking ligands used in drug design never meet initial expectations because the protein targets are found not to be druggable [6]. To address this issue, we determined the residues and druggable binding pocket in NS3/NS2B based on five properties; hydrophobicity, the percentage of charged residues, volume, and depth using computational approach. The Computational approach played an important role to modeled the DENV protein. It can give us the better insight about the protein active site and way to inference the protein activity [7]. We also performed *in silico* molecular docking using 4-hydroxypanduratin A and panduratin A to model the interaction between ligands and protein [8].

2. Methods

2.1. Preparation of dengue virus NS2B-NS3 protease structure and standard ligand

The complex three-dimensional structure of the DENV NS2B-NS3 protease was obtained from the Protein Data Bank (PDB) database using the specific enzyme ID of 2FOM (http://www.rcsb.org/pdb/). The geometry optimization and energy minimization of DENV NS2B-NS3 protease were performed using *MOE2008.10 software*. We chose ligands 4-hydroxypanduratin A and panduratin A as a standard ligand. The structure of the ligands 4-hydroxypanduratin A and panduratin A were drawn using *ACD/Chemsketch* software version 12.01. *MOE2008.10* software was used for optimization of the ligands.

2.2 Binding site analysis

The binding sites on the surface protein of the NS3/NS2B of DENV were predicted using *Metapocket 2.0, DoGSiteScorer,* and *DEPTH* server. In *Metapocket 2.0* server, the structure of the protein used for prediction was in PDB format, and the amino acid residues will be shown as the result of binding site residues [9–11]. *DoGSiteScorer* was used to determine the 3D structure of the pocket of the protein using the calculation of size, shape and chemical features of the predicted (sub) pockets. PDB and SDF formats were used in this prediction [12,13]. *DEPTH* server calculated cavity volumes and predicted the location of small molecule-binding sites by measuring the closest distance of a residue/atom to the bulk solvent [14–16].

2.3 Molecular docking

Molecular docking is a method which predicts the preferred orientation of one molecule to a second when bound to each other to form a stable complex. This process was done using the *MOE 2008.10* software through the MOE-dock tools. The method for placement, rescoring and refinement used a matching triangle with 2000000 iterations, London dG, and force field with a population of 1000 repetitions respectively [17].

3. Result and Discussion

3.1 Binding site prediction

Modeling and characterization of the active site of NS3/NS2B is an important step for screening the potential reactivity of drug candidates [7]. The application *Metapocket* 2.0 calculates the geometric and energy interaction between the proteins and target ligands to locate favorable binding sites. The most favorable site is ranked first. NS2B/ NS3 protease of DENV structure contained two pockets. The first site consisted of 37 residues in a chain (7 polar residues, 22 nonpolar residues, and 8 charge residues) and the second binding site consisted of 27 residues (3 polar residues and 11 nonpolar residues and 13 charge residues) in chain A and B (Table 1).

Table 1. The amino acid residues in binding pocket 1 and 2 based on residue properties

Pocket	Hydrophobicity		Charge	
	Hydrophilic	Hydrophobic	(+ve)	(-ve)
1	THR_B118;THR_B120 THR_B122;**THR_B156** ASN_B119 **ASN_B152** ASN_B167	MET_B49;**TRP_B69**;VAL_B72 **LEU_B76;TRP_B83;**LEU_B85 GLY_B87;TRP_B89;GLY_B121 ILE_B123;GLY_B124;VAL_B146 VAL_B147;**GLY_B148;**LEU_B149;GLY_B151;GLY_B153;VAL_B154;AL_B155;**ALA_B164;ILE_B165;**ALA_B166	LYS_B117;LYS_B73;LYS_B74;LYS_B84	ASP_B71;ASP_B75;GLU_B86;GLU_B88
2	SER_A75;THR_A94 ASN_B105	LEU_A53;**ALA_A56;**ALA_A57;ILE_A73;**LEU_A74;**ILE_A76;**LEU_A95;**GLY_A96;TYR_B23;PRO_B106 **LEU_B18**	**ARG_B24**;ARG_A55;**ARG_A60**;LYS_A87;LYS_B61; HIS_B60	GLU_A52;GLU_A54;ASP_A58;GLU_B19;GLU_A62;GLU_A89 GLU_A90

[a]Black = residues detected by *Metapocket* 2.0; **Black bold=** residues detected by both *Metapocket* 2.0 and *DEPTH*

DEPTH server, an application that calculates cavity volumes and predicts the location of small molecule-binding sites, confirmed the *Metapocket* result that there were two cavities suitable for ligand binding. The small molecule predicted can bind to 8 amino acid residues in pocket 1 and 6 amino acid residues in pocket 2 (Table 1). The three amino acid residues with the highest probability of binding were, GLY_B148, followed by ALA_B164, and LEU_B76, are located in pocket 1 as determined by *Metapocket* 2.0.

DoGSiteScorer determined the druggable pocket based on physical properties; pocket volume, pocket surface, lipophilic surfaces, and depth [12,13]. Measurement of physical properties has important implications for ligand design. Residues that have a volume ≥ 500 Å^3, depth ≥ 10.4 Å, enclosure ≥0.28, percentage of charge residue < 26.3 and hydrophobicity ≥ -1.12 generally have the highest probability of forming a druggable pocket [18]. The 2FOM protein has 13 residues which have more than 10 Å of depth in a β chain. The total depth of each pocket is shown in Table 2.

Table 2. Drug score of pocket protein by assessment of the physical properties of the pocket using *DoGSiteScorer*

Pocket	Volume [Å^3]	Surface [Å^2]	Lipophilic surfaces [Å^2]	Depth [Å]	Drug Score
P1	600.00	976.78	576.15	18.92	0.81
P2	573.31	843.74	551.08	16.42	0.78
P3	236.54	457.92	295.86	13.18	0.54
P4	208.45	373.51	202.56	14.27	0.53
P5	154.69	338.81	195.60	9.34	0.32
P6	140.74	336.88	205.71	8.45	0.25
P7	117.70	258.98	184.42	6.34	0.16

legend: undruggable => druggable

The polarity influences the hydrogen bond form because it occurs due to a bond being polar. Hydrogen bonds confer directionality and specificity to the intramolecular interactions in the structures. This interaction is particularly important for proteins, where the hydrogen bond provides the organization for distinct folds and also provides the selectivity in the protein-ligand interactions [19]. So, the number of polar and nonpolar residues will affect the degree of hydrogen bonding in ligand-NS3/NS2B protease complexes. Since the small compounds or ligands will easily find their binding sites at the hydrophobic pocket, hydrophobic pockets are preferable as drug targets. Based on Table 1, both Pocket 1 and 2 have high percentages of hydrophobicity, 59% and 41%, which indicated both pockets are good for drug target.Furthermore, potential binding sites with a lower frequency of amino acid residues in a charged state are favored as drug targets. The preferable range of charged residues is 10 to ≤ 26.3% [18]. Since the percentage of charged residues in Pocket 1 and 2 was 21.6% and 48% respectively, only Pocket 1 is indicated as a potentially good drug target. The ligand binding pocket had a wide range of sizes from 100 to 1000Å^3. If the pocket volume is lower than 700 Å^3, the size of potential ligands needs to be considered[20]. Since Pocket 1 and 2 have the volume 600 Å^3 and 573 Å^3 respectively, the small ligand is suitable for this protein. The pocket cavity of NS2B/NS3 Protease is shown in Figure 1.

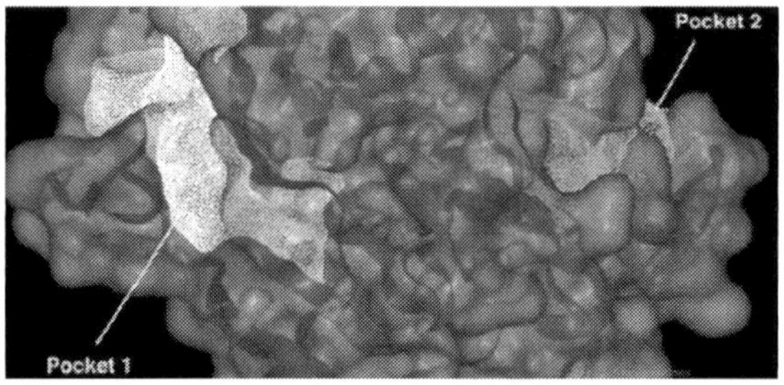

Figure 1. Two drugable pockets of NS3/NS2B protease DENV

3.2 Molecular docking

In order to explore the enzyme-substrate interaction and to determine the key residues responsible for interaction, two substrates (4-hydroxypanduratin A and panduratin A) were used as model ligands for the NS3/NS2B protease using the *MOE2008.10* software. Table 3 shows that in all cases there was a negative and low value of free energy of binding ($\Delta G_{binding}$) interaction between the NS3/NS2B protease pockets and the 2 ligands indicating a potential strong bond in the most favorable conformations. The noncovalent binding will take place as a spontaneous process if it is correlated with a negative binding free energy [21]. Table 3 showed that the interaction between ligand 4-hydroxypanduratin A and panduratin A and NS2B /NS3 protein should occur spontaneously. In Pocket 1 the ligand 4-hydroxypanduratin A formed 2 hydrogen bonds and panduratin A formed one hydrogen bond. In Pocket 2, 4-hydroxypanduratin A is predicted to form 3 hydrogen bonds and panduratin A 4 hydrogen bonds. As mentioned previously hydrophobic pockets are good candidates for druggable pockets even though spectroscopic evidence demonstrates that hydrogen bonds at hydrophobic sites are weaker [19,21].

Even though the $\Delta G_{binding}$ for the ligand-protein complex in Pocket 1 is more negative than Pocket 2 (Table 3), the number of hydrogen bonds in Pocket 1 is lower than Pocket 2. There are many factors influencing the stabilization of the ligand-enzyme complex, for example, the distance and angle of hydrogen bonds [22]. In this research, the length of hydrogen bonds ranged from 1.59 to 2.47 (Table 3) which fits the accepted (and most frequently observed) geometry for a hydrogen bond, a length of less than 2.5Å.

Table 3. Result of Molecular docking simulation

Pocket	Ligand	$\Delta G_{binding}$ (kcal/mol)	Hydrogen bonds	Bond length (Å)
1	4-hydroxypanduratin A	-13.00	Leu_B140 (66%); Lys_B74(45%)	1.79; 1.61
1	Panduratin A	-11.49	Lys_B74(21%)	2.47
2	4-hydroxypanduratin A	-10.77	His_B60(60%); Leu_A53(44%) Lys_B61(48%)	1.75; 1.84; 1.59
2	Panduratin A	-10.80	Glu_B19(90%); Leu_B19(17%); His_B60(63%); Arg_A55(19%)	1.78; 2.3;1.66 2.08

4. Conclusion

This article described the analysis of amino acid residues for potential binding sites of DENV NS2B/NS3 Protease using *Metapocket*, *Depth,* and *DoGSiteScorer* to perform molecular docking as a part of efforts to discover the potential binding site and new candidates for drug design. Our results clearly identify Pocket 1 as the best candidate as it has a lower overall hydrophobicity and a more spontaneous binding energy ($\Delta G_{binding}$) than pocket 2.

Conflict of Interest

We declare that there is no conflict of interest in this research nor publication

Acknowledgements

We thank the Indonesian government for funding this research, Hibah Pekerti – DIKTI No: 237/UN35.2/PG/2015 and Hibah Penelitian Unggulan Perguruan Tinggi (PUPT) 2017 No: 2716/UN2.R3.1/HKP.05.00/2017. In this research; Usman Sumo Friend Tambunan, Arli Aditya Parikesit, Ridla Bakri and Fitri Amelia were supervising the research. Iryani and Prima Yulia Sari conducted the experiments, while Erwin Prasetya Toepak helped to proofread the paper

References

[1] Nitsche C, Holloway S, Schirmeister T and Klein C D 2014 Biochemistry and Medicinal Chemistry of the Dengue Virus Protease *Chem. Rev.* **114** 11348–81

[2] Frecer V and Miertus S 2010 Design, structure-based focusing and in silico screening of combinatorial library of peptidomimetic inhibitors of Dengue virus NS2B-NS3 protease. *J. Comput. Aided. Mol. Des.* **24** 195–212

[3] de Almeida H, Bastos I M D, Ribeiro B M, Maigret B and Santana J M 2013 New binding site conformations of the dengue virus NS3 protease accessed by molecular dynamics simulation. ed J Zheng *PLoS One* **8** e72402

[4] Verma R, Jatav V K and Sharma S 2015 Identification Inhibitors of Dengue Virus (DENV1, DENV2 & DENV3) NS2B/NS3 Serine Protease: A Molecular Docking and Simulation Approach *Asian J. Pharm. Clin. Res.* **8** 4–9

[5] Schüller A, Yin Z, Brian Chia C S, Doan D N P, Kim H-K, Shang L, Loh T P, Hill J and Vasudevan S G 2011 Tripeptide inhibitors of dengue and West Nile virus NS2B–NS3 protease *Antiviral Res.* **92** 96–101

[6] Halgren T A 2009 Identifying and Characterizing Binding Sites and Assessing Druggability *J. Chem. Inf. Model.* **49** 377–89

[7] U.S.F. Tambunan A A P 2011 In silico Design of Drugs and Vaccines for Dengue Disease *Trends Bioinforma.* **4**

[8] Kiat T S, Pippen R, Yusof R, Ibrahim H, Khalid N and Rahman N A 2006 Inhibitory activity of cyclohexenyl chalcone derivatives and flavonoids of fingerroot, Boesenbergia rotunda (L.), towards dengue-2 virus NS3 protease *Bioorg. Med. Chem. Lett.* **16** 3337–40

[9] Anon http://project.biotec.tu-dresden.de/metapocket/index.php

[10] Zhang Z, Li Y, Lin B, Schroeder M and Huang B 2011 Identification of cavities on protein surface using multiple computational approaches for drug binding site prediction *Bioinformatics* **27** 2083–8

[11] Huang B 2009 MetaPocket: A Meta Approach to Improve Protein Ligand Binding Site Prediction *Omi. A J. Integr. Biol.* **13** 325–30

[12] Volkamer A, Kuhn D, Grombacher T, Rippmann F and Rarey M 2012 Combining Global and Local Measures for Structure-Based Druggability Predictions *J. Chem. Inf. Model.* **52** 360–72
[13] Anon http://proteinsplus.zbh.uni-hamburg.de/.
[14] Tan K P, Varadarajan R and Madhusudhan M S 2011 DEPTH: a web server to compute depth and predict small-molecule binding cavities in proteins *Nucleic Acids Res.* **39** W242–8
[15] Chakravarty S and Varadarajan R 1999 Residue depth: a novel parameter for the analysis of protein structure and stability *Structure* **7** 723–32
[16] Anon http://cospi.iiserpune.ac.in/depth/htdocs/run_depth.html.
[17] C.C.G. Inc 2013 Molecular Operating Environment (MOE)
[18] Perola E, Herman L and Weiss J 2012 Development of a Rule-Based Method for the Assessment of Protein Druggability *J. Chem. Inf. Model.* **52** 1027–38
[19] Hubbard R E and Kamran Haider M 2010 Hydrogen Bonds in Proteins: Role and Strength *Encyclopedia of Life Sciences* (Wiley-Blackwell)
[20] Liang J, Woodward C and Edelsbrunner H 1998 Anatomy of protein pockets and cavities: Measurement of binding site geometry and implications for ligand design *Protein Sci.* **7** 1884–97
[21] Bissantz C, Kuhn B and Stahl M 2010 A Medicinal Chemist's Guide to Molecular Interactions *J. Med. Chem.* **53** 5061–84
[22] Chikalov I, Yao P, Moshkov M and Latombe J-C 2011 Learning probabilistic models of hydrogen bond stability from molecular dynamics simulation trajectories *BMC Bioinformatics* **12** S34

Dynamic adsorption of mixtures of Rhodamine B, Pb (II), Cu (II) and Zn(II) ions on composites chitosan-silica-polyethylene glycol membrane

F W Mahatmanti[1*], W D P Rengga[2], E Kusumastuti[1], Nuryono[3]

[1] Department of Chemistry, Faculty of Mathematics and Natural Sciences, Universitas Negeri Semarang, Indonesia 50229
[2] Department of Chemical Engineering, Faculty of Engineering, Universitas Negeri Semarang, Indonesia 50229
[3] Department of Chemistry, Faculty of Mathematics and Natural Sciences, Universitas Gadjah Mada, Sekip Utara, Yogyakarta, Indonesia 55281

*widhimahatmanti17@gmail.com,

Abstract. The adsorption of a solution mixture of Rhodamine B, Pb (II), Cu (II) and Zn(II) was studied using dynamic methods employing chitosan-silica-polyethylene glycol (Ch/Si/P) composite membrane as an adsorptive membrane. The composite Ch/Si/P membrane was prepared by mixing a chitosan-based membrane with silica isolated from rice husk ash (ASP) and polyethylene glycol (PEG) as a plasticizer. The resultant composite membrane was a stronger and more flexible membrane than the original chitosan-based membrane as indicated by the maximum percentage of elongation (20.5 %) and minimum Young's Modulus (80.5 MPa). The composite membrane also showed increased mechanical and hydrophilic properties compared to the chitosan membranes. The membrane was used as adsorption membrane for Pb (II), Cu (II), Cd (II) ions and Rhodamine B dyes in a dynamic system where the permeation and selectivity were determined. The permeation of the components was observed to be in the following order: Rhodamine B > Cd (II) > Pb (II) > Cu (II) whereas the selectivity was shown to decrease the order of Cu (II) > Pb (II) > Cd (II) > Rhodamine B.

Introduction

The textile industry is one of the few developed industrial sectors in Indonesia. One aspect that needs to be considered in the textile industry is the waste that can pollute the environment, especially water. Textile industry waste is dominated by dyestuff pollution due to the use of both natural and synthetic dyes in the production process. The commonly used colors in the textile industry are Rhodamine B dyes to color silk, wool, and synthetic materials. Rhodamine B dyes are basic dyes which are chloride or oxalic salts of organic bases. These dyes can be Zwitterionic, having both positive and negative electrical charges in the same structures. Regularly found metal ions in textile industry waste are mainly As (III), Cd (II), Cr (III), Pb (II), Cu (II), and Zn (II).

The dye intensity and the content of metal ions in the textile waste can be reduced physically, chemically and biologically. Chemically, several methods have been reported to be used by researchers to reduce pollution in textile industry waste including coagulation-flocculation, electrochemical oxidation, adsorption and membrane filtration [1]. Membrane technology is considered an effective chemical way to reduce dye pollutants and metal ions in waste. Therefore, this paper reports the investigation of the effectiveness of a composite membrane Ch/Si/P for the permeation and selectivity of dyes and cations for potential use in the textile waste treatment.

Materials and Methods

Materials

Materials used were NaOH, HCl, Rhodamine B and nitrate salts of Pb (II), Cu (II) and Cd (II); purchased from Merck and used as received without purification., Pearson Panke equipment Ltd for measuring the mechanical parameters and Thermogravimetric Analysis (TGA) (Perkin-Elmer TGA-7) for evaluating thermal stability. Spectrophotometry visible (Shimadzu 1240) was applied for the determination of Rhodamine B concentration in the sample solution. Atomic absorption spectroscopy, AAS (Perkin-Elmer AA 200), was applied for the determination of Pb (II), Cd (II) and Cu (II) concentration in the solution.

Methods

Preparation of composites Ch/Si/P membrane

Chitosan solution 1 wt % was prepared by dissolving 1 g of chitosan in 100 mL 2 % (v/v) acetic acid solution. An appropriate amount of sodium silicate from isolated rice hull ask with the different mass ratio (0.1-1.5 g) were mixed with 100 mL chitosan solution, and the mixture was thoroughly stirred until there was no air bubble in the solution. The membrane was obtained by casting the mixture on a glass plate and was allowed to evaporate the solvent under an ambient temperature for three days. The membrane was neutralized with a 5 % (w/v) NaOH solution after drying. Afterward, the membrane was washed with water to remove the remaining NaOH. Finally, the membrane was stored in a desiccator before use. The optimum ratio mass of chitosan to silica is obtained from the membrane giving the highest tensile strength. The same work was carried out by adding varying the mass of PEG (0.05-0.50 g) on chitosan-silica to evaluate the effect of PEG content on the membrane properties.

Characterization of composites Ch/Si/P membrane

The mechanical properties of the membrane in dry state were performed at 25°C using a universal material testing instrument with a crosshead speed of 10 mm/min. The results were presented in mean values of three independent measurements. Thermo-gravimetric (TG) data was performed for a 10 mg sample from room temperature to 400 °C at a heating rate of 10 °C/min in a dynamic (30 mL/min) synthetic air atmosphere. For determination of swelling index (SI), a procedure reported by [2] was used. The pH-dependent on swelling properties of the film, pre- weighed dry samples was immersed in buffer solutions at pH of 2.0-11.0. After 24 h, the swollen films were removed, excess water was bottled from the surface, and then were weighted. The following equation was used to determine SI.

$$SI\ (\%) = [(w_w - w_d) / w_d] \times 100\ \%, \quad (1)$$

with w_w and w_d are the sample weights after swelling for 24 h and dry state, respectively. In each specimen, three samples were tested.

Preparation of solution mixture:

An aqueous mixture of Rhodamine B, Pb (II), Cu (II) and Cd (II) was prepared by mixing two mmol of each chemical into a 1000 mL volumetric flask and adding distilled water to the mark. The solution was shaken to ensure a complete dissolution and mixing.

Test the ability of Ch/Si/P membrane as an adsorptive membrane of metal ions and dye. Test of membrane capability as metal ions separating membrane was carried out on a solution containing

Rhodamine B and Pb (II), Cu (II) and Cd (II) metal ion by dynamic system. The sample used is sample of metal ion solution and artificial dye. The metal ions solution and the dye with a concentration of 2 mmol/L are included in the diffusion cell consisting of the chamber A and B. The chamber A contains a solution which contains no metal ions, whereas chamber B contains a solution containing both metal ions and dyes. The diffusion test was carried out using the cell set up presented schematically in Figure 1.

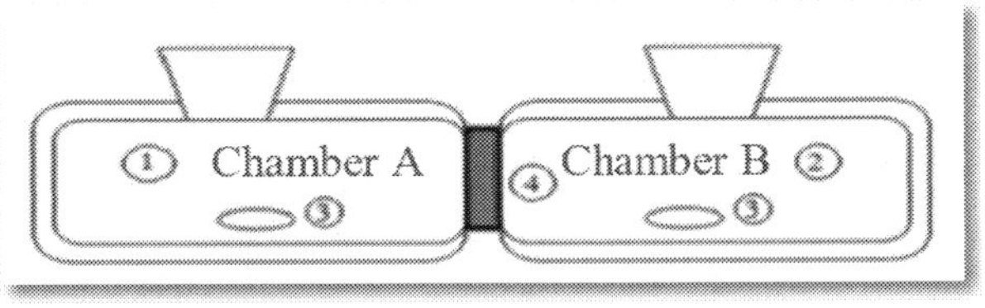

Figure 1. Diffusion cells for dynamic system adsorption test where (1) Chamber A containing the solution of metal ions/dye, (2) Chamber B containing distilled water, (3) Magnetic stirrers, (4) Ch/Si/P composites membrane

Chambers A and B were filled with the solution mixture (50 mL) and distilled water (50 mL), respectively, with constant stirring. The two chambers were separated with the Ch/Si/P membrane of 0.16 mm thickness. Every 2 hours for 24 hours a sample of solution in both chambers A and B was collected and tested in an Atomic Absorption Spectrometer (AAS) and UV-Visible spectrophotometer to determine the concentrations of metal cations and the Rhodamine B. The tests for the ability of the membrane as an adsorbent were carried out for each of the metal ion solutions separately, as mixture of metal ions and as mixture of dyestuff with metal ions.

Results and Discussion

Membrane characterization

The membranes were tested for mechanical properties to determine their tensile strength, stress, strain and Young's Modulus following ASTM D638-03 using Pearson Panke Equipment Ltd. The tensile strength was determined by measuring the maximum voltage that can be retained by a material when stretched or withdrawn before the material is broken. Tensile strength is commonly used in designing the part of a ductile and brittle structure that is not static. In all materials, in the early stages of the tensile test, the relationship between the load or force is proportional to the length of the material, in which the ratio of stress and strain is constant. Stress is the load divided by the cross-sectional area of the material (F/A), and the strain is the length increase divided by the initial length of the material ($\Delta L/L$). Stress and strain comparisons are always fixed, formulated as Young's Modulus. The test results of mechanical properties are presented in Figures 2(a) and 2(b).

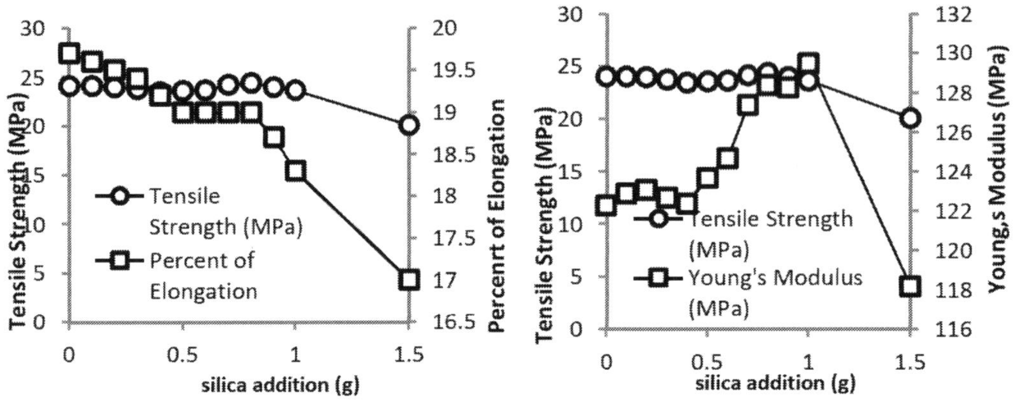

2(a) 2(b)

Figure 2. Effect of adding silica to (a) Tensile strength and Percentage of elongation, (b) Tensile stress and Young's Modulus

The tensile strength of membrane tends to decrease with the increase of the chitosan-silica ratio up to 1.0:0.8; at the higher content of silica, the tensile strength drastically decreases. The optimum ratio showing the highest tensile strength is 1.0:0.8. The addition of silica more than 0.8 g ratio of chitosan, both properties above deteriorate supporting a conclusion that silica may harden and toughen chitosan. The percentage of elongation decreases with increasing the silica content. The high silica content would bring chitosan-silica membranes to high inorganic characters such as the brittleness and makes low elongation at break. The test results revealed that formation chitosan-silica membranes did not bring positive effect on improving the mechanical properties of the membrane. The addition of silica-enhanced the membrane character of being more inorganic, harder and more fragile [3]. According to [4] increasing silica content would result in a negative effect on the mechanical properties.

The effect on the mechanical properties of the addition of PEG is presented in Figure 3(a) and 3(b). In general, the addition of PEG increased dramatically the percentage of elongation up to a maximum of 21 % at 0.2 g of PEG addition followed by a slight decrease beyond the maximum, but the tensile strength of the membrane decreased slightly. The addition of PEG resulted in a significant drop of Young's Modulus indicating that the viscosity of the membrane increased significantly. Considering the minimum Young's Modulus, the maximum percentage of elongation, no significant drop of tensile strength indicating elastic and strong membrane, the optimum composition of Chitosan/silica/ PEG was indicated to be 1: 0.8: 05. This is in agreement with the report by [5]. PEG as a plasticizer could improve mechanical properties, maintaining wholeness and avoiding holes and cracks. In the membrane matrix, the addition of a plasticizer is expected to reduce the intermolecular bonds between the chitosan molecules thereby weakening the intermolecular forces [6-7].

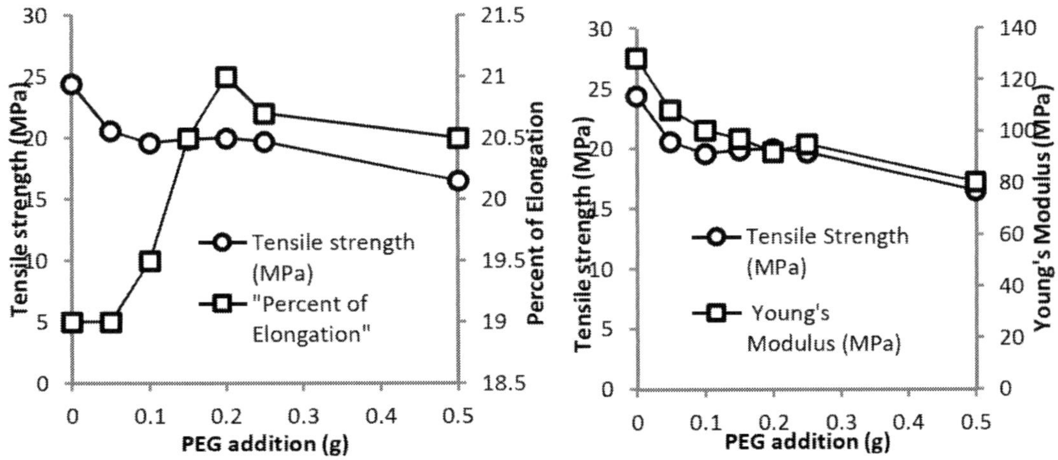

3 (a) 3(b)

Figure 3. Effect of polyethylene glycol (PEG) addition to (a) Tensile strength and percentage of elongation, (b) Tensile strength and Young's Modulus

The addition of PEG plasticizer in membrane material has been reported by several researchers. [6] added PEG with 6000 and 8000 weight molecular on the manufacture of chitosan membranes. In the addition of PEG with 6000 weight molecular with chitosan ratio: PEG (2: 1), the value of tensile strength decreased from 5.5 to 4.2 MPa. Addition of PEG with 6000 weight molecular and chitosan ratio: PEG (4: 1) increase tensile strength value to 7 MPa. Also PEG with 8000 weigth molecular produces membranes with mechanical properties equal to the membrane with the addition of PEG 6000 weight molecular. [8] use PEG because of its non-toxic, biocompatible and biodegradable nature. Through mixing technique, PEG is commonly added to chitosan as an effort to improve membrane elasticity. According to [9], the addition of PEG to the chitosan-silica membrane with PDMS as a source of silica tends to increase chitosan dispersion with PDMS. This indicates that increase in PEG increases drastically the membrane flexibility, due to the formation of hydrogen bonds between chitosan molecules with PEG [10].

The thermal properties of the membranes were measured by TG presented in Figure 4. The weight loss takes place in two stages. The first stage starts at 60 °C with a weight loss of 10 %. The second stage begins at 240 °C and reaches a maximum at 380 °C with a weight loss of 41.4 % [11]. For chitosan membrane, chitosan-silica membrane and chitosan-silica-PEG, the first stage of decomposition occurs between ambient and 200 °C being accounted for around 14 % weight loss, corresponding to dehydration of the membranes. The second stage of decomposition which starts from around 200 to 550 °C is attributed to a weight loss due to the decomposition of the polymeric network [11-13]. For the chitosan-silica film, the second stage starts from 240 to 550 °C with the weight loss of 60.0 %. Blending chitosan-silica with PEG, the second stage starts from 240 to 550 °C with the weight loss of 62.2 %. As the amount of PEG is increased, the intra-molecular interactions between the terminal hydroxyl groups of PEG predominate, the attractive bonds are inhibited, the blend becomes incompatible, and as a consequence, the thermal stability of the chitosan and chitosan-silica membranes is lower than that of chitosan-silica-PEG one. This result is contradicted with result reported by [14]. It was observed that thermal stability of chitosan blending PEG is higher than that of chitosan, but thermal stability decreases at a too high mass ratio of chitosan to PEG.

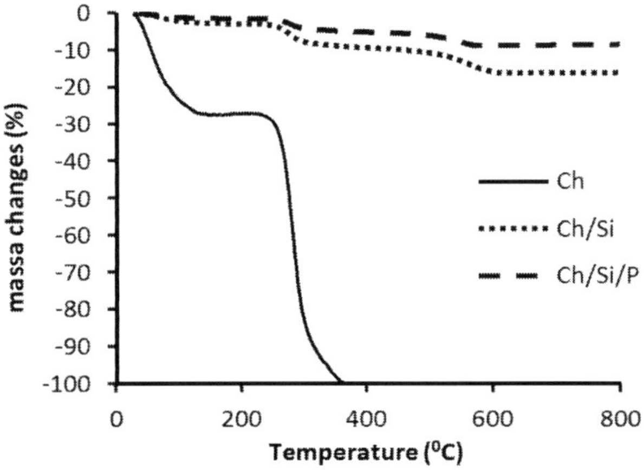

Figure 4. TGA thermograms of the membranes

The hydrophilic properties of the membranes were tested by immersing the membrane in a buffer solution with a pH of 2.0-11.0 for 24 hours. Test results on membrane hydrophilicity indicated by the values of the swelling index (SI) are presented in Figure 5. In Figure 5 it can be seen that in a strongly acidic solution, the value of the Swelling Index was high. The high Swelling Index value results in a *hydrophilic* membrane at pH 2.0 to pH 3.0. In an acidic solution, the protonation of the $-NH_2$ groups on chitosan produces $-NH_3^+$ which is a weak base. The dissolution mechanism occurring in the membrane is thought to be due to the acid reaction in the solvent with the weak base in the $-NH_3^+$ group. In neutral and slightly alkaline solutions, the Swelling Index value decreased. In general, it can be concluded that the increase of silica concentration decreased the hydrophilicity and the increase of PEG concentration increased the hydrophilicity characteristic along with the increase of pH value. PEG is a polymer having a sufficiently large number of -OH and is hydrophilic [15], explaining this observation.

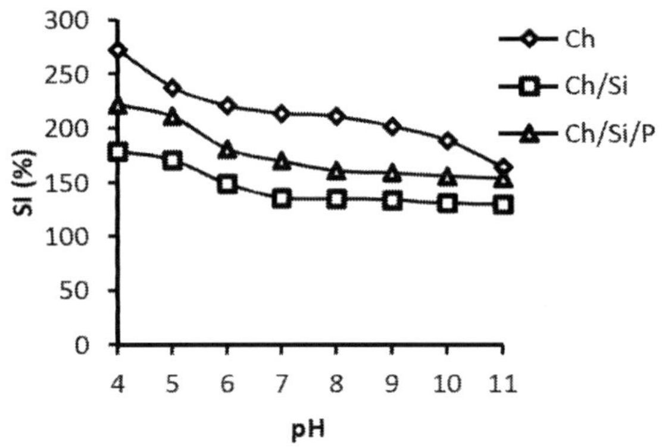

Figure 5. The Swelling Index (SI) of membranes at various pH

The separation process using a membrane is a separation process at a very small molecular or particle level. The membrane can move one component faster than the other based on the physical and chemical properties of the membrane and the separated component. The displacement may occur due to differences in pressure (ΔP), concentration (ΔC), electrical potential (ΔE) and temperature (ΔT).

Membrane selectivity is the ability of the membrane to pass a component relative to the other, expressed by rejection coefficient (R) [16]. The transport of ions and molecules through the membrane can occur through passive or active transportation. A passive transport does not require energy such as diffusion, facilitated diffusion and osmosis. The basic principles of transport through the membrane are based on the tendency of molecules to occupy space evenly. Molecules at high concentrations have a greater pressure, and each molecule tends to always move because of its kinetic energy content. Thus naturally there is a tendency of molecules at high concentrations to move to low concentrations. An active transport uses energy as it is transported through the membrane that resists the natural tendency of concentration difference.

In this study, the Ch/Si/P membrane was studied for its transport properties of Cu (II), Pb (II), Cd (II) and Rhodamine B dyes by calculating the rejection coefficient (R) to determine the selectivity. The membrane capability test as a separating membrane is performed for each of the metal ion solutions, metal ion mixture and dyestuff mixture with metal ions. The amount of absorbed metal ions and dissolved metal ions (permeated) were determined using AAS and UV-Vis.

Membrane selectivity is expressed by rejection coefficient (R), also called % removal, a fraction of the concentration of a non-penetrating membrane or. Rejection coefficient (R) is defined as:

$$R = (1-\frac{Cp}{Cf}) \qquad (2)$$

where: C_p is the permeated concentration and C_f is the feed concentration.

The results of the study on the transport properties of Rhodamine B and all metal ions on the Ch/Si/P membrane are presented in Tables 2-5.

Table 2 Percentage of permeated Rhodamine B for individual and mixed solutions

Contact time (h)	Permeated Rhodamine B (%)	
	Individual	Mixed
0	0.0	0.0
8	96.0	78.0
16	95.3	78.0
24	85.0	75.0

Table 3 Percentage of permeated Pb (II) for individual and mixed solutions

Contact time (h)	Permeated Pb (II) ion (%)	
	individually	mixed
0	0.0	0.0
8	2.4	5.0
16	6.1	3.7
24	6.03	3.5

Table 4 Percentage of permeated Cu (II) for individual and mixed solutions

Contact time (h)	permeated Cu (II) ion (%)	
	individually	mixed
0	0	0
8	13.7	0.5
16	24.1	0.31
24	24.9	0.27

Table 5 Percentage of permeated Cd (II) for individual and mixed solutions

Contact time (h)	permeated Cd (II) ion (%)	
	individually	mixed
0	0	0
8	24.5	90.2
16	18.6	39.7
24	18.4	39.7

Tables 2-5 revealed that the order of permeability of the dye and metal cations is as follows: Rhodamine B > Cd (II) > Pb (II) > Cu (II). Permeated ions are the ions that pass through the membrane as a result of the equilibrium of concentration, and the size of the molecule is smaller than the pore size of the membrane. The Rhodamin B molecule successfully penetrates the Ch / Si / P membrane before it is attached to the pore wall of the membrane. Membrane selectivity can be illustrated as rejection coefficient (R). The Rejection coefficient (R) can be calculated by using equation 2. Table 2-5 is calculated the value of R can be written in Table 6-9.

Table 6 Percentage of removed Pb (II) ion for individual and mixed solutions

Contact time (h)	Removed Pb (II) ion (%)	
	Individual	Mixed
0	0.0	0.0
8	97.6	95.0
16	93.9	96.4
24	93.97	96.5

Table 7 Percentage of removed Cu (II) ion for individual and mixed solutions

Contact time (h)	Removed Cu (II) ion (%)	
	individually	mixed
0	0	0
8	86.3	99.5
16	75.9	99.7
24	75.1	99.6

Table 8 Percentage of removed Cd (II) ion for individual and mixed solutions

Contact time (h)	Removed Cd (II) ion (%)	
	individually	mixed
0	0	0
8	75,5	90.2
16	81,4	60.3
24	81,6	60.3

Table 9 Percentage of removed Rhodamine B for individual and mixed solutions

Contact time (h)	Removed Rhodamine B (%)	
	individually	mixed
0	0	0
8	4	22
16	4.7	22
24	15	25

The results displayed in Tables 6-9 showed that Cu (II) ions were the most adsorbed ions compared to Rhodamine B dyes or other ions. The adsorbed portion as a result of equilibrium concentrations in the two membrane chambers may then be ion bound by an active group present on the membrane surface as well as on the inside of the membrane. Cu (II) ions are easily bonded to membrane walls compared to Rhodamine B dyes or other metal ions. Table 6-9 also indicates that Cu (II) ions are the largest adsorbed ions compared to Rhodamin B dyes or other ions. The adsorbed portion as a result of the equilibrium concentration in the two membrane chambers then the ion may be bound by the active group present on the membrane surface as well as on the inside of the membrane. Cu (II) ions are easily bonded to membrane walls compared to Rhodamine B dyes or other metal ions.

In general, metal ions were hydrated by the presence of water to form metallic aquo, $M(H_2O)_n$ which increased the size of the moiety. By the concept of HSAB, Cu (II) ion is middle acids that tend to bind to the $-NH_2$ group which is a hard base compared to Cd (II) ion which is a softer acid. This result is consistent with previous research reported by [17] that the selectivity of the composite membrane which is capable of forming complex compounds with metal ions, was in the order of Cu (II) > Hg(II) > Zn(II) > Cd (II) > Ni(II) > Co(II), Ca(II). According to the HSAB concept, the ability to form complexes depends on the size and hardness of the metal ions. Disadvantages of membrane technology include: flux and selectivity because the membrane process occurs flux phenomenon inversely proportional to the selectivity. The higher the flux often results in decreased selectivity and vice versa.

Conclusion

In the study of adsorption of Rhodamine B dyes, Pb (II), Cu (II), and Cd (II) ions using Ch/Si/P membrane, the permeability of the membrane increases following the order of Rhodamine B > Cd (II) > Pb (II) > Cu (II). The selectivity of the membrane increases following the order of Cu (II) > Pb (II) > Cd (II) > Rhodamine B.

References

[1] Min, M., She, L., Hong, G., Zhu, M., Zhang, Y., and Wang, X., 2012, Micro-nano structure poy(ether sulfones)/poly(ethyleneimine) nano fibrous affinity membranes for adsorption of anionic dyes and heavy metal ions in aqueous solution, *Chem. Eng. J.*, 197, 88-100.

[2] Zeng, M dan Fang, Z., 2004, Preparation of a submicrometer porous membrane from chitosan/polyethylene glycol semi-IPN, *J. Membr. Sci.*, 245, 95-102.

[3] Yeh, J-T., Chin-Lai, C., dan Kuo-Shien, H., 2007, Synthesis and properties of chitosan/SiO_2 hybrid materials, *Mater. Lett.*, 61, 1292-1295.

[4] Pandis, C., Sara, M., Joana, M., Apostolos, K., Joao, F.M., and Jose, L.G.R., 2014, Chitosan-silica hybrid porous membranes, *Mater. Sci. Eng.* C 42, 553-561.

[5] Nadarajah, K., 2005, Development and Characterization of Antimicrobial Edible Film from Crawfish Chitosan. *Ph.D. dissertation,* Louisiana State University, Baton Rouge, LA.

[6] Zhang, M., Li, X.H., Gong, Y.D., Zhao, N.M., and Zhang, X.F., 2002, Properties and biocompatibility of chitosan films modified by blending with PEG, *Biomaterials.*, 23, 2641-2648.

[7] He, L-H., Xue, R., De-bin, Y., Liu, Y., and Song, R., 2009, Effects of blending chitosan with PEG on surface morphology, crystallization and thermal properties, *Chinese J. Polym. Sci.*, 27(4), 501-510.

[8] Rodkate, N., Uthai, W., Boonjira, B., dan Meta, R, 2010, semi-interpenetrating polymer network hydrogels between polydimethylsiloxane /polyethylene glycol and chitosan, *Carbohydr. Polym.*, 81, 617-625.

[9] Rutnakornpituk, M and Paradorn, N, 2006, Surface and mechanical properties of microporous membranes of poly (ethylene glycol)–poly dimethylsiloxane copolymer/chitosan, *Polym.* 47, 7909-7917.

IOP Conf. Series: Materials Science and Engineering **349** (2018) 012022 doi:10.1088/1757-899X/349/1/012022

[10] Rodriguez-Nunez, J.R., Madera-Santana, T.J., Sanchez-Machado, D.I., Lopez-Cervantes, J., and Valdez, H.S, 2014, Chitosan/hydrophilic plasticizer-based films: preparation, physicochemical and antimicrobial properties, *J. Polym. Environ.*, 22, 41-51.
[11] Neto, C.G.T., Giacometti, J.a. job, A.E., Ferreira, F.C., Fonseca, J.L.C., and Pereira, M.R, 2005, Thermal analysis of chitosan-based networks, *Carbohydr. Polym.*, 62, 97-103.
[12] Ma, G., Yang, D., Zhou, Y., Xiao, M., Kennedy, J.F., and Nie, J., 2008, Preparation and characterization of water-soluble N-alkylated chitosan, *Carbohydr. Polym.*, 74, 121-126.
[13] Martinez-Camacho, A.P., Cortez-Rocha, M.O., Ezquerra-Brauer, J.M., Graciano-Verdugo, A.Z., Rodriguez-Felix, F., Castillo-Ortega, M.M., Yepiz-Gomez, M.S., and Plascencia-Jatomea, M, 2010, Chitosan composite films: Thermal, structural, mechanical and antifungal properties, *Carbohydr. Polym.*, 82, 305-315.
[14] Reiad, N.A., Salam, O.E.A., Abadir, E.F., and Harraz, F.A, 2012, Adsorptive removal of iron and manganese ions from aqueous solutions with microporous chitosan/polyethylene glycol blend membrane, *J.Environ. Sci.*, 24(8), 1425-1432.
[15] Salehi, E and S.S. Madaeni, 2014, Influence of poly(ethylene glycol) as pore-generator on morphology and performance of chitosan/poly(vinyl alcohol) membrane adsorbents, *Appl. Surf. Sci.*, 288, 537-541.
[16] Hwang, S-T, 2011, Fundamentals of membrane transport, *Korean J. Chem. Eng.* 28(1), 1-15.
[17] Rhazi, M., J. Desbrieres., A. Tolaimate., M. Rinaudo., P. Vottero., A. Alagui., and M. El Meray, 2002, Influence of the nature of the metal ions on the complexation with chitosan. Application to the treatment of liquid waste, *Europ. Polym. J.*, 38, 1523-1530.

Supramolecular assembly of group 11 phosphorescent metal complexes for chemosensors of alcohol derivatives

H O Lintang[1*,2,3], N F Ghazalli[4,5] and L Yuliati[1,2,3]

[1] Ma Chung Research Center for Photosynthetic Pigments, Universitas Ma Chung, Malang 65151, East Java, Indonesia
[2] Department of Chemistry, Faculty of Science and Technology, Universitas Ma Chung, Malang 65151, East Java, Indonesia
[3] Centre for Sustainable Nanomaterials, Ibnu Sina Institute for Scientific and Industrial Research, Universiti Teknologi Malaysia, 81310 UTM, Johor Bahru, Johor, Malaysia
[4] Department of Chemistry, Faculty of Science, Universiti Teknologi Malaysia, 81310 UTM, Johor Bahru, Johor, Malaysia
[5] School of Fundamental Science, Universiti Malaysia Terengganu, 21030 Kuala Terengganu, Terengganu, Malaysia

E-mail: hendrik.lintang@machung.ac.id

Abstract. We report on systematic study on vapochromic sensing of ethanol by using phosphorescent trinuclear metal pyrazolate complexes with supramolecular assembly of weak intermolecular metal-metal interactions using 4-(3,5-dimethoxybenzyl)-3,5-dimethyl pyrazole ligand (**1**) and group 11 metal ions (Cu(I), Ag(I), Au(I)). Upon excitation at 284, the resulting complexes showed emission bands with a peak centered at 616, 473 and 612 nm for **2(Cu)**, **2(Ag)** and **2(Au)**, respectively. Chemosensor **2(Cu)** showed positive response to ethanol vapors in 5 mins by blue-shifting its emission band from 616 to 555 nm and emitting bright orange to green. Otherwise, **2(Au)** gave shifting from its emission band centered at 612 to 587 nm with $\Delta\lambda$ of 25 nm (41%) and color changes from red-orange to light green-orange while 2(Ag) showed quenching in its original emission intensity at 473 nm in 40% with color changes from dark green to less emissive. These results demonstrate that sensing capability of chemosensor **2(Cu)** with suitable molecular design of ligand and metal ion in the complex is due to the formation of a weak intermolecular hydrogen bonding interaction of O atom at the methoxy of the benzyl ring with the OH of the vapors at the outside of the molecules.

1. Introduction

Over the last few years, many research groups have been focused on the development of vapochromic sensor materials with high sensing capability to detect the presence of volatile organic compounds (VOCs), which are considered to be important issues due to their significant contribution to pollution [1-3]. In particular, phosphorescent d^6, d^8 and d^{10} metal compounds with extended π or coordination donor systems have been widely developed as vapochromic chemosensors with several advantages from their photophysical properties such as sensitivity of emission, Stokes shifts, excitation of single photon and longer lifetime [4-6]. Considering luminescent changes such as shifting, quenching, positive response and photoinduced energy transfer, phosphorescent metal complexes have shown great attention as vapochromic chemosensors with high tunability and rigidity of supramolecular metal-solvent, metal-

metal, π–π, hydrogen bonding and host-guest interactions [4, 7-9]. For example, Nagel and co-workers in 1988 [10] synthesized palladium (Pd) and platinum (Pt) double-complex or mixed salt for the detection of vapors. This type of complex was fabricated into fiber optics for the detection of VOCs by Lancaster 1995 [11]. In 1998, Maan and co-workers had also successfully synthesized Pt-Pt double-complex salt for vapochromic sensors of organic vapors with color, absorption and emission changes [12].

Group 11 complexes from d^{10} metal ions and various types of ligands have been investigated as phosphorescent chemosensors such as monometallic gold and bimetallic gold-silver complexes [13]. Although copper complexes can be synthesized using less expensive precursors and showed a strong luminescent intensity and longer lifetime compared other group 11 metal complexes [14,15], they have received less attention as chemosensors so far. Previously, we have reported that phosphorescent trinuclear copper(I) pyrazolate complexes from 3,5-dimethyl and 4-(3,5-dimethoxybenzyl)-3,5-dimethyl pyrazole ligands were successfully synthesized and then used for chemosensors of ethanol vapors [16]. Interestingly, compared to other non-side chain, 3,5-dimethyl, 3,5-bis(trifluoromethyl) and 3,5-diphenyl pyrazole ligands, the resulting trinuclear copper(I) 4-(3,5-dimethoxybenzyl)-3,5-dimethyl pyrazolate complex (**2(Cu)**) only showed a positive response to ethanol by shifting the emission band from 604 to 554 nm where the original intensity can be recovered without external stimuli at room temperature [17]. Considering the different metal ions from group 11 have produced metal complexes with different metal-metal distance as a sensing site [18], we report the sensing capability of trinuclear group 11 metal trinuclear pyrazolate complexes from the same ligands for detection of ethanol vapors. Indeed, we found that sensing capability of **2(Cu)** was the best chemosensors for detection of ethanol vapors compared to another group 11 metal complexes. This finding suggests that there is no significant effect of metal-metal distance on accessibility of vapors to sensing site where the vapors are more likely to interact with O atom at the methoxy of the benzyl ring with at the outside of the molecules through a weak intermolecular hydrogen interaction.

2. Experimental

2.1 Instrumentations

Fourier Transform Infrared Spectroscopy (FT–IR) was used on Thermo Scientific model a Nicolet iS50 using potassium bromide (KBr). Mass spectra were recorded by electrospray ionization (ESI) source for ionization using JEOL model a JMS–T100LP (AccuTOF LC–plus) with dichloromethane (DCM) and addition of methanol (MeOH) or acetone for direct injection in 10 μL/min at needle voltage of 2000 V. Absorption spectra were measured using Shimadzu Diffuse Reflectance (DR) Ultraviolet-Visible (UV-Vis) spectrophotometer (UV-2600). Luminescent properties were monitored on a JASCO model of FP-8500 spectrofluorophotometer. The photography images were taken using Panasonic Lumix digital camera (DMC–FZ38) under a UV lamp (Vilber Lourmat, 8 watts) in daylight or in the dark room. Photon Technology International (PTI) with a model of QuantaMaster 400 containing of near infrared (NIR) PMT detector model 914 under liquid N_2 cooling system and a Xenon lamp (75 W) was used as steady state photoluminescent spectrophotometer for the determination of emission and excitation maxima in the lifetime measurement.

2.2 Synthesis of Group 11 Trinuclear Metal Pyrazolate Complexes

Trinuclear copper(I) pyrazolate complex **2(Cu)** was prepared according to our previous reports [14-17]. As shown in Figure 1, at room temperature, the mixture of 4-(3,5-dimethoxybenzyl)-3,5-dimethyl pyrazole ligand (**1**) and tetrakis (acetonitrile) copper (I) hexafluorophosphate ($[Cu(MeCN)_4]PF_6$) in dry tetrahydrofuran (THF) was reacted for 5 mins using a Schlenk technique and followed by adding the distilled dry Et_3N under an inert condition. After overnight reaction, THF was removed from the reaction mixture under reduced pressure, and the remaining residues were dried-off to isolate the desired complexes. In–situ recrystallization was carried out using dry DCM and MeOH at room temperature to give solid powder of **2(Cu)** after dried under reducing of vacuum pressure in 85% yields. The same procedure was repeated for synthesis other metal complexes using silver(I)hexafluorophosphate ($AgPF_6$; 1.02 g, 4.06 mmol) and chloro(dimethyl sulfide) gold(I) ([$Au(SMe_2)$]Cl; 1.19 g, 4.06 mmol) to give **2(Ag)** and **2(Au)** as white powder in 75% and 95% yields [19-25].

2.3 Evaluation of Sensing Capability

Chemosensors complexes **2(Cu, Ag, Au)** in 22 mg were used for evaluation of sensing capability by directly expose of ethanol vapors (200 μL) from 0 to 5 mins. The study was carried out by arranging cell holder for spectrofluorophotometer containing chemosensors into the sealed beaker for desired exposure time and followed by evaluation of sensing capability using spectrofluorophotometer and photographs.

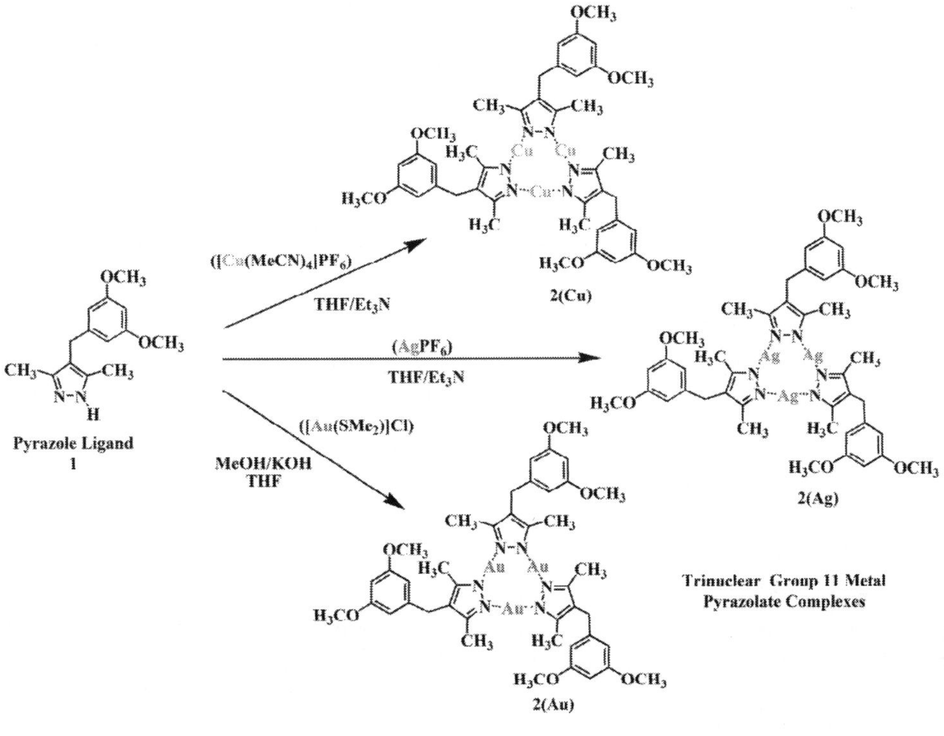

Figure 1. Schematic and molecular structures of trinuclear pyrazolate complexes **2(Cu, Ag, Au)** from pyrazole ligand **1** with different metal ions.

3. Results and Discussion

3.1 Structure elucidation of Group 11 Trinuclear Metal Pyrazolate Complexes

To confirm the synthesis of complexes **2(Cu, Ag, Au)**, FT-IR and mass spectroscopy measurements were carried out to observe characteristic vibration bands of functional groups and molecular weight compared to ligands **1**. Figure 2(i) shows FT-IR spectra of ligand **1** with characteristic vibration bands at 3201-3058 and 1750-1595 cm^{-1} for N–H stretching and bending from 1° or 2° amine groups. These vibration bands of **1** did not observe in that of **2(Cu, Ag, Au)** (Figure 2(ii-iv)) due to the formation of N–Cu–N with vibration bands at 560-832 cm^{-1}. As the results, **2(Cu, Ag, Au)** still preserved C–H, C–N and N–N stretchings of **1** with vibration bands at 2835, 1371-1289 and 1289-1151 cm^{-1} for **2(Cu)**, 2834, 1339-1291 and 1205-1156 cm^{-1} for **2(Ag)** and 2991, 1376-1289 and 1207-1155 cm^{-1} for **2(Au)**. The mass spectra showed that the observed molecular weights were 927.60, 1057.12 and 1326.29 Da for $[M+H]^+$ with molecular formula of $C_{42}H_{51}M_3N_6O_6$ (M is group 11 metal ions). Moreover, monoisotopic patterns and observed molecular weights for all complexes were almost similar to the predicted ones. Hence, all complexes were successfully synthesized.

3.2 Optical Properties of Group 11 Trinuclear Metal Pyrazolate Complexes

The luminescent properties of d^{10} complexes have been found to show phosphorescent properties with unique molecular geometries. The absorption spectra were measured to find the maximum energy for excitation of electrons from ground to excited state in the emission measurement. By using the peak top of the absorption spectra at 270, 273 and 280 for **2(Cu, Ag, Au)**, respectively as excitation wavelength (λ_{ext}), emission spectra were observed with a peak centered at 616, 473 and 612 nm for **2(Cu, Ag, Au)** as shown in Figure 2. Monitoring at maxima emission wavelength (λ_{em}), it provided the excitation spectra for maximum absorption of light with a peak centered at 274, 280 and 280 nm for **2(Cu, Ag, Au)**, which were closed to the peak top of absorption of luminophores. These results were in good agreement with their colors as green, orange and red under exposure to a UV hand lamp at 254 nm in the dark room (Figure 3d), respectively. The calculated Stokes shift ($\Delta\lambda$) was 338, 193 and 332 nm with phosphorescent lifetime (τ) of 8.16±1.8, 7.84±0.7 and 8.91±0.7 μs and at room temperature, indicating the characteristics of phosphorescent compounds from metal-metal interaction [14-18].

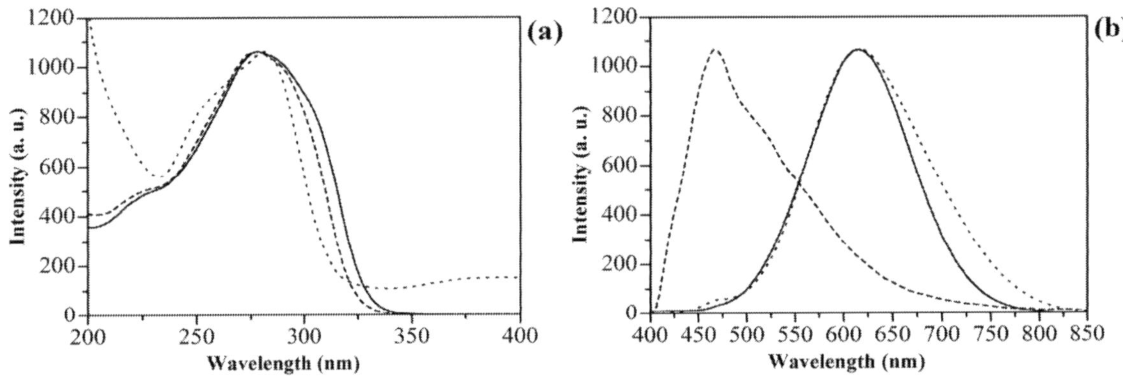

Figure 2. (a) Normalized excitation and (b) emission spectra of **2(Cu)** (solid line), **2(Ag)** (dash line) and **2(Au)** (dot line).

3.3 Evaluation of Sensing Capability

Sensing capability of group 11 pyrazolate complexes was studied using ethanol vapors in the closed system for vapochromic chemosensors of alcohol derivatives. As reported in our paper [17], upon exposure to ethanol vapors in 5 mins until no significant shifting in emission wavelength and color, chemosensor **2(Cu)** was found to show shifting phenomenon of the emission band centered at 616 to 554 nm with $\Delta\lambda$ of 62 nm with excitation at 270 nm (Figure 3a) which was in good agreement with its changes in color from bright orange to green (Figure 3e). Interestingly, when chemosensor **2(Ag)** was utilized for the same VOC, the emission band centered at 473 nm was significantly quenched in 40% upon excitation at 273 nm (Figure 3b) of its original intensity. This phenomenon was well supported with its color changes from dark green to less emissive (Figure 3f). On the other hand, an emission band centered at 612 nm for chemosensor **2(Au)** was shifted to 587 nm upon excitation at 280 (Figure 3c) with $\Delta\lambda$ of 25 nm. The color of chemosensor **2(Au)** was changed from red-orange to light green-orange (Figure 3g). Hence, chemosensor **2(Cu)** provides the best sensing capability for detection of ethanol vapors compared to others. The proposed mechanism for the shifting phenomenon was discussed in the previous report [17], in which the vapors is more likely to use a weak intermolecular hydrogen bonding for interaction with electronegative O atom at the methoxy of the benzyl ring at the outside of the molecules. This shifting phenomenon was also observed for **2(Au)** with small $\Delta\lambda$ due to a little bit shorter metal-metal distance compared to **2(Cu)**. On the other hand, **2(Ag)** with longer metal-metal distance tends to provide accessibility of vapors to its sensing site for quenching of emission intensity until certain level. Such sensing capability of chemosensor **2(Ag)** might possible to occur from the competition of vapors to Ag(I)–Ag(I) interaction and π–π interactions.

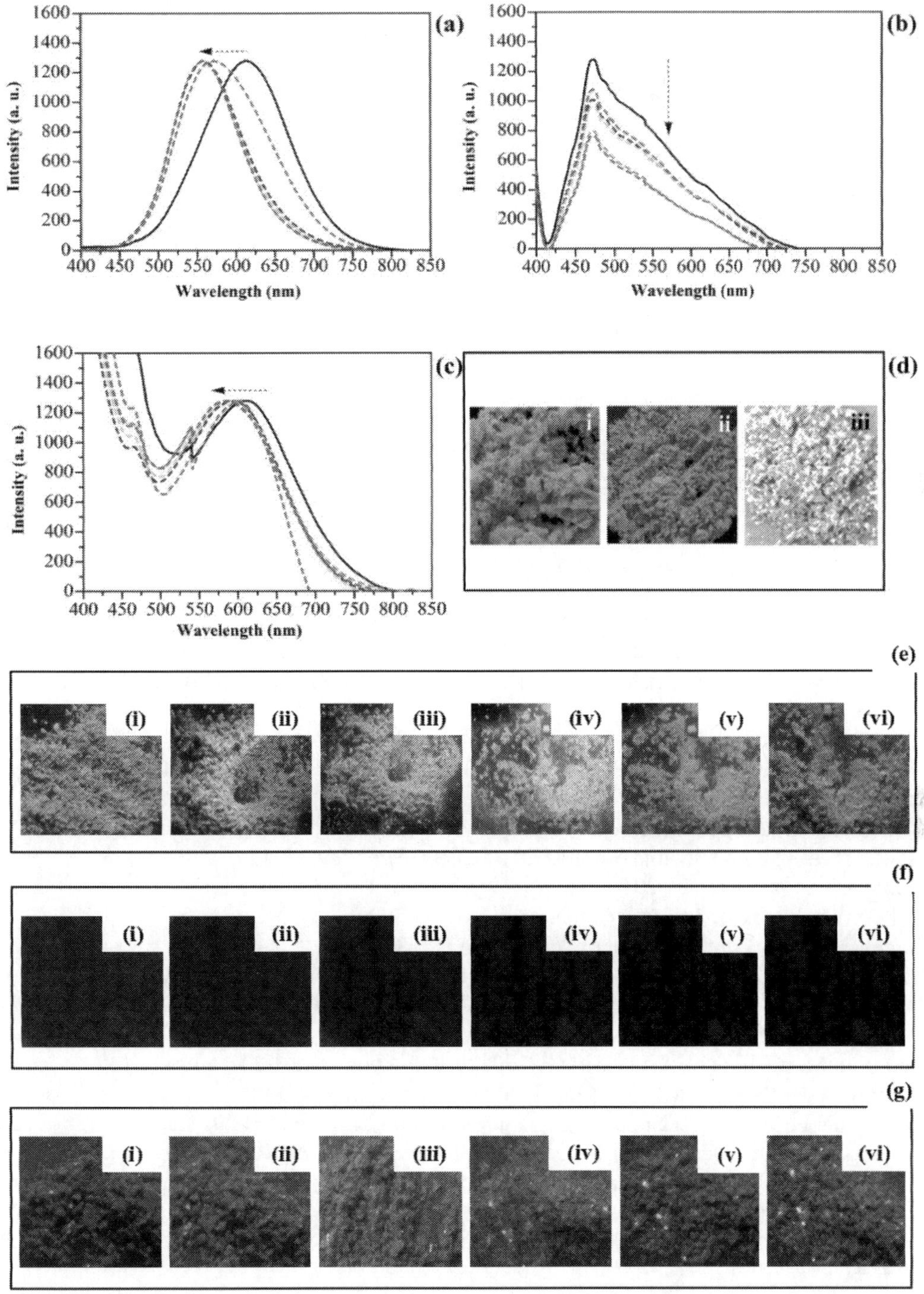

Figure 3. Emission spectral changes of (a) chemosensors **2(Cu)**, (b) **2(Ag)** and (c) **2(Au)** upon exposure to ethanol from 0 (black solid line) to 5 mins (red, blue, green, orange and pink dash line, respectively) and (d) their photography images of chemosensors **2(Cu)** (i), **2(Ag)** (ii) and **2(Au)** (iii) before exposure to ethanol. The photography images of (e) chemosensors **2(Cu)**, (f) **2(Ag)** and (g) **2(Au)** upon exposure

to ethanol for (i) 0, (ii) 1, (iii) 2, (iv) 3, (v) 4 and (vi) 5 mins. The photographs were taken at room temperature on exposure to a UV hand lamp at 254 nm in the dark room.

4. Conclusion

In conclusion, we have successfully synthesized trinuclear metal pyrazolate complexes **2(Cu**, **Ag**, **Au)** from 4-(3,5-dimethoxybenzyl)-3,5-dimethyl pyrazole ligand **1** in 85%, 75%, and 95% yields. All complexes showed characteristic phosphorescent properties with emission bands centered at 616, 473 and 612 nm, large Stokes shifts and luminescent lifetime decay in the range of microseconds, which were in good agreement with their colors from green, orange and red, respectively. Upon exposure to ethanol vapors, chemosensors **2(Cu)** showed the best sensing capability with shifting of its emission center compared to others due to the formation of a weak intermolecular hydrogen bonding interaction with vapors at the O atom of the methoxy at the benzyl ring. By using molecular design of ligand and metal ion in the synthesis of complexes, the phosphorescent chemosensors with high sensing capability can be prepared for the detection of VOCs in next future.

5. Acknowledgements

The authors thank Centre for Sustainable Nanomaterials, Ibnu Sina Institute for Scientific and Industrial Research (ISI-SIR), Universiti Teknologi Malaysia (UTM), Malaysia for characterization facilities. The authors also thank consultancy project from RGS Corporation Sdn. Bhd. through Uni-Technologies Sdn. Bhd. with vote no. 1079 for the financial supports.

References

[1] Jiang H, Wang Y, Ye Q, Zou G and Zhang Q 2010 *Sens. Actuators B Chem.* **143** 789
[2] Yu L, Song H, Tang Y, Zhang L and Lv Y 2014 *Sens. Actuators B Chem.* **203** 726
[3] Dubar A D F, Brittle S, Richardson T H, Hutchinson J and Hunter C A 2010 *J. Phys. Chem. B* **114** 11697
[4] Zhao Q, Li F, and Huang C 2010 *Chem. Soc. Rev.* **39** 3007
[5] Ma D-L, Ma V P-Y, Chan D S-H, Leung K-H, He H-Z and Leung C-H 2012 *Coord. Chem. Rev* **256** 3087
[6] You L, Zha D, and Anslyn E V 2015 *Chem. Rev.* **115** 7840
[7] Xu Z, Chen X, Kim H N and Yoon J 2010 *Chem. Soc. Rev.* **39** 127
[8] Kobayashi A and Kato M 2014 *Eur. J. Inorg. Chem.* **27** 4469
[9] Bell T W and Hext N M 2004 *Chem. Soc. Rev.* **33** 589
[10] Nagel C C 1998 European Patent Application EP, 277003.
[11] Lancaster G D, Moore G A, Stone M L and Reagen W K 1995 U.S. Patent 5, 445795.
[12] Mann K R, Daws C A, Exstrom C L , Janzen D E and Pomije M 1998 U.S. Patent 5, 766952.
[13] Wenger O S 2013 *Chem. Rev.* **113** 3686
[14] Enomoto M, Kishimura A and Aida T 2001 *J. Am. Chem. Soc.* **123** 5608
[15] Kishimura A, Yamashita T and Aida T 2005 *Nature Mater.* **4** 546
[16] Ghazalli N F, Yuliati L, Endud S, Shamsuddin M and Lintang H O 2014 *Adv. Mater. Res.* **970** 44
[17] Lintang H O, Ghazalli N F and Yuliati L 2017 *Indones. J. Chem.* **17(2)** 191
[18] Omary M A, Rawashdeh-Omary M A, Gonser M W A, Elbjeirami O, Grimes T and Cundari T R 2005 *Inorg. Chem.* **44** 8200
[19] Lintang H O, Kinbara K, Tanaka K, Yamashita T and Aida T 2010 *Angew. Chem. Int. End.* **49** 4241
[20] Lintang H O, Kinbara K, Yamashita T and Aida T 2010 *Int. Conf. on ESciNano 2010 – Proceedings 2010, IEEE proceedings (Kuala Lumpur, Malaysia, 2010),* Article number 5700970
[21] Lintang H O, Kinbara K, Tanaka K, Yamashita T and Aida T 2012 *Chem. Asian. J.* **7** 2068

[22] Lintang H O, Kinbara K, Yamashita T and Aida T 2012 *Int. Conf. on ESciNano 2012 – Proceedings 2012, IEEE proceedings (Johor, Malaysia, 2012),* Article number 6149684
[23] Lintang H O, Yuliati Y and Endud S 2014 *Mater. Res. Innov.* **18**, S6-444
[24] Jalani M A, Yuliati Y, Endud S and Lintang H O 2014 *Adv. Mater. Res.* **925**, 233
[25] Jalani M A, Yuliati Y and Lintang H O 2014 *Jurnal Teknologi* **70**, 131

IOP Conf. Series: Materials Science and Engineering **349** (2018) 012024 doi:10.1088/1757-899X/349/1/012024

Chemical characteristics and fatty acid profile of butterfly tree seed oil (*Bauhinia purpurea* L)

H Soetjipto*, C A Riyanto* T Victoria*

Universitas Kristen Satya Wacana, Salatiga, Indonesia

E-mail: hartati.sucipto@staff.uksw.edu

Abstract. Butterfly tree (Kachnar) in Indonesia is only used as ornamental plants in garden, park, and roadsides. The seed of Butterfly tree was extracted with n-hexane and physicochemical properties were determined based on Standard Nasional Indonesia (SNI) 01-3555-1998 while the oil chemical composition was determined using GC-MS. The result showed that yield of the oil as 57.33±1.14 % (w/w) and the chemical characteristic of seed oil include acid value (13.7.8±0.23 mg KOH/g) saponification value (153.32±1.85 mg KOH/g), peroxide value (43.51±0.57. mg KOH/g). The butterfly tree seed oil showed that linoleic acid (28.11 %), palmitic acid (29.2%), oleic acid (19.82%) and stearic acid (10.7.4 %) were the main fatty acids in the crude seed oils. Minor amounts of neophytadiena and arachidic acid were also identified.

1 Introduction

To fill the need of world vegetable oil and fat which is continually increased. Moreover is predicted reach 236 million of ton at 2020 [1] while for this moment palm oil is still used as an important source of vegetable oil especially in Indonesia. ***Indonesian Palm Oil Producers Association (Gapki) & Indonesian Ministry of Agriculture*** said that around 81% of Indonesian palm oil products in 2015 was exported to other countries. As we know that palm oil plantation destroys the tropical forest in Indonesian national park and gives a lot of bad influence to the environmental. So that very important for Indonesia to look for new alternative plants for support as vegetable oil sources.

Butterfly tree (*Bauhinia purpurea L.*) is a small evergreen ornamental plant in Indonesia, but this plant has already cultivated in Pakistan [2] The mature seeds and young pods also can be eaten in Indian native countries [3]

Dewi et al (2014) analyzed the composition of butterfly tree seed oil, in the total lipids, free fatty acids were present [4]. The lipid composition including neutral lipids (94.91%), phospholipids (1.86%) and glycolipids (3.23%) . The major fatty acid composition of neutral lipids were linoleic acid (51.32%), palmitic acid (29.31%) [3-4].

Purification oil was needed to reduce unlikely aroma, taste, and color, especially for vegetable oil because of the presence of mucous, gum, ash, minerals, free fatty acids, sterol, hydrocarbon, mono and diglyceride and pigment. [5-6]

De-gumming, neutralization, bleaching, and deodorization were purification steps which were done to vegetable oil. Except for food industrial, usually only de-gumming and neutralization were done. Phosphoric acid (H_3PO_4) was used for washing vegetable oil in de-gumming step because phosphoric acid can change nonhydratable fosfatide to be hydratable and follow by water washing. Neutralization is important to do and using sodium hydroxide which is easy to find and cheap. The alkali concentration was used influence to the amount of emulsion in saponification process [6]. The objective of the present study was to obtain knowledge about the chemical characteristic and fatty acid profile of butterfly seed oil before and after purification method.

2 Material and method

2.1. Sample preparation and Extraction process (modified from [4])

The dry Butterfly tree seeds (100 g) was extracted by Soxhlet method for 6 hours in n-hexane (300 ml). The butterfly tree seed oil was obtained after all the solvent removed by rotary-evaporation.

2.2. Physicochemical Property Assays for the crude Butterfly tree seed oil [7]

Important physicochemical properties of the crude oil comprises moisture, density, acid value saponification value, peroxide value was determined according to SNI 01-3555-1998. While the aroma and color of the oil were determined as descriptive.

2.3. Moisture and Density of Butterfly tree seed oil

Moisture of Butterfly tree seed oil (1.0 g) was measured by using Moisturizer balance (Ohaus, MB 150) while the density (1.0 ml) was measured by analytical balance (Mettler H80. Mettler Instrument Corp., USA).

2.4. Acid value and acidity (SNI 01-3555-1998).

The amount of 2.0 g oil was added 50 ml ethanol 95% and 3 -5 drops phenolphthalein indicator, then was titrated by KOH 0.1 N until pale pink color.

$$\text{Acid value} = \frac{V \times T \times 56.1}{m} \qquad \text{FFA (\%)} = \frac{M \times V \times T}{10\ m}$$

V = Volume in ml standard KOH used

T = Normality of standard KOH

m = weight in gram of the sample

M = Molecular mass of dominant fatty acid (expressed as oleic acid)

2.5. Saponification value (SNI-3555-1998)

The amount of 2.0 g oil was added 25 ml KOH 0.5M, refluxed for 1 hour. After the sample have cooled the solution was added 0.5 ml phenolphthalein indicator and titrated by HCl 0.5M until the pink color disappeared

$$\text{Saponification value } \left({}^{\text{mg KOH}}/_{\text{g oil}}\right) = \frac{56.1 \times T \times (V0 - V1)}{m}$$

V_0 = Volume in standard ml HCl 0.5 M used as a blank.

V_1 = Volume in standard ml HCl 0.5 M used as sample

T = Normality of standard HCl 0.5 M

m = weight in gram of the sample

2.6. Peroxide Value (SNI-3555-1998)

The amount of 0.3 g oil was added mixture of 55ml chloroform, 20 ml glacial acetic acid and 25 ml ethanol 95 %. Then 1 g KI was added to the solution and kept in the dark for 30 minutes. The next step 50 ml water-distillate free CO_2 was added to the mixture. The excess of Iod in the solution was titrated by $Na_2S_2O_3$ 0.02 N with a starch solution as indicator.

$$\text{Peroxyde value (mgrek/kg)} = \frac{(V1\text{-}V0) \times T \times 1000}{m}$$

V0 = Volume in ml standard sodium thiosulfate for blank

V1 = Volume in ml standard sodium thiosulfate for sample

T = Normality of standard sodium thiosulfate solution

m = weight in gram of the sample

2.7. *Fatty Acid Composition*

The fatty acid composition of butterfly tree seed oil was analyzed according to IUPAC method 2.302 [8] And the analysis of methyl esters was performed on *Gas Chromatography-Mass Spectrometry* (GCMS-QP2010 SE Shimadzu) equipped with a flame ionization detector (FID) capillary column AGILENTTJ%W DB-(30 m x 0.25mm.). The column, injector, and detector temperatures were set at 80°C, 300 °C, and 300°C respectively. The flow rate of carrier gas Helium with split ratio 7.3.0was set as 3.0 mL/min. The fatty acids were identified concerning the retention times of standard fatty acid methyl ester performed at the same condition.

2.8. *Purification*

2.8.1. *Degumming process*

Butterfly tree seed oil was heated at 7.0°C then added with 0.2% (v/w) solution of H_3PO_4 20%, followed by stirring for 10 minutes at constant temperature. By using separating funnel, the impurities can be separated from the oil. Then the oil was washed with warm water (60°C) until pH of the washing water was neutral.

2.8.2. *Neutralization Process*

After the degumming process, oil is going to neutralization step. Oil was heated at 7.0-7.5°C, then add with 0.2% NaOH solution 0.3 N (0.87. ml), followed by stirring for 15 minutes then washed to neutral by warm water 60°C.

2.8.3. *Statistical analysis*

All determinations were carried out in triplicate and their mean value (± standard deviation) is given.

3 Result and Discussion

3.1. *The physicochemical properties of butterfly tree seed oil*

Table 1 showed the physicochemical properties of butterfly tree seed oil before and after purification.

Table 1. Physicochemical properties of butterfly tree seed oil before and after purification

Parameter	Crude oil	After Purification
Rendemen	14.18 %	8.12 %
Density (g/cm$^{3)}$	0.86 ±0.02	0.82 ±0.006
Color	Brownish yellow	Yellow
Aroma	specific	Specific
Acid value (mg KOH/g)	13.7.8±0.23	3.87.±0.19
Peroxide value (mequiv O2/kg)	43.51±0.57.	15.62±0.22
Saponification value (mg KOH/g)	153.32±1.85	187.,39±0.47.

The refining process for vegetable oil is necessary to do to eliminate the impurities from crude oils especially the phosphatides or so-called gums (degumming process). It was started with degumming then follow by neutralization, bleaching, and deodorization. For nonfood industrial consumption, the third and fourth steps (bleaching and deodorization) is not done. This process aims to neutralize water degummed or partially purified oils by using NaOH (alkali neutralization).

Seeds of *B. purpurea* (100 g) contained 14.18 % (w/w) of oil. This seed oil content is not quite different with previously reported [4]. The density of this oil 0.86 ±0.02 g/cm^3 similar to other edible oil [9] The acid value is an important variable for the oil quality because it showed the content of free fatty acid inside the oil. The lower the free fatty acid, the better the quality of oil [10] Peroxide value is indicated by the oxidative state of oils. The peroxide value of butterfly tree seed oil is relatively high 43.51±0.57. (mequiv O2/kg). Oils easy become rancid when the peroxide value ranges from 20 to 40 mequiv O_2/kg of oil. Saponification value is expressed as the number of mg KOH required to saponify 1 g of sample [11]. Saponification value of Butterfly tree seed oil indicates that the oil contains high molecular weight fatty acids on average 153.32±1.85 (mg KOH/g) or very high proportion of low molecular weight triacylglycerols.

The result showed that purification step increases the quality of the oil. The colour of the oil to be clean, bright yellow and reduced acid value (13.7.8±0.23 to 3.87.±0.19 mg KOH/g) and peroxide value (43.51±0.57. to 15.62±0.22 mequiv O_2/kg) whereas increased saponification value (153.32±1.85 to 187.,39±0.47. mg KOH/g). As a comparison with another species of Bauhinia, *B.variegata* seed oil showed the level of Free Fatty Acid (FFA) *B.variegata* 0.6 mgKOH/g while 0.9 mgKOH/g sample for *B.linnaei* [3]. The lower acid value, the more stable of oil. A low acidity and peroxide value indicate a good quality and stability. The oils which have high peroxide values are not so stable and easily become rancid with an undesirable odor.

3.2 Fatty acid profile of butterfly tree seed oil

The fatty acid composition is presented in Table 2. The 4 major fatty acids of crude butterfly tree seed oil were linoleic acid (28.11%), palmitic acid (25.42%), oleic acid (19.82%), stearic acid (10.7.4%) and palmitic acid (3.7.8%). Another compound (6 compounds) which are each compounds less than 3%.

After purification, the formation of fatty acid profile of butterfly tree seed oil changed. Linoleic acid, palmitic acid, and stearic acid were reduced to be 16.00%, 22.60%, and 7.23% respectively, whereas oleic acid increased to 25.53%. As a comparison, the major fatty acids in butterfly tree seed oil obtained is similar to Arain et al., [3] A very low acidity in oil indicate is good quality and stability [12]. The difference between species, or the same species but come from different places which have a different climate, soil or environment will influence the quality of the plants. The phytochemical variations and content of the compounds of species is dependent on geographical location. Purification process also influenced the fatty acid profile but only in content not in various kind of fatty acid.

Table 2. Fatty acid composition of crude and purified butterfly tree seed oil

No	Fatty acids	Content (%)		
		This study		Arain et al. [3]
		Crude oil *B.purpurea*	Purified oil *B.purpurea*	*B.variegata*
1	Linoleic acid (C18:2)	28.11	16.00	42.1±1.8
2	Palmitic acid (C16:0)	29.20	22.60	22.1±1.5
3	Oleic acid (C18:1)	19.82	25.53	0.5±0.1
4	Stearic acid (C18:0)	10.7.4	7.32	17.5±1.7.
5	Arachidic acid	1.17.	-	

IOP Conf. Series: Materials Science and Engineering **349** (2018) 012024 doi:10.1088/1757-899X/349/1/012024

4 Conclusions

The physicochemical of butterfly seed oil and its fatty acid profile were influenced by degumming and neutralization process (purification). The moisture, density, acid value, saponification value and peroxide value of bauhinia tree seed oil were changed after purification. This oil could be a good source of natural oil rich in linoleic acid and oleic acid. Purification method reduced the content of linoleic acid but increase the oleic acid

References

[1] Amri Q. (2014). 2020 Kebutuhan Minyak Nabati Dunia bergantung kepada CPO Indonesia [online]. 20 September 2014.

[2] Arain, S.T.H . Sherazi, M. I. Bhanger, F. N. talpur and S. A. Mahesar, *Thermochemica Acta*, 484 (2009)3.

[3] Arain S., Najma M, Muhammad TR, Syed THS. Muhammad I B and Sarfaraz A M. 2012. Physico-chemical Characteristics of Oil and Seed Residues of *Bauhinia variegata* and *Bauhinia linnaei.* Pak. J. Anal. Environ. Chem. Vol.13. No.1 (2012) 16-21.

[3] N. Rajaram and K. Janardhanan, J. Sci. Food. Agric., 55 9 1991) 431.

[4] Dewi, E.M.K (2014). Karakterisasi dan Komposisi Kimia Minyak Biji Tumbuhan Kupu-kupu (*Bauhinia purpurea* L) Bunga Merah Muda. Prosiding Seminar Nasional Sains dan Pendidikan Sains IX. Fakultas Sains dan Matematika, UKSW. Vol 5. No 1. hal 11-17. Salatiga. 21 Juni 2014

[5] Djatmiko dan Ketaren,1985. Pemurnian Minyak Makan . Bogor. Jurusan Teknologi Industri Pertanian., FATETA, IPB: Agroindustri Press.

[6] Ketaren, S. (1986). Pengantar Teknologi Minyak dan lemak Pangan. Jakarta; UI – Press

[7] BSN. Cara Uji Minyak dan Lemak : SNI 01-3555-1998. Jakarta: Badan Standardisasi Nasional. 1998

[8]Paquot C, Hauntfenne A, 1987. IUPAC Standard methods for the analysis of oils, fat and derivatives. Blackwell, London.

[9] Bello EI, Agg M. 2012 Biodiesel production grom ground nut oil. J. Emerging Trends in Engineering and Appl. Sci. **3.** 27.6-280.

[10] Ibeto CN, Okoye COB, Ofoefule AU. 2012. Comparative Study of the Physicochemical Characterization of some oils as Potential Feedstock for Biodiesel Production. *Renew Energy Article* ID 621518.

[11] Nayak BS, Patel KN. 2010. Physicochemical characterization of seed and seed oil of *Jatropha curcas* 1 collected from Bardoli (South Gujarat) (Ciri-ciri Fizikokimia Biji dan Minyak Biji *Jatropha curcas* L. Bardoli (Selatan Gujarat) Sains Malays 39. 951-955

[12] Norman OVS, Composition and Characteristic of individual fats and Oils, 1997.in Baileys Industrial Oil and Fats Products (ed. D. Swern). 4 th, Wiely, New York, pp 289-459

IOP Conf. Series: Materials Science and Engineering **349** (2018) 012025 doi:10.1088/1757-899X/349/1/012025

Synthesis and Characterization of Diranitidinecopper(II) Sulfate Dihydrate

H Syaima[1], S B Rahardjo[2], and I M Zein[2]

[1]Chemistry Graduate Program, Graduate School, Universitas Sebelas Maret, Surakarta, 57126, Indonesia
[2]Department of Chemistry, Faculty of Mathematics and Natural Sciences, Universitas Sebelas Maret, Surakarta, 57126, Indonesia

E-mail: husna_syaima@yahoo.com

Abstract. The complex of ranitidine with Cu(II) has been synthesized in 1:2-mole ratio of metal to the ligand in water. The forming of the complex was indicated by shifting of maximum wavelength from 816 nm ($CuSO_4{\cdot}5H_2O$) to 626 nm (the complex). Infrared spectra indicated NO_2 and NH functional group were coordinated to Cu(II). The percentage of copper in the complex measured by Atomic Absorption Spectroscopy (AAS) analysis was 7.5% indicating that formula of the complex was $Cu(ranitidine)_2SO_4(H_2O)_n$ (n=2, 3 or 4). The electrical conductivity of Cu(II) complex in water was 71.0 Scm^2mol^{-1} corresponding to 1:1 electrolytes. Thermogravimetric/Differential Thermal Analysis (TG/DTA) showed the presence of two molecules of H_2O in the complex. UV-Vis spectra showed a transition peak on 15974 cm^{-1} indicating square planar geometry. The complex was paramagnetic with μ_{eff} 1.77 BM. The proposed formula of the complex was $[Cu(ranitidine)_2]SO_4{\cdot}2H_2O$.

1. Introduction

Study and synthesis of a metal complex with various drugs as ligands become considerable research field due to the synergistic action of both ligands and metals to improve the activity of the drugs [1-11]. The work and activity of the drug an enhancement after converted to chelate transition metals that turned out to be better than using only organic compounds [12-16]. Famotidine (fam) and cimetidine (cim) are anti-ulcer drugs and able to function as ligands because they have several free electron pairs. Famotidine and cimetidine have been reported to form complexes with Cu(II) ion [17-18]. In the tetragonal [Cu(fam)] complexes, N-guanidine (N3), N-thiazole (N9), N-amidine (N16) and S groups in the thioether are coordinated on Cu^{2+} ions [17]. Thus, the Cu^{2+} ion is coordinated by the cimetidine ligand through the -C=N group and the -CS group forming the square planar $[Cu(cim)_2]$ complex [18].

Another anti-ulcer compound that has a potentiality to form complexes with a metal ion is ranitidine. Ranitidine hydrochloride is a histamine H_2-receptor antagonist with a furan ring structure that increases its potency to inhibit gastric acid secretion induced by various stimuli while lacking the anti-androgenic and hepatic microsomal enzyme inhibiting effects [19]. The ranitidine structure showed in figure 1, has five donor atomic groups, i.e. -CN, -CS, -NH-, -CO, and $-NO_2$ which can be coordinated on metal ions.

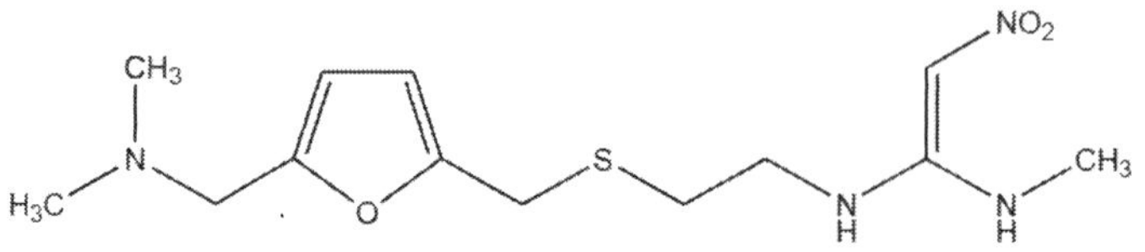

Figure 1. Structure of ranitidine

It is reported that ranitidine (ran) and glycine (gly) forming a complex with zinc. The complex of $[Zn(ran)(gly)]NO_3$ is shown in figure 2. The -NH- and $-NO_2$ groups of ranitidine together with the -COO- and $-NH_2$ groups of glycine are coordinated to Zn^{2+} ions form tetrahedral geometry [20].

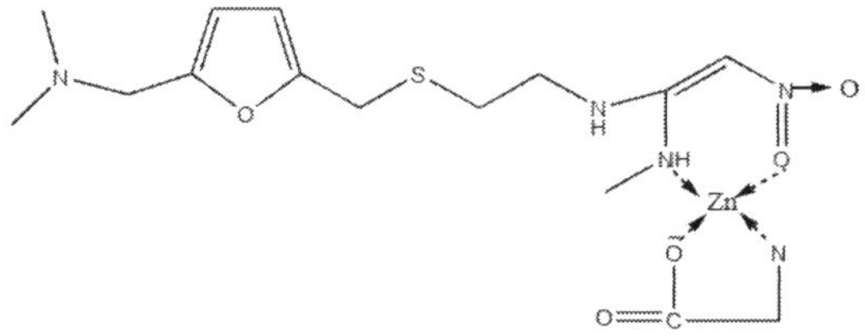

Figure 2. Structure of $[Zn(ran)(gly)]NO_3$ complex

The metal complex is one of the compounds that can be used for various applications such as science in biology, clinics, analytics and renewable science. In this paper, we use Cu(II) as metal central ions due to the effect of increasing of work activity as anti-ulcer drug compared to its free drug as ligand when complexed with Cu(II) such as in $[Cu(cim)_2]$ [21]. The Cu^{2+} ion has a $3d^9$ configuration with one unpaired electron, able to bind to a ligand that has a free electron pair. The new complex of copper with ranitidine ligand and its properties are then characterized by UV-Visible (UV-Vis) spectroscopy, Atomic Absorption Spectroscopy (AAS), molar conductivity, Thermogravimetric/Differential Thermal Analysis (TG/DTA), infrared spectroscopy, and magnetic susceptibility measurement.

2. Experimental

2.1. Materials

The chemicals and solvents were of reagent grade and used without further purification. All chemicals such as $CuSO_4{\cdot}5H_2O$, ranitidine hydrochloride, NH_4OH, $NiSO_4{\cdot}6H_2O$, $FeSO_4{\cdot}7H_2O$, $CuCl_2{\cdot}2H_2O$, $FeCl_2{\cdot}4H_2O$, $FeCl_3{\cdot}6H_2O$, $AlCl_3{\cdot}6H_2O$, and nitric acid were purchased from E. Merck.

2.2. Physical measurements

Spectra UV-VIS of metal complex and ligand was recorded in methanol solvent using UV-Vis Double Beam Shimadzu PC 1601 spectrophotometer. The copper content was determined by Atomic Absorption Spectrometer (AAS) Shimadzu AA-6650. Infrared spectra were recorded on Prestige-21 Shimadzu spectrophotometers as KBr pellets in the frequency range of 4000–450 cm^{-1}. Molar conductivity (Λ*m) of 1 mM solution in methanol was measured on Jenway CE 4071 conductivity meter at 25 °C. The presence or absence of H_2O molecules in the complex was estimated from the results of thermal analysis using Differential Thermal Analyzer Shimadzu 50. The magnetic moment was measured using Auto Sherwood Scientific 10169 Magnetic Susceptibility Balance.

2.3. Synthesis of Cu(II) complex

Ranitidine hydrochloride (2.10 g, 6 mmol) was dissolved in 15 ml of distilled water plus NH_4OH to reach pH 7.5 then mixed with 5 mL distilled water solution of $CuSO_4{\cdot}5H_2O$ (0.249 g, 3 mmol). The mixture was then refluxed for 2 hours while stirring. After 24 hours, obtained precipitation was filtered with filter paper and dried in a desiccator.

3. Result and Discussion

3.1. Formation of the complex

Figure 3 shows the shift of the maximum wavelength of the Cu(II)-ranitidine complex from 810 nm ($CuSO_4$ solution) to 626 nm (Cu-ranitidine complex). This large wavelength shift (185 nm) shows that ranitidine is a strong ligand.

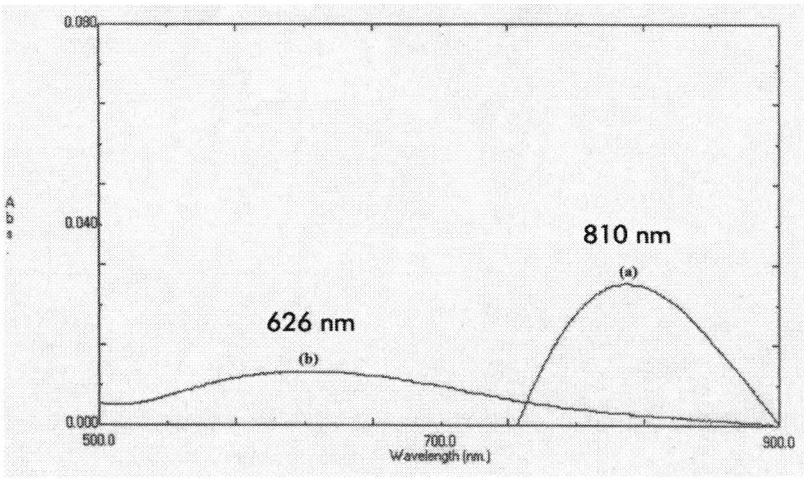

Figure 3. Electronic spectra (a) $CuSO_4{\cdot}5H_2O$ and (b) Cu(II)–ranitidine complex in water with a mole ratio of metals and ligands 1: 2.

3.2. Analysis of the amount of copper with Atomic Absorption Spectroscopy (AAS)

The copper content analyzed by AAS in the Cu(II)-ranitidine complex was 7.54 ± 0.45%. If the result of the measurement is compared with theoretical analysis shown in table 1, it can predict the formula of Cu(II)-ranitidine. The different value of copper content based on theoretic and experiment generally happen [22-24]. It needs more methods to confirm the number of presence of water molecules, such as thermogravimetry and differential thermal analysis. Therefore, the formula of the formed complex is $Cu(ranitidine)_2SO_4(H_2O)_n$ (n=2, 3 or 4).

Table 1. Percentage of copper in the complex theoretically.

Empirical Formula	Mr	% Cu
$Cu(ran)_2(SO_4)(H_2O)_2$	824.5	7.70
$Cu(ran)_2(SO_4)(H_2O)_3$	842.5	7.53
$Cu(ran)_2(SO_4)(H_2O)_4$	860.5	7.37

3.3. Analysis of electrical conductivity

The results of standard electrical conductivity measurements and complex in water are shown in table 2. If the electrical conductivity results of the Cu(II)-ranitidine complex compared to the standard electrical conductivity of the solution, the conductivity of the Cu(II)-ranitidine complex is close to the electrical conductivity of $CuSO_4{\cdot}5H_2O$ and $NiSO_4{\cdot}6H_2O$. The value of the measurement is consistent with the 1:1 electrolytic nature of the complex, indicating that the SO_4^{2-} ions are not coordinated to Cu^{2+}.

Table 2. Molar conductivity of metal salts and Cu(II)-ranitidine in water

Solution	Λ_m (S cm^2 mol^{-1})	Cation:anion Charge
Water	0	-
$CuSO_4 \cdot 5H_2O$	87 ± 1	1:1
$NiSO_4 \cdot 6H_2O$	81 ± 3	1:1
$FeSO_4 \cdot 7H_2O$	103 ± 1	1:1
$CuCl_2 \cdot 2H_2O$	134 ± 0.5	2:1
$FeCl_2 \cdot 4H_2O$	173 ± 1	2:1
$FeCl_3 \cdot 6H_2O$	286 ± 1	3:1
$AlCl_3 \cdot 6H_2O$	225 ± 0.3	3:1
Cu(II)-ranitidine	**71.04 ± 3**	**1:1**

3.4. Thermal analysis

TG/DTA showed the peak of endotherm at temperature 99.08 °C, shown in figure 4. Thus, at that temperature also happened the reduction of 4% compound mass at 80.05-99.72 °C showing the evaporation of two molecules of lattice water. It is reported that decomposition step at 60-100 °C is assignable to the removal of co-crystallized water molecules [25-26]. Therefore, the possible structure is $[Cu(ranitidine)_2]SO_4 \cdot 2H_2O$.

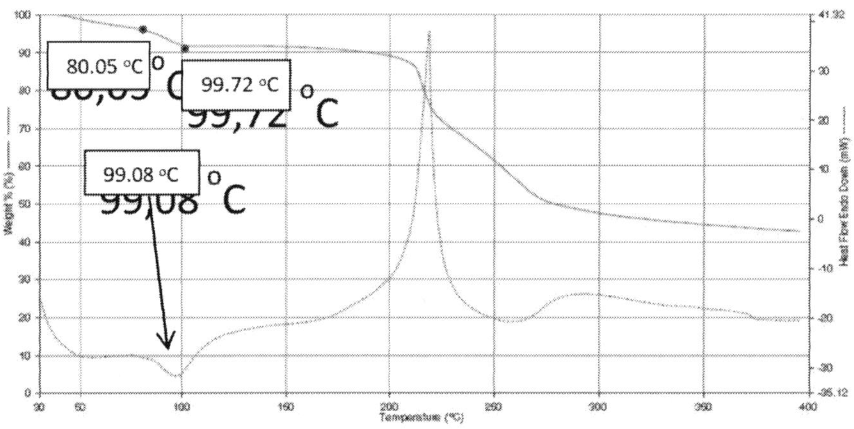

Figure 4. Thermogravimetric/differential thermal analysis spectra

3.5. Infrared analysis

Figure 5 and table 3 shows that the uptake of the -NH groups in the complex (3157 and 1519 cm^{-1}) has shifted toward a smaller wave number than the free ligand absorption of 3255 and 1568 cm^{-1}. It indicates coordination of amine group to Cu(II). The uptake of the NO_2 group in the complex (1620 and 1381 cm^{-1}) undergoes a shift from the free ligand (1631 cm-1 and 1346 cm^{-1}), indicating the presence of the coordinated NO_2 group with the metal. This case is similar to the $[Zn(ran)(gly)]NO_3$ complex which ranitidine is coordinated to Cu(II) through -NH and -NO_2 group [20]. Cu-O vibration appears at wave numbers 436 cm^{-1} and Cu-N appears at wave number 599 cm^{-1}. It is same with the complex of Cu(2-hydroxy-4-butoxy aniline) with Cu-O bonding vibrations occur in 447 cm^{-1} while Cu-N (617 cm^{-1}) [27]. No absorption band of -OH was a presence due to the overlapping of -OH and N-H absorption.

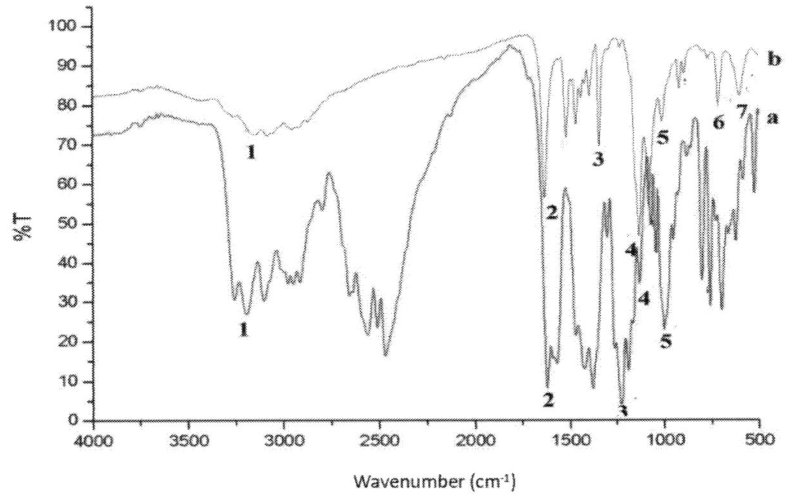

Figure 5. Infrared spectra of ranitidine hydrochloride (a) and the complex (b)

Table 3. Infrared absorption of ranitidine hydrochloride and the complex

Numbering	Functional Group	Ligand (cm^{-1}) (a)	The complex (cm^{-1}) (b)
1	-NH-	3255 1568	3157 1519
2	$-NO_2$	1620 1381	1631 1346
3	-CO	1224	1134
4	$-C-N_{ter}$	1070	1012
5	-C-S	698	700
6	Cu-O	-	599
7	Cu-N	-	435

3.6. Magnetic properties (μ_{eff})

The measurement of the effective magnetic moment (μ_{eff}) of the $[Cu(ranitidine)_2]SO_4{\cdot}2H_2O$ is 1.77±0.03 BM. The effective magnetic moment value shows the $[Cu(ranitidine)_2]SO_4{\cdot}2H_2O$ is paramagnetic with one unpaired electron. The value of the effective magnetic moment (μ_{eff}) also shows no Cu-Cu bonding, because if the Cu-Cu bond is formed the unpaired electron will be paired and the effective magnetic moment value (μ_{eff}) becomes smaller than the spin only magnetic moment value (μs) [28]. The effective magnetic moment (μ_{eff}) value of the complex is also the normal value for Cu^{2+} with an unpaired electron, which is the effective moment magnet value (μeff) of 1.70-2.20 BM [29].

3.7. Electronic spectra

The solution UV–Vis spectrum of the formed complex showed a low-intensity band at 626 nm (15974 cm^{-1}) shown in figure 3. This absorption indicates the presence of d-d transition as occurs in the Cu(II)-2-hydroxy-4-butoxy aniline complex which has a 17637 cm^{-1} uptake showing 2Eg → $2T_2g$ transition indicating the planar square geometry complex [27].

From all the characterization analysis above, the proposed structure of the complex is shown in figure 6.

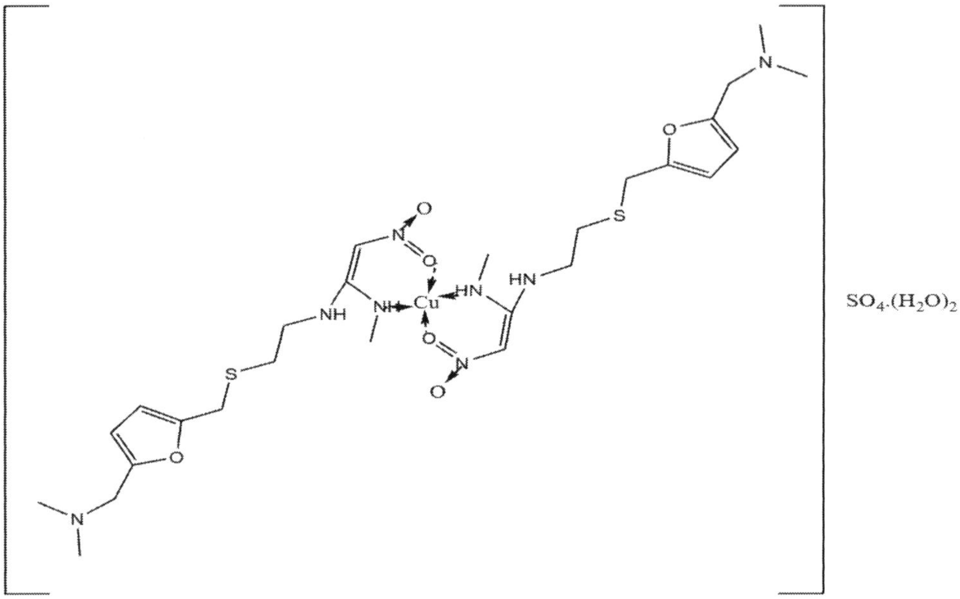

Figure 6. Suggested structure $[Cu(ranitidine)_2]SO_4{\cdot}2H_2O$

4. Conclusion

The Cu(II)-ranitidine complex can be synthesized from $CuSO_4{\cdot}5H_2O$ and ranitidine in refluxing water with mole ratio of metal to ligand = 1:2. The complex formula is $[Cu(ranitidine)_2]SO_4{\cdot}2H_2O$, namely diranitidinecopper(II) sulfate dihydrate. The coordinated groups are the O atom of $-NO_2$ and -NH. The complex is paramagnetic and has one peak at maximum UV-Vis absorption of 626 nm. It is indicated that the complex forming square planar geometry.

Acknowledgement

The authors gratefully express acknowledgments to Ministry of Research, Technology and Higher Education (RISTEKDIKTI) of the Republic of Indonesia for financial support.

References

[1] McCaffrey L J, Henderson W, Nicholson B.K, Mackay J.E & Dinger M B 1977 *J Chem Soc, Dalton Trans* 2577.
[2] Cavaglioni A, Cini R 1997 *J Chem Soc, Dalton Trans* 1149.
[3] Farrell N (Ed) 1989 *Transition Metal Complexes as Drugs and Chemotherapeutic Agents* (Kluwer, Dordrecht),
[4] Bloemink M J, Reedijk J 1996 *Metal Ions in Biological Systems edited by Sigel H & Sigel A vol. 32* (New York: Marcel Decker).
[5] Roat R M, Jerardi M J, Kopay C B, Heath D C, Clark J A, DeMars J A, Weaver J M, Bezemer E, Reedijk J 1997 *J Chem Soc, Dalton Trans* 3615.
[6] Kenawi I M, Barsoum B N, Youssef M A 2005 *J Pharm Biomed Anal* **37** 655.
[7] Soreson J R J 1976 *Inflammation* **1** 317.
[8] Singh M M, Basin DK 1984 *Indian, J. Pharmac* **16** 136.
[9] Kozlowaski H, Kowalik J, Jainkowska J 1992 *Inorganic Biochem* **48** 233.
[10] Mc. Colm A A, Mc. Laren A, Klinkejt G, Francis M R, Collnolly P C, Brinham C J, Comptde C J, Selway S, Williamson R 1996 *Elementary Pharmacology and Therapeutics* **10** 241.
[11] Martini N, Parente J E, Toledo M E, Escudero G E, Laino C H, Medina J J M, Echeverría G A,

Piro O E, Lezama L, Williams P A M, Ferrer E G 2017 *Journal of Inorganic Biochemistry* **174** 76-79.
[12] Xie J, Li Y, Song L, Pan Z, Ye S, and Hou Z 2017 *Drug Delivery* **24(1)** 707-719.
[13] Tarushi A, Perontsis S, Hatzidimitriou A G, Papadopoulos A N, Kessissoglou D P, Psomas G 2015 *Journal of Inorganic Biochemistry* **149** 68–79
[14] Sembiring Z, Illim 2008 *Sintesis dan Karakterisasi Kompleks Cu(II) dan Mn(II) dengan Derivat Ligan Basa Schift 1,5 Dimethylcarbazone dan aniline* (Lampung: Jurusan Kimia FMIPA Universitas Lampung).
[15] Loboda D, Rowińska-Żyrek M 2017 *Journal of Inorganic Biochemistry* **174** 150-155.
[16] Loganathan R, Ganeshpandian M, Nattamai SP, Bhuvanesh, Palaniandavar M, Muruganantham A, Ghosh SK, Riyasdeen A, Akbarsha MA 2017 *Journal of Inorganic Biochemistry* **174** 1-13.
[17] Kubiak M A, Duda M, Ganadu M L, Kozlowski H 1996 *J. Chem. Shock. Dalton Trans* 905-1908.
[18] Reedijk J 2012 *Journal of Inorganic Biochemistry* 182-185.
[19] Nakai M, Sekiguchi F, Obata M, Ohtsuki C, Adachi Y, Sakurai H, Orvig C, Rehder D Yano S 2005 *J Inorg Biochem* **99** 1275.
[20] Arya, P S, Singh S N, Chandra S 2010 *J. Chem. Pharm. Res.* **2(3)** 626-630.
[21] Baranska M, Kontecka E G, Kozlowski H, Proniewicz 2002 *Journal of Inorganic Biochemistry* **92** 112-120.
[22] Ranskiy A, Didenko N, Gordienko O 2017 *Chem. Chem. Technol.* **11** 11-18.
[23] Yousif E, Majeed A, Al-Sammarrae K, Salih N, Salimon J, Abdullah B 2017 *Arabian Journal of Chemistry* **10** 1639–1644.
[24] Refat M S, Mohamed G G, El-Sayed M Y, Killa H M A, Fetooh H 2017 *Arabian Journal of Chemistry* **10** 2376–2387.
[25] Abdolmaleki S, Ghadermazi M 2017 *Inorganica Chimica Acta* **461** 221–232.
[26] Kavitha N, Lakshmi P V A 2015 *Journal of Saudi Chemical Society* **21** S457-S466.
[27] Sultan K M 2011 *Diyala Journal for Pure Sciences* **7** 47-57.
[28] Szafran Z, Pie R, Singh M 1991 *Microscale Inorganic Chemistry* (Canada:John Willey).
[29] Huheey E, James A, Keiter E, Keiter R L 1993 *Inorganic Chemistry Fourth edition* (New York: Harper Collins College Publisher).

IOP Conf. Series: Materials Science and Engineering **349** (2018) 012026 doi:10.1088/1757-899X/349/1/012026

Photocatalyst of Perovskite $CaTiO_3$ Nanopowder Synthesized from CaO derived from Snail Shell in Comparison with The Use of CaO and $CaCO_3$

I Fatimah, Y Rahmadianti, R A Pudiasari

Chemistry Department, Universitas Islam Indonesia, Kampus Terpadu UII, Jl. Kaliurang KM 14, Sleman, Yogyakartam Indonesia 555584.

E-mail : isfatimah@uii.ac.id

Abstract. Calcium titanate belongs to the important group of compounds with a perovskite structure having high dielectric loss for various applications including photocatalysis mechanism. Refer to the principles of green chemistry, in this work preparation of $CaTiO_3$ was conducted by using CaO derived from snail shell. Aim of this research are to study the physicochemical character of perovskite derived from snail shell and its comparison with CaO and $CaCO_3$ as Ca sources. Material preparation was performed by solid reaction of Ca sources with TiO_2 under comparison with CaO and $CaCO_3$ precursors. Mixture of Ca sources with TiO_2 in certain proportion were ground and calcined at the temperature of 200 °C for 2 hs. Materials were characterized by using X-ray diffractometer (XRD), Fourier Transform-Infra Red (FTIR) and the photocatalytic activity was tested by using methylene blue photooxidation. Perovskite synthesized using CaO derived from snail shell exhibits the similar XRD pattern with that were prepared by using CaO and $CaCO_3$. From the photooxidation activity test, it is proven that $CaTiO_3$ shows similar photocatalytic activity correspond to that were prepared by CaO and $CaCO_3$ Utilazation of shell as agricultural waste of the synthesis of $CaTiO_3$ perovskite is the novelty of this work. Furthermore, the study on material structure and photoactivity is the main focuses for the application in industry and environment.

1. Introduction

During last two decades, photocatalysis becomes developing technology in wastewater treatment and intensively studied. Some advantageous of photocatalysis utilization are laid on the reusability of the material in further process and the more economist and efficient process. [1-3]. Conventional methods for the dye removal from waste water and textile industries viz. physical methods, chemical methods, adsorption, absorption, incineration and biological. Each method has some merits and drawbacks. For example due to large number of dyes molecules and the stability of modern dyes, conventional biological methods are ineffective for decolorization and degradation.

Some previous works explore the utilization of semiconductor materials instead of TiO_2 and ZnO that has been an interest for photocatalysis application. One of these materials is perovskite. Perovskites are the class of compounds presenting the general formula ABO_3[1]. The perovskite crystal structure has corner connected BO_6 octahedra and 12 oxygen coordinated A cations, located in between the eight BO6 octahedra (Figure 1). Some of perovskite material are $CaTiO_3$, $PbZrO_3$, $BaTiO_3$, $PbTiO_3$ commonly used piezoelectric compounds[2,3].

Perovskite of $CaTiO_3$ is the economist material for photocatalysis application and can be synthesized from the reaction of CaO and TiO_2. The high content of $CaCO_3$ in snail shell is a high potential commodity for precursors in the synthesis of $CaTiO_3$ material[4–6]. Aim of study is to evaluate chemical properties of $CaTiO_3$ derived from sail shell in comparison with $CaTiO_3$ synthesized from CaO and $CaCO_3$.

2. Materials and Method

2.1. Materials:

Snail shell (*Philla ampulacea*) was obtained from Paddy field in Bantul District, Special Region of Yogyakarta Indonesia. The determination of snail was performed by Faculty of Biology, Gadjah Mada University, Indonesia. TiO_2, methylene blue (MB), H_2O_2, CaO and $CaCO_3$ were purchased from Merck.

2.2. Synthesis of $CaTiO_3$

$CaTiO_3$ nanocrystalline powder was prepared by using stoichiometric mixture of Ca and TiO_2 at mole ratio of 1:1. The mixture was grinded until 20 minutes by using mortar and pastle and followed by calcination at 200°C for 2 hours. The syntheses were at varied source of Ca consist of snail shell, CaO and $CaCO_3$, prepared $CaTiO_3$ were encoded as $CaTiO_3$-ss, $CaTiO_3$-CaO and $CaTiO_3$-$CaCO_3$ respectively

2.3. Characterization of $CaTiO_3$

Materials were characterized by using x-ray diffractometer (XRD). XRD Shimadzu X6000 was utilized for the analyses. The corresponding X-ray diffraction patterns recorded agreement with the reported values of Joint Commitee on Powder Difraction Standards (JCPDS). FT-IR spectra of a powder sample was recorded using Perkin Elmer Spectrum 10™. Spectrophotometer and SEM profile of materials were recorded by JEOL JSM-7001F instrument.

2.4. Photodegradation experiments

Photodegradation experiment of MB using prepared $CaTiO_3$ was conducted in a photocatalytic reactor equipped with Philips UV Lamp 360nm at 40 watt of power (Figure 1). A solution containing the proper concentration of the dye, e.g., 1, 2, 3 and 5 ppm was transferred into the reactor batch photocatalyst, and then 0.5 g of $CaTiO_3$ nanocrystalline powder and 1 mL H_2O_2 was added. This mixture was irradiated under ultraviolet light with variations time 5; 10; 15; 30; 45; 60 and 120 minutes, which induced the photochemical reaction to proceed. The samples of the test were monitored on UV-visible spectrophotometer (Shimadzu 1800) at wavelength 664 nm. Co and C are the initial and sample concentration obtained by photometric method using calibration standard method.

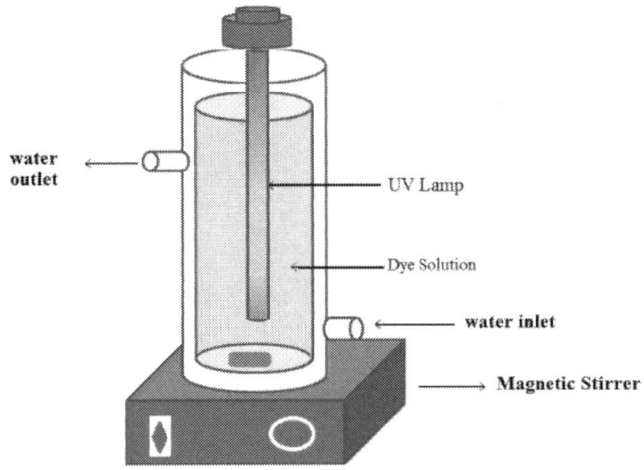

Figure 1. Schematic photoreactor

3. Results and Discussion

The crystallinity of prepared materials was identified by using XRD and the patterns are presented in Fig.2. The patterns indicate that all samples exhibit the formation of $CaTiO_3$ in mixture with TiO_2, CaO and $CaCO_3$ phases. The spectra of $CaTiO_3$ shows the reflections as indication of (012), (006), (202), (116), (018) and (214) refered to the JCPDS (Card No.22-0153)[7].

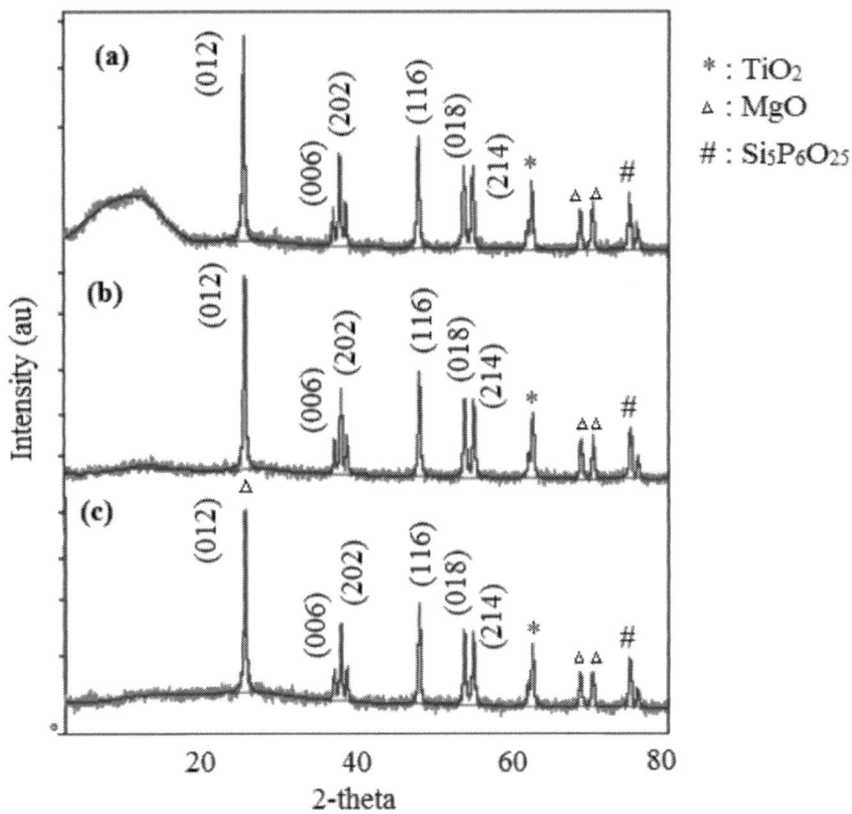

Figure 2. XRD pattern of (a) $CaTiO_3$-CaO, (b) $CaTiO_3$-$CaCO_3$,(c) $CaTiO_3$-ss,

However, some reflections are appeared as an indication of the other minerals than the perovskite phases in the solids . The indentified minerals are silica (SiO_2), MgO and $Si_5P_6O_{25}$. The presence of impurities related to the incomplete reaction during calcination at 200°C.
Comparison on surface profile of materials is described by SEM-EDX profile in Figure 3. It can be seen that there is a surface evolution of snail shell powder to $CaTiO_3$ in which the smaller grain obtained after modification to $CaTiO_3$. There is no significant difference between $CaTiO_3$ from different CaO source indicating that the chemical interaction in the synthesis is not affected much by the impurities in snail shell content (Table 1).

Table 1. Composition of snail shell

Component	Percentange (% wt.)
Na_2O	0.12
C	14.2
CaO	83.9
Si	0.32
Mg	0.21

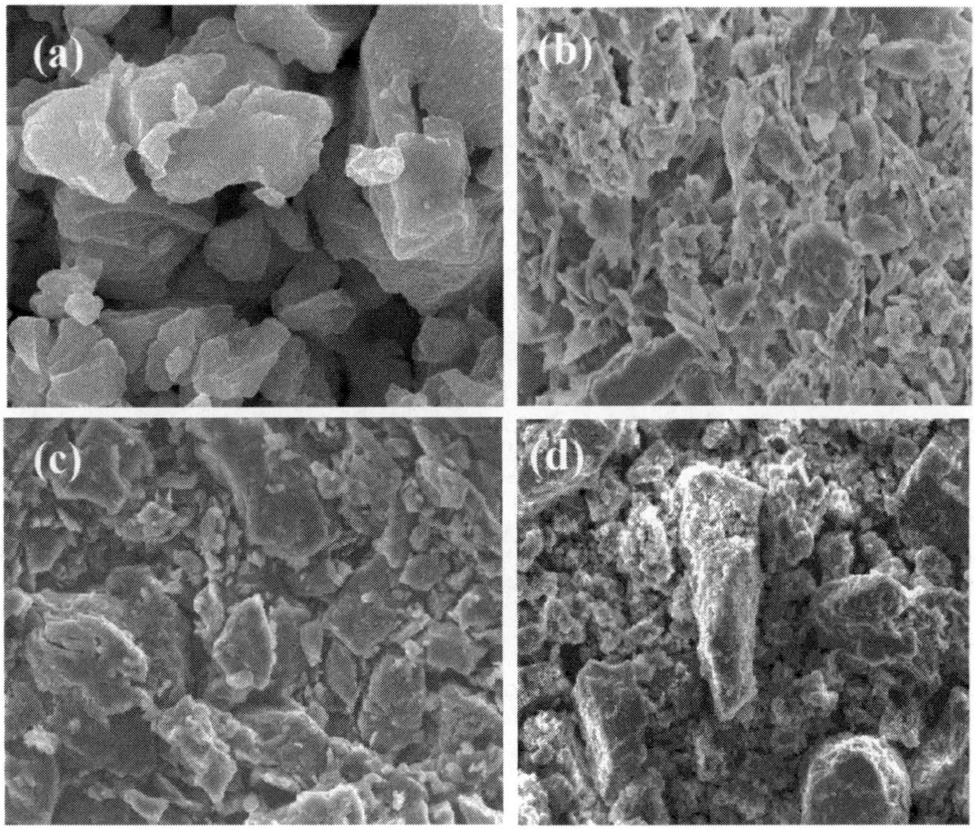

Fig.3. SEM profile of (a) snail shell powder (b) $CaTiO_3$-CaO, (c) $CaTiO_3$-$CaCO_3$,(d) $CaTiO_3$-ss

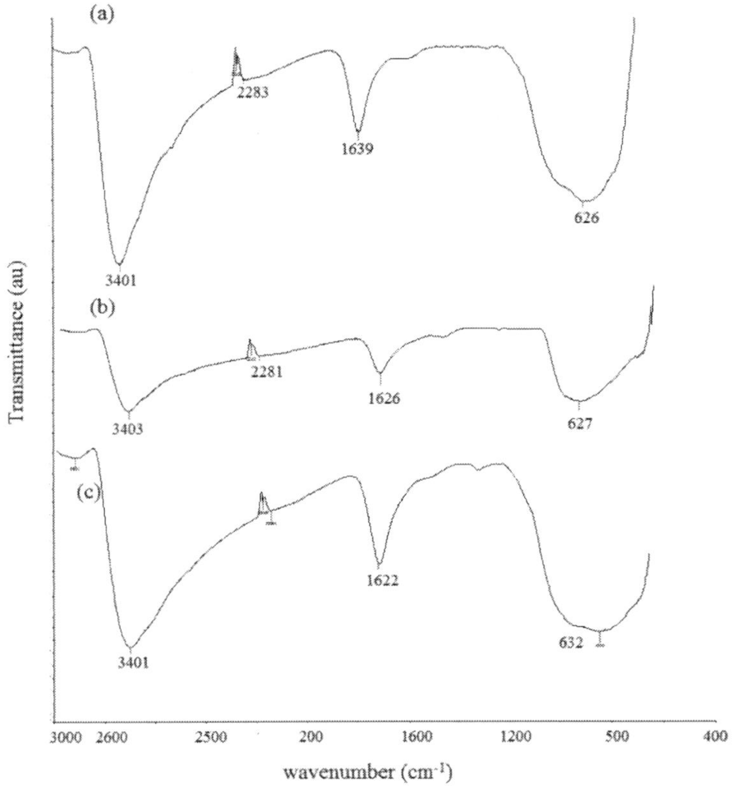

Fig. 4. FTIR spectra of (a) $CaTiO_3$-CaO, (b) $CaTiO_3$-$CaCO_3$,(c) $CaTiO_3$-ss

The FTIR spectra of materials presented in Fig.4 demonstrate some important peaks at around 632cm-cm^{-1}, 1626 cm-1 and 3403 cm^{-1}. The peaks at 1626 cm^{-1} can be ascribed to the bending vibration of O-H and the peak in the range of 500–900 cm^{-1} was assigned to the Ti-O stretching and Ti–O–Ti bridging stretching modes. The bands at 3400 cm^{-1} are clearly shows the presence of moisture and water molecular.

Photocatalytic activity of materials was evaluated in MB photodegradation and the kinetics are presented in Fig.5. The photodegradation was conducted in varied MB concentration. It is seen that all $CaTiO_3$ demonstrate the photoactive properties as decreasing MB concentration along increasing time of treatment is found.

IOP Conf. Series: Materials Science and Engineering **349** (2018) 012026 doi:10.1088/1757-899X/349/1/012026

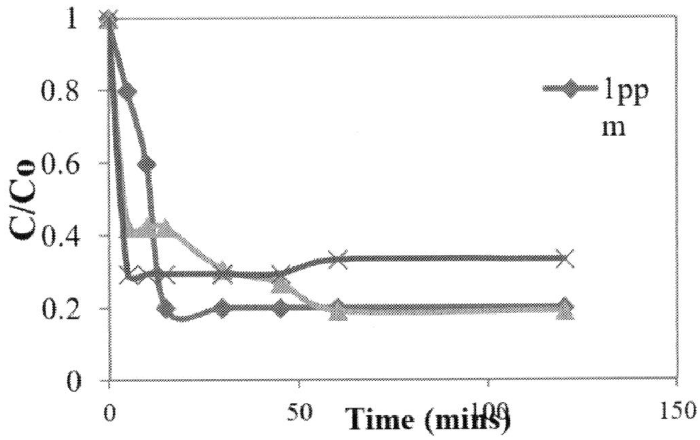

(a)

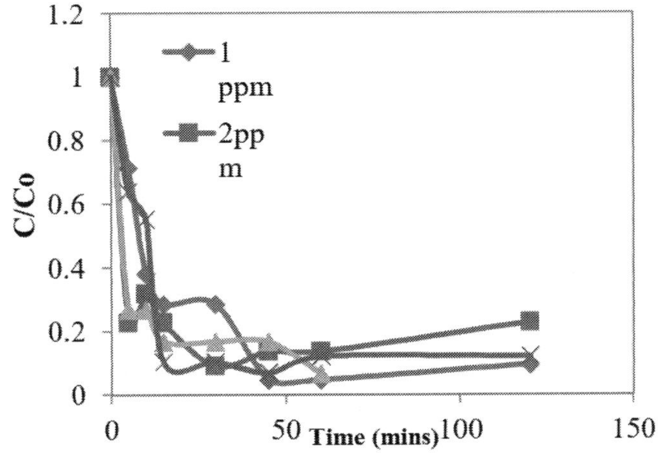

(b)

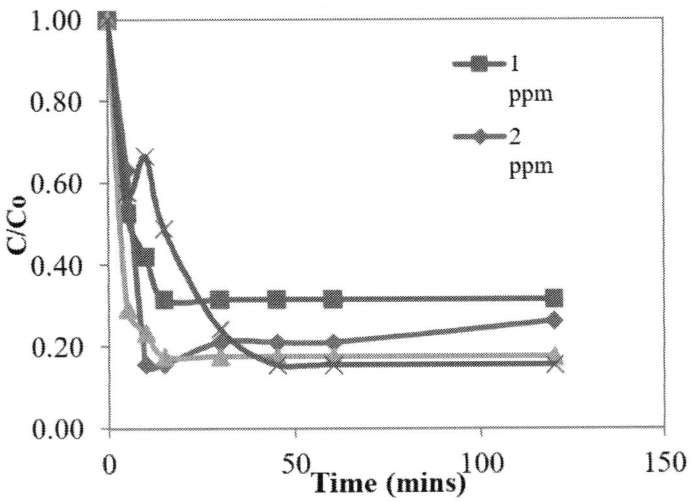

(c)

Fig 5. Kinetics of MB over (a) $CaTiO_3$-CaO, (b) $CaTiO_3$-$CaCO_3$,(c) $CaTiO_3$-ss,

From photooxidation test, kinetics of MB degradation shows that $CaTiO_3$ derived snail shell is in similarly values with the kinetics degradation over $CaTiO_3$ synthesized from CaO and $CaCO_3$. The initial rate data of MB degradation over varied $CaTiO_3$ are presented in Table 2. From the data it is found that photocatalytic activity of $CaTiO_3$-ss is lower compared to other $CaTiO_3$ samples.
The kinetics of MB photooxidation occurs in random trend but the in general lowest rate over $CaTiO_3$-ss. The possible reason for the photoactivity is from the presence of impurities in the sample.

Table 2. Initial rate of MB photooxidation over prepared materials

[MB]/ppm	**Initial rate (ppm/mins)**		
	$CaTiO_3$-CaO	**$CaTiO_3$-$CaCO_3$**	**$CaTiO_3$-ss**
1	0.0533	0.047	0.045
2	0.022	0.103	0.112
3	0.115	0.167	0.164
5	0.235	0.299	0.170

From the kinetic simulation, it is found that the MB photodegradation obeys second order reaction.

4. Conclusion

The $CaTiO_3$ synthesis was carried out under ecofriendly, easily and cheap solid state mechanochemical method. Synthesis of $CaTiO_3$ exhibits photocatalytic activity in MB photodegradation.

References

[1] Heydari M, Fazaeli R and Yousefi M 2012 Preparation of Perovskite Nanocomposites and Photochemical Degradation Kinetics of Acid Yellow 199 **15** 7–15

[2] Seo S S A, Lee H N and Noh T W 2005 Infrared spectroscopy of CaTiO3, SrTiO3, BaTiO 3, Ba0.5Sr0.5TiO3 thin films, and (BaTiO3)5/(SrTiO3)5 superlattice grown on SrRuO3/SrTiO3(001) substrates *Thin Solid Films* **486** 94–7

[3] Liotta L F, Gruttadauria M, Di Carlo G, Perrini G and Librando V 2009 Heterogeneous catalytic degradation of phenolic substrates: catalysts activity. *Journal of hazardous materials* **162** 588–606

[4] Liu H, Guo H shuang, Wang X jing, Jiang J zhong, Lin H, Han S and Pei S peng 2016 Mixed and ground KBr-impregnated calcined snail shell and kaolin as solid base catalysts for biodiesel production *Renewable Energy* **93** 648–57

[5] Roschat W, Siritanon T, Kaewpuang T, Yoosuk B and Promarak V 2016 Economical and green biodiesel production process using river snail shells-derived heterogeneous catalyst and co-solvent method *Bioresource Technology* **209** 343–50

[6] Viriya-Empikul N, Krasae P, Nualpaeng W, Yoosuk B and Faungnawakij K 2012 Biodiesel production over Ca-based solid catalysts derived from industrial wastes *Fuel* **92** 239–44

[7] Gaikwad S S, Borhade A V. and Gaikwad V B 2012 A green chemistry approach for synthesis of CaTiO3 photocatalyst: Its effects on degradation of methylene blue, phytotoxicity and microbial study *Der Pharma Chemica* **4** 184–93

ZrO_2/bamboo leaves ash (BLA) Catalyst in Biodiesel Conversion of Rice Bran Oil

Is Fatimah, Ana Taushiyah, Fitri Badriatun Najah, Ulil Azmi

Chemistry Department, Universitas Islam Indonesia, Kampus Terpadu UII, Jl. Kaliurang KM 14, Sleman, Yogyakarta Indonesia 555584.

E-mail : isfatimah@uii.ac.id

Abstract. Preparation, characterization and catalytic activity of ZrO_2/bamboo leaves ash (BLA) catalyst for conversion of rice bran oil to biodiesel have been investigated. The catalyst was prepared by impregnation method of $ZrOCl_2$ as ZrO_2 precursor with BLA at a theoretical content of 20% wt. followed by calcination. The physicochemical properties of the catalyst material were characterized by x-ray diffraction (XRD), FTIR and surface acidity measurement. Activity test of materials in biodiesel conversion of rice bran oil was used by reflux method and microwave (MW) assisted method. Reaction variables studied in the investigation were the effect of catalyst weight and time of MW irradiation compared with the use reflux method. The results showed that ZrO2/BLA catalyst exhibited competitively effective and efficient processes for the production of biodiesel. The reflux method demonstrated an higher conversion (%) compared to MW method, however MW method showed the better reusable properties.

1. Introduction

Alternative energy is still the focus of human need due to the limitation of fossil fuel availability. In this scheme, biodiesel is one important fuel alternative having a potency to be developed in Asian region since it is synthesized by transesterification of vegetable/animal oils which are abundantly present. The conversion reaction of vegetable/animal oil via transesterification reaction can be carried out using both acidic and basic (homogeneous or heterogeneous) catalysts and usually the NaOH or KOH homogeneous base catalyst were employed. The use of both kinds of basic solution has drawback in non-reusability as well as consumable material for production. Referring to some principles of green chemistry, the use of a heterogeneous catalyst is an important consideration. Beside the catalyst election as a concern, the use of rapid procedure including the use of microwave irradiation procedure is also widely reported. Many previous works reported the use of active metal oxide catalyst to replace homogeneous catalyst[1–3]. They consist of silica-alumina, clay, zeolite-based materials instead of a well-known metal oxide of ZrO_2. In another scheme, seeking low-cost catalyst is an interesting concern[4]. For this purpose, many investigations reported the catalytic activity of natural materials containing silica as a composite builder with a metal oxide active catalyst. The use of rice husk ash as the material of catalyst composite with ZrO_2, $KHCO_3$, and other metal demonstrated activity in biodiesel production has been reported (references??). As other alternative for low-cost catalyst in biodiesel production, in this research the use of Bamboo leaves ash for the composite builder with ZrO_2 was attempted. Bamboos plant is grown well in south-east Asia including Indonesia. It's approximately 1,000 kinds of bamboo in 80 countries, about 200 species of 20 countries found in Southeast Asia. In Indonesia, bamboo plants grow well in the lowlands to the mountains with a height of about 3,000 m above sea level and generally found in open places and their regions are free from flooding. Bamboo leave is a part of bamboo plant which is not utilized well, and from the previous report it was known that silica content in bamboo leaves was high. By this reason, this work took advantage by utilizing bamboo leave ash as a silica source for supporting zirconia catalyst for biodiesel production in the form of ZrO_2-SiO_2. Research aimed to evaluate the physicochemical character of the composite and its activity in biodiesel production. Another consideration of MW irradiation utilization was to evaluate the economical aspect of the conversion which also intensively studied[5,6].

2. Materials and Methods

2.1. Materials

The materials used in this study were bamboo leaves obtained from village plantation in Sleman, Yogyakarta Province, Indonesia. Zirconium (IV) oxide chloride ($ZrOCl_2$), tetra ethyl ortho silicate, NaOH, methanol (CH_3OH), HCl, n-hexane, (H_3PO_4) the Phenol phthalin ($C_{20}H_{14}O_4$) were purchased from Merck. For biodiesel production, the rice bran oil produced by PT. Indofood was utilized.

The equipment used for physicochemical characterization in this study consisted of a furnace, the analytic balance, mortar pestle, a porcelain, glass clock, stone melting, stat and the clamps, the mouthpiece of divorce, the erlenmeyer mouthpiece, beaker glasses, a peck, a spray bottle, magnetic stirrer. The physicochemical character of the samples was performed using FTIR (Shimadzu Type IRPrestige21), SEM-EDX (JEOL JED-2300), XRD (Shimadzu X6000).

2.2. Preparation of ZrO_2/bamboo leaves ash (ZrO_2/BLA) Catalyst.

The catalyst of ZrO_2/BLA was prepared by two main steps: Bamboo leaves ashing and ZrO_2 immobilization. First, the leaves of bamboo were washed with clean water to remove the polluter, followed by sun drying before calcined in the furnace at 500 °C for 4 hours. The ash was milled using a mortar. The aim of calcination in this step was to remove the organic content of bamboo leaves and the ash was rich in silica. The bamboo leaves ash was then mixed with the precursor of ZrO_2 by zirconium (IV) Oxide chloride. The solution was added dropwise into an ash suspension followed by stirring for 24 hours before drying in the oven. The dry solid was then calcined to obtain the catalyst. For comparison purpose, the synthesis of ZrO_2-SiO_2 was also conducted. A similar procedure with the synthesis of ZrO_2/BLA was applied however the Bamboo ash suspension was replaced by tetra ethyl ortho silicate (TEOS). The schematic procedure of the synthesis is described in Fig. 1.

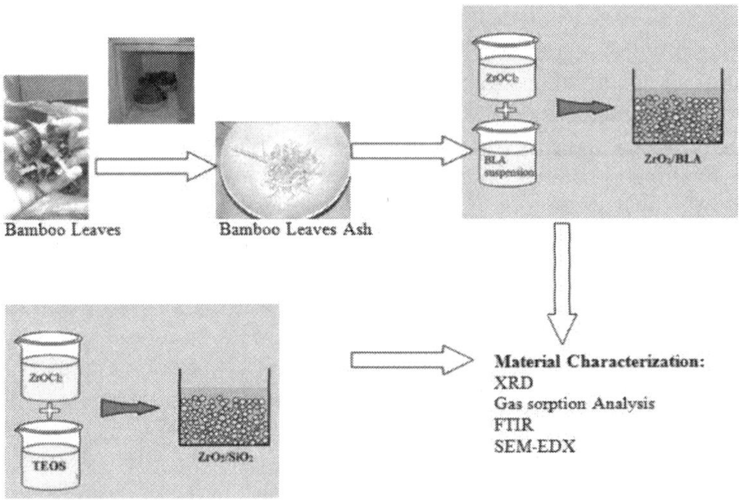

Fig. 1. Schematic representation of ZrO_2/BLA synthesis

The material was characterized using XRD, gas sorption analyzer and SEM-EDX. A Ni-filtered Cu-Kα radiation was utilized for x-ray analysis in the 2θ of 5-40° with the ramp rate of 2°/min.

2.3 Catalytic activity testing

The catalytic activity of ZrO_2/BLA in the biodiesel conversion was tested using rice bran oil as a vegetable oil source. The biodiesel conversion was conducted by methanol to oil ratio of 1:5 under a varied time of reflux; 2, 3 and 5 hours. The conversion was also conducted using microwave irradiation method for 30 minutes. The results of each reaction were separated into the separating funnel and the biodiesel fraction was examined using GCMS analysis (using GCMS (Shimadzu Type GCMS QP2010 Plus).

3. Results & Discussion

Initial evaluation of BLA content was performed by SEM-EDX analysis and the result is presented in Fig.2. From the EDX analysis results (Table 1) it can be seen that Si presented at more than 78% and was the major component of BLA with some impurities consist of Ca, C, P, Na and Mg. Based on the result it can be concluded that BLA had potential as a support for metal oxide. This assumption was referred to the silica content of rice husk ash and bagasse ash reported by previous reports [7,8].

Table 1. Result of EDX analysis of BLA

Component	**Percentage (% wt.)**
SiO_2	78.98
C	5.09
MgO	4.99
Na_2O	2.80
K_2O	1.58
P_2O_5	2.77
CaO	3.08

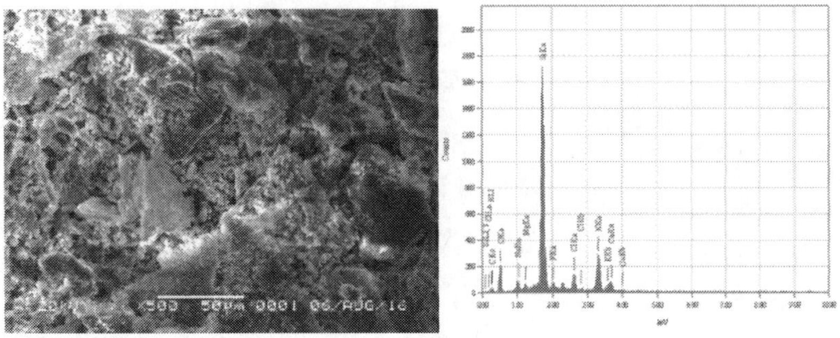

Fig.2. SEM profile and EDX analysis result of BLA

XRD pattern of the ZrO_2/BLA compared with the pattern of bamboo leaves ash (BLA) and ZrO_2/BLA in comparison with ZrO_2-SiO_2 is presented by Fig. 3. XRD pattern of ZrO_2/SiO_2 as standard showed the reflections of ZrO_2 in mixture phase of monoclinic and tetragonal phases. BLA did not reflect any crystalline phase as an indication of amorphous silica content. The BLA pattern showed the amorphous structure of silica as presented by a broad peak at around 20°. After composite formation with ZrO_2, the amorphous structure appeared as a broad peak corresponding to silica content which there was a reflection match with the peak of ZrO_2 in tetragonal phase as represented by the reflection at around 30°

[9–11]. The surface profile of ZrO_2/BLA in comparison with ZrO_2-SiO_2 is presented in Fig. 4. A rougher surface is appeared by ZrO_2/BLA representing the distribution of ZrO_2 on the surface. The profile is in line with the surface analysis of presented by adsorption-desorption profile and calculated parameters (Table 2).

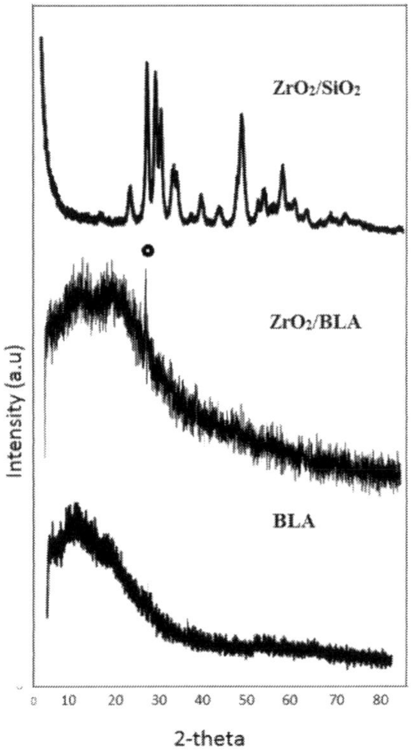

Fig. 3. XRD pattern of materials (• = *t*-ZrO_2)

Fig. 4. SEM profile of (a) ZrO_2/SiO_2 (b) ZrO_2/BLA

Table 2. Composition of ZrO_2/SiO_2 and ZrO_2/BLA

Component	ZrO_2/SiO_2 **(% wt.)**	ZrO_2/BLA **(% wt.)**
SiO_2	58.98	38.88
ZrO_2	11.03	6.58
MgO	-	1.24
Al_2O_3	-	0.3

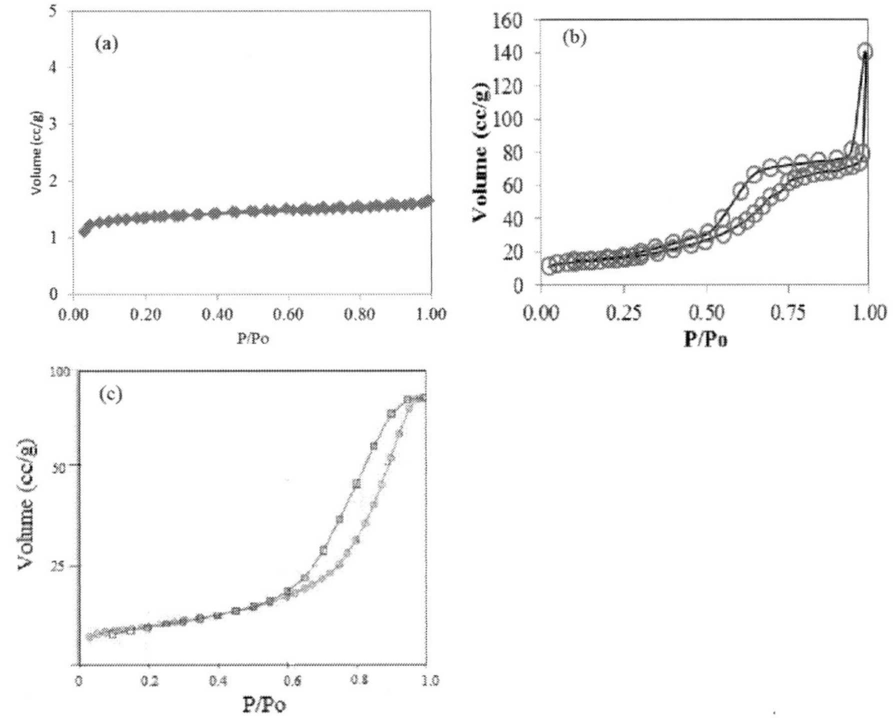

Fig. 5. Adsorption-desorption pattern of (a) BLA (b) ZrO_2/BLA (c) ZrO_2/SiO_2

The adsorption-desorption pattern of ZrO_2/BLA represented the micropore structure of material which is referred to the IUPAC type III – adsorption-desorption curve classification with hysteresis loop indicating the mesoporous formation in the structure (Fig. 5). The pattern was commonly found in the amorphous structure and silica-based materials. Different with the pattern, BLA expressed the micropore structure while ZrO_2/SiO_2 profile suggested the hysteresis loop meant the mesoporous content in the structure. The specific surface area, pore volume, and pore radius data measured based on the adsorption-desorption pattern are listed in Table 3.

Table 3. Specific surface area, pore volume and pore radius data of materials

Parameter	ZrO_2**/BLA**	**BLA**	ZrO_2/SiO_2
BET Specific surface area (m^2/g)	103.09	14.45	58.02
Pore Radius (Å)	18.45	14.05	14.34
Pore Volume (cc/g)	0.568	0.036	0.126

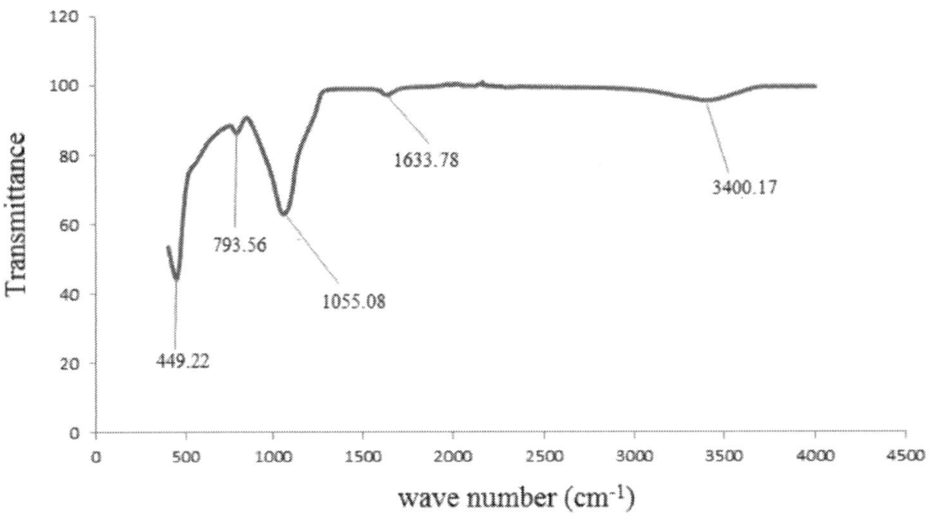

Fig. 6. FT-IR spectra of ZrO2/BLA

From spectra of FTIR analysis (Fig.6), it can be seen that Si-O-Si asymmetric stretching at 1055.08 cm^{-1} corresponding to SiO_2 is observed. The characteristic peaks of Si–OH at 793.50 cm^{-1} may be attributed to the silanol groups on the SiO_2 particles. In both spectra, the peaks which appear at about 3400,17 cm^{-1} are due to the bending vibration of OH from the water. The presence of Si-O-Zr interaction can be assigned from the spectra at 1055 cm^{-1} which are attributed to the silica network forming the Si-O-Zr band[10,12]. These reports are consistent with our findings and verified the oxygen bridge between Si and Zr. This means that there is a composite formation of ZrO_2 with SiO_2.

Catalytic activity

The catalytic activity of material was tested by biodiesel conversion. Yield and percentage of fatty acid methyl ester (FAME) content is listed in Table 4. The yield of biodiesel fraction is calculated by the following equation:

$$\%\text{Yield} = \frac{Vol.Biodiesel\ fraction}{Vol.of\ Oil}.100\% \quad (1)$$

and the percentage of fatty acid methyl ester (FAME) was measured based on GCMS data interpretation.

Table 4. Yield of biodiesel conversion

Reaction condition	**Catalyst**	**Yield (%)**	**FAME(%)**
Reflux-2h	ZrO_2/BLA	50.15	100
Reflux-3h	ZrO_2/BLA	100	100
Reflux-5h	ZrO_2/BLA	100	100
MW- 30 mins	ZrO_2/BLA	80.5	100
Reflux-3h	BLA	43.22	89.03
Reflux-3h	ZrO_2/SiO_2	100	100

From the data in Table 3 it is concluded that both reflux and microwave irradiation method of biodiesel conversion give 100% of FAME composition in the biodiesel fraction. Based on the variation of reaction condition it can be concluded that ZrO_2/BLA exhibits the same activity with ZrO_2/SiO_2 in 3 h reflux method. This data suggests that the activity of prepared material is comparative. By the use of MW irradiation it is also seen that shorter time of reaction may be applied to give comparative results.

4. Conclusion

The results showed that physicochemical characterization of BLA could be applied for preparing the heterogeneous catalyst of ZrO_2/BLA in biodiesel conversion application. The material exhibited the microporous in mixture with mesoporous structure and accommodating ZrO_2 in tetrahedral phase in form. The catalytic activity of material was comparable respect to the use of ZrO_2/SiO_2 in which the composite give the maximum yield over reflux for 2h and 3h as well as over MW irradiation for 30 mins.

References

[1] Xie W, Huang X and Li H 2007 Soybean oil methyl esters preparation using NaX zeolites loaded with KOH as a heterogeneous catalyst *Bioresource Technology* **98** 936–9

[2] do Nascimento L A S, Angélica R S, da Costa C E F, Zamian J R and da Rocha Filho G N 2011 Comparative study between catalysts for esterification prepared from kaolins *Applied Clay Science* **51** 267–73

[3] Guzmán-Vargas A, Santos-Guti??rrez T, Lima E, Flores-Moreno J L, Oliver-Tolentino M A and Mart??nez-Ortiz M D J 2015 Efficient KF loaded on MgCaAl hydrotalcite-like compounds in the transesterification of Jatropha curcas oil *Journal of Alloys and Compounds* **643** S159–64

[4] Turapan S, Yotkamchornkun C and Nuithitikul K 2010 Esterification of Free Fatty Acids in Crude Palm Oil with Sulfated Zirconia : Effect of Calcination Temperature **4** 310–4

[5] Hincapié G M, Valange S, Barrault J, Moreno J a. and López D P 2014 Effect of microwave-assisted system on transesterification of castor oil with ethanol *Universitas Scientiarum* **19** 193–200

[6] El Sherbiny S A, Refaat A A and El Sheltawy S T 2010 Production of biodiesel using the microwave technique *Journal of Advanced Research* **1** 309–14

[7] Alves R H, Vitória T, Rovani S and Fungaro D A 2017 Green Synthesis and Characterization of Biosilica Produced from Sugarcane Waste Ash **2017**

[8] Chen G-Y, Shan R, Shi J-F and Yan B-B 2015 Transesterification of palm oil to biodiesel using rice husk ash-based catalysts *Fuel Processing Technology* **133** 8–13

[9] Zhang P and Choy K 2015 The Synthesis of Single Tetragonal Phase Zirconia by Sol-Gel Route 18–24

[10] Chandradass J, Balasubramanian M and Hyeon Ki. K 2011 Solution phase synthesis of t-ZrO2 nanoparticles in ZrO2-SiO2 mixed oxide. *Journal of Experimental Nanoscience* **6** 38–48

[11] Hexin Zhang C H and Y Q 2011 Synthesis, Structural and Thermal Properties of Nano-porous SiO2-based Aerogels *Advances in Nanocomposites - Synthesis, Characterization and Industrial Applications* pp 40–60

[12] Tarafdar A, Panda A B and Pramanik P 2005 Synthesis of ZrO2-SiO2 mesocomposite with high ZrO2 content via a novel sol-gel method *Microporous and Mesoporous Materials* **84** 223–8

Utilization of hydrotalcite modified with 3,4,5-trihydroxy-benzoic acid for the treatment of silver-containing wastewater

I Yanti[1*], W F Winata[2], M Anugrahwati[1]

[1] Chemistry Department, Universitas Islam Indonesia, Yogyakarta, Indonesia
[2] Environmental Engineering, Universitas Proklamasi 45, Yogyakarta, Indonesia

*E-mail: ika.yanti@uii.ac.id

Abstract. Utilization of hydrotalcite modified with 3,4,5-trihydroxybenzoic acid as a medium for the treatment of silver-containing wastewater has been done by using adsorption method. The modified hydrotalcite was prepared by direct precipitation using Mg and Al metals precursor and an organic compound of 3,4,5-trihydroxybenzoic acid. The modified material was then used for the treatment of silver-containing wastewater through adsorption method together with the determination of adsorption capacity, kinetics study as well as the material characterization after the treatment process. It could be observed from the results that the adsorption of silver onto the modified hydrotalcite was in accordance with Langmuir adsorption model with the adsorption capacity (q_{max}) of 400 $mg.g^{-1}$. Moreover, the adsorption kinetics of the treatment showed that the process followed the kinetics model of pseudo-second-order of Ho with the velocity constant (k) of $1.15x10^{-4} g.mg^{-1}.minute^{-1}$. The adsorption process exhibited an exergonic property with the adsorption energy of $28.10x10^3$ $J.mol^{-1}$. All of the adsorption processes were conducted in the condition of $pH_{adsorption} = 3$. Meanwhile, results from the characterization using FTIR showed the functional groups of C=C, -OH and NO_3^- on the modified hydrotalcite underwent some alteration after the adsorption process of Ag(I). Therefore it can be observed that the functional groups of COOH and OH had an important role in the adsorption of silver from wastewater.

1. Introduction

Silver-containing wastes are presented in the environment either in the form of solid or liquid waste. The waste in the form of liquid is generated for example from the photography industry because the process of photo printing using a washing solution and fixer which each of them contains silver solution [1]. Silver-containing waste in a solid form is produced from electronic material that contain various metals such as Au 0.1%; Ag 0.2%; Pd 0.005%; Cu 20%; Ni 2%; Sn 4% and Pb 2% [2].

The treatment process of industrial waste particularly the silver-containing one can be done by using materials as adsorbent through adsorption process. Adsorption method has been used for metal recovery from water [3]. The election of adsorbent in this process is tremendously important. The chosen adsorbent must provide a high adsorption capacity, low cost, and an effective capability in adsorption process [4].

Clay material is a layered hydroxide material which is commonly applied in many fields. One of the clay which numerously applied as an adsorbent is hydrotalcite (HT) [5]. Utilization of HT is also applied in the adsorption process of Cu(II), Zn(II) dan Cd(II) [6,7], Pb(II) [8], Cr, As, Se and Hg in the form of species; CrO_4^{2-}, $Cr_2O_7^{2-}$, $H_2AsO_3^-$, AsO_4^{3-}, SeO_3^{2-}, SeO_4^{2-}, $HgCl_4^{2-}$[9], respectively. HT which is modified with organic compound such as decanoic acid, can increase the adsorption capacity for adsorption process of dyes in a solution [10]. Therefore, in this research, HT modified with 3,4,5-

trihydroxybenzoic acid was prepared to be used in the process of adsorption of silver in a solution, together with the observation of the kinetics studies, adsorption and material characterization for the initial and the final adsorption process.

2. Experimental

2.1. Materials

Materials which were used in this research acquired in pro analysis (p.a.) quality, i.e. $Mg(NO_3)_2{\cdot}6H_2O$, $Al(NO_3)_3{\cdot}9H_2O$, HCl, NaOH, HNO_3, $Ag(NO_3)$ and 3,4,5-trihydroxybenzoic acid ($C_7H_6O_5{\cdot}H_2O$), H_2O, N_2 gas and Whatman filter paper no. 42.

2.2. Instruments

Some laboratory equipments were used in the research, i.e. magnetic stirrer, analytical balance, pH-meter, oven, mortar and pestle. Moreover, analytical instruments which were used i.e. *Fourier Transform Infrared* (FTIR), *Atomic Absorption Spectroscopy* (AAS), *shaker*, glassware and plastic apparatuses.

2.3. Synthesis of Hydrotalcite Modified with 3,4,5-trihydroxybenzoic acid

Some amount of $Mg(NO_3)_2{\cdot}6H_2O$, $Al(NO_3)_3{\cdot}9H_2O$ dan ($C_7H_6O_5{\cdot}H_2O$) in a mole ratio of 2:1:0.5 were mixed and dissolved in a 100 mL volumetric flask with CO_2 –free aquadest. The mixture was then added with 0,5 M NaOH solution dropwise until the solution pH was 9 and continued with stirring for ±30 min under the N_2 flow. The mixture was then heated at 120 °C for 5 h, and cooled until the precipitation process was complete. The precipitate was separated by centrifugation then washed using CO_2–free aquadest until the pH was neutral. The colloidal precipitate still contained with water was then filtered by using Whatman filter paper no. 42 and dried at 70 °C for 48 h in an oven. The dried solid was then grinned using porcelain mortar and pestle. Finally the solid was characterized using FTIR in order to observe the vibration of some functional groups contained in the adsorbent.

2.4. Kinetics study of Ag(I) adsorption using the synthesized adsorbent.

As much as 10 mg hyrotalcite modified with 3,4,5-trihydroxybenzoate was interacted with 10 mL solution of Ag(I) 500 $mg.L^{-1}$ in the pH of 3, the mixture was then shaken in the various time i.e. 0, 10, 20, 30, 60, 90, 120, 150, 200, 250, 300 and 350 min., afterwards, it was filtered by using Whatman filter paper no. 42. At last, the concentration of Ag(I) was measured using AAS.

2.5. Study of adsorption isotherm of Ag(I) onto hydrotalcite modified with 3,4,5-trihydroxybenzoic acid

As much as 10 mg of the modified hydrotalcite was interacted with 10 mL of Ag(I) solution in the concentration variation of 0, 100, 200, 400, 600, 800 and 1000 $mg.L^{-1}$ under the condition of pH of 3. Then the mixture was stirred for 120 min. and filtered with Whatman filter paper no. 42. Concentration of Ag (I) was analyzed using AAS.

3. Result and Discussion

Study of Ag(I) adsorption was conducted via two methods i.e. (1) determination of adsorption kinetics of Ag(I) by adsorbent of hydrotalcite modified with 3,4,5-trihydroxybenzoic acid in the variation of time and constant concentration of Ag(I), constant pH system, and constant mass of the adsorbent i.e. 500 $mg.L^{-1}$, pH=3 and 10 mg, respectively, and (2) determination of the model of adsorption isotherm of

Ag(I) by the modified hydrotalcite to determine the adsorption capacity of Ag(I) by the adsorbent in a various concentration of Ag(I) in the constant mass of the adsorbent and the constant pH i.e. 10 mg and pH=3, respectively. Under the condition of pH=3 (adsorption process), Ag ion was present in the form of Ag(I) [11].

Adsorbtion kinetics of Ag(I) onto the modified hyrotalcite which was synthesized in the mole ratio of Mg^{2+}:Al^{3+}:AG (2:1:0,5) and one which was synthesized in the pH of 9 was observed by interacting as much as 10 mg of the modified hydrotalcite with 10 mL of Ag(I) 500 mg.L^{-1} solution in a time variation of 0, 15, 30, 60, 90, 120, 150, 200, 250 and 300 min. so that the adsorption equilibrium could be obtained. Active sites on the modified adsorbent would be more actively interact with Ag(I) along with the contact time until the adsorption equilibrium was achieved between the adsorbent and the adsorbate.

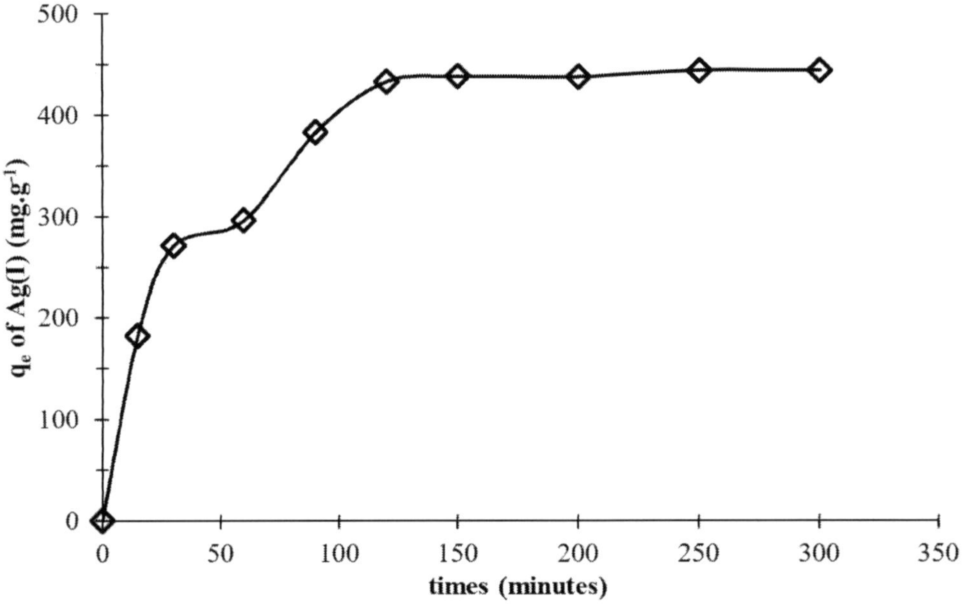

Figure 1. Effect of time on the adsorption of Ag(I) by hydrotalcite modified with 3,4,5-trihydroxybenzoic acid

Based on the results which is shown in Figure 1 adsorption equilibrium of Ag(I) (q_e) by the modified hyrotalcite occurred at the minute of 120 (2 h). After the equilibrium was reached, the adsorption of Ag(I) increased along with the time, however, after 120 min the increase was less significant. Therefore it can be concluded whether the adsorption equilibrium was occurred at the minute of 120. Model of adsorption kinetics of Ag(I) by the adsorbent was observed using kinetics models of Langmuir-Hinshelwood, Lagergreen, Santosa and Ho. Thus, from the kinetics models were obtained some data as it is exhibited in the Table 1.

The model of adsorption kinetics from Langmuir-Hinshelwood is shown an equation in Equation 1.

$$\frac{\ln\frac{C_0}{C_A}}{C_0-C_A}+K=\frac{k.t}{C_0-C_A} \quad (1)$$

In which C_0 is an initial concentration, C_A is solution concentration at t, t is adsorption time, k is adsorption rate constant and K is equilibrium constant.

The kinetics model of Langmuir-Hinshelwood was obtained by using linier curve between $\frac{\ln \frac{C_0}{C_A}}{C_0 - C_A}$ vs $\frac{t}{C_0 - C_A}$ so that the slope of the resulting linier equation was assumed as k and the intercept as K. In the kinetics model, the obtaining K value was negative and it was irrational for the adsorption, therefore the other equation of kinetics model was used. In which, this new model assumed that the adsorption rate of the process involved an adsorption in the order one [12]. This model is called as Santosa kinetics model of adsorption, which is written as Equation 2.

$$\frac{\ln \frac{C_0}{C_A}}{C_A} = \frac{k.t}{C_A} + K \tag{2}$$

The Santosa model was obtained by plotting a linier curve between $\frac{\ln \frac{C_0}{C_A}}{C_A}$ vs $\frac{t}{C_A}$ so that the obtaining slope was assumed as k and the intercept as K. Equation of Santosa kinetics model is more realistic because it gives k and K as a positive value. Another kinetics model which is known as Lagergreen (pseudo-first-order) kinetics model, has an equation of kinetics model shown in Equation 3.

$$\log(q_e - q_t) = \log q_e - \left(\frac{k}{2.303}\right) \tag{3}$$

Where, q_e is an adsorption capacity in equilibrium and q_t is an adsorption capacity at a certain t. Lagergreen kinetics model was obtained by plotting a linier curve between log q_e–q_t vs t so that the obtaining slope was assumed as (k/2.303) and the intercept was log q_e value. The adsorption kinetics was based on the adsorption capacity in solid phases which was mentioned as pseudo-second-order [13]. In the kinetics model of adsorption based on Ho, it was stated as Equation 4.

$$\frac{t}{q_t} = \frac{1}{h} + \frac{1}{q_e} t \tag{4}$$

Initial adsorption rate was stated in a symbol of h (min^{-1}) and the adsorption rate constant in a symbol of k ($g.mg^{-1}.min^{-1}$). Kinetics model of pseudo-second-order which was expressed by Ho was obtained by plotting a linier curve between $\frac{t}{q_t}$ vs t so that the obtaining slope was assumed as $\frac{1}{q_e}$ and the intercept as $\frac{1}{h}$, in the equation of $h = kq_e^2$ [13].

Table 1. Calculation results of adsorption kinetics of Ag(I) by hydrotalcite which was modified with 3,4,5-trihydroxybenzoic acid by using several kinetics model

Kinetics model	Kinetics parameters

	R^2	k	K	h
Langmuir-Hinshelwood	0.0167	$6x10^{-7}$ min^{-1}	0.2519 $L.mol^{-1}$	-
Santosa	0.8873	0.01 min^{-1}	904.92 $L.mol^{-1}$	-
Lagergreen	0.9412	0.0094 min^{-1}	-	-
Ho	0.9916	$1.151x10^{-4}$ $g.mg^{-1}.min^{-1}$	-	26.109 $mg.g^{-1}.min^{-1}$

Based on Table 1 concerning the parameters calculation of adsorption kinetics of Ag(I) onto the hydrotalcite, from the four models, it was seen that Ho kinetics model gave the best R^2 value in which $R^2 \approx 1$, compared to the other models. Thus, the adsorption process of Ag(I) by the modified adsorbent was more likely to follow the kinetics model of pseudo-second-order from Ho, with the adsorption rate constant of $1.151x10^{-4}$ $g.mg^{-1}.min^{-1}$. It meant that $1.151x10^{-4}$ g of the adsorbent could adsorb 1.0 mg of the adsorbate in 1 min. It could be indirectly assumed that there are two factors which affected adsorption rate of Ag(I) onto the modified hydrotalcite. Moreover, the adsorption constant of Ag(I) could reach $3.798x10^{-4}$ $g.mg^{-1}.min^{-1}$ by using the Ho, kinetics model.

Model of adsorption isotherm which could figure the adsorption pattern of adsorbate onto the adsorbent surface usually was observed by using isotherm model of Langmuir dan Freundlich. The model of adsorption isotherm of Langmuir in the Equation (5) was based on the assumption that adsorption energy depends on the closure of the active sites of the adsorbent in which the sites have a homogenous property so that if the sites have been contained by the adsorbate, it will cause the formation of an only monolayer of the adsorbate on the adsorbent's surface [14-17].

$$\frac{C_e}{q_e} = \frac{1}{q_{maks} K_L} + \frac{C_e}{q_{maks}} \quad (5)$$

C_e is the adsorbate concentration at the equilibrium in a solution ($mg.L^{-1}$), q_e is the amount of adsorbate which was adsorbed at the equilibrium by each g of adsorbent ($mg.g^{-1}$), K_L is Langmuir constant ($L.mg^{-1}$) dan q_{max} is the maximum capacity of the adsorbent ($mg.g^{-1}$). Determination of the maximum adsorption capacity (q_{max}) was conducted by the model of Langmuir isotherm, i.e. by plotting a linear curve between C_e/q_e vs C_e.

$$\log q_e = \log K_F - \frac{1}{n} \log C_e \quad (6)$$

Model of adsorption isotherm of Langmuir in the Equation 6 was based on the assumption that the adsorption energy depends on the interaction of adsorbate with the adsorbent's surface which has a heterogeneous property [14-17]. The surface heterogeneity could cause different interaction among one active site and the others with the adsorbate molecule so that the adsorbate's surface which was bonded on the adsorbent's surface could form more than one layer (*multilayers*).

Table 2. Parameters of adsorption isotherm of Freundlich and Langmuir in the adsorption process of Ag(I) by hydrotalcite modified with 3,4,5-trihydroxybenzoic acid

Isoterm Model	**Parameters**			
	R^2	K	n	q_{max}

Freundlich	0.5294	366.7 mg.g^{-1} (K_F)	50.761	-
Langmuir	0.9992	8.4273x10^4 L.mg^{-1} (K_L)	-	400 mg.g^{-1}

In the Tabel 2, it is known that the adsorption process of Ag(I) by the modified hydrotalcite followed the model of Langmuir isotherm with the R^2 value was 0.9992, and the adsorption capacity was 400 mg.g-^{1}. Based on the model, it was assumed that the adsorbent's surface was homogenous which caused the formation of the adsorbate monolayer from the Ag(I) adsorption onto the adsorbent's active sites. It occurred as the active sites had the same adsorption energy to the adsorbate so that it was more likely to form the adsorbate monolayer. It went to the same way for the adsorption isotherm of Ag(I) onto the modified hydrotalcite which was known to follow the isotherm model of Langmuir (Figure 2).

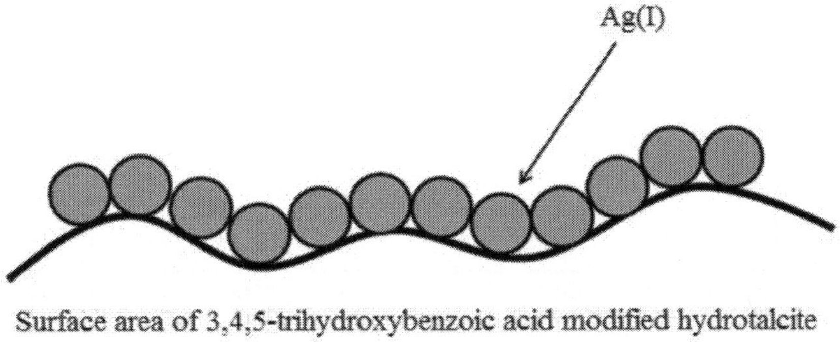

Figure 2. Illustration of the adsorption model of Langmuir in the adsorption peocess of Ag(I) onto the hydrotalcite modified with 3,4,5-trihydroxybenzoic acid

The adsorption energy was calculated by the equation of Gibbs free energy as mentioned in the Equation 1. According to the equation, adsorption energy could be calculated from the Langmuir constant so that it could be obtained value of the adsorption energy of Ag(I) by the 3,4,5-trihydroxybenzoic acid modified hydrotalcite of -28.100x10^3 J.mol^{-1} which was shown that the adsorption reaction had an *exergonic* property.

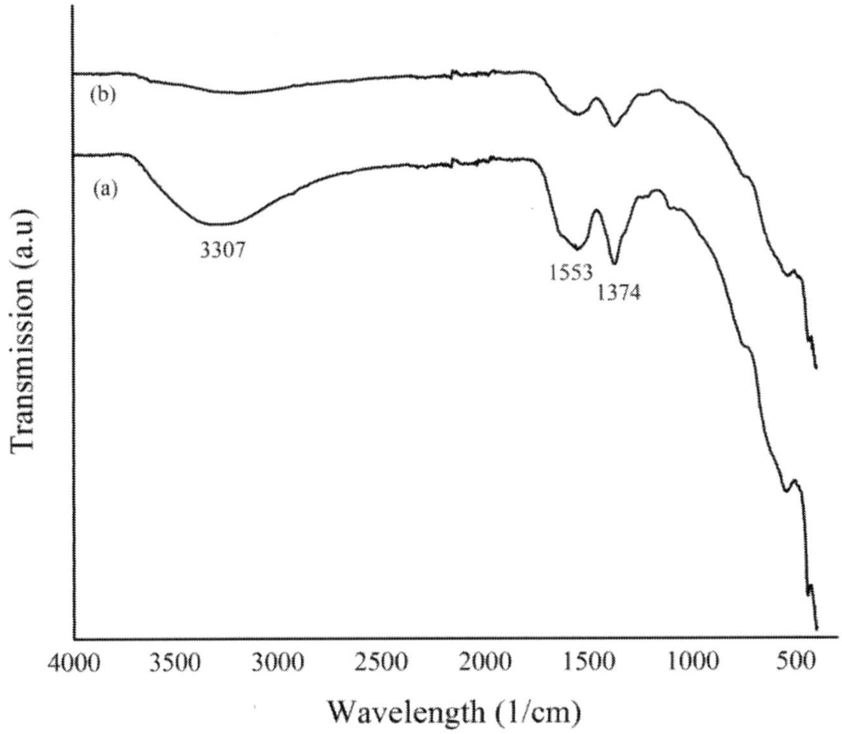

Figure 3. FT-IR spectra of 3,4,5-trihydroxybenzoic acid modified hydrotalcite (a) before and (b) after being used to remove Ag(I)

Based on the Figure 3., it could be seen that the functional groups in the 3,4,5-trihydroxybenzoic acid modified hydrotalcite had an important roles in the adsorption process of Ag(I) in the solution. It was observed from the alteration in the wavenumber of 3307 cm^{-1} which showed vibration of –OH either from the groups of 3,4,5-trihydroxybenzoic acid or from the hydroxyl group in the hydrotalcite itself. The alteration in the wavenumber of 1553 cm^{-1} showed the vibration of aromatic C=C owned by 3,4,5-trihydroxybenzoic acid. Finally, the wavenumber of 1374 cm^{-1} illustrated the vibration of N–O from the NO_3^- group which filled the interlayer area of hydrotalcite.

4. Conclusion

Adsorption model of the Ag(I) containing wastewater by using 3,4,5-trihydroxybenzoic acid modified hydrotalcite followed the adsorption model of Langmuir with the adsorption capacity (q_{max}) of 400 $mg.g^{-1}$. The results of the kinetics study from the adsorption process showed that it followed the Ho kinetics model of pseudo-second-order with the rate constant (k) of $1.15x10^{-4}g.mg^{-1}.min^{-}1$. The adsorption process was exergonic with the adsorption energy value of $28.10x10^3$ $J.mol^{-1}$. All of the adsorption conditions was set at $pH_{adsorption}$=3. The results from the characterization using FTIR figured that the functional groups of C=C, –OH dan NO_3^- in the modified hydrotalcite underwent some alterations after the Ag(I) adsorption process. Therefore, it could be assumed that the groups of COOH and OH had an important role in the adsorption of Ag(I) in a solution.

References

[1] Songkroah C, Nakbanpote W, Thiravetyan P 2004 Process Biochem. **39** 1553-9
[2] Groot D R and van der Linde J A N 2009 *The Journal of The Southern African Institute of Mining and Metallurgy* **109** 697-700.
[3] Sounthararajah D P, Loganathan P, Kandasamy J, and Vigneswaran S 2015 *Int. J. Environ. Res. Public Health* **12** 10475-89
[4] Thomas W J and Crittenden B D 1998 *Adsorption technology and design* Butterworth – Hienemann Oxford 7
[5] Cavani F, Trifiri F and Vaccari A 1991 *Catal. Today* **11** 173-301
[6] Anirudhan T S, Jalajamony S and Sreekumari S S 2013 *J. Chem. Eng. Data* **58** 24-31
[7] Anirudhan T S and Suchithra P S 2010 *Indian Journal of Chemical Technologyi* **17** 247-259
[8] Setshedi K, Ren J, Aoyi O and Onyango M S 2012 *International Journal of the Physical Sciences* **7(1)** 63-72
[9] Mitsuo Y 2002 *J. Ecotechnology Research* **8(2)** 248-9
[10] Nong L, Xiao C and Jiang W 2011 *Korean J. Chem. Eng.* **28(3)** 933-8
[11] Alonso A R, Lapidus G T, Gonzalez I 2007 *Hydrometallurgy* **85(2)** 144-153
[12] Santosa S J, Siswanta D, Kurniawan A and Rahmanto W H 2007 *Appl. Surf. Sci.* **601** 5155-61
[13] Ho Y S 2006 *J. Hazard. Mater.* **36** 681-9
[14] Dada A O, Olaleka A P, Olatunya A M and Dada O 2012 *IOSR Journal of Applied Chemistry (IOSR-JAC)* **3** 38-45
[15] Hossein S, Nafiseh B, Sohrab A G, Ahmad H and Sara P 2013 *World Jornal of Modelling and Simulation* **9** 243-254
[16] Qing C, Yuanyuan T, Peng L, Changhui Y, Yu P, Li Z, Hucheng D, Wen Z and Xianghao M 2017 *Journal of Chemsitry* **2017** 1-11
[17] Mohamed E, Fatima Al-T, Ragwan M and Khadija A 2015 *American Journal of Analytical Chemistry* **6** 1-10

IOP Conf. Series: Materials Science and Engineering **349** (2018) 012029 doi:10.1088/1757-899X/349/1/012029

Identification and control of unspecified impurity in trimetazidine dihydrochloride tablet formulation

Jefri[1], A D Puspitasari[2], J S R Talpaneni[1], R R Tjandrawinata[1,2,3,*]

[1]Analytical Development, Dexa Development Center, Dexa Medica, Cikarang, Indonesia
[2]Formulation Development, Dexa Development Center, Dexa Medica, Cikarang, Indonesia
[3]Dexa Laboratories of Biomolecular Sciences, Dexa Medica, Cikarang, Indonesia

Email: raymond@dexa-medica.com*

Abstract: Trimetazidine dihydrochloride is an anti-ischemic metabolic agent which is used as drug for angina pectoris treatment. The drug substance monograph is available in European Pharmacopoeia and British Pharmacopoeia, while the drug product monograph is not available in any of the pharmacopoeias. During development of trimetazidine dihydrochloride tablet formulation, we found increase of an unspecified impurity during preliminary stability study. The unspecified impurity was identified by high performance liquid chromatography coupled with mass spectrometry (LC-MS) and the molecular weight obtained was matching with the molecular weight of *N*-formyl trimetazidine (m/z 295). Further experiments were performed to confirm the suspected result by injecting the impurity standard and spiking formic acid into the drug substance. The retention time of *N*-formyl trimetazidine was similar to the unspecified impurity in drug product. Even spiking of formic acid into drug substance showed that the suspected impurity increased with increasing concentration of formic acid. The proposed mechanism of impurity formation is via amidation of piperazine moiety of trimetazidine by formic acid which present as residual solvent in tablet binder used in the formulation. Subsequently, the impurity in our product was controlled by choosing the primary packaging which could minimize the formation of impurity.

1 Introduction

Trimetazidine (1-(2,3,4-trimethoxybenzyl)piperazine) dihydrochloride (Figure 1) is an anti-ischemic metabolic agent used for treatment of angina pectoris with no vasodilator properties[1]. The drug's mechanism of action is through inhibition of long-chain 3-ketoacyl CoA thiolase activity, thus inhibiting the fatty acid oxidation and stimulating glucose utilization[2]. Clinical studies show that anti-ischemic effect of trimetazidine is not associated with myocardial oxygen consumption which affects the heart rate and systolic blood pressure[1].

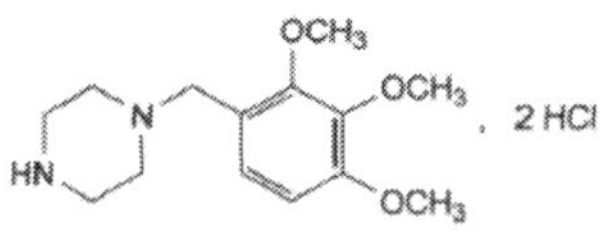

Figure 1. Chemical structure of trimetazidine dihydrochloride

The monograph of trimetazidine dihydrochloride drug substance is available in European Pharmacopoeia (EP) and British Pharmacopoeia (BP). However, currently there are no monographs available for the drug tablet formulation; hence it is considered as non-compendial product. In the drug substance monographs, there are nine specified impurities (impurity A – I)[3,4]. None of the impurities specified in drug substance monograph were observed during drug product development as most of them were process-related impurities owing to the route of synthesis. Preliminary stability studies of the drug product were performed at 30 °C, 75% RH and 40 °C, 75% RH for one month and showed that one unspecified impurity at relative retention time (RRT) 1.2 with respect to trimetazidine was increasing with time. The impurity level increased up to 0.24% in 4 weeks under accelerated stability condition (40 °C, 75% RH). This value has exceeded the identification threshold as per International Conference on Harmonisation (ICH) guideline (0.2% for maximum daily dose of >10 mg – 2 g)[5]. Hence, the identification of this unspecified impurity using LC-MS was initiated and measures were taken to control it in our drug product.

2 Experimental

2.1 Chemicals, reagents and samples

Trimetazidine dihydrochloride drug substance was purchased from Bachem SA (Vionnaz, Switzerland) and the tablet formulation (35 mg/ tablet) was developed in-house by Dexa Medica. Trifluoroacetic acid spectroscopy grade, methanol HPLC grade, ortho-phosphoric acid 85% AR grade and formic acid 98-100% AR grade were commercially obtained from Merck (Germany). Ion-pairing agent sodium 1-heptanesulfonate was procured from Tokyo Chemical Industry (TCI) Co. Ltd. Water used for chromatography was prepared using Milli-Q water purification system from Merck Millipore. *N*-formyl trimetazidine impurity standard was procured from Simson Pharma (Mumbai, India).

2.2 LC-MS conditions

Identification of the unspecified impurity was performed on Shimadzu LCMS-2010 series equipped with Photo Diode Array (PDA) detector. The method was based on the publication of Y Jiao and co-workers[6] with modifications in column length and flow rate. Zodiac-C_{18} column with dimensions of 250 mm x 4.6 mm; 5 μm particle size was utilized at 40 °C. The mobile phase consisted of water adjusted to pH 2 with trifluoroacetic acid : methanol = 6:4 (v/v). Flow rate was 0.8 mL/min in an isocratic mode and injection volume was 50 μL. The UV chromatogram was extracted at 240 nm. MS spectrum was scanned from *m/z* 100-450 using Atmospheric Pressure Chemical Ionization (APCI) probe in positive ion mode.

2.3 HPLC-UV conditions

Confirmation of identification result using HPLC-UV was performed on Waters alliance e2695 separation module equipped with Waters 2489 UV/Vis detector and *Empower 2* data handling system. The method was directly adopted from monograph of trimetazidine dihydrochloride drug substance[3,4] utilizing Atlantis dC-18 column (150 mm x 4.6 mm; 5 μm particle size).

2.4 Sample preparation

For LC-MS analysis, one tablet of trimetazidine dihydrochloride was transferred into 50 mL volumetric flask (previously rinsed with methanol and dried), added 15 mL of methanol and sonicated for 15 minutes to disintegrate the tablet, diluted to volume with water and centrifuged at 6000 rpm for 15 minutes. Later, 7.5 mL of the supernatant was pipetted and transferred into 50 mL volumetric flask, diluted to volume with water and filtered through 0.2 μm PTFE membrane filter. In HPLC-UV analysis, 200 mg of drug substance was weighed and transferred into 50 mL volumetric flask, added 30 mL of water and sonicated for 5 minutes until dissolved. Formic acid solution was spiked into three sample solutions separately to obtain three concentration levels (50, 100, 150 ppm), then each solution was diluted to volume with water and mixed homogenously.

3 Results and discussion

3.1 Identification of unspecified impurity by LC-MS

The stability sample of trimetazidine dihydrochloride tablet was analyzed using LC-MS with conditions as previously mentioned to identify the unspecified impurity by its molecular weight. The target impurity eluted after the main peak with retention time (RT) at about 5.5 min. Figure 2 and 3 shows the typical chromatogram of sample and mass spectrum of unspecified impurity obtained in the analysis.

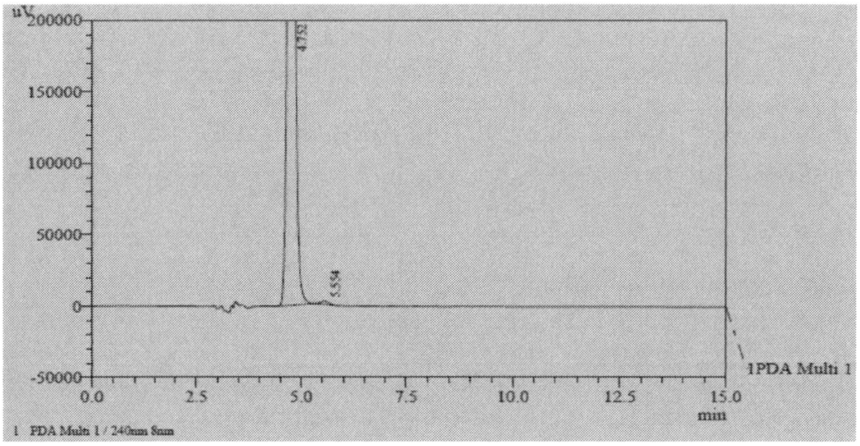

Figure 2. Typical chromatogram of sample in LC-MS analysis

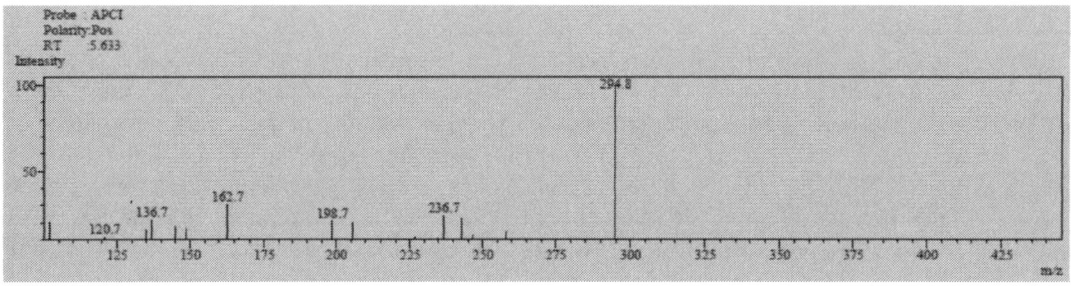

Figure 3. Mass spectrum of unspecified impurity

The mass spectrum, recorded in positive ion mode, of the unspecified impurity exhibited a base peak at *m/z* about 295 (M+H) which suggested that the molecular weight of impurity was 294. Compared to trimetazidine (MW 266), the molecular weight of the impurity is 28 amu more than the main compound. This mass difference corresponds to carbonyl (C=O) functional group which can attach to the piperazine moiety of trimetazidine forming the *N*-formyl impurity whose structure is shown in Figure 4 below.

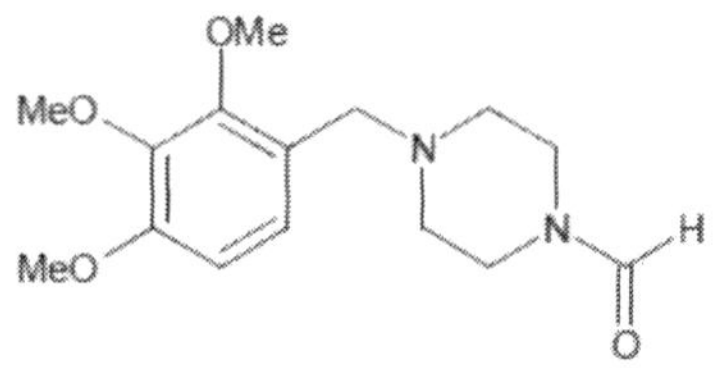

Figure 4. Chemical structure of *N*-formyl trimetazidine

3.2 Confirmation of identification result by HPLC-UV

Based on its molecular weight, the unspecified impurity was suspected as *N*-formyl trimetazidine. To further confirm the identity of impurity, we purchased the impurity standard and compared the retention time of impurity with the standard in HPLC. Moreover, three concentration levels (50, 100, 150 ppm) of formic acid was spiked into the sample to observe the correlation between impurity formation and formic acid concentration as the acid could react with trimetazidine to form *N*-formyl trimetazidine[7]. The chromatogram in Figure 5a showed that the retention time of the unspecified impurity was matching with *N*-formyl trimetazidine standard at around 10 min. The other experiment (Figure 5b) demonstrated that the impurity increased along with increasing concentration of formic acid spiked into the drug substance.

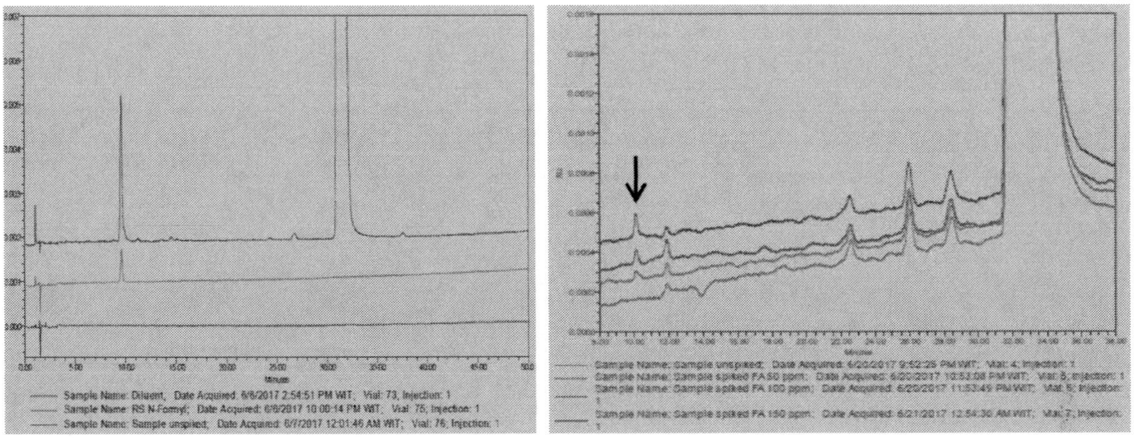

(a) (b)

Figure 5. Typical chromatograms of (a) sample and *N*-formyl trimetazidine standard, (b) spiking of formic acid into trimetazidine dihydrochloride drug substance in HPLC-UV analysis

3.3 Proposed mechanism of impurity formation

Increase of impurity level by spiking of formic acid suggested that the formation of *N*-formyl trimetazidine involved reaction between trimetazidine in the drug product with formic acid[7]. After reviewing the list of excipients used in the formula, the tablet binder contains formic acid as residual solvent with limit of 0.5% (reported value 0.2%). The presence of formic acid in the tablet binder could react with the drug substance via amidation of piperazine moiety of trimetazidine forming the *N*-formyl impurity as shown in Figure 6.

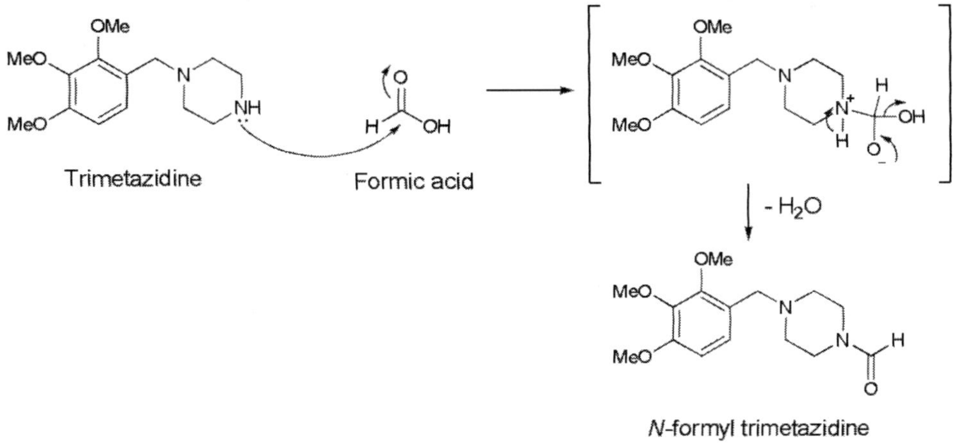

Figure 6. Proposed mechanism of *N*-formyl trimetazidine formation

3.4 Control of impurity in finish product

Although the root cause of the impurity formation had been identified, the use of the tablet binder which contained formic acid was inevitable due to its critical function in the tablet formulation. The other solution was to control the impurity by minimizing its formation. In one of the optimization trials, we observed that primary packaging material selection might have impact on the impurity formation. Based on the data presented in Table 1, we concluded that polycellonium strip was the best primary packaging for the drug product.

Table 1. Impurity result of trimetazidine dihydrochloride tablet in three different packaging materials after stored in climatic chamber at 60 °C, 75% RH for one week

Impurities	**Primary packaging materials**		
	PVC blister	**PVDC blister**	**Polycellonium strip**
Impurity A	ND	ND	ND
Impurity B	ND	ND	ND
Impurity C	ND	ND	ND
Impurity D	ND	ND	ND
Impurity E	ND	ND	ND
Impurity F	ND	ND	ND
Impurity G	ND	ND	ND
Impurity H	ND	ND	ND
Impurity I	ND	ND	ND
N-formyl trimetazidine	0.28%	0.26%	0.22%
Total impurities	0.28%	0.26%	0.22%

*ND = not detected

4 Conclusion

The unspecified impurity in trimetazidine dihydrochloride tablet formulation was identified as *N*-formyl trimetazidine by means of LC-MS. Further experiments confirmed the identity of the impurity. The possible mechanism of the impurity formation was also proposed which involved reaction of the drug with residual formic acid present in tablet binder used in the formula. Finally, the impurity formation was controlled by choosing the best primary packaging material used for the drug product.

Acknowledgments

The authors would like to acknowledge Simson Pharma/ Simson Labs for the help in carrying out the LC-MS analysis as well as the management of Dexa Medica for supporting this work. Authors also wish to thank the colleagues of Analytical, Formulation, and Packaging Development of Dexa Development Center for their support and co-operation during this work.

References

[1] McKellan K J and Plosker G L 1999 *Drugs* **58(1)** 143-57
[2] Kantor P F, Lucien A, Kozak R and Lopaschuk G D 2000 *Circ. Res.* **86** 580-88
[3] Council of Europe 2016 *European Pharmacopoeia 9.0* vol III (Strasbourg: EDQM) pp 3843-45
[4] British Pharmacopoeia Commission 2016 *British Pharmacopoeia 2017* vol II (London: The Stationery Office) pp 1150-52
[5] International Conferences on Harmonization 2006 *Q3B (R2): Impurities in New Drug Products*
[6] Jiao Y, Su M, Chen M, Jia W, Chou Y, Huang Z, Yang N and Tong W 2007 *J. Pharm. Biomed. Anal.* **43** 1804-07
[7] Baertschi S W (Ed.) 2005 *Pharmaceutical Stress Testing: Predicting Drug Degradation* (Indianapolis: Taylor & Francis) p 73

IOP Conf. Series: Materials Science and Engineering **349** (2018) 012030 doi:10.1088/1757-899X/349/1/012030

The catalytic activity of CoMo/USY on deoxygenation reaction of anisole in a batch reactor

K D Nugrahaningtyas*, I F Putri, E Heraldy, Y Hidayat
Department of Chemistry, Faculty of Mathematics and Natural Sciences, University of Sebelas Maret, Surakarta, Indonesia

*E-mail: khoirina@mipa.uns.ac.id

Abstract. The catalytic hydrodeoxigenation of the bio oil model compounds (biomass pyrolysis results) typically uses sulphide catalysts. In this study, we studied the activity of non-sulphide catalyst, the effect of temperature and reaction time on anisole deoxygenation. The catalytic activity was performed in a batch reactor, using N_2 gas at 1 bar of pressure. The product was analyzed by gas chromatography (GC) and gas chromatography-mass spectrometry (GC-MS). The result showed that the Co-Mo/USY catalyst perform a highest activity and produce pentamethylbenzene, an oxygen free products, when reaction time is 2 hours. The Co-Mo/USY catalysts has the value of the total yield of the product increased with time increase drastically.

1. Introduction

Bio oil is an alternate energy resource for replacing Crude oils [1]. It is able to minimize air pollution, by mean of replacing the CO_2 emissions released. But the bio oil cannot be directly used as a fuel, because it contains various compounds in the form of alcohols, aldehydes, sulphur, and oxygen. The highest content in the bio oil is oxygenated compounds that will generate undesirable properties, such as high thermal instability, high viscosity, and low heating value [2]. Therefore, to earn quality fuel products, deoxygenation reaction is one alternative to improve the characteristics [3]. The composition of bio oil is very complex, so there is required models compounds such as anisole as reactant, which is capable to representing the actual compounds[4–12].

The deoxygenation process for removing carbonyls group in aromatic compounds need catalyst [11]. According to Mora et al. [13], the catalyst CoMo/Al_2O_3 capable to produce the highest benzene yield from deoxygenation of hydroxyl anisole. Jongerius et al. [14], explained that the catalysts CoMo/Al_2O_3 provide a high deoxygenation activity of lignin. Anisole is derivative of lignin, that it has an oxygen atom at methoxy group [15].

However, Al_2O_3 support has a lower catalytic activity than the Ultra Stable Y-Zeolite (USY) on desulphurization reaction of thiophene [16]. This study was investigated the used of the CoMo/USY catalyst for deoxygenation reaction of anisole. The Co metal has a moderate energy for adsorption and desorption mechanism, so it is suitable for the hydrogenation reaction [17]. The Mo metal can activate C-O bonds [18,19], that is making it suitable for deoxygenation reaction [18]. The acid sites of USY serve the active site so that a combination of metal with USY will produce bifunctional catalysts [20–22].

Temperature and time of reaction influent to activate catalyst and to remove the C-O bond. The product of hydroxy anisole using a Pd/Al_2O_3 catalyst increase with reaction temperatures from 200°C until 300°C[23]. But according to Payormhorm et al. [24], at higher temperature (300°C-420°C), the deoxygenation product is declined. Meanwhile, Lee et al. [25] said thatthe raising time (30-180 minutes) of hydroxy anisole deoxygenation causes cyclohexane product increased. The catalytic reaction with batch systems [25,26], has a lower cost and simple operation than the flow system. So, this research aimed to test the catalytic activity of CoMo/USY on the anisole deoxygenation in a batch reactor with varied of catalyst types, temperature, and reaction time.

For captions not placed at the side of the figure, captions should be set to the width of the figure for wider figures, centred across the width of the figure, or, for narrow figures with wide captions, slightly extended beyond the width of the figure. The caption should finish with a full stop (period).

2. Experimental

The catalyst support used in this study was NH4Y zeolite, that was purchased from Tosoh Inc. (Japan). Catalyst metal precursors for deoxygenation reactions (Co and Mo) were ammonium heptamolybdate and cobalt nitrate hexahydrate purchased from Merck (Germany). Anisole (methoxy benzene) was 99% from sigma Aldrich.

The catalyst that used in this study was prepared by using Rachmadani methods [27]. Deoxygenation reaction was first conducted by oven of catalyst before used for 3 hours at temperature of 110°C. Amount of dry CoMo/USY catalysts (0.5 grams) and 5 mL anisole are added to columns autoclave. The next column autoclave injected into the furnace. Reactor was prepared, as shown as Figure 1. Then the reactor cleaned using N_2 gas (all valves opened) for 5 minutes. After 5 minutes, the valve no 2 are closed.The reactor is filled N_2 gas until reach 1 bar pressure. Then the first valve was closed or locked, N_2 gas removed from the reactor. The reactor heated until 200°C, then left to 60 minutes. After that, the liquid product obtained, analyzed by GC and GC-MS. The same procedure for the reactor temperature 250°C and 300°C, the time on stream for 120 minutes, and the catalyst sample are Co-Mo/USY and Mo-Co/USY.

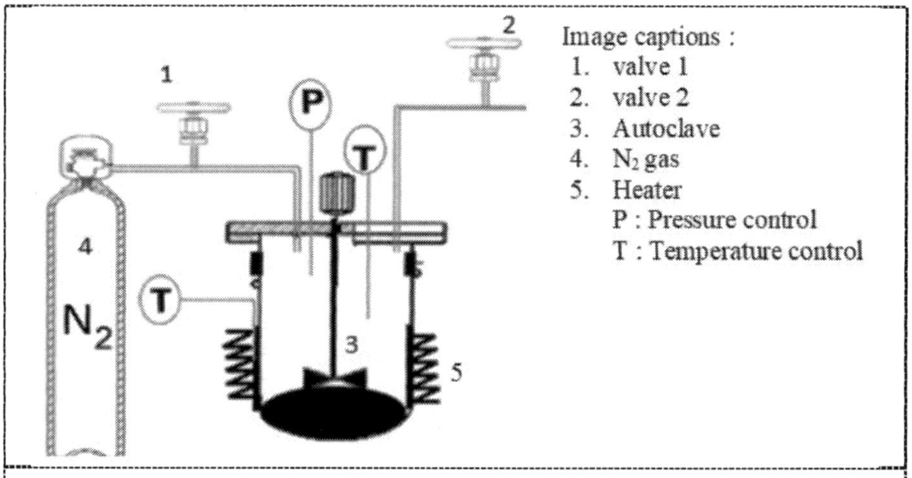

Figure 1. The batch reactor for anisole deoxygenation [26].

The conversion of anisole (total yield) and the selectivity of component i are calculated based on the following equation [25]:

$$Conversion\ (\%) = \frac{(Weight\ of\ reactant)_{initial} - (Weight\ of\ reactant)_{final}}{(Weight\ of\ reactant)_{initial}}\ x\ 100 \quad (1)$$

$$\text{Selectivity of component i } (\%) = \frac{Weight\ of\ product\ i}{Weight\ of\ total\ products}\ x\ 100 \quad (2)$$

3. Result and Discussion

The catalytic activity tests of CoMo/USY, Co-Mo/USY, and Mo-Co/USY on deoxygenation reaction of anisole at temperature from 200°C until 300°C. The total yield is the yield of products containing oxygen and without oxygen. Correlation between the types of catalysts and total yield anisole at various temperatures with a reaction time of 1 hour is presented in Figure 2.

Based on the graph in Figure 2, the highest total yield is provided by CoMo/USY catalyst at 300°C is 24.64%. It was estimated that the total acidity of CoMo/USY catalyst highest compared with other catalysts, is 11.206 mmol/g (showed in Table 2). This will indicate the strength of the Brönsted and Lewis acid sites is high , so that the catalytic activity is maximal [25].

The product yield increases with rising reaction temperature (from 200 to 300 ° C) for all types of catalysts. This is according with research that conducted by Hong et al. [23]. Hong research [15] found that rising reaction temperatures from 200 to 300 °C resulted in increased deoxygenation products of hydroxyanisole.

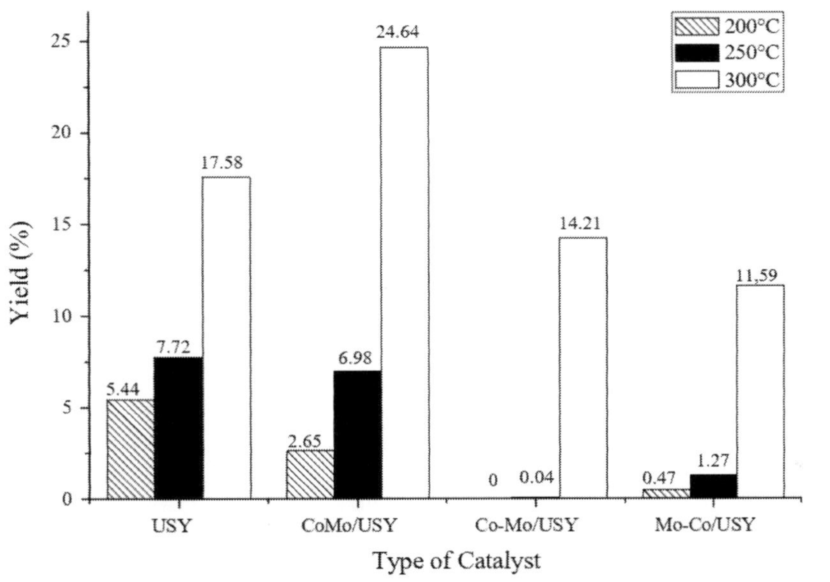

Figure 2. The correlation of catalyst type and total yield on various temperatures at 1 hour reaction time.

For each type of catalyst, the increased temperatures will generally rise deoxygenation activity, which is characterized by high conversion of anisole. In addition, at higher temperatures, the reactants are more actively moving, so the probability of collisions between the reactants and the catalyst becomes larger. The greater of collision probability, the more rapid the reaction and the more and more the product is formed. This phenomenon applies equally to anisole deoxygenation reactions using CoMo / USY, Co-Mo / USY, Mo-Co / USY or USY catalysts. Based on the data of variation temperature at 1 hour reaction time (Figure 2), it was found that the best reaction temperature was 300 °C.

Meanwhile, the Co-Mo / USY catalyst has the highest surface area, total acidity of the medium, and many particle types, as shown in Table 1. Therefore, the highest total yield of the anisole deoxygenation reaction is obtained in reaction with Co-Mo / USY catalyst. This is consistent with research Osakoo et al. [28], that a sequentiale impregnation catalyst that has a higher surface area (473 m^2/g) were able to produce more product than co-impregnation catalyst (397 m^2/g). The higher surface area indicates that the distribution of active sites on the catalyst is more evenly distributed and becomes homogeneous[28]. The availability of an evenly active site allows more opportunities to interact between reactants and catalysts. The interaction causes the reactants to become active, so that the reaction is more effective and the more products are formed. It makes the role of Co-Mo / USY as a catalyst getting better [29], product yields are increasing, and stable over a long reaction period.

Table 1. The total acidity and surface area spesific of catalyst [27].

Type of Catalyst	Total Acidity (mmol/g)	Surface area spesific (m^2/g)
Mo-Co/USY	8.304	437.782
Co-Mo/USY	10.849	557.172
CoMo/USY	11.206	497.004

Table 2 shows the deoxygenation reaction anisole using USY, CoMo/USY, Co-Mo/USY, Mo-Co/USY catalyst compared with HY zeolite. HY zeolite uses H_2 gas in the catalytic reaction [30]. Based on the analysis of table 2, it was found that all of the catalysts have almost the same ability in terms selectivity of deoxygenation and hydrogenation.

Table 2. Total Product Selectivity of anisole deoxygenation at temperature 300°C and 1 hours.

Product	Type of Catalysts				
	HY zeolite*	USY	CoMo/USY	Co-Mo/USY	Mo-Co/USY
Phenol	0.60	0.56	0.48	0.71	0.17
2-methylanisole	0.13	–	–	–	0.25
4-methylanisole	0.15	0.40	0.17	0.28	0.54
2-methylphenol	0.11	–	0.19	–	–
2,3-dimethylanisole	–	0.03	0.02	–	–
2,6-dimethylphenol	0.01	–	–	–	–
4-methylphenol	0.04	–	–	–	–

*Runnebaum, *et al.* (2011) [30]

Co and Mo metal content for each type of catalyst has the same amount, so the reaction occurs in the impregnation catalyst and co-impregnation produce almost the same product. The results showed that the reaction product that always formed was phenol and 4-methylanisole. In addition, also obtained that all products produced still contain oxygen. This is possible because the supply of hydrogen is only anchored from a USY carrier, resulting in a hydrogen source deficiency reaction. The lack of hydrogen source causes the hydrogenation reaction not to occur maximally. Although hydrogen is slightly, there is still a methyl transfer reaction at the Brönsted acid site of USY.

According to Ruiz *et al.* [31], the catalyst activity for 2-hydroxy anisole becomes more optimal in the event of a relatively long interaction between the reactants with the catalyst. They found that the longer reaction time (0-4.15 hours) of deoxygenation of 2-hydroxy anisole, the more liquid product of phenol and cyclohexene was formed. Therefore, in this study deoxygenation reaction applied with various types of catalyst at reaction temperature of 300 ° C for 2 hours. Correlation types of catalysts with a total yield of the product at 300°C for 1 and 2 hours is presented in Figure 3.

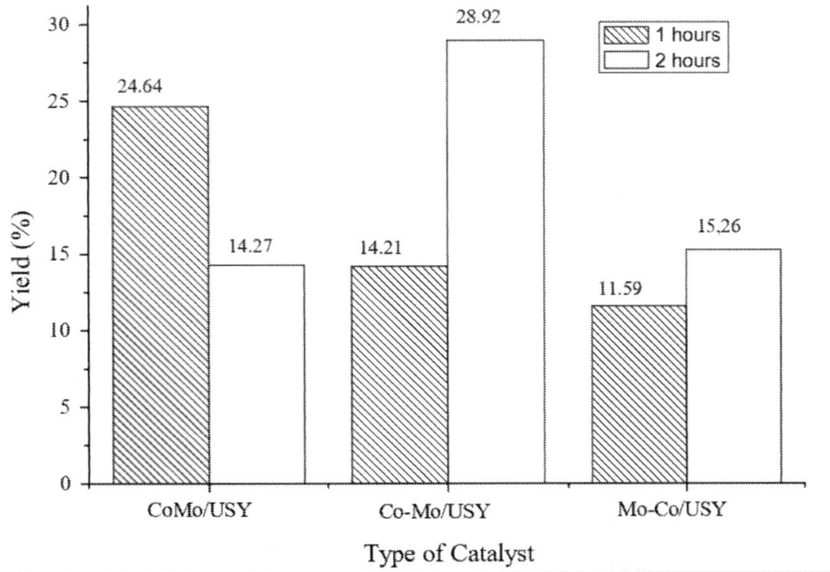

Figure 3. Correlation types of catalysts with a total yield of the product at 300°C for 1 and 2 hours

After 2 hours reaction time, CoMo / USY catalyst becomes unstable and total product yield decreases. This is possible because of the relatively smaller surface area of CoMo / USY and the highest acidity compared to other catalysts, causing the CoMo / USY catalyst become rapidly faster deactivated. The catalyst deactivation can be characterized by a decreased conversion reaction, although the deoxygenation of anisole has only been running for 1 hour. These results are consistent with Runnebaum, *et al.* [32]. According to Hellinger, *et al.* [33], deactivation of the catalyst occurs rapidly at elevated temperatures (250 - 380 ° C). However, the total yield is not proportional to the anisole conversion value. Whereas, the catalyst is said to be good when it produced more free oxygen products.

Table 3. Total product selectivity of Anisole at temperature 300°C and 2 hours.

Product	**Product selectivity (%)**		
	CoMo/USY	Co-Mo/USY	Mo-Co/USY
Phenol	31	20	18
4-methylphenol	2	3	1
2-methylphenol	–	18	12
2-methylanisole	4.4	35	40
4-methylanisole	1.8	19	23
2,3-dimethylanisole	–	6	3
1,2,3,4,5-pentamethyl benzene	–	1	–

Product selectivity at 2 hours reaction time are presented in Table 3. Catalysts Co-Mo/USY produce one compound deoxygenation oxygen-free products i.e. 1,2,3,4,5-pentamethyl benzene. Large surface area capable to make the catalyst remains stable in absorb reactants, although the reaction time become longer. This is in accordance with Messou *et al.* [34], that a sequential impregnation of catalyst is able to produce oxygen free (C6) compared co-impregnation catalyst in the transformation of sorbitol. Meanwhile the anisole deoxygenation used CoMo/USY and Mo-Co/USY catalyst are not produce oxygen-free products. The main reaction all of catalysts are transferring of a methyl group.

The Oxygen-free product of anisol deoxygenation only owned by Co-Mo/USY catalyst that produce 1,2,3,4,5-pentamethyl benzene. The yield product of anisole deoxygenation is 17%. Although the few supply of hydrogen from the carrier USY, but the catalyst Co-Mo/USY capable of breaking the hydrogen to react with anisole compound, thus forming an oxygen-free product. The formation of 1,2,3,4,5-pentamethylbenzene with demethylation and methyl transfer reaction.

Based on these results and in agreement with previous studies [30,35], that phenol is the product intermediate. The yield of the product intermediate is presented in Figure 4. The intermediate product yield is shown in Figure 4. The CoMo / USY catalyst has the best phenol yield, ie 11.774% at 1 hour reaction time. Meanwhile, at 2 hours reaction time, the best result is owned by Co-Mo / USY catalyst which is 5,718%. The result of the greatest 4-methylanisol product obtained by Mo-Co / USY catalyst was 6,2474% at reaction time 1 hour and for reaction time 2 hours, the biggest result was owned by Co-Mo / USY catalyst which was 5,387%.

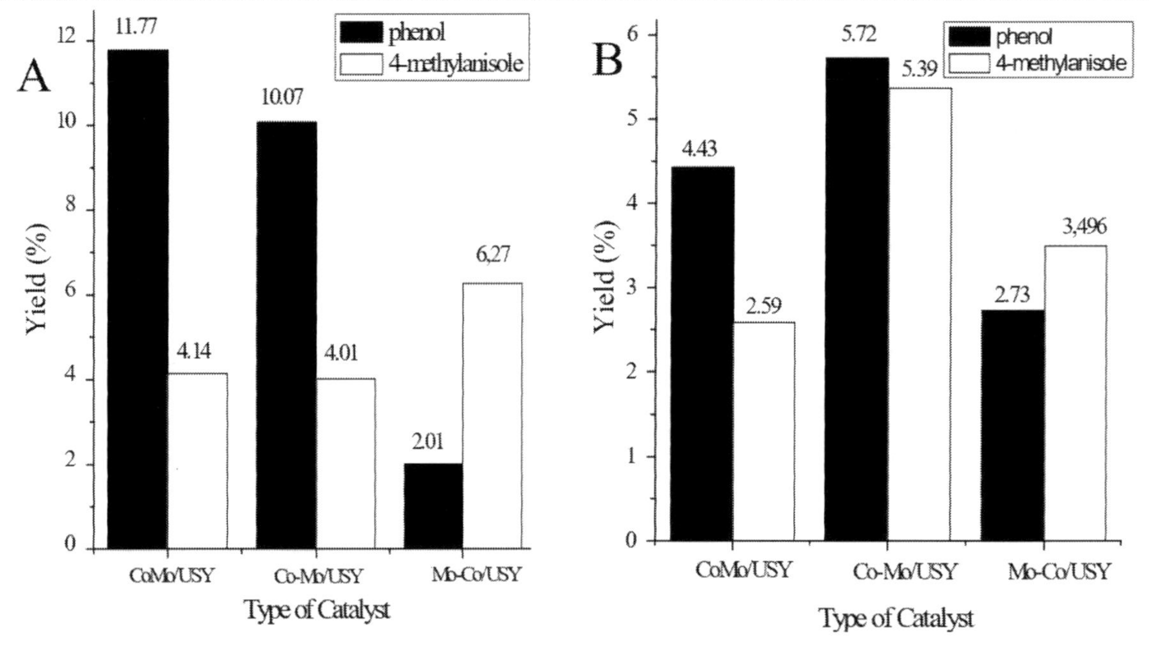

Figure 4. The yield of the product intermediate phenol (black) and 4-methylanisole (white) at (A) 1 hours and (B) 2 hours.

4. Conclusion

The product yield rises with increasing reaction temperature (200 ° -300 ° C). Meanwhile, the Co-Mo / USY catalyst has the best activity, that it is shown from the increase of the yield value (14.21% to 28.92% from reaction time 1 h to 2 h) and obtained oxygen-free products (1,2,3,4,5-pentamethyl benzene). The highest yield towards phenol products when 1 hours is CoMo/USY catalyst (11.774%) and at 2 hours is Co-Mo/USY catalyst (5.718%). While the highest yield to 4-methylanisole product when 1 hour is the catalyst Mo-Co/USY (6.274%) and time of 2 hours is the catalyst Co-Mo/USY (5.387%).

Acknowledgements
This research is funded by The Ministry of Research, Technology and Higher Education through The Directorate of general of research and community service (DP2M DIKTI) under research competent Grant (2017) with contract number **873/UN2.21/PP/2017**. Therefore, I would like to thank the Ministry of Research, Technology and Higher Education, especially DP2M DIKTI, for funding this research.

References

[1] C. González, P. Marín, F. V. Díez, S. Ordóñez, Ind. Eng. Chem. Res. 55 (2016) 2319–2327.
[2] P.M. Mortensen, J.D. Grunwaldt, P.A. Jensen, K.G. Knudsen, A.D. Jensen, Appl. Catal. A Gen. 407 (2011) 1–19.
[3] C.A. Fisk, T. Morgan, Y. Ji, M. Crocker, C. Crofcheck, S.A. Lewis, 358 (2009) 150–156.
[4] X. Zhu, R.G. Mallinson, D.E. Resasco, Appl. Catal. A Gen. 379 (2010) 172–181.
[5] D.E. Resasco, 25 (2011) 4155–4162.
[6] Y. Yang, C. Ochoa-Hernández, V.A. de la Peña O'Shea, P. Pizarro, J.M. Coronado, D.P. Serrano, Appl. Catal. B Environ. 145 (2014).
[7] E. Barker Hemings, G. Bozzano, M. Dente, E. Ranzi, Chem. Eng. Trans. 24 (2011) 61–66.
[8] T. Prasomsri, A.T. To, S. Crossley, W.E. Alvarez, D.E. Resasco, Appl. Catal. B Environ. 106 (2011) 204–211.
[9] T.M. Sankaranarayanan, A. Berenguer, C. Ochoa-Hernández, I. Moreno, P. Jana, J.M. Coronado, D.P. Serrano, P. Pizarro, Catal. Today 243 (2015) 163–172.
[10] X. Zhu, L.L. Lobban, R.G. Mallinson, D.E. Resasco, J. Catal. 281 (2011) 21–29.
[11] S. Jin, Z. Xiao, C. Li, X. Chen, L. Wang, J. Xing, W. Li, C. Liang, Catal. Today 234 (2014).
[12] S. Wang, Q. Cai, J. Chen, L. Zhang, X. Wang, C. Yu, Ind. Eng. Chem. Res. 53 (2014) 13935–13944.
[13] I.D. Mora, E. Méndez, L.J. Duarte, S.A. Giraldo, Appl. Catal. A Gen. 474 (2014).
[14] A.L. Jongerius, R. Jastrzebski, P.C.A. Bruijnincx, B.M. Weckhuysen, J. Catal. 285 (2012) 315–323.
[15] W.-S. Lee, Z. Wang, R.J. Wu, A. Bhan, J. Catal. 319 (2014) 44–53.
[16] K. Rawat, M. Kumar, J. Gupta, R. Bal, G. Dhar, A. Data, in:, 20th Natl. Symp. Catal. Energy Conserv. Env., 2010, pp. 19–22.
[17] Z. Pan, R. Wang, Z. Nie, J. Chen, J. Energy Chem. 25 (2016) 418–426.
[18] D. Kubička, L. Kaluža, Appl. Catal. A Gen. 372 (2010) 199–208.
[19] W. Wang, Y. Yang, H. Luo, T. Hu, W. Liu, React. Kinet. Mech. Catal. 102 (2011) 207–217.
[20] K.D. Nugrahaningtyas, Y. Hidayat, T.D. Saputri, Alchemy J. Penelit. Kim. 10 (2014) 40–48.
[21] K.D. Nugrahaningtyas, Y. Hidayat, Patiha, N. Prihastuti, B. Yelvi, R.U.N. Kalimah, IOP Conf. Ser. Mater. Sci. Eng. 176 (2017) 12024.
[22] R. Sesario, K. Dwi, F. Rahmawati, E. Heraldy, Rachmadani, in:, 2017, p. 20002.
[23] Y. Hong, D. Lee, H. Eom, K. Lee, "Applied Catal. B, Environ. 150–151 (2014) 438–445.
[24] J. Payormhorm, K. Kangvansaichol, P. Reubroycharoen, P. Kuchonthara, N. Hinchiranan, Bioresour. Technol. 139 (2013) 128–135.
[25] H.W. Lee, B.R. Jun, H. Kim, D.H. Kim, J.K. Jeon, S.H. Park, C.H. Ko, T.W. Kim, Y.K. Park, Energy 81 (2015) 33–40.
[26] C. Larabi, W. Al Maksoud, K.C. Szeto, A. Garron, P.P. Arquilliere, J.J. Walter, C.C. Santini, Fuel Process. Technol. 142 (2016) 192–200.
[27] Rachmadhani, Pengaruh Proses Kalsinasi Dan Reduksi Terhadap Karakter Katalis Sistem Logam-Pengemban (Co Dan Mo)/Ultra Stable Y Zeolite, Universitas Sebelas Maret, 2016.
[28] N. Osakoo, R. Henkel, S. Loiha, F. Roessner, J. Wittayakun, CATCOM 66 (2015) 73–78.
[29] M.S. Rana, J. Ancheyta, P. Rayo, S.K. Maity, 98 (2004) 151–160.
[30] R.C. Runnebaum, T. Nimmanwudipong, D.E. Block, B.C. Gates, (2011) 817–820.

[31] P.E. Ruiz, B.G. Frederick, W.J. De Sisto, R.N. Austin, L.R. Radovic, K. Leiva, R. García, N. Escalona, M.C. Wheeler, 27 (2012) 44–48.
[32] T. Nimmanwudipong, R.C. Runnebaum, D.E. Block, B.C. Gates, (2011).
[33] M. Hellinger, H.W.P. Carvalho, S. Baier, D. Wang, W. Kleist, J.D. Grunwaldt, Appl. Catal. A Gen. 490 (2015).
[34] D. Messou, L. Vivier, C. Especel, Energy Convers. Manag. 127 (2016) 55–65.
[35] T.R. Viljava, R.S. Komulainen, A.O.I. Krause, Catal. Today 60 (2000) 83–92.

IOP Conf. Series: Materials Science and Engineering **349** (2018) 012031 doi:10.1088/1757-899X/349/1/012031

Intertextual learning strategy with guided inquiry on solubility equilibrium concept to improve the student's scientific processing skills

K U Wardani[1], S Mulyani[2], Wiji[2]

[1]Students of Chemistry Education School of Postgraduate Studies, Indonesia Education University
[2]Lecturer of Chemistry Education in Indonesia Education University

Email : [1]qqdani20@gmail.com

Abstract: The aim of this study was to develop intertextual learning strategy with guided inquiry on solubility equilibrium concept to enhance student's scientific processing skills. This study was conducted with consideration of some various studies which found that lack of student's process skills in learning chemistry was caused by learning chemistry is just a concept. The method used in this study is a Research and Development to generate the intertextual learning strategy with guided inquiry. The instruments used in the form of sheets validation are used to determine the congruence of learning activities by step guided inquiry learning and scientific processing skills with aspects of learning activities. Validation results obtained that the learning activities conducted in line with aspects of indicators of the scientific processing skills.

1. Introduction

There are two things related to inseparable chemistry, chemistry as a product (chemical knowledge in the form of facts, concepts, principles, laws, and theories) the findings of scientists and chemists as a process (scientific work). Therefore, chemistry learning and chemical learning outcomes should take into account the chemical characteristics as products and processes [1]. Chemistry as part of Science, is inseparable from the process. Tawil and Liliasari (2014) declared the process of doing activities related to science commonly called the scientific processing skills [2]. In science teaching and learning activities, in addition it is important to develop scientific processing skills to ensure that students master the concepts taught well [3].

Scientific processing skills is progressing in line with various studies of science process skills. Karsli and Şahin (2009) stated that the scientific processing skills are the skills used by scientists to construct knowledge, solve problems, and make decisions [4]. Each process skill is a typical intellectual skill used by all scientists, and can be used to understand any phenomenon as well. In line with Özgelen (2012) which states that the skills of the process of science is a skill used by scientists to build knowledge, solve problems and produce decisions [5]. Meanwhile, according to Çepni *et al* define the skills of the science process as a fundamental activity that facilitates in learning science, linking methods and research techniques, helping students to be active in learning [4].

One of the learning strategies that can be used to improve scientific processing skills is the intertextual learning strategy. Intertextual learning strategies are chosen because intertekstual learning is a learning strategy that links the conceptual relationship between macroscopic, microscopic, and symbolic representation [6]. Gilbert and Treagust (2009) suggest that the linkage between the three chemical representations is one of the keys used in chemical education [7]. The linkage between the three chemical representations proposed by Adadan (2014) plays an important role in learning science to build and communicate abstract understanding [8]. When student given a phenomenon from macroscopic level, student must be observe and make decision with linkage three level representation in chemistry learning, so student doing some indicator from scientific processing skill. As stated by Tawil and Liliasari (2014) several indicators of scientific processing skills include observing, classifying, interpreting, concluding, and designing experiments, conducting experiments. Intertextual learning strategies include learning activities that connect chemical representations, students' real experiences in everyday life and classroom events [9]. Wu (2003) states that learning environments, including teachers, chemistry and learning tools, should explicitly show the relationship between macroscopic, submicroscopic and symbolic in the inquiry context [9].

In the preparation of intertextual learning strategies can be combined with the stages contained in inquiry. Through the inquiry stage students are expected to build the concept through observing activities, making hypotheses, designing experiments to conclude (Some indicators of scientific processing skills). Several studies suggesting a link between inquiry and scientific processing skills are found in studies conducted by Özgelen (2012) and Basaga, Gebban, Tekayya (1994). Özgelen argues that the student's scientific processing skills can be bridged using the inquiry model [5] and Basaga et al states that through inquiry the student's scientific processing skills can improve [10].

In general, the inquiry is divided into two, namely open inquiry and guided inquiry, as proposed by Hassard [11]. Open inquiry requires a high-level thinking as proposed by Orlich, *et al* [11]. Guided inquiry can be provided for students who do not have experience in doing inquiry [11]. In addition to that Moore, Herzogb and Perkinsa (2013) state that guided inquiry helps students specifically for investigations guided by facilitators or with guidance to make the investigation process more effective [12].

Magnusson and Palinscar suggest that guided inquiry can integrate scientific facts and principles in science [11]. If connected with intertextuals have the same characteristics to relate students' real experiences, chemical representations, and learning in the classroom. The students' real experiences can be facts that students find in everyday life. In addition to the students' experience, students can find scientific facts and principles in science when conducting classroom learning and chemical concepts using macroscopic, submicroscopic, symbolic representations.

In guided inquiry according to Bilgin (2009) teachers and students play an important role in compiling questions, building answers, and composing materials and issues, activities in guided inquiry also help students to build responsibility, cognitive student, report, solve problems and skills understand [11]. Based on the role of students in the inquiry activities proposed by Bilgin in line with the indicators contained in the scientific processing skills proposed by Tawil and Liliasari. Guided inquiry is chosen because in the inquiry phase is consistent with the indicators contained in the scientific processing skills. As the observing stage, put forward the problem, making hypotheses, designing and experimenting and conclucing. Through these stages can prove students' scientific processing skills.

The concept of solubility equllibrium is chosen because in the concept of solubility equilibrium can be developed activities to display student's scientific processing skills. Solubility equilibrium is a complex topic since it integrates solubility, molarity, physical, chemical equilibrium and Le Chatelier's principles concepts. Also, it interacts with biology concept such as osmotic pressure and osmosis [13]. For example in solubility concept, student must know to determining the saturation level of the solution. Students directed to observe the phenomenon about saturation level so that students are expected to be able to formulate the problem through asking then making hypotheses and testing hypotheses through experimental activities. So it is expected that through the stages can improve student's scientific processing skills.

The purpose of this research is to know the effectiveness of intertextual learning strategy with guided inquiry to improve student's scientific processing skills on solubility equilibrium concept. Based

IOP Conf. Series: Materials Science and Engineering **349** (2018) 012031 doi:10.1088/1757-899X/349/1/012031

on this objective, the research question states "How do student's scientific processing skills on solubility equilibrium concept using intertextual learning strategies with guided inquiry?"

2. Method

The method used in this research is Research and Development (R & D) to generate intertextual learning strategy with guidance to improve student's scientific processing skills. The object of this study is an intertextual learning strategy with guided inquiry on the concept of solubility. Subjects In this study were 21 of 11th graders in one high school in West Bandung. Instrument used to know how far it can improve students's scientific processing skills validated to decide the accuracy of measurement of the student's scientific processing skills indicators. The research was conducted up to a limited trial stage. In the early stages of development, a study of basic competencies, solubility equilibrium concepts to be developed, and various contexts about the scientific processing skills. All of them are done to obtain the indicator that will be achieved in developing the scientific processing skills.

3. Results and Discusion

3.1 Development of intertextual learning strategies with guided inquiry is based on the concept of solubility equilibrium

Development of intertextual learning strategy begins with the preparation of indicators of scientific processing skills. The scientific processing skill indicators are derived from the basic competencies of the skills. After determining scientific processing skills on solubility equilibrium concept, intertextual learning strategy with guided inquiry for improving student's scientific processing skills are designed before being validated. The intertextual learning strategy composed consists of three activities. Activities undertaken are based on concepts to be conveyed to students. Stages used in each activity have similarities, because the strategy is structured based on the stages contained in guided inquiry.

First activity is based on the concept of solubility. In the first activity the concept of solubility prepared is related to the concept of the saturation level of the solution. This is done because the concept of solubility is closely related to the concept of the saturation level of the solution, especially the saturated solution. The second activity deals with the concept of solubility equilibrium. In the second activity the students are guided to be able to determine the solubility product constant. The third activity is based on the concept of predicting the formation of sediment.

3.2 Validation of intertextual learning strategies with guided inquiry is based on the concept of solubility equilibrium

Sheet validation of intertextual learning strategy with guided inquiry to improve student's scientific processing skills with guided inquiry steps, teacher learning activities, student learning activities, student's scientific processing skill aspect, suitability of learning activities with guided inquiry steps, suitability of learning step with student's scientific processing skill aspect and suggestions for improvement.

Validation is done by experts in the field of lecturers in the department of chemical education. Validation results obtained that the learning activities conducted in line with aspects of indicators of the scientific processing skills. Based on the validation done for the concept of solubility some indicators that can be used to demonstrate students' scientific processing skills are observing, asking questions, making hypotheses, interpreting.

3.3 Scientific processing skills on solubility equilibrium concept

The student's scientific processing skills are measured using 6 test items to measure the scientific processing skills. Students are given pretest and posttest questions to find out whether student's scientific processing skills are improving. Among the pretest and posttest activities, students follow the learning activities using an intertextual learning strategy with guided inquiry. The research results of intertextual learning strategies that learning environments, including teachers, chemistry and learning tools, should explicitly show the relationship between macroscopic, microscopic and symbolic in the inquiry context [9]. The learning activities used help students to relate the three chemical representations. Students may also link prior knowledge with their new knowledge. Group discussions that are widely used in the process, can facilitate students to work together in conducting investigations, students also have the opportunity to express opinions during the learning process.

The most common skill of the students' scientific processing skills are inferring skill and observing skill, as shown in figure 1. Student's scientific processing skills in conclude skill based on posttest results have the lowest value. This can happen because in the instrument used for conclude skill it involves determining the solubility of some slightly soluble compounds. Students are still having trouble in determining the solubility value from the solubility-product constant (K_{sp}). In learning activities, students can make conclusions from an experience. For example, when students are asked to inferring from the issues raised about the maximum limit of a solute the student can make the conclusion as expected. This is because in the learning activities of the students directly observe the phenomenon about the solubility of substances associated with the maximum amount of substances that can dissolve in a particular solvent.

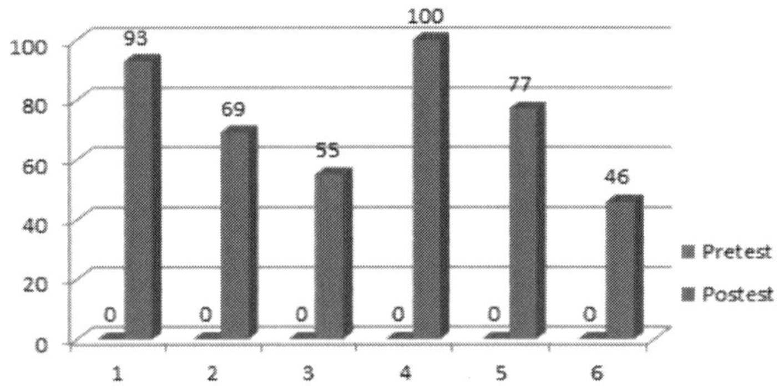

Caption: X = Indicator, Y = Persentage
1. Observe 4. Inferring
2. Thinking Question 5. Designing Experiments
3. Hypothesis 6. Conclude

Figure 1. Graph of the percentage of scientific processing skills

Based on intertextual learning strategy with guided inquiry, students get chemistry learning experience especially on the solubility equilibrium concept by connecting the macroscopic level through direct observation, microscopic level by describing the condition in solution with certain solubility and know the chemical formula of the compound used as the symbolic level.

4. Conclusion

Based on the data obtained, student's scientific processing skills through intertextual learning strategies with guided inquiry are based on the concept solubility equilibrium to be increased in terms of observing and inferring skills. Through observation activities and data interpretation students are given some data to be able to obtain information. From the information provided students can easily make observations and interpretation of data so that based on the results of tests conducted by students easily able to answer the problems given. Activities conducted through discussion can bridge the students' difficulties in understanding the given phenomenon. Students also got opportunity to express their opinions and communicate during learning process.

Acknowledgments

We thank to Dr. Heli Siti Halimatul, M.Si, Dr. Ratnaningsih Eko, M.Si, and Dr. Soja Siti Fatimah for their support and advices.

References

[1] Badan Standar Nasional Pendidikan 2006 *Standar Isi untuk Satuan Pendidikan Dasar dan Menengah : Standar Kompetensi dan Kompetensi Dasar SMA/MA* (Jakarta : BSNP)
[2] Tawil M and Liliasari 2014 *Keterampilan-Keterampilan Sains dan Implementasinya dalam Pembelajaran IPA* (Makassar : Badan Penerbit UNM)
[3] Sukarno Pemanasari A and Hamidah I 2013 The profile of science process skill (SPS) student at secondary high school (case study in Jambi) *International Journal of Scientific Engineering and Research (IJSER)*. **1** 79
[4] Karsli F Şahin Ç 2009 Developing worksheet based on science process skills : Factors affecting solubility *Asia-Pacific Forum on Science Learning and Teaching* **10** 1
[5] Özgelen S 2012 Student's science process skills within a cognitive domain framework *Eurasia Journal of Mathematics, Science & Technology Education* **8** 283.
[6] Heli S Mulyani S Dwiyanti G and Wahyu, W 2010 Pengembangan pembelajaran berbasis intertekstual untuk meningkatkan pemahaman konsep mahasiswa pada topik hormon *UPI Jurusan Pendidikan Kimia*
[7] Gilbert J K Treagust D F 2009 *Multiple Representations in Chemical Education* (United Kingdom : Springer)
[8] Adadan E 2014 Investigating the influence of pre-service chemistry teache's understanding of the particle natue of matter on their conceptual understanding of solution chemistry *Chemistry Education Research and Practice* **15** 219
[9] Wu K H 2003 Linking the microscopic view of chemistry to real-life experiences: Intertextuality in a high-school science classroom *Wiley Periodicals Inc* **87** 868
[10] Basağa H Geban Ö and Tekkaya C 1994 The effect of the inquiry teaching method on biochemistry and science process skill achievements *Biochemical Education* **22** 29
[11] Bilgin I 2009 The effects of guided inkuiri instruction incorporating a cooperative learning approach on university students' achievement of acid and bases concepts and attitude toward guided inkuiry instruction *Scientific Research and Essay* **4** 1038
[12] Moore E Herzog T and Perkins K 2013 Interactive simulations as implicit support for guided-inkuiri. *Chemistry Education Research and Practice* **14** 257
[13] Çam A and Geban Ö 2013 Effectiveness of case-based learning instruction on students' understanding of solubility equilibrium concepts *H UJournal of Education* **44** 97

Synthesis and characterization of $NaCo_{(1-x)}Mn_xO_2$ solid electrolyte using sol-gel method: the effect of milling speed variations

L Suyati, O A Widyayanti, M Qushoyyi, A Darmawan, R Nuryanto

Physical Chemistry Laboratory, Chemistry Department, Faculty of Sciences and Mathematics, Diponegoro University, Jl. Prof. Sudharto SH, Tembalang, Semarang 50275 (Telp/Fax 024- 76480824)

*Corresponding Author: linda_suyati@live.undip.ac.id

Abstract. Battery is a device that converts chemical energy into electrical energy through electrochemical process. Further research on the synthesis of cathode of Na-ion battery that has good conductivity to maximize the battery performance needs to be conducted. One of the production steps of the $NaCo_{(1-x)}NaCo$ cathode synthesis in the Na-Ion battery was a ball-milling process, in which by the ball-milling process, the crystal size of $NaCo_{(1-x)}Mn_xO_2$ cathode can be minimized. The purpose of this study was to determine the effect of variation of ball-milling speed to the characteristics of resulting product including the oxide types composing $NaCo_{(1-x)}Mn_xO_2$ cathode, surface morphology, and conductivity. The main ingredients used were sodium acetate, manganese acetate, cobalt acetate with molar ratio of 0.7: 0.66: 0.22, respectively and citric acid as chelating agent with the M/CA ratio of 1: 1. The variations of milling speed were 0, 300, 400, 500, 600 and 700 rpm. Characterization of the product was conducted using XRD, SEM-EDS, and conductivity meter (LCR-meter). The result showed that a solid electrolyte of $NaCo_{(1-x)}Mn_xO_2$ consisting of $NaMnO_2$, NaO_2, CoO, Co_2O_3, MnO_2 components was successfully synthesized. The observation on the milling speed at 400 rpm showed that the solid electrolyte produced had the highest conductivity i.e. 4.08 x 10-6 Scm^{-1} with a homogeneous surface morphology and had a spinel formula $NaCo_{0,65}Mn_{0,35}O_2$.
Keywords: Ball milling, Sodium Battery, Sol-gel, Solid Electrolytes

1. Introduction

Battery is a device that can generate electrical energy. Battery performance is transferring electrons from negative electrode (anode) to positive electrode (cathode) as to generate electric current and potential difference. Battery based on the operating principle is categorized into 2 types, namely, primary and secondary batteries. Primary battery production can only be applied for single use hence once discharged, it is discarded. On the other hand, secondary battery can be used to convert chemical energy into electrical energy through an electrochemical process. This electrochemical process takes place in reversible way therefore this battery is also called rechargeable battery [1].

The electrolyte or ionic conductor is the medium for transferring ions, which are between the anodes and the cathode in the cell. Currently the use of electrolyte tends to switch to solid electrolyte form. This trend is based on safety, leak-free media, easy to use, and can be made with smaller dimensions in sheet form [1].

The research on Na-ion batteries was conducted by Bucher *et.al.* [2] which reported that $Na_{0.7}Co_{0.11}Mn_{0.89}O_{2+z}$ cathode material was synthesized by combustion synthesis method that was in accordance with the research conducted by Cheng and colleagues [3] which the product was $Na_{2/3}Co_{2/3}Mn_{1/3}O_2$ by co-precipitation method. According to Datta *et.al.* [4], there was an effect of milling time on the character of $NaMn_2O_4$ cathode with milling condition using ball weight and sample ratio 10:1. According to research conducted by Lu *et.al.* [5], the ball milling process helped to reduce the crystal size of $Li_{1.2}Ni_{0.2}Mn_{0.6}O_2$ cathode material, which increased the conductivity of the material. While Doubaji *et.al.* [6] reported that $Na_xCo_{2/3}Mn_{2/9}Ni_{1/9}O_2$ cathode could be synthesised by sol-gel method followed by ball milling process and calcination at 800°C.

Panabière *et.al.* [7] performed variation on milling speeds i.e. 200, 500 and 700 rpm for 1 h and generated the best retention capacity in the negative Li_7MnN_4 electrode for Lithium ion batteries, producing the best capacity of 250 $mAhg^{-1}$ in ball-milled with the speed of 500 rpm. The stable capacity was also obtained at 200 and 700 rpm; however, the retention capacity was smaller which were of 220 $mAhg^{-1}$ and 205 $mAhg^{-1}$ respectively. The resulting product ($NaCo_{(1-x)}Mn_xO_2$) was then characterised using LCR, XRD and SEM-EDS.

This study investigated the effect of milling speed variations conducted at 0, 300, 400, 500, 600 and 700 rpm in producing electrode material that had the best electrical conductivity. The electrodes were synthesised via sol-gel method using sodium acetate, manganese acetate, cobalt acetate and citric acid as chelating agent and continued by characterization using XRD and SEM-EDS.

2. Experimental Method

2.1. Material.
Sodium acetate trihydrate p.a. [$CH_3COONa.3H_2O$ (Merck)], cobalt acetate tetrahydrate p.a. [$Co(CH_3COO)_2.4H_2O$ (Merck)], manganese acetate tetrahydrate p.a. [Mn $(CH_3COO)_2.4H_2O$ (Sigma Aldrich)], aquabidest p.a., and citric acid monohydrate [$C_6H_8O_7.H_2O$ (Merck)].

2.2. Tool
Glass, porcelain, spatula, digital balance, Thermolyne furnace 47900, High Energy Milling (HEM) E3D, X-Ray Diffraction (XRD), Scanning Electron Microscopy (SEM) - Energy Dispersive X-ray Spectroscopy (EDS), LCR HI HIOKI Tester 3532-50.

2.3. Synthesis of $NaCo_{(1-x)}Mn_xO_2$ *Solid Electrolyte*
25 mL of CH_3COONa 0.7 M, $Co(CH_3COO)$ 0.66 M and $Mn(CH_3COO)_2$ 0,22 M were mixed into $C_6H_8O_7$ citric acid solution, then stirred constantly with magnetic stirrer for 2 hours continued by evaporation at 80°C until the gel formed. The formed gel was dried in an oven at 110°C. Further, the result was milled by a ball milling with the speed variations of 300, 400, 500, 600 and 700 rpm for 3 hours at room temperature. The ratio of powder and ball milling used was 1:10. Then, the result in the form of powder was calcined at 800°C for 12 hours.

2.4. Characterization of results
The powder was made into pellet and performed its conductivity using LCR meter to know the ability of solid in conducting electricity. The powder obtained from the calcination was also characterised using XRD and SEM-EDS.

3. RESULTS AND DISCUSSION

3.1. Product Characterization using LCR meter
Figure 1 shows the solid electrolyte conductivity of $NaCo_{(1-x)}Mn_xO_2$ as function of milling speed. It can be seen that milling speed of 400 rpm, the solid electrolyte had the highest conductivity of 4.08 x 10^{-6} Scm^{-1}, whilst the lowest conductivity owed by the $NaCo_{(1-x)}Mn_xO_2$ solid electrolyte with a milling speed

of 300 rpm which was 0.963 x 10^{-6} Scm^{-1}. This indicates that variations in milling speed produced the best electrical conductivity at milling speed of 400 rpm. The electrical conductivity increased 4-fold at a milling rate of 300 rpm to 400 rpm and decreased gradually thereafter. Thus, the milling process of a solid electrolyte should be applied at a given milling speed to produce the highest conductivity. This result may be due to at a given milling speed produced the highest number of holes hence the conductivity increases.

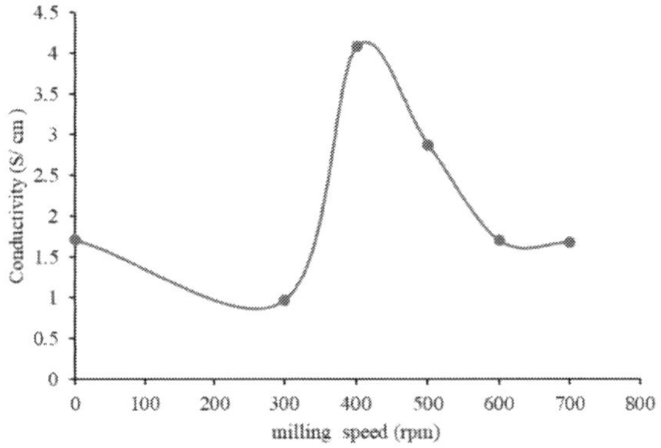

Figure 1. Graph of effect of milling speed on conductivity

3.2. Product Characterization using XRD

To confirm the effects of milling speed on the crystal structure of $NaCo_{(1-x)}Mn_xO_2$, XRD observations were carried out and displayed in Figure 2. All the XRD patterns of $NaCo_{(1-x)}Mn_xO_2$ samples display a set of narrow peaks, which reflect good crystallinity of these materials and all samples have been well prepared. However, the XRD pattern shows the noise that may be caused by by-products on the spinel making process or the formation of low-concentration amorphous materials mixed with the crystal structure. In general, all XRD diffractograms produced similar peak patterns that indicated that the formed solids had the same structure. However, the relative intensity of one peak with another peak were different. These results indicate that the speed of milling does not affect the resulting crystal structure and crystal of $NaCo_{(1-x)}Mn_xO_2$ structure was not destroyed during milling process.

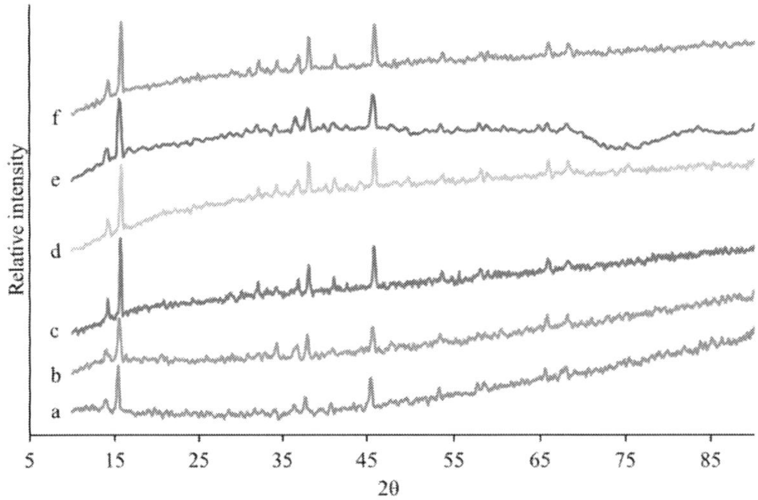

Figure 2. XRD diffractogram of $NaCo_{(1-x)}Mn_xO_2$ solid electrolyte (a) un-milled (b) 300rpm (c) 400rpm (d) 500rpm (e) 600rpm (f) 700rpm

According to the Scherrer formula, the average crystal size for each sample was calculated to be 19.42, 1.04, 1.02, 3.47, 0.74 and 0.84 nm for un-milled sample, samples milled at 300, 400, 500, 600 and 700 rpm respectively. This indicated that the particle size decreased with the increasing milling speed. This result is in accordance with Kar *et.al.* [8], reporting that the widening peak in diffractogram was associated with small crystal size or lattice strain caused by structural defects such as dislocations or defective arrays.

The peak intensity followed the same pattern as the conductivity pattern presented in Fig. 1. For non-milling samples and samples milled at 300 rpm, the peak of $2\theta = 15.7$ shows a low intensity. The 2θ peak of 15.7° from the sample with 400 rpm milling speed displays the highest intensity. Furthermore, the peak of $2\theta = 15.7°$ decreases gradually with increasing milling speed. The intensity rise pattern of this diffractogram is closely correlated with the electrical conductivity pattern. These results indicate a correlation between the crystallinity of the battery samples and their electrical conductivity properties. This result is in line with what were reported [9-11] that there was a close relationship between the crystallinity of the material and the electrical conductivity. The conductivity increases with increase of crystallinity may be due to the increased crystallinity reduced the energy barrier to facilitate the diffusion of ions in the electrolyte.

3.3. Product Characterization using SEM-EDS

Further, the products were characterized using SEM-EDS. The three samples observed were non-milled sample, sample having the highest conductivity (milled at 400 rpm) and having the lowest conductivity (milled at 300 rpm) as shown in Figure 3. The electrode materials exhibit exclusively a worm-like structure for all samples. All samples have dense structures and show uniform distribution.

A b c

Figure 3. The surface morphology of product (a) un-milled sample, samples milled at (b) 300 rpm (c) 400 rpm

Figure 3a shows that the particle size of the $NaCo_{(1-x)}Mn_xO_2$ sample was still large and the surface morphology was not homogeneous. This is because the two synthesis materials were simply mixed hence the mixing was uneven. The powder particles were not perfect to produce smaller crystalline sizes so uniform particle size has not yet taken place.

The material treated with a 300 rpm milling speed (Fig. 4b) shows a fairly high porosity. It can be seen that there was a large cavity over the distance between the particles and the presence of agglomerates. For materials treated with 400 rpm milling speed (Fig. 4c), the particle arrangement is finer and denser compared to the 300 rpm milling speed. This may be caused by repeat collision energy during higher grinding hence the particles broke up more frequently.

The compositions of solid electrolyte constituents were analysed using Energy Dispersive Spectroscopy (EDS). EDS results are presented in Table 1.

Table 1. Composition of solid electrolyte constituent elements

Variation of speed milling	Content of Element (% Mass)			
	Na	Co	Mn	O
300 rpm (I)	20.90	48.72	30.37	22.65
400 rpm (II)	25.44	49.68	24.89	22.79
0 rpm (III)	26.64	51.63	21.73	22.80

EDS results can be used to find out the value of x using the formula

$$x=\frac{(Ar_{Co}\times \text{mass Mn})}{((Ar_{Mn}\times \text{mass Co}+ Ar_{Co}\times \text{mass Mn}))}$$

The values of x obtained from the calculation for samples treated by milling speed of 300, 400 rpm and un-milled samples were 0.29; 0.40 and 0.35 respectively. Thus, the spinel formulas for sample I, sample II and standard sample were $NaCo_{0.6}Mn_{0.4}O_2$, $NaCo_{0.65}Mn_{0.35}O_2$ and $NaCo_{0.71}Mn_{0.29}O_2$, respectively.

4. Conclusion

$NaCo_{(1-x)}Mn_xO_2$ solid electrolyte was successfully synthesised using several mineral composers consisting of $NaMnO_2$, CoO, Na_2O, Co_2O_3 and MnO_2. Treatment of sample by milling speed at 400 rpm showed the highest conductivity of 4.08 x 10^{-6} S/cm with homogeneous surface morphology and has a spinel formula $NaCo_{0,65}Mn_{0,35}O_2$.

5. References

[1] David Lindenand Thomas Reddy, Linden's Handbook of Batteries, 3rd ed., McGraw-Hill Education, 2010.

[2] Nicolas Bucher, Steffen Hartung, Irina Gocheva, Yan L. Cheah, Madhavi Srinivasanand Harry E. Hoster, Combustion-synthesized sodium manganese (cobalt) oxides as cathodes for sodium ion batteries, *Journal of Solid State Electrochemistry*, 17, 7, (2013) 1923-1929 https://doi.org/10.1007/s10008-013-2047-x

[3] Ju-hsiang Cheng, Chun-Jern Pan, Jyh-Fu Lee, J. M. Chen, Marie Guignard, Delmas Claude, Dany Carlierand Bing Joe Hwang, Simultaneous Reduction of Co3+ and Mn4+ in P2-Na2/3Co2/3Mn1/3O2 As Evidenced by X-ray Absorption Spectroscopy during Electrochemical Sodium Intercalation, *Chemistry of Materials*, 26, 2, (2014) 1219–1225 http://dx.doi.org/10.1021/cm403597h

[4] Moni Kanchan Datta, Ramalinga Kuruba, Prashanth H. Jampani, Sung Jae Chung, Partha Saha, Rigved Epur, Karan Kadakia, Prasad Patel, Bharat Gattu, Ayyakkannu Manivannanand Prashant N. Kumta, Electrochemical properties of a new nanocrystalline NaMn2O4 cathode for rechargeable sodium ion batteries, *Materials Science and Engineering: B*, 188, (2014) 1-7 https://doi.org/10.1016/j.mseb.2014.05.007

[5] Chi-Yuan Lu, Jui-Yeh Rau, Jyh-Cherng Chen, Shih-Teng Huangand Ming-Yen Wey, Removal of NO and fly ash over a carbon supported catalyst: Effects of fly ash concentration and operating time, *Powder Technology*, 239, (2013) 239-247 https://doi.org/10.1016/j.powtec.2013.02.002

[6] Siham Doubaji, Mario Valvo, Ismael Saadoune, Mohammed Dahbiand Kristina Edström, Synthesis and characterization of a new layered cathode material for sodium ion batteries, *Journal of Power Sources*, 266, (2014) 275-281 https://doi.org/10.1016/j.jpowsour.2014.05.042

[7] E. Panabière, N. Emery, S. Bach, J. P. Pereira-Ramosand P. Willmann, Ball-milled Li7MnN4: An attractive negative electrode material for lithium-ion batteries, *Electrochimica Acta*, 97, (2013) 393-397 https://doi.org/10.1016/j.electacta.2013.03.012

[8] Sujan Kar Kar, Shweta Logad, Om P. Choudhary, Chiranjit Debnath, Sunil Vermaand Kunwar S. Bartwal, Preparation of Lithium Niobate Nanoparticles by High Energy Ball Milling and their Characterization, *Universal Journal of Materials Science*, 1, 2, (2013) 18-24 http://dx.doi.or/10.13189/ujms.2013.010202.

[9] Erin M. Sullivan, Yun Ju Oh, Rosario A. Gerhardt, Ben Wangand Kyriaki Kalaitzidou, Understanding the effect of polymer crystallinity on the electrical conductivity of exfoliated graphite nanoplatelet/polylactic acid composite films, *Journal of Polymer Research*, 21, 10, (2014) 563 http://dx.doi.or/10.1007/s10965-014-0563-8

[10] Angela S. Wochnik, Matthias Handloser, Dajana Durach, Achim Hartschuhand Christina Scheu, Increasing Crystallinity for Improved Electrical Conductivity of TiO2 Blocking Layers, *ACS Applied Materials & Interfaces*, 5, 12, (2013) 5696-5699 http://dx.doi.org/10.1021/am401110n

[11] M. Diantoro, I. N. Fitriana, F. Parasmayanti, Nasikhudin, A. Taufiq, Sunaryono, N. Muftiand H. Nur, Crystallinity and Electrical Conductivity of PANI-Ag/Ni Film: The Role of Ultrasonic and Silver Doped, *IOP Conference Series: Materials Science and Engineering*, 202, 1, (2017) 012005

IOP Conf. Series: Materials Science and Engineering **349** (2018) 012033 doi:10.1088/1757-899X/349/1/012033

Effect of preparation methods on the activity of titanium dioxide-carbon nitride composites for photocatalytic degradation of salicylic acid

L Yuliati[1-3]*, A M Salleh[4], M H M Hatta[4], H O Lintang[1-3]

[1] Ma Chung Research Center for Photosynthetic Pigments, Universitas Ma Chung, Villa Puncak Tidar N-01, Malang 65151, East Java, Indonesia
[2] Department of Chemistry, Faculty of Science and Engineering, Universitas Ma Chung, Villa Puncak Tidar N-01, Malang 65151, East Java, Indonesia
[3] Centre for Sustainable Nanomaterials, Ibnu Sina Institute for Scientific and Industrial Research, Universiti Teknologi Malaysia, 81310 UTM Johor Bahru, Johor, Malaysia
[4] Department of Chemistry, Faculty of Science, Universiti Teknologi Malaysia, 81310 UTM Johor Bahru, Johor, Malaysia

leny.yuliati@machung.ac.id

Abstract. In this study, titanium dioxide-carbon nitride (TiO_2-CN) composites were prepared by three methods, which were one pot oxidation, impregnation, and physical mixing. Each series of the photocatalysts was prepared with different ratios of titanium to carbon (Ti/C), *i.e.*, 1, 5, 10, 20, and 50 mol%. All samples were characterized by X-ray diffraction (XRD) and diffuse reflectance ultraviolet-visible (DR UV-Vis) spectroscopies. The characterization results confirmed the successful preparation of TiO_2, CN, and the TiO_2-CN composites. Photocatalytic activity tests were carried out for degradation of salicylic acid at room temperature for 6 h under UV and visible light irradiations. It was confirmed that all the prepared TiO_2-CN composites showed better photocatalytic activities than the bare TiO_2 and the bare CN. Under UV light irradiation, 90.6% of salicylic acid degradation was achieved on the best composite prepared by one pot oxidation with 5 mol% of titanium to carbon (Ti/C) ratio. On the other hand, the highest degradation under visible light irradiation was 94.3%, observed on the composite that was prepared also by one pot oxidation method with the Ti/C ratio of 10 mol%. Therefore, among the investigated methods, the best method to prepare the titanium dioxide-carbon nitride composites with high photocatalytic activity was one pot oxidation method.

1. Introduction

Photocatalysis by titanium dioxide (TiO_2) has been discovered more than 90 years ago [1]. During the last three decades, many researches have been explored to understand its possible applications in the remediation of air and water. Owing to its high stability and non-toxicity, TiO_2 has been the most used and researched photocatalyst [2]. In general, pure anatase TiO_2 showed better photocatalytic activity than the pure rutile TiO_2. However, the high charge recombination rate in the TiO_2 leads to the low efficiency in the photocatalytic reactions [1]. Moreover, due to their large band gap energy (3.0–3.2 eV), the TiO_2 in anatase form can be mainly activated by UV light while TiO_2 in rutile form can only extend its absorption near to visible light irradiation.

Since sunlight energy has a greater portion of visible light than the UV light, it is most beneficial to develop visible light-driven photocatalysts. In order to make TiO_2 become active in visible light irradiation, the material needs to be incorporated with other materials, which able to absorb visible light and also can act as active compounds, such as carbon nitride (CN). Such strategy was used in some semiconductors, such as zinc oxide [3] and gallium oxide [4]. CN itself has been recognized as a good photocatalyst for degradation of organic pollutants [5-7]. In the present study, the TiO_2-CN composites were prepared by three methods, namely one pot oxidation, impregnation, and physical mixing. Investigating the effect of these preparation methods is an important approach to develop highly active photocatalysts. The composites were evaluated for degradation of salicylic acid as one of the organic pollutants.

2. Experimental

2.1. *Preparation of TiO_2 and CN*
Titanium chloride ($TiCl_4$, > 97%, Merck) was used as a precursor to prepare TiO_2. The first step was the reaction between $TiCl_4$ precursor and water to form titanium hydroxide, $Ti(OH)_4$. The precipitate was then dried and calcined at 823 K for 4 h to obtain TiO_2. As for the preparation of CN, urea was used as the precursor [5]. The thermal polymerization of urea to CN was carried out by calcination process at 823 K for 4 h.

2.2. *Preparation of TiO_2-CN composites*

2.2.1. One pot oxidation. Certain amounts of urea and $TiCl_4$ with different mol ratios of titanium to carbon (Ti/C, x = 1, 5, 10, 20, and 50 mol%) were dissolved in ethanol. The mixture was then heated to dry by stirring it on a hot plate at 348–353 K. The obtained solid sample was then calcined at 823 K for 4 h. The final product was denoted as $TiO_2(x)$-CN OPO composite, with x shows the mol ratio of Ti/C.

2.2.2. Impregnation Method. The $TiO_2(x)$-CN IM composites were synthesized by an impregnation method where a certain amount of $TiCl_4$ was impregnated onto the prepared CN with various mol ratios of Ti/C (x = 1, 5, 10, 20, and 50 mol%). The $TiCl_4$ was first dissolved in ethanol and the prepared CN was added to the solution. The mixture was stirred on a hot plate at 348–353 K to evaporate the solvent. The obtained solid was subjected to calcination process for 4 h at 823 K to produce the $TiO_2(x)$-CN IM composite.

2.2.3. Physical Mixing. The prepared TiO_2 was mixed physically with the prepared CN by grinding process. Various mol ratios of Ti/C (x = 1, 5, 10, 20, and 50 mol%) were used and the obtained samples were labeled as $TiO_2(x)$-CN PM composites.

2.3. *Characterizations of TiO_2, CN, and TiO_2-CN composites*
In order to study the structural and optical properties of the prepared TiO_2, CN, and TiO_2-CN composites, all the samples were characterized by X-ray diffractometer and diffuse reflectance ultraviolet-visible spectrophotometer (DR UV-Vis), respectively. The XRD patterns were recorded on a D8 Advance (Bruker) at room temperature using radiation of Cu-Kα at 40 kV and 40 mA. The DR UV-Vis spectra were measured on a Lambda 900 (Perkin Elmer) by using barium sulfate as the reference.

2.4. *Photocatalytic degradation of salicylic acid*
In order to determine the required time for the sample to achieve adsorption equilibrium, the sample (0.05 g) was dispersed in a beaker containing 50 mL of salicylic acid (100 μM). The beaker was placed on the stirring plate in a closed box to ensure that there would be no activation by light. The dark

condition tests were conducted for 30 min, 1, 2, and 3 h. After each time, the photocatalyst was separated from the solution by centrifugation. The concentration of salicylic acid was then measured by using a UV-Vis spectrophotometer (Genesys 10S, Thermo Scientific). Since the required time to achieve the equilibrium was 1 h, such dark condition was conducted before the photocatalytic degradation tests were carried out. Similar to dark condition test, the sample (00.5 g) was dispersed in a beaker containing 50 mL of salicylic acid (100 μM). After 1 h adsorption process in the dark, the photocatalytic degradation of salicylic acid was carried out for 6 hours under 8 W UV lamp or 150 W halogen lamp as the visible light source. After the sample was separated from the solution, the concentration of remaining salicylic acid was determined by the UV-Vis spectrophotometer.

3. Results and Discussion

3.1. Properties of the prepared TiO_2, CN, and TiO_2-CN composites

The structural and optical properties of the prepared samples were investigated by XRD and DR UV-Vis spectroscopies. Figure 1 [A](a) shows the XRD pattern of the prepared TiO_2. As indicated in the figure, the prepared TiO_2 has the crystal planes of anatase crystal structure [8]. On the other hand, the successful formation of graphitic CN could be shown in Figure 1 [A](b), where one peak around 28° was observed due to the interlayer spacing of the CN [5-7]. The XRD patterns of the $TiO_2(x)$-CN OPO composites are displayed in Figure 1 [A](c)-(g). It was obvious that when the ratio of Ti/C was low, the composite samples only showed the main peak of CN. On the other hand, with the increase of the Ti/C ratio, the peak of CN decreased and the peak of anatase TiO_2 became the dominant diffraction peaks. The disappearance of the CN peak might suggest the possibility that the CN was oxidized easier in the presence of TiO_2 during the heating process in the synthesis part.

Figure 1 [B] shows the DR UV-Vis spectra of all samples. In Figure 1 [B](a) TiO_2 showed a significant increase in the absorption at a wavelength shorter than 400 nm, which could be assigned to the intrinsic band gap absorption of pure anatase TiO_2 (about 3.2 eV) [2]. On the other hand, the CN gave a strong absorption peak at around 370 nm, which was the characteristic of CN [5-7]. The DR UV-Vis spectra of $TiO_2(x)$-CN OPO composites are shown in Figure 1 [B](c)-(g). All samples prepared by one pot oxidation method showed absorption in the visible region above 400 nm, suggesting that they have good ability to absorb visible light and might show good activity under visible light irradiation.

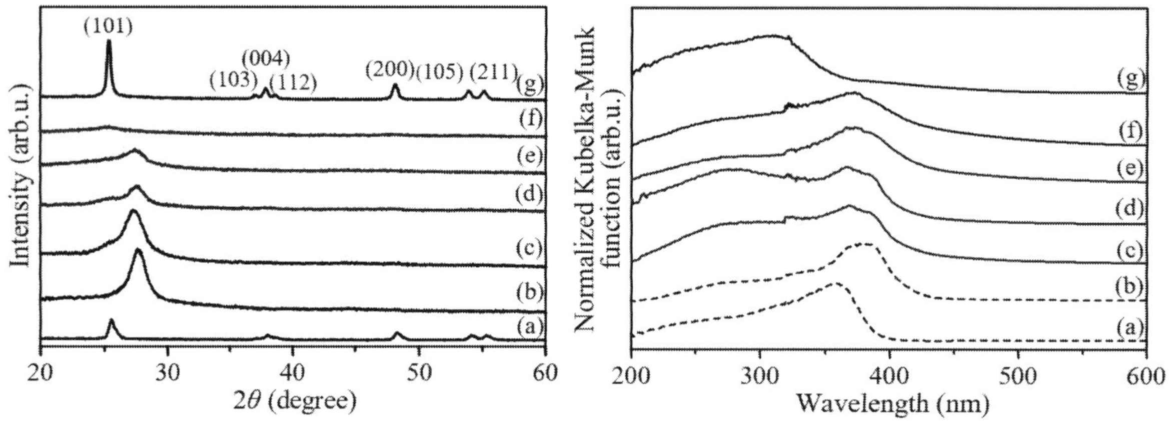

Figure 1. [A] XRD patterns and [B] DR UV-Vis spectra of (a) TiO_2, (b) CN, (c) $TiO_2(1)$-CN OPO, (d) $TiO_2(5)$-CN OPO, (e) $TiO_2(10)$-CN OPO, (f) $TiO_2(20)$-CN OPO, and (g) $TiO_2(50)$-CN OPO.

Figure 2 [A] shows the XRD patterns of $TiO_2(x)$-CN IM composites. Similar to the trend observed on the $TiO_2(x)$-CN OPO composites, the $TiO_2(x)$-CN IM composites with low Ti/C ratio (1-5%) were mainly characterized by the presence of CN, while high Ti/C ratio ($\geq$ 10%) were dominated by the diffraction peaks of TiO_2. Again, the result suggested that the CN might be decomposed easier when heated in the presence of TiO_2. The DR UV-Vis spectra of the TiO_2(*x*)-CN IM composites are illustrated in Figure 2 [B]. Depicted in Figure 2 [B](b), only TiO_2(*1*)-CN IM composite showed absorption in the visible light region above 400 nm. On the other hand, with the increasing amount of Ti/C ratio, the absorption in the visible light region decreased and shifted to a lower wavelength. These DR UV-Vis spectra are in good agreement with their respective XRD patterns.

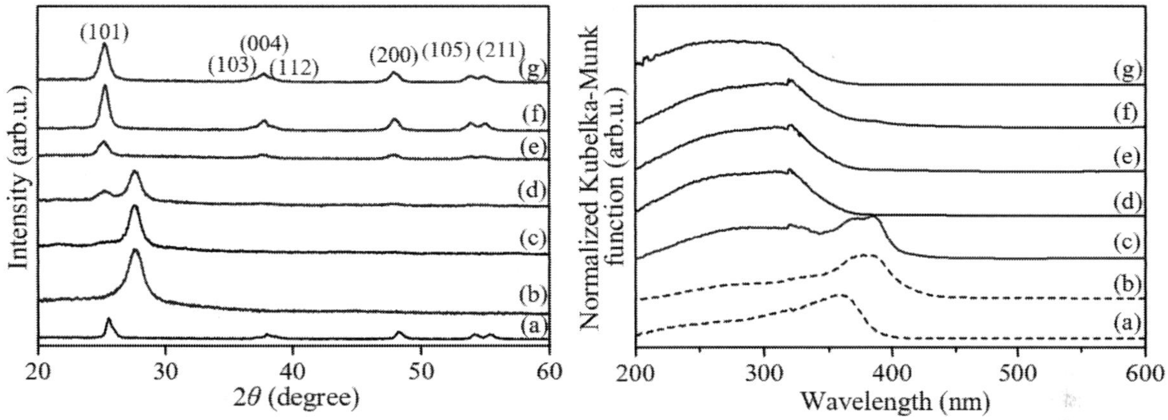

Figure 2. [A] XRD patterns and [B] DR UV-Vis spectra of (a) TiO_2, (b) CN, (c) TiO_2(*1*)-CN IM, (d) TiO_2(*5*)-CN IM, (e) TiO_2(*10*)-CN IM, (f) TiO_2(*20*)-CN IM, and (g) TiO_2(*50*)-CN IM.

The XRD patterns of the TiO_2(*x*)-CN PM composites are shown in Figure 3 [A]. Similar to the composites prepared by one pot oxidation and impregnation methods, the composite with low Ti/C ratio, *i.e.*, TiO_2(*1*)-CN PM only exhibited the CN peak. However, the diffraction peak of CN was not disappeared even when the Ti/C ratio was increased to 50%. Even though the diffraction peaks of TiO_2 increased with the increase of the Ti/C ratio, all composites with the Ti/C ratio of more than 5% showed the diffraction peaks of both TiO_2 and CN, suggesting that both TiO_2 and CN existed in these samples. This is reasonable since physical mixing only included grinding process at room temperature. Shown in Figure 3 [B] is the DR UV-Vis spectra of the TiO_2(*x*)-CN PM composites. In contrast to the composites prepared by one pot oxidation and impregnation methods, all composites exhibited a similar absorption spectrum to each other regardless the difference in the Ti/C ratio. All composites showed the characteristics of light absorption for both TiO_2 and CN, and thus, able to absorb both UV and visible light.

3.2. Photocatalytic activity of the prepared TiO_2-CN composites

The photocatalytic performance of the TiO_2(*x*)-CN OPO composites prepared by the one pot oxidation method was evaluated for degradation of salicylic acid under UV or visible light irradiation. As shown in Table 1, under UV light irradiation, the CN only gave degradation of 33.7%. The addition of TiO_2 up to 5 mol% increased the photocatalytic activity to 90.6%. The activity decreased with further increase of the Ti/C ratio and the TiO_2(*50*)-CN OPO showed the lowest degradation percentage among other composites, which was 79.6%. Compared to the bare TiO_2, all composites still gave better photocatalytic activity. Under visible light irradiation, the CN only showed 43% degradation. On the other hand, the composite with a Ti/C ratio of 10 mol% gave an optimum enhanced photocatalytic activity of 94.3%. When the Ti/C ratio increased to 50%, the degradation percentage decreased to 77.7%, but this activity was still much higher than the bare TiO_2. Since all the TiO_2(*x*)-CN OPO composites gave higher photocatalytic activity than the bare TiO_2 and bare CN both under UV and visible light, it was suggested that there was a synergic effect between TiO_2 and CN to give high activity for the degradation of salicylic acid regardless the type of light source used in the reaction.

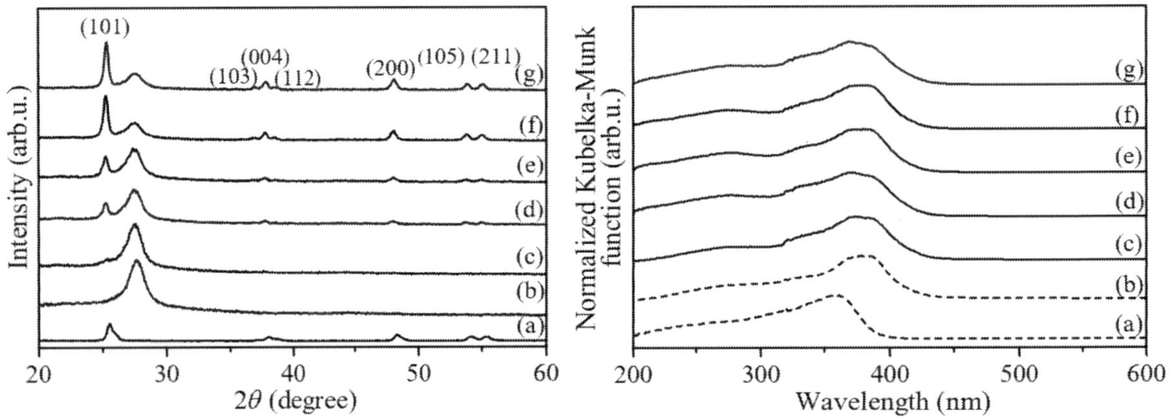

Figure 3. [A] XRD patterns and [B] DR UV-Vis spectra of (a) TiO_2, (b) CN, (c) TiO_2(*1*)-CN PM, (d) TiO_2(*5*)-CN PM, (e) TiO_2(*10*)-CN PM, (f) TiO_2(*20*)-CN PM, and (g) TiO_2(*50*)-CN PM.

Table 1. Photocatalytic degradation of salicylic acid on the TiO_2-CN composites prepared by one pot oxidation, impregnation, and physical mixing methods under UV or visible light irradiation

Preparation Method	Sample	Percentage of Degradation (%)	
		Under UV light	Under visible light
One Pot Oxidation	CN	33.7	43.0
	TiO_2(*1*)-CN OPO	67.3	60.6
	TiO_2(*5*)-CN OPO	90.6	91.6
	TiO_2(*10*)-CN OPO	77.4	94.3
	TiO_2(*20*)-CN OPO	82.6	84.1
	TiO_2(*50*)-CN OPO	79.6	77.7
	TiO_2	60.9	21.4
Impregnation	CN	33.7	43.0
	TiO_2(*1*)-CN IM	70.7	65.1
	TiO_2(*5*)-CN IM	89.7	82.6
	TiO_2(*10*)-CN IM	77.0	86.7
	TiO_2(*20*)-CN IM	67.3	81.0
	TiO_2(*50*)-CN IM	66.4	78.0
	TiO_2	60.9	21.4
Physical Mixing	CN	33.7	43.0
	TiO_2(*1*)-CN PM	80.0	71.4
	TiO_2(*5*)-CN PM	79.7	56.7
	TiO_2(*10*)-CN PM	78.1	56.3
	TiO_2(*20*)-CN PM	76.1	46.3
	TiO_2(*50*)-CN PM	81.0	62.4
	TiO_2	60.9	21.4

As for the TiO_2(*x*)-CN IM composites, the photocatalytic activity under UV light irradiation increased first to 89.7% when the ratio of Ti/C increased up to 5 mol%, then the activity decreased to 66.4% with a further increase of the Ti/C ratio. However, it was obvious that even though the activity decreased, compared to the bare TiO_2 and CN, all the TiO_2(*x*)-CN IM composites showed much higher activities. Similar to the case of reactions under UV light irradiation, the reactions under visible light irradiation also gave the same trend. The activity increased to 86.7% when the ratio of Ti/C increased up to 10 mol%, but then the activity decreased with further increase of the Ti/C ratio. Again, all the TiO_2(*x*)-CN IM showed much higher activities than the bare TiO_2 and bare CN.

The photocatalytic activities of the $TiO_2(x)$-CN PM composites were also tested under both UV and visible light irradiation. Under UV light irradiation, it was obtained that the photocatalytic activities of the samples were not much influenced by the amount of Ti/C ratio. All samples showed a similar level of activity with salicylic acid degradation was *ca.* 80%. It is worthy noted here that all samples prepared by physical mixing showed much higher activity than the bare TiO_2 and the bare CN. As for the reactions under visible light irradiation, the highest activity was obtained on the $TiO_2(1)$-CN PM that gave the salicylic acid degradation of 71.4%. While there was no clear relationship between the Ti/C ratio to the photocatalytic activity, all composites showed the percentage removals of 40–60%. Even though the activity of the composite was less enhanced compared to when it was prepared by impregnation or one pot oxidation method, the activity of $TiO_2(1)$-CN PM was much higher than the activity of bare TiO_2 or bare CN. This result again showed the synergic effect between TiO_2 and CN.

Among all composites prepared by three methods above, the $TiO_2(5)$-CN OPO and the $TiO_2(10)$-CN OPO showed the highest photocatalytic activity towards degradation of salicylic acid under UV and visible light, respectively. These results suggested that one pot oxidation method was the best approach to obtain TiO_2-CN composites with high activity. Under light irradiation with enough energy, the electrons in the valence bands of both TiO_2 and CN in the composite will be excited to the conduction bands. Since the conduction band of TiO_2 is higher than the conduction band of CN, the excited electrons will be transferred from the conduction band of TiO_2 to the conduction band of CN. Thus, the electron recombination rate of photoexcited TiO_2 will be reduced, giving the increased activity under UV light irradiation. The excited electrons would react with the dissolved oxygen to produce superoxide radical that could oxidize salicylic acid. In this study, the formation of hydroxyl and hydroperoxyl radicals could be excluded since water was not used as the solvent. Holes that were photogenerated in the valence band of both TiO_2 and CN would also react with the salicylic acid to finally produce water and carbon dioxide.

4. Conclusions

TiO_2-CN composites could be prepared by one pot oxidation, impregnation, and physical mixing. Among the three methods, the composites prepared by one pot oxidation method showed the highest photocatalytic activity for degradation of salicylic acid both under UV and visible light irradiation. The synergic effect between TiO_2 and CN in the composite was proposed to reduce the electron-hole recombination in the TiO_2, resulting in the better photocatalytic activity of the composite than the bare TiO_2 and the bare CN.

Acknowledgements

The academy of sciences for the developing world (TWAS) via COMSTECH-TWAS joint research grants programme for young scientists (RGA No: 10-231 RG/MSN/AS_C-UNESCO FR: 3240246318) and Directorate General of Strengthening Research and Development, Ministry of Research, Technology and Higher Education of the Republic of Indonesia via the Higher Education Excellent Applied Research Grant (PTUPT 2018) are greatly acknowledged.

References

[1] Fujishima A, Zhang X and Tryk D A 2008 *Surf. Sci. Rep.* **63** 515
[2] Luttrell T, Halpegamage S, Tao J, Kramer A, Sutter E and Batzil M 2014 *Sci. Rep.* **4** 4043
[3] Hussin F, Lintang H O and Yuliati L 2016 *Mal. J. Anal. Sci.* **20** 102
[4] Lee S C, Chew W S, Lintang H O and Yuliati L 2015 *Mal. J. Fund. App. Sci.* **11** 98
[5] Lee S C, Lintang H O and Yuliati L 2012 *Chem. Asian J.* **7** 2139
[6] Zhang Y, Liu J, Wu G and Chen W 2012 *Nanoscale* **4** 5300
[7] Wang X, Blechert S and Antonietti M 2012 *ACS Catal.* **2** 1596
[8] Thamaphat K, Limsuwan P and Ngotawornchai B 2008 *Kasetsart J. (Nat. Sci.)* **42** 357

IOP Conf. Series: Materials Science and Engineering **349** (2018) 012034 doi:10.1088/1757-899X/349/1/012034

The Impact of Template Types on Polyeugenol to the Adsorption Selectivity of Ionic Imprinted Polymer (IIP) Fe Metal Ion

M C Djunaidi, A Haris, Pardoyo, Rosdiana K

Chemistry Department, Faculty of Sciences and Mathematics
Diponegoro University

choliddjunaidi@live.undip.ac.id

Abstract. The synthesis of IIP was carried out by variation of Fe(III) ion templates from $Fe(NO_3)_3$, $K_3[Fe(CN)_6]$ and $NH_4Fe(SO_4)_2$ compounds which then tested IIP selectivity to the Fe metal ions through adsorption process. Ionic Imprinted Polymer (IIP) is a method of printing metal ions bound in a polymer, subsequently released from the polymer matrix to produce a suitable imprint for the target ion. The purposes of this study were to produce IIP from $Fe(NO_3)_3$, $K_3[Fe(CN)_6]$ and $NH_4Fe(SO_4)_2$ templates, to know the effect of templates on adsorption selectivity of IIP involving imprint cavity, and to know the impact of metal competitor on the selectivity adsorption of IIP to the Fe metals. The results obtained showed that IIP synthesized by variations of $Fe(NO_3)_3$, $K_3[Fe(CN)_6]$ and $NH_4Fe(SO_4)_2$ templates were successfully synthesized. The adsorption selectivity of Fe (III) metal ion in the $Fe(NO_3)_3$ template was greater than that of in the $K_3[Fe(CN)_6]$ and $NH_4Fe(SO_4)_2$ templates. The adsorption selectivity of Fe was greater on Fe-Cr compared to on Fe-Cd and Fe-Pb.

1 Introduction

Heavy metal pollution has become a global problem along with the increasing of industrialization, mining, laboratory and daily activities. Heavy metals have detrimental effects on the environment even in very low concentration. High level of Fe (iron) contained in the water influences its use, for example, concentrations of Fe content in water exceeding 1 mg/L will cause the color of the water becomes reddish, stains on the equipment and materials becomes white. World Health Organization, WHO (World Health Organization) limits the content of Fe in drinking water 0.3 mg/L [1].

One of the latest techniques developed for the preparation of highly effective adsorbents is the use of Ionic Imprinted Polymer (IIP), in which the synthesis material contains very specific receptor sites to the target compounds [2]. The advantages of IIP are easy and effective preparations for making polymer media having a specific molecular identifier for target compounds [3]. IIP can recognize and bind the desired target ions with high affinity and selectivity, due to the high cross-linking properties of the polymer in IIP material, it is stable and robust. The selectivity of IIP is determined by several factors, namely, the presence of template ion and crosslinking agent [4].

Eugenol as a natural ingredient of Indonesia has been utilized for the separation of metal ions. Eugenol can also be used as a starting material for the synthesis of a compound because of the presence of three functional groups attached to it, i.e., alkyl groups, hydroxy and methoxy. Through the allyl group, eugenol can be polymerized into polyeugenol which is the starting material for the synthesis of IIP [5-8].

IOP Conf. Series: Materials Science and Engineering **349** (2018) 012034 doi:10.1088/1757-899X/349/1/012034

In this research, the synthesis of IIP was conducted by using polyeugenol with variation of sources of Fe (III) template, namely, $Fe(NO_3)_3$, $K_3[Fe(CN)_6]$ and $NH_4Fe(SO_4)_2$, which was then crosslinked with PEGDE (Polyethylene glycol diglycidyl either) for selectivity test to Fe metal.

2 Experimental Method

The first stage of the research was the synthesis of polyeugenol from eugenol. Furthermore, IIP synthesis with Fe (III) ion templates from $Fe(NO_3)_3$, $K_3[Fe(CN)_6]$ and $NH_4Fe(SO_4)_2$ cross-linked using PEGDE. Subsequently, Fe ion contained was then released from the polymer formed using acid, the acid filtrate was added KSCN complexing agent which then tested using UV-Vis spectrophotometer. The final stage was the adsorption test of the product in which the adsorbent had been contacted with the binary Fe metal mixtures for 24 hours. The adsorption filtrate was analyzed by Atomic Absorption Spectrometer (AAS).

Tools: Laboratory glassware, reflux apparatus, analytical balance, magnetic stirrer, spatula, pH meter, fine filter paper, UV-Vis Spectrophotometer, Atomic Absorption Spectroscopy (AAS), and FTIR.

Materials: Eugenol p.a. (SIGMA Aldrich), anhydrous Na_2SO_4 technical grade (Merck), BF_3-diethyl ether (SIGMA Aldrich), HCl (Merck), NaOH p.a. (Merck), chloroform p.a. (Merck), methanol p.a. (SIGMA Aldrich), Poly Ethylene Glycol Diglycidyl Ether (PEGDE) (SIGMA Aldrich), KSCN (Merck), Aquabidest, $K_3[Fe(CN)_6$ (Merck), $NH_4Fe(SO_4)_2$ (Merck), $Fe(NO_3)_3$ (Merck), $Cr(NO_3)_3$ (Merck), $Pb(NO_3)_2$ (Merck), $Cd(NO_3)_2$ (Merck).

Experimental procedures

Material Preparation

Preparation of 50 ppm Fe (III) solution from 1000 ppm $Fe(NO_3)_3$ solution. 5 ml of $Fe(NO_3)_3$ 1000 ppm was diluted in 100 mL measuring flask with distilled water up to meniscus. The preparation of 50 ppm Fe (III) solution from $K_3[Fe(CN)_6]$ that was 29.4 mg of $K_3[Fe(CN)_6]$ was dissolved in 100 mL of distilled water. The preparation of 50 ppm Fe (III) solution from $NH_4Fe(SO_4)_2$ that was 43.05 mg of $NH_4Fe(SO_4)_2$ was dissolved in 100 mL of distilled water.

Synthesis of Polyeugenol [5-8].

Synthesis of IIP with Template Variation

0.5 g (0.003 mol) of poly eugenol was stirred with 50 ppm of Fe (III) ion which was conditioned at pH 3 for 24 h, then the result was filtered with filter paper and dried at room temperature. At this stage, the Fe (III) templates used were from $Fe(NO_3)_3$, $K_3[Fe(CN)_6]$ and $NH_4Fe(SO_4)_2$. After forming of Polyeugenol-Fe powder then each was crosslinked with 0.96 g (1 mol) of PEGDE and catalyzed using 20 mL of NaOH 1 M. The mixture had been refluxed at 80-90 °C for 15 minutes, the results were filtered and neutralized with aquabidest. The results in the form of precipitate (resin) were then heated at 110 °C for 6 hours. The resulting resin was then added HCl to release Fe (III) ion bound for 24 hours and the acid filtrate was subsequently added by KSCN complexing agent and then tested using UV-Vis spectrophotometer. After the filtrate did not form red color (6 days) when KSCN was added, it was filtered, and the result (IIP) was dried.

NIP Synthesis

NIP was synthesized in the same procedure as IIP; however, without the binding of Fe (III) ions. Characterization of IIP and NIP was undertaken using FTIR.

Synthesis of NIP Acid

The procedure of the NIP acid preparation was the same as NIP preparation, but at the end of the procedure, the polymer was immersed in 0.5 M HCl and stirred for 24 hours by 6 days.

Adsorption Experiment
50 mg of adsorbent was contacted with a mixture of 10 ml of 10 ppm Fe (III) ion and 10 ppm of competitor ion for 24 hours at a constant velocity. The mixture was filtered through fine filter paper, and the Fe (III) concentration in the filtrate was determined using AAS.

Adsorption Selectivity Test
It was performed on competitive adsorption of binary solution consisting of Fe (III)/Cd (II), Fe (III)/Cr (III), Fe (III)/Pb (II) and compared with NIP.

3 Results and discussion

3.1 Synthesis of Polyeugenol [5-8]
The mechanism of formation of polyeugenol and characterization was published by Djunaidi *et al.* 2015)

3.2 Synthesis of IIP with Template Variation
The synthesized polyeugenol was bound to 50 ppm of Fe (III) ion (pH 3) using contacting them for 24 hours through stirring. The templates used were variation of $Fe(NO_3)_3$, $K_3[Fe(CN)_6]$ and $NH_4Fe(SO_4)_2$. The resulting polyeugenol-Fe was filtered and dried, then crosslinked with 0.96 g (1 mol) of PEGDE (Polyethylene glycol diglycidylether) (Mr 500); the resulting product was Polyeugenol-Fe-PEGDE. Further, the bounded Fe (III) ion was released by washing using acid (HCl) to produce imprint cavity. The binding reaction of Fe (III) template (in this case $Fe(NO_3)_3$ in example) to polyeugenol is shown in Figure 1a-d.

The test of Fe content released was carried out by adding KSCN as a complexing agent, the filtrate changed color to red indicating the presence of Fe metal that was released from the polymer. The reaction is [9]:

$$3SCN^- + Fe^{3+} \rightleftharpoons Fe(SCN)_3$$

The test was undertaken until the filtrate did not form a red complex when KSCN was added. The washing filtrate was analyzed using UV-Vis spectrophotometer at a wavelength of 460 nm for 6 days. The graph of Fe template release was shown in Figure 2.

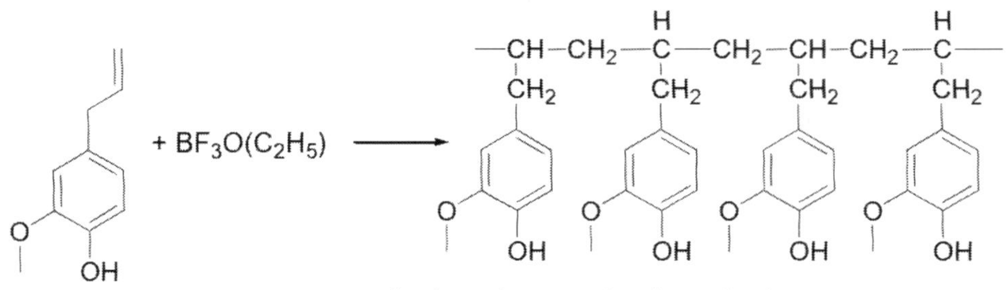

Figure 1a. IIP Synthesis: Polyeugenol Polymerization

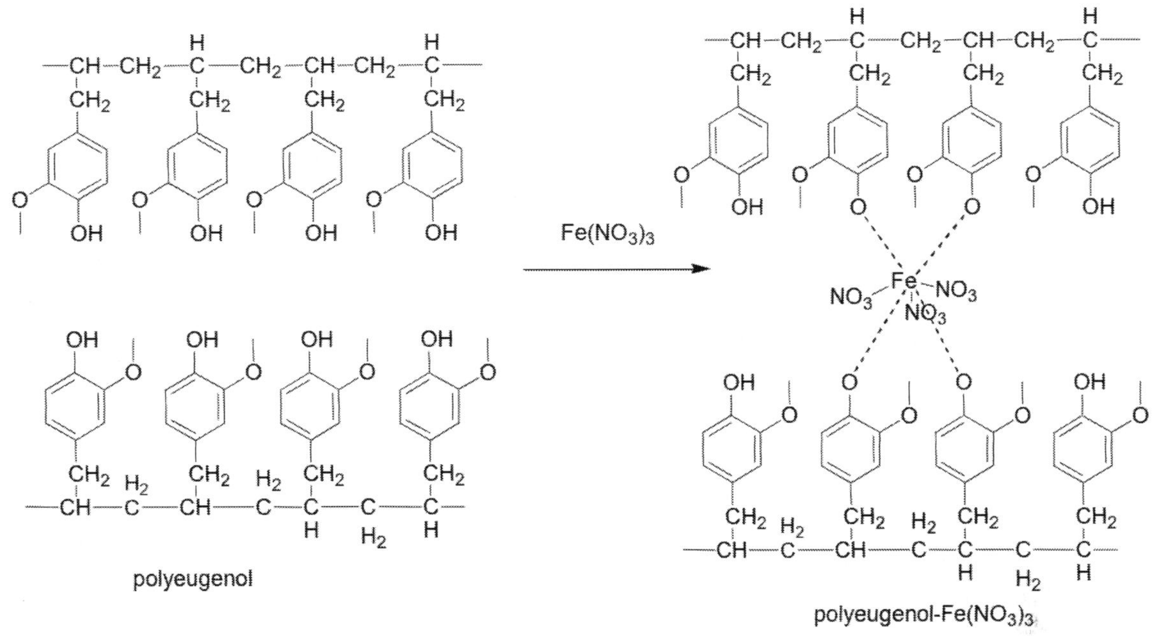

Figure 1b. IIP Synthesis: Binding of Ion Imprint, Fe (III)

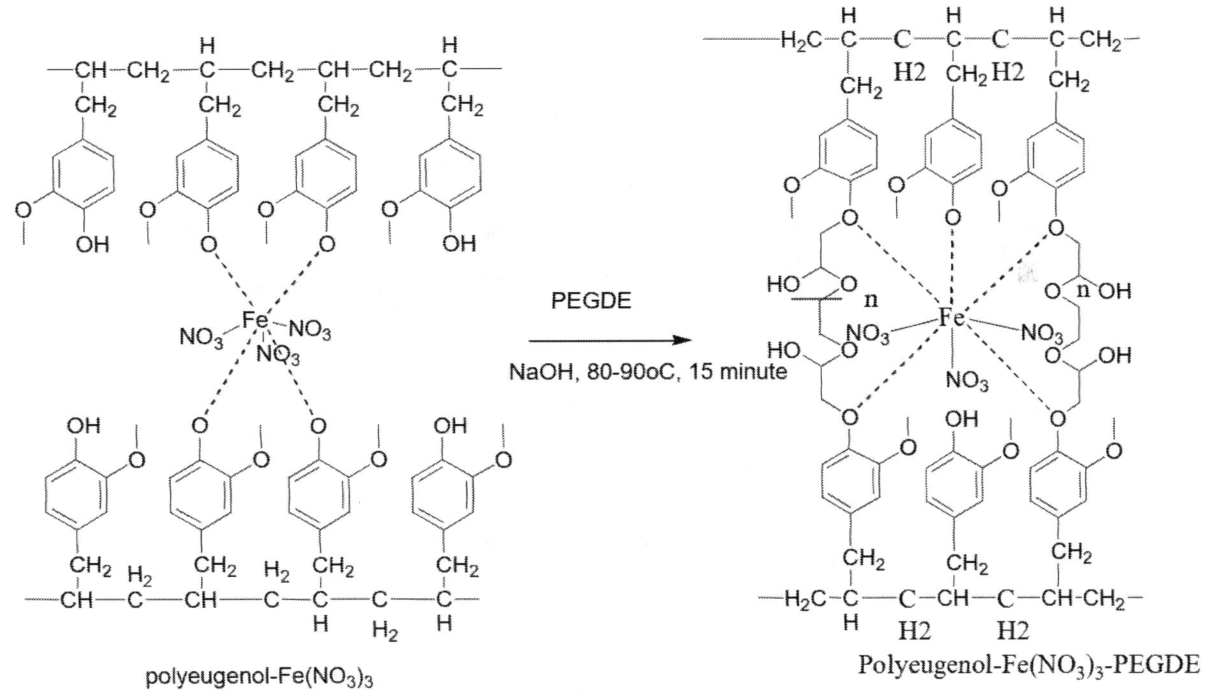

Figure 1c. IIP Synthesis: Crosslinker, PEGDE

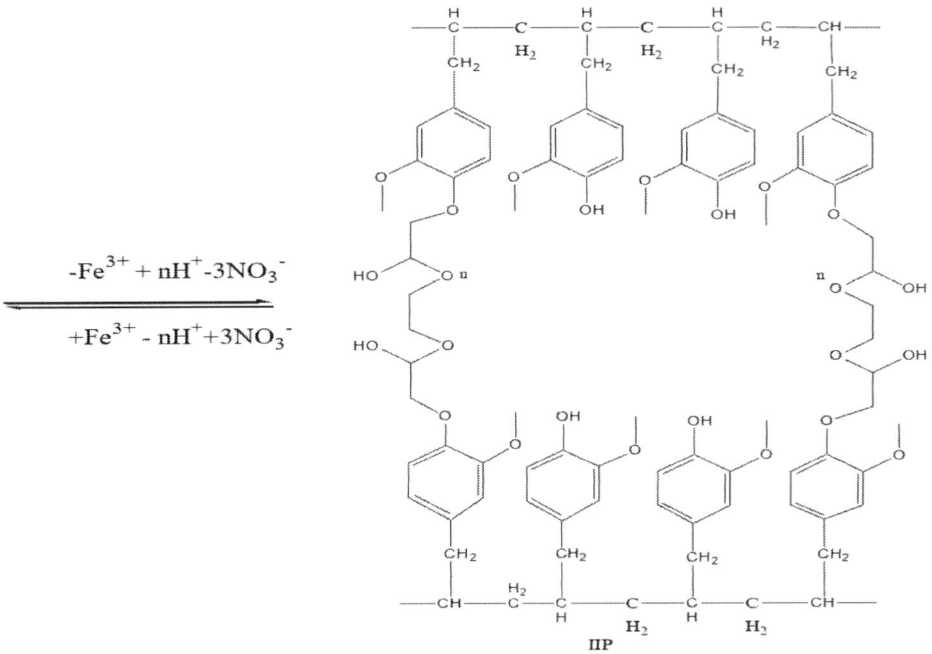

Figure 1d. IIP Synthesis: $Fe(NO_3)_3$ release

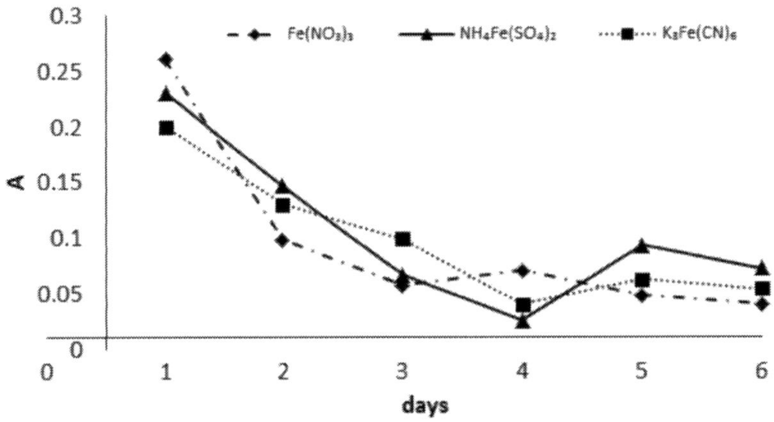

Figure 2. The graph of Fe template release

Polyeugenol synthesized using $Fe(NO_3)_3$ as template was able to bind Fe (III) higher than that of other templates; this was proved by generating high absorbance value in the acid release process. The resin which has been removed from the template is then dried and can be used as an adsorbent to adsorb metals.

3.3 Synthesis of NIP (Non-Imprinted Polymer)

The procedure of the NIP synthesis was similar to that of IIP synthesis but without binding of Fe (III) ion. Here is the reaction of NIP formation (Figure 3):

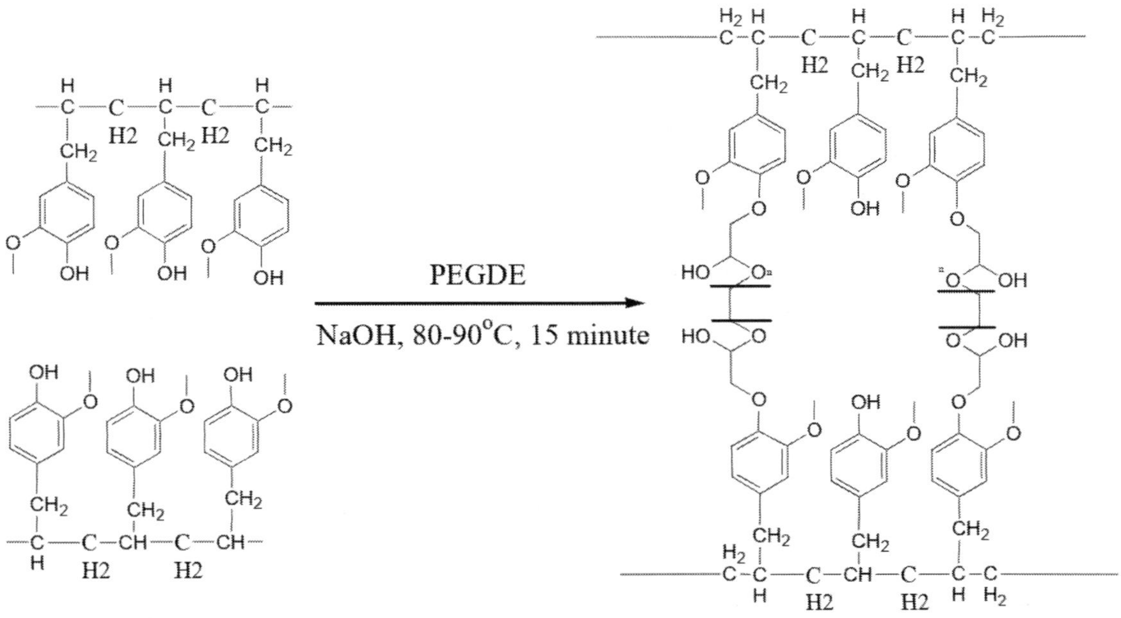

Polyeugenol IIP

Figure 3. NIP Synthesis

The result obtained was further used as an adsorbent for metal adsorption to be compared with the IIP adsorbent.

3.4 FTIR analysis

From the analysis using FTIR, it can be seen that the -OH spectrum in polyeugenol decreased in intensity when Fe (III) was added because it bound Fe (III) and the intensity decreased when it was cross-linked using PEGDE (Polyeugenol-Fe-PEGDE). This intensity then rose again after the Polyeugenol-Fe-PEGDE polymer released Fe-containing ion using acid (HCl) to produce IIP. The IIP spectrum is sharper than NIP and NIP acid, indicating the role of -OH in IIP work. This can be seen in Figure 4 until 7. FTIR analysis results can be seen in the Figures 3-5, and The comparison of FTIR spectra of template variation can be seen in Figure 6.:

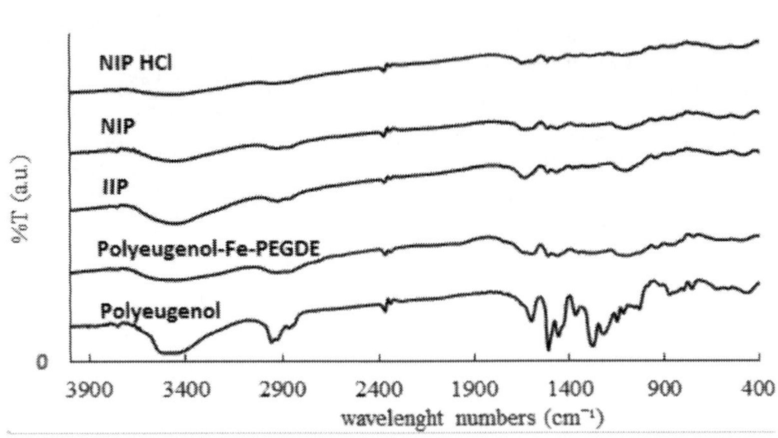

Figure 4. The graph comparison of FTIR spectra of $Fe(NO_3)_3$ template

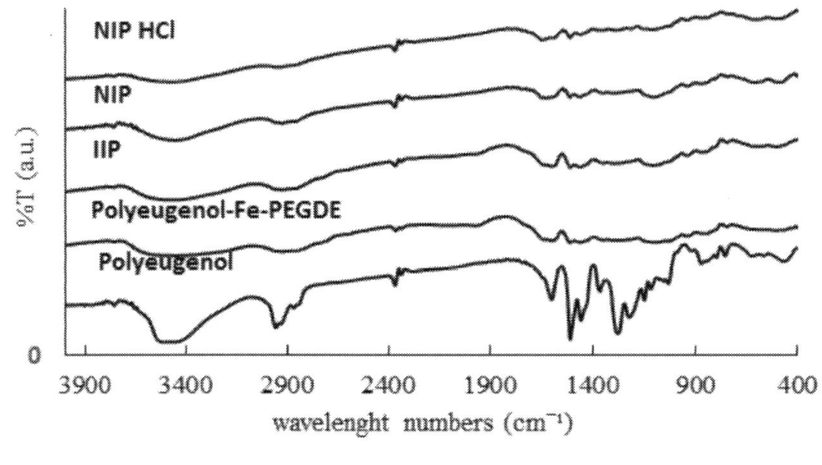

Figure 5. The graph comparison of FTIR spectra of $K_3[Fe(CN)_6]$ template

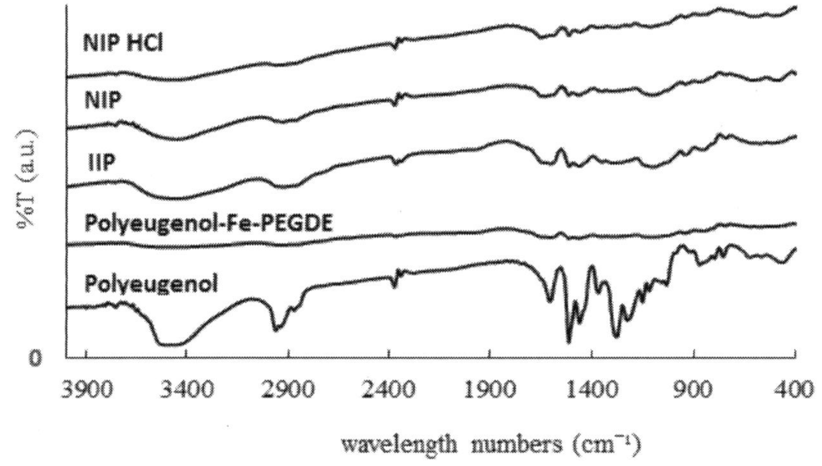

Figure 6. The graph comparison of FTIR spectra of $NH_4Fe(SO_4)_2$

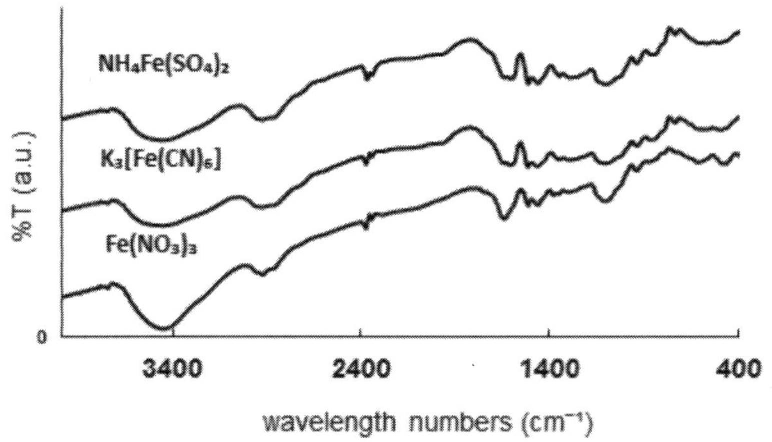

Figure 7. The comparison of FTIR spectra of template variation

3.5 SEM (Scanning Electron Microscopy) Analysis

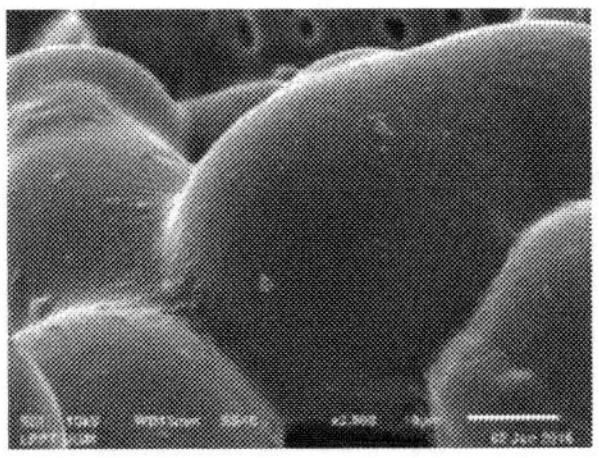

Polyeugenol-Fe-PEGDE IIP

Figure 8. SEM analysis

Analysis of the products using SEM as presented in Figure 8 shows that the pores of Polyeugenol-Fe-PEGDE are bigger relative than that of IIP. It may be due to the presence of Fe (III) bound while IIP is fairly free from Fe, it could be due to the role of Fe bound and then released from the adsorbent. Djunaidi *et al.* 2015 were found the same phenomena.

3.6 Adsorption Selectivity of Metal Ion

The selectivity of Fe metal ion adsorption based on template variation can be seen in Table 1.

Table 1. Selectivity of Fe metal ion adsorption based on template

Selectivity	IIP of $Fe(NO_3)_3$	IIP of $K_3[Fe(CN)_6]$	IIP of $NH_4Fe(SO_4)_2$	NIP	NIP of acid
Fe-Cr	2.51	1.08	0	0	0
Fe-Cd	1.47	1.51	0.78	0.38	0.07
Fe-Pb	0.75	1.24	1.32	1.31	0

Based on Table 1 it shows that the IIP using $Fe(NO_3)_3$ template produced the greatest selectivity compared to other templates. This is because the Fe (III) solution used for adsorption contained nitrate. so that the adsorption of binary Fe metal was more easily bound to IIP synthesized using the same template. i.e. $Fe(NO_3)_3$. The result of the highest Fe selectivity on Fe-Cr shows that the imprint appropriate with the template was formed. that is the adsorbent was more suitable for Fe metal. although Fe and Cr are in the same group which have similar properties; however. the imprint of the adsorbent was highly selective to Fe metal.

The adsorption selectivity of Fe metal on Fe-Cd binary metal mixture was greater than Fe-Pb since Fe is categorized as a hard acid with small hydrated atomic radius compared to Cd in the soft acid group and Pb in the medium acid group (borderline) which has the greater hydrated atomic radius [10]. The greater the size of the hydrated radius of the competitor metal. the lower of the selectivity of the adsorbent for Fe metal because it inhibited Fe metal to enter the imprint of the template. The template-based selectivity sequence on polyeugenol is $Fe(NO_3)_3 > K_3[Fe(CN)_6] > NH_4Fe(SO_4)_2$.

4 Conclusion

IIP synthesized from polyeugenol with variation of Fe (III) templates $Fe(NO_3)_3$. $K_3[Fe(CN)_6]$ and $NH_4Fe(SO_4)_2$ were successfully synthesized. The adsorption selectivity of Fe (III) metal ion in the $Fe(NO_3)_3$ template was greater than that of in the $K_3[Fe(CN)_6)$ and $NH_4Fe(SO_4)_2$ templates.The adsorption selectivity of Fe was greater in Fe-Cr binary metal compared to in the Fe-Cd and Fe-Pb.

References

1. Fan. H.-T. dan T. Sun. 2012. Selective removal of iron from aqueous solution using ion imprinted thiocyanato-functionalized silica gel sorbents. *Korean Journal of Chemical Engineering* 29(6): 798-803.
2. Parmpi. P. dan P. Kofinas. 2004. Biomimetic glucose recognition using molecularly imprinted polymer hydrogels. *Biomaterials* **25**(10): 1969-1973.
3. Harera. R.. L. S. Tety dan W. Meyliana. 2015. Sintesis Cu(II)-Imprinted Polymers untuk Ekstraksi Fasa Padat dan Prakonsentrasi Ion Tembaga(II) dengan Ligan Pengkhelat 4-(2-Pyridylazo) Recorcinol. *Al Kimiya*. 2 (1): 30-39.
4. Lee. L. L.. 2006. Synthesis And Application Of Molecularly Imprinted Solid-Phase Extraction For The Determination Of Terbutaline In Biological Matrices *[QD382. I43 L732 2006 f rb]. Universiti Sains Malaysia.*
5. Djunaidi. M.C. Jumina. Siswanta. D. Ulbricht. M. 2015. Selective Transport of Fe(III) Using Polyeugenol as Functional Polymer with IIP (IonicImprinted Polymer) Membrane Method. *Asian J. Chem 27 (12) p. 4553-4562*
6. Djunaidi. M.C. Jumina. Siswanta. D. Ulbricht. M. 2015. Synthesis of Fe(III)-Imprinted Polyeugenol using Polyethylene Glycol Diglycidilether (PEGDE) as Cross-Linking Agent and Its Application as Fe(III) Sorbent. *Indonesia. J. Chem. 15 (3). p 305-314.*
7. Djunaidi. M.C. Jumina. Siswanta. D. Eluent Influences on Synthesis of Fe(III)-imprinted Polyeugenol using Polyethylene Glycol Diglycidilether (PEGDE) as Cross-Linking Agent and its Application as Fe(III) Sorbent. 2015. *Oriental Journal of Chemistry Vol. 31. No. (4). p 2223-2229*
8. Djunaidi. M.C. Jumina. Siswanta. D. Ulbricht. M. 2016. Synthesis of Ionic Imprinted Polymer Particles for Selective MembraneTransport of Fe(III) using Polyeugenol as the Functional Polymer. *Oriental Journal of Chemistry Vol. 32. No. (1). pp. 77-84*
9. Svehla. G. 1985. Text Book of Macro and Semimicro Qualitative Inorganic Analysis. *ab L. Setiono and Hadyana Pudjaatmaka. part II. Fifth edition. PT. Kalman Media Pusaka. Jakarta* **369**.
10. Persson. I.. 2010. Hydrated metal ions in aqueous solution: how regular are their structures. *Pure and Applied Chemistry* 82(10): 1901-1917.

Characterization of Bio-Oil from Fast Pyrolysis of Palm Frond and Empty Fruit Bunch

M D Solikhah[1], F T Pratiwi[1], Y Heryana[1], A R Wimada[1], F Karuana[1], AA Raksodewanto[1], and A Kismanto[1]

[1]Agency for the Assessment and Application of Technology, Build. 480 PUSPIPTEK, Tangerang Selatan 15314 Indonesia

E-mail: maharani.dewi@bppt.go.id

Abstract. As the world's biggest producer of palm oil, 109 million tons of palm frond and 46 million tons of empty fruit bunch (EFB) were produced annually in Indonesia. These two kinds of palm biomass were still in low-application and could be potentially used as future energy resources such as biofuel. One of the promising methods to convert palm frond and EFB into biofuel, as a dense and easy to transport material, is fast pyrolysis. Before pyrolysis, biomass feedstock was characterized their component and elemental compositions, moisture content and higher heating value (HHV). Fast pyrolysis processes were conducted at a temperature of 350°C using thermal oil heater as a heat carrier. The gas phase from pyrolysis was condensed and produced a dark color and water soluble liquid called bio-oil. As GC-MS data shows, the bio-oil from both feed stocks was dominated by acetic acid, furans, phenols, aldehydes, and ketones. The HHV was reported 12.19 and 26.49 MJ/kg, while water content was 41.91 and 11.54 wt% for bio-oil from palm frond and EFB, respectively. The high content of lignin in EFB effects to the low content of water, high content of phenolic compound, and high calorific value in the bio-oil from EFB.

1 Introduction

The increasing of global energy demand and the depletion of fossil fuel reserves has forces study on new energy resources. Biomass has become favorable as a potential renewable energy source because of its abundance, making it easy to obtain, especially in countries with high agricultural activity. Waste generated in general has not been well utilizedto address problems in the environment [1].

Indonesia is the largest palm oil producing country in the world has a palm oil plant of more than 10.5 million hectares [2] which can produce 4.4 tons$\cdot$ha$^{-1}\cdot$y^{-1} of palm empty bunches and 10.4 tons$\cdot$ha$^{-1}\cdot$y^{-1} of palm frond [3]. Currently, empty bunches and palm frond is partly utilized for composting, and animal feed and the rest was left on the ground. Instead of being piled up as waste, it can be converted into biofuel that more dense and easy to transport material.

The process of converting biomass into fuel can be through two paths, namely biochemical pathways and thermochemistry. The biochemical process is done through anaerobic fermentation or digestion process using microbes [4] while the thermochemical process is considered more efficient because it

IOP Conf. Series: Materials Science and Engineering **349** (2018) 012035 doi:10.1088/1757-899X/349/1/012035

produces solid, liquid, and gas products under thermal conditions. This process can be pursued through pyrolysis pathways, gasification, liquefaction, and combustion [5].

Pyrolysis is one of the most promising processes for converting biomass that produces bio-oil, biochar and noncondensable gas [5]. This process generally occurs at high temperatures and in the absence of oxygen. Both exothermic and endothermic reactions involved in pyrolysis process. Exothermic reactions occur at low temperatures in the early stages when the pyrolysis process begins, including the breaking of the bonds of organic compounds into small fractions. As temperatures increase, some primary products vaporize and break into secondary products; this stage is called endothermic reaction [6].

Bio-oil, as the main product of biomass pyrolysis, is a high-density oxygenated compound that forms of thick dark brown or black liquid that iseasy flowing and smelling sharp [5]. The chemical composition in bio-oil depends on the biomass feedstock used. Because it comes from decomposition and depolymerization of cellulose, hemicellulose, and lignin, the main composition of the bio-oil consists of water, acids, alcohols, ketones, aldehydes, phenols, ether, esters, sugars, furans, nitrogen compounds, and various other compounds. These components make bio oil corrosive, unstable, has high water content, and has a low calorific value for direct use as fuel [7]. Therefore, characterizations of bio-oil are required to determine further upgrading process to improve its quality.

In this research, the raw material of palm empty fruit bunches (EFB) and palm frond has been converted into bio-oil through fast pyrolysis process at 350°C. Characterization of bio-oil was conducted to determine the water content, pH, calorific value, and characterization of its components using GC-MS.

2 Materials and Methods

2.1 Materials

Palm Frond used in this experiment was collected from a palm tree in PUSPIPTEK, South Tangerang, Indonesia. The leaves were removed from the frondbefore chopping and powdering until the size of 16 – 30 mesh. Palm EFB was collected from a local palm oil mill in Lebak Regency, Indonesia. Before pyrolysis process, EFB was chopped and powdered into the same size as palm frond. The moisture content of palm frond and EFB was 13.06 and 12.73%, respectively.

2.2 Bio-oil Productions

Bio-oil productions were conducted using a reactor describes in Solikhah et al.[8] operated at 350°C using thermal oil as a heat carrier. Bio-oil as the liquid product was collected from the condenser in the form of dark color and water soluble liquid.

2.3 Characterizations

Analysis of cellulose, hemicellulose, and lignin in the palm frond and EFB was conducted using an HPLC (Waters Corp, USA) according to the method developed by Sluiter et al. [9].

The bio-oils were qualitatively analyzed using a gas chromatograph mass spectrometer of Shimadzu GC/MS QP 2010 (Shimadzu Corp, Tokyo, Japan). In the analysis, a column of DB1 (dimethyl polysiloxane, 30 m x 0.32 mm x 1 μm capillary column) obtained from Agilent, USA was used. Helium gas on GCMS grade (Air Liquide, Indonesia) was used for gas carrier. A 0.8 μL sample was injected on the column that was started at 40 °C then hold for 4 minutes, after that, ramped at 5 °C/minute until 250 °C and then hold for 10 minutes.

Water content on bio-oil was analyzed using Karl Fisher KF200 (Mitsubishi Chemical Analytech Co.,Ltd., Kanagawa, Japan). The heating value of bio-oil was measured using Automatic Calorimeter

AC500 (Leco Corp., Saint Joseph, USA).The pH analysis was performed on bio-oil using 848 Titrino plus (Metrohm AG, Switzerland).

3 Results and Discussions

3.1 Celluloses and Lignin

The analysis shows that the cellulose, hemicellulose, and lignin composition in palm frond were 36.67, 18.71, and 27.38% while in EFB were 23.83, 15.90, and 39.51%, respectively. This result shows that palm frond is dominated by cellulose which is in accordance previous studies by Abnisa et al. and Kabir et al. [3,10]. On the other hand, EFB is dominated by lignin. This result differs from the study of Abnisa et al. [7] but similar to the analysis of softwood mentioned by Kan et al. [11].

3.2 Water Content

Based on Table 1, it can be seen that the bio-oil from palm frond (41.91%) has higher water content than the bio-oil from EFB (11.54%). The water content of bio-oil is influenced by the water content of raw materials used and water formation generated during pyrolysis process [11]. The lower water content in bio-oil from EFB also affected by higher lignin content in the biomass, as mentioned by Fahmi et al.[12].

3.3 pH

The identification of pH is required because low pH can cause corrosion at the time of storage. The measurement results in Table 1 show that pH value in both bio-oil from palm frond and EFB was low/acidic. This indicates that the decomposition of lignin and hemicellulose compounds was occurred at the pyrolysis to form acid compounds, as mentioned by Kanet al.[11].

Table 1. Water content, pH, and Heating Value of Bio-oil

Parameter	Bio-oil from EFB	Bio-oil from Palm Frond
Water content, % mass	11.54	41.91
pH	2.8	3.1
Heating Value (HHV), MJ/kg	26.49	12.19

3.4 GC-MS Analysis

Details compositions of bio-oil from palm frond and EFB can be seen in Table 2 and 3. Based on the table, both bio-oils are dominated by acetic acid, furan, phenol, aldehyde, and ketone compounds. Bio-oil from palm frond shows a high content of acetic acid and phenolic compounds at 35 and 40%, respectively. On the other hand, acetic acid in bio-oil from EFB is less than that from a palm frond. This result is in accordance with the result by Wang et al. [13] that acetic acid and 2-furfural are the major products obtained from hemicellulose and that the presence of lignin in EFB becomes an inhibitory effect on the formation of acetic acid. Table 3 shows that bio-oil from EFB contains phenolic compounds as much as 47%. The high content of the phenolic compound in bio-oil from EFB is affected by the high content of lignin that decomposed into the phenolic compound, as mentioned by Wang et al.[14].

Table 2. Major Components of Bio-oil from Palm Frond

Component	Area, %
Acetic acid	35.79
Chloroform	2.26
2-Propanone, 1-hydroxy-	3.93
2-Propanone	0.90
2-Furancarboxaldehyde	2.68
Phenol	5.78
Phenol, 2,6-bis(1,1-dimethylethyl)-4-methyl-	34.18

Table 3. Major Components of Bio-oil from EFB

Component	Area, %
Acetic acid	4.07
2-Propanone, 1-hydroxy-	1.95
2-Furancarboxaldehyde	1.19
2-Furanmethanol	1.64
Phenol	24.64
2-Furanmethanol, tetrahydro-	1.37
2-Cyclopenten-1-one, 2-hydroxy-3-methyl	1.44
Phenol, 2-methoxy-	3.25
2-Methoxy-4-methylphenol	0.99
Phenol, 2-(1,1-dimethylethyl)-	3.97
Phenol, 4-ethenyl-2-methoxy-	1.51
Phenol, 2,6-dimethoxy-	7.76
Benzaldehyde, 4-hydroxy-3-methoxy	0.95
Benzene, 1,2,3-trimethoxy-	1.50
Benzoic acid, 4-hydroxy-, methyl ester	0.93
Phenol, 2-methoxy-4-(2-propenyl)-	1.84
Benzene, 1,2,3-trimethoxy-5-methyl	1.90
Benzaldehyde, 6-hydroxy-4-methoxy-2,3-dimethyl	1.76
Phenol, 2,6-dimethoxy-4-(2-propenyl)-	3.18
1H-Purin-6-amine, [(2-fluorophenyl)methyl]	1.40

3.5 Calorific Value

Calorific value is an important parameter in fuel characterization. The calorific value of bio-oil is represented by higher heating value (HHV). The result on Table 1 shows that bio-oil from palm frond has a lower heating value than the one of the EFB. This result is affected by the water content in both bio-oil. The higher moisture content of the bio-oils causes the lower heating value since the presence of water will influence the combustion process. As mentioned above, the water content in bio-oil is affected by lignin content in biomass so that higher content of lignin in EFB results lower water content. On the other hand, calorific value also affected by chemical composition of the bio-oil as mentioned by Zhao et al. [15]. Pyrolysis reaction of lignin produced phenolic compound that gives bio-oil with heating value of 20.4 to 24.5 MJ/kg [16]. Therefore, higher content of lignin in biomass produced a higher calorific value of its bio-oil.

4 Conclusions

Characterization of bio-oil from fast pyrolysis of palmfrond and empty fruit bunch has been conducted to determine the water content, pH, calorific value, and characterization of its components using GC-MS. As GC-MS data shows, the bio-oil from both feed stocks was dominated by acetic acid, furans, phenols, aldehydes, and ketones. The HHV was reported 12.19 and 26.49 MJ/kg, while moisture content was 41.91 and 11.54 wt% for bio-oil from palm frond and EFB, respectively. The high content of lignin in EFB effects to the low content of water, high content of phenolic compound, and high calorific value in the bio-oil from EFB.

Acknowledgements

This research project is funded by Indonesia Estate Crop Fund for Palm Oil (in Indonesian: Badan Pengelola Dana Perkebunan Kelapa Sawit) No. PRJ-30/DPKS/2015.

References

[1] Bridgwater A. Fast pyrolysis processes for biomass. Renew Sustain Energy Rev 2000;4:1–73. doi:10.1016/S1364-0321(99)00007-6.
[2] BPS-Statistics Indonesia. Indonesian Oil Palm Statistics 2015. BPS-Statistics Indonesia; 2016.
[3] Abnisa F, Arami-Niya A, Wan Daud WMA, Sahu JN, Noor IM. Utilization of oil palm tree residues to produce bio-oil and bio-char via pyrolysis. Energy Convers Manag 2013;76:1073–82. doi:10.1016/j.enconman.2013.08.038.
[4] Bridgwater A V. Review of fast pyrolysis of biomass and product upgrading. Biomass and Bioenergy 2012;38:68–94. doi:10.1016/j.biombioe.2011.01.048.
[5] Cardoso CAL, Machado ME, Caramão EB. Characterization of bio-oils obtained from pyrolysis of bocaiuva residues. Renew Energy 2016;91:21–31. doi:10.1016/j.renene.2015.11.086.
[6] Xu Y, Hu X, Li W, Shi Y. Preparation and Characterization of Bio-oil from Biomass. Prog. Biomass Bioenergy Prod., InTech; 2011. doi:10.5772/16466.
[7] Abnisa F, Arami-Niya A, Daud WMAW, Sahu JN. Characterization of Bio-oil and Bio-char from Pyrolysis of Palm Oil Wastes. BioEnergy Res 2013;6:830–40. doi:10.1007/s12155-013-9313-8.
[8] Solikhah MD, Raksodewanto AA, Kismanto A, Karuana F, Heryana Y, Riza, et al. Development of bio-fuel from palm frond via fast pyrolysis. IOP Conf Ser Earth Environ Sci 2017;65:12014. doi:10.1088/1755-1315/65/1/012014.
[9] Sluiter A, Hames B, Ruiz R, Scarlata C, Sluiter J, Templeton D, et al. Determination of Structural Carbohydrates and Lignin in Biomass. vol. 2011. National Renewable Energy Laboratory; 2012.
[10] Kabir G, Mohd Din AT, Hameed BH. Pyrolysis of oil palm mesocarp fiber and palm frond in a slow-heating fixed-bed reactor: A comparative study. Bioresour Technol 2017;241:563–72. doi:10.1016/j.biortech.2017.05.180.
[11] Kan T, Strezov V, Evans TJ. Lignocellulosic biomass pyrolysis: A review of product properties and effects of pyrolysis parameters. Renew Sustain Energy Rev 2016;57:1126–40. doi:10.1016/j.rser.2015.12.185.
[12] Fahmi R, Bridgwater AV, Donnison I, Yates N, Jones JM. The effect of lignin and inorganic species in biomass on pyrolysis oil yields, quality and stability. Fuel 2008;87:1230–40. doi:10.1016/j.fuel.2007.07.026.
[13] Wang S, Guo X, Wang K, Luo Z. Influence of the interaction of components on the pyrolysis behavior of biomass. J Anal Appl Pyrolysis 2011;91:183–9. doi:10.1016/j.jaap.2011.02.006.
[14] Wang S, Dai G, Yang H, Luo Z. Lignocellulosic biomass pyrolysis mechanism: A state-of-the-art review. Prog Energy Combust Sci 2017;62:33–86. doi:10.1016/j.pecs.2017.05.004.
[15] Zhao C, Jiang E, Chen A. Volatile production from pyrolysis of cellulose, hemicellulose and lignin. J Energy Inst 2017;90:902–13. doi:10.1016/j.joei.2016.08.004.
[16] Bu Q, Lei H, Wang L, Wei Y, Zhu L, Zhang X, et al. Bio-based phenols and fuel production from catalytic microwave pyrolysis of lignin by activated carbons. Bioresour Technol 2014;162:142–7. doi:10.1016/j.biortech.2014.03.103.

IOP Conf. Series: Materials Science and Engineering **349** (2018) 012036 doi:10.1088/1757-899X/349/1/012036

Synthesis of 4-hydroxy-3-methylchalcone from Reimer-Tiemann reaction product and its antibacterial activity test

M Hapsari, T Windarti, Purbowatiningrum, Ngadiwiyana, Ismiyarto*

Organic Chemistry Laboratorium, Chemistry Departement,Faculty of Sciences and Mathematics Diponegoro University

*Email: Ismiyarto@live.undip.ac.id

Abstract. A 4-hydroxy-3-methylchalcone has been synthesized from 4-hydroxy-3-methylbenzaldehyde as the Reimer-Tiemann reaction product. This research consists of three steps involve synthesize of 4-hydroxy-3-methylbenzaldehyde from ortho-cresol, synthesize of chalcone derivatives from 4-hydroxy-3-methylbenzaldehyde and 4-hydroxy-3-methoxybenzaldehyde or vanillin for the comparison, the last is antibacterial activity test of both chalcone derivatives against *Escherichia coli* (negative gram) and *Staphylococcus aureus* (positive gram) bacteria using disc diffusion method. Results of Reimer-Tiemann reaction is 4-hydroxy-3-methylbenzaldehyde compound in an orange colour solid form which has 43% yields and melting point 110-114°C. A 4-hydroxy-3-methylbenzaldehyde then reacted with acetophenone in a base condition and form 4-hydroxy-3-methylchalcone compound in a yellow colour solid form which has 40% yields and melting point 83-86°C. The antibacterial activity of the 4-hydroxy-3-methylchalcone against gram-positive bacteria *Staphylococcus aureus* is better than the 4-hydroxy-3-methoxychalcone.

1 INTRODUCTION

Chalcone is an open chain flavonoid group which has two aromatic ring containing unsaturated α,β – carbonyl three carbons system as a connector[12]. During this time, chalcone is synthesized from acetophenone with benzaldehyde and its derivatives by Claisen-Schmidt condensation reaction[6]. According to researches[2][8] a 4-hydroxy-3-methoxychalcone have been synthesized from the reaction of acetophenone and 4-hydroxy-3-methoxybenzaldehyde or vanillin. Chalcone derived from vanillin has a better antibacterial activity due to the influence of its two substituents. In this study, synthesis of 4-hydroxy-3-methylchalcone will be done by reacting acetophenone and 4-hydroxy-3-methylbenzaldehyde. The 4-hydroxy-3-methylbenzaldehyde compound used is obtained from Reimer-Tiemann formylation reaction of ortho-cresol and chloroform in a base condition. Then the antibacterial activity of this compound will be compared with the 4-hydroxy-3-methoxychalcone obtained from the reaction of acetophenone and vanillin. The 4-hydroxy-3-methylchalcone has a lipophilic methyl group, so it is easier to bind and form complexes with cell membranes that can cause damage to the membrane. Therefore, the antibacterial activity of 4-hydroxy-3-methylchalcone would be greater than 4-hydroxy-3-methoxychalcone.

2 PROCEDURE

2.1 Synthesis of 4-hydroxy-3-methylbenzaldehyde

A total of 15 g of sodium hydroxide (NaOH) dissolved in 75 mL of distilled water was added to 5.15 mL ortho-cresol (50 mmol) in 100 mL of absolute ethanol. The reaction mixture was heated at 60°C. After the reaction lasts 5 minutes, the addition of 20 mmol of chloroform was carefully started. Chloroform was added excess because it is easily evaporated. Stirring was continued for 1 hour after the addition of chloroform was completed. Ethanol was removed, and 1 M of HCl was added to neutralize the excess NaOH to pH = 2-3. Then the solution was separated by extraction method using ethyl acetate for 3 times. After that, ethyl acetate fraction was added by anhydrous Na_2SO_4 and filtered before evaporated by reducing the pressure. The reaction mixture is further purified by column chromatography with silica gel. The product obtained were analyzed using spectrophotometer FT-IR, GC-MS, and H-NMR[5][10][11].

2.2 Synthesis of Chalcone Derivative

A total of 1.7 mL of Acetophenone (14.68 mmol) and 9 mL of NaOH 60% were added to 2 g of 4-hydroxy-3-methylbenzaldehyde (14.68 mmol) in 15 mL absolute ethanol. The mixture was heated at 70°C for 3 hours and every hour monitored by TLC. Ethanol was removed and 1 M HCl was added to pH = 1. The reaction mixture was separated by extraction method using ethyl acetate. Then the organic phase obtained was added by anhydrous Na_2SO_4 to bind water that still exists in the organic phase. Solvents were distilled off and then the reaction mixture was further purified by column chromatography on silica gel. The formed solids were then analyzed using spectrophotometer UV-VIS, FT-IR, and LC-MS. Under the same procedure, acetophenone was reacted with vanillin to obtain a 4-hydroxy-3-methoxychalcone [2][8].

2.3 Antibacterial Activity Test

Nutrient agar media as much as 15 mL was poured into a petri dish and left to solid. All bacterial planting process was done in Laminar Air Flow. A total of 0.1 mL of diluted bacterial suspension was inoculated on the nutrient agar medium which had been solidified. Furthermore, the disc paper (6 mm diameter) was inserted into the dilute synthesized product solution as a test solution into various concentrations for 1 minute until it was fully diffused on the disc paper. The disc paper containing the test solution was placed in the agar medium. The test medium which has been ready, was incubated for 24 hours in an incubator at 37°C. The clear zone around the disc were a zone of inhibition which measured using a vernier caliper in millimeters.

3 RESULT AND DISCUSSION

3.1 Synthesis of 4-hydroxy-3-methylbenzaldehyde

The 4-hydroxy-3-methylbenzaldehyde has been synthesized from ortho-cresol through Reimer-Tiemann formylation reaction. The reaction is shown in ***Figure 1*** below.

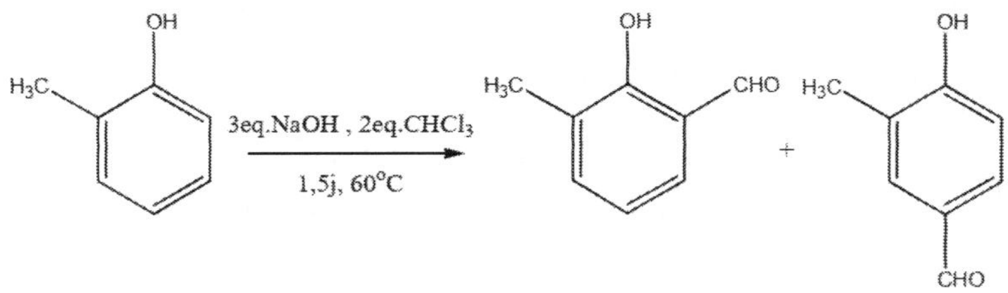

Figure 1. *Reimer-Tiemann Formylation Reaction*

IOP Conf. Series: Materials Science and Engineering **349** (2018) 012036 doi:10.1088/1757-899X/349/1/012036

Chromatogram of the product mixture developed by GC-MS method showed the presence of 3-methylsalicylaldehyde (m/z 136) at the retention time (RT) of 12.658 minutes with an abundance of 2% and 4-hydroxy-3-methylbenzaldehyde (m/z 136) at RT of 25,068 minutes as much as 53%.

The product mixture which has been purified using column chromatography then analyzed using FT-IR and H-NMR. The presence of 4-hydroxy-3-methylbenzaldehyde in the compound with a retention time of 25.068 minutes was proven by FT-IR spectrum. The spectrum showed there were five peaks gained at 3190.03 cm^{-1} (O-H), 2850.58 cm^{-1} (C-H Aldehyde), 1662.52 cm^{-1} (-C=O, carbonyl), 3055,02 cm^{-1} (sp^2 C-H, aromatic), and 1377.07 cm^{-1} (CH_3, methyl).

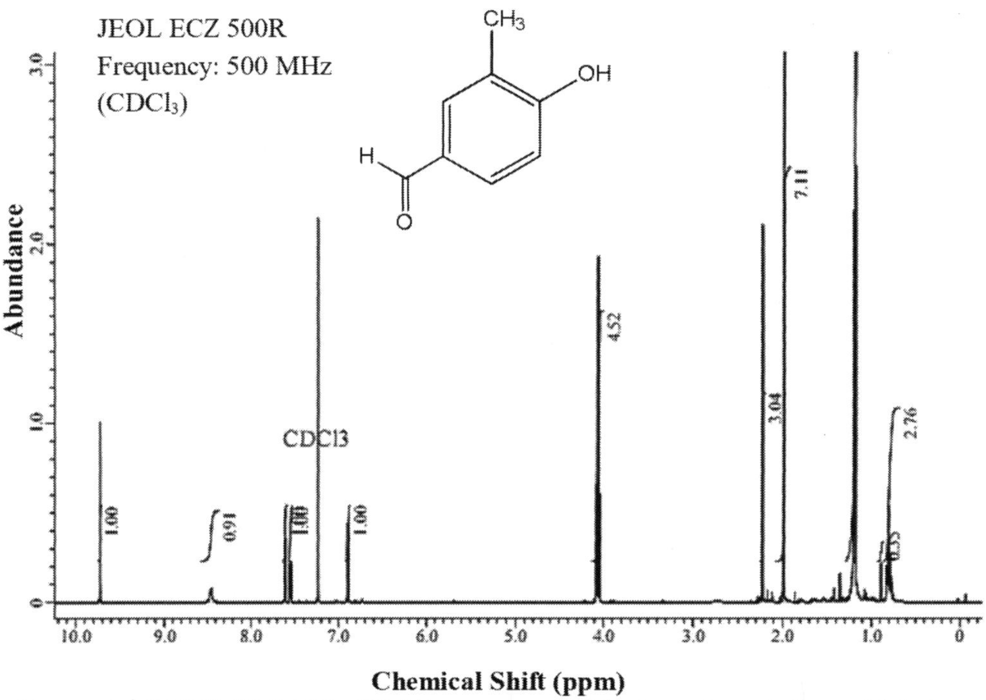

Figure 2. ¹H-NMR (500 MHz, $CDCl_3$) spectrum of 4-hydroxy-3-methylbenzaldehyde

A typical ¹H-NMR spectrum of 4-hydroxy-3-methylbenzaldehyde can be seen in *Figure 2*. There are 6 signals of a proton which indicated the presence of 4-hydroxy-3-methylbenzaldehyde. Proton from aldehyde group is shown at a chemical shift (δ) 9.726 ppm (s, 1H), proton from hydroxyl group (–OH) at 8.457 ppm (*broad*, 1H), protons from aromatic ring at 7.612 ppm (s, 1H), 7.562 ppm (d, 1H, J = 8 Hz), and 6,886-6,903 ppm (d, 1H, J = 8,5 Hz), then proton from methyl group ($-CH_3$) is shown at 2..233 ppm (s, 3H). The solid form is obtained in an orange color, has a yield of 43% with an abundance of 53%, and has a melting point of 110-114°C.

3.2 Synthesis of Chalcone Derivatives
The 4-hydroxy-3-methylchalcone and 4-hydroxy-3-methoxychalcone as its comparative compound have been synthesized from the reaction of acetophenone and 4-hydroxy-3-methylbenzaldehyde while its comparative compound from 4-hydroxy-3-methoxybenzaldehyde or vanillin through Claisen-Schmidt condensation reaction. The reaction is shown in *Figure 3* below.

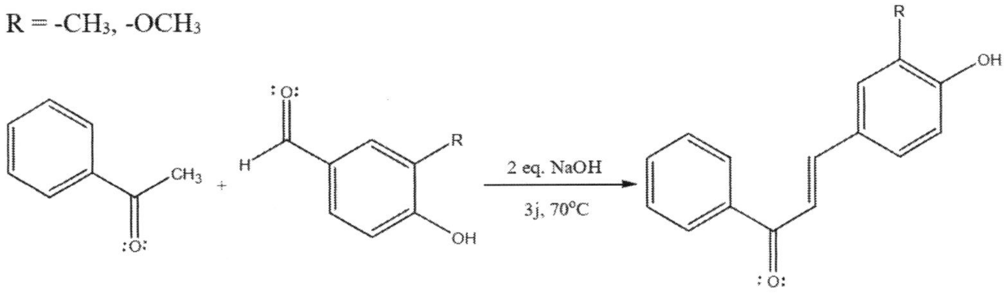

Figure 3. Claisen-Schmidt Condensation Reaction

The reaction product obtained has been purified by column chromatography. It was then analyzed using spectrophotometer UV-VIS, FT-IR, and LC-MS. The presence of 4-hydroxy-3-methylchalcone was proven by UV-VIS spectrum which showed two peaks gained at a wavelength 353 nm (1st Band) and 253 nm (2nd Band). These results prove that the product obtained is a group of flavonoid chalcone. The UV-VIS spectrum of chalcone has a characteristic that consists of two bands in the 340-390 nm range (1st band) and 230-270 nm (weak, as the 2nd band) (Markham, 1988). The UV-VIS spectrum of 4-hydroxy-3-methylchalcone is shown in *Figure 4* below.

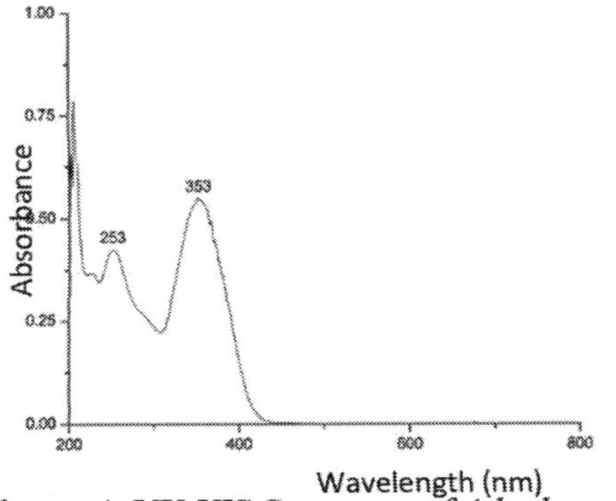

Figure 4. *UV-VIS Spectrum of 4-hydroxy-3-methylchalcone*

The FT-IR spectrum *(Figure 5)* of the reaction product also indicated the presence of 4-hydroxy-3-methylchalcone. These peaks are gained at 3305.75 cm^{-1} (O-H), 3024.16 cm^{-1} (sp^2 C-H, aromatic), 1666.37 cm^{-1} (C=O, carbonyl), 1581.51 cm^{-1} (C=C, aromatic), and 1380.35 cm^{-1} (methyl/ CH_3).

A dominant peak at the chromatogram *(Figure 6)* with a retention time (RT) of 2,6 minutes developed by LC-MS method was indicating that there was only one constituent in the product. From the mass spectrum of the product, that constituent was identified and characterized as 4-hydroxy-3-metylchalcone (m/z 238). The 4-hydroxy-3-methylchalcone obtained in a yellow color solid form at room temperature, has a yield of 40%, abundance of >95%, and has a melting point of 83-86°C.

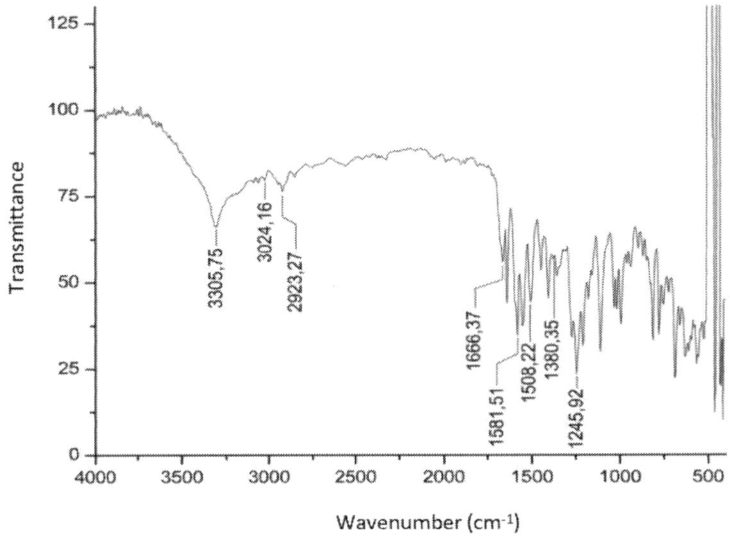

Figure 5. *FT-IR Spectrum of 4-hydroxy-3-methylchalcone*

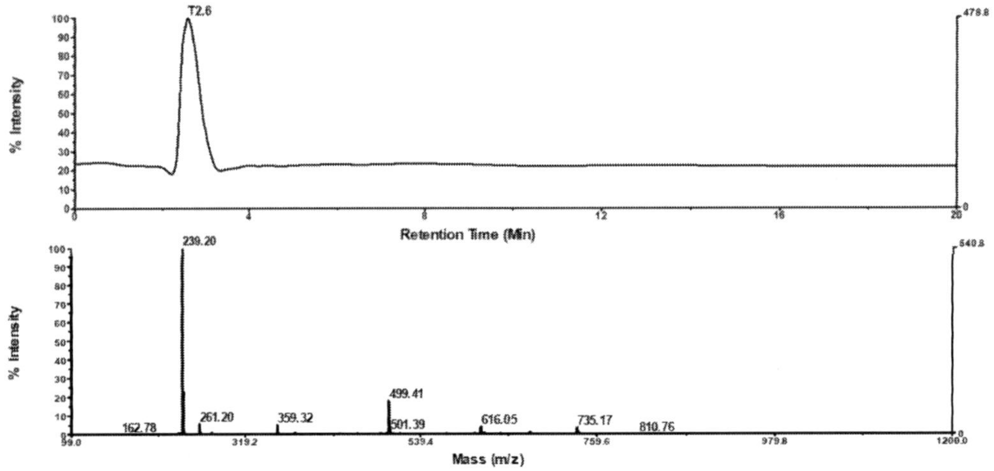

Figure 6. *Chromatogram and Mass Spectrum of 4-hydroxy-3-methylchalcone*

Meanwhile, as a comparison, the product obtained from the reaction of acetophenone and vanillin also gave similar results to the 4-hydroxy-3-methylchalcone. The UV-VIS spectrum of the reaction products of acetophenone and vanillin compounds showed two peaks at the wavelength of 363 nm (1st Band) and 261 nm (2nd Band). Moreover, a dominant peak at the retention time of 3.8 minutes on the chromatogram and mass spectrum developed by LC-MS method was indicating that the product was 4-hydroxy-3-methoxychalcone (m/z 254). The 4-hydroxy-3-methoxychalcone obtained in a yellow color solid form at room temperature, has a yield of 76%, abundance of >95%, and has a melting point of 78°C.

If the yield of 4-hydroxy-3-methylchalcone and 4-hydroxy-3-methoxychalcone were compared, there was a significant difference. This occurs due to its differences in substituents $-CH_3$ dan $-OCH_3$ and their relation to each substituent character (σ). The value of σ on each substituent shows the resultant between induction and mesomery influence on a compound[4]. In the two chalcone derivatives, position of the $-CH_3$ and $-OCH_3$ substituents is meta to the aldehyde group. At the meta position, the $-CH_3$ group has a value of σ_m = -0.07 and the $-OCH_3$ group has a value of σ_m = +0.12. It should be noted that if the value of σ is positive, it means the substituent is electron withdrawing group. Conversely, if the value of σ is negative then the substituent is electron releasing group[4].

Electron withdrawing substituents also can be called as a ring activator group. This is because it can increase the reactivity of the aromatic ring. Since the aromatic ring of vanillin is more active than the aromatic ring in the 4-hydroxy-3-methylbenzaldehyde, the vanillin is more reactive with the enolate ions derived from the acetophenone. Because it is easy to react, the yield is much more than a compound with an electron releasing substituent like 4-hydroxy-3-methylbenzaldehyde.

3.3 Antibacterial Activity Test

The antibacterial activity test of 4-hydroxy-3-methylchalcone against *Escherichia coli* (negative gram) and *Staphylococcus aureus* (Positive gram) bacteria have been done by disc diffusion method. In this test, peptone and yeast extract were used as bacterial growth medium. Before the antibacterial test, the bacterial stock was incubated in a nutrient broth medium for about 3 hours to obtain an equivalent of 0.5 Mc Farland turbidity. Results of antibacterial activity test of **chalcone derivative are shown in table 1 below.**

Table 1: Data of the Antibacterial Activity Test

No.	Sample	Zone (mm)	
		E.coli	*S.aureus*
1.	Chalcone	1,2	2,1
2.	4-hydroxy-3-methoxychalcone	2,4	2,2
3.	4-hydroxy-3-methylchalcone	1,7	2,4
4.	DMSO	0	0
5.	Amoxicillin	12,2	4,5

The antibacterial mechanism of flavonoid can occur through the formation of complexes with proteins by forming hydrogen bonds, covalent bonds, and hydrophobic effects. This is related to the ability of flavonoid antibacterial compounds in disabling microbial solvents, enzymes, and envelope cell proteins[7].

When the antibacterial activity of the 4-hydroxy-3-methylchalcone is compared with the 4-hydroxy-3-methoxychalcone and the unsubstituted chalcone, antibacterial activity of the 4-hydroxy-3-methylchalcone against gram-positive bacteria *Staphylococcus* aureus is better than the comparable compounds. This is because besides the simpler cell wall structure, the thick layer of peptidoglycan contains lipoteichoic acid that is hydrophobic. This lipoteichoic acid connects to the outer layers of the bacterial cell wall and the cell membrane layer that lies beneath a thick layer of peptidoglycan[3]. The presence of lipoteichoic acid causes the lipophilic 4-hydroxy-3-methylcchalcone will more easily bond and form complexes with cell membranes that can cause damage to the membrane. Nevertheless, the increase in antibacterial activity of flavonoid compounds especially chalcone is not only influenced by the type of substituent or functional group attached to the aromatic ring. However, the functional group position in the aromatic ring also influences the increase in antibacterial activity of chalcone[1].

4 CONCLUSION

Results of Reimer-Tiemann reaction is 4-hydroxy-3-methylbenzaldehyde compound in an orange color solid form which has 43% yields and melting point 110-114°C. A 4-hydroxy-3-methylbenzaldehyde then reacted with acetophenone in a base condition and form 4-hydroxy-3-methylchalcone compound in a yellow colour solid form which has 40% yields and melting point 83-86°C. The 4-hydroxy-3-methoxychalcone compound obtained in 76% yields. The antibacterial activity of the 4-hydroxy-3-methylchalcone against gram-positive bacteria *Staphylococcus aureus* is better than the 4-hydroxy-3-methoxychalcone.

ACKNOWLEDGEMENT

We are grateful to financial support from Diponegoro University (No: SP DIPA-042.01.2.400898/2016, December, 7th 2015) second year.

REFERENCES

[1] Avila, H.P., Smania, E.F.A., Monache, F.D., and Junior, A.S., 2008, Structure-activity Relationship of Antibacterial Chalcones, *Bioorganic & Medicinal Chemistry*, **16** 9790-9794.
[2] Azizah, Y., Hanapi, A., and Adi, T.K., 2015, Synthesis of 3-(4-Hydroxy-3-Metoxyphenyl)-1-Phenyl-2-Propen-1-On and its Antioxidant Activity Assay using DPPH, *ALCHEMY*, **4** 67-72.
[3] Brown, L., Wolf, J.M., Prados-Rosales, R., and Casadevall, A., 2015, Through Cell Wall: Extracellular Vesicles in Gram-positive Bacteria, Mycobacteria, and Fungi, *Nature Reviews Mictrobiology*.
[4] Cahyono, B., 2004, Kimia Organik Fisik, Diponegoro University Publisher, Semarang.
[5] Crisan, R., and Modra, D., 2013, The Synthesis of Salicylaldehyde Varying Different Parameters, *Series of Chemistry*, **22** 57-66.
[6] Ismiyarto, Mastjeh, S., and Anwar, C., 1998, Sintesis Senyawa Kalkon dan Flavanon Menggunakan Bahan Dasar Senyawa Turunan Asetofenon dan Benzaldehida, *Indonesian Journal of Chemistry*.
[7] Kumar, S. and Pandey, A.K, 2013, Chemistry and Biological Activities of Flavonoids: An Overview, *The Scientific World Journal*.
[8] Madiyono, M., 2002, Sintesis Senyawa 4-hidroksi-3-metoksikalkon dari Vanilin dan Asetofenon, Undergraduate Thesis, FMIPA UNDIP, Semarang.
[9] Markham, K.R., 1988, Cara Mengidentifikasi Flavonoid, Translated by Padmawinata, K., ITB Publisher, Bandung.
[10] Mokle, S., Vibhute, Y.B., and Sayyed, M., 2006, An Improved Procedure for the Synthesis of 2-hydroxybenzaldehyde and 2-hydroxynaphthalene-1-carbaldehyde, *Journal of Chemical Research*, **5** 3236.
[11] Thoer, A., Denis, G., Delmas, M., and Gaset, A., 1988, The Reimer-Tiemann Reaction in Slightly Hydrated Solid-liquid Medium: A New Method for the Synthesis of Formyl and Diformyl Phenols, *Synthetic Communications*, **18** 2095-2101.
[12] Tran, T., Do, T., Tran, N., Ngo, T., Huynh, T., Tran, C., and Thai, K., Synthesis and Anti Methicillin Resistant *Staphylococcus aureus* Activity of Substituted Chalcones Alone and in Combination with non-beta-lactam Antibiotics, *Bioorganic and Medicinal Chemistry Letters*, **22** 4555-4560.

Toxicity tests, antioxidant activity, and antimicrobial activity of chitosan

M Kurniasih*[1], Purwati[1], R S Dewi[2]

[1]Chemistry Department, Faculty of Mathematics and Natural Sciences, Jenderal Soedirman University
[2]Mycology & Phytopathology Laboratory, Faculty of Biology, Jenderal Soedirman University,

E-mail: m_kurniasih@yahoo.com

Abstract. Chitosan is a naturally occurring cationic biopolymer, obtained by alkaline deacetylation of chitin. This research aims to investigate the toxicity, antioxidant activity and antibacterial activity of chitosan from shrimp chitin. In this study, chitin extracted from shrimp waste material. Chitin is then deacetylation with 60% NaOH so that chitosan produced. Degrees of deacetylation, molecular weight, toxicity test, antioxidant activity and antimicrobial activity of chitosan then evaluated. Toxicity test using Brine Shrimp Lethality Test. The antioxidant analysis was performed using DPPH method (2,2-diphenyl-1-picrylhydrazyl) and FTC method (ferric thiocyanate) in which the radical formed will reduce Ferro to Ferri resulting in a complex with thiocyanate. To determine the antibacterial activity of *Staphylococcus aureus*, antifungal in *Candida albicans* and *Aspergillus niger* by measuring antimicrobial effects and minimum inhibitory concentrations (MIC). Based on the result of research, the value of degrees of deacetylation, molecular weight, and LC_{50} values of chitosan synthesis was 94,32, 1052.93 g/mol and 1364.41 ppm, respectively. In general, the antioxidative activities increased as the concentration of chitosan increased. MIC value of chitosan against *S. aureus*, *C. albicans,* and *A. niger* was 10 ppm, 15.6 ppm, and 5 ppm, respectively.

1. Introduction

Nowadays people use only a small portion of the waste of shrimp shells as flavorings in the process of making chips and shrimp paste [1]. Therefore, the waste needs a treatment, mainly because the waste contains a substance, chitin, which further can have potential use. One of the derivatives of chitin is chitosan. The main commercial sources of chitin are the shell wastes of shrimp, lobster, krill, and crab [2].

Chitosan is attained through the process of chitin deacetylation, in which the acetyl groups in chitin changed into amino groups by adding high concentration alkali [3]. Chitosan is bio-active, bio-degradation and non-toxic [4] polymer from nature. Chitin and chitosan are often applied in industry as well as health. Chitosan consists of three reactive and functional groups, namely amino group, a primary hydroxyl group and secondary hydroxyl group [5]. Chitosan has a free positively-charged amino group so that it can bind with negatively-charged one. Chitosan and its derivatives had some biological activities, like the anti-microbial feature [6] because chitosan has an active group inside it that reacts with microbes so that it can inhibit their growth.

Free radicals are relatively unstable species, have unpaired electron on their outside orbit, so they are reactive in finding another electron. Free radicals found in environments, several metals (iron, copper), smoke from cigarettes, air pollution, medicines, toxic substances, packaged food, additives, and UV rays from both the sun and radiation [7]. Antioxidants can prevent the presumed deleterious effects of free radicals in the human body, and prevent the deterioration of fats and other constituents of foodstuffs [8]. According to a study conducted by [9], chitosan can inhibit DPPH radical (1,1-diphenyl-2-picrylhydrazyl) in the concentration ranging 0.125 – 1.0 mg/mL with the inhibition of 28.37 – 38.03%. According to the study by [10], hexanoyl chitin could trap peroxide radicals in the organic solvent when the chain reaction of radicals is initiated 2,20-azobis (2,4-dimethylvaleronitrile). Based on the above background, a study was conducted to investigate the toxicity, antioxidant activity and antibacterial activity of chitosan from shrimp chitin.

2. Experimental

2.1. A Chitin Isolation [11]

Shrimp shells are cleaned, dried, and mashed until smooth. The powder, which is 100 mesh, is used for demineralization process using HCl 1.5 M solvent with the ratio of 1:15 (b/v) at 60-70°C for 4 hours while stirred once in a while. The result, which is in the form of solids, is then cleaned using distilled water for several times until neutral. The solids dried at 80°C. The obtained powder cooled then it weighed. The protein from the demineralized solids is then removed using NaOH 3.5% (1:10) (b/v) at 60-70°C for 4 hours while stirred once in a while. The resulted solids are cleansed using distilled water for several times until neutral. The resulted solids dried in an oven at 80°C then they are cooled inside a desiccator and are weighed. The resulted solids are then conducted depigmentation using NaOCl 4% (1:10) while stirred at room temperature for1 hour. The result is then cleaned using distilled water and then dried at 60°C, so obtained chitin.

2.2. Chitosan Synthesis [3]

Chitin is dissolved using NaOH 60% with the ratio of 1:10 (b/v). The mixture is stirred and heated at 120°C for 3 hours. The resulted solids are cleansed using distilled water for several times until neutral. The solids dried at 80°C then they are cooled inside a desiccator and weighed until they reach the constant weight.

2.3. Chitosan Characterizing

Chitosan is characterized using FITR to count the degree of deacetylation (DD). DD is calculated using baseline method [12]. Measurement of water content and ash content chitosan refers to [13].

2.4. Molecule weight determined

The molecule weight of chitosan is measured based on the intrinsic viscosity (η). Chitosan solution is made in a concentration ranging 0.02-0.1% in acetic acid solution 0.1 M and sodium chloride 0.2 M (1:2) and put into an Ostwald viscometer. The data mapped in a graph ηsp/C toward C. Intrinsic viscosity is a point in the graph that shows C=0. The molecular weight is determined based on Mark-Houwink equation:

$$[\eta]=Km(Mw)\alpha.$$

[η] is intrinsic viscosity, Km is solvent constant (1.8×10^{-3}), α is constant (0.93), and Mw is the molecule weight [2].

2.5. Antioxidant activity test

2.5.1 DPPH Method (1,1-diphenyl-2- picrylhydrazyl radical). Four mL of 0.05mM DPPH solution added to one mL methanol. After that, it left for 30 minutes in a dark place. The uptake measured at 400-600 nm of wavelength using spectrophotometer UV-Vis. Four mL of 0.05mM DPPH solution is added with one mL chitosan 100 ppm. The solution is measured for its absorbance with a 5-minute interval to determine the operating time of chitosan solution test. Four mL of 0.05mM DPPH solution is added to 1 mL chitosan solution with different concentration 500, 1000, 2000, 3000 and 4000 ppm. Then incubated for 300 minutes at room temperature; then it is measured for its absorbance at the maximum wavelength. The percentage of inhibition calculated as follows:

$$\text{Inhibiting percentage (\%)} = 1 - \frac{sample\ absorbance}{control\ absorbance} \text{ x } 100$$

2.5.2 FTC (ferric thiocyanate) method. The sample about to test is made by adding 8 mL buffer phosphate 0.1 M pH 7 and 10 mL linoleic acid 0.05% (b/v) into 2 mL chitosan solution 500, 1000, 2000, 3000 and 4000 ppm. Linoleic acid with the same treatment without adding chitosan solution used as a controller. As the comparison to the sample used antioxidant BHT 500 ppm and α-tocopherol 500 ppm. This solution is then kept in an oven at 60°C for seven days [14].

Two mL of BHT solution which has been kept in an oven for 24 hours added with 4.7 mL of 75% ethanol, 0.1 mL of 30% NH_4CNS and 0.02 M $FeCl_2H_2O$ to 0.1 mL of 3.5% HCl. The solution measured for its absorbance at 470 – 570 nm of wavelength. The highest absorbance tells the maximum wavelength of the solution. The solution is measured for its absorbance with a 5-minute interval to determine the balance time of chitosan solution test.

The sample is taken out after being kept in an oven, let it reach the room temperature. As many as 0.1 mL of 30% NH_4CNS and 0.02 M $FeCl_2H_20$ in 0.1 mL of 3.5% HCl is allowed to reach the balance time after the mixing; it then measured for its absorbance using a spectrophotometer UV-Vis at the maximum wavelength. The measurement is conducted every 24 hours for seven days [14]. Antioxidant activity is based on the oxidation inhibition of linoleic acid toward a sample that has antioxidant trait based on the calculation,

$$\text{\% of inhibition} = \frac{control\ absorbance - sample\ absorbance}{sample\ absorbance} \text{ x100\%}$$

2.6. Toxicity test using Brine Shrimp Lethality Test (BSLT) method

The first step is breeding *A. salina Leach* larvae. Seawater put into a small container that has been divided into two chambers using a net partition. Some shrimp eggs of *A. salina Leach* put into one of the chambers; then it is closed. The other chamber is kept open and given light to attract hatched shrimps through the partition so that they can separate from the eggs and the unhealthy shrimps. The eggs will hatch into little shrimps after two days and are ready for testing.

As many as 0.1 grams of chitosan is taken and added with two drops of tween 80, then dissolved in 5 mL of seawater. The adding of tween 80 is to help dissolve chitosan in the seawater to obtained 10000 ppm of chitosan solution. Some main solution of chitosan is taken using micropipette and diluted into several test solutions with 1400, 1200, 1000, and 800 ppm at the final concentration.

Each tube is filled with ten shrimp larvae and seawater, is given two drops of yeast (3 mg/mL sea water) as their nutrition than the tube is given test sample in a certain amount and diluted with seawater until five mL. Control tube is the same content as the test tube but without a sample. The test tube and control tube is incubated openly and given light for 24 hours. The number of living larvae for each concentration calculated and recorded to determine the percentage of shrimp mortality as well as LC_{50}. If there is a dead larva in the control solution, the number of larvae killed in control solution subtracted by the number of larvae killed in the test solution. Base on [15], the percentage of the mortality calculated as follows:

$$\%\ of\ mortality = \frac{\Sigma\ controlling\ living\ larvae - \Sigma\ treated\ living\ larvae}{\Sigma\ controlling\ living\ larvae}\ x\ 100\%$$

2.7. Antimicrobial activity test

S. aureus (ATCC 25923), *Candida albicans* and *Aspergillus niger* used for the measurement of antimicrobial activity. *C. albicans* is a clinical blood isolate of the Candida species collected in our laboratory were used at an inoculum size of 1-5 x 10^5 CFU/ml. *Aspergillus niger* strain used in the present study was previously isolated and identified [16]. Bacteria cultures incubated in Luria Broth and agar Medium at room temperature. The concentration of the bacteria controlled from 10^7 to 10^8 CFU/mL. Yeast and the filamentous fungi maintained on potato dextrose agar (PDA).

2.7.1. *Measurement of antibacterial activity and Antifungal susceptibility testing.* Antibacterial activity measurement was performed by the agar dilution method using 90mm × 15mm Petri dishes. Luria agar medium prepared with the bacteria cultures which have a hole in the center (10 mm diameter). Ten microliters (10 ppm) of chitosan added in the hole of Luria agar medium. The LB medium incubated for 24 h for *S. aureus*, PDB medium during 24 h for *C. albicans*, and 48 h for *A. niger* at room temperature. The bacterial and fungal growth checked for a visible growth, control plates prepared without any chitosan. The antibacterial effects of chitosan measured by growth suppression exchange = total growth suppression exchange−diameter of the hole (8 mm). Inoculums without treatment as negative control also included.

2.7.2. *Measurement of minimum inhibitory concentration (MIC).* Different concentration of chitosan (0, 5, 10, 50, 100 ppm) and added in Luria Broth medium. The bacterium culture (*S. aureus*) controlled at 10^5–10^6 CFU/mL. The minimum inhibitory concentration of chitosan for *S. aureus*, *C. albicans*, and *A. niger* determined by the optical density of the bacterial culture solution containing different concentration after 24 h.

2.7.3. Measurement of the colony-forming unit (CFU). A bacterial culture medium containing chitosan solution spared on LB medium. The bacterium (*S. aureus*) was incubated its cultivation temperature for 48 h after that colony-forming unit (CFU) was measured. The LB medium incubated the same method of above measurement of the MIC. The decrease of bacterial at each LB medium was measured. MIC defined as the lowest concentration of extract that inhibited visible growth on agar.

3. Results and discussion

3.1 Chitosan

Chitosan is deacetylation product of chitin which is a polymer long chain of glucosamine (2-amino-2-deoksi-β-(1-4)-D-glucose) [17]. Chitin, however, is a linear polymer most of which comprises of β-(1→4)-2-acetamide-2-deoksi-β-D-glucopyranose units [18] and can be isolated from shrimp shells. Chitin isolation usually conducted in three stages, namely demineralization, deproteinization, and depigmentation. Demineralization is to eliminate minerals or inorganic compounds in the shells. Calcium will react with hydrochloric acid and result in calcium chloride and phosphoric acid dissolved in water [3]. CO_2 bubbles from the demineralization process show that there has been a reaction between HCl with minerals in the shells [19]. Deproteinization is to eliminate protein using alkali solution and proper heating [3]. The last stage of chitin isolation is depigmentation. This stages to eliminate color so that whiter chitosan with the higher degree of deacetylation compared to that without depigmentation obtained. Pigments in the shells are astaxanthin pigment which can be eliminated using NaOCl solution. NaOCl acts as the bleaching agent that can produce white chitin [20]. This process will produce chitin. Based on the results, % of the yield of chitin is 31.92 %.

Chitosan is obtained by chitin deacetylation using high concentrate alkali solution [3]. Deacetylation process is the eliminating process of acetyl groups ($-COCH_3$) of chitin using alkali solution to change it into amino groups (-NH2). Chitin has a long crystalline structure with a strong bond between the nitrogen atom and acetyl groups. Therefore, in the deacetylation process, it takes a long time, a high concentrate of NaOH solution and high temperature to break the bond between acetyl groups and the nitrogen atom to produce amino groups [21]. The final product from chitin deacetylation is chitosan and sodium acetate as the byproducts. Based on the results, % of the yield of chitosan is 16.63 %.

3.2. Chitosan characteristics

The chitosan from the synthesis process is characterized using FTIR instruments. The result of chitosan IR spectrum shown in Figure 1. There is an expanding uptake in the wave area of 3452 cm^{-1} where there is overlapping on vibration uptake of –OH groups. The moderate uptake in 2877 cm^{-1} shows that there is vibration range C-H on $–CH_2–$ aliphatic. The uptake in the area of 1419 cm^{-1} is the bending vibration of $–CH_2–$. The range of C=O ($-NHCOCH_3$) occurs in 1651 cm^{-1}. The uptake in 1597 cm^{-1} shows that there is a bending vibration N-H from NH_2. The wavelength of 1381 cm^{-1} indicates that there is bending uptake in $–CH_3$ with the weaker intensity; this happens because of the deacetylation process causing the loss of most of the methyl groups, $-CH_3$. The range of C-O of the C-O-C bond detected in 1033 and 1087 cm^{-1}. The 894 cm^{-1} shows that there is moderate uptake as a result of the vibration change of NH_2 from primary amide.

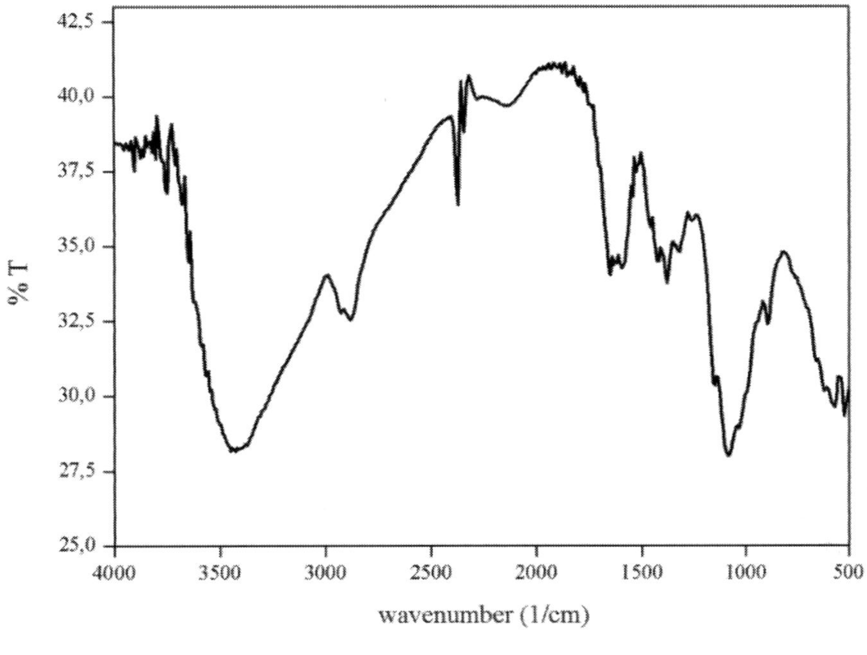

Figure 1. IR spectrum of chitosan

The obtained IR spectrum can be used to calculate the degree of deacetylation of chitosan. The level of deacetylation shows the diminishing of acetyl groups in chitin to become amine groups in chitosan [12]. The value is affected by several factors, like the alkali concentration, temperature, the ratio of chitin and alkali solution, and the particle size [21] and the reaction time [22]. The use of high concentration of alkali solution as well as high temperature during the deacetylation process can affect the results. The more NaOH to use, the more substances that will react and there is a high possibility that impact will occur. However, if the NaOH concentration is more than 60%, the solution will become thick. Consequently, a part of chitin that imperfectly reacts with NaOH solution so the resulted amino groups are little or the deacetylation level decreases. The longer deacetylation time, resulting in more reduction of acetyl groups in chitin. This results in the less output of chitosan but better quality of chitosan. The calculation degree of deacetylation chitosan using *baseline b* [12] is 94.32%

Water content shows how much water in the chitosan in percent. It obtained by heating several grams of samples into the oven at 100-105°C until constant weight reached. The result of the study is 0.33%. The obtained chitosan has a proper water content according to the quality standard of chitosan water content, which is ≤10%. Water content also depends on the humidity level around the storage area of chitosan [23]. The ash content shows that there is an inorganic compound in chitosan. How high and low the level indicates how pure the chitosan is. If the content is high, then chitosan is less pure, and vice versa. The ash content determines how successful the demineralization process in chitin isolation. The study result shows 1.53%.

The use of a very high temperature in the chitosan synthesis causes the breaking in polymer bond (depolymerization) of chitosan molecule chains so that their mass loses. The mass can be determined using Ostwald viscosity method. The measurement conducted by determining the time needed by a certain amount of solution to flow between two calibrating signs. The flow time then compared to the flow time of the pure solvent, so specific viscosity values obtained. The mass of chitosan molecule is determined using 0.1 M acetic acid and 0.2 M sodium chloride solvent. The varied chitosan samples used are 0.02; 0.04; 0.06; 0.08; and 0.1%. From the measurement, the relative viscosity and specific viscosity can calculate (Table 1). From the chart of the correlation between reduced viscosity (ηsp/C) and various concentration (C) of chitosan, there is a regression equation y = 8.7412x + 1.1646. The molecule mass is calculated using Km value (solvent constant) and α (constant) as much as 1.81×10^{-3} and 0.9 [2]. Based on the calculation, the molecule mass (Mw) of chitosan is 1052.93 g/mol.

Table 1. Molecule mass determination of chitosan

Concentration	Flow time (seconds)	ηintrinsic	ηspecific (ηsp)	ηreduction $\left(\frac{\eta sp}{C}\right)$
Solvent	25.51	-	-	-
0.02%	26.12	1.0239	0.0239	1.195
0.04%	27.21	1.0666	0.1043	1.665
0.06%	28.17	1.1043	0.1043	1.738
0.08%	29.37	1.1513	0.1513	1.891
0.1%	30.50	1.1956	0.1956	1.956

3.3 Antioxidant activity using DPPH method

The maximum wavelength (λmax) measurement shows the substance absorption or maximum absorbance at a specific wavelength. The analysis uses DPPH 0.05 mM; then the solution is measured for its absorbance at 400-600 nm wavelength using spectrophotometer UV-Vis. The result shows that 515.6 nm wavelength with 0.560 absorbances is λmax. Determining the operating time is conducted by reacting sample solution with 0.05 mM DPPH solution, and then it is observed for its absorbance at λmax 515.6 nm from 5 until 60 minutes with a 5-minute interval. Operating time is chosen when the decreasing absorbance is relatively stable [8]. The result obtained at minute 30 – 40.

Antioxidant activity of chitosan solution is conducted using DPPH method. The measurement of antioxidant activity from chitosan conducted by reacting four mL of 0.05 mM DPPH solution with one mL chitosan solution with varying concentrations: 500, 1000, 2000, 3000, and 4000 ppm. It is then incubated for 30 minutes and measured for its absorbance at λmax. The absorbance is used to know the antioxidant activity by calculating the damping percentage. The free radical inhibition percentage by chitosan solution can see in Figure 2. According to Figure 2, the solution inhibition percentage chitosan increases from 27% to 36.9%. The bigger the concentration, the bigger the percentage.

The higher the concentration of chitosan solution is, the lower the uptake is. The principle of DPPH method that antioxidant compound will release its hydrogen atom in DPPH radical so that DPPH becomes a non-radical reduced form. A non-radical DPPH will lose its purple intensity proportionally to the number of electrons it captures. The fading color marked with the decreasing DPPH absorbance at the maximum wavelength [8].

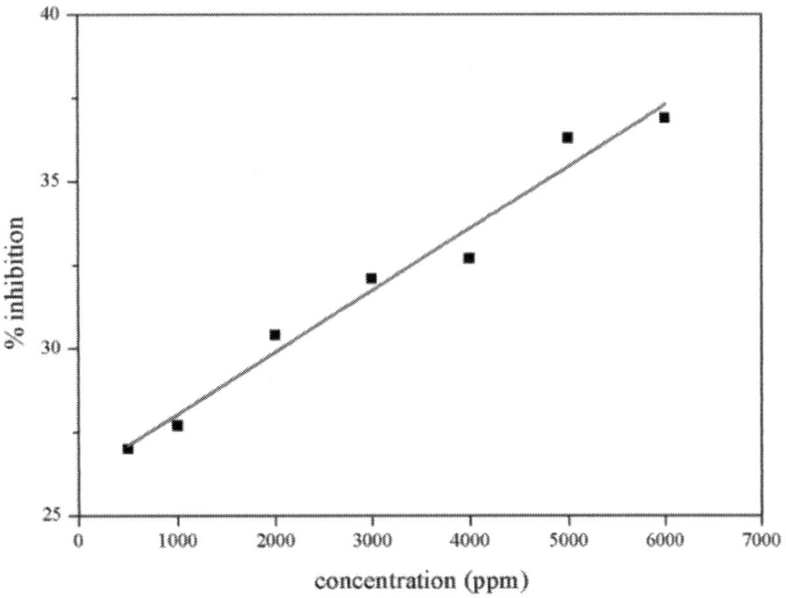

Figure 2. Free radical inhibition percentage by chitosan solution

3.4 Antioxidant activity using Thiocyanate method

Determination of maximum wavelength (λmax) aims to obtain maximum absorption. Measurement of sample uptake at wavelength 470 nm to 570 nm due to the color of the solution orange. The results show the λmax is 481 nm. The measurement of complex stability time uses 500 ppm BHT solution, and it conducted during the minute 5 until 60 with 5-minute interval using 481 nm wavelength. The time showing the stable absorbance in the minute 35 – 40.

Thiocyanate method is a method that measures the number of free radicals based on lipid peroxidation, which is the forming of the alkoxy radical [24]. This method uses linoleic acid, which is unsaturated fatty acids acting as free radicals [25]. The linoleic acid used as the substrate, and oxygen will oxidize it. The linoleic acid oxidation accelerated by the existence of light, temperature, pH, oxygen, metal ions and lipid radicals. Therefore, linoleic acid incubation is set at 60°C so that the high temperature can catalyze linoleic acid oxidation. After being incubated for 24 hours, a measurement of peroxide number conducted by reacting sample with ammonium thiocyanate and Ferro (II) chloride to produce reddish orange Ferri thiocyanate using spectrophotometer UV-Vis at λmax 481 nm.

The resulted of absorbance 500, 1000, 2000, 3000, 4000 ppm chitosan; 500 ppm BHT; 500 ppm α-tocopherol shows that the uptake for control (linoleic acid) with the incubation time is quite big. For chitosan samples has a less uptake than the control. α-tocopherol and BHT as the comparates have a little uptake compared to the control or the sample. According to [26], the absorbance value is directly proportional to the amount of peroxide in samples because of the oxidation. The bigger the peroxide values, the bigger the absorbance values and the more reddish orange intensity. The ability of thiocyanate method can see from the absorbance values compared to the values of control absorbance. The obtained result of absorbance is used to calculate the inhibition percentage of linoleic acid oxidation so that the activity of the antioxidant sample is known. The percentage shown in Table 2.

Table 2. The inhibition percentage of linoleic acid oxidation

Days to	% inhibition						
	Cs 500 ppm	Cs 1000 ppm	Cs 2000 ppm	Cs 3000 ppm	Cs 4000 ppm	a-tokoferol	BHT
1	2.02	27.77	30.95	35.08	35.45	55.25	63.80
2	13.33	45.43	53.95	51.00	47.77	65.42	62.63
3	26.48	46.31	46.45	55.57	48.48	76.63	70.69
4	39.46	60.83	57.26	57.69	58.55	75.06	74.81
5	57.67	77.41	77.07	75.95	75.26	70.54	79.97
6	55.50	75.70	74.60	76.90	74.00	55.11	77.22
7	32.28	50.72	54.99	64.82	57.84	73.95	80.37

Table 2 shows that chitosan (Cs) has inhibition percentage ranging from 2.02% to 77.41%, while BHT and α-tocopherol range from 55.11% to 80.37%. On day one there is a correlation between the increasing concentration. This is that the inhibitory power is getting bigger by the concentration adding. On day two until day 7, the inhibitory power fluctuates when there is concentration adding. This is because of the uncontrollable oxidation process. The correlation of adding the days is seen in 3000 ppm concentration which reaches the 6th day and shows that the longer the storage time, the bigger the inhibitory power and it reaches its peak then goes down on day 7. This means that on day seven peroxide begins to experience the decomposition to become another secondary compound. Peroxide is an unstable product of primary oxidation and will be decomposed gradually to become a secondary product (e.g., alkane, alcohol, and aldehyde) [26]. The inhibitory power of BHT and α-tocopherol fluctuates from day 1 to day seven caused by uncontrollable oxidation reaction that influences ferric thiocyanate complex [27].

BHT and α-tocopherol have the antioxidant ability to inhibit linoleic acid oxidation process. BHT is phenol compounds that are substituted in the ortho position (tertiary butyl group) and can expand the ability as an antioxidant. On the other hand, α-tocopherol has methyl group substituents bound to aromatic rings in the meta position. This causes the reactivity as the antioxidant to weaken. Chitosan can inhibit linoleic acid oxidation process although its inhibitory power is not as strong as BHT or α-tocopherol. The inhibition process varies depending on the chemical structures and mechanisms. In this mechanism, the most important thing is the reaction with free-lipid radicals which form non-active products.

3.5. Chitosan toxicity

Toxicity test is a test conducted to know the toxicity level of a compound. The study uses *Brine Shrimp Lethality Test* (BSLT) method [15] because this method is one method that mostly used, cheap, fast, easy (does not require aseptic conditions) and reliable. The result can be known from the mortality *Artemia salina* Leach larvae because of the effect of certain compounds with a certain dosage that has been specified previously. This method conducted by determining the value of LC_{50} for 24 hours. The data are analyzed using Probit Analysis on a computer to determine the value of LC_{50}. The result from the 24-hour observation of *Artemia salina* Leach larvae toward chitosan is shown in Table 3.

Table 3. The 24-hour observation of *Artemia salina* Leach larvae toward chitosan

Concentration (ppm)	Living larvae			The average of living larvae	Mortality percentage	Concentration Log
	T1	T2	T3			
K	10	10	10	10.00	0.00	-
800	9	8	9	8.67	13.33	2.90
1000	7	8	7	7.33	26.67	3.00
1200	7	6	6	6.33	36.67	3.08
1400	4	5	5	4.67	53.33	3.15

Based on Table 3, a curve is then made to show the correlation between the mortality percentage of larvae and chitosan concentration log (Figure 3). The linear regression equation from the above chart is then used to determine LC_{50}. The obtained line equation is $y = 4.8323x - 10.149$, $R^2 = 0.992$. Based on the result, the value of LC_{50} at probit (Y) = 5, the concentration (X) 3.1349 is $10^{3.1349} = 1364.41$ ppm.

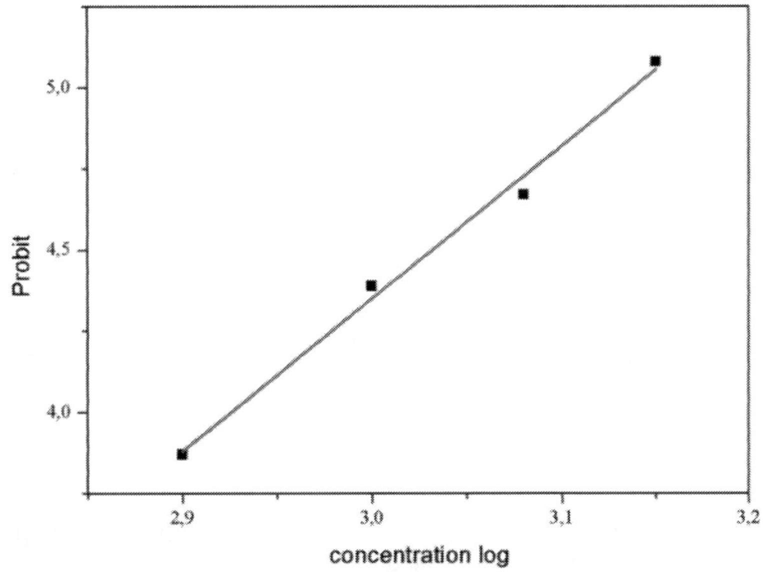

Figure 3. Curve of the correlation between the mortality percentage of larvae and chitosan concentration log

As it can see in Figure 3, the value of chitosan concentration is directly proportional to the mortality percentage of *Artemia Salina* Leach larvae. The existence of larvae test in control is to find out the mortality percentage of *Artemia Salina* Leach larvae caused by natural death. According to [28], a compound shows its toxicity activity in BSLT if the compound can cause 50% mortality of the animal tests on the concentration less than 1000 ppm. Based on the statement, it can conclude that chitosan is not toxic because the LC_{50} value is more than 1000 ppm. Chitosan reaches LC_{50} at the 1364.41 ppm concentration. The result proves that chitosan is not toxic.

3.6.Antimicrobial activity

The bacteria test used in the antimicrobial activity test is *S. aureus* ATCC 25923 (for bacteria) and *Candida albicans* and *Aspergillus niger* (for fungi). Antibacterial and antifungal tests are conducted aseptically to avoid contaminants. The antibacterial activities of chitosan show for S. *aureus*. Figure 4, figure 5 and figure 6 showed the antibacterial activities of chitosan for S. *aureus*.

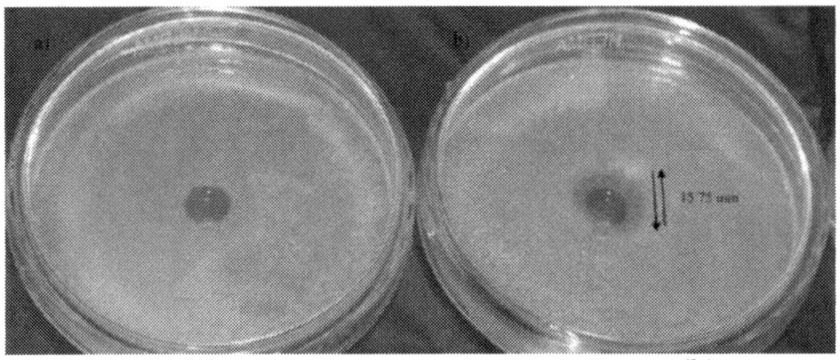

(a) (b)

Figure 4. Antibacterial activity of (a) control and (b) chitosan

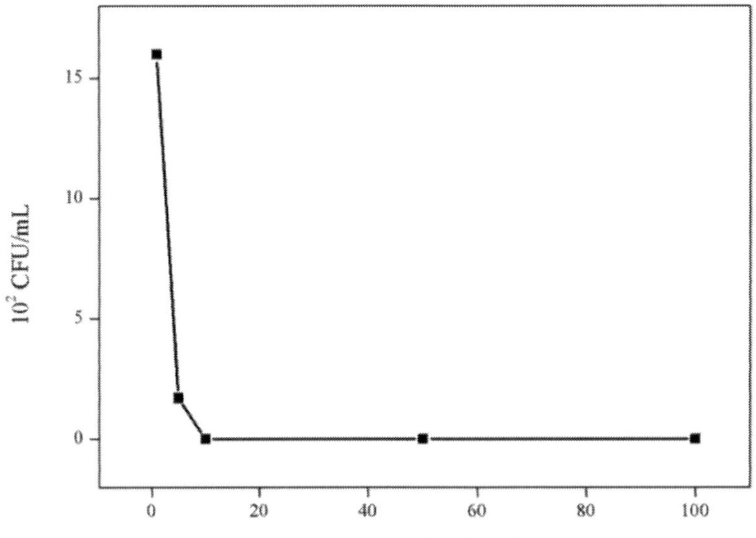

Figure 5. curves of *S. Aureus* Growth inhibition of chitosan with different concentrations in LB medium

Table 4 showed *antibacterial activity and antifungal susceptibility testing* of chitosan. According to [29], acetic acid has antibacterial activity. Therefore, when it uses as chitosan solvent, the acetic acid cannot be ignored. Based on the previous studies, the result should use chitosan. In the bacteria test acetic acid solvent with a concentration less than 100 ppm is used [30]. In this study, used acetic acid with 50 ppm concentration.

Table 4. *Antibacterial activity and Antifungal susceptibility testing* of chitosan

Sample	Antifungal activity MIC (ppm)		
	S. aureus	*C. albicans*	*A. niger*
Chitosan	10	15.6	5

4. Conclusion

The synthesized chitosan from *vanamei* shrimp shells has the output as many as 17.83% and the deacetylation 94.32%. The chitosan from synthesis has weak antioxidant activity. The chitosan gives antimicrobial activity toward *S. aureus* bacterium as well as *C. albicans* and *A. niger* fungi.

Acknowledgments
This research supported by *Penelitian Produk Terapan* Research Program funded by the Ministry of Research, Technology and Higher Education of the Republic of Indonesia in 2017.

References

[1] Natsir, H., Dali, S., Jawahir, B., dan Aziz, F., (2007). Konversi Kitin dari Limbah Udang Api-Api (Metapenaeus monoceros) Menjadi Senyawa Kitosan Secara Enzimatis. *Jurnal Marina Cemica Acta, Edisi Khusus Seminar Nasional FK3TI*: 82-89.

[2] Cervera, M. F.; Heinämäki, J.; Räsänen, M.; Maunu, S.L.; Karjalainen, M.; Acosta, O.M.; Nieto; Colartea, A. I. dan Yliruusi, J., (2004). Solid-state characterization of chitosans derived from lobster chitin. *Carbohydrate Polymers*, **58**, 401–408.

[3] Kurniasih, M, Kartika, D. (2011) Sintesis dan Karakterisasi Fisika-Kimia Kitosan, *Jurnal Inovasi,* Vol. **5** No. 1: 42-48.

[4] Mohd Zaid Arqam Z. A., Julkapli N. M., Juahir H., Azaman F., Sulaiman N. H., Abidin I. Z., (2015). Fabrication and properties of chitosan with starch for packaging application. *Malaysian Journal of Analytical Sciences,* Vol. **19** No 5: 1032 – 1042

[5] El-Fattah Hma, Abdel-Kaderzm, Hassnin Ea, El-Rahman Mka and Hassan Le., (2013). Chitosan as A Hepato-Protective Agent Against Single Oral Dose of Dioxin. *Journal Of Environmental Science, Toxicology, and Food Technology*. **7**, 3: 11–17.

[6] Lin. H.Y., and Cheng-Chun Chou. (2004). Antioxidative Activities of Water-Soluble Disaccharide Chitosan Derivatives. *Food Research International* **37**: 883–889.

[7] Kothari, S., Thompson, A., Agarwal, A., and Plessis, S. S du., (2010). Free Radicals: Their Beneficial and Detrimental Effects on Sperm Function. *Indian Journal of Experimental Biology*. Vol. **48**.

[8] Molyneux, P., (2004). The use of the stable free radical diphenylpicrylhydrazyl (DPPH) for estimating antioxidant activity. *Songklanakarin Journal of Science and Technology*. Vol. **26**, 2: 211-219.

[9] Kuppusamy, S. R and Karuppaiah, J. (2013). In Vitro evaluation of free radical scavenging activity of chitosan. *International Journal Of Pharmacy & Life Sciences (IJPLS)*. Vol. **4** Issue 5: 2685-2690.

[10] Xie W., Xu P., and Liu Q. (2001). Antioxidative Activity of Watersoluble Chitosan Derivatives. *Bioorganic and Medicinal Chemistry Letters*. Vol **1**. 1699–1701.

[11] Agustina, S., Swantara, I.M.D., dan Suartha, I.N., (2015). Isolasi Kitin, Karakterisasi, dan Sintesis Kitosan dari Kulit Udang. *Jurnal Kimia*. Vol. **9** (2): 271-278.

[12] Khan T. A., Peh, K.K., dan Ch'ng, H.S. (2002). Reporting Degree of Deacetylation, Values of Chitosan: The Influence of Analytical Methods. *J Pharm Pharmaceut Sci*. Vol. **5**, No. 3: 201-212.

[13] AOAC. (2005). *Official Methods of Analysis*. 15th Edition. Washington, D.C: Association of Official Analytical Chemists.

[14] Boer, Y. (2000). Uji Aktivitas Antioksidan Ekstrak Kulit Buah Kandis (*Garcinia parvifolia Miq*), *Jurnal Matematika dan IPA*. **1**: 26-33.

[15] McLaughlin, Roger, and Anderson. (1998). The use of biological assays to evaluate botanicals. *Drug Information Journal*., Vol. **32**: 513-524.

[16] Dewi, R.S. & Hana. 2014. Formulation of decoloring and heavy metal agent in batik effluent with fungi active ingredients, and study of enzyme activity and toxicity test. *Research Report of Inter-University Cooperation (Pekerti)*. Unsoed

[17] Shaji J., Jain V., and Lodha S., (2010). Chitosan: A Novel Pharmaceutical Excipient, *International Journal of Pharmaceutical Excipient and Applied Sciences*, Vol. **1**, No. 1, pp. 11-28.

[18] Kumirska J., Czerwicka M., Kaczyński Z., Bychowska A., Brzozowski K., Thöming J., Stepnowski P., (2010). Application of Spectroscopic Methods for Structural Analysis of Chitin and Chitosan. *Mar. Drugs*, **8**(5), 1567-1636; doi:10.3390/md8051567

[19] Harjanti, R.S. (2014). Kitosan dari Limbah Udang sebagai Bahan Pengawet Ayam Goreng, *Jurnal Rekayasa Proses*, Vol. **8**, No. 1.

[20] Zulfikar dan Ratnadewi, A. A. I. (2006). Isolasi dan Karakterisasi Fisikokimia-Fungsional Kitosan Udang Air Tawar (*Macrobrachium sintangense de Man.*). *Jurnal Teknologi Proses*. Vol. **5**, No. 2: 129-137.

[21] Odete, P.M.O., Struszczyk, M.K., dan Peter, M.G. (2005). Characterization of Chitosan from Blowfly Larvae and Some Crustacean Species from Kenyan Marine Waters Prepared Under Different Conditions, *Western Indian Ocean J Sci.* **4**, 1: 99-107

[22] Wahyuni, Ridhay. A Nurakhirawati. (2016). Pengaruh Waktu Proses Deasetilasi Kitin dari Cangkang Bekicot (*Achatina Fulica*) terhadap Derajat Deasetilasi. *Jurnal Kovalen*. Vol. **2**, No 1: 1–7.

[23] Kurniasih M., Purwati., and Zaki M. (2014). Optimum Conditions for The Synthesis of High Solubility Carboxymethyl Chitosan. *Malaysian Journal of Fundamental and Applied Sciences*. Vol **10**, No. 4. 189-195.

[24] Hanani E, Mun'im A, Sekarini R, dan Wiryowidagdo S. (2006). Uji Aktivitas Antioksidan Beberapa Spons Laut dari Kepulauan Seribu. *Jurnal Bahan Alam Indonesia*. **6**: 1-3.

[25] Behera BC, Verna N, Sonone A., Makhija U. (2006). Determination of Antioxidative Potential of *Usnea ghattensis* L. In Vitro. *LWT-Food Sci Tech*. **36**: 80-5

[26] Chayati, I., Isnatin, M. (2014). Kandungan Komponen Fenolat, Kadar Fenolat Total dan Aktivitas Antioksidan Madu dari Beberapa Daerah di Jawa dan Sumatera. *MGMI*. Vol. **6**, No. 1: 11-24.

[27] Suparmi. Hady, A, Niche, D. (2012). Uji Aktivitas Antioksidan Ekstrak Etanol Kulit Buah Rambutan (*Nephelium Lappaceum*, L.) dengan Metode Linoleat-Tiosianat. *Jurnal Ilmiah Farmasi*. Vol. **9**, No. 1.

[28] Meyer, B.N., Ferrigni, N.R., Putnam, J.E., Jacobsen, L.B., Nichols, D.E., dan McLaughin, J.L., 1982, Brine Shrimp: A Convenient General Bioassay for Active Plant Constituent, *Planta Medica*. **45** (5):31-34.

[29] Liu N., Xi-Guang, C., Hyun-Jin P., Chen-Guang, L., Cheng-Sheng, L, Xiang-Hong, M. and Le-Jun, Y.,(2006), Effect of MW and Concentration of Chitosan On Antibacterial Activity of Escherichia coli, *Carbohydr. Polym*., **64**, 60-65.

[30] Kurniasih M., Kartika D., 2009, aktivitas antibakteri kitosan terhadap bakteri *S.aureus*, *Molekul*, Vol. **4.** No. 1: 1 - 5

IOP Conf. Series: Materials Science and Engineering **349** (2018) 012038 doi:10.1088/1757-899X/349/1/012038

Sorption Isotherm Modelling Of Fermented Cassava Flour by Red Yeast Rice

M N Cahyanti[1], M N Alfiah[2], S Hartini[3]

[1,2,3]Satya Wacana Christian University, Salatiga, Indonesia

Email: margareta.cahyanti@staff.uksw.edu

Abstract. The objective of the study is to determine the characteristic of moisture sorption isotherm from fermented cassava flour by red yeast rice using various modeling. This research used seven salt solutions and storage temperature of 298K, 303K, and 308K. The models used were Brunauer-Emmet-Teller (BET), Guggenheim-Anderson-de Boer (GAB) and Caurie model. The monolayer moisture content was around 4.51 – 5.99% db. Constant related to absorption heat in the multilayer area of [GAB model was around 0.86-0,91. Constant related to absorption heat in the monolayer area of GAB model was around 4.67-5.97. Constant related to absorption heat in the monolayer area of BET model was around 4.83-7.04. Caurie constant was around 1.25-1.59. The equilibrium and monolayer moisture content on fermented cassava flour by red yeast rice was decreasing as increasing temperature. GAB constant value indicated that the process of moisture absorption on the fermented cassava flour by red yeast rice categorized in type II.

1 Introduction

The fermented cassava flour by red yeast rice is flour made from cassava made by the fermentation process by red yeast rice. In the process of making fermented cassava flour, cassava undergoes the drying process. Dehydrated food products are susceptible to damage due to rising moisture content. The moisture content in the food product will form the equilibrium with relative humidity. The relationship between equilibrium moisture content and relative humidity is illustrated in a moisture sorption isotherm curve. This curve is then modeled for further study. Earlier research suggests that GAB modeling is the most suitable model to illustrate the relationship between equilibrium moisture content and the relative humidity of cassava flour fermented by red yeast rice [1]. The purpose of this research is to determine the characteristics of cassava flour fermented by red yeast rice using GAB, BET, and Caurie models in depth.

2 Material and method

2.1 Material

The materials used were cassava and red yeast rice originally from Salatiga, aqua dest, NaOH, $MgCl_2$, K_2CO_3, $Mg(NO_3)_2$, KI, NaCl, and KCl. The tools used were drying cabinet, grinder, 61 mesh strainer, incubator, sorption container, and moisture analyzer Ohaus BM 25.

2.2 Method

The flour was made based on the previous research conducted by earlier study [2]. The cassava was peeled and then washed. Cassava is steamed for approximately 60 minutes, then drained and cooled until cassava reaches room temperature. Cassava is cut and weighed as much as 1 kg then inoculated with red rice yeast with a concentration of 12%, then fermented for three days at room temperature. The fermentation product was dried at 55 ° C until dry and then mashed and sieved with mesh sieve 61.

The measurement of equilibrium moisture content was done based on the method by Budijanto et al. [3] which had been modifying the type of salt and the storage temperature. This research used seven kinds of saturated salt such as NaOH, $MgCl_2$, K_2CO_3, $Mg(NO_3)_2$, KI, NaCl, and KCl. It was done to form certain RH and storage temperature of 303K, 308K, and 313K.

The models used in this research were Brunauer-Emmet-Teller (BET), Guggenheim-Anderson-de Boer (GAB) and Caurie. The equation for each model can be seen in Table 1.

Table 1. Equation of GAB, BET, and Caurie model

Model	Equation	Description
GAB	$m = \frac{m_0 k_b c a_w}{(1 - k_b a_w)(1 - k_b a_w + c k_b a_w)}$ (a)	a_w : water activity m : equilibrium moisture content (% db) m_0 : monolayer moisture content((% db) c and k_b : constant
BET	$\frac{a_w}{(1 - a_w)m} = \frac{1}{m_0 c} + \left(\frac{c - 1}{m_0 c}\right) a_w$ (b)	a_w : water activity m : equilibrium moisture content (% db) m_0 : monolayer moisture content (% db) c : constant
Caurie	$\ln \frac{1}{m} = - \ln (C m_0) + \frac{2C}{M_0} \ln \left(\frac{1 - a_w}{a_w}\right)$	m : equilibrium moisture content (% db) C :Caurie Constant m_0 : monolayer moisture content (% db) a_w : water activity

Monolayer moisture content can be determined from the equilibrium sorption isotherm data by fitting BET, GAB, and Caurie model. Monolayer moisture content is one of the important parameters. Estimates of monolayer moisture content are related to the stability of food products physically and chemically. In the range below the monolayer moisture content, various food damage reactions are infrequent. This is due to a very strong bound water with food products, and water does not participate in the reaction. Determination of monolayer moisture content can be used to calculate the surface area of solid from food product using BET and GAB model based on Equation 1.

$$\text{Surface area of solid} = m_0 \times 35.53 \quad (1)$$

Monolayer moisture content in Caurie model and Caurie constant can be used to determine the properties of sorbed water. Number of the adsorbed monolayer can be calculated by Equation 2 and bound water content can be calculated by Equation 3. The density of adsorbed water in the monolayer calculation is done by Equation 4 and surface area by Equation 5.

$$\text{Number of adsorbed monolayer} = \frac{m_0}{C} \quad (2)$$

$$\text{Bound water content} = m_0 \times \text{Number of adsorbed monolayer} \quad (3)$$

$$\text{Density of adsorbed water in the monolayer} = \text{Caurie Constant} \quad (4)$$

$$\text{Surface area of adsorption} = \frac{m_0}{C\,x\;diameter\;of\;a\;water\;molecule\;x10^8} \quad (5)$$

3 Result and Discussion

The equilibrium moisture content of fermented cassava flour by red yeast rice has been published earlier [1]. It is known that the equilibrium moisture content shows a decrease as the storage temperature increases. At high temperatures, the molecules are in a state of excitation, thereby increasing the distance between the water and the surface of the product that causes a decrease in attractiveness between them [4]. Another reason is that higher temperatures give higher energy to water molecules that cause water molecules tend to move and not bind to food products [5]. It was precisely correlated with the previous research about the equilibrium moisture content of dry merunggai leaves in several temperatures [6] and the research about moisture sorption isotherm on dry mung beans [5].

The data of equilibrium moisture content and water activity was then applied in the GAB, BET, and Caurie equations to get description about absorption process on the sample. The valuable constants from GAB, BET, and Caurie equations were shown in Table 2.

Table 2. GAB, BET, and Caurie constants on several storage temperatures.

Model	GAB			BET		Caurie	
Constant	m_0	C	k_b	m_0	c	m_0	C
Temperature	Article	Article II	Article I	Article	Article V	Article	Article
303K	5.56	5.2641	0.8835	4.53	6.4751	5.98	1.2905
308K	5.54	5.9704	0.8593	4.51	7.0381	5.99	1.2494
313K	5.53	4.6685	0.9117	4.99	4.8345	5.06	1.5859
Dried Cassava (303K)[7]	6.16	12.9004	0.8939	5.55	15.1463	-	-
Cassava Flour (303K)[8]	7.28	15.550	0.8329	-	-	-	-
Cassava Flour (308K)[8]	6.32	30.390	0.8677	-	-	-	-
Gari (303K)[9]	5.9	38.64	0.811	-	-	-	-
Enriched Gari (303K)[9]	3.2	5.76	0.900	-	-	-	-

Monolayer moisture content for GAB model was around 4.51-5.99% db and for BET model was around 4.51-4.99% db. Monolayer moisture content of fermented cassava flour by red yeast rice lower than another cassava product except enriched gari using soy melon. The monolayer moisture content changed as the composition changed as well. Commonly, monolayer moisture content of starchy food supply in GAB model was around 3.2-16.3% [10]. In areas of low water activity, proteins and carbohydrates are preferred sites of absorption at lower temperatures than higher temperatures. Proteins and carbohydrates are preferred sites of absorption at lower temperatures is because proteins and carbohydrates have a polar hydrophilic part with a high water binding capacity in it. Water also forms hydrogen bonds with certain groups causing hydrophobic hydration [11]. But with rising temperatures, there is a decrease in the number of hydrophilic groups, and hydrophobic hydrogen biopolymers also begin to break at higher temperatures. It may cause a decrease in monolayer moisture content.

Constant (c) on GAB and BET models were connected to absorption heat in monolayer area. BET constants were related to absorption heat in accordance with an algorithm in a monolayer area. Whereas in accordance to algorithm BET constants were related to chemical potential differences of the water molecule in the form of pure liquid in a monolayer area. GAB constants were related to variation in the energy value in the monolayer area and the layers above [12]. BET constants score tended to be higher than the score of GAB constants. It was the same with the research of Koua et al. [7] which used samples from dry cassava. The decrease of c was an indication of more enthalpy and a gain in kinetic energy resulting in the loss of more moisture at higher temperature [13].

The k_b constant was a constant related to absorption heat in the multilayer area [14]. The k_b score was around 0.8593-0.9117. Increase in k_b was an indication that at higher temperature, the multilayer molecules became more entropic [13].

The k_b score was around 0.8593-0.9117, while the c score was around 4.6685-5.9704. Those two constants indicated that sorption isotherm on "gaplek" flour fortified by protein from red bead tree flour was following isotherm type II. Blahovec[15] said that if $0<k<1$ and $c>2$, the sorption isotherm followed type II. Type II had a graphic about the relation of aw and m which is similar to letter S.

Once the monolayer moisture content was known, the solid surface area of the samples could be determined. The solid surface based on GAB and BET model for each temperature shown in Table 3.

Table 3. The Solid Surface Based on GAB and BET Model

Model	GAB	BET	Cassava
Temperature	Article IX.	Article X.	Article XI.
303K	196.27	159.91	217.55
308K	195.56	159.20	-
313K	195.21	176.15	-

Monolayer moisture content in Caurie model and Caurie constant can be used to determine the properties of sorbed water. The properties of sorbed water based on Caurie model shown in Table 4.

Table 4. The Properties of Sorbed Water Based on Caurie Model

Constant	m_0	C	N	Density	Surface area of Adsorption	Bound water
303K	5.98	1.2905	4.6339	1.2905	126.1602	27.7107
308K	5.99	1.2494	4.7943	1.2494	130.5282	28.7179
313K	5.06	1.5859	3.1906	1.5859	86.8668	16.1444

A larger surface area means a large number of exposed polar groups, resulting in increased water absorption. Table 3 and Table 4 show the higher the temperature, the lower absorption surface area. This is in line with previous research conducted by Koua et al. [7] using cassava as a sample.

The highest density is achieved at the highest temperature. Possibly at higher temperatures, the rate of chemical reactions at the multimolar phase and capillary water condensation in region III also increased which may have led to the increased solubilization of sugars present in considerable amounts in food. This intermingling of the sugars with the water molecules may have caused an increase in the density of the sorbed water as adding sugar or salt to water increases its density [7].

4 Conclusions

The equilibrium and monolayer moisture content on fermented cassava flour by red yeast rice was decreasing as increasing temperature. GAB constant value indicated that the process of moisture absorption on the fermented cassava flour by red yeast rice categorized in type II.

References

[1] Alfiah, M. N., Hartini, S., and Cahyanti, M. N.. Mathematical Models and Thermodynamic Properties of Moisture Sorption Isotherms of Fermented Cassava Flour by Red Yeast Rice. *Alchemy*. 2017. 13(1), 29-40.

[2]Lakahina, O., Saputri, Y. L. I. D., and Hartanto, B. D., *Mocaf Merah – Pangan Kaya Antioksidan Berbasis Kearifan Lokal*. [Research report]. Salatiga : Universitas Kristen Satya Wacana. 2015.

[3] Budijanto, S., Sitanggang, A.B.andKartika, Y.D.Shelf Life Study of Tortilla Using Accelerated Shelf Life Testing (ASLT) Method and its Mathematical Modeling of Moisture. *Jurnal Teknologi dan Industri Pangan*. 2010. 21(2), 165-170.

[4] Kaya A, Aydin O. An experimental study on drying kinetics of some herbal leaves. *Energy Convers Manag*. 2009. 50,118–124

[5]Chowdhury, M.M.I, Huda, M.D, Hosain, M.A., and Hassan, M.S.. Moisture Sorption Isotherms for Mungbean (Vigna radiata L). *Journal of Food Engineering*, 2006. 74 (4), 462-467.

[6] Sobolawe, S.S., Oke, M.O., Odunmbaku, L.A., and Adebo, O.A.. Equilibrium Sorption Isotherms of Moringa Oleifera Leaves at different temperatures. *African Journal of Science, Technology, Innovation and Development*, 2017. 9(1), 61-68.

[7]Koua, B.K, Koffi, P.M, Gbaha, P, and Toure,S. Thermodynamic analysis of sorption isotherms of cassava (Manihot esculenta). *Journal of Food Science and Technology*, 2014. 51(9), 1711–1723.

[8] Ayala-Aponte, A. A., Thermodynamic Properties of Moisture Sorption in Cassava Flour. *Dyna* 2015. 83 (197), 138-144.

[9] Oluwamukomi, M O.Adsorption isotherm modeling of soy-melon-enriched and un-enriched 'gari' using GAB equation. *African Journal of Food Science* 2009. 3(5), 117-124

[10]Lomauro, C. J., Bakshi, A. S., and Labuza, T. P. Evaluation of food moisture sorption isotherm equations. Part I: Fruit, vegetable and meat products. *Lebensmittel-Wissenschaft und Technologie*, 1985. 18(2), 111–117.

[11] Kinsella JF, Fox PF. Water sorption by milk proteins. *Bull Int Dairy Fed*. 1987. 209,12–40.

[12]Erbaş, M., Aykın, E., Arslan, S., and Durak, A. N. Adsorption behaviour of bulgur. *Food Chemistry*, 2016. 195, 87-90.

[13]Diosady LL, Risvi SSH, Cai W, Jagdeo DJ Moisture sorption Isotherms of Canola meals, and application to packaging. *J. Food Sci*. 1996. 61(1), 204-208.

[14]Basu, S., Shivhare, U.S., & Mujumdar, A.S.. Model for Sorption Isotherm for Food: A Review. *Drying Technology* 2006. 24(8), 917-930

[15]Blahovec, J. Sorption isotherms in materials of biological origin mathematical and physical approach. *Journal of Food Engineering* 2004. 65(4), 489-495.

IOP Conf. Series: Materials Science and Engineering **349** (2018) 012039 doi:10.1088/1757-899X/349/1/012039

2-Thiophenecarboxylic acid hydrazide Derivatives: Synthesis and Anti-Tuberculosis Studies

M R G Fahmi[1], L Khumaidah[1], T K Ilmiah[1], A Fadlan[1], and M Santoso[1*]

1Department of Chemistry, Institut Teknologi Sepuluh Nopember, Kampus ITS Sukolilo, Surabaya 60111, Indonesia

*E-mail : tsv09@chem.its.ac.id

Abstract. One of the most frequent and widespread infectious diseases especially in developing countries is tuberculosis (TB). The number of TB drug resistant tend to increase, and there has been no new TB drug introduce since the 1960s. Six 2-Thiophenecarboxylic acid hydrazide derivatives were synthesized in 90-97% yields, and 2-thiophenecarbonylhydrazone-5,7-dibromoisatin showed the highest activity in inhibiting *M. tuberculosis* $H_{37}Rv$.

1. Introduction

Tuberculosis (TB) is an airborne infectious disease caused by *Mycobacterium tuberculosis* [1,2]. TB is one of top ten cause of death which usually attacks the lungs (pulmonary TB), but also can affect other parts of body (extra pulmonary TB) [3,4]. World Health Organization (2016) reported that there were 6.1 million people infected TB disease, including 330.279 people in Indonesia [3]. First-line and second-line of anti-TB drugs have been used for TB treatment [1,4,5]. However TB bacteria have been resistant to some of anti-TB drugs in the forms of MDR-TB and XDR-TB [6]. MDR TB resistant to the first line drugs such as isoniazid and rifampicin, whilst XDR TB resistant to isoniazid, rifampicin and at least one of the second line TB drugs such as fluoroquinolones groups or even one of injected drugs such as amikacin [2,7,8]. World Health Organization (2016) reported that 2.135 and 28 people infected MDR-TB and XDR-TB in Indonesia respectively [3]. In general TB treatment with the second line drugs are not effective enough, more costly and even leads to another side effect [1,9]. Recently, totally drug resistance tuberculosis (TDR) has been reported which resistance to all first-line and second-line TB drugs. TDR-TB were reported in Italy, Iran, South Africa and India [9–12]. There has been no new TB drug introduce since the 1960s, therefore development of novel TB drugs is urgently needed.

Isatin derivatives have showed board spectrum bioactivity such as anti-tuberculosis [13,14]. Schiff bases and hydrazones from isatins showed good activities against *M. tuberculosis* [14–16]. The structure-activity relationship of the isatin derivatives showed that 5-halogenation or *N*-alkylation gave *N*-Schiff base and hydrazones which increase inhibition against TB bacteria [6,7,12,17–24]. This paper reported synthesis and anti-tubercular activity against *Mycobacterium tuberculosis* H37Rv strains of some hydrazones derived from 2-thiophenecarboxilic acid hydrazide.

IOP Conf. Series: Materials Science and Engineering **349** (2018) 012039 doi:10.1088/1757-899X/349/1/012039

2. Result and Discussion

2.1. Chemistry

5,7-Dibromoisatin was prepared by bromination of isatin under reflux conditions following Vine *et al.* method [25]. The reaction gave precipitate which on recrystallization in ethyl acetate resulted in 5,7-dibromoisatin as an orange solid. The ^{1}H NMR spectrum of 5,7-dibromoisatin showed two doublet signals for two aromatic protons at δ 7.68 (J = 0.9 Hz) and 8.04 (J = 0.9 Hz) ppm, and one singlet signal for NH proton at δ 11.44 ppm. The ^{13}C NMR spectrum exhibited eight signals for eight 5,7-dibromoisatin carbons.

2-Thiophenecarboxylic acid hydrazide derivatives **(1a-f)** were prepared by the condensation reactions of 5-haloisatins with 2-thiophenecarboxilic acid hydrazide catalyzed by sulfuric acid under reflux conditions [19]. After cooling the resulting precipitated was filtered off, washed with cold ethanol, and dried to yield the derivatives **(1a-f)** in 90-97 % yields. IR spectra of compound **(1a-f)** showed sharp bands at around 1658-1743 and 1533-1620 due to the amide carbonyl and C=N frequency, respectively. In the ^{1}H-NMR spectra gave singlets proton for NH hydrazide protons at 11.35-13.32 ppm. The ^{13}C NMR spectrum showed the absence of carbon carbonyl in C-3 position signal, and the presence of NH hydrazide proton at 11.35-13.32 ppm for the ^{1}H NMR spectra of 2-thiophenecarboxylic acid hydrazide derivatives **(1a-f)** confirmed that hydrazone derivatives have formed.

2.2. Anti-tubercular assay

All the synthesized compounds **(1a-f)** were evaluated against *Mycobacterium tuberculosis* H37Rv strain by dilution method in order to determine minimum inhibitory concentrations (MIC) which is the lowest concentration of compound required to inhibit the bacteria growth completely [14,26,27]. As can be seen in Table 1 that all derivatives active against the bacteria. The presence of *N*-methyl group in compound **(1b)** slightly increase the activity compare with compound **(1a)**. Compound **(1d)** showed the highest anti-tubercular activity due to the presence of dibromo groups at C-5 and C-7.

Table 1. Anti-tubercular activities of 2-thiophenecarboxylic acid hydrazide derivatives **(1a-f)**

Compounds	Concentrations (in ppm)				
	31.25	62.5	125	250	500
(1a)	2+,2+	2+,2+	2+,2+	2+,2+	2+,2+
(1b)	2+,2+	2+,2+	2+,2+	2+,2+	1+,2+
(1c)	2+,2+	2+,2+	2+,2+	2+,2+	2+,2+
(1d)	2+,2+	2+,2+	2+,2+	2+,2+	2+,1+
(1e)	2+,2+	2+,2+	2+,2+	2+,2+	2+,2+
(1f)	2+,1+	1+,1+	1+,1+	1+,1+	5,3
rifampicin[a]	-	-	-	-	-

Note : +2 = 100-200 colony, +1 = 20-100 colony, negative (-) = no growth
[a] rifampicin concentration at $8x10^{-4}$ ppm

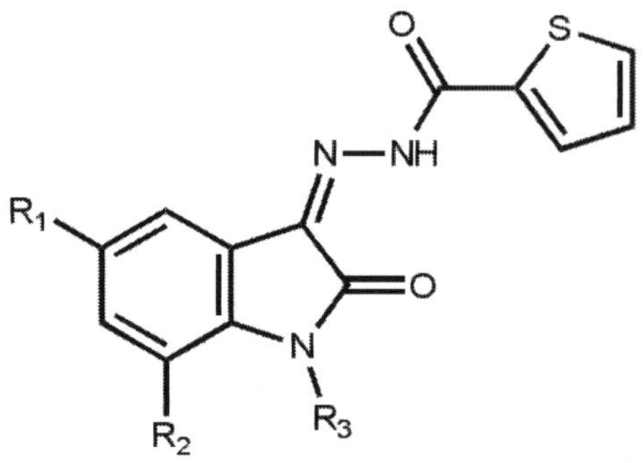

(1a) $R_1 = H$ $R_2 = H$ $R_3 = H$
(1b) $R_1 = H$ $R_2 = H$ $R_3 = CH_3$
(1c) $R_1 = F$ $R_2 = H$ $R_3 = H$
(1d) $R_1 = Cl$ $R_2 = H$ $R_3 = H$
(1e) $R_1 = Br$ $R_2 = H$ $R_3 = H$
(1f) $R_1 = Br$ $R_2 = Br$ $R_3 = H$

3. Experimental

3.1. General

All chemicals and solvents were purchased from Sigma-Aldrich and Merck and used without further purification. Melting points were measured using a Fisher-John melting point apparatus, and are uncorrected. ^{1}H and ^{13}C NMR spectra were obtained in the designated solvents on a Jeol Resonance (500 MHz) spectrometer or Agilent (500 MHz) spectrometer. Mass spectra were recorded on a LC-MS (ESI) Mariner Bio spectrometry Hitachi L 6200 with column C18 (RP 18) Phenomenex, spectrometer masa ESI Waters LCP Premier XE. The anti-tubercular assay was carried out at the Faculty of Medicine, Gadjah Mada University, Indonesia.

3.2. Synthesis of 5,7-dibromoisatin

A solution of isatin (0.74 g, 4.92 mmol) in ethanol (15 mL) was refluxed, and added dropwise bromine (0.8 mL, 14.76 mmol). The solution was refluxed further for 60 minutes. After cooling, the resulting precipitated was filtered off, washed with cold ethanol and recrystallized in ethyl acetate to yied 5,7-dibromoisatin as a yellow powder (1.10 g, 73%), m.p. 205-206°C. ^{1}H-NMR (DMSO-d_6, 500 MHz): δ 7.68 (d, J = 0.9 Hz, 1H, ArH), 8.04 (d, J = 0.9 Hz, 1H, ArH), (11.44 (s, 1H, NH). ^{13}C-NMR (DMSO-d_6, 125 MHz): δ 106.14, 115.03, 12.73, 126.39, 141.59, 149.02, 159.86, 182.93.

3.3. General method for synthesis of 2-thiophenecarboxylic acid hydrazide derivatives

A solution of isatin in ethanol containing a catalytical amount of sulfuric acid was refluxed, and added with a solution 2-thiophenecarboxilic acid hydrazide in ethanol. The solution was refluxed for additional 120 minutes. The resulting precipitated was filtered off after cooling, washed with cold ethanol, and dried to yield the derivatives.

*3.3.1. 2-Thiophenecarbonylhydrazone-3-isatin **(1a)*** as an orange powder (0.24 g, 97%), m.p. 222-223. IR (KBr, cm^{-1}) 1658 and 1712 (C=O), 3190 (N-H), 1620 (C=N), 1076 (C-O). ^{1}H NMR (DMSO-d_6, 500 MHz) 6.96 (d, 1H, ArH), 7.12 (t, 1H, ArH), 7.29 (t, 1H, ArH), 7.40 (t, 1H, ArH), 7.61 (d, 1H, ArH), 8.04 (d, 1H, ArH), 7.39 (s, 1H, NH), 11.33 (s, 1H, NH). *m/z* (ESI) 272 $[M+H]^+$. -1H

3.3.2. 2-Thiophenecarbonylhydrazone-3-(N-methyl)isatin ***(1b)*** as an orange powder (0.49 g, 96%), m.p. 235-236 °C. IR (KBr, cm^{-1}) 1647 and 1743 (C=O), 3518 (N-H), 2924 (CH_3), 1608 (C=N), 1103 (C-O). ^{1}H NMR ($CDCl_3$, 500 MHz) δ 3.29 (s, 3H), 6.87 (d, 1H, ArH), 7.17 (m, 2H, ArH), 7.40 (dd, 1H, ArH), 7.67 (s, 1H, ArH), 7.80 (s, 1H, ArH), 13.99 (s, 1H, NH). *m/z* (ESI) 286 $[M+H]^+$. -1H

3.3.3. 2-Thiophenecarbonylhydrazone-5-fluoroisatin ***(1c)*** as an orange powder (0.24 g, 92 %), m.p. 242-243 °C. IR (KBr, cm^{-1}) 1658 and 1716 (C=O), 3211 (N-H), 1589 (C=N), 1074 (C-O). ^{1}H NMR (DMSO-d_6, 500 MHz) δ 6.96 (m, 1H, ArH), 7.24 (m, 1H, ArH), 7.30 (t, 1H, ArH), 7.45 (dd, 1H, ArH), 7.95 (s, 1H, ArH), 8.05 (d, 1H, ArH), 11.00 (s, 1H, NH), 11.35 (s, 1H, NH). ^{13}C NMR (DMSO-d_6, 125 MHz) δ 108.41, 108.62, 112.84, 118.47, 118.67, 121.42, 121.49, 128.63, 135.34, 139.24, 157.88, 159.78, 163.38. *m/z* (ESI) 290 $[M+H]^+$.

3.3.4. 2-Thiophenecarbonylhydrazone-5-chloroisatin ***(1d)*** as an orange powder (0.25 g, 92 %), m.p. 230-231°C. IR (KBr, cm^{-1}) 1670 and 1716 (C=O), 3230 (N-H), 1599 (C=N), 1072 (C-O). ^{1}H NMR (DMSO-d_6, 500 MHz) δ 6.98 (d, 1H, ArH), 7.30 (t, 1H, ArH), 7.44 (d, 1H, ArH), 7.61 (m, 1H, ArH), 7.97 (s, 1H, ArH), 8.06 (d, 1H, ArH), 11.45 (s, 1H, NH), 13.32 (s, 1H, NH). *m/z* (ESI) 308 $[M(^{37}Cl)+H]^+$, 306 $[M(^{35}Cl)+H]^+$.

3.3.5. 2-Thiophenecarbonylhydrazone-5-bromoisatin ***(1e)*** as a yellow powder (0.22 g, 90%), m.p. 299-300°C. IR (KBr, cm^{-1}) 1688 and 1716 (C=O), 3234 (N-H), 1533 (C=N), 1058 (C-O). ^{1}H NMR ($CDCl_3$, 500 MHz) δ 6.76-7.52 (m, 6H, ArH), 7.68 (s, 1H, NH), 7.86 (s, 1H, NH). *m/z* (ESI) 350 $[M(^{79}Br)+H]^+$, 352 $[M(^{81}Br)+H]^+$.

3.3.6. 2-Thiophenecarbonylhydrazone-5,7-dibromoisatin ***(1f)*** as a yellow powder (0,23 g, 91%), m.p. 269-270°C. IR (KBr, cm^{-1}) 1664 and 1708 (C=O), 3192 (N-H), 1614 (C=N), 1043 (C-O). ^{1}H NMR (DMSO-d_6, 500 MHz) δ 7.30 (t, 1H, ArH), 7.71 (d, 1H, ArH), 7.84 (d, 1H, ArH), 8.07 (d, 1H, ArH), 7.99 (s, 1H, NH), 11.77 (s, 1H, NH). ^{13}C NMR (DMSO-d_6, 125 MHz) δ 104.95, 110.55, 115.18, 118.66, 122.67, 123,09, 123.75, 127.07, 135.87; 141.39; 163.06. -1H

3.4. Anti-tubercular assay

The anti-tubercular assay was carried out by dilution method. Suspension of TB bacteria was prepared by dropped 1 ose of it into screw cap tube containing 10 glassbeads, 2 drops of tween 0.05%, and aquadest. Then 0,8 ml of those solution was placed into MB DS media to make a 10^6 CFU/mL suspension of TB bacteria in 80mL MB DS. The 10^6 CFU/mL of suspension with tested compounds, positive control, solvent control, and antibiotic control then were inoculated and incubated at 37 C for 10 days. Furthermore 100 μL test compounds were taken and incubate on the with Lowenstein-Jensen (LJ) media for 3-4 weeks, and observed the growth.

4. Conclusion

Bromination of isatin gave 5,7-dibromoisatin in 73% yield. 2-Thiophenecarboxylic acid hydrazide derivatives have been prepared in 90-97% yields, and *in vitro* evaluated against *M. tuberculosis* $H_{37}Rv$ strain.

Acknowledgments

We would like to thank Indonesian Government and ITS Research Institute for their financial support.

References

[1] Mhiri C, Boudriga S, Askri M, Knorr M, Sriram D, Yogeeswari P, Nana F, Golz C and Strohmann C 2015 Design of novel dispirooxindolopyrrolidine and

dispirooxindolopyrrolothiazole derivatives as potential antitubercular agents *Bioorganic Med. Chem. Lett.* **25** 4308–13
[2] He C X, Meng H, Zhang X, Cui H Q and Yin D L 2015 Synthesis and bio-evaluation of phenothiazine derivatives as new anti-tuberculosis agents *Chinese Chem. Lett.* **26** 951–4
[3] WHO 2016 *WHO Global tuberculosis report 2016*
[4] Elhakeem M A, Taher A T and Abuel-Maaty S M 2015 Synthesis and anti-mycobacterial evaluation of some new isonicotinylhydrazide analogues *Bull. Fac. Pharmacy, Cairo Univ.* **53** 45–52
[5] Naveen Kumar H S, Parumasivam T, Jumaat F, Ibrahim P, Asmawi M Z and Sadikun A 2014 Synthesis and evaluation of isonicotinoyl hydrazone derivatives as antimycobacterial and anticancer agents *Med. Chem. Res.* **23** 269–79
[6] Feng L S, Liu M L, Zhang S, Chai Y, Wang B, Zhang Y Bin, Lv K, Guan Y, Guo H Y and Xiao C L 2011 Synthesis and in vitro antimycobacterial activity of 8-OCH3 ciprofloxacin methylene and ethylene isatin derivatives *Eur. J. Med. Chem.* **46** 341–8
[7] Xu Z, Zhang S, Gao C, Fan J, Zhao F, Lv Z S and Feng L S 2017 Isatin hybrids and their anti-tuberculosis activity *Chinese Chem. Lett.* **28** 159–67
[8] Ganihigama D U, Sureram S, Sangher S, Hongmanee P, Aree T, Mahidol C, Ruchirawat S and Kittakoop P 2014 Antimycobacterial activity of natural products and synthetic agents: Pyrrolodiquinolines and vermelhotin as anti-tubercular leads against clinical multidrug resistant isolates of Mycobacterium tuberculosis *Eur. J. Med. Chem.* **89** 1–12
[9] Kumar D, Khare G, Kidwai S, Tyagi A K, Singh R and Rawat D S 2014 Synthesis of novel 1 , 2 , 3-triazole derivatives of isoniazid and their in vitro and in vivo antimycobacterial activity evaluation *Eur. J. Med. Chem.* **81** 301–13
[10] Parida S K, Axelsson-Robertson R, Rao M V., Singh N, Master I, Lutckii A, Keshavjee S, Andersson J, Zumla A and Maeurer M J 2015 Totally drug-resistant tuberculosis and adjunct therapies *J. Intern. Med.* **277** 388–405
[11] Klopper M, Warren R M, Hayes C, van Pittius N C G, Streicher E M, M??ller B, Sirgel F A, Chabula-Nxiweni M, Hoosain E, Coetzee G, van Helden P D, Victor T C and Trollip A P 2013 Emergence and spread of extensively and totally drug-resistant tuberculosis, South Africa *Emerg. Infect. Dis.* **19** 449–55
[12] Eldehna W, Fares M, Abdel-Aziz M and Abdel-Aziz H 2015 Design, Synthesis and Antitubercular Activity of Certain Nicotinic Acid Hydrazides *Molecules* **20** 8800–15
[13] Pal M, Sharma N K, Priyanka and K.K.Jha 2011 Synthetic and Biological Multiplicity of Isatin: a Review *J. Adv. Sci. Res.* **2** 35–44
[14] Saxena A, Sharma S and Husain A 2015 Antimicrobial and anti-tubercular activity of some new indole derivatives *World J. Pharm. Sci.* **3** 1877–82
[15] Pakravan P, Kashanian S, Khodaei M M and Harding F J 2013 Biochemical and pharmacological characterization of isatin and its derivatives: From structure to activity *Pharmacol. Reports* **65** 313–35
[16] Rollas S and Küçükgüzel Ş G 2007 Biological activities of hydrazone derivatives *Molecules* **12** 1910–39
[17] Aboul-Fadl T, Bin-Jubair F A S and Aboul-Wafa O 2010 Schiff bases of indoline-2,3-dione (isatin) derivatives and nalidixic acid carbohydrazide, synthesis, antitubercular activity and pharmacophoric model building *Eur. J. Med. Chem.* **45** 4578–86
[18] Tehrani K H M E, Maryam H, Maryam H, Kobarfard F and Shohreh M 2016 Synthesis and antibacterial activity of Schiff bases of 5-substituted isatins *Chinese Chem. Lett.* **27** 221–5
[19] Aboul-fadl T, Abdel-hamid M K and Youssef A F 2015 Schiff bases of indoline-2 , 3-dione (isatin) derivatives as efficient agents against resistant strains of Mycobacterium tuberculosis § *Dhe Pharma Chem.* **7** 217–25
[20] Fadlan A, Santoso M and Ediati R 2014 Kompleks Tembaga (II) Dari 2-Tiofenakarbonilhidrazon-3-(N-metil)-5-bromoisatin: Sintesis dan Uji Antibakteri *J. Sains dan*

Terap. Kim. **8** 103–11

[21] Kumar S B, Ravinder M, Kishore G, Jayathirtha Rao V, Yogeeswari P and Sriram D 2014 Synthesis, antitubercular and anticancer activity of new Baylis-Hillman adduct-derived N-cinnamyl-substituted isatin derivatives *Med. Chem. Res.* **23** 1934–40

[22] Rodríguez-Argüelles M C, Cao R, García-Deibe A M, Pelizzi C, Sanmartín-Matalobos J and Zani F 2009 Antibacterial and antifungal activity of metal(II) complexes of acylhydrazones of 3-isatin and 3-(N-methyl)isatin *Polyhedron* **28** 2187–95

[23] Rodríguez-Argüelles M C, Mosquera-Vázquez S, Tourón-Touceda P, Sanmartín-Matalobos J, García-Deibe A M, Belicchi-Ferrari M, Pelosi G, Pelizzi C and Zani F 2007 Complexes of 2-thiophenecarbonyl and isonicotinoyl hydrazones of 3-(N-methyl)isatin. A study of their antimicrobial activity *J. Inorg. Biochem.* **101** 138–47

[24] Zhang X M, Guo H, Li Z S, Song F H, Wang W M, Dai H Q, Zhang L X and Wang J G 2015 Synthesis and evaluation of isatin-??-thiosemicarbazones as novel agents against antibiotic-resistant Gram-positive bacterial species *Eur. J. Med. Chem.* **101** 419–30

[25] Vine K L, Locke J M, Ranson M, Pyne S G and Bremner J B 2007 An Investigation into the Cytotoxicity and Mode of Action of Some Novel N-Alkyl-Sub Isatin *J. Med. Chem.* **50** 5109–17

[26] Pahlavani E, Hadi K and Rad N S 2015 A Study on Antitubercular and Antimicrobial Activity of Isoniazid Derivative *Zahedan J. Res. Med. Sci.* **17** 7–10

[27] Saxena A, Sharma S, Husain A, Chandanwadi P O, Nagar P and Delhi N 2015 STUDIES ON ANTIMICROBIAL POTENTIAL OF NEW ISATIN *World J. Pharm. Pharm. Sci.* **4** 497–505

Improvement the Yoghurt Nutritional Value, Organoleptic Properties and Preferences by Spirulina (*Spirulina platensis*) Supplementation

M Suzery[1], Hadiyanto[2], H Sutanto[3], Y Widiastuti[4], Judiono[5]

[1]Department of Chemistry, Faculty of Science and Mathematics, Diponegoro University, Jl. Prof Soedarto, SH, Tembalang, Semarang 50275
[2]Center of Biomass and Renewable Energy, Department of Chemical Engineering, Faculty of Engineering, Diponegoro University, Jl. Prof Soedarto, SH, Tembalang, Semarang 50275,
[3]Department of Physics, Faculty of Science and Mathematic, Diponegoro University, Jl. Prof Soedarto, SH, Tembalang, Semarang 50275,
[4]Head of Study Programme on Nutrition Department, The Health Polytechnic of Bandung, MOH R.I-Bandung 40514, Indonesia
[5]Head of Study Programme on Nutrition Department, The Bandung Health Polytechnic, MOH R.I. Jl. Babakan Loa, Kelurahan Pasirkaliki, Cimahi Utara, West Java, Indonesia 40514.

Corresponding Author: judi.fkundip@gmail.com

Abstract Spirulina sp has been identified as potential source food functional such as protein, amino acids and other high added value compounds from microalgae. One of the compounds is phycocyanin as also known for antioxidant use. This research was aimed to increase the nutritional value and organoleptic properties and preferences yoghurt by *Spirulina platensis* supplementation. Completely randomized controlled group design conducted by 31 respondents. Spirulina Research accomplished in Food Technology, Microbiology and Chemistry Laboratory at Bandung Health Polytechnic in Bandung. Samples divided randomly into three groups: (1) yoghurt standard supplemented 50 ml, (2) yoghurt and spirulina 1% supplemented 50 ml, (3) yoghurt and spirulina 1,2% supplemented 50 ml. Spirulina was added concentration by 0, 1, 1,2 % calculated by dx trial. Proximate analysis revealed the highest protein in enriched by spirulina 1%. However fat was lower 3.48 – 3.56 compare to control. All products found growing Lactobacillus acidophilus with a pH of 4.0 range. There was no microbial contamination such as E coli, Salmonella sp, Pseudomonas sp. Supplementation of spirulina to yoghurt products can be accepted by organoleptic, chemical and microbiological tests of concentrations of 1 and 1.2%. The material makes the blue colours more attractive and can be accepted by the expert panellist.

Introduction

Nowadays there is a tendency of people consume the foodour beverages, it's not only judged by nutritional content and delicious but also consider the influence of these foods on the health of the body is called functional properties [1].

The development of the current dimension of the disease tends toward degenerative diseases, where the main trigger is free radical involvement and poor eating behaviour. The presence of free radicals can damage the integrity of cells in the body and decrease the body's antioxidant status. Functional utilization becomes an alternative to avoid the substance.

One food that has a functional substance is *Spirulina platensis. Spirulina platensis* application is usually for food colouring, because of its phycocyanin content. The study has found the presence of antioxidant and anti-inflammatory properties of phycocyanin[2] the stability of phycocyanin extract [2-4]. *Spirulina platensis* can be used for the production of functional foods because its bioactive compounds increase the nutritional value. However, some of the *Spirulina platensis* bioactive compounds are heat sensitive, so processing techniques need to be considered strictly. Spirulina contains chlorophyll and carotenoids as well as phenolic compounds and flavonoids as natural substances [5]. *Spirulina platensis* in powder form contains phenolic compounds and potential flavonoids. Beta-carotenee in *Spirulina platensis* is one of the antioxidants neutralize free radical substances in the human body. Beta-carotenee is one of the simplest forms of carotenoids with the molecular formula C40 H56 [6,7]. *Spirulina platensis* is one of the microalgae used by people living around water reservoirs, or water with high salt concentrations. Spirulina plants harvested from seawater have a higher mineral content than from freshwater and brackish water areas. Seawater contains high salt such as NaCl, KCl, MgCl. Spirulina processed on salted medium also contains phycocyanin, polysaccharide, inositol higher than fresh water.

Technical fermentation technology can be defined as an anaerobic or partial anaerobic carbohydrate oxidation process that produces alcohol and some acids. The result of fermentation is obtained as a result of metabolism of microbes in a food substance in an anaerobic state. Microbes that perform fermentation require energy generally obtained from glucose [9]. Food processing by fermentation process is a preservation method that depends on the production of certain microorganisms, chemical and physical changes that alter the original body shape and flavour of the foodstuff. These changes can improve the nutrients of the product and generally inhibit the growth of microorganisms which is not desirable. One of the fermented products is fermented milk.

Yoghurt is a dairy product that is pasteurized, fermented with bacteria mixture of *Lactobacillus bulgaricus* and *Streptococcus thermophillu*s, or *Lactobacillus acidophillus* that can break down the milk sugar (lactose) into lactic acid [10] to obtain the acidity, smell, and taste Which is typical [11]. The use of *Streptococcus thermophilus* as starter culture is its ability to ferment lactose and lowers the pH of the product quickly [12], while *L. bulgaricus* is an important characteristic in yoghurt that stimulates the growth of *Streptococcus thermophilus*, also produces exopolysaccaride Which improves the texture and rheology of fermented milk products [12]. Some factors that may affect the fermentation process is the influence of nutrients, temperature, water activity, pH, and oxygen.

Yoghurt has the advantage of being proven to lose weight, low cholesterol, and blood sugar levels. Milk has a high enough antioxidant and 10 times ascorbic acid (vitamin C) [13]. Previous research Milk fermented into low-symbiotic yoghurt cholesterol [14]

This study aimed to observe the implementation of *Spirulina platensis* powder into different yoghurt products *Spirulina platensis* which the application is at a relatively low temperature to protect the bioactive compounds of Amage. According to Agustini et al., the nutritional value of fresh *Spirulina platensis* has a significant difference compared with that of *Spirulina platensis* dry [15]. In addition, here is very different quantitatively for its bioactive compounds. The maximum concentration of *Spirulina platensis* that can be added to the product is determined, thus, will be accepted for consumption. Protein content and fat content were analyzed for both products.[16, 17] The water content, the Ash content, and the β-carotene analysis are performed gently. Based on this research, this research was aimed to increase the nutritional value and organoleptic properties and preferences yoghurt by *Spirulina platensis* supplementation.

Material and Metods

This study used an experimental design of Completely Randomized Design with three replicates for each treatment[18]. Research subjects 31 panels of organoleptic test experts who are in the Health Polytechnic Bandung, Ministry of Health, Department of Nutrition Bandung. The research was conducted in several stages: (1) first step to get the most appropriate spirulina concentration, this activity was done with the aid of DX Trial software. This tool helps choose the most appropriate concentration of spirulina. In the selection test the concentration of *Spirulina platensis* was 0.5, 0.7, 0.9, 1.0[23], 1,2. The experimental results obtained that are organoleptically feasible are concentrations 1.0, 1.2. (2) The test was continued to measure the parameters of organoleptic properties[19], (3) stage performed to measure parameters of proximate analysis including Protein, Fat, Minerals, and pH. Yoghurt is made from milk fermented with a commercial inoculum of Lactobacillus acidophilus. Samples divided randomly into three groups: (1) yoghurt standard supplemented 50 ml, (2) yoghurt and spirulina 1% supplemented 50 ml, (3) yoghurt and spirulina 1,2% supplemented 50 ml. Creating *Spirulina platensis* yoghurt, prepare milk, pour it into a saucepan and boil at a temperature of about 70°C by letting the pot open. Add spirulina as percentage that has been designed (1.0, 1,2%). Add 8% sugar, spirulina, and boiling lasts 15 minutes while stirring for whole cream milk not burnt. Finish boiling milk juice cooled to 45oC. Prepare a stater in the form of culture of Lactobacillus acidophilus bacteria as much as 5%. Dairy milk with a temperature of 42°C was inoculated with the stater. Stater mixing with milk should be evenly distributed and done cleanly so as not to be contaminated by other bacteria. Store a mixture of milk and stater juice in an incubator at 42°C. Duration of curing is 6 - 8 hours. Incubation is considered sufficient when the presence of thickened coagulation of milk, slowly pH 4 - 4.5 and forming yoghurt. Test data characteristic of yoghurt probiotic microorganism was measured by Total Plate Count (TPC) method, microbial contamination, total amount of probiotics. Statistical analysis presented bivariate (Wilcoxon, Pairs t-test), bivariate (Kruskal Wallis, Anova / Post Hoc Duncan Multi Range Test, multivariate Anacova with significance level <0.05.

Results

1. Organoleptic Test

The organoleptic test of 31 trained panellists consisting The Bandung Health Polytechnics Department of Nutrition to provide hedonic quality assessment of colour, aroma, taste, and consistency. The picture of the level of acceptance is done by organoleptic test which includes colour parameters, taste, aroma, and consistency. The results showed on the table as mentioned below.

Table 1. Organoleptic test of Yoghurt with addition of *Spirulina platensis* powder

Parameters	**Concentration of S. platensis powder (5w/w)**			**p**
	0	1,0	1,2	
Colour	4.80 ± 0.913	5.96 ± 0.730	6.0 ± 0.957	0,001
Taste	4.40 ± 1.258	5.0 ± 1.258	5.40 ± 1.155	0,018
Aroma	5.08 ± 1.37	5.36 ± 0.907	5.52 ± 0.770	0,275
Consistency	3.76 ± 1.615	5.52 ± 1.005	5.40 ± 0.913	0,001

Significance, p<0,05

The results of organoleptic test showed that there was a significant difference in the colour (p <0.001). Different turning products were white control group (4.80 ± 0.913), while the most preferred product group colour was 1.2% spirulina addition group. However the product formula was 1.0 (5.96 ± 0.730), and 1, 2% (6.0 ± 0.957) did not differ significantly.

The result of organoleptic test showed that there was a significant difference to taste (p <0,018). The differentiated product flavours were 1.2% (5.40 ± 1.155), whereas most preferred was the 1.2% spirulina addition group compared with controls (4.40 ± 1.258) and 1% spirulina (5.0 ± 1.258).

The results of the organoleptic test show that there is no difference between the three formulas; Control group (5.08 ± 1.37), spirulina group 1% (5.36 ± 0.907) and 1.2% spirulina group (5.52 ± 0.770) significantly to the aroma (p> 0.275), so it can be said Panellists favor the fragrance of the product even when spirulina is added.

The results of organoleptic test showed that there was a significant difference to consistency (p <0.001). Consistency of different turning products was the control group (3.76 ± 1.615), while most preferred was the 1.0% spirulina addition group (5.52 ± 1.005) and the 1.2% (5.40 ± 0.913) group of products Conditions are not significantly different.

2. Proximate Analysis

The result of proximate analysis obtained the data as follows; protein content showed results in the range 10,13 - 10,35, the biggest result was group with addition of 1.0% (10,30 gr). Fat content decreased 3.46 - 4.45. The lowest fat content was at a concentration of 1.2% of 3.48, and the highest was 4.45 in the control group. Mineral showed in the range of 2.21 - 2.34, the lowest group of 2.21 in the control group, while the increased spirulina content also affected the mineral content, as evidenced by the 1.2% group. Based on observations pH levels are in acidic conditions (pH 4.0).

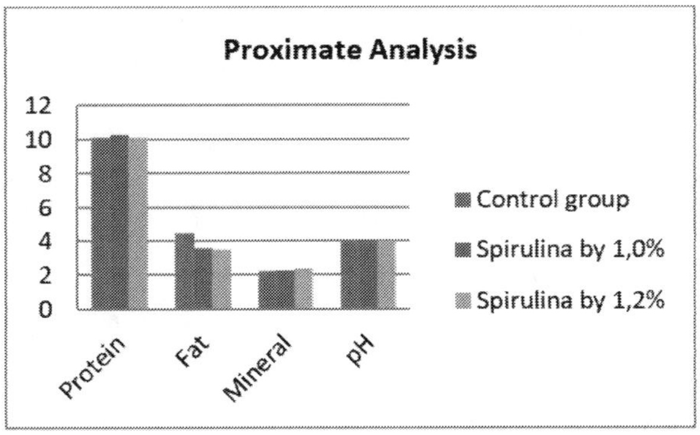

Graph 1. Proximate analysis of Yoghurt with addition of *Spirulina platensis* powder

3. Microbiological Analysis

Microbiological analysis was performed on total plate number parameter (TPCN) obtained the result as follows.

Table 2. Microbiology parameter test of Yoghurt with addition of *Spirulina platensis* powder

No	Parameters / Groups	Results
1	Total Plate Count	
	Control	> 10^5
	Spirulina 1,0% Group	669 X 10^4
	Spirulina 1,2% Group	> 10^5
2	MPN Coliform	
	Control	0 MPN/100 mL
	Spirulina 1,0% Group	0 MPN/100 mL
	Spirulina 1,2% Group	0 MPN/100 mL
3	Microorganism Contamination	
	a. *Escherichia coli*	Negative
	b. Salmonella sp.	Negative
	c. Pseudomonas sp.	Negative

Based on the observation and analysis of types of microbes with the help of ManRogosa Agar (MRSA) Media grown in the yoghurt is Lactobacillus acidophilus.

Discussion

Organoleptic properties are also called subjective traits of Food our sensory properties because of measurement using organs of the human senses, sometimes also called the sensory nature because the assessment is based on sensory stimulation of the sense organs. Organoleptic properties are the most decisive factor in determining the level of consumer preference because it can be done quickly and directly. The test of organoleptic properties includes tests on colour, taste, aroma, and consistency. [19]

The colour attribute involves the physical and psychological phenomena of perception by the visual system caused by the presence of light in the visible wave range. In the colour notation system, the colour assessment includes three attributes: the chromatic colour (the type of colour actually seen visually) called hue, value or brightness indicating the achromatic colour parameter and the sharpness or intensity of the colour called chroma purity or saturation. [19, 22]

Colour uniformity is also an important parameter of appearance. Damage and degradation of food product quality are often indicated by the change of colour. [19, 22] The smell or smell of food much determines the delicacy of these foods. In case the smell is more related to the means of the five sense sensors. In general, the smell received by the nose and the brain is more a variety of herbs or a mixture of four major odours are fragrant, sour, rancid and charred. For the food industry, the assessment using the odour test is very important because it quickly gives the results of the assessment of a preferred product or not. [19,20,22] Taste is a mixture of tastes, smells and touches that are combined with other impressions such as sight, touch, and hearing. Pain involves the taste senses of the tongue. Speech sensation can be grouped into four main ingredients, namely acid, salty, sweet and bitter. [20] Consistency or texture is a sensation of pressure that can be observed by mouth (when bitten, chewed and swallowed) or touch with the fingers. The boundaries of tactile nature are closely related to the type of material being observed, so it is very difficult to draw conclusions or to provide general restrictions. [21]

Based on the proximate analysis such as of protein content showed results in the range 10.13 - 10.35, the largest result was the group with the addition of 1.0% (10.30 gr). In addition, fat content decreased 3.46 - 4.45. The lowest fat content was at a concentration of 1.2% of 3.48, and the highest was 4.45 by the control group. Mineral showed in the range of 2.21 - 2.34, the lowest group of 2.21 in the control group, while the increased spirulina content also affected the mineral content, as evidenced by the 1.2% group. Those result fat, ash content were closely other research product in spirullina.[23] Based on observations pH levels are in acidic conditions (pH 4.0). The results of Judiono study on the potential of probiotics as anti-diabetic and the results significantly affected the decrease of blood fasting glucose and postprandial blood glucose, HbA1c, increase c-peptide and insulin[24] proinflammatory cytokines such as IL_1, IL_6 and lipid peroxidation (MDA) and increase endogen antioxidants (SOD, Catalase, GPx) anti-proliferative cytokines IL_{10} and pancreatic beta cell function.[25]

Analysis of probiotic levels in general spirulina yoghurt products is at a level of more than 10^5, with a degree of acidity ranging from 4.0.

Conclusions

Supplementation of spirulina to yoghurt products can be accepted by organoleptic, chemical and microbiological tests of concentrations of 1 and 1.2%. The material makes the blue colours more attractive and can be accepted by expert panellist. The result of organoleptic test on the colour, flavour, and consistency expected in the manufacture of spirulina yoghurt is at concentration of 1.2%, the aroma expected in the manufacture of spirulina yoghurt is equal to the control means there is no significant difference. Protein content in the highest 1% group of 10.30 gr, fat content with increased spirulina showed a low decrease. Microbial contamination was not found to be very good. It is recommended that the addition of concentrations of 1.0 and 1.2% is very potential and excellent in the yoghurt industry.

Acknowledgement
This project was supported by Non-Tax Nation Income of Diponegoro University through "Riset Unggulan Univeritas" 2017

Reference

1. Goldberg I, 1994. *Functional Food*. New York: Chapman & Hall.
2. Romay, C., J. Armesto, D. Remirez, R. Gonzales, N. Ledon, and I. Garcia. 1998. Antioxidant and Anti-inflammatory Properties of C-phycocyanine from Blue-green Algae. *Inflam. Res*. 47: 36–41.
3. Sarada, R., M. G. Pillai, and G. A. Ravishankar. 1999. Phycocyanin from *Spirulina* sp. Influence of Processing of Biomass on Phycocyanin Yield, Analysis of Efficacy of Extraction Methods and Stability Studies on Phycocyanin. *Process Biochem*. 34: 795–801.
4. Duangsee, R., N. Phoopat and S. Ningsanond. 2009. Phycocyanin Extraction from *Spirulina platensis* and Extract Stability under Various pH and Temperature. *As.J.Food and Agro-Industry*. 2(4): 819–826.
5. Martelli, G., C. Folli, L. Visai, M. Daglia and D. Ferrari. 2014. Thermal Stability Improvement of Blue Colourant C- phycocyanin from *Spirulina platensis* for Food Industry Applications. *Process Bichemistry*. 49: 154–159.
6. Minea, R., M. Brasoveanu, M. N. Grecu, and M. R. Nemtamu. 2006. Preliminary Studies on Irradiated Spirulina. *Rom. J. Phys*. 51(1–2): 141–145.
7. Khoo, H.E., K.N. Prasad, K.W. Kong, Y. Jiang and A. Ismail. 2011. Carotenoids and theirs Isomer : Fruit and Vegetable. Moleculles 16: 1710=1738.
8. Codex Alimentarius 2011. Milk and Milk Product, 2 nd Edition. Rome : FAO and WHO.
9. Muchtadi, Deddy. 2012. *Pangan Fungsional dan Senyawa Bioaktif*. Bandung: Alfabeta.
10. Belitz H.D., Peter Schieberle dan W. Grosch. 1987. *Food Chemistry*. Munchen: Springer.
11. Surajudin, Kusuma, dan Fauzi R. 2008. *Yoghurt Susu Fermentasi yang Menyehatkan*. Jakarta: Agromedia Pustaka.
12. Bastomi, Riska R. 2009. *Pengembangan Yoghurt Menggunakan Bakteri Asam Laktat Isolat Air Susu Ibu. Skripi*. Fakultas Teknologi Pertanian. Institut Pertanian Bogor.
13. Suharjanto Toto. 2010. Respon Hasil Susu Terhadap Intensitas Cekaman Kekeringan. *Agrika*. Malang: Universitas Widyagama Malang.
14. Winarti S. 2010. *Yoghurt Sinbiotik Rendah Kolesterol dari Susu*. Fakultas Teknologi Pertanian. Institut Pertanian Bogor.
15. Agustini T.W., **M. Suzery**, D. Sustrisnanto, W.F. Ma'ruf, and Hadiyanto. 2015. Comapartive Study of Biactive Substances Extracted from Fresh and Dried Spirulina sp. In Hadiyanto, H., H. Susanto, and O.K Radjasa. Proceding Environmental Science. 24(2015): 282-289
16. Badan Standarisasi Nasional. 2006. Standar Nasional Indonesia (SNI). SNI-01-2346-2006. Petunjuk Pengujian Organoleptik dan/atau Sensori pada Produk Perikanan [Instructions of Organoleptic and/or Sensory Testing for Fishery Products]. Dewan Standarisasi Indonesia. Jakarta. [Bahasa Indonesia].
17. Feldsine, P., C. Abeyta, and W. H. Andrew. 2007. INTERNATIONAL Methods Committee guidelines for validation of qualitative and quantitative food microbiological official methods of analysis. J . AOAC. 85(5):1187-1200.
18. Donald T. Campbell, Julian C. Stanley. Experimental and Quasi-experimental designs for research. Rand Mc Nally College Publishing Company, Chicago. 1963. Page 145 – 170.
19. Soekarto Soewarno T. 1985. *Penilaian Organoleptik*. Jakarta:Bhratara Karya Aksara.
20. Dede R. Adawiyah, , dan Waysima. 2009. *Evaluasi Sensori Produk Pangan*. Bogor: Institut Pertanian Bogor.
21. FG. Winarno. 1991. *Kimia Pangan dan Gizi*. Cet. Ke-8. Jakarta: PT Gramedia Pustaka.
22. Kartika Bambang, Pudji Hastuti, dan Wahyu Supartono. 1988. *Pedoman Uji Inderawi Bahan Pangan*. Yogyakarta:Universitas Gadjah Mada.

23. Tri Winarni Agustini, Widodo Farid Ma'ruf, Widayat, Meiny Suzery, Hadiyanto, Soottawat Benjakul. 2016. Application Of *Spirulina platensis* On Ice Cream And Soft Cheese With Respect To Their Nutritional And Sensory Perspectives. Jurnal Teknologi 78:4–2 (2016) 245–251 |www.jurnalteknologi.utm.my. eISSN 2180–3722
24. Judiono, Suharyo Hadisaputro, Indranila KS, Bambang Cahyono, Meiny Suzery, Yuliati Widiastuti, Asep Iwan Purnawan, 2014. Effects of Clear Kefir on Biomolecular Aspects of Glycemic Status of Type 2 Diabetes Mellitus (T2DM) Patients in Bandung, West Java [Study on Human Blood Glucose, c Peptide and Insulin]. Functional Foods in Health and Disease 2014; 4(8):340-348
25. Judiono, RRJ Djokomoeljanto, Soeharyo Hadisaputro, 2011. POTENSI KEFIR BENING SEBAGAI ANTIDIABETES Dikaji dari Status Glikemik, Antioksidan, Respon imun dan Regenerasi sel β pankreas Pada Tikus Strain Wistar Hiperglikemia Diinduksi Streptozotocin. Thesis Program Doktor Kedokteran Universitas Diponegoro Semarang.

Effect of Temperature to Adsorption Capacity and Coefficient Distribution on Rare Earth Elements Adsorption (Y, Gd, Dy) Using SIR

N Aziz[1], A Mindaryani[1], Supranto[1], A Taftazani[2], D Biyantoro[2]

[1] Chemical Engineering - Universitas Gadjah Mada

[2] Centre for Science and Accelerator Technology - National Nuclear Energy Agency of Indonesia (PSTA – BATAN), Yogyakarta, Indonesia

nofriadyaziz@gmail.com

Abstract. The use of REE like element of Yttrium (Y) as a superconducting material requires a purity of more than 90%, so it needs to increase the purity of Y from the settling process. The purpose of this research is to study the separation process of REE that is Y, Gd, Dy elements from REE hydroxide ($REE(OH)_3$) using SIR method are consisting of Amberlite XAD-16 resin impregnated with Di-(2-ethylhexyl) phosphate (D2EHPA) and Tributyl Phosphate (TBP) and determine the isotherm model on REE adsorption and determine the kinetic model of pseudo adsorption reaction. This research was started by activating XAD -16 resin and is mixed with TBP-D2EHPA solvents so it will form SIR, then it is conducted on variation of SIR composition, temperature variation of adsorption process, determination of equilibrium equation and kinetic sorption occurring in SIR adsorption based on experimental data of liquid concentration as function of time. Based on the calculation result, the most effective SIR composition for REE separation is 0.75 g, the equilibrium equation for Y, Gd and Dy follows the Henry equilibrium model and the pseudo kinetic model of the reaction order Y, Gd, and Dy is followed by the pseudo reaction of order 2 The result of separation of LTJ with SIR is said to be effective from another method because purity is obtained that is 96.73% and qualify as a super conductor material.

1 Introduction

Rare earth metals (REE) are very rare, in nature can be complex compounds, generally complex phosphate compounds and carbonates [1]. One of them is xenotime which is a rare earth elements phosphate compounds and is usually written with $REEPO_4$ or YPO_4 and containing the main elements that are Yttrium (Y), Gadolinium (Gd), and Dysprosium (Dy) with each content is 60%, 3.6%, and 7.5%. In Indonesia, xenotime is found on the islands of Bangka, Belitung and Singkep which is a by-product of tin mining of PT Timah, Tbk in Bangka [2].

Because of the prospect, potential and economic value of REE is very large, especially the use of a single REE like Y then it is necessary to separate the REE to obtain elements Y with high purity (more than 90%) to be used as a super conductor material. While processing xenotime before the purification process, Y only has content 79-80%, so it needs to be increased purity. However, the problems experienced by the REE elements have similar chemical properties and physical properties that are difficult to separate. REE separation is usually done by solvent extraction or ion exchange (resin).

However, solvent extraction and ion exchange techniques have some limitations. To solve the limitations of extraction and ion-exchange resins, Warshawsky in 1974 introduced a method called solvent impregnated resins (SIR) method for the separation of metal ions [3]. This study purpose to study the separation process of REE (Y, Gd, Dy) from REE hydroxide ($REE(OH)_3$) using SIR method and are consisting of Amberlite XAD-16 resin impregnated with Di-(2-ethylhexyl) phosphate (D2EHPA) and Tributyl Phosphate (TBP), to determine the isotherm model on REE adsorption and determine the kinetic model of pseudo adsorption reaction.

2 Methods

2.1. Tools and Materials

The tools to be used in this research are beaker glass, erlenmeyer, magnetic stirrer, hot plate, stirrer bar, pumpkin neck 3, measuring, bulpet, plastic bottle, XRF tool and the device where sample for XRF analysis, measuring cylinder, measuring pipette, volume pipette, porcelain plate, analytical balance, filter paper oven, glass funnel, thermometer. Materials to be used in this study are Tributyl Phosphate (TBP), Bis (2-ethylhexyl) Phosphate Acid (D2EHPA), $REE(OH)_3$, Aquadest, XAD-16 Resin, Nitric Acid (HNO_3) and Acetone.

2.2 Procedure

2.2.1 Activation of XAD – 16

The process begins with activating XAD - 16 as much as 50 g XAD - 16 soaked HNO_3 2 M for 2 hours then filtered. Then soaked again with 2 M NaOH for 2 hours then filtered. After it was rinsed with aquadest to neutral, soaked with acetone for 2 hours, and dried with oven for 50°C.

2.2.2 Preparation of D2EHPA / TBP Solution (3: 1) and Resin Impregnation

Prepared D2EHPA and TBP solution, then mixing to homogeneous with a ratio of 3: 1. Then 20 g solvent soaked with 100 mL of acetone and add 20 g of resin (50% / 50%) was stirred at 350 rpm at room temperature for 2 Hour and dried in 50°C oven for 24 hours.

2.2.3 Making Feed Solution $REE(OH)_3$

The $REE(OH)_3$ feedstock standard solution was prepared by weighing as much as 25 g of $REE(OH)_3$ dissolved in concentrated HNO_3 and then heated to 120°C until the acidity of the depleted HNO_3 vapor was then dissolved in 0.5 M.

2.2.4 *Determination of Adsorption Optimization Y, Gd, Dy*

The SIR variation is made by a certain ratio with a fixed feed amount of 10 mL as in Table 1.

Table 1. Variation of SIR

Solution	SIR
Feed	0
A	0.25 g
B	0.5 g
C	0.75 g
D	1 g
E	1.5 g

In each variation, the SIR was then mixed with feed at 30 °C and atmospheric pressure stirred at 350 rpm for 30 minutes. Then separated by filter paper. The filtrate was taken as much as 5 mL to be analyzed using XRF. This process is also carried out at temperatures of 30°C, 40°C, 50°C, and 60°C.

2.2.5 Kinetics of sorption

Comparison of 0.75 g SIR composition: 10 mL of feed then mixed at 30°C and atmospheric pressure stirred at 350 rpm for 30 minutes. Then separated by filter paper. The filtrate was taken as much as 5 mL to be analyzed using XRF. This adsorption action is carried out for temperature variations of 30°C, 40°C, 50°C, 60°C with mixing time variation of 2 minutes, 5 minutes, 10 minutes, 20 minutes, 30 minutes.

3. Result and Discussion

3.1. Rare Earth Separation Mechanisms

After processing from leaching to settling obtained levels of Y, Gd and Dy of 79.4%; 12.1% and 8.5%. However, the use of Y as a superconductor material requires a purity of Y of 90%, so it needs to be increased Y purity. After the calculation is done the results Table 2.

Table 2 Type of Separation based on% purity and % taken

Type of Separation	% purity			% taken		
	Y	Gd	Dy	Y	Gd	Dy
Adsorption						
Activated Carbon	77.15	10.08	12.77	7.27	4.85	31.76
Silica	81.98	7.09	10.93	6.19	2.73	21.78
Extraction						
Solvent D2EHPA	91.11	6.33	2.57	60.27	39.45	55.88
Solvent TBP	90.57	6.91	2.53	23.27	16.72	21.36
Solvent Mixture TBP-D2EHPA	91.87	5.26	2.87	66.38	35.85	68.23
Resin XAD -16	92.20	6.30	1.50	17.29	6.03	7.42
SIR						
SIR Solvent D2EHPA	93.91	3.62	2.47	22.69	4.46	15.73
SIR Solvent TBP	90.95	7.21	1.84	14.20	5.75	7.57
SIR Solvent TBP - D2EHPA	96.733	2.743	0.524	31.57	4.57	4.51

In Table 2, the separation by using SIR obtained by Y purity over 90% with Gd and Dy of 2.743% and 0.524% compared with adsorption using activated carbon or silica, extraction and ion exchange resin and qualified to be used as the superconductor material is Y 89.76%. The analysis used XRF with 99.95% accuracy and 90.43% precision using Merck Yttrium Oxide standard with 99.99%. In addition, based on the results of previous research, it can be seen that the separation by using SIR is much greater than using liquid-liquid extraction and ion exchange resin. The results of this study were compared using the extraction method, ion-exchange resin, and SIR and presented in Table 3. Table 3 shows the synergistic effect of the solvent mix of D2EHPA and TBP that is greater than 1.

Table 3 Separation Method of REE

Type of Separation	Kd			β		Effect of Synergism
	Y	Gd	Dy	Gd	Dy	
Adsorption						
Activated Carbon	0.08	0.05	0.47	1.54	0.17	
Silica	0.07	0.03	0.28	2.35	0.24	
Extraction						
Solvent D2EHPA	1.52	0.65	1.27	2.33	1.20	
Solvent TBP	0.30	0.20	0.27	1.51	1.12	
Solvent MixtureTBP-D2EHPA	1.97	0.56	2.15	3.53	0.92	1.08
Resin	0.21	0.06	0.08	3.26	2.61	
SIR						
SIR Solvent D2EHPA	0.29	0.05	0.19	6.28	1.57	
SIR Solvent TBP	0.17	0.06	0.08	2.72	2.02	
SIR Solvent TBP - D2EHPA	0.46	0.048	0.047	10.00	9.76	1.22

3.2 Effect of Composition Variation SIR

3.2.1 Effect of Weight SIR against Kd and separation factor (β)

Based on the calculations, we obtain the weight relationship SIR (g) Vs. Kd is presented in Figure 1, and relationship the weight of SIR (g) Vs separation factor (β) is presented in Figure 2. Based on Figure 1 and 2 it can be seen that with the increasing weight of SIR used in the separation of REE the distribution coefficients of each element (Kd_Y, Kd_{Gd}, Kd_{Dy}) will also increase. The largest distribution coefficient is owned by Y indicating that the Y^{3+} ion is absorbed larger than Gd and Dy.

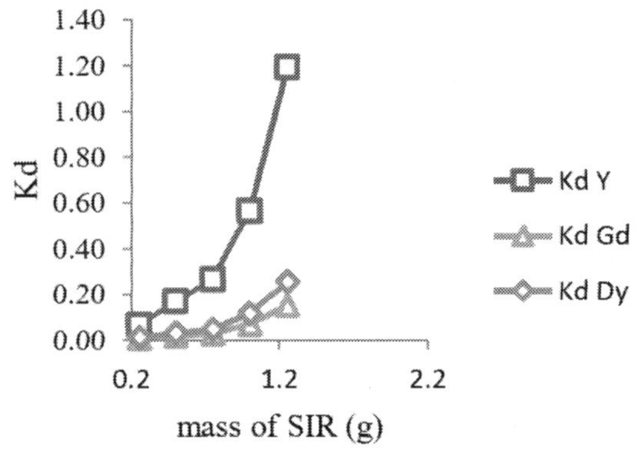

Figure 1. Relations mass of SIR (g) vs Kd

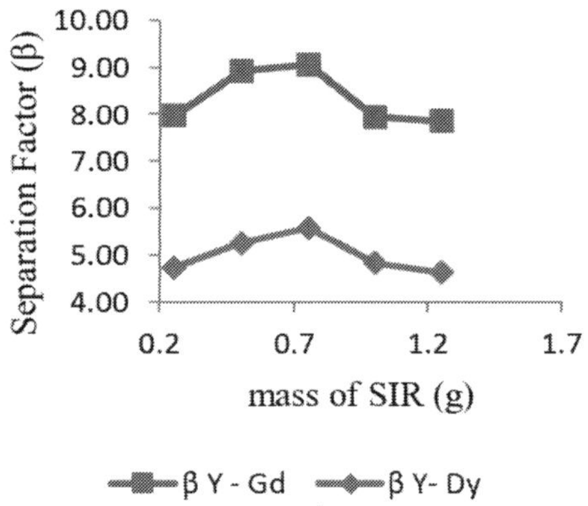

Figure 2. Relations mass of SIR (g) vs Separation factor

3.2.2 The Effect of Temperature on Kd and separation factor (β)

Plot of temperature (g) Vs Kd is presented in Figure 1, and temperature SIR (g) Vs separation (β) is presented in Figure 2. Based on Figure 3 it can be seen that with the decreasing temperature used in the separation of REE the distribution coefficients of each element (Kd_Y, Kd_{Gd}, Kd_{Dy}) will also increase. In

Figure 4 it can be seen that with the decreasing temperature used in the separation of the REE, the separation factor (β) of Y-Gd and Y-Dy increases, so that the optimum temperature is 30°C.

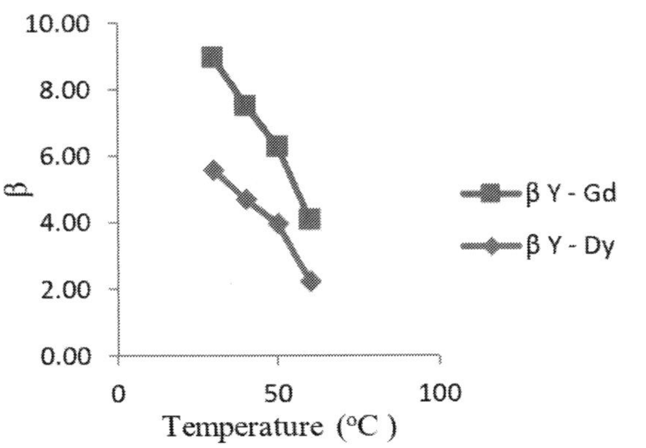

Figure 3. Relations Temperature (°C) vs. Kd

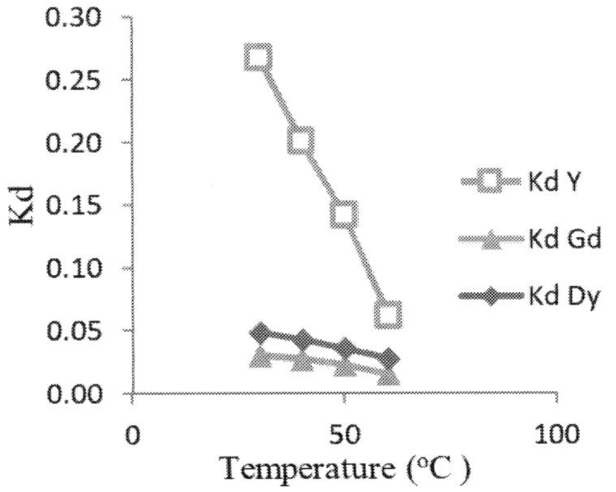

Figure 4. Relations Temperature (°C) Vs Separation Factor (*β*)

3.3 Adsorption Equilibrium Equation

3.3.1 Determination of Equilibrium Time

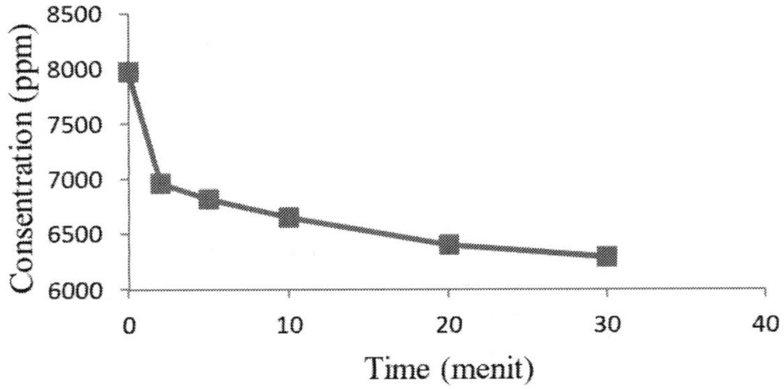

Figure 5. Equilibrium Time (min) for Yttrium ion sorption

Figure 5 shows that there is a decrease in concentration and the equilibrium time has been reached in 30 minutes.

3.3.2 Isotherms Adsorption

Through this research, SIR adsorption power will be exposed to REE to formulate the adsorption equilibrium behavior of Y, Gd and Dy ions in SIR.

Table 4 Adsorption Isotherm Equations

	Y			
Temperature	30	40	50	60
Henry				
r^2	0.9824	0.9829	0.9977	0.9812
error (%)	3.35	3.22	1.71	7.05
Freundlich				
r^2	0.9	0.8654	0.9515	0.9946
Error (%)	5.59	7.75	7.22	3.25
Langmuir				
r^2	0.9626	0.9356	0.9107	0.9538
Error (%)	11.28	13.00	14.50	10.04

	Gd				Dy			
Temperature	30	40	50	60	30	40	50	60
Henry								
r^2	0.9926	0.9946	0.9967	0.9887	0.9894	0.9931	0.9953	0.9969
error (%)	5.02	4.91	4.92	8.70	5.83	5.25	5.31	5.73
Freundlich								
r^2	0.8953	0.9106	0.9669	0.9872	0.8789	0.9064	0.9221	0.9261
Error (%)	13.05	13.13	8.99	7.05	13.49	12.42	13.57	32.61
Langmuir								
r^2	0.846	0.8534	0.8499	0.8904	0.8422	0.848	0.8437	0.8132
Error (%)	19.99	19.08	19.68	16.31	20.37	20.02	19.62	21.18

Based on Table 4 it can be stated that the adsorption of Y, Gd and Dy metals follows the Henry isotherm model because r^2 is close to 1 and the isotherm error of Henry is less than that of Langmuir isotherms and Freundlich isotherms.

3.3.3 Isotherms Adsorption of temperature function and van't Hoff equation

The van't Hoff equation can be done to find out the dependence of the equilibrium concentration of temperature and the results obtained in Table 5.

Table 5 Reaction Heat of REE

Element	Ho	ΔH (kcal/mol)
Y	1646.65	-3.05
Gd	38368.80	-0.87
Dy	48194.49	-0.74

In Table 4.4 it can be seen that the largest ΔH is owned by Yttrium that is 3.05 kcal/mol and ΔH is negative, i.e., the exothermic reaction due to K decreases with temperature so that the metal sorption of Y in SIR is faster than Gd or Dy metal.

3.4 Kinetics of sorption

3.4.1 Compare Model

Result of rate constant adsorption speed based on some models can be seen in Table 6.

Table 6 Reaction Model of REE Pesudo Kinetics

Pseudo Model/Temperature	Y				
	30	40	50	60	Average
Order 1					
r^2	0.7	0.6723	0.9557	0.8353	0.7908
error (%)	25.58	31.45	16.75	29.51	25.82
Order 2					
r^2	0.99996	0.9999	0.9996	0.9985	0.9995
error (%)	1.89	0.96	2.23	5.69	2.69

Pseudo Model/Temperature	Gd				
	30	40	50	60	Average
Order 1					
r^2	0.965	0.9955	0.9983	0.9959	0.9887
error (%)	31.45	14.96	-7.04	-9.95	7.36
Order 2					
r^2	0.9378	0.8626	0.7868	0.6433	0.8076
error (%)	6.07	12.96	15.30	17.09	12.86

Pseudo Model/Temperature	Dy				
	30	40	50	60	Average
Order 1					
r^2	0.9497	0.987	0.9821	0.9569	0.9689
error (%)	18.72	20.17	13.01	17.74	17.41
Order 2					
r^2	0.9206	0.9088	0.799	0.5481	0.7941
error (%)	10.140	9.346	14.738	13.943	12.04

Table 6 shows that the REE adsorption kinetics of the SIRs are more suitably explained by the second order pseudo adsorption kinetics model because r^2 is close to 1 and the error by the 2nd order reaction kinetics is smaller than the order 1.

3.4.2 Effect of temperature on sorption kinetics

The effect of temperature on the value of the velocity constant (k) on the pseudo adsorption kinetics model 2 can be seen in Table 7, and Figure 6. It can be seen that the higher the temperature the constant value of the speed will be smaller because of the SIR resistance at temperatures below 60°C so that the REE will be bound complex and stable in the SIR at a temperature of 30°C. The largest reaction rate constant is owned by Y so that Y is more chemically stable in the SIR.

Table 7 Kinetic Constants Reaction of Order 2 Pseudo REE

Temperature (°C)	k_2 (mg/g adsorbent. min)		
	Y	Gd	Dy
30	0.3872	0.3434	0.3533
40	0.3005	0.2660	0.2894
50	0.2344	0.2127	0.2188
60	0.2158	0.1911	0.2092

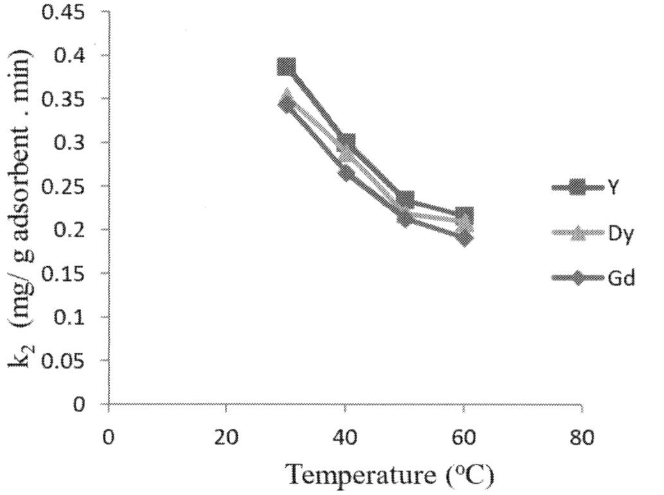

Figure 6. Kinetic Constants Reaction of Order 2 Pseudo REE Temperature Relation (°C) and Order Rate Reaction Pseudo 2 (mg / g adsorbent.min)

4. Conclusion

SIR method is more effective than extraction or adsorption. The most effective SIR composition for rare earth metal separation is 0.75 g with a temperature of 30 °C. Equilibrium equations for Y, Gd, Dy approach the Henry, equilibrium model. Y, Gd, Dy kinetic sorption follows the pseudo reaction kinetics of order 2 and the reaction kinetics constant belongs to Y that is 0.3872 mg / g of minute adsorbent at a temperature of 30 ° C. The larger the temperature the absorption will be smaller and the greater the temperature then constant rate pseudo reaction order 2 will also be smaller.

Reference

[1] Suprapto, S.J., 2009. *Tinjauan Tentang Unsur Tanah Jarang, Buletin Sumber Daya Geologi,* Volume 4, 36-47

[2] Atmawinata, A., *Pengembangan Industri REE di Indonesia*, Kementerian Perindustrian, Jakarta, 2011.

[3] Khaldun, I. Buchari, Amran, M. B. dan Sulaeman, A. 2009. *Pengaruh Komposisi Asam Bis(2-Etilheksil) Fosfat (D2EHPA) dan Tributil Fosfat (TBP) dalam Resin Amberlite Xad-16 terhadap Sorpsion-Ion La(III), Nd(III) dan Gd(III)*. Bandung. ITB.

[4] de Haan AB. 2004. Separation Technology at University of Twente. Di dalam *EFCE-WP Distillation, Absorption Extraction Meeting*; Huelva; 16-18 Sept 2004. Separation Technology Group.

IOP Conf. Series: Materials Science and Engineering **349** (2018) 012041 doi:10.1088/1757-899X/349/1/012041

[5] Cortina JL, Warshwasky A. 1997. Developments in Solid-Liquid Extraction by Solvent-Impregnated Resins. Di dalam: Marinsky JA, Marcus Y, editor. *Ion Exchange and Solvent Extraction*. Vol. 13 New York: Marcel Dekker. hlm. 195-287.
[6] Jorderns, A., Cheng, Y.P., Waters, K.E., 2013. A review of benefication of rare earth elements bearing minerals. Miner. Eng, 41, 97-114.
[7] Aziz, N. 2014. *Pemisahan Zr – Hf dengan Metode SIR.* Yogyakarta. STTN - BATAN
[8] Do,D.D, 1998, Adsorption Analysis : *Equilibria and Kinetics*, Vol 1, 64-103, Imperial, College Press, London

Separation of Gadolinium (Gd) using Synergic Solvent Mixed Topo-D2EHPA with Extraction Method.

N Effendy[1], K T Basuki[1], D Biyantoro[2], N K Perwira[1]
[1] Polytechnic Institute of Nuclear Technology, BATAN Yogyakarta
[2] Centre for Science and Accelerator Technology - National Nuclear Energy Agency of Indonesia (PSTA – BATAN), Yogyakarta, Indonesia

nuradame@gmail.com

Abstract. The main problem to obtain Gd with high purity is the similarity of chemical properties and physical properties with the other rare earth elements (REE) such as Y and Dy, it is necessary to do separation by the extraction process. The purpose of this research to determine the best solvent type, amount of solvent, feed and solvent ratio in the Gd extraction process, to determine the rate order and the value of the rate constant of Gd concentration based on experimental data of aqueous phase concentration as a function of time and to know the effect of temperature on the reaction speed constant. This research was conducted on variation of solvent, amount of solvent, feed and solvent ratio in the extraction process of Gd separation, extraction time to determine the order value and the rate constant of Gd concentration in extraction process based on the aqueous phase concentration data as a function of time, to the rate constant of decreasing concentration of Gd. Based on the calculation results, the solvent composition was obtained with the best feed to separate the rare earth elements Gd in the extraction process is 1 : 4 with 15% concentration of TOPO and 10% concentration of D2EHPA. The separation process of Gd using extraction method by solvent TOPO-D2EHPA 2 : 1 comparison is better than single solvent D2EHPA / TOPO because of the synergistic effect. The rate order of separation process of Gd follows order 1. The Arrhenius Gd equation becomes k = 1.46 x 10-7 exp (-6.96 kcal / mol / RT).

1. Introduction

One of the rare earth elements (REE) in senotim that have potential in the development of the nuclear industry is Gadolinium (Gd) with content 3.6% [1]. Rare Earth Elements in the form of Gd in the form of oxide (Gd2O3) is used as the main component of a nuclear reactor control rods because Gd has a neutron absorption cross section (σ) is very large [2]. The required Gd level as control rod is 99%. The main problem for obtaining high purity Gd is the similarity of chemical properties and physical characteristics with other Rare Earth elements such as Y and Dy, it is necessary to separate with the most technologically feasible and efficient to obtain Gd in REE [3].

Separation of Gd in the Rare Earth elements can be done with a liquid-liquid extraction process. Methods of liquid-liquid extraction is used to separate the elements or compounds from low levels to high levels, but it has advantages such as more efficient, effective, selective against certain ions, the solvent can be reused, the energy requirements are lower, and more equipment small [4]. So liquid-liquid extraction methods are considered feasible and efficient to obtain Gd from REE.
In the process the selection of liquid-liquid extraction solvent, the amount of solvent, and the ratio of

the amount of feed and the solvent is very important, this is because the solvent play a role in determining the distribution coefficient value, increased efficiency, and the separation factor [5]. Solvents used in this study were Acid di- (2-ethylhexyl) Posfat (D2EHPA), Trioctylphosphine Oxide (TOPO), and solvent mixture of both. Mixed solvents need to be used to elicit synergistic effects that have the effect of strengthening the extraction process against organic solvents. Other factors that influence the extraction process are time and temperature extraction. Temperature and extraction time are used in determining the kinetics of reduction of Gd concentration (rGd) [6]. The variation of extraction temperature is carried out to determine the activation energy (E) and the frequency factor (A) by using the Arrhenius Equation. By doing research on the kinetics of the extraction process some important data such as the order of the reaction (n), the reaction rate constant (k), the activation energy (E),and frequency factor (A) can be known and useful as data and scientific information useful for research and design in the future.

So the purpose of this research is to determine the type, number of solvent, and the best ratio of feed and solvent in the Gd extraction process, and to determine the order and the value of the rate constant of the reduction of Gd concentration based on experimental data of aqueous phase concentration as a function of time and to know the effect of temperature on the reaction speed constant.

2. Method

2.1 Tools and Materials

The tools to be used in this research are beker, erlenmeyer, magnetic stirrer, hot plate, stirrer bar, pumpkin neck 3, measuring, bulpet, plastic bottle, XRF tool and the device where sample for xrf analysis, measuring cylinder, measuring pipette, volume pipette, porcelein plate, analytical balance, filter paper oven, glass funnel, thermometer. Materials to be used in this study are Tributyl Phosphate (TBP), Bis (2-ethylhexyl) Phosphate Acid (D2EHPA), $REE(OH)_3$, Aquadest, XAD-16 Resin, Nitric Acid (HNO_3) and Acetone.

The materials used in this study were TOPO (TOPO), Bis (2-ethylhexyl) Phosphate Acid (D2EHPA), Concentrated REE senotim (result of enhanced senotim cracker concentration of Rare Earth Metal with Gd concentration 15.45%), Aquadest , HNO3 65% pa (ex. Merck), Kerosene. The tool used in this research is glass beker as reactor that used as place of extraction process that is beaker. This tool is coupled with hot plate, stirrer, water batch and thermometer in kinetic study. XRF device for measuring quantitative and qualitative analysis, measuring glass, watch glass, stirrer, magnetic strirrer, measuring flask, dropper pipette, measuring pipette, spatula, analytical balance, purple spoon, erlenmeyer, bulpet, hotplate stirrer.

2.2 Procedures

Making of Feed Solution

20 g of REE $(OH)_3$ was dissolved into 50 mL of 65% HNO3 then heated at 120 °C until the gas from HNO_3 was depleted. Then the solution was diluted to 200 mL HNO_3 0.5 M. The mother liquor was diluted to 900 ppm Gd.

Making of Solvent Solution

TOPO

TOPO 50% is prepared by weighing 1 g of pure TOPO solid then is added kerosene up to 200 mL. For 10 mL TOPO 5% is prepared by diluting 1 mL TOPO 50% then is added with kerosene up to 10 mL. This step is repeated for the preparation of TOPO solutions of 10%, 15%, and 30%.

D2EHPA

D2EHPA 50% is made by taking 100 mL of pure D2EHPA then diluted with kerosene to 200 mL. For 10 mL D2EHPA 5% was prepared by diluting 1 mL of D2EHPA 50% and then it is added with kerosene up to 10 mL. This step is repeated for the preparation of 10%, 20%, 30%, and 40% D2EHPA solution.

Extraction

a) The feed prepared on point 1 (water phase) was extracted using 10% D2EHPA in kerosene (v/-v) (organic phase). The ratio of the volume of the aqueous phase (FA) to the organic phase (FO) is 1: 1. The mixture is stirred at 350 rpm for 30 minutes and then is allowed to stand for 10 minutes until between the water phase and the organic phase separately. The two phases are separated by a separation funnel. Then the filtrate results is analyzed using XRF to know the concentration of Gd.

b) Process a) is performed for TOPO 5%, 10%, 15%, 20%, 30%, D2EHPA 5%, 10%, 15%, 20%, 30%, 40%, mix D2EHPA- TOPO 3: 1, 1 : 3, 2: 1, 1: 2, 1: 1 and the composition between the solvent and the feed with a ratio of 1: 1, 1: 2, 1: 3, 1: 4, 2: 1 with a mixture of D2EHPA - TOPO.

3. Result and Discussion

3.1 Variation of Solvent

Research conducted on the separation of Gd in the concentrate REE senotim by extraction process, has the dominant element content that is Y, Gd and Dy with the content 80.89%; 15.49%, and 3.62%. Solvents used in the extraction process consist of solvent solvents and mixed solvents. Single solven ie TOPO or D2EHPA and mixed solvent is a mixture of TOPO and D2EHPA. Solvent is diluted using kerosene to reduce the density of D2EHPA or TOPO so that separation can be done due to the formation of 2 perfect phases.

3.1.1 TOPO

From Table 1 can be seen that the greater the concentration of TOPO solves, the greater the distribution coefficient of each element due to the solvents used more and more so that the complex formed to bind each of the larger REE. However, to determine the optimum condition of TOPO on gadolinium it is necessary to note the separation factor against other elements, so that 15% concentration has the best disadvantage for gadolinium by considering the separation factor which shows the greatest value to Gd-Y is 0.538 and to Gd-Dy 0.548.

Table 1. Kd and β Variation of Solvent TOPO

	Kd			β	
TOPO	Y	Gd	Dy	Gd-Y	Gd-Dy
5%	0.180	0.197	0.461	1.096	0.427
10%	0.297	0.139	0.577	0.467	0.240
15%	0.554	0.298	0.544	0.538	0.548
30%	0.881	0.269	1.359	0.305	0.198
50%	1.695	0.695	3.475	0.410	0.200

3.1.2 D2EHPA

From Table 2 it can be seen that the greater the concentration of the D2EHPA solvent the greater the distribution coefficient of each element due to the solvent being used the more so that the complex formed to bind each of the REE is greater, especially in the Ittrium indicating the compatibility of D2EHPA solvent to Ittrium. However, to determine the optimum state of D2EHPA against gadolinium it is necessary to note the separation factor against other elements, so that with 10% concentration has the best disadvantage for gadolinium by considering the distribution coefficient and separation factor. This optimum condition was chosen from the value of distribution coefficient to obtain more gadolinium quantity by considering good separation factor (β) to Gd-Y that is 0.281 and to Gd-Dy that is 0.061.

Table 2. Kd and β Variation of Solvent D2EHPA

	Kd			β	
D2EHPA	Y	Gd	Dy	Gd-Y	Gd-Dy
5%	0.477	0.276	10.034	0.572	0.027
10%	1.191	0.334	5.476	0.281	0.061
20%	24.577	0.633	11.019	0.026	0.057
30%	128.633	0.606	10.472	0.005	0.058
40%	843.967	0.567	243.564	0.001	0.002

3.1.3 Solvent TOPO – D2EHPA

The optimum concentration of TOPO - D2EHPA mixture for gadolinium is at 2: 1 with the highest distribution coefficient (Kd) value is 0.636 and the separation factor of Gd to Y and Dy is greater than other concentrations respectively is 1.014 and 1.181 (Table 3).

Table 3. Kd and β Variation of Solvent TOPO-D2EHPA

TOPO :	Kd			β	
D2EHPA	Y	Gd	Dy	Gd-Y	Gd-Dy
1;3	0.989	0.435	13.870	0.440	0.031
3;1	0.885	0.505	1.209	0.570	0.417
2;1	0.627	0.636	0.538	1.014	1.181
1;2	0.995	0.253	3.665	0.254	0.069
1;1	0.998	0.244	0.241	0.244	1.012

The value of Kd elements Y and Dy on D2EHPA is greater than Gd due to sole D2EHPA solvent more selective against elements Y and Dy shown by Table 4 so that Y and Dy elements are extracted more stable in solvent solvent extraction process D2EHPA, this is because what happens is ion exchange reactions, so that elements with greater electronegativity will be easier to bind to D2EHPA solvents whereas in the TOPO solvent is a Solvation reaction. Separation using a mixed solvent yields a larger Kd than a single solvent for Gd so that the 2: 1 mix ratio is optimized for the separation of gadolinium from the senotime REE concentrate. This optimum state is influenced by the synergistic effect of the TOPO and D2EHPA mixtures, in accordance with the theory that solvent solvent and ion exchange mixtures will produce a synergistic effect that will strengthen the bonding complex in the organic phase. The synergistic effect on Gd metal extraction is due to the role of TOPO in distributing the metal, the presence of this TOPO compound can increase the complex of metal formed between metals by D2EHPA. The synergistic effect of the TOPO and D2EHPA solvent mixtures is 1.01 indicating that there is a synergistic effect between the two solvents.

Table 4. REE Separation with variation solvent

	Kd			B	
Type of Solvent	Y	Gd	Dy	Gd-Y	Gd-Dy
Solvent D2EHPA	1.191	0.334	5.476	0.281	0.061
Solvent TOPO	0.554	0.298	0.544	0.538	0.548
Mixing Solvent TOPO-D2EHPA	0.627	0.636	0.538	1.014	1.181

3.2 Composition Feed and Solvent in Extraction process in Separation of REE

Based on Table 5 it can be seen that the greater the ratio of the solvent to the feed, the greater the REE is drawn, but it affects the distribution coefficients of each element (KdY, KdGd, KdDy) will vary depending on the number of mixed solvents. For gadolinium, the solvent mixture concentration: the optimum feed is 1: 4 because gadolinium is more competitive with TOPO-D2EHPA solvents in a 2: 1 ratio so that gadolinium is relatively stable compared to Y and Dy with 0.609 distribution coefficient and Gd to Y and Dy are 3.452 and 19.239 respectively.

Table 5. Kd and β Variation of Variasi Solvent : Feed

	Kd			β	
Solvent : Feed	Y	Gd	Dy	Gd-Y	Gd-Dy
2;1	2.376	0.845	138.72	0.356	0.007
1;1	0.610	0.636	0.551	1.043	1.815
1;2	0.384	0.611	0.257	1.590	3.893
1;3	0.286	0.576	0.187	2.011	5.337
1;4	0.176	0.609	0.052	3.452	19.239

3.3 Rate Process

3.3.1 Determination of Time Equilibrium

The extraction process occurs because of the separation of a particular component from a fluid to the organic phase.

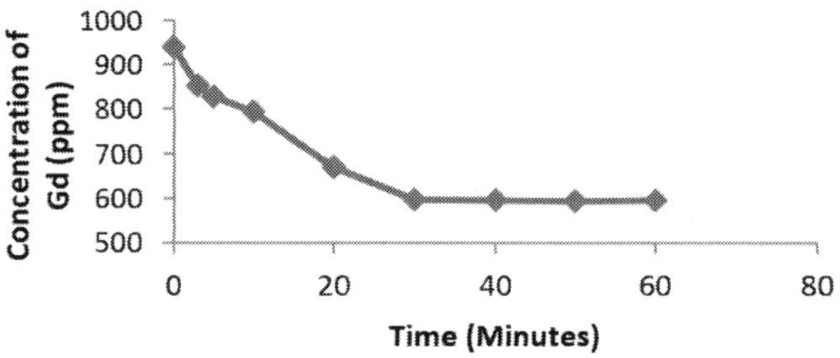

Figure 1. Time Equiblirium (minutes) foor Gd Extraction

Figure 1 shows that there is a decrease in the concentration of Gd in the liquid phase and the equilibrium time has been reached in the 30 minute meaning that the concentration reached in the time range is relatively constant, since the ability to bind the mixed solvent to Gd is maximum in other words the maximum solvent extract capacity has been reached. In addition, the decrease of Gd concentration in the extraction process proves that the solvent can be used to extract the Gd.

IOP Conf. Series: Materials Science and Engineering **349** (2018) 012042 doi:10.1088/1757-899X/349/1/012042

3.3.2 Reaction rate order and Constant Rate Reaction of Separation REE

The result of determination of extraction rate constant based on some models can be seen From Table 6. From Table 6 shows that the kinetics of extraction of Gd more in proportional with the model described first order reaction kinetics for r^2 close to 1 and error by a first-order reaction kinetics smaller than the order of 2.

Effect of temperature on extraction kinetics and the rate constant value (k) on the extraction kinetics model order 1 can be seen From Table 7. From Table 7 shows that the higher the temperature the separation factor will be smaller, this is not in accordance with the theory. This situation is due to a significant volume change in the process of taking the solution at any time on the temperature and time variations, so that the solvent becomes damaged due to rising temperature, causing the separation becomes less optimal.

Table 6. Kinetic Extraction Model of REE

Kinetic Model/Temperature	Gd			
	30°C	40 °C	50 °C	60 °C
Orde 1				
r^2	0.983	0.978	0.979	0.991
error (%)	4.51	4.07	4.24	0.9
Orde 2				
r^2	0.990	0.976	0.988	0.995
error (%)	7.037	6.526	7.402	6.325

Table 7. Effect of Temperature on Kinetic Extraction of Gd Order 1

Temperature (°C)	Gd Consentration (ppm)
30	340.24
40	321.20
50	228.23
60	135.61

From Table 8 shows that the higher the temperature of the reaction rate constants value will be smaller because of changes in the volume, which means the process of determining the kinetics causes the solvent resistance is reduced due to increased temperatures and increased impurities D2EHPA decomposed into M2HPA, causing the separation selectivity also reduced.

Table 8. Rate Constant First Order Reaction of Gd

Temperature (°C)	k (1/minutes) Gd
30	0.014
40	0.012
50	0.008
60	0.005

3.3.3 Arrhenius Equation

The equilibrium constant will decrease with decreasing temperature as shown in Figure 2. This does not correspond to the kinetic theory due to damage of the solvents as described previously. The Arrhenius equation is performed to find out the energy dependence of activation (E) and the collision factor (A) on temperature and the results obtained as in Figure 2.

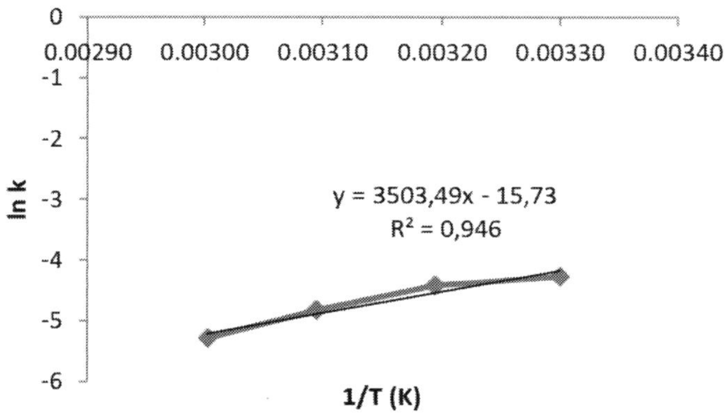

Figure 2. Curve ln k Vs 1/T

In Figure 2 it can be seen that the activation energy possessed by Gd is -6.96 kcal / mol and the collision factor owned by Gd is 1.46 x 10^{-7} so that the Arrhenius Gd equation becomes k = 1.46 x 10^{-7} exp (-6.96 kcal / mol / RT). Activation energy is negative due to the value of k and T is not in accordance with the theory. This event is experienced because of the significant volume changes in every solution taking during the kinetic determination process, so that the solvent capacity is reduced and the decomposition of D2EHPA into M2EHPA causes the separation selectivity also to decrease.

4 Conclusion

Based on this research, it can be concluded that the mixed solvent composition with the best feed to separate the rare earth metal Gd in the extraction process is 1: 4 with TOPO concentration of 15% and D2EHPA by 10%. The process of separation of the rare earth metals Gd with a solvent extraction method using a mixture of D2EHPA - TOPO better than a single solvent D2EHPA or TOPO with a mixture of TOPO-D2EHPA ratio is 2: 1.Orde reaction rate approaching Gd separation order 1 and Gd Arrhenius equation becomes k = 1.46 x 10^{-7} exp (-6.96 kcal / mol / RT).

References

1. Biyantoro, D., Tri, H., Moch. Setyadji., 2016, *Ekstraksi-Stripping Y, Dy, Gd, Ce, La, Nd dari Hasil Olah Pasir Senotim*, Pusat Sains dan Teknologi Akselerator –BATAN, Prosiding Seminar Nasional Teknologi Energi Nuklir 2016, Yogyakarta.
2. Jerabek K, Hankova L, Strikovsky AG, Warshwasky A. 1996. *Solvent impregnate*resins: relation between impregnation process and polymer support morphology I. Di-(2-ethylhexyl)ditiophosphoric acid.React Func Polym. 28:201-207.
   Aziz, N. 2014. *Pemisahan Zr – Hf dengan Metode SIR.* Yogyakarta. STTN - BATAN
3. Biyantoro, D dan Muhadi AW. 2006. Pemisahan Ltj (Y, La, Ce, Nd) Dari Hasil Olah Pasir Xenotim Dengan Cara Ekstraksi Cair-Cair Memakai Organofosfor. Yogyakarta. PTAPB BATAN.
   Moris,B., 2006, *Rare Earths*. PIRSA Minerals-Mineral Resource Potential Rare Earth Elements
4. Basuki, Kris Tri dan Dwi Biyantoro. "Kinetika Reaksi Pemisahan Zr – Hf pada Ekstraksi Cair – Cair dalam Media Asam Nitrat." *J. Tek. Bhn. Nukl. Vol. 7 No. 1* , Januari 2011: 1-73.
5. Biyantoro, D. 2016. *Kajian pemisahan mineral senotim.* Bogor. BATAN.
6. Biyantoro, D dan Muhadi AW. 2006. Ekstraksi Itrium Dari Konsentrat Senotim Menggunakan D2EHPA-Dodekan. Yogyakarta. PTAPB BATAN.

IOP Conf. Series: Materials Science and Engineering **349** (2018) 012042 doi:10.1088/1757-899X/349/1/012042

7. Khaldun, I. Buchari, Amran, M. B. dan Sulaeman, A. 2009. Pengaruh Komposisi Asam Bis(2-Etilheksil) Fosfat (D2EHPA) dan Tributil Fosfat (TBP) dalam Resin Amberlite Xad-16 terhadap Sorpsion-Ion La(III), Nd(III) dan Gd(III). Bandung. ITB.
8. Hartati, P. Indrati, T. Sunarjo dan Sulistyo. B. 2009. *Ekstraksi pemisahan zirkonium dan hafnium dalam suasana asam khlorida.* Yogyakarta. PTAPB BATAN.
9. Rachmi, W. 2014. Pemisahan Logam Tanah Jarang dengan Proses Ekstraksi. Yogyakarta. PTAPB BATAN.
10. Ali, M. M. "Synergistic extraction of uranium from acidic sulfate leach liquor using D2EHPA mixed with TOPO." *J Radioanal Nucl Chem (2014) 300*, 16 Maret 2014: 963–967.
11. Alamdari, D. Drvishi, D. Sadrnezhaad, S. Shabestari, Z. O'hadizadeh, A. Akbari, M. 2002. Effect of TBP as a Modifier for Extraction of Zinc and Cadmium with a Mixture of D2EHPA and MEHPA, Int. Conf. Solvent Extrancion Conference. S. Afr. Inst. Min and Metal, Johannesburg.
12. Soepriyanto, S. Dan Buchari. 2010. Ekstraksi Logam Tanah Jarang di Indonesia: *Permasalahan dan Prospek Pemanfaatannya.* Seminar nasional Pertambangan dan Metalurgi, ITB, 2010.
13. Sulistyani, R. 2016. Pemisahan Y, Gd, Dy Hasil Ekstraksi dari Konsentrat Itrium Menggunakan Kolom Penukar Ion, PSTA. BATAN

Application of Sodium Ligno Sulphonate as Surfactant in Enhanced Oil Recovery and Its Feasibility Test for TPN 008 Oil

N I Prakoso[1], Rochmadi[2] and S Purwono[2]

[1]Chemistry Department, Universitas Islam Indonesia, Yogyakarta-55584, Indonesia
[2]Chemical Engineering Department, Universitas Gadjah Mada, Yogyakarta-55281, Indonesia

E-mail: nurcahyo.ip@uii.ac.id

Abstract. One of enhanced oil recovery (EOR) methods is using surfactants to reduce the interfacial tension between the injected fluid and the oil in old reservoir. The most important principle in enhanced oil recovery process is the dynamic interaction of surfactants with crude oil. Sodium ligno sulphonate (SLS) is a commercial surfactant and already synthesized from palm solid waste by another researcher. This work aimed to apply SLS as a surfactant for EOR especially in TPN 008 oil from Pertamina Indonesia. In its application as an EOR's surfactant, SLS shall be passed feasibility test like IFT, thermal stability, compatibility, filtration, molecular weight, density, viscosity and pH tests. The feasibility test is very important for a preliminary test prior to another advanced test. The results demonstrated that 1% SLS solution in formation water (TPN 008) had 0.254 mN/M IFT value and was also great in thermal stability, compatibility, filtration, molecular weight, viscosity and pH test.

1 Introduction

The development of chemicals for the enhanced oil recovery (EOR) in Indonesia has been done by many researchers [1]. A surfactant can be used to reduce the interfacial tension between the injected fluid and the oil in the reservoir. However, Indonesian national oil industries are still using surfactants which are imported from other countries that in fact, the price is very high. Therefore, cheap and well-performing surfactants are urgently needed[1].

The dynamic interaction between the surfactants and the oil in the reservoir is the basic principle in EOR process [2]. The surfactant should interact very well with the oil, and it can be seen from the interfacial tension values. A surfactant with low IFT value and low adsorption on the reservoir rock material is a good candidate before continuing to core flooding process [3]. Interfacial tension analysis is very important as the initial analysis prior to preliminary tests such as core flooding and another advanced test.

Sodium ligno sulphonate is a well known commercial surfactant that can be synthesized from lignin [4] which is isolated from the waste of crude palm oil industry [5][6][7][8]. This surfactant is very potential if applied in the Indonesian oil field considering that Indonesia have big palm oil industry. However, sodium ligno sulphonate has not been tested for the feasibility to Indonesian crude oil. This study is very important for tested the feasibility as an EOR agent through IFT, compatibility, phase behavior, thermal stability, filtration, molecular weight, viscosity and pH test. Surfactants with low prices and sustainable raw material can be obtained. This study can be a novelty in the field of surfactants in Indonesia, especially for TPN 008 oil.

2 Experimental

All chemicals used were of analytical grade from Merck and Co. Inc (Ethanol, Methanol, $NaHSO_3$, NaOH, and H_2SO_4). Chemicals with technical grade are oil palm empty fruit bunches from PT Asiatic Persada Sungai Bahar, Jambi, aquadest from CV Progomulyo, dead oil TPN-008 and formation water sample with code TPN-008 from Pertamina Indonesia. The instruments used in this study included infrared spectrophotometer (FT-IR, Shimadzu Prestige-21), IFT Meter TX500D Spinning Drop Tensiometer, and Viscosity kinematic P 906 size 150.

Lignin isolation from oil palm empty fruit bunches isolated by using batch method with reaction temperature, system pressure, concentration of NaOH, the ratio of fibers to solvents and reaction time of 170 °C, 15 atm, 1% (w/v), 9% (w/v), respectively, and the reaction was carried out for 5 hours. The black liquor was acidified until the pH reached 1, and the formed solid lignin was filtered and dried [5].

Synthesis of SLS was carried out by refluxing lignin to $NaHSO_3$ solution for 4 hours. The reaction was conducted under reflux conditions at 97 °C and the mass ratio of lignin to the $NaHSO_3$ solution, concentration of $NaHSO_3$ solution, reaction temperature, and reaction time were 0.3 M, 0.1 M 97 °C, respectively [4]. The 1% of SLS in formation water TPN 008 tested for feasibility as an EOR agent through IFT (temperature of test of 60 °C and root speed of 6000 r/min), compatibility (temperature of test of 60 °C), phase behavior (temperature of test of 60 °C), thermal stability test (temperature of test of 60 °C), filtration rate (0.45 μm), molecular weight (viscometer), viscosity (ASTM D 445-07) and pH test.

3 Results and Discussion

3.1 FT-IR for determination of lignin and SLS structure

To determine the key functional groups from lignin, the FTIR spectrophotometer was used. The spectra are presented in Figure 1. The existence of OH-stretch (3425 cm^{-1}), CH-stretch (2924 and 2854 cm^{-1}), C=C aromatics (1512 and 1604 cm^{-1}), asymmetry C-H stretch (1458 cm^{-1}), syringyl rings (1327 cm^{-1}), guaiacyl rings (1219 cm^{-1}), C-O either stretch (1033 cm^{-1}), and C-H aromatics (840 cm^{-1}) groups were a sign that the isolation of lignin was successful.

To determine the key functional groups of SLS, determination was conducted by using FTIR spectrophotometer (Figure 2). The existence of OH-stretch (3433 cm^{-1}), CH-stretch (2916 and 2846 cm^{-1}), C=C aromatics (1512 and 1604 cm^{-1}), $-SO_3$ (1465 cm^{-1}), S=O asymmetric stretch (1118 cm^{-1}), S-O stretch (1) (1026 cm^{-1}), and S-O stretch (2) (825 cm^{-1}) groups were a sign that the synthesis of SLS was successful.

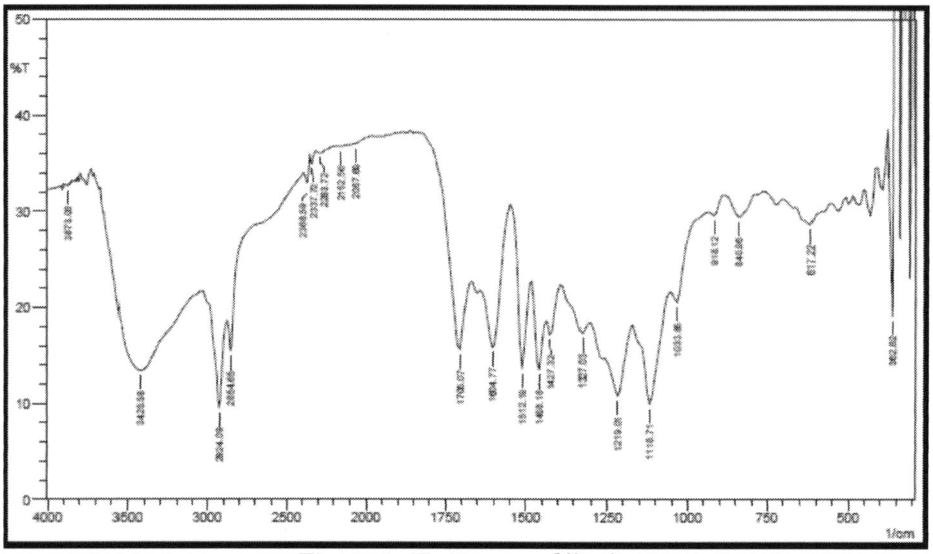

Figure 1. IR Spectra of lignin

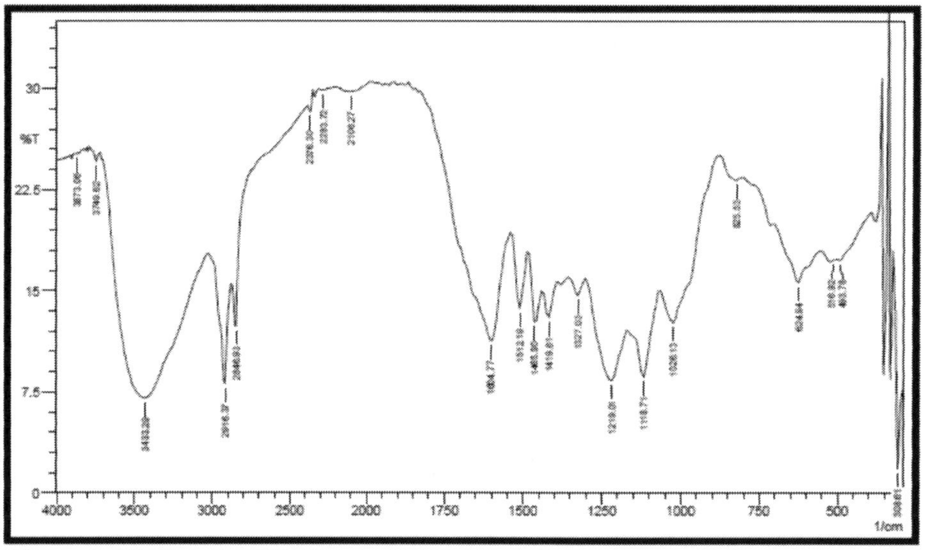

Figure 2. IR Spectra of SLS

3.2 Analysis of interfacial tension (IFT), phase behaviour, compatibility and thermal stability test

Water and oil are immiscible. However, in the presence of surfactants, they are miscible as the oil can be dissolved in the surfactant micelle nuclei and the surfactant is soluble in water. Combination between oil-water-surfactant mixture can form a lower-phase (water-soluble) emulsion, middle-phase emulsion (called microemulsion, soluble in oil and water), and upper phase emulsion (soluble in oil).

These tests are needed to see the performance of the surfactant as an EOR agent. All test have been done using dead oil and formation water. The results are shown in Table 1. Sodium lignosulphonate will be confirmed to be good as surfactant for an EOR agent when IFT value was obtained on a scale of 10^{-3}, no precipitate in the compatibility test for 1 to 24 hours, and formed stable microemulsion in the phase behavior test for 24 hours. From IFT, compatibility and phase behavior analysis showed that SLS well performed as an EOR agent. It can be seen by the compatibility phase behavior and thermal stability tests from SLS. There are no precipitate formed in the solution, formed a stable microemulsion and stable in reservoir temperature as surfactant (Figure 3). Nevertheless, IFT value from sodium ligno sulphonate was very high (0.254 mN/M).

Table 1. IFT value of SLS between formation water and dead oil TPN 008

Test	Results	Explanation
IFT (interfacial tension)	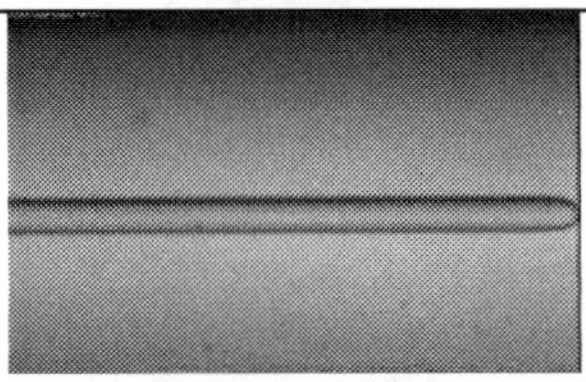 0.254 mN/M	not achieved
Compatibility	 1 hour	no precipitate
	 24 hours	no precipitate
Phase behavior	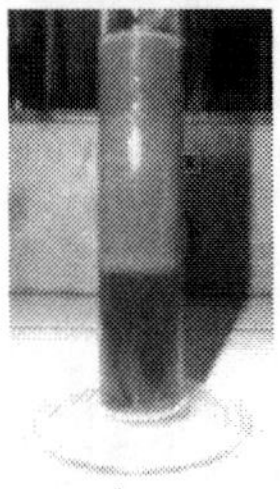 1 hour	Middle layer have formed like a white froth wherein thick and clear. The oil and surfactant are still visibly emulsified after shaking process.
	 24 hours	The microemulsion (middle phase) was formed with 0.3 mL thick (10 mL measuring glass scale). Oil layer and water layer are connected by the third phase layer in the middle.
Thermal stability	0.254-0.0809 mN/M	very stable for 3 months

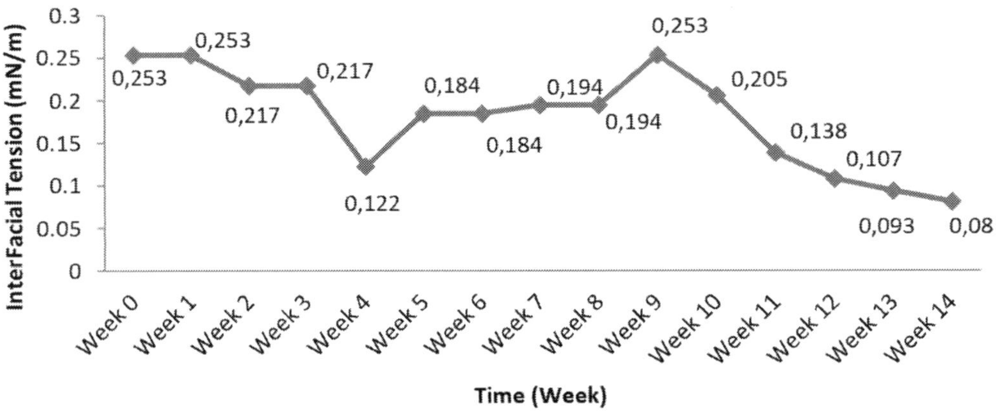

Figure 3. Thermal stability test of 1% SLS solution at reservoir temperature (60 °C) for 14 weeks.

3.3 Analysis of filtration ratio of 1% SLS in 0.45μm membrane

The filtration ratio test was performed to identify the presence of sediment in the surfactant solution. When the surfactant injection process in the reservoir, it is expected that the injected surfactant would not produce precipitates. If the injected surfactant produces precipitates, blocking of the reservoir inlet could occur and inhibit the production process of the oil. Filtration was done using a filtration tool which is designed as Figure 4.

Figure 4. Filtration rate tes equipment.

In this filtration process, the flow rate of the surfactant was observed. The filtration test was carried out by passing a fluid (a surfactant liquid) through a 0.45 μm filter (filter membrane) at a constant pressure and calculating the time required by the fluid to reach a given volume. The rate was determined by recording the time per 50 mL of the solution accommodated in the sample container. From this analysis, the streamed sample contained no impurities, the flow rate would tend to be constant so that the filtration ratio (the ratio of filtration time per given volume) would be equal to or close to 1. The results of filtration test on 1% surfactant was given in Table 2 and Figure 5.

Table 2. Filtration ratio of 1% SLS solution to 0.45 μm membrane.

Volume (mL)	1st test time (s)	2nd test time (s)	3rd test time (s)
50	142.91	37.56	85.13
100	326.44	77.06	179.56
150	523.2	116.71	273.13
200	721.26	155.95	366.71
250	922.75	195	460.71
300	1123.14	234.24	554.33
350	1325.25	273.91	649.53
400	1525.5	313.23	744.9
450	1727.55	354.18	839.54
500	1928.6	394.73	933.9
Filtration Ratio	1.10	1.03	1.00

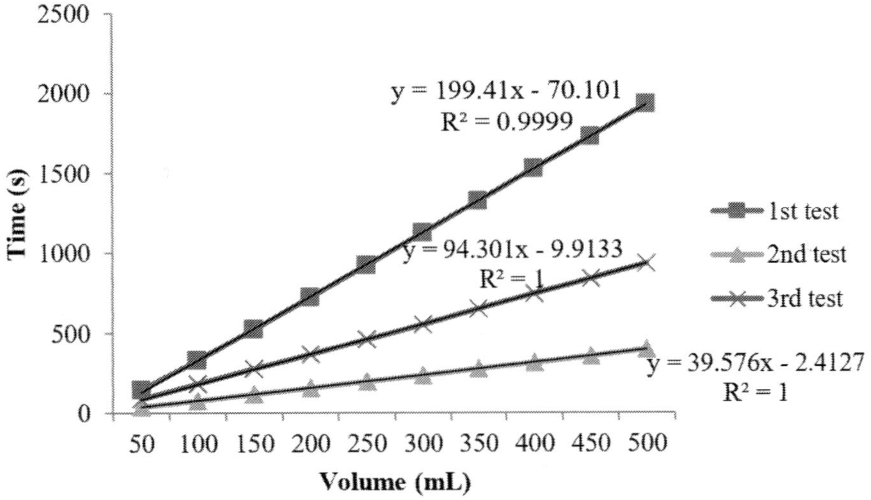

Figure 5. Graphic of filtration ratio of 1% SLS solution

3.4 Analysis of molecular weight of SLS

Molecular weight measurements of the SLS are according the Mark-Houwink-Sakurada equationwhere pure solvent of 0.1 M NaCl was used. Then, the SLS solution was applied to the viscometer column until t_0, t_1, t_2, t_3, dan t_4 were obtained. Through calculations, relative viscosity (η_r) and specific viscosity(η_{sp}) were obtained. Then the graph was η_{sp}/C against C was made and by using Huggin's equation. $[\eta]$ value can be obtained as an intercept of the graph. The average molecular weight of the polymer can be obtained by following the Mark-Houwink-Sakurada equation: $[\eta] = K\,(M_v)^{\alpha}$ [9]. The result of SLS molecular weight was 21.351 $g.mol^{-1}$ and the calculation are given in Table 3 and Figure 6.

Table 3. Viscosity data of the SLS solution

C (g/mL)	time (s) I	II	III	Average	ηr	ηsp	ηsp/C
0.0000	67.80	67.80	67.80	67.80			
0.0002	68.75	68.76	68.85	68.79	1.01	0.01	72.76
0.0004	69.30	69.33	69.35	69.33	1.02	0.02	56.29
0.0006	69.02	69.04	69.05	69.04	1.02	0.02	30.40
0.0008	68.44	68.42	68.43	68.43	1.01	0.01	11.62
0.0010	68.61	68.63	68.63	68.62	1.01	0.01	12.14

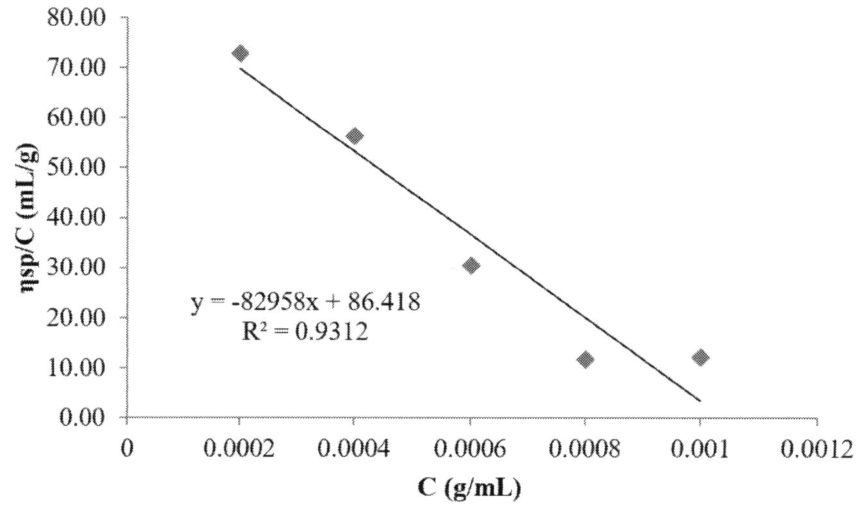

Figure 6. Graphic η_{sp}/C versus C

Refering to the data presented in Figure 4, calculation of molecular weight of SLS could be calculated by using the following equations .

The relative viscosity (η_r) is expressed as $(\eta_r) = \frac{\eta}{\eta_0} = \frac{t}{t_0}$ (1)

The specific viscosity (η_{sp}) is expressed as $\eta_{sp} = \eta_r - 1$ (2)

The reduced viscosity (η_{red}) is expressed as $\eta_{red} = \eta_{sp}/C$ (3)

Where C is the concentration of surfactant (g/mL)

Mark-Houwink-Sakurada equation is: $[\eta] = K\,(M_v)^{\alpha}$.

K value and α for sodium lignosulfonate in 0.1 M NaCl solution is 0.12 mL/g and 0,66. Therefore the (M_v) value is 21.351 $g.mol^{-1}$

3.5 Analysis of pH and viscosity of 1% SLS solution

This analysis was conducted to determine the pH and viscosity values of the surfactant (SLS). In common, surfactants should not be alkaline or have a pH above 10 and preferably not too concentrated because it will clog the pores of the rocks. If it happens, the oil cannot be mined. The results of pH and viscosity tests are given in Table 4 and Table 5.

Table 4. pH value of surfactant and formation water TPN 008

Sample	pH
Formation water TPN 008	8,3
1% SLS solution in formation water TPN 008	9,5

Table 5. Kinematic viscosity of 1% SLS solution in formation water TPN 008.

Temperature (°C)	Kinematic viscosity (mm^2/s)
30	0,844
40	0,705
50	0,673
60	0,545

From this analysis, it can be concluded that 1% SLS solution in formation water TPN 008 could be applied at an oil reservoir. It can be seen by the pH value of 1% SLS solution in formation water TPN 008 very closed with pH value of formation water TPN 008 and still below 10. For kinematic viscosity at reservoir temperature of 60 °C also gave a small value (below 1 mm^2/s).

4 Conclusion

In conclusion, we have reported the results of EOR agent feasibility test demonstrated that 1% surfactant solution has 0,254 mN/M IFT value and also great in thermal stability, compatibility, filtration, molecular weight, viscosity and pH test. From this research, it is found that SLS cannot be applied as EOR agent although it fulfill some requirements. It is necessary to modify the SLS either by adding some additive or altering some functional groups to produce a suitable surfactant as EOR agent.

References

[1] Suryo P, Murachman B. Development of non-petroleum base chemicals for improving oil recovery in Indonesia. InSPE Asia Pacific Oil and Gas Conference and Exhibition 2001 Jan 1. Society of Petroleum Engineers.
[2] Porzer M, Bujok P, Klempa M, Pánek P. Evaluation Of Surfactant Performance For Enhanced Oil Recovery/Posouzení Vhodnosti Aplikace Povrchově Aktivních Látek (Pal) Pro Potřeby Zvýšení Vytěžitelnosti Ložisek Uhlovodíků. GeoScience Engineering. 2013 Dec 1;59(4):32-8.
[3] Iglauer S, Wu Y, Shuler P, Tang Y, Goddard WA. New surfactant classes for enhanced oil recovery and their tertiary oil recovery potential. Journal of Petroleum Science and Engineering. 2010 Mar 31;71(1):23-9.
[4] Prakoso NI, Purwono S, Rochmadi. Synthesis of sodium lignosulphonate from oil palm empty fruit bunches's lignin. InAIP Conference Proceedings 2017 Mar 17 (Vol. 1823, No. 1, p. 020037). AIP Publishing.

[5] Prakoso NI, Purwono S, Rochmadi R. Study on Lignin Isolation from Oil Palm Empty Fruit Bunches. Jurnal Eksakta. 2016 May 22;16(1):46-54.
[6] Harmaja Simatupang AN, Herlina N. Studi isolasi dan rendemen lignin dari tandan kosong kelapa sawit (TKKS). Jurnal Teknik Kimia USU. 2012 Sep 20;1(1).
[7] Hassan NS, Badri KH. Lignin recovery from alkaline hydrolysis and glycerolysis of oil palm fiber. In Razak FA, Maideen HM, Sidek HM, Noorani MS, Ibrahim N, Ibrahim NB, Simon N, Maseran N, Karim NH, editors.AIP Conference Proceedings 2014 Sep 3 (Vol. 1614, No. 1, pp. 433-438). AIP.
[8] Ma'ruf A, Pramudono B, Aryanti N. Lignin isolation process from rice husk by alkaline hydrogen peroxide: Lignin and silica extracted. InAIP Conference Proceedings 2017 Mar 17 (Vol. 1823, No. 1, p. 020013). AIP Publishing.
[9] Paul, Hiemenz C., and Lodge P. Timothy. *Polymer Chemistry*. Second ed. Boca Raton: CRC P, 2007. 336, 338–339.

IOP Conf. Series: Materials Science and Engineering **349** (2018) 012044 doi:10.1088/1757-899X/349/1/012044

Optimization Recovery of Yttrium Oxide in Precipitation, Extraction, and Stripping Process

N I Perwira[1], K T Basuki[1], D Biyantoro[2], N Effendy[1]

[1] *Polytechnic Institute of Nuclear Technology, BATAN Yogyakarta, Indonesia*

[2] *Centre for Science and Accelerator Technology - National Nuclear Energy Agency of Indonesia (PSTA – BATAN), Yogyakarta, Indonesia*

Abstract. Yttrium oxide can be used as a dopant control rod of nuclear reactors in YSH material and superconductors. Yttrium oxide is obtained in the Xenotime mineral derived from byproduct of tin mining PT Timah Bangka which contain rare earth elements (REE) dominant Y, Dy, and Gd whose content respectively about 29.53%, 7.76%, and 2.58%. Both usage in the field of nuclear and non-nuclear science and technology is need to pure from the impurities. The presence of impurities in the yttrium oxide may affect the characteristic of the material and the efficiency of its use. Thus it needs to be separated by precipitation and extraction-stripping and calcination in the making of the oxide. However, to obtain higher levels of Yttrium oxide, it is necessary to determine the optimum conditions for its separation. The purpose of this research was to determine the optimum pH of precipitation, determine acid media and concentration optimum in extraction and stripping process and determine the efficiency of the separation of Y from REE concentrate. This research was conducted with pH variation in the precipitation process that pHs were 4 - 8, the difference of acid media for the extraction process, i.e., HNO_3, HCl and H_2SO_4 with each concentration of 0,5 M; 1 M; 1,5 M; and 2 M and for stripping process were HNO_3, HCl, and H_2SO_4 with each concentration of 1 M; 2M; and 3 M. Based on the result, the optimum pH of precipitation process was 6,5, the optimumacid media was HNO_3 0,5 M, and for stripping process media was HNO_3 3 M. The efficiency of precipitation process at pH 6,5 was 69,53 %, extraction process was 96.39% and stripping process was 4,50%. The separation process from precipitation to extraction had increased the purity and the highest efficiency recovery of Y was in the extraction process and obtained Y_2O_3 purer compared to the feed with the Y_2O_3 content of 92.87%.

1 Introduction

Xenotime is a phosphate rare earth elements compound (Y, Dy, Gd) PO_4 obtained from byproducts of tin mining in PT Timah Bangka island which has not been widely developed for its separation [1]. The rare earth metal elements (REE) contained in xenotime were Gadolinium (Gd), Dysprosium (Dy), Yttrium (Y), Samarium (Sm), Lanthanum (La), Cerium (Ce), and Neodymium (Nd) (1,2,3). Total of rare earth elements ranged from 55% to 70%. The Rare Earth Elements Oxides (REE_2O_3) with dominant elements Y, Dy, and Gd contained in Xenotime were 29.53%, 7.76%, and 2.58%.[1].

REE has been widely used especially in the field of advanced technology. In nuclear science and technology, REE such as Y in the form of an oxide (Y_2O_3) was used as a control rod dopant material at a nuclear reactor made from Yttria-stabilized hafnia[4]. The permitted impurities on Y_2O_3 with maximum

Dy and Gd level was 500 µg / g of powder[5]. In the non-nuclear field, Y_2O_3 could be used as a superconducting material that is highly potent in the manufacture of fast trains[6].

Both usage in the field of nuclear and non-nuclear science and technology, Y_2O_3 must be in the pure state of its impurity. The presence of impurities or additional compounds may modify the behavior and efficiency in its use a separation is necessary before the preparation of Y_2O_3. Yttrium that has been separated from the impurities, then calcined to obtain Y_2O_3. In this research, it was used the concentrate of xenotime in the form of REE hydroxide. The Y element contained in the concentrate still had a purity of 85.17%. However, these content could not be used for nuclear and non-nuclear science and technology application such as dopant material of Yttria-stabilized hafnia (YSH) and control rods for nuclear application and for non-nuclear application such as especially a superconducting material with purity requirement of $Y_2O_3 > 90\%$ thus it needed to be increased in purity. In this research, the study of REE purification including precipitation, extraction, and stripping procces.

2 Research Method

2.1 Reagent

The Reagents used in this research were rare earth element hydroxide concentrate [$REE(OH)_3$] which had composition: Y = 85.17%, La= 0.71 %, Ce = 5.81%, Gd = 2.33%, Dy = 5.98%, HNO_3 65% p.a, NH_4OH 25%, HCl 32% p.a, H_2SO_4 97% p.a, $H_2C_2O_4$ 15%, *Bis(2-etilheksil) Phosphate Acid* (D2EHPA) 97% and kerosene.

2.2 Tool

The tools used in this research were glass beaker as a reactor in the process of precipitation, extraction, and stripping, pH meter as an instrument for measurement pH of solution in settling process, and X-ray spectrophotometry as an instrument for quantitative and qualitative analysis,hot plate, volumetric flask, Volumetric pippete magnetic stirrer, Graduated cylinder, pipette, spatula, analytical balance, Erlenmeyer, bullet, oven, desiccator, and furnace.

2.3 Procedure

2.3.1 Optimization pH of Precipitation Process

An amount of 20 g $REE(OH)_3$ was dissolved in 250 mL HNO_3 0.5 M. Then it was settled by adding 15% NH_4OH solution to pH = 4. A Filtrate I and a precipitate I were separated. The precipitate was dried at 110°C and stored in the desiccator for 30 minutes. The precipitate and the filter paper were weighed until had a constant weight. Process a) was repeated for simultaneous precipitation, starting pH = 5 until Y did not re-form the precipitate. Each of the precipitation process results in different pH (filtrate (n) and precipitate (n) was separated. The filtrate (n) and precipitate (n) at various pH were analyzed using XRF. After analyzed, it would be known the optimum pHs to recovery concentrate $Y(OH)_3$.

2.3.2 Optimization Acidity of Feed Extraction

An amount og 1 g precipitate of $Y(OH)_3$ in the precipitation process was dissolved in the acidity media of HNO_3 0.5 M, 25 mL. Preparations for variation acidity on extraction process were performed by making of HNO_3, H_2SO_4 and HCl acidic media at concentrations of 0.5 M; 1 M; 1.5M; and 2M. Then, solution $Y(NO_3)_3$ was extracted using 30% D2EHPA in Kerosene (organic phase v/v) with a water phases on volume (FA) and an organic phase (FO) ratio was 1: 1. The mixture was stirred on a hot plate stirrer at 30 ± 1 ° C for 30 minutes then allowed for 15 minutes until it formed two phases that was the water phase and the organic phase and could be separated using a separation funnel. The aqueous phase were analyzed using XRF.

The processes were repeated for variations of the acidity media of HNO_3, H_2SO_4, and HCl with each concentration of 0.5 M; 1 M; 1.5M; and 2M.

2.3.3 Optimization Acidity of Stripping Process

An amouny of 5 g precipitate of $Y(OH)_3$ in the precipitation process was dissolved in the optimum acidity at 100 mL extraction process. Then, solution Y was extracted using 30% D2EHPA in kerosene (organic phase v/v) with a water phase (FA) and an organic phase (FO) volume was 1: 1. The mixture was stirred on a hotplate for 30 minutes then allowed for 15 minutes until the water phase and the organic phase was separate. Two phases which were formed separated by a separation funnel.

The extracted organic phase (FO) was stripped with 3M H_2SO_4 with the volume ratio of the water phase (FA) and the organic phase (FO) was 1: 1. The mixture was stirred using a stirrer hot plate at 30 °C ± 1°C at 4 °C for 30 minutes, then allowed until it formed two phases that were the water phase and the organic phase and could be separated using a separation funnel to obtain a stripping phase (FSwater) and an organic phase (FO). Process b) was repeated for variations of the acidity media of stripping HNO_3, H_2SO_4 and HCl with each concentration of 1 M; 2M; and 3M.

The stripping phase (FSwater) was added 15% $H_2C_2O_4$ (b/v) hence, forming precipitate Yttrium Oxalate ($Y_2(C_2O_4)_3$). FSwater in each variation was analyzed using XRF to find the type of acidity media and the optimum concentration. The precipitate $Y_2(C_2O_4)_3$ was filtered and dried at 110°C then stored in the desiccator.

3 Result and Discussion

3.1 Optimization of Precipitation pH

The occurrence of element precipitation could be predicted by calculating from the pKsp data of each element hence it could be seen that each element would settle to a certain pH.

Table 1. Precipitation of REE in Deposition Results of Variation pH

Type	Content (%)					Mass of precipitate (g)	Total Efficiency (%)
	Y	La	Ce	Gd	Dy		
Feed	85.17	0.71	5.81	2.33	5.98	20.3102	-
pH 4	0	0	0	0	0	0	0
pH 5	0	0	0	0	0	0	0
pH 6	37.69	0.75	55.63	0	5.93	0.1537	0.76
pH 6.5	89.57	0	0.03	1.09	9.31	14.1209	69.53
pH 6.8	89.77	0.42	1.59	5.31	2.91	2.4207	11.92
pH 7	97.35	2.65	0	0	0	0.0088	0.04
pH 7.5	70.59	0.27	25.12	4.02	0	1.2611	6.21
pH 8	0	0	0	0	0	0	0

Note: Feed Volume was 250 mL, Temperature was 30 ± 1 °C

Based on Table 1, it was known that the largest efficiency of precipitation was at pH 6.5 with a total efficiency was 69.53% and Y content was 89.57%.

The precipitation results in each pH 4 to 8 differed from the estimates in pKsp calculations. According to calculations, the precipitation for $Y(OH)_3$ would begin to settle at pH 8.14. However, this experiment which was performed indicated that the yttrium was started to settle at pH 6 and the optimum precipitation for $Y(OH)_3$ was pH 6.5. This may occur because the solution deposited had content another rare earth metal elements which will cause a decrease in the solubility of Y in the solution.

3.2 Optimization of acid aqueous media extraction

The precipitate of $Y(OH)_3$ at pH 6.5 was selected as a feed for the variation of acidity media in the extraction process. The selected of acidity media types were HNO_3, HCl and H_2SO_4 with concentrations of 0.5 M, 1M, 1.5M and 2M respectively. The water phase as the extraction media was selected from acid solution because D2EHPA can extract REE at low pH state [14].

Table 2. Efficiency and Purity of Extraction Process

Type of acidity media	Efficiency Y (%)	Purity of Y (%)
HNO_3 0.5 M	96.39	90.09
HNO_3 1 M	85.88	90.96
HNO_3 1.5 M	*73.19*	91.98
HNO_3 2 M	75.05	96.15
H_2SO_4 0.5 M	63.72	84.01
H_2SO_4 1 M	67.95	90.15
H_2SO_4 1.5 M	53.05	77.30
H_2SO_4 2 M	29.50	87.85
HCl 0.5 M	95.62	89.68
HCl 1 M	82.45	82.68
HCl 1.5 M	70.03	92.57
HCl 2 M	56.75	96.84

Note: temperature is 30°C ± 1, stirring time is 30 minute, D2EHPA: Kerosin 30%, FO: FA 1:1

Based on Table 2, the highest efficiency of recovery Y was obtained on the acidity medium of HNO_3 0,5 M and the lowest efficiency was on H_2SO_4 2M. The lower the acid concentration, it will get the higher the efficiency of recovery Y. This was occured according to the following reaction equation,

$$HNO_{3(aq)} + H_2A_{2(org)} \leftrightarrow HNO_3.H_2A_2 \quad (1)$$

$$Y^{3+}{}_{(aq)} + 3HNO.H_2A_2 \leftrightarrow Ln(H_2A_2)_{3org} + 3H^+ + 3NO_3^- \quad (2)$$

The first reaction was the exchange of cations between $H^+{}_{(aq)}$ water phase with $H^+{}_{(org)}$ organic phase, so that would equally donate H^+ which means no significant change in the reaction. Furthermore, the second reaction was the exchange of $H^+{}_{(org)}$ with Y^{3+} which would cause Y^{3+} to react complex with D2EHPA ligands and $3H^+$ donated into the water phase. In the case of H_2SO_4, it was the most less effective because of the first reaction. H_2SO_4 had $2H^+$ which would cause D2EHPA prefer exchange $H^+{}_{(org)}$ in H_2SO_4 to binding complex at Y^{3+}.

HCl had a lower efficiency compared to HNO_3 because HNO_3 had a lower acidity level than HCl. It was easier D2EHPA to react complex bonds with Y^{3+} at lower acidity than in high acidity. thus, the acidity media in the extraction of Y for the extraction process from the highest to the lowest which was according to the efficiency ie HNO_3> HCl> H_2SO_4.

3.3 Optimization of acid aqueous media stripping

Feed of stripping process used organic phase which was extracted from HNO_3 0.5 M. In the stripping experiment used variations of acidity type of HNO_3, H_2SO_4, and HCl with concentrations of 1 M, 2 M, and 3 M respectively. The stripping process results were obtained in Table 3.

Table 3. Efficiency and Purity of Stripping Process

Type of acidity media	Efficiency Y (%)	Purity of Y (%)
HNO_3 1 M	0.00	0.00
HNO_3 2 M	18.59	83.01
HNO_3 3 M	40.50	92.64
H_2SO_4 1 M	5.17	73.53
H_2SO_4 2 M	9.55	90.95
H_2SO_4 3 M	15.79	94.13
HCl 1 M	0.88	80.14
HCl 2 M	6.92	92.74
HCl 3 M	21.57	91.08

Note: temperature is 30°C ± 1, stirring time is 30 minute, FO: FA 1:1

Based on Table 3, the largest efficiency recovery of Y was obtained from the 3M HNO_3 stripping agent of 40.50% with a Y purity of 92.64%. The efficiency of stripping was still low. The higher acid concentration, the higher efficiency recovery of the Y in the stripping process. It was occurred because it wouldl be easier to break down the complex at high acid concentration. If the acid concentration was high, the efficiency of the Y taken would be high in the stripping process. It occured because process would be easier to break down the complex at high acid concentration. If D2EHPA was expected to form a complex in extraction process with metal by minimizing H^+_{org} reacts with H^+_{aq} hence H^+_{org} was more to bind metal complex, then on stripping process it was expected metal complex that has been bound would be split or decomplexation. At high concentrations, H^+ dissociates in water was easier than low concentrations, hence the potential of Y^{3+} in the organic would be bind H^+ higher in water phase than the low concentration. Thus in this study, if concentration was higher, so stripping efficiency also will be higher .

However, in the type of acid medium, HNO_3 was better than H_2SO_4 and HCl. This occurred because in the extraction process, it occured the next solvent reaction was NO_3^- also slightly into the organic phase. When there was the addition of an acid with a similar anion of HNO_3, presence NO_3^- was in the organic phase also becomes a ligand in the complex Y bonds, hence it was easier to release into the water phase by carrying the Y metal than the acid with the unlike anions.

In this research, H_2SO_4 and HCl have low efficiency for Y so that agent strriping for metal Y was from high to low HNO_3> HCl> H_2SO_4.

3.4 The Making of Yttrium Oxide

After the stripping process had performed, simultaneous precipitation with oxalic acid should be performed because all rare earth metals would settle by oxalic acid without treatment of pH variation. The results of simultaneous precipitation was shown in Table 4.

Table 4. Precipitation with Oxalate Acid

Sample	ppm	
	Y	Dy
Aqueous Stripping HNO_3 3M	5821	462
Filtrate	0	0

In Table 4, it was shown that Y and Dy had simultaneous precipitation which was shown no concentration in the filtrate. Then, the oxalate precipitate formed was dried a temperature of 110°C before calcination was performed. The temperature of calcination was performed at 1100°C for 2 hours.

Table 5. Result of Yttrium Oxide from Calcination Process at 1100°C for 2 hours

Sample	Content (%)	
	Y	Dy
Product 1 stripping HNO_3 3 M	92,87	7,13
Product 2 stripping H_2SO_4 3 M	94,19	5,81

Ket : Temperature at 1100°C, t = 2 hours

In Table 5, it could be seen that the result of calcination product 1 had 92.87% content of Y and product had 94.19 content of Y. In Figure 1, product 2 had a white oxide color compared to product 1. It was occured because Dy in product 1 was more than product 2 which caused a brownish-white color.

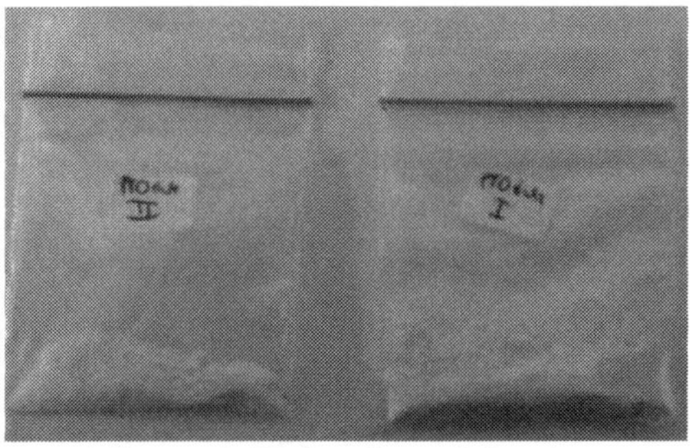

Figure 1. Y_2O_3 after calcination

3.5 Efficiency of Each Process

The separation of Y from the rare earth metal concentrate in the experiments was performed by the precipitation process at various pH, extraction-stripping, and calcination process to obtain Y_2O_3. Efficiency and purity of the concentrate from the precipitation process to extraction - stripping with the optimum conditions were shown in Table 6.

Based on Table 6, each process from precipitation to extraction had increased the purity and the highest efficiency recovery of Y was in the extraction process and it can be obtained that Y_2O_3 purer than the feed with the Y_2O_3 content of 92.87%.

Table 6. Efficiency and Purity of Each Process

Process	Efficiency Y (%)	Purity of Y (%)
Feed	-	85.17
Precipitation (pH 6.5)	69.53	89.57
Extraction (HNO_3 0.5 M)	96.39	90.09
Stripping (HNO_3 3 M)	40.50	92.64

4 Conclusion

Based on this research, it could be concluded that the result of optimum precipitation of yttrium was pH 6.5, extraction with the highest effect that was on acidity type HNO_3 0.5 M and stripping on acidity type HNO_3 3 M. Efficiency of precipitation process at pH 6.5 was 69.53%, extraction process was 96.39%, stripping process was 40.50% and yield Y_2O_3 with purity 92.87%.

Acknowledgements
The authors wish to thank Mr. Sukmajaya, Nofriady, Imam, Bambang, Suhardi and Mrs. Prihartini, Tri Handini, Racmi, Suyanti who helped in this research.

References

[1] Atmawinata, A., Ferry, Y., Sakri, W., Roosmariharso, Drajadm, I., Adriano, A., Yus, S., Wartam R., Massarudin, Denny, N., Ari, I.S., Shinta, V., Sih, W., Bayu, P.W., Ardhana, 2014, *Telaah Penguatan Struktur Industri Pemetaan Potensi Logam Tanah Jarang di Indonesia*, The Ministry of Industry of Indonesia; Jakarta. (in Indonesia).
[2] Jordens. A., Ying. P.C., Kristian. E.W. 2013. *J Minerals Engineering* **41**. 97–114
[3] Rodliyah, I. 2015. *J. M&E*. **13**. (1). 71 – 80. (in Indonesia)
[4] Campo. D.K., Domingos. D.S.M., Annie. B., et al. 2011. *J. The American Ceramic Society*. **94**. (6). 1859 – 1864.
[3] American Standard Testing and Material (ASTM), 2012, *Standard Specification for Nuclear-Grade Hafnium Oxide Powder*, ASTM international; United States.
[4] Savitri. M., Sini. N.K., Yati. M., Kiki. W.V., Wahyu. A.S., Togar. S., Risdiana. 2015. *J. Material dan Energi Indonesia*. **05**. 35 – 38.
[5] Biyantoro. D., Tri. H., and Moch. Setyadji. 2016. *Proc. Nat. Conf. on Nuclear Energy and Technology (Indonesia). (Yogyakarta: Indonesia)* p 1-9.
[6] Patel. M. J., Agrawal. D. H., and Pathan. A.M. 2011. *Proc. Nat. Conference on Recent Trends in Engineering & Technology. (Gujarat:India)*
[7] Soemarsono, M and Biyantoro, D., 2005. *Proc. Nat. Conf. on Center of Research and Development of Advanced Technology – National Nuclear Energy Agency of Indonesia*. p 120-126
[8] Rouco. R, R. Córdoba, J. M. De Teresa, L. A. Rodríguez, C. Navau, N. Del-Valle, G. Via, A. Sánchez, C. Monton, F. Kronast, X. Obradors, T. Puig, Palau. 2017. *J Scientific Report*. **7**. 5663.
[9] Prakash. S., 1975. *Advanced Chemistry of Rare Earth Elements*. (Ram Nagar, New Delhi: S.Chamc. & Co (PVT) (TI))
[10] Gupta, C.K., and Krishnamurthy, N., 2005, *Extractive Metallurgy of Rare-Earths*. London: CRC Press.
[11] Ahmed. Sherien H, Osama. S. Helaly, and Mohamed. S. Abd. El-Ghany. 2014. *J Chemical, Molecular, Nuclear, Materials and Metallurgical Engineering*. **8**. 866 – 872
[12] Al. Wadeeah M, Che Nor Aniza Che Zainul Bahri, Amran Ab. Majid, Sukiman Sarmani. 2016. *Malaysian. J. Analytical Science*. **20**. 770 – 776
[13] Jha Manis K, Archana K, Rekha P, Jyothi R.K, Kyoungkeun Y, Jin Y.L. 2016. *J. Hydrometallurgy*. **165**. 2-26

[14] Prameswara. G. 2016. Preliminary Design of ThO_2 Plant from Bangka's Monazite Processing 10 Ton/Year. Bachelors Thesis. Polytechnic Institute of Nuclear Technology, BATAN Yogyakarta, Indonesia.

[15] Bahri. C.N.A.C.Z, Wadeeah M. A, Amran Ab. Majid and Mohd Izzat Fahmi Mohd Ruf. 2016. *Malaysian. J. Analytical Science*. **20.** 44 – 50

[16] Suyanti, M.V.Purwani., and Muhadi. 2008. *Proc. Nat. Conf. on Human Resources of Nuclear Technology IV*. p 429 – 438.

[17] Bongaerts. J.C. 2013. *J. The IMRE*. **7**. 1-9. TU Bergakademie Freiberg.

[18] Alberts, E. 2011. Stripping Rare Earth Elements and Iron From D2EHPA During Zinc Solvent Organic. Masters Thesis. Faculty of Engineering, Stellenbosch University.

[19] Skoog D.A., West D.M., and Holler F.J. 1996. *Fundamental of Analytical Chemistry, 7th edition*, (Florida: Saunders College Publishing)

[20] International Union of Pure and Applied Chemistry (IUPAC). 1992. *IUPAC Solubility Data Project Solubility Data Series*. Oxford: Pergamon Press.

[21] Kislik, V.S., 2012. *Solvent Extraction: Classical and Novel Approaches*. Netherlands: Elsevier.

[22] Xie. F, Ting. A.Z. David. D, and Fiona. D. 2014. *J. Minerals Engineering*. **56**. 10 – 28.

[23] Mohammadi. M, Kerstin. F, Lars. K, Joaquin. M.D.L.C, and Ake. R. 2015, *j Hydrometallurgy*, Article in-press.

[24] Sato,T. 1984. *Proc . Int. Conf. on of On Actinide/Lanthanide Separations*. Honolulu, USA.

[25] Yoon, H.S., Chul, J.K., Kyeong, W.C., Sung, D.K., Jin, Y.L., dan Jyothi, R.K., 2016, *j Hydromet.* **4281**

[26] Handini. T, Bambang. EHB, Sri Sukmajaya. 2017. *Ganendra. J. Nuclear Science and Technology*. 20. 49 – 55. (in Indonesia)

[27] Jorjani E and Malek Shahbazi. 2016. *Arabian. J. Chemistry,* **9**. S1352 – 1539

Effect of SrO content on Zeolite Structure

N Widiarti*, U S Sari*, F W Mahatmanti*, Harjito*, C Kurniawan*, D Prasetyoko, Suprapto****

*Department of Chemistry FMIPA Universitas Negeri Semarang
D6 Building Sekaran Gunungpati Telp. (024) 8508112 Semarang 50229
**Department of Chemistry FMIPA Sepuluh Nopember Institute of Technology Surabaya

E-mail: wiwid_mgl_78@yahoo.comee

Abstract. The aims of current studies is to investigate the effect of strontium oxide content (SrO) on synthesized zeolite. Zeolite was synthesized from *Tetraethyl orthosilicate* (TEOS) as precursors of SiO_2 and *aluminum isopropoxide* (AIP) precursors. The mixture was aged for 3 days and hydrothermally treated for 6 days. The SrO content was added by impregnation method. The products were then characterized using X-Ray Diffraction (XRD), Fourier Transform Infrared (FTIR), and Surface Area Analyzer (SAA). The diffractogram confirmed the formation of Faujasite-like zeolite. However, after the addition of SrO, the crystallinity of zeolite was deformed. The diffractograms shows the amorphous phase of zeolite were decrease as the SrO content is increase. The structural changes was also observed from FTIR spectra which shows the shifting and peak formation. The surface area analysis showed that the increasing loading of SrO/Zeolites reduced the catalyst surface area.

1. Introduction

Zeolite Y is a large-pore high-silica material first synthesized in 1967 [1]. Its structure has been recently reported to be an intergrowth of two or three polymorphs having a 3-dimensional system of interconnected 12-membered ring channels. Its crystallization kinetics and mechanism have been investigated in a Na, K, TEA-containing system with tetraethylorthosilicate (TEOS) as source of silicon in static conditions [2-3]. Amorphous and colloidal silica have also been used as raw materials [4-7]. In these studies, the effect of the relative amounts of Na and K in the parent gel was also systematically investigated. However, the nature and concentration of alkali metals have been shown to influence the crystallization rate and the crystal size and morphology of other high-silica zeolites [7].

In the case of zeolite Y, the possibility of synthesizing a pure sample of one of the polymorphs by control of their relative nucleation and crystallization rates provides an additional strong reason for studying the effect of the synthesis parameters. Furthermore, both the crystal intergrowth and the crystal size and morphology are known to influence the catalytic behavior of zeolites.

In the present work, the effect of Sr^{2+} cation on the crystallization of zeolite Y is reported. Special attention has been given to the chemical changes in the liquid and solid phases, as well as to the functional group and diffractogram pattern of the products.

2. Experimental

The tools used in the study were equipment glass (glass beaker, erlenmeyer, funnel glass, funnel, measuring cup, pipette, pipette measuring, pipette volume, flask, a stirrer, a spatula), hotplate magnetic stirrer (IKAMAG), magnetic stirrer, oven (Memmert), furnaces (Barnstead Thermolyne 1400), analytical balance (AND GR-200), the reactor autoclave stainless steel, thermometer, instrument X-Ray Diffraction Zhimadzu, FTIR Shimadzu instruments Spectrum One 8400S, a set of tools Reflux, fine filter paper, GC-MS (Shimadzu QP2010S).

The materials needed in this research are Strontium Nitrate $Sr(NO_3)_2$. (99%) (Merck), tetraethyl orthosilicate (TEOS) 99% ρ = 0.94 kg/L, sodium hydroxide (NaOH) (99%), tetramethylammonium hydroxide (TMAOH) (acidimetric 24.0 to 27.0% (m)) ρ = 1.02 kg/L, $C_9H_{21}AlO_3$Aluminium isopropoxide (ALP) 98%, deionized water (Aqua Demin), Methanol Merck (Mr 32.04 g / mol, about 99% of ρ 0.792 g/cm^3), used cooking oil, alcohol 95%, Potassium Hydroxide (KOH) Merck (Mr 56.11 g/mol, purity 85%), Merck phenolphthalein indicator (Mr 318.33 g/mol, about 99%), natural zeolites bayah Banten, aqueous HCl Merck (Mr. 36.46 g/mol, 32% purity, ρ 1.16 g/cm^3).

2.1. Synthesis of Zeolite

Zeolite were synthesized in accordance with the methods conducted by Taufiqurrahmi *et al*, (2011) with some modifications [8]. Preparation of zeolite begins with making a 0.065 mol NaOH solution. The next stage added 17.744 grams of tetramethylammonium hydroxide (TMAOH) and 16.34 grams of aluminum isopropoxide (AIP) to a solution of NaOH while stirring using a magnetic stirrer until a homogeneous or formed sol. At the same time TEOS of 30,796 mL was added to the mixture dropwise until the solution is white as milk. The mixture was stirred using a magnetic stirrer for 3 days while continuously stirring at room temperature. In this preparation obtained a molar composition ratio of 4.964 TMAOH: 0.065 NaOH: 2 AIP: 3.448 TEOS: 125 H_2O. The next stage was hydrothermal by incorporating the mixture into a stainless steel autoclave heated in an oven at 100°C for 6 days. The solid obtained at this stage was further separated by centrifuge at a rate of 3,000 rpm for 20 min. The solid was washed with distilled water until the pH of neutral laundry, then dried in an oven at 100°C for 2 hours. The obtained solids were calcined using a furnace with a temperature of 550°C for 8 hours. The zeolite solids were characterized by Fourier Transform Infrared (FTIR) and X-Ray Diffraction (XRD)

2.2 Preparation of SrO-modified Zeolite

In this research, the Sr content was calculated as SrO. Preparation of zeolite with 1% of Sr (w/w) was carried out by first immersing of zeolite within NH_4OH solutions, followed by putting zeolite solids into 10 mL of $Sr(NO_3)_2$ solution. The solution mixture was then stirred for 30 minutes, while heated until the mixture was shaped like a slurry. The mixture was dried in an oven at a temperature of 100°C for 2 hours. The obtained solids were then calcined in the furnace at 550°C for 5 hours. The SrO/Zeolite catalysts were characterized by Fourier Transform Infrared (FTIR), X-RayDiffraction (XRD) and Surface Area Analyzer (SAA). The similar treatment was also done for different concentration of SrO.

3. Result And Discussion

3.1 Synthesis of Zeolite

3.1.1 Crystallinity Analysis X-Ray Diffraction (XRD)

Crystallinity Analysis Using X-Ray Diffraction (XRD) synthesis of Zeolite was matched with a standard faujasite zeolite diffractogram according to Treacy and Higgins [9].

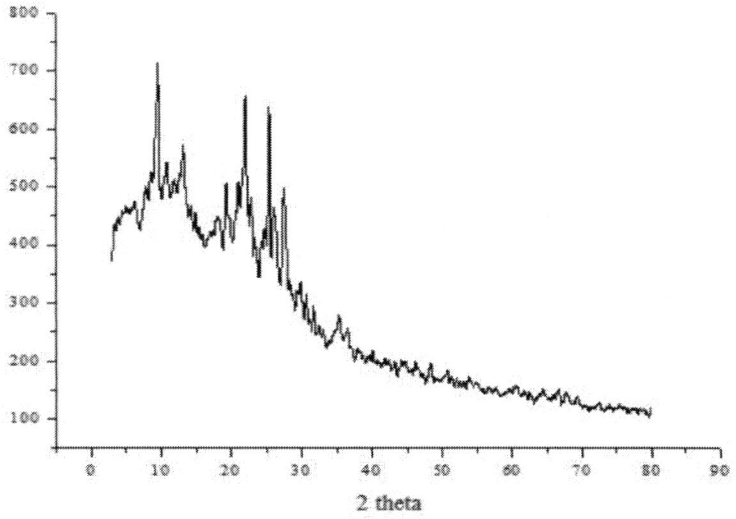

Figure 1. Diffragtogram Synthesis of Zeolites Using X-Ray Diffraction (XRD)

The diffractogram of synthesized zeolite Y shows a characteristic peak at $2\theta = 10.84°; 21.98°$; 25.40. However, the zeolite is still containing high amorphous phase. Zeolites had the best performance since the channel was good and the range of 3-15 Å cavity [10]. Because of the zeolite Y type faujasite standard first peak appears at an angle 2θ smaller than 10, the Zeolite also have the distance d which is equal to the zeolite comparison, possible arrangement of polyhedral cage sodalite on this product is perfect [9].

3.1.2 Zeolite Function Analysis of Fourier Transform Infrared Nanoparticles (FTIR)

FTIR spectra results of synthesis of Zeolite can be seen in Figure 2. It can bee seen that the Zeolite had formed the zeolite framework. The band at 3435.98 cm^{-1} indicate -OH group and 1639.82 cm^{-1} was overtone vibration of -OH group. While the band at 1052.08 cm^{-1} was a vibration asymmetric span. The presence of a curved peak in an area of about 1000 cm^{-1} arose because of the vibration of the Al-OH cage formed by a cation vacancy, and 722.25 cm^{-1} was a symmetric spacer vibration. The presence of vibration and bending indicates the formation of aluminosilicate framework [11]. This results was in accordance with Zahro *et al* [12] which states that in structure of zeolite Y was known character of aluminosilicate had internal and external network, which was internal network of zeolite shown at absorption area about 500-420 cm^{-1}, 820-650 cm^{-1}, and 1250-950 cm^{-1}. The absorption area of about 820-650 cm^{-1} represents the vibration of O-Si-O and O-Al-O symmetry, while the absorption region of about 1250-950 cm^{-1} represents the asymmetric stretching vibration.

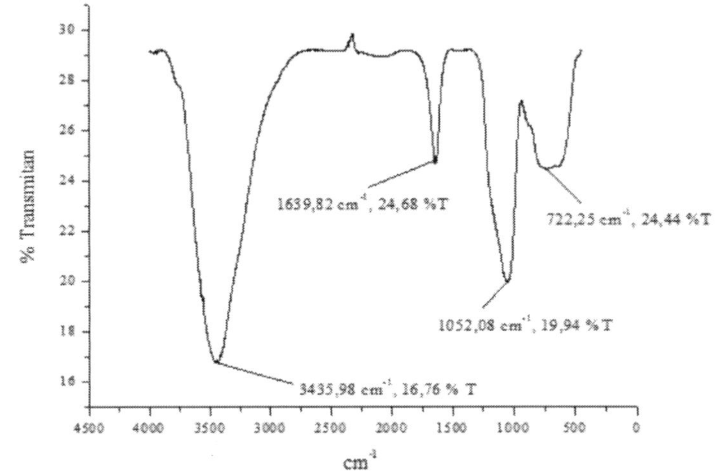

Figure 2. Clustered Spectrum Function of Zeolite Using FTIR

3.2 Synthesis of SrO/Zeolites

SrO metal bonding in Zeolite was done by impregnation method. The SrO/zeolite solids of the formed nanoparticles are separated from the solvent by being filtered. The obtained solid is dried to remove any residual demineralized water solvent present in SrO/Zeolites. The result of impregnation was obtained by slurry solids.

3.2.1 Analysis of Crystallinity SrO/Zeolites Using X-Ray Diffraction (XRD)

Diffractogram of crystallinity SrO/Zeolites can be shown in Figure 3.

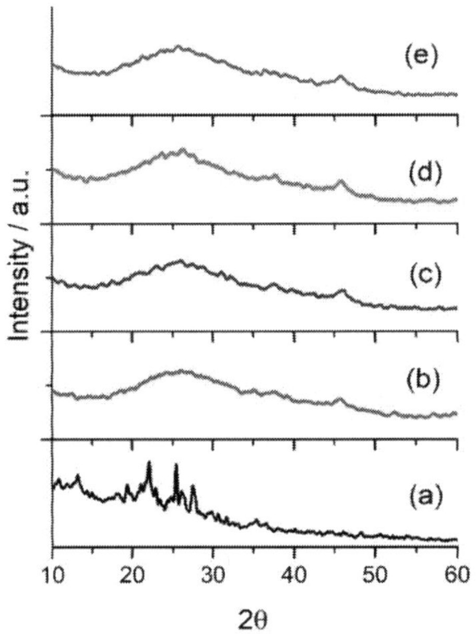

Figure 3. Diffractogram of (a) H-zeolite, (b) 1% SrO/Zeolite, (c) 2% SrO/Zeolite, (d) 4% SrO/Zeolite, and (e) 8% SrO/Zeolite

XRD data analysis results show that SrO metal cannot be embedded in Zeolite. This was known by the peak loss of SrO/zeolite. The loss of zeolite peak was due to SrO formed on the pore of zeolite, urging the aluminosilicate skeleton on the zeolite. New emerging peaks were characterized as strontium oxide, which was presented in zeolites [12]. After the impregnation process, the zeolite structure becomes amorphous, indicating a pure silica composition. The diffraction pattern of FIG. 3 shows a peak width at 2θ between 21°-25° for all types of catalysts. The peak was similar to the peak of the amorphous silica [13], so it can be said that all synthesized materials were mostly amorphous materials.

Based on this, the zeolite of the impregnated nanoparticles was transformed into a silicalite zeolite, a zeolite having high Si and it was hydrophobic and had an affinity for hydrocarbons. Alunimium in the zeolite framework was a weak point of zeolite crystals. Al easily knocked out of the framework caused damage to the zeolite structure and lost its activity [1]. According Kasmui *et al* [14] stated that the silicon zeolite faujasite was derived from zeolite Y and it was caused the formation of the faujasite framework can experience a metastable state.

The impregnated catalyst with SrO had a pattern similar to that of H-Zeolite but with different intensities. It shows that the six catalysts still have the same minerals as the nanoparticle zeolite but with different amounts. The decrease in XRD peak intensity was also experienced by [14] which developed CuO in Titanium Silicalite (TS-1). They stated that the decrease in intensity in TS-1 was due to the agglomeration of CuO oxide that closed the pores of TS-1.

3.2.2 Analysis of Cluster Function SrO/Zeolites Using Fourier Transform Infrared (FTIR)

Comparison of FTIR spectra of catalyst H-Zeolite, SrO, SrO/Zeolites 1%, 2%, 4% and 8% can be seen in Figure 4.

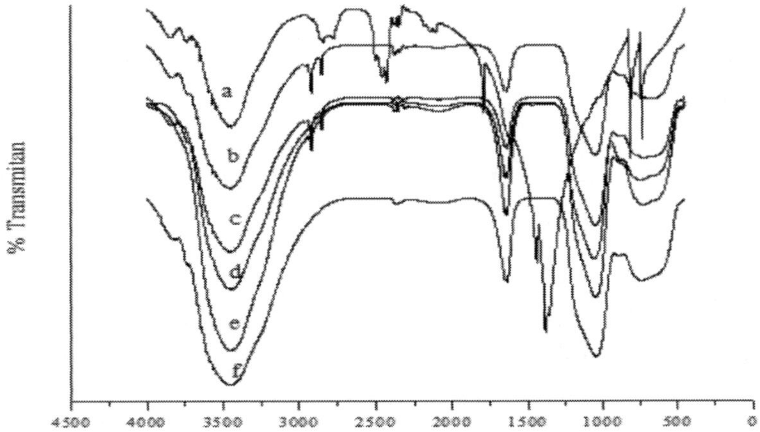

Figure 4. a) SrO/Zeolites 8% b) H-Zeolite c) SrO d) SrO/Zeolites 4% e) SrO/Zeolites 1% f) SrO/Zeolites 2%

According to Figure 4, the emerging absorption can be seen that the zeolite of the SrO/zeolite spectrum of nanoparticles had been formed. The absorption peak that appeared in the area around 400-1300 cm^{-1} was the typical peak of zeolite faujasit [5]Typical vibration band zeolite Y at wave number 1055.5; 1052.14; 1052.04; 1051,78; And 1043,91 cm^{-1}. According Sudiyono (2016) was typical of zeolite Y vibrational band at wave number 1020 cm^{-1} (internal asymmetric stretch vibration) after bonding experiencing widening, so that indicated the zeolite structure changes. This was in accordance with the XRD results data showing the increment made causing zeolite Y to shift the value of 2θ [15]. The new absorption peaks after loading were present at wave numbers at about 2360-2380 cm^{-1}, which assumed the uptake of the hydrocarbon impurities [16].

3.2.3 Surface Area Analysis of SrO/Zeolites Using Surface Area Analyzer (SAA)

The results of catalyst surface area measurement SrO/Zeolites are synthesized and H-Zeolite and SrO are presented in Table 1.

Table 1. Surface Surface Results

No	Sample	Surface Area (m^2/g)
1.	H-Zeolite	46,969
2.	SrO	386,395
3.	SrO/Zeolites 1%	382,818
4.	SrO/Zeolites 2%	148,851
5.	SrO/Zeolites 4%	81,008
6.	SrO/Zeolites 8%	279,394

Table 1 shows that on the stand alone SrO catalyst and SrO the synthesis results decreased the specific surface area along with the percent increase loading. This decrease in surface area indicated that the formation of smaller particle size and the breakdown of crystal structure of SrO/zeolite synthesized nanoparticle catalyst. The smaller the particle crystal size, the larger the surface area of the catalyst [17-18].

In addition to the SrO/zeolite catalyst nanoparticles produced large particle size crystals and small surface area. This was inversely related to the synthesis results of SrO/Zeolites 1%, 2%, and 4% catalysts. However, in this case it can be reinforced with the crystallinity yield resulting from an 8% SrO/Zeolite catalyst that there is a better 2θ peak than the H-Zeolite catalyst, SrO/Zeolites of 1%, 2%, and 4%.

4. Conclusion

SrO/Zeolites 8% shows a shift in the wave number 1043.91 cm^{-1} and the emergence of new absorption bands at 2359.97 cm^{-1} indicating the interaction between SrO and H-Zeolite. The highest surface area was obtained at 8% of SrO was added into zeolite.

References

[1] Ayoup, M., Ghrair, J.I, & Thilo, S. 2009. Preparation and Characterization, Water Air Soil Pollut. *Journal of Nanoparticulate Zeolitic Tuff for Immobilizing Heavy Metals in Soil,* (203):155 168.

[2] Chen, C., Chien-Chang, H., Dang-Thuan, T, & Jo-Shu, C. 2012. Biodiesel Synthesis Via Heterogeneous Catalysis Using Modified Strontium Oxides As The Catalys. *Bioresource Technology*, (113): 8-13.

[3] Hartanto, Djoko., Tri Esti Purbaningtias., Hamzah Fansuri., Didik Prasetyoko. 2011. Karakterisasi Struktur Pori dan Morfologi ZSM-2 Mesopori yang Disintesis dengan Variasi Waktu Aging. *Jurnal Ilmu Dasar,* 12(1): 80-90

[4] Htun, Mie Mie Han., Mu Mu Htay, & May Zin Lwin. 2012. Preparation of Zeolite (NaX, Faujasite) from Pure Silica and Alumina Sources. *International Conference on Chemical Processes and Environmental*, 15-16.

[5] Kasmui, M. M. Z, & S.Woro. 2008. Kajian Pengaruh Variasi Rasio Si/Al dan Variasi Kation Terhadap Perubahan Ukuran Pori Zeolit Y dengan Menggunakan Metode Mekanika Molekuler. *Jurnal Kimia.*

[6] Nikmah, R. A. Syukuri., Widiastuti, N, & Fansuri, H. 2008. Pengaruh Waktu dan Perbandingan Si/Al Terhadap Pembentukan Zeolit A dari Abu Dasar Bebas Karbon dari PLTU PT. IPMOMI dengan Metode Hidrotermal. *Journal of Indonesia Zeolites*, 7(1): 1411-6723.

IOP Conf. Series: Materials Science and Engineering **349** (2018) 012045 doi:10.1088/1757-899X/349/1/012045

[7] Raida, Ariyanti. 2012. *Pengaruh garam prekursor terhadap aktivitas katalis CuO/γ-Al2O3 Yang Digunakan Dalam Reaksi Hidrogenasi Minyak Jarak*. Skripsi. Universitas Indonesia.
[8] Socrates, G. 1994. *Infrared Spectroscopy*. Chicester: John Willey & Sons Ltd.
[9] Sriatun. 2004. *Sintesis Zeolit A dan Kemungkinan Penggunaannya sebagai Penukar Kation*, 7(3): 66-72.
[10] Subagjo. 2008. Pembuatan Zeolit Y dan USY Untuk Komponen Aktif Katalis Perengkahan. *Jurnal Zeolit Indonesia*, 7(1): 1411-6723.
[11] Sudiyono, Dinda Atikah. 2016. *Sintesis Dan Karakterisasi Zeolit Nay Sebagai Pengemban Senyawa Antikanker Hasil Ekstrak Etanol Akar Rumput Bambu (Lophatherum Gracile Brongn)*. Skripsi. Jurusan Kimia Fakultas Sains Dan Teknologi Universitas Islam Negeri Maulana Malik Ibrahim Malang.
[12] Taufiqurrahmi, N., Abdul, R.M, & Subhash, B. 2011. Nanocrystalline zeolite beta and zeolite Y as catalysts in used palm oil cracking for the production of biofuel. *J Nanopart Res*, 13: 3177-3189.
[13] Treacy, M.M.J. & Higgins, J.B., 2007, *Collection of Simulated XRD Powder Patterns for Zeolites 5th edition*, Structure Commission of the International Zeolite Association, Elsevier.
[14] Widiarti, N. & D. Prasetyoko. 2011. *Sintesis dan Karakterisasi Katalis Cu/Ts-1*. Kimia FMIPA Institut Teknologi Sepuluh Nopember Surabaya.
[15] Widjajanti, Endang. 2007. *Kinetika Kimia*. Makalah disampaikan dalam kegiatan Pengabdian Kepada Masyarakat. FMIPA UNY.
[16] Xia, Y. & Mokaya, R. 2006. Molecularly Ordered Layered Aluminosilicate-Surfactant Mesophases and Their Conversion to Hydrothermally Stable Mesoporous Aluminosilicates, *Microporous and Mesoporous Materials*, 94: 295–303.
[17] Yee, Maxine. & Yaacob, Iskandar. I. 2004. Synthesis and characterization of iron oxide nanostructured particles in Na–Y zeolite matrix. *Journal of materials research*, 19(3): 930-936.
[18] Zahro, A. 2014. *Sintesis dan Karakterisasi Zeolit Y dari Abu Ampas Tebu Variasi Rasio Molar* SiO_2/Al_2O_3 *dengan Metode Sol-Gel Hidrotermal*. Skripsi. Jurusan Kimia Fakultas Sains dan Teknologi. UIN Malang.

Isomerization of α-pinene in the terpentin oil with TCA/ Natural Zeolite using microwave irradiation

N. Wijayati[1], Supartono[1] and E. Kusumastuti[1]

[1]Semarang State University

Email: nanikanang@gmail.com

Abstract: The catalytic potensial of trichloroacetic acid (TCA)//Natural Zeolite in the isomerization of α -pinene in the terpentin oil was investigated. The purpose of this study is to investigate the influence of the power of microvawe on activity and selectivity of catalyst. The main product were champhene, terpinene, limonene, p-cymene, and terpinolene. The highest selectivity was 28.26% with a conversion of 23.25%, whereas the higher conversion was 98.99% with selectivity of 16.90% at room temperature using power of microwave 640 W.

Keywords: α-pinene, isomerization reaction, microwave

1 Introduction

Turpentine oil are mobile liquids, usually non-colored or slightly colored, with characteristic pleasant odor. The boiling point of the oils varies between 154-170°C, and not soluble in water, but are soluble in ethers, dichloromethane and in other oils. Chemically, turpentine oil is mixture of cyclyc monoterpene hydrocarbons, $C_{10}H_{16}$, such as α-pinene, β-pinene, 3-carene and d-longifolene. Alpha-pinene compound is a major component of turpentine oil (>80%). Synthesis of some compounds based α-pinene produce terpineol, camphor, bornyl chloride, and camphene an industrial scale has a high selling price. Alpha-pinena can be converted chemically into the basic material for the manufacture of compound that is more useful through addition reaction, hydration reaction, and isomerization [1-4].

The isomerization α-pinena can produce bicyclic compound, monocyclic compound, or other products. Bicyclic product such as camphene, while products monocyclic such as limonene, p- cimene, and terpinolene [5]. The result isomerized α-pinene are used as fragrance ingredients, cosmetics, food, pharmaceuticals, flavorings, and solvent [6-7].

In chemical reactions, homogeneous catalysts can't be reused because it is unstable, besides the catalyst separation process of its products are widely experienced problems because both are in one phase. The homogeneous catalysts lead to the decomposition of the reaction products because of the homogeneous catalysts participate in the reaction. Heterogeneous catalysts can be used as an alternative with positive opportunities related to increased yields and selectivity through the isomerization reaction of α-pinene [8].

The chemical reactions using microwave is already done. Method of heating by microwaves has many advantages, such faster reaction time, the product is cleaner, higher selectivity and better results. This is the main alternative to obtain a synthesis of various organic of intermediates the isomerization of compounds more efficiently, with operational simplicity and mild reaction conditions [9].

In this study the effect of power microwave to isomerization of α -pinene in The turpentin oil using TCA/ Natural Zeolite as heterogeneous catalyst

2 Experimental

2.1 Catalyst Preparation

Natural zeolites are dried and then activated. Natural zeolite has been activated, calcined at temperatures of 500 °C for 4 h in a calcining furnace. Furthermore, natural zeolites was impregnated with trichloroacetic acid at a concentration of 35% followed calcination at a temperature of 500 °C for 4 h. The catalyst was analyzed by X-Ray Diffraction (XRD) and acidity test by gravimetric method.

2.2 Catalytic test

The isomerization reaction α-pinene using turpentine oil as a raw material (α-pinene (87%), β-pinene (2%), and 3-carene (11%). The isomerization α-pinene in the turpentine oil 10 mL was performed in the flask by adding 1 g TCA/Natural zeolite catalyst. The mixture was reacted by using microwave at 320, 480, 640 and 800 W power for 15, 30, 45, 60 and 90 min. The mixture is separated by centrifugation for 15 min to separate catalyst and results, the reaction product is then filtered and analyzed by FTIR and GC-MS.

3 Result and Discusion

Trichloroacetic acid/natural zeolite is found as a good catalyst for the production of camphene from α-pinene. Impregnation is done by adding TCA to natural zeolite. The purpose of this process is that the active side of the TCA will be bound to the pore structure of zeolite solids. The good results are due to strong acidity and high affinity with organic and aqueous phases during reaction. The crystallinity structure of the catalyst was studied by XRD. The powder diffractograms of natural zeolite (ZA), and H-Natural Zeolite H-(ZA) showed reflection lines at 17°, 20°, 22°, and 25°, which were assigned to TCA (JPDC 42.1605), while in this case of TCA/ZA, the diffractogram did not show refflection lines corresponding to TCA. This was attributed to TCA dispersion due to the high surface area of the natural zeolite as the support [1]. (Figure 1).

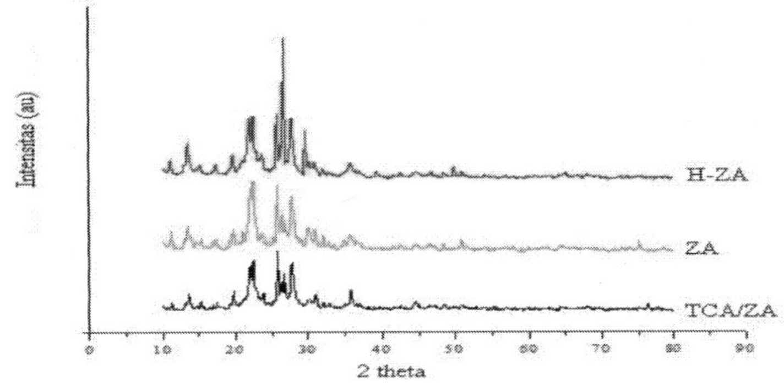

Figure 1. Diffractograms natural zeolite (ZA), H-Natural Zeolite H-(ZA), and TCA/Natural zeolite (TCA/ZA).

The acidity of the catalysts in this study using the gravimetric method by using ammonia as a base adsorbate to determine the acidity in the zeolite cavities (Table 1). The use of ammonia as adsorbate alkaline due to the size smaller than pyridine then ammonia gas can be adsorbed on the outer surface and within the zeolite while pyridine only adsorbed on the outer surface of zeolite [10].

Table 1. The acidity of catalysts

Sample	Surface acidity (mol/g)
Natural zeolite	2.22×10^{-3}
H-Natural zeolite	2.42×10^{-3}
TCA/H-Natural zeolite	3.23×10^{-3}

Acidity measurement results indicate acidity TCA/Natural zeolite is greater than the H-Natural zeolite and acidity of H-Natural zeolite is greater than the Natural zeolite Trichloroacetic acid has a pair of free electrons that can accept an electron pair. Increased acid sites is possible because impregnated of trichloroacetic acid on the zeolite.

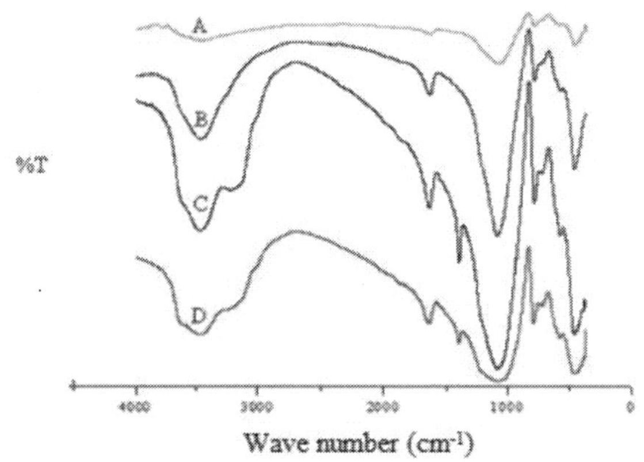

Figure 2. IR spectra (A) TCA/Natural zeolite (B) H-Natural zeolite (C) TCA/Natural zeolite+ammonia (D) H-Natural zeolite+ammonia

The presence of TCA in the catalysts was studied by FT-IR. Figure 2 shows the spectra of Natural zeolite, and TCA/natural zeolite. The carbonyl group stretching signal appeared at 1660 cm^{-1}. the The bands assigned to stretching and bending due Cl-C bonds were observed at 830 cm^{-1} and 680 cm^{-1}[1].

3.2 Isomerization Reaction

The isomerization reaction of α-pinene was studied in heterogeneous phase by TCA/Natural zeolite as catalyst. Conversion of α-pinene to some compouds e.g camphene as major product and limonene, p-cimene, and terpinolene as minor product. Conversion of α-pinene and selectivity to camphene was determined in the following ways. Conversion (X) was defined here as moles of α-pinene converted per 100 moles of α-pinene feed. The selectivities to camphene (S) was defined as moles of camphene formed per 100 moles of α-pinene converted.

Table 2. Conversion α-pinene and selectivity to camphene with TCA/Natural zeolite catalyst

Catalysts	Reaction time (min)	320 W		480 W		640 W		800 W	
		X (%)	S (%)	X (%)	S (%)	X (%)	S (%)	X (%)	S (%)
TCA/ H-Natural zeolite	15	3,47	33,97	12,26	27,73	23,25	28,26	29,83	22,90
	30	3,11	41,73	18,13	26,86	35,60	27,78	70,85	23,60
	45	5,40	33,88	29,75	26,35	66,88	26,30	88,48	19,43
	60	8,68	26,27	42,37	26,22	84,97	24,88	97,05	13,73
	90	13,46	22,43	76,01	24,26	98,99	16,90	98,85	6,59

X =Conversion of α-pinene, S = selectivity to camphene

Table 2 presents α-pinene conversion values (%X) and selectivities to camphene for the different treatments performed. When the TCA/Natural zeolite is exposed to a more severe power of microwave treatment, 320, 480, 640, 800 W the conversion increases but selectivities to camphene is decreases. The main product were camphene, α-terpinene, limonene, p-cimene, γ-terpinene, and terpinolene (Scheme 1).

The α-pinene isomerization can pass two parallel routes, leading to the formation of monocyclic products such as limonene, terpinolene, terpinene, p-cymene and bicyclic products, such as camphene. The first step in The reaction involves protonation of the doble bond giving pinanyl cation. The second involves a Wagner-Meerwein type rearragements of isoborrnyls and p-menthenyl cations giving camphene and others [11].

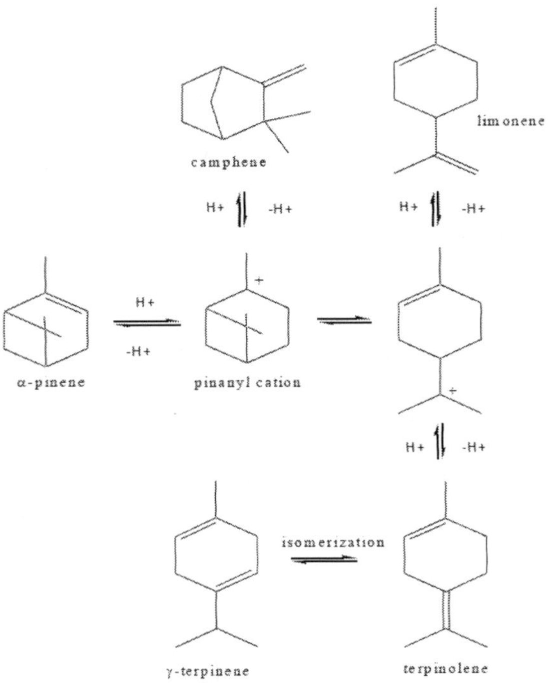

Scheme 1. Mechanism of the α-pinene isomerization with TCA/Natural zeolite

Heating with radiation microwave unlike conventional heating methods, heating with radiation microwave provides uniform heating throughout the reaction mixture so as to increase the temperature of the entire volume of the resulting reaction can be run faster [12].

Trichloroacetic acid (TCA) impregnated in natural zeolite can be used as a heterogeneous catalyst that can convert the α-pinene compound into its derivative compound. It is shown that α-pinene isomerization reaction with natural TCA / zeolite catalyst can produce camphene compound as main product with α-pinene conversion of 98.99% with selectivity 16.90%. The longer isomerization reaction time can increase the α-pinene conversion [1, 13].

4 Conclusion

TCA/Natural Zeolite was found a good catalyst for a production of camphene from α-pinene. The treatment microwave power has an important effect on the reaction activity on the selectivity to camphene, α-terpinene, limonene, p-cymene, terpinene, and terpinolene. The study on the effect microwave power (320 - 800 W) on conversion showed that 640W gives higher conversion (98.99%) and selectivity for camphene (16.90%) at room temperature.

Acknowledgment

The authors would like to thank Directorate General of Higher Education (DGHE), Department of National Education Republic Indonesia for to support.

References

[1] Avila M.C.,ComelliN.A.,CastellonE.R.,LopezA.J.,FloresR.C.,PonziE.N., PonziM.I., 2010.Study of Solid Acid Catalysis For The Hydration of α-pinene. *Journal of Molecular Catalysis*, 322(1-2): 106-112.

[2] Bruno S.M., Pillinger M., Kuhn F.E., Goncalves II.S., Valente A.A., 2013. Isomerization of α-pinene oxide in the presence of methyltrioxorhenium (VII). *Catalysis Communications*. 35: 40-44.

[3] Wijayati, N., Pranowo, dan Triyono, P., 2013. The Acid Catalyzed Reaction Of α -Pinene Over Y-Zeolite. *Indonesian Journal of Chemistry*, 13(1): 59-65.

[4] Wijayati, N., T. Handayani, Supartono. 2017. Isomerization of α-pinene using Zirconiua/Natural Zeolite Catalysts. *Asian Journal of Chemistry*. 29 (8) : 1705-1708.

[5] Comelli N. A., PonziE. N., dan PonziM.I., 2006. α-Pinene Isomerization to Camphene Effect of Thermal Treatment on Sulfated Zirconia. *Chemical Engineering Journal*, 117(2): 93-99.

[6] Wang J., W. Hua, Y. Yue, Z. Gao, 2010. MSU-S Mesoporous Material : An Efficient Catalys For Isomerization of α-Pinene. Bioresource Technology,101: 7224-7230.

[7] Avila M.C., Ponzi M.I., dan Comelli N.A., 2015. Hydration of α-Pinene over Heteropoly Acid H3PW12O40 and H3PMo12O40. *Journal Chromatogr Sep Tech*, 6(7) : 1-6.

[8] Reddy B.M., P.M. Sreekanth, V.R.Reddy, 2005. Modified Zirconia Solid Acid Catalysts For Organic Synthesis and Transformations. *Journal of Molecular Catalysis*, 225: 17-78.

[9] Bhuiyan, Hossain, Mahmud and Al-Amin, 2011. Microwave-assisted Efficient Synthesis of Chalcones as Probes for Antimicrobial Activities. *Chemistry Journal*, 3(2): 2465-2479.

[10] Flores-Holguin, N., A. Aguilar-Elguezabal, L.M. Rodriguez-Valdez, 2008. Theoretical Study Of Chemical Reactivity Of The Main Species In The α-Pinene Isomerization Reaction. Journal of Molecular Structure, 854: 81-88.

[11] Rachwalik, R., Olejniczak, Z., Jiao, J., Huang, J., Hunger, M., dan Sulikowski, B., 2007. Isomerization of α-Pinene Over Dealuminated Ferrite-Type Zeolites. *Journal Catal*, 252: 161-170.

[12] Lidström, P., J. Tierney, B. Wathey, dan J. Westman, 2001. Microwave Assisted Organic Synthesis - a Review. *Tetrahedron*, 57: 9225-9283.

[13] Wijayati, N.,Pranowo, Jumina and Triyono, 2011. Synthesis Of Terpineol From α-pinene Catalyzed By TCA/Y-Zeolite. *Indonesian Journal of Chemistry*, 11(3): 234-237.

Effectiveness Study of Drinking Water Treatment Using Clays/Andisol Adsorbent in Lariat Heavy Metal Cadmium (Cd) and Bacterial Pathogens

Pranoto[1]*, Inayati[2], Fathoni Firmansyah[3]
[1]Department of Chemistry, Faculty of Mathematics and Natural Sciences, Sebelas Maret University, Jl. Ir. Sutami 36A Surakarta, 57126 Indonesia
[2]Department of Chemical Engineering, Faculty of Engineering, Sebelas Maret University, Jl. Ir. Sutami 36A Surakarta, 57126 Indonesia
[3]Master Program of Environment Science, Graduate School, Sebelas Maret University, Surakarta, Indonesia.

*Corresponding author, Email: pakpranotomipa@staff.uns.ac.id

Abstract. Water is a natural resource that is essential for all living creatures. In addition, water also caused of disease affecting humans. The existence of one of heavy metal pollutants cadmium (Cd) in the body of water is an environmental problem having a negative impact on the quality of water resources. Adsorption is one of the ways or methods that are often used for the treatment of wastewater. Clay and allophanic soil were used as Cd adsorbent by batch method. Ceramic filter was used to reduce Cd concentration in the ground water. This study aims to determine the effect of the composition of clay and Allophane, activation temperature and contact time on the adsorption capacity of Cd in the model solution. The optimum adsorption condition and the effectiveness of drinking water treatment in accordance with Regulation of the Minister of Health using clay/Andisol adsorbents in ensnare heavy metals Cd and bacterial pathogens. Identification and characterization of adsorbent is done by using NaF, Infrared Spectroscopy (FTIR), X-ray diffraction (XRD), specific surface area and total acidity specific. The Cd metal concentrations were analysed by atomic absorption spectroscopy. Adsorption isotherms determined by Freundlich and Langmuir equations. Modified water purification technology using ceramic filters are made with a mixture of clay and Andisol composition. The results showed samples of clay and Andisol containing minerals. The optimum condition of adsorption was achieved at 200 °C of activation temperature, 60 minutes of contact time and the 60:40 of clay:Andisol adsorbent composition. Freundlich isotherm represented Cd adsorption on the clay/Andisol adsorbent with a coefficient of determination (R^2=0.99) and constant (k=1.59), higher than Langmuir (R^2=0.89). The measurement results show the water purification technology using ceramic filters effectively reduce *E. coli* bacterial and Cd content in the water.

1. Introduction

Water is a vital necessity for human sustainability. Heavy metals are one of the toxic and dangerous pollutants since it is not degraded in nature [1, 2]. Cadmium (Cd) is a heavy metal resulted from rocks and soil erosion, volcanic eruptions, mining activities and industrial waste (leather, textiles, batteries, paint and pesticides) [3, 4]. At high concentrations and in the long term, heavy metals can cause health risks for humans and ecosystems.

IOP Conf. Series: Materials Science and Engineering **349** (2018) 012047 doi:10.1088/1757-899X/349/1/012047

Various methods have been developed by the researchers to improve water quality [4-6]. Adsorption method has been developed as a method of reducing heavy metals due to it is more effective, simpler and cheaper than other methods [4, 7]. Sorption is the process of centralization of adsorbate molecules or ions on the adsorbent surfaces, either physics or chemistry.

Andisol soil is soil that occurs from the weathering of volcanic rocks that are scattered throughout its existence Indonesia, especially in Lawu, Dieng, Merapi, Merbabu and Wilis Mountains [8]. Allophane adsorbent has good characteristics, such as porosity, absorption and high cation exchange. Many studies have utilized allophane as adsorbent with relatively high effectiveness [9, 10]. Allophane and clays can be combined since they have similar characteristics as adsorbent. Clay is an aggregate of minerals that form of land consisting mainly of hydrous aluminium silicates, having active sites on the surface, hard and stiff when dry, and stable at high temperatures [8,11-12].

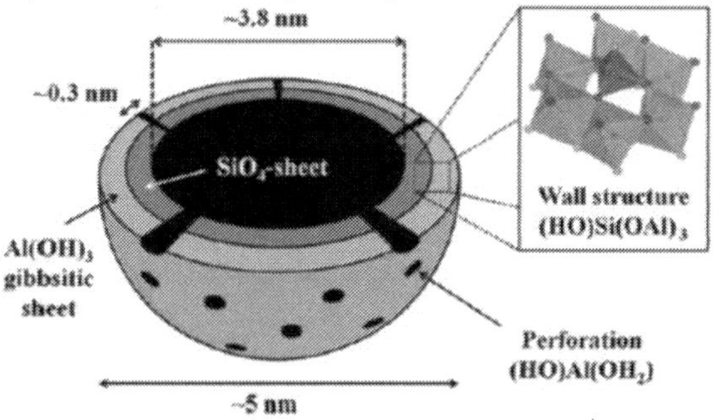

Figure 1. Structure Allophane [13]

Clay samples can be obtained from Bekonang, Mojolaban, and Sukoharjo, meanwhile Andisol can be obtained from Cemoro Sewu, Lawu Karanganyar. Andisol and clays further developed for the manufacture of ceramic filters to reduce the cadmium levels in water. Development of the main ceramic filter made from clay, is expected to result in a reduction of pollutant physical, biological and chemical that more effectively in order to obtain clean water that can be tolerated for drinking water; simple because its operation does not require any special skills; materials are used locally available and easy to obtain and inexpensive [1, 12, 14-16]. Muhdarina et al. [17] examined the manufacture of ceramic membranes as a water purifier. Porous ceramic membrane plays a major role in water purification because of its properties, which is stable at high temperatures, high mechanical strength and easy regeneration. Agmalini et al. [1] used a ceramic membrane made from clay and coal fly ash to improve the water quality of the swamp. The ceramic filter can be made by mixing clay with sawdust, bark rice [21], leaf tea, powdered coffee, or organic materials other combustible [16] to increase the flow rate. In this study studied the physical and chemical parameters such as pH, TDS, cadmium concentration and flow rate, as well as reduce *E. coli* bacteria.

The purpose of this study was to determine the effect of the composition of clay and Andisol, activation temperature and contact time on the capacity of adsorption of metal ions Cd in the solution of the model, determine the optimum conditions adsorbent mixture of clay and Andisol as adsorbent metal ions Cd in the solution of the model and determine the effectiveness of the filter ceramics for the reduction of Cd metal ions in water, and reduce coli bacteria.

2. Materials and Methods

2.1. *Materials*

Clay was obtained from Bekonang, Sukoharjo, Central Java. Andisol soil was obtained from Cemoro Kandang, Lawu Mountain, Central Java. Distilled water was obtained from the Chemistry Laboratory of Universitas Sebelas Maret. The Cd standard solution, HNO_3 (Merck), NaF (Merck), Ammonia (Merck) were used in this research.

IOP Conf. Series: Materials Science and Engineering **349** (2018) 012047 doi:10.1088/1757-899X/349/1/012047

2.2. *Preparation of Adsorbent*

2.2.1. *Clay*

The clay used in this study came from the area of Bekonang, Sukoharjo, Central Java. The clay obtained is cleaned, dried with aerated at room temperature, and crushed until smooth. Clay was then sieved with a 150 mesh sieve. The powder that passed 150 mesh was soaked in distilled water, filtered, and dried at a temperature of 105 °C for 4 hours [42].

2.2.2. *Andisol Soil*

Andisol soil used in this study came from Cemoro Sewu, Lawu, East Java. Andisol soil obtained was cleaned, washed with distilled water and dried with aerated in the open air, then crushed until smooth. Furthermore, Andisol soil was sieved with a 150 mesh sieve. The powder that passed 150 mesh was soaked in distilled water and filtered, then dried at a temperature of 105 °C for 4 hours [42].

2.2.3. *Identification*

Identification and characterization of Adsorbent were done with NaF test, XRD, FTIR, SAA, AAS, total acidity test specific.

2.3. *Adsorbent Activation*

Activation adsorbent was done chemically and physically. Chemical activation was only performed for Andisol soil, as many as 50 grams of Andisol soil is added to 250 mL of NaOH with a concentration of 3 M. Then the mixture was stirred at 70 °C with stirring time for 5 hours, then cooled. Once the mixture is cool then filtered and washed with distilled water until the neutral pH condition or equal to pH solvent. After that, the Andisol soil was dried in an oven for 4 hours at a temperature of 105 °C [42]. Subsequently, adsorbent was made with composition variation between clay and soil Andisol, namely 0: 100, 20:80, 40:60, 50:50, 60:40, 80:20 and 100: 0 wherein the mixing of the clay and the soil is done in a way Andisol stirring for 1 hour. After that, the solid phase is filtered and washed with distilled water several times, followed by drying in an oven for 4 hours at 105 °C. Each composition of the mixture was physically activated at temperature variations of 100, 200 and 400 °C for 3 hours.

2.4. *Performance Test of Adsorbent*

Cd metal ion adsorption process was carried out by the batch method of immersion. About 0.05 g of a mixture of clay and Andisol soil was added into 15 mL of 8 ppm Cd solution. Then, it was stirred at a constant speed (150 rpm) at room temperature for 30, 60 and 90 minutes. Afterward, it was filtered and measured by Atomic Absorption Spectroscopy (AAS) to determine the concentration of Cd metal ion that is not adsorbed by adsorbent.

2.5. *Determination of Sorption Isotherm*

A total of 0.05 g of optimum adsorbent was added into 15 mL of Cd solution with variations concentration by 15, 20, 25 and 30 ppm and then stirred at the optimum time. The sorption results was filtered with Whatman paper #40, then measured by AAS to determine Cd metal ion that is not adsorbed. The results are then analysed by the Langmuir and Freundlich isotherm models.

2.6. *Making Ceramic Filter*

The process of making ceramic filter was made by craftsmen of pottery in Bayat, Klaten by adjusting the composition of raw materials. As for the stages of manufacture are that dry mixing is performed between clay, Andisol soil and flour powder with a weight ratio of 2:3:1. The mixture was added with water and stirred to form a clay mixture. The mixture was printed in gypsum cylindrical mold with a inside diameter of 4 cm, the outside diameter of 5 cm, thickness of 0.5 cm and a length of 20 cm. Material was removed from the mold and dried at room temperature for 7 days. The, it was burnt on a wood stove for 12 hours.

2.7. *Ceramic Filter Performance Test*

The ground water was drained from the tank through a PVC pipe with the aid of a pump. The water was passed through ceramic filter, activated carbon granules, carbon block, then to osmosis membranes and the last pass of activated carbon powder. The flow rate was measured at intervals of 10 minutes. Water filtrate is collected and measured the TDS and pH. The Cd metal ion content was measured by AAS and compared with commercial water filter and control (untreated well water).

3. Result and discussion

3.1. *Identification of Adsorbent*

3.1.1. *The NaF Test*

The existence of allophane soil can be identified by NaF test. NaF test was done by measuring the pH of 1 g of soil in 50 mL of 1 N NaF solution for 2 minutes. The soil pH value greater than 9.4 indicates a high allophane content in soil samples [30]. The NaF can react quickly when added to the soil sample containing allophane. The fluoride (F) group can react with aluminum (Al) and break down the structure of soil aggregates so that it will release OH^- ions [31]. The NaF test results of Andisol soil samples showed that the pH value of 10.88 indicates that the Andisol soil contains allophane adequate.

3.1.2. *FTIR Characterization*

The FTIR characterization was employed to determine and evaluate as well as compare the functional group each raw materials. The FTIR spectra of Andisol Soil and Clay reveal in Fig. 2 and 3. The characteristic of each functional group shows in Table 1.

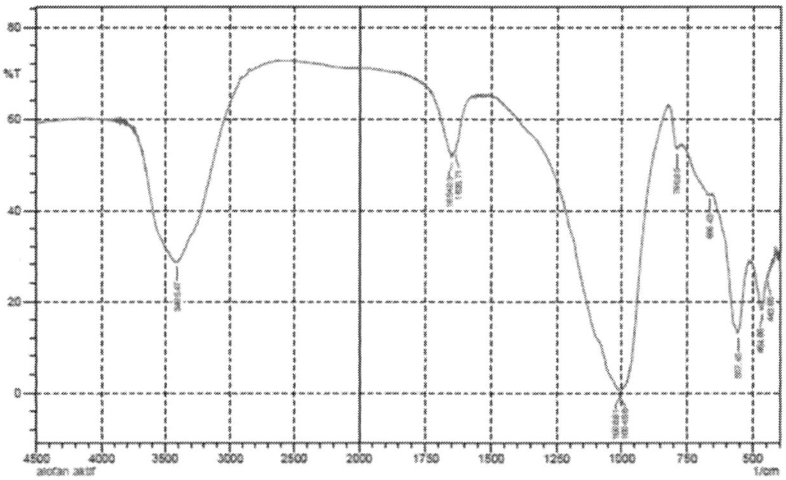

Figure 2. FT-IR spectra of Andisol Soil

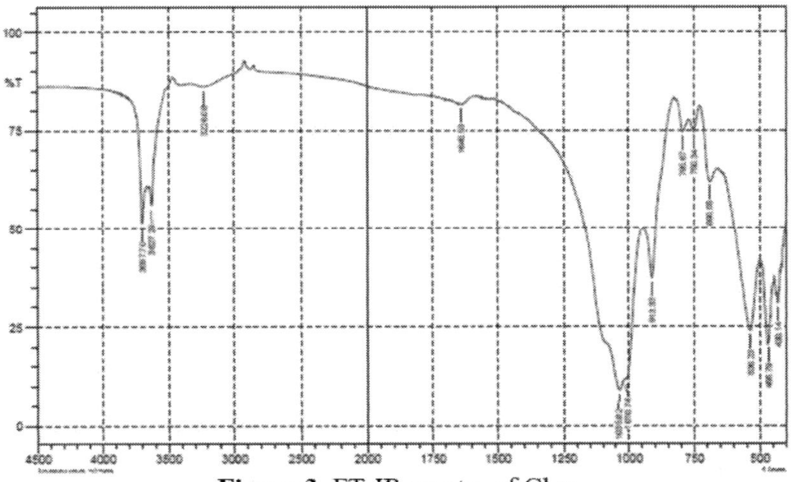

Figure 3. FT-IR spectra of Clay

Table 1. Data from the analysis of the functional group clays and Andisols

Functional Group	Wave number (cm^{-1})		
	reference	**Clay**	**Andisol**
Stretch–OH	3700-3000 (1) 3455(2)	3697.71-3226.08	3405.47
Bending Vibration H-O-H	1640 (3)	1640.53	1654.03
O-Si-O or O-Al-O	973. 11 (2); 1039.60 (1)	1035.82 1010.74	1008.81 1004.96
Bending Vibration Si-O or Al-O	470.60 (1) 485. 58 (2)	795.67 750.34	790.85
Kaolinit	3600-3800(5) 1030 (4)	3627.29; 3697.7; 1010.74; 1035.82	1004.96; 1008.81
Gibsit	3400-3500 (5); 1030; 974 (7)	1010.74; 1035.82	3405.47; 1004.96; 1008.81
Felspar	647 (4)	690.55	666.43
Allophane	3400; 1640; 1040; 670; 470(2); 430 (7)	1640.53; 1035.82; 690.55 430.14; 466.79;	3405.47; 1635.71;1654.03; 1008.81; 1004.96 443.65; 464.86;

3.1.3. *XRD Characterization*

The XRD characterization is employed to evaluate the presence of chemical and minerals in the Andisol or Clay samples revealed from the specific 2θ showed in Fig. 3. Based on the analysis of Fig. 2, 3 and 4, as well as Tables 1 and 2, it can be concluded that adsorbent Andisol and clay contain minerals identified by the diffraction peaks at d /(A) the characteristics. Minerals in clays and Andisol adsorbent are allophane, feldspar, gibsit, kaolinite and montmorillonite. The presence of those minerals was revealed by the presence of their characteristic of both functional group and basal spacing at specific 2θ which it was already explain in Table 1 and Table 2.

IOP Conf. Series: Materials Science and Engineering **349** (2018) 012047 doi:10.1088/1757-899X/349/1/012047

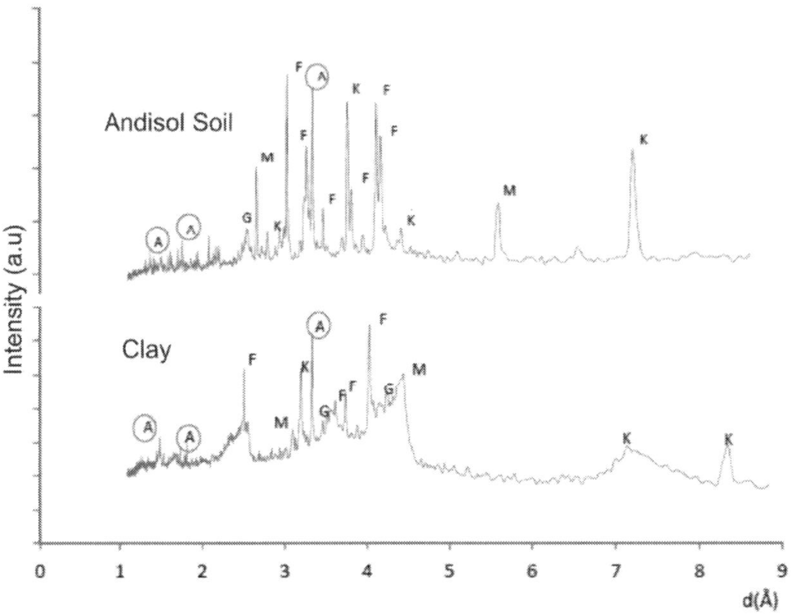

Figure 4. XRD diffractograms of Andisol Soil and Clay

Table 2. Results of XRD analysis clay soil and Andisol

Materials	**d (Å)**		
	Reference	**Clay Sample**	**Andisol Sample**
Allophane	3.30; 2.25; 1.80; 1.40 (1)	3.33; 1.81	3.29; 1.74; 1.36
Feldspar	4.00-4.20; 6.30-6.45; 3.80-3.90; 3.73-3.75; 3.64-3.67; 3.44-3.48; 3.00-3.25 (2) 3.27; 3.23; 3.59; 3.76; 3.44; (3)	4.03; 4.22; 3.74; 3.61; 3.20	4.04; 3.71; 3.70; 6.47; 3.22; 3.19; 2.98
Gibbsite	4.34; 4.83; 3.30 (2) 4.85; 4.36; 2.45; 2.38 (4)	3.45;	2.53; 3.41
Kaolinite	7.10-7.20; 4.45-4.46; 4.35-4.36; 4.17; 4.12; 3.84; 3.56-3.58 (2) 8.91; 4.65; 4.14; 3.81; 3.51 (5)	8.34; 7.14;	7.09; 3.76; 3.89; 4.34
Montmorillonite	12.00-15.00; 5.90 (2) 15.00; 1.49; 2.53; 1.29; 4.05 (6)	2.51; 1.49	5.48; 2.62

3.2. *Performance Test of Adsorbent*

Andisol/clay adsorbent has a high affinity for metal ions because they have active sites of Si-OH, Al-OH and -OH capable to provide electronegative charge on their surfaces allowing the exchange of cations in the process of adsorption of metal ions in the Cd solution [25, 37]. Sorption ability of silicate minerals derived from the number of negative charge on the structure of silicate minerals. The negative charge is neutralized by the positive charge of adsorbate, for example heavy metal cations [46]. Adsorption data results in Fig. 5 show that the adsorption of metal ions decreases with increasing of clay percentage in the mixture of clay and Andisol. The Reduction of Andisol/clay adsorbent ability can be explained by the data of surface area and acidity number of each adsorbent.

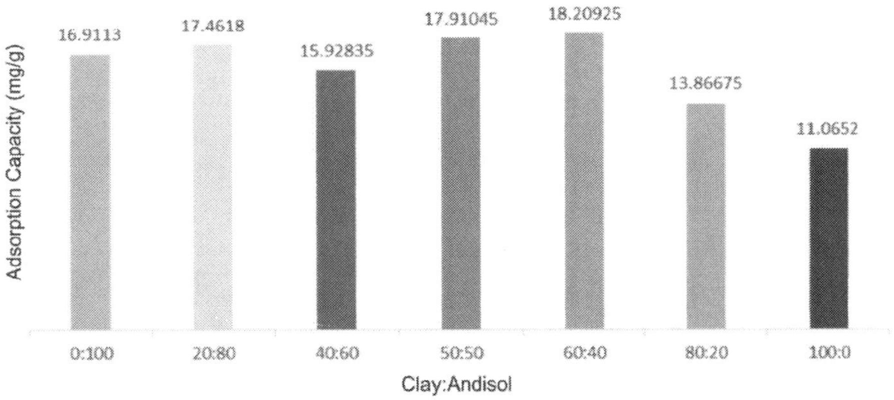

Figure 5. Comparison of adsorption capacity between various adsorbent.

The surface area is one of the decisive factors in the sorption process since the surface area provides the size of the area on the surface of adsorbent in the adsorption process of the Cd metal ion. So, the larger of the adsorbent surface area will produce greater adsorption capacity [36]. Clay has a surface area of 56.54 m^2/g, while the Andisol soil has a much larger surface area, which is 257.84 m^2/g. The surface area data of both Andisol and clay show in Table 3. Adsorbent with composition ratio 0:100, or 100 percent of the Andisol shows the highest sorption capacity compared to other adsorbent composition.

The measurement results show Andisol soil acidity has a larger value, i.e. 4.18 mmol/g compared to clay and the mixture, that are 2.65 and 3.47 mmol/g, respectively (see Table 3). The greater the number of acidity means adsorbent able to provide more active sites on the surface as a medium in the process of adsorption of the metal ion Cd. [13].

Table 3. Surface area and acidity of Andisol and Clay

Sample	**Surface Area (m^2/g)**	**Acidity (mmol/g)**
Andisol	257.84	4.18
Clay	56.54	2.65

Adsorption capacity of adsorbent calcined at 100 °C and 200 °C higher than the adsorbent calcined at 400 °C as shown in Fig. 6. The Increasing of calcination temperature is able to free the water molecules trapped in the matrix of clay and Andisol leaving a porous cavity structure and increase the surface area of adsorption. The calcination temperature that is higher than the optimum temperature damaged the adsorbent structure and consequently decreased the surface area so that the adsorbent media limited [27]. The results of this study are also consistent with the results of Wicaksono et al. [49] that they calcined bentonite at 120 and 400 °C. Bentonite adsorption ability was tested by conducting adsorption process against methylene blue solution. The results showed that the bentonite with a calcination of 120 °C was able to absorb the methylene blue and better than the bentonite which is calcined by 400 °C.

Adsorption capacity of Andisol/clay adsorbent increased with the long of contact time. When adsorption has reached the maximum, adsorption capacity was reduced by increasing contact time. At a temperature of 100 and 200 °C, all compositions of adsorbent adsorption capacity increased with increasing of contact time by 30 to 60 minutes and decreased when the contact time of 90 minutes.

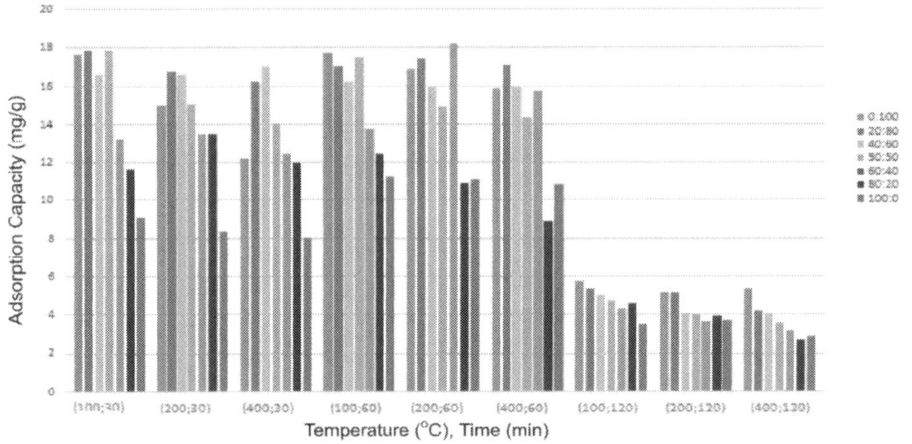

Figure 6. Effect of temperature and contact time toward adsorption capacity of various composition of Andsiol/clay adsorbent

At the contact time of 60 minutes, most of the active sites on the surface of clay and Andisol has been occupied by metal ions Cd and has been occured repulsion between adsorbed metal ions Cd and metal ions in solution to occupy the active sites. The repulsive force and the competition among the Cd metal ions caused reduction of metal ion adsorption capacity of Cd^{2+} [9, 14]. With increasing contact time, the more hydrated cations formed with a radius larger than the radius of Cd metal ions thus blocking the process of adsorption [29]. Longer contact time is not always followed by a rise adsorption capacity. When the contact time has reached saturation point, there will be events desorption, the metal ions Cd will be released back into solution due to an excessive contact between metal ions Cd in solution and adsorbed ion [44, 20].

3.3. *Adsorption Isotherm*

Langmuir isotherm equation is determined using the equation $^{c}/_{Q} = {}^{1}/_{(k.Q_{max})} + {}^{c}/_{Q_{max}}$ and Freundlich isotherm using equation $\log Q = \log K + {}^{1}/_{n} \log c$ with each is made curve c vs. c/Q and log c vs log Q as shown in Fig 7a and 7b, respectively. Based on the fitting linier, the linearity of Freundlich isotherm has a value of $R^2 = 0.9914$, higher than the Langmuir isotherm, i.e. $R^2 = 0.8872$. Thus, it can be concluded that the type of isotherm in this study followed the Freundlich model.

Freundlich isotherm model explains that the adsorption occurs on more than one surface (multilayer) and adsorbent have heterogeneous surfaces with a different binding energy [3; 18]. Cd metal ion adsorbed into Andisol/clay adsorbent is physical sorption. This sorption type is suitable for sorption mechanisms that require regeneration process.

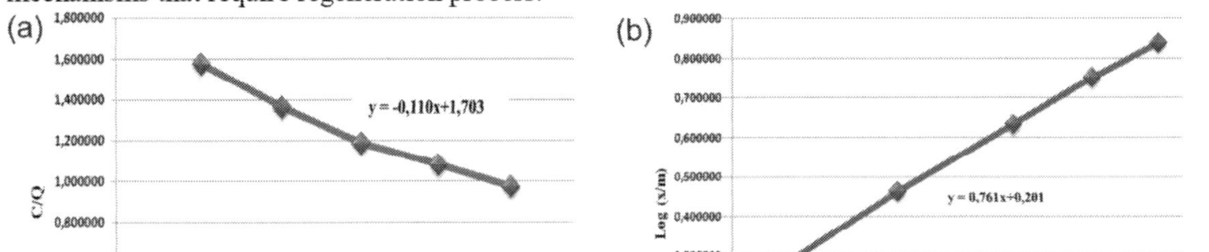

Figure 7. (a) Langmuir and (b) Freundlich isotherm adsorption models.

3.4. *Performance of Filter Purifier*
One of the drinking water treatment processes is utility of a ceramic filter. The ceramic filter is able to reduce physical, chemical and biological pollutants in order to obtain clean water that can be used for drinking water [43, 19, 47, 21, 26, 17, 51, 35, 11]. Muhdarina et al. [26] explains in their research on special properties of porous ceramic membranes, such as the high temperature stability, mechanical strength and easy in terms of regeneration.

The ingredients for making a ceramic filter can be vary, but as a main ingredient is clay because of its ability to be shaped and resistant to high temperatures. Other ingredients combine to improve the effectiveness of the performance of the ceramic filter, such as Andisol soil and flour. Andisol soil is added as a mixture for the manufacture of filters since its cationic exchange capacity is higher to adsorb heavy metals [33, 23, 10], but this material is difficult to set up and less resistant to high temperatures. Starch powder is used to increase the flow rate or discharge of water. At high combustion temperatures flour powder will be burned and left the room in a filter so that these spaces will be used as a water molecule. The measurement results in Table 5 shows that the ceramic filter is effective in reducing dissolved solids (94%) and lower Cd concentration in water wells (98.9%), higher than a commercial filter, ie 93.1% (TDS) and 93.8 % (Cd).

4. Conclusion
Composition of clay and Andisol, calcination temperature and contact time affect the adsorption capacity of Cd metal ions in solution models. The optimum conditions of Andisol/clay adsorbent to adsorb Cd metal ions in solution models are the composition of Andisol:Clay (60:40), the calcination temperature of 100 °C and a contact time of 60 minutes. The ceramic filter of Andisol/Clay mixture is effective to reduce the metal ion content of cadmium in water by 98.9% and bacterial pathogens that produce potable water.

References
[1] Al-Jill S A, Latif M S 2013 *MTAEC9*, **47(4)** 481.
[2] Alhawas M, Alwabel M, Ghoneim A, Alfarraj A, Sallam A 2013 *Proc. Int. Acad. Eco. Environ. Sci.* **3(2)** 160.
[3] ASTDR, 2012. *Toxicological Profile for Manganese. U.S. Department of Health and Human Services. Public Health Service.* Agency for Toxic Substances and Disease Registry. Division of Toxicology and Human Health Sciences. Atlanta.
[4] Sunaryo W W 2010 *Spektra.* **10(2)** 119.
[5] Schulze D G 2005 *Clay Minerals. Purdue University, West Lafayette, IN, USA* http://www.geoinfo.amu.edu.pl/geoinf/m/GL11. accessed on 16 April 2014
[6] Agmalini S, Lingga N N, Nasir S 2013 *Jurnal Teknik Kimia* **19(2)** 59.
[7] Pranoto, Suranto, Sugiyarto K H, Ashadi 2013 *J. Environ. Earth Sci.* **3(5)** 48.
[8] Munir 1996 *Main Soils Indonesia.* Pustaka Jaya. Jakarta, 71-75
[9] Hariyadi S, Margono, Purwanto D S 2013 *Gema Kesehatan Lingkungan* **X(1)**, ISSN 1693-3761
[10] Prakash N, Latha S, Sudha P N, Renganathan N G 2013 *Chem Sci Trans.* **2(S1)** S239.
[11] Sulistyarini A 2012 *Identification, activation, and Characterization of Soil Volcanic Mount Arjuna as Penjerap Metal Copper (Cu).* Fakultas Matematika dan Ilmu Pengetahuan Alam Universitas Sebelas Maret
[12] Van-Halem D 2006 *Ceramic Silver Impregnated Pot Filters for Household Drinking Water Treatment in Developing Countries.* Thesis. Sanitary Engineering Section, Department of Water Management, Faculty of Civil Engineering, Delft University of Technology.
[13] Flynn T 2005 New Filter Promises Clean Water for Millions. The Australian National University. *http://info.anu.edu.au/ovc/assets/claypotfil terfinalweb.pdf.* accessed on 31 April 2014
[14] Connors S, Lanza R, Sirocki A 2013 *Removal of Ibuprofen from Drinking Water Using Adsorption.* Worcester Polytechnic Institute, Worcester

[15] Dewi Y S 2011 *Lingkungan Tropis*. **5(2)** 93.
[16] Diantariani N P 2010 *Jurnal Kimia*. **4(1)** 91
[17] Haryati S, Supraptiah E, Bustan M 2011 *J. Appl. Eng. Chem.* **1** 18.
[18] Alemayehu D D, Sing S J, Tessema D A 2012 *Universal J. Environ. Res. Technol.* **2(5)** 411.
[19] Ariffeni 2011. *Metal biosorption of Cd (VII) with Fungi Using Saccharomyces Sp isolated from the Solid Waste Coca Cola. Essay*. Department of Chemistry. Fakultas Matematika dan Ilmu Pengetahuan Alam. Universitas Andalas. Padang.
[20] Arpa C, Say R, Satiroglu, Bektas S, Yurum Y, Geng O 2000 *Turk. J. Chem.* **24** 209.
[21] Auliah A 2009 *Jurnal Chemica*. **10(2)** 14.
[22] Bhattacharyya K G, Gupta S S 2008 *Adv. Colloid Interface Sci.* **140** 114.
[23] Clark C J, McBride M B 1984 *Clays and Clay Minerals*. **32(4)** 300.
[24] Eba F, Gueu S, Eya'A-Mvongbote A, Ondo J A, Yao B K, Nlo J N, Biboutou R K 2010 *Int. J. Eng. Sci. Technol.* **2(10)** 5001.
[25] Hagan J M, Harley N, Pointing D, Sampson M, Smith K, Soam V 2009 *Resource Development International -Cambodia Ceramic Water Filter Handbook-*, Version 1.1, Phnom Penh, Cambodia
[26] Handayani M, Sulistyono E 2009 *Prosiding Seminar Nasional Sains dan Teknologi Nuklir*, PTNBR- Batan, Bandung, 3 June 2009.
[27] Henry M, Maley S, Mehta K 2013 *Int. J. Service Learn. Eng.* **8(1)** 62
[28] Heraldy E, Pranoto, Prowida D 2004 *Journal Alchemy*. **3(1)** 32.
[29] Iyoda F, Hayashi S, Arakawa S, Okamoto M 2011 *Nanostructure and Adsorption Behaviour of Natural/Synthetic Allophanes*. PPS-27, 27th World Congress of the Polymer Processing Society, May 10-14, Marrakech, Morocco.
[30] Itou Y, Shozugawa K, Matsuo M 2009 *XAFS Speciation of Adsorbed Zinc Ion on Allophane in Presence of Humic Acid*. Photon Factory Activity Report. 2009#27 Part B (2010). Graduate School of Arts and Sciences, The University of Tokyo, Tokyo
[31] Li L, Lee R 2009 *Sep. Sci. Technol.* **44** 3455.
[32] Mortland M M 1954 *Soil Sci.* **78** 343
[33] Muhdarina, Muhammad A B, Muchtar A 2010 *Reaktor*. **13(2)** 81.
[34] Munir 1996 *Main Soils Indonesia*. Pustaka Jaya. Jakarta, 71-75
[35] Parfitt R L 2009 *Clay Mineral.* **44** 135.
[36] Sajidu S M I, Person I, Masamba W R L, Henry E M T, Kayambazinthu D 2006 *Water SA*. **32(4)** 519.
[37] Sapana M M, Sonal G C, Raut P D 2012 *Res. J. Recent Sci.* **1(3)** 31.
[38] Sanchez A G, Ayuso E A, De Blas J 1999 *Clay Mineral.* **34**, 469.
[39] Hartopo P S, Pranoto 2014 *Kajian Efektifitas Campuran Lempung Bekonang dan Alofan Gunung Lawu Sebagai Adsorben Mn*. Tesis Prodi Ilmu Lingkungan, Program Pascasarjana UNS.
[40] *EB/1b%20Clay%20minerals_EncSoilEnv_SCHULZE%2005.pdf.* accessed on 26 Maret 2014.
[41] Sinha A, Khare S K 2013 *Int. J. Appl. Sci. Biotechnol.* **1(4)** 162.
[42] Sulastri S, Kristianingrum S 2007 *Proceedings of the National Seminar of Chemistry and Chemical Education*. ISBN: 978-979-98063-1-4.
[43] Talaat H A, El Delrawy N M, Abulnour A G, Hani H A 2011. *Evaluation of Heavy Metals Removing Using Some Egyptian Clays*. 2nd International Conference on Environmental Science and Technology. IPCBEE. Vol. 6. IACSIT Press
[44] Tan K H 1982 *Dasar-Dasar Kimia Tanah* University Gajah Mada Press Yogyakarta.
[45] Visekruna A, Strkalj A, Pajc L M 2011 *The Holistic Approach to Environment*. **1(1)** 29.
[46] Wicakso D R, Mirwan A, Abdullah 2012 *Conversion*. **1(1)** 7.
[47] Widjonarko D W, Pranoto, Christina Y 2003 *Journal Alchemy*. **2**(2).
[48] Yakup I 2010 *Micro- and Nano- Porous Adsorptive Materials for Removal of Contaminants from Water at Point-of- Use*. Disertasi. Department Of Mechanical and Aerospace Engineering. Princeton University.
[49] Yuan P, Fan M, Yang D, He H, Lui D, Yuan A, Zhu J, Chen T 2009 *J. Hazard Mater.* **166** 821

IOP Conf. Series: Materials Science and Engineering **349** (2018) 012048 doi:10.1088/1757-899X/349/1/012048

Antidiabetic activity from cinnamaldydhe encapsulated by nanochitosan

Purbowatingrum[1], Ngadiwiyana[1], E Fachriyah, Ismiyarto, B Ariestiani, Khikmah

Chemistry Department, Science and Mathematics Faculty, Diponegoro University, Jl. Prof. H. Soedarto, SH, Tembalang, Semarang, Indonesia, Post code 50275.

purbowatining@gmail.com

Abstract. Diabetes mellitus (DM) is a disease characterized by chronic hyperglycemia and metabolic disorders of carbohydrates, proteins, and fats due to reduced function of insulin. Treatment of diabetes can be done by insulin therapy or hypoglycemic drugs. Hypoglycemic drugs usually contain compounds that can inhibit the action of α-glucosidase enzymes that play a role in breaking carbohydrates into blood sugar. Cinnamaldehyde has α-glucosidase inhibit activity because it has a functional group of alkene that is conjugated with a benzene ring and a carbonyl group. However, the use of this compound still provides unsatisfactory results due to its degradation during the absorption process. The solution offered to solve the problem is by encapsulated it within chitosan nanoparticles that serve to protect the bioactive compound from degradation, increases of solubility and delivery of a bioactive compound to the target site by using freeze-drying technique. The value of encapsulation efficiency (EE) of cinnamaldyhde which encapsulated within chitosan nanoparticles is about 74%. Inhibition test result showed that cinnamaldehyde-chitosan nanoparticles at 100 ppm could inhibit α-glucosidase activity in 23.9% with 134,13 in IC_{50}. So it can be concluded that cinnamaldehyde can be encapsulated in nanoparticles of chitosan and proved that it could inhibit α-glucosidase.

1. Introduction

The worldwide estimation of diabetic patients around 2030 will be more than double from that of 2005, and most of these will be dominated by those suffering from type 2 diabetes [1]. So diabetes has become the third largest health killer, which is only second to cancer and cardiovascular diseases. At present, the drugs for diabetes include sulfonylureas and biguanides. Both of them have some beneficial effects on hyperglycemia and also have some adverse effects. So it was an urgent need to have some new kinds of drugs to solve this problem until α-glucosidase inhibitors turned up. The α-glucosidase inhibitors which inhibit the enzyme at the brush border of small intestine are effective for delaying glucose absorption and preventing postprandial blood glucose level elevation, therefore they play a significant role in the therapy as chemotherapeutic agents for Noninsulin-dependent diabetes mellitus (NIDDM) [2]. Such inhibitors, including acarbose and voglibose, are currently used clinically in combination with either diet or other anti-diabetic agents to control blood glucose levels of patients [3]. Natural resources provide a huge and highly diversified chemical bank from which we can explore for potential therapeutic agents by bioactivity-targeted screenings [2]. Recently, some of the natural phenolic compounds and polysaccharides (such as tea, grapefruit, strawberries) have been shown to inhibit the α - glucosidase enzyme and potential as the natural therapeutic agent for the treatment of *diabetes mellitus* [4].

Chitosan is prepared by the alkaline deacetylation of chitin [5]. Chitosan uses one as a means of delivering drugs and genes and biomaterials for immobilization. Chitosan is a natural biopolymer of interest because of the presence of reactive amino groups and hydroxyl functional groups. Chitosan has the desired biocompatibility characteristics and its ability to increase membrane permeability so that chitosan is very promising as it has the ability to form membranes [6]. Physical modification of chitosan involves changing the particle size or chitosan granules to be smaller for wider use. Utilization of physical modification produces nanoparticle size [7]. Biological activities of chitosan, such as its antitumor effect [8], cholesterol-lowering effects [9] and antibacterial effects [10] are well known. However, few reports are available on the antidiabetic effects of chitosan.

Cinnamaldehyde has an alkene functional group which is conjugated with a benzene ring and a carbonyl group. Research Subash et al [11] mentioned that sinamaldehid has an antidiabetic effect in diabetic rats induced streptozotocin with a decrease in glucose levels of 63.29%. Sinamaldehid contained in cinnamon oil is useful as an antidiabetic with IC50 value of 27.96 ppm [12], because sinamaldehid has the most functional group similar to caffein and cinnamic compounds compared to other cinnamon oil components. This advantage causes sinamaldehid to have potential as an antidiabetic

This research investigated the characterization of chitosan structure, and anti-diabetic potential of encapsulation cinnamaldyhde-chitosan nanoparticles. Firstly, synthesized chitosan nanoparticles, nanoemulsion cinnamaldyhde, and encapsulated cinnamladyhde in chitosan nanoparticles. To determine the encapsulation of cinnamaldehyde-chitosan nanoparticles as anti-diabetic was test by inhibition *α-glucosidase* enzyme.

2. Methods

2.1 Materials and Instrumentation

Materials; Chitosan, Cinnamaldehyde (Sigma aldrich), Aquades, Na-TPP (Sigma Aldrich), 1% glacial acetic acid, Ethanol 70%, Tween 80, ethanol, ether, bovine serum albumin, ρ-nitrophenyl α-D-glucopyranoside.

Research tools used are glass that commonly used in the laboratory, Hot Plate Stirrer (IKA C-MAG HS 7), Centrifuge (Corning LSE Compact), Freeze drying (LLI500), PSA (Microtrac-Particle Size Analyzer), FTIR (Shimadzu-Fourier Transform Infrared) and Spectrophotometer Ultraviolet-Visible (Spectroquant).

2.2 Eksperimental

2.2.1. Preparation of chitosan nanoparticles.

Chitosan of 0.2 g dissolved in 100ml glacial acetic acid 1% (v / v). Na-TPP 0.5 g dissolved in 100ml aquades. The Na-TPP solution was added dropwise into the chitosan solution, stirred using a magnetic stirrer for 1 hour. Subsequently the chitosan nanoparticle suspension was generated at centrifugation at 6000 rpm for 10 min. The centrifugation results were then dried using freeze drying and PSA characterization was performed to detect the nano size formed.

2.2.2 Preparation of cinnamaldehyde emulsion

Preparation of nanoemulsion using spontaneous emulsification technique. The emulsion system consists of organic phases (sinamaldehid and ethanol 70%) and water phases (aquades and Tween 80). The organic phase was prepared by mixing the cyclamaldehyde and 70% ethanol solvent to reach the total dissolved solids of 20° brix. The spontaneous emulsification technique is done by adding the organic phase into the water phase through drip (dropwise). At the time of dripping the organic phase into the water phase, the water phase is stirred using a magnetic stirrer, the surfactant used is Tween 80. The best emulsion formula is determined by the best characteristics (particle size, polydispersity index value, zeta potential) using light-scattering Particle Size Analyzer DelsaTM Nano C (Beckman Coulter, France).

2.2.3 Encapsulation of cinnamaldehyde nanoemulsion

After the manufacture of nanoemulsion, a coating material of nanocitosan was added with a ratio of 1: 1 (v / v). Then stirring for 1 hour using a magnetic stirrer. A slightly thicker solution is produced, then centrifuged at 6000 rpm for 30 minutes to produce gel suspension and supernatant. The gel suspension is dried using a freeze dryer to produce a cinnamaldehyde encapsulation powder. The sinamaldehyde encapsulation powder was then calculated by encapsulation efficiency (% EE) using a UV-Visible spectrophotometer.

2.2.4 Alpha-glucosidase inhibition test.

An enzyme solution was prepared by dissolving 1.0 mg α-glucosidase in 100 ml of phosphate buffer (pH 7.0) containing 200 mg bovine serum albumin. Before use as much as 250μL 20 mM ρ-nitrophenyl α-D-glucopyranoside as substrate, then added 490μL 100mM phosphate buffer (pH 7.0) and 10 μL solution of the synthesis compound in DMSO. The reaction mixture was incubated at the optimum temperature for 5 min, after which 250 μL of enzyme solution was added and incubated. The enzymatic reaction was discontinued by the addition of 1000 μl of 200 mM sodium carbonate. The reaction product is a ρ-nitrophenol compound which can be measured using a UV-Visible spectrophotometer at a 400nm wavelength. The sample used in the α-glucosidase inhibition test was sinamaldehyde with various concentrations of 6.25 ppm, 12.5 ppm, 25 ppm, and 50 ppm with DMS solvent. Percent of inhibition can be calculated by the equation:

$$\% = \frac{C - S}{C} x\ 100\%$$

with S = absorbance of the sample (obtained from the S1 - So, where S1 = absorbance of the sample with enzyme addition and So = the absorbance of the sample without enzyme addition) and C = absorbance of control (DMSO), without sample (control-blank). IC50 values was obtained from: a linear regression equation Y = a + bx.

IC50 ⟶ Y = 50, then it was substituted into the linear regression equation becomes:

IC50 = (50-a)/b

3. Results and Discussion

3.1 Charaterization

Nanoparticles have very specific properties and surface area that multiply, usually can increase the chances of the occurrence of more chemical reactions. A substance can be absorbed directly into the bloodstream where it is needed so that it is more effective than broken down during the digestive system [13]. Factors that affect the work of substances such as repeated doses and income [12]. The application of nanotechnology to chitosan that is in the formation of nanoparticles allows chitosan to be a conductor of functional compounds or drugs to be more effective. The use of nanoparticles in the coating process can increase the absorption of active substances. The nanoencapsulation process means that various nutrients can be absorbed directly into the bloodstream where the nutrients are needed, it will be much more effective than broken down during processing or breakdown with enzymes in the digestive system. New mechanisms can be developed with nanotechnology in drug delivery issues. The foundation used in the drug delivery system is to stimulate the effectiveness of the drug, through specific targeting and specific cells, the acceleration of delivery time, and the prevention of digestive enzymes in breaking down the drugs being consumed [17].

Modern particle calculations generally use image analysis or some kind of particle counting such as Particle Size Analyzer (PSA) analysis. The average particle size distribution obtained by the size of the sinamaldehid emulsion has a particle size of 336.3 nm while nanokitosan has size 2249, 7 nm.

The value of encapsulation efficiency (EE) of cinnamaldehyde which encapsulated within chitosan nanoparticles is about 74%. This shows that the coated sinamaldehid in chitosan nanoparticles is 74%

3.2 Inhibition α-glukosidase test

The antidiabetic activity test of cinnamaldehyde-encapsulated with nanochitosan was performed by inhibition of glucosidase enzyme activity with glucobay tablet as standard. On this test, the alpha-glucosidase enzyme will hydrolyze p-nitrophenyl-α-D-glukopyranoside substrate into the yellow p-nitrophenol and glucose with the following the reaction:

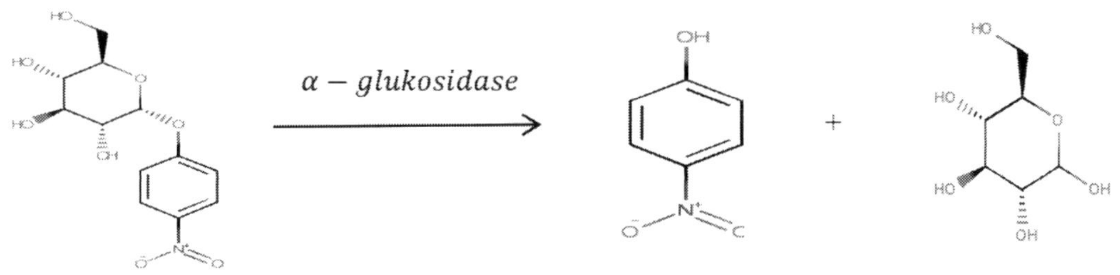

Fig 1. An enzymatic reaction of α-glucosidase and *p*-nitrophenyl-α-D-glucopiranoside

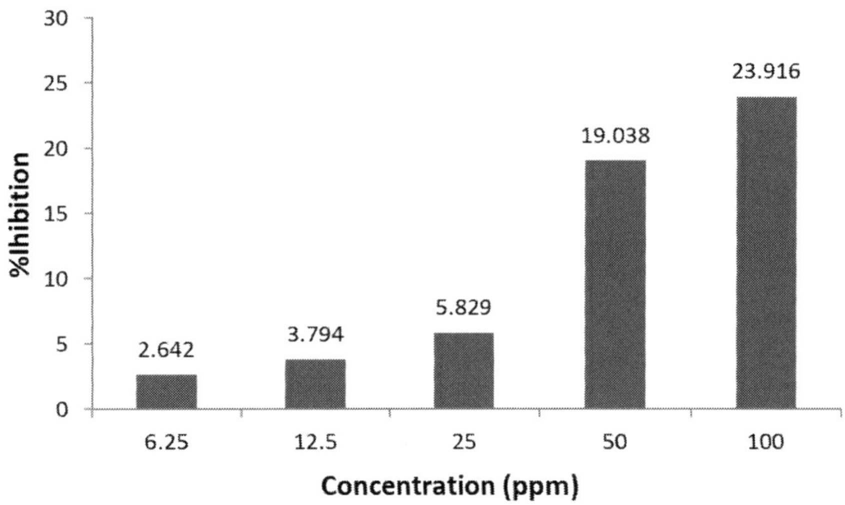

Fig 2. Percent inhibition of α-glucosidase enzyme from cinnamaldehyde-encpasulated

Enzyme activity was measured based on the absorbance of the p-nitrophenol which has yellow color. In the presence of Gallic acid-chitosan nanoparticles which acts as an alpha-glucosidase inhibitor so the p-nitrophenol produced will reduce which was characterized by a lowering in the intensity of yellow color. The p-nitrophenol is a product of the enzymatic reaction of alpha-glucosidase with p-nitrophenyl- α -

IOP Conf. Series: Materials Science and Engineering **349** (2018) 012048 doi:10.1088/1757-899X/349/1/012048

D-glucopyranose substrate. Gallic acid-chitosan nanoparticles can inhibit the action of the alpha-glucosidase enzyme which was characterized by a reduced concentration of p-nitrophenol.

In Figure 2 shows a percentage of inhibition of α- glucosidase enzyme from sinamaldehyde-encapsulated at a concentration of 6.25 ppm, 12.5ppm, 25ppm, 50ppm, 100ppm. From Figure 2 it can be seen that cinnamaldehyde-encapsulated at 100 ppm can inhibit α- glucosidase enzyme by 23,916% with IC_{50} value were 134,13ppm. IC_{50} value is a number indicating the concentration of sample (ppm) capable of inhibiting 50% of α-glucosidase enzyme activity

REFERENCES

[1] Gershell, L., 2005. Type 2 diabetes market. Nat. Rev. Drug Discov. 4,367–368.

[2] H.M. Chen, X.J. Yan, W. Lin,. 2005. Progress in α-glycosidase inhibitors, Mar. Sci.73–76.

[3] Van de Laar, F.A., Lucassen, P.L., Akkermans, R.P., Van de Lisdonk, E.H., Rutten, G.E., Van Weel, C., 2005. alpha-Glucosidase inhibitors for type 2 diabetes mellitus. Cochrane Database Syst. Rev. 18, CD003639.

[4] Dewi UK, Tyas RS. 2009. Efek rebusan daun tapak dara pada dosis dan frekuensi yang berbeda terhadap kerusakan dan akumulasi glikogen pada hepar mencit (*Mus musculuc*). *BIOMA* 11(1): 1-5.

[5] Furda I., 1983. Unconventional Sources of Dietary Fiber, ACS symposium series 214, 1983, pp. 105—122.

[6] Gallaher C. M., Munion J., Hesslink R., Jr., Wise J., Gallaher D. D., J. Nutr., 130, 2753—2759 (2000).

[7] Irianto et all. 2011. Proses dan aplikasi nanopartikel kitosan sebagai penghantar obat. Squalen Vol. 6 No. 1

[8] Jeon Y. J., Kim S. K., J. 2001. Microbiol. Biotechnol., 11, 281—286

[9] Liu G S Xu and L Chen 2012 Chemical composition and bioactivities of a water-soluble polysaccharide from the endodermis of shaddock. Int. J. Biol. Macromol 51 763–766

[10] Suzuki, Ko. Mikami, Takeshi. Okawa, Yoshi. Tokoro, Akio. Suzuki, Shigeo. Suzuki, Masuko. 1986. Antitumor effect of hexa-N-acetylchitohexaose and chitohexaose. Carbohydrate Research. 403-408

[11] Wahyono. 2010. Ciri nanopartikel kitosan dan pengaruhnya pada ukuran partikel dan efisiensi penyalutan ketoprofen. Institut Pertanian Bogor

[12] Ngadiwiyana, Ismiyarto, Nor Basid, A.P, Purbowatingrum, R.S. 2011. Potensi Sinamaldehid Hasil Isolasi Minyak Kayu Manis sebagai Senyawa Antidiabetes. Majalah Farmasi Indonesia, Vol 22, No. I, PP. 9-14

[13] Winarno FG, Fernandez IE. 2010. *Nanoteknologi bagi Industri Pangan dan Kemasan.* Bogor (ID): M-Brio Press.

IOP Conf. Series: Materials Science and Engineering **349** (2018) 012049 doi:10.1088/1757-899X/349/1/012049

Ab initio computational study of –N-C and –O-C bonding formation : functional group modification reaction based chitosan

P Siahaan[1], S N M Salimah[1], M J Sipangkar[1], D Hudiyanti[1], M C Djunaidi[1] and M D Laksitorini[2]

[1]Department of Chemistry, Diponegoro University, Semarang, Indonesia,
[2]Department of Pharmaceutics, Faculty of Pharmacy, Gadjah Mada University, Sekip Utara, Yogjakarta, Indonesia

E-mail : parsaoran_s@undip.ac.id

Abstract. Chitosan application in pharmaceutics and cosmeceutics industries is limited by its solubility issue. Modification of -NH_2 and -OH fuctional groups of chitosan by adding carboxyl group has been shown to improve its solubility and application. Attempt to synthesize carboxymethyl chitosan (CMC) from monocloroacetic acid (MCAA) has been done prior this report. However no information is available wether –OH (-O-C bonding formation) or -NH_2 (-N-C bonding formation) is the preference for - CH_2COOH to attach. In the current study, the reaction mechanism between chitosan and MCAA reactants into carboxymethyl chitosan (CMC) was examined by computational approach. Dimer from of chitosan used as a molecular model in calculation All the molecular structure involved in the reaction mechanism was optimized by ab initio computational on the theory and basis set HF/6-31G(d,p). The results showed that the -N-C bonding formation via S_N2 than the -O-C bonding formation via S_N2 which have activation energy 469.437 kJ/mol and 533.219 kJ/mol respectively. However, the -O-C bonding formation more spontaneous than the -N-C bonding formation because ΔG the formation of O-CMC-2 reaction is more negative than ΔG of formation N-CMC-2 reaction is -4.353 kJ/mol and -1.095 kJ/mol respectively. The synthesis of N,O-CMC first forms -O-CH_2COOH, then continues to form -NH-CH_2COOH. This information is valuable to further optimize the reaction codition for CMC synthesis.

1. Introduction

Chitosan is a copolymer of glucosamine and N-acetyl glucosamine monomer linked by 1,4-glucoside bonds[1]. It is commonly obtained from deacetylation of chitin produced by crustacean. Chitosan has high biocompatibility, relatively non toxic and biodegradable, promoting wide use of this polymer in drug delivery, antibacterial, cosmetic, food film and water purifier [2,3]. This polymer has reactive groups -NH_2 and -OH that been associated with its insolubility at neutral and alkaline solutions [3]. Some strategies has been performed to increase the solubility of chitosan. One of them by reacting chitosan with monocloro acetic acid (MCAA) to produce carboxymethyl chitosan (CMC) [4]. The hydrophilic group (-CH_2COOH) in MCAA group increased chitosan solubility. Modification of chitosan into CMC occurs via nucleophilic substitution reactions with the formation of -N-C and -O-C bonds resulting in N-CMC (-NHCH_2COOH) and O-CMC (-O-CH_2COOH) [1]. The O-CMC synthesis occurs at 30 °C with 40-50% NaOH concentration, while N-CMC synthesis occurs at 90 °C with 25% NaOH and at 60 °C yields N,O-CMC product [5]. Previous studies done in our lab shown that modification of chitosan to CMC improved its solubility and encapsulation efficency [6]. However from those studies it is not known if the substitution present on -NH_2 and -OH groups. In real experimental setting it is costly to study intermediate. Thus, in this studies, the formation preference between -N-C with -O-C bond will be examined through ab-initio computational method.

The term ab-initio means "from first principle" which means the total wave function, molecular orbital, and energy are calculated respectively without empirical parameters [7]. The ab initio computational method has been efficiently predicted the formation mechanism of-N-C(= O)- on the cadherin peptide synthesis [8]. Computational methods are also able to determine the mechanism of glycine formation of CO_2, H_2 and H_2C=NH which lead to a conclusion thet glycine formation was via joint reaction pathway where CO_2, H_2 and H_2C=NH react together [9].

This research used ab-initio computational method with theory and basis set RHF / 6-31G (d, p) to determine the stable structure and energy of each molecule (R: reactant, TS-m-n: transition state, I-m-n: intermediate and P-n : Product) involved in CMC formation reaction mechanism. The energy obtained becomes the basis for determining the most likely mechanisms to occur. The molecular model used is chitosan dimer. Analysis of the mechanism of the reaction was carried out on six mechanism models, two via S_N1 (monomolecular nucleofilic substitution) that produced P_O-1 (O-CMC-1) and P_N-1 (N-CMC-1), and four via S_N2 (bimolecular nucleofilic substitution) produced P_O-2 (O-CMC-2), P_N-2 (N-CMC-2), $P_{N,O}$-1 (N, O-CMC-1) and $P_{N,O}$-2 (N, O-CMC -2). The proposed reaction mechanism is as follows:

Mechanism 1 : $ClCH_2COOH$ → $^+CH_2COOH$ + Cl^- + chitosan-OH → chitosan-$OHCH_2COOH$ → chitosan-O-CH_2COOH + HCl

Mechanism 2 : $ClCH_2COOH$ → $^+CH_2COOH$ + Cl^- + chitosan -NH_2 → chitosan-NH_2CH_2COOH → chitosan -NH-CH_2COOH + HCl

Mechanism 3: chitosan-OH + $ClCH_2COOH$ → chitosan-OH-CH_2COOH + Cl^-→ chitosan-O-CH_2COOH + HCl

Mechanism 4 : chitosan-NH_2 + $ClCH_2COOH$ → chitosan-NH_2-CH_2COOH + Cl^-→ chitosan-NH-CH_2COOH + HCl

2. Computational Metods

All molecular calculations are performed with linux-based Gaussian 03 applications and all molecules are visualized with chemcraft software with input files created using a notepad++. In this research all calculations using the theory and basis set RHF / 6-31G (d, p). The stable structure at the stationary point in PES (Potential Energy Surface) as well as the minimum energy is determined by the "opt (CalcFC)" command, while the TS structure is determined by "opt (calcfc, ts, noeigentest)" command. Frequency of harmonic vibration is calculated by the "freq" command to determine the correction of zero-point vibration energy (ZPVE) for all molecules. Frequency is also used to determine the TS structure by ensuring that there is only one imaginary frequency, whereas the molecule at the stationary point has no imaginary frequency. In addition, the frequency calculation is also to know the free energy of Gibs of each molecule involved in the reaction mechanism [10,11].

3. Results and Discussion

Before studying the mechanism, computational study of reactants was done to determine bond characteristic. Reactant in this synthesis was chitosan dimer and MCAA. Chitosan dimer has two types of -NH_2 bound to C4 and C9 and two types of -OH bound to C17 and C22 (**Figure 1.a**). We found that substitution of carboxymethyl groups occurred at -NH_2 bound to C4 and -OH bound to C17. Substitution reactions did not occur in -C9-NH_2 and -C22-OH, although N and O atoms have a more negative charge than -C4-NH_2 and -C17-OH but the steric blockages in C9-NH_2 and C22-OH are larger. Large steric hindrances make substitution reactions difficult because the chitosan molecule undergoes a flip that makes the transition state structure difficult to form.

A stable structure of MCAA, a hydrophil group donor for CMC formation, was depicted at **Figure 1.b**. The minimum energy structures of chitosan dimer and MCAA are -328.35 x 10^4 kJ/mol and -180.28 x10^4 kJ/mol respectively. From the frequency calculation, the enthalpy energy (H) and free energy Gibs (ΔG) for chitosan dimer molecule are -328.34 x 10^4 kJ/mol and -328.36 x10^4 kJ/mol. The enthalpy energy and ΔG for MCAA molecules are -180.277 x 10^4 kJ/mol and -180.287 x 10^4 kJ/mol. The energy of chitosan enthalpy is negative which signifies exotherm [12].

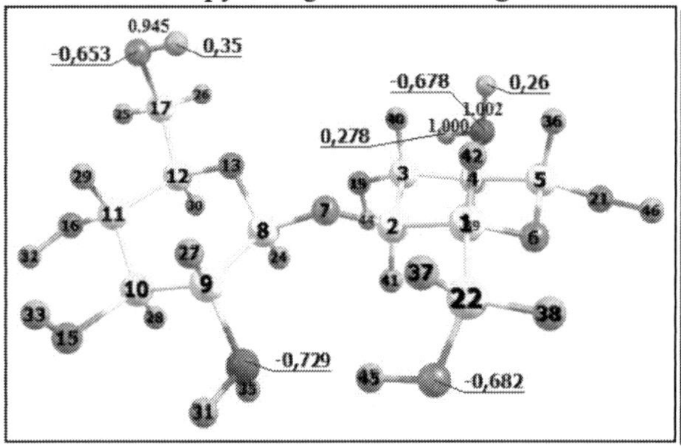

a. The Stable structure of chitosan dimer

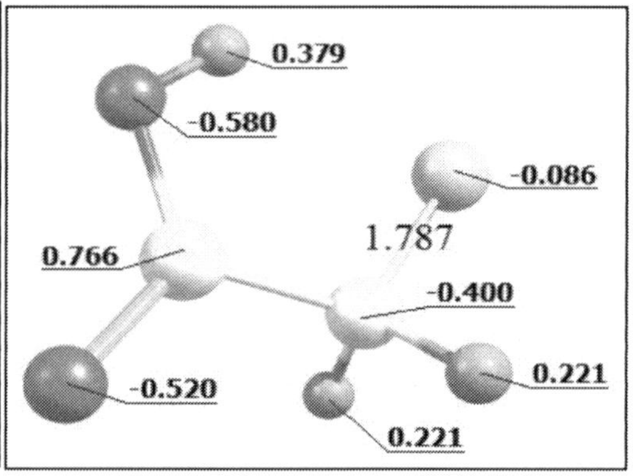

b. The Stable structure of MCAA

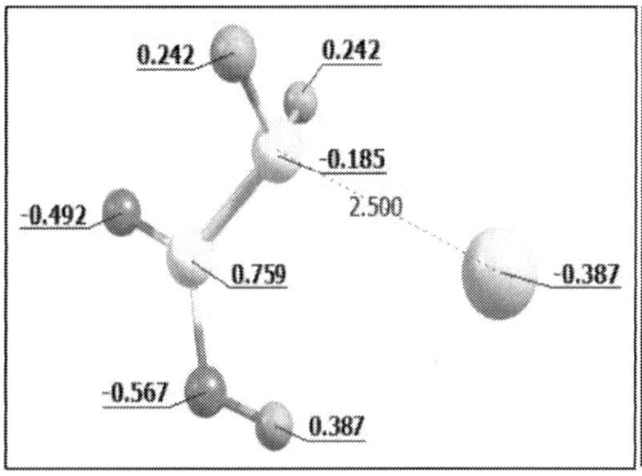

c. The stable tructure of TS_{1-1}
imaginary frequency = -394,64 cm^{-1}

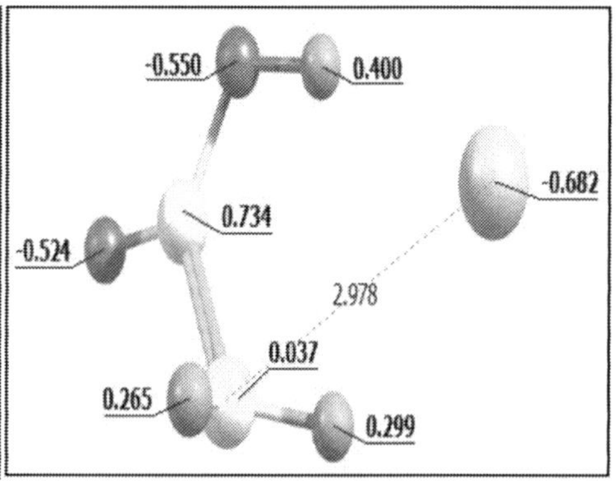

d. The stasle Structure of I_{O-1-1}

Figure 1. The stable structure of reactant, TS_{1-1} and I_{O-1-1}

C Cl H N O

3.1. The mechanism of synthesis reaction of O-CMC

The mechanism of synthesis reaction of O-CMC is a CMC synthesis substituted through the -OH group. This synthesis passes through two paths of reaction mechanism is S_N1 path and S_N2 path. The mechanism of synthesis reaction of O-CMC-1 is the reaction of S_N1 pathway. The mechanism of this reaction begins with a weakened C-Cl bond which is indicated by the longer bonding distance of the TS_{O-1-1} molecule the bond length of C-Cl is 2.5 Å (**Figure 1.c**), compared to in MCAA(1.787 Å). This result is supported by an imaginary vibrational frequency on the center reaction area (υ = -394.64 cm^{-1}). Weak bonds cause the C-Cl bond to break into carboxymethyl ions (+CH_2COOH) and Cl ions, which are the first intermediates (I_{O-1-1}).

The second stage is the attack of carbocation by the nucleophile group of chitosan (-OH) through the transition state (TS_{O-2-1}) with the bond length between –O-C is 1.8 Å. The transition state structure is confirmed by the presence of an imaginary vibrational frequency of the freq calculation (υ = -288.53 cm^{-1}). After the Ea from the second state is passed the molecule will formed I_{O-2-1} with the formation of the -OH-CH_2COOH bonds with the bond length –O-C to 1.437 Å, and the length of the oxygen bonds with hydrogen weakened from 0.947 Å to 0.98 Å (**Figure 2.a**). This is because the molecules I_{O-2-1} is charged +1 where the oxygen atom binds 3 atoms [13].

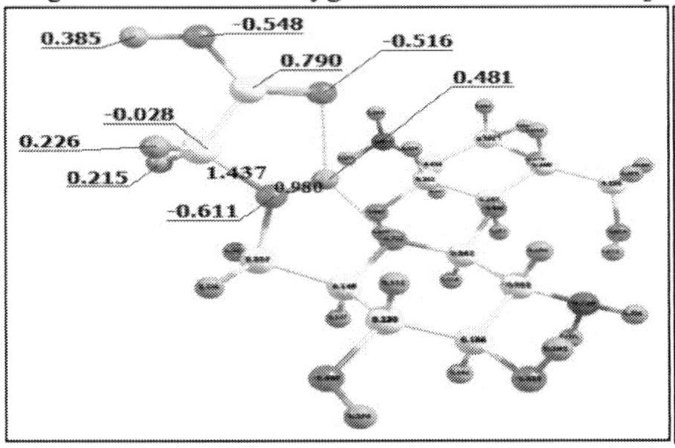

a. The stable structure of I_{O-2-1}

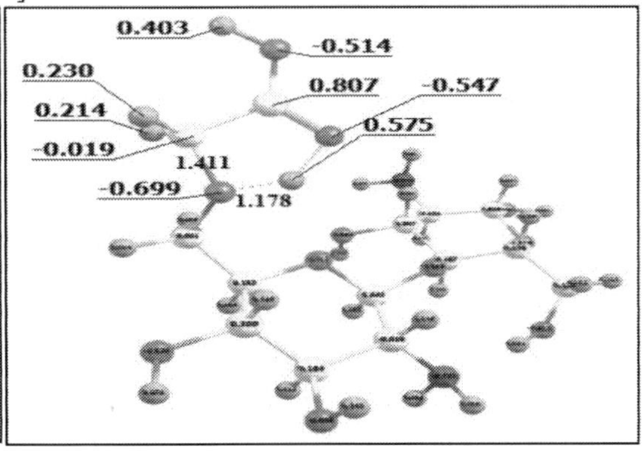

b. The stable structure of TS_{O-3-1}

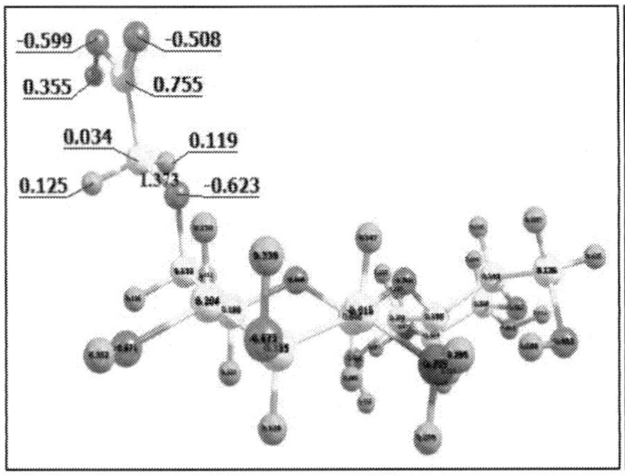

c. The stable structure of P_{O-1}

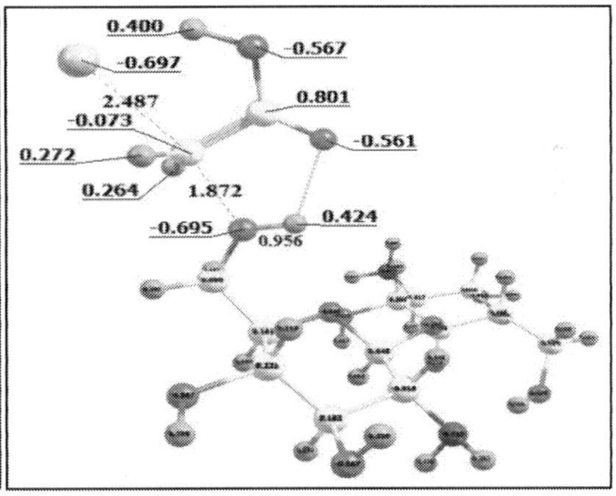

d. The stable structure of TS_{O-1-2}

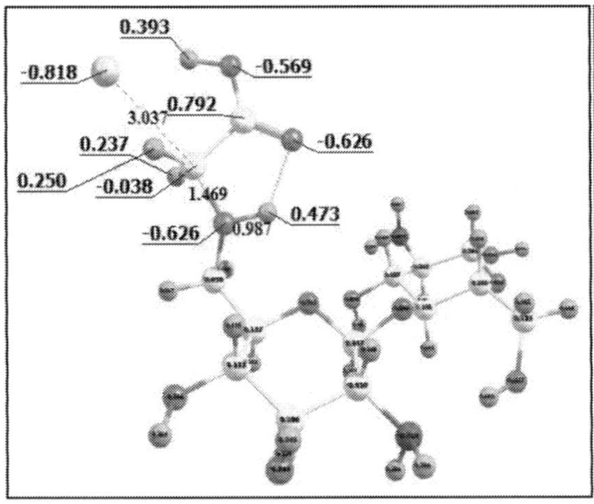

e. The stable structure of I_{O-1-2}

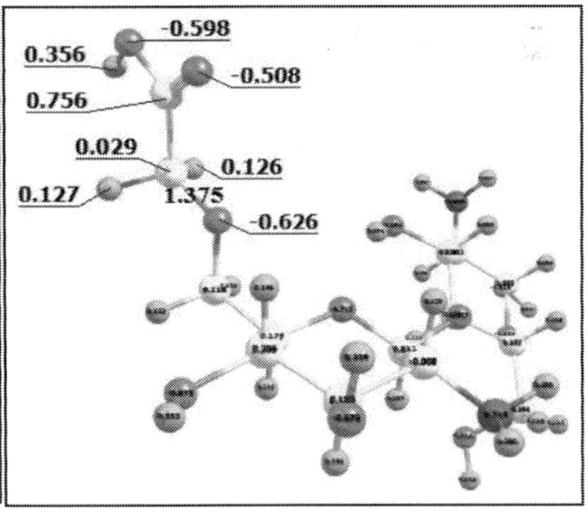

f. The stable structure of P_{O-2}

Figure 2. The stable structure of the O-CMC synthesis via S_N1 and S_N2 pathway

The I_{O-2-1} molecules is highly reactive and unstable, so they must release an atom is hydrogen atom, this is the third reaction step. The third reaction stage begins with the formation of TS_{O-3-1} which H atoms begin to move away from O with bond lengths to 1.178 Å (**Figure 2.b**). The transition state is confirmed by an imaginary vibrational frequency of -1550.36 cm^{-1}. The length of the H bond with O will be longer and finally H is released forming O-CMC-1and H^+ ions. H^+ ions reacts with Cl^- ion to form HCl, as a by product. The length of the O-C bond becomes 1.373 Å (**Figure 2.c**). The minimum energy of all molecules involved in the reaction mechanism of the O-CMC bond formation of S_N1 path is presented in the energy energy potential (**Figure 3**) [14].

The mechanism of synthesis reaction of O-CMC-2 is the reaction of S_N2 pathway. The mechanism of this reaction begins with the Cl^- ion starting away and the nucleophile groups of chitosan (-OH) begin to approaching simultaneously, forming TS_{O-1-2}. As a result of this process there is a change in the length of bond and charge shown in **Figure 2.d**. The most significant changes are in the reaction areas (-OH-C-Cl). The Cl^- charge on the MCAA molecule is -0.086, while in the TS_{O-1-2} molecule to -0.765, which makes the Cl^- ion more easily released. It is supported by an imaginary vibrational frequency on the center reaction area υ = -506,29 cm^{-1}. TS_{O-1-2} through the formation of intermediates I_{O-2} which is the +1-charged molecules at the reaction center that are formed -OH-CH_2COOH ions (**Figure 2.e**). In this stage the bonds between -O-H is getting weaker with each bond length to 0.987 Å.

The intermediate substance is not a stable molecule so release its hydrogen forming TS_{O-2-2}. The second stage transition state (TS_{O-2-2}) was confirmed by an imaginary vibrational frequency is -1408.61 cm^{-1}. The weak hydrogen bonds with bond length O-H to 1.161 Å make the hydrogen easily loose to form O-CMC-2 (**Figure 2.f**) with a by-product of HCl. The minimum energy of all molecules involved in the reaction mechanism of the O-CMC bond formation of S_N2 path is presented in the energy energy potential (**Figure 3**) [14].

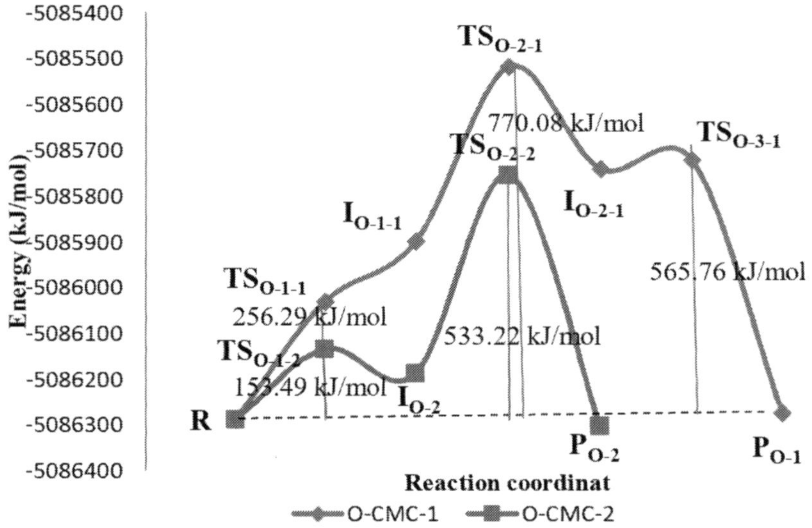

Figure 3. Potential energy surface of the reaction mechanism of the O-CMC synthesis

Figure 3 shows the reaction rate determinant for O-CMC-1 and O-CMC-2 is the second stage, because it has the greatest activation energy. According to the Arrhenius equation ($k = Ae^{-Ea/RT}$), the greater the value of the activation energy(Ea) the value of the rate constant (k) will be smaller and the reaction rate will be slow at that stage. The value of the rate constant (k) obtained is presented in **Table 1**. From **Table 1,** it can be seen that the mechanism of the reaction of O-CMC formation most preferred is through the path S_N2. This is because The Ea2 of S_N2 pathway is lower than S_N1 pathway, It is also because the value of the k of S_N2 pathway is greater than the S_N1 pathway, so the reaction is faster. Abou-zied et al., (2013) stated that carboxymethyl groups are formed by the reaction of S_N2 [1]. The S_N2 reaction path is preferable because the positive C atoms of the MCAA are primary C-type in the reaction path S_N2 has the fastest reaction rate [15, 16]. From the data of gibbs free energy (ΔG) known that the O-CMC-1 reaction is not spontaneous and the O-CMC-2 reaction is spontaneous [14]. The O-CMC-1 formation reaction is an endothermic reaction, whereas the O-CMC-2 formation reaction is an exothermic reaction.

3.2. The mechanism of synthesis reaction of N-CMC

The mechanism of synthesis reaction of N-CMC is a CMC synthesis substituted through the $-NH_2$ group. This synthesis passes through two paths of reaction mechanism is S_N1 path and S_N2 path. The mechanism of synthesis reaction of N-CMC-1 is the reaction of S_N1 pathway. The first stage of this mechanism is the same as the first stage of O-CMC-1 synthesis via the S_N1 pathway.

The second stage is the attack of carbocation by the nucleophile group of chitosan ($-NH_2$) through the transition state (TS_{N-2-1}) with the bond length between –N-C is 2.162 Å. The transition state structure is confirmed by the presence of an imaginary vibrational frequency of the freq calculation (υ = -338.657 cm^{-1}). After the Ea from the second state is passed the molecule will formed I_{N-2-1} with the formation of the –NH2-CH_2COOH bonds with the bond length –N-C to 1.487 Å, and the length of the –N-H bond weakened from 0.947 Å to 0.98 Å (**Figure 4.a**). This is because the molecules I_{N-2-1} is charged +1 where the nitrogen atom binds 4 atoms.

The I_{N-2-1} molecules is highly reactive and unstable, so they must release an atom is hydrogen atom, this is the third reaction step. The third reaction stage begins with the formation of TS_{N-3-1} which H atoms begin to move away from N with bond lengths to 1.42 Å (**Figure 4.b**). The transition state is confirmed by an imaginary vibrational frequency of -1295.77 cm^{-1}. The length of the H bond with N will be longer and finally H is released forming N-CMC-1and H^+ ions. H^+ ions reacts with Cl^- ion to form HCl, as a by product. The length of the N-C bond becomes 1.434 Å (**Figure 4.c**). The minimum energy of all molecules involved in the reaction mechanism of the N-CMC bond formation of S_N1 path is presented in the energy energy potential (**Figure 5**).

The mechanism of synthesis reaction of N-CMC-2 is the reaction of S_N2 pathway. The mechanism of this reaction begins with the Cl^- ion starting away and the nucleophile groups of chitosan ($-NH_2$) begin to approaching simultaneously, forming TS_{N-1-2}. As a result of this process there is a change in the length of bond and charge shown in **Figure 4.d**. The most significant changes are in the reaction areas ($-NH_2$-C-Cl). The Cl^- charge on the MCAA molecule is -0.086, while in the TS_{N-1-2} molecule to -0.768, which makes the Cl^- ion more easily released. It is supported by an imaginary vibrational frequency on the center reaction area υ = -563.23 cm^{-1}. TS_{N-1-2} through the formation of intermediates I_{N-2} which is the +1-charged molecules at the reaction center that are formed $-NH_2$-CH_2COOH ions (**Figure 4.e**). In this stage the bonds between -N-H is getting weaker with each bond length to 1.012 Å.

The intermediate substance is not a stable molecule so release its hydrogen forming TS_{N-2-2}. The second stage transition state (TS_{N-2-2}) was confirmed by an imaginary vibrational frequency is -1558.81 cm^{-1}. The weak hydrogen bonds with bond length N-H to 1.378 Å make the hydrogen easily loose to form N-CMC-2 (**Figure 4.f**) with a by-product of HCl. The minimum energy of all molecules involved in the reaction mechanism of the O-CMC bond formation of S_N2 path is presented in the energy energy potential (**Figure 5**).

Figure 5 shows the reaction rate determinant for N-CMC-1 and N-CMC-2 is the second stage, because it has the greatest activation energy. According to the Arrhenius equation ($k = Ae^{-Ea/RT}$), the greater the value of the activation energy(Ea) the value of the rate constant (k) will be smaller and the reaction rate will be slow at that stage. The value of the rate constant (k) obtained is presented in **Table 1**. From **Table 1,** it can be seen that the mechanism of the reaction of N-CMC formation most preferred is through the path S_N2. This is because The Ea2 of S_N2 pathway is lower than S_N1 pathway, It is also because the value of the k of S_N2 pathway is greater than the S_N1 pathway, so the reaction is faster. The S_N2 reaction path is preferable because the positive C atoms of the MCAA are primary C-type in the reaction path S_N2 has the fastest reaction rate [15, 16]. From the enthalpy energy (ΔH) and gibs free energy, both the N-CMC-1 and N-CMC-2 forming reactions are exothermic and spontaneous.

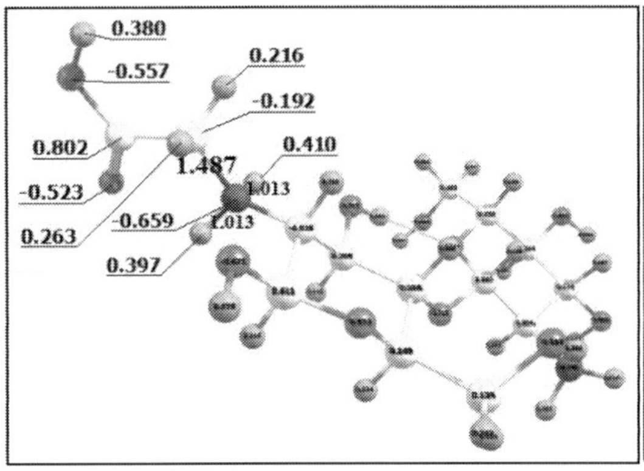

a. The stable structure of I_{N-2-1}

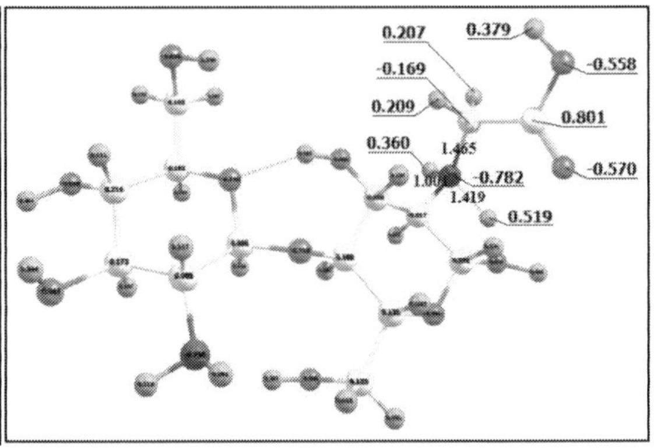

b. The stable structure of TS_{N-3-1}

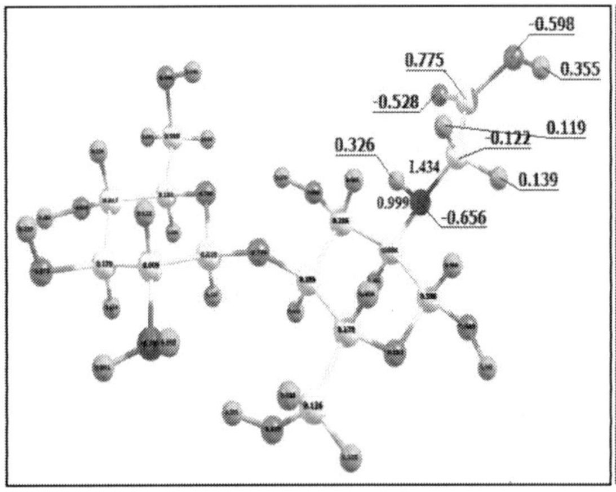

c. The stable structure of P_{N-1}

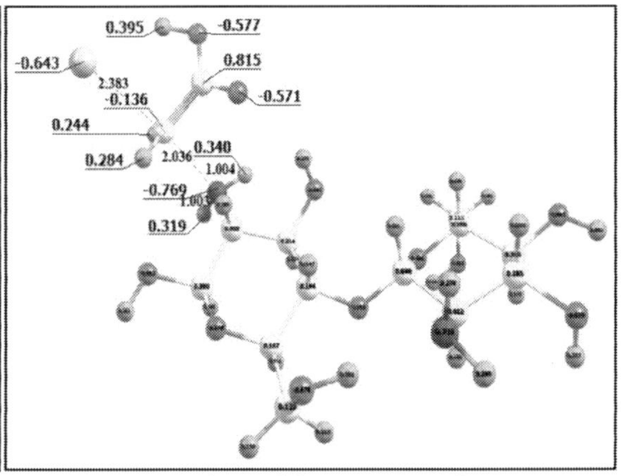

d. The stable structure of TS_{N-1-2}

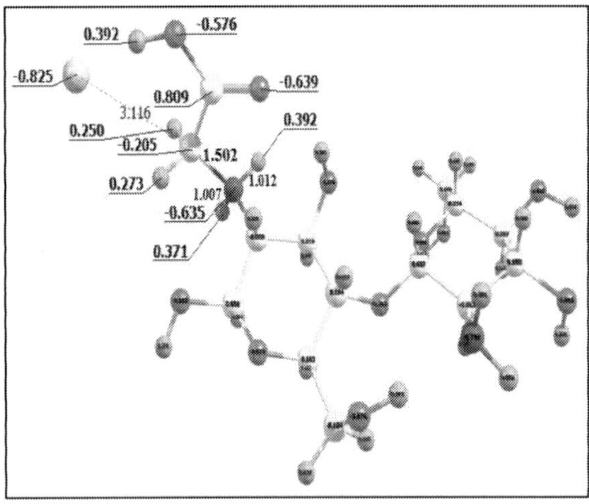

e. The stable structure of I_{N-1-2}

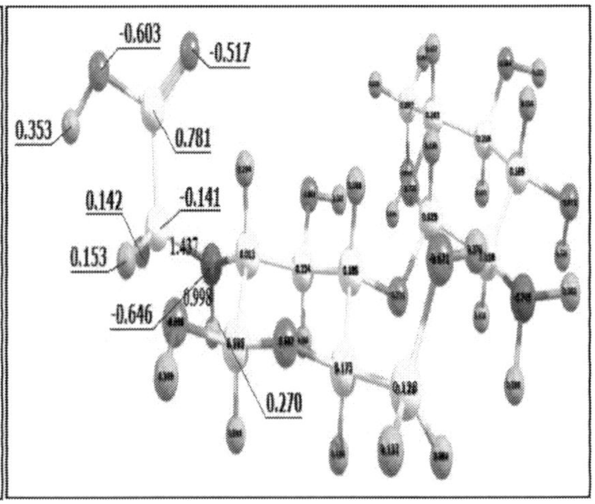

f. The stable structure of P_{N-2}

Figure 4. The stable structure of the N-CMC synthesis via S_N1 and S_N2 pathway

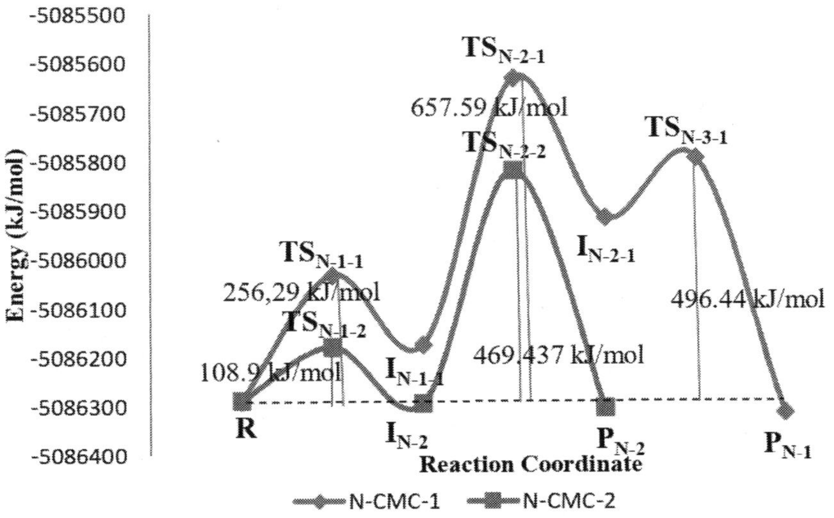

Figure 5. Potential energy surface of the reaction mechanism of the O-CMC synthesis

Table 1. Parameters determining the most likely reaction mechanism.

Parameter	O-CMC		N-CMC		N,O-CMC	
	SN1 (O-CMC-1)	SN2 (O-CMC-2)	SN1 (N-CMC-1)	SN2 (N-CMC-2)	(N,O-CMC-1	(N,O-CMC-2)
Ea (kJ/mol)	770.08	533.22	657.59	469.437	533.22	469.437
k	$1.203x10^{-135}$	$3.79x10^{-94}$	$6.15x10^{-116}$	$5.67x10^{-83}$	$3.79x10^{-94}$	$5.67x10^{-83}$
ΔH (kJ/mol)	17.35	-9.341	-19.707	-11.628	-33.47	-17.48
ΔG (kJ/mol)	24.629	-4.353	-8.097	-1.948	-4.95	15.47

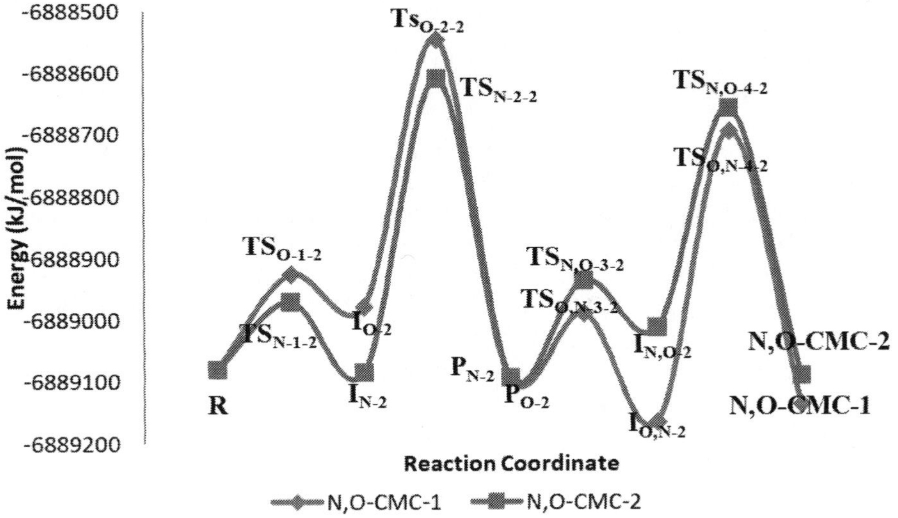

Figure 6. Potential energy surface of the reaction mechanism of the N,O-CMC synthesis

The CMC synthesis not only produces O-CMC and N-CMC but also can produce N,O-CMC, substituted carboxymethyl ions in both groups -OH and $-NH_2$. The mechanism of the N,O-CMC synthesis reaction is formed by continuing the reaction mechanism through the S_N2 reaction pathway. N,O-CMC-1 is a synthesis of N,O-CMC which begins with substitution of -OH group then substituted in the -NH_2 group, while N,O-CMC-2 begins substitution in the -NH_2 group then substituted in the -OH group. The results obtained are presented in **Figure 6**. In the N,O-CMC-1 synthesis, the third and fourth phase Ea is smaller than the second stage, so the reaction rate determinant stage is the second stage such as O-CMC-2. As well as the N,O-CMC-2, Ea3 and Ea4 are smaller than Ea1 and Ea2. It's means substitution via the -NH_2 group is easier after the chitosan is substituted in the -OH group, because the atomic charge N of -NH_2 is initially -0.678 become more negative that is to be -0.685. As well as the N,O-CMC-2 synthesis with the stage of reaction rate at the second stage. It's means substitution of -OH groups is easier to occur after substitution of the -NH_2 group.

ΔH and ΔG the synthesis reaction of N,O-CMC-1 is -33.47 kJ/mol and -17.48 kJ/mol, whereas ΔH and ΔG the synthesis reaction of N,O-CMC-2 is -4.95 kJ/mol and 15.47 kJ/mol. From the ΔG data it suggest that the N,O-CMC-2 synthesis is positive so that the reaction is not spontaneous and difficult to occur so that the reaction mechanism of N,O-CMC occurs through substitution -OH first, then new -NH_2 substituted. Oleh karena itu, untuk mensintesis N, O-CMC secara eksperimental dengan membuat kondisi reaksi untuk O-CMC terbentuk terlebih dahulu, kemudian lanjutkan untuk mengendalikan kondisi reaksi yang akan diganti pada kelompok -NH_2.

4. Conclusion

A mechanism for the carboxymethyl chitosan synthesis by dimer chitosan and monocloro acetic acid by HF calculation was presented. The results showed that the reaction of CMC synthesis via S_N2 reaction pathway, based on the kinitecs data, the N-CMC-2 synthesis is most easily formed with Ea determinant of reaction rate is 469.44 kJ/mol, while from the thermodynamic data, ΔG of the O-CMC-2 synthesis is -4.353 kJ/mol more spontaneous than the N-CMC-2 synthesis. Both N-CMC-2 and O-CMC-2 are exothermic reactions. The synthesis of N,O-CMC occurs through the formation of -O-CH_2COOH the first and then proceeds to form -NH-CH_2COOH, which is indicated by low and negative ΔH and ΔG values of -33.47 kJ/mol and -17.48 kJ/mol.

Acknowledgements

Thanks to Prof. Teruna J. Siahaan, Ph.D. (Pharmacy Department, University of Kansas, US), for the valuable direction and discussion of encapsulation and drug delivery systems. Thanks to the Directorate General of Higher Education. Last but not least Faculty of Science and Mathematics, which has funded this research through the Research Funding Scheme 2016.

References

[1] Abou-Zied N Y, Waly A I, Kandile N G, Rushdy A A, El-Sheikh M A, and Ibrahim H M 2013 *J. Mater. Sci. Eng. with Adv. Technol.* 7 93–123
[2] Mohanasrinivasan V, Mishra M, Paliwal J S, Singh S K, Selvarajan E, Suganti V and Devi Subathra 2013 *biotech* 1-9
[3] Muzzarelli R A A and Muzzarelli C 2005 *Adv Polym Sci* **186** 151-209
[4] Upadhyaya L, Singh J, Agarwal Vl, and Tewari R P 2014 *J. Control. Release* **186** 54–87
[5] Mourya V K, Inamdar N N, and Tiwari A 2010 *Adv. Mat. Lett.* 111–33
[6] Siahaan P, Mentari N C, Wiedyanto U O, Hudiyanti D, Hildayani S Z, and Laksitorini M D 2017 Indones. J. Chem. **17** 291–300
[7] Anslyn E V and Dougherty D A 2006 *Modern Physical Organic Chemistry* (United States of America: Wilsted & Taylor Publishing Services)
[8] Siahaan P, Lalita M N T, Cahyono B, Laksitorini M D, and Hildayani S Z 2017 *IOP Conf. Ser. Mater. Sci. Eng.* **172** 012040
[9] Nhlabatsi Z P, Bhasi P, and Sitha S 2016 *Physical Chemistry Chemical Physics*
[10] Foresman J B and Frisch A 1996 *Exploring Chemistry With Electronic Structure Methods second edition*, Pittsburgh, 3-297.
[11] Cramer C J 2004 *Essentials of Computational Chemistry Theories and Models* (England: JohnWiley&Sons)

[12] Amorim M L, Max G, Ferreira D, and Soares L D S 2016 *Food Biophysics* 1-12
[13] Jonest P G and Kirby A J 1984 *J. Am. Soc.* **106** 6207–12
[14] Højer M and Hashmi A S K 2014 *Chem. Eur. J.* **20** 1901–08
[15] Grabowski S J 2014 *phys. Chem. Chem. Phys* **16** 1824–34
[16] Fessenden R J and Fessenden J S 1997 *Kimia Organik* (Jakarta: Erlangga)

IOP Conf. Series: Materials Science and Engineering **349** (2018) 012050 doi:10.1088/1757-899X/349/1/012050

Probing the Interaction between Cyclic ADTC1 Ac-CADTPPVC-NH_2) Peptide with EC1-EC2 domain of E-cadherin using Molecular Docking Approach

P Siahaan[1], S Wuning[1], A Manna[1], V D Prasasty[2], D Hudiyanti[1]

1Departement of Chemistry, Faculty of Science and Mathematics, Diponegoro University
2Faculty of Biotechnology, Atma Jaya Catholic University of Indonesia

E-mail : parsaoran_s@undip.ac.id; siahaan_parsaoran@yahoo.com

Abstract. Deeply understanding that intermolecular interaction between molecules on the paracellular pathway has given insight to its microscopic and macroscopic properties. In the paracellular pathway, synthetic cyclic ADTC1 (Ac-CADTPPVC-NH_2) peptide has been studied to modulate EC1-EC2 domain, computationally using molecular docking method. The aim of this research is to probe the effect of amino acid alanine (A) of ADTC1 on its interaction properties. The study carried out in two steps: 1. the optimization using GROMACS v4.6.5 program and; 2. Determination of the interaction properties using AutoDock 4.2 program. The interaction was done for A-J box, and the best position of the binding site and binding energy on the OC and CC ADTC1 peptides against the EC1-EC2 domain of E-cadherin was selected. The result showed that the CC of the F box ADTC1 has the best interaction with binding energy of -26.36 kJ/mol and its energy was lower than ADTC5 without alanine amino acid. ADTC1 interacted with EC1 of EC1-EC2 on Asp1, Trp2, Val3, Ile4, Ile24, Lys25, Ser26, Asn27, and Met92 residues.

1. Introduction

Intermolecular interaction is a very important study in understanding the nature of a molecule macroscopically and microscopically [1]. This interaction study has been studied computationally using the molecular docking method between cadherin peptides against EC1 and EC5 domains [2-5]. The cadherin peptide is a derivative synthesis peptide of the bulge and groove region sequences in the cadherin, where the bulge region is an ADT structure and the groove region is a HAV structure [6]. This cadherin peptide can be used to modulate intercellular junction by inhibiting cadherin-cadherin interactions [7-9].

In this research has been studied the interaction study between ADT structure with EC1 domain on E-cadherin [10]. Where the ADT structure has several derivatives, one of which is the ADTC1 (Ac-CADTPPVC-NH_2) peptide. This peptide is thought to increase its biological activity in inhibiting cadherin-cadherin interactions. Moreover, previous study has shown that ADTC1 peptide can inhibit cadherin-cadherin interaction with ADT peptide binding on groove region the EC1 domain [11]. However, the biological activity is affected by the structure form of the open-cyclic (OC) and close-cyclic (CC) ADTC1 peptide, aside from that also influenced by changes in the structure of each time using Molecular Dynamics (MD) simulation for 20 ns (20,000 ps) as shown in Table 1.

Table 1. Generated OC and CC forms of the ADTC1 peptide by molecular dynamics approach

Code	OC		
	Distance constraint of $S_{14}...S_{105}$ (nm)	force restraint (kJ/mol^1.nm^2)	Total Energy (kJ/mol)
A1	All bond freely rotatable	None	-82,504.52
A2	0.2-0.3	None	-82,830.28
A3	0.3-0.4	None	-82,685.92
B1	All bond freely rotatable	4,000	-82,567.45
B2	0.2-0.3	4,000	-82,964.15
B3	0.3-0.4	4,000	-82,869.02
C1	All bond freely rotatable	12,000	-82,638.36
C2	0.2-0.3	12,000	-82,823.03
C3	0.3-0.4	12,000	-82,690.93
	CC		
D1	All bond freely rotatable	None	-59,769.39
D2	All bond freely rotatable	4,000	-59,807.41
D3	All bond freely rotatable	12,000	-59,851.81

Based on Table 1 the formation of CC peptide using OC peptide (code B2) and on the optimization result of CC obtained stable structure CC (code D3) that is by giving force restraint for 12,000 kJ/mol^1.nm^2 which result then used for optimization of DM simulation for 120 ns (120,000 ps). Previous studies have performed DM simulations for 120 ns for the ADTC5 peptide and have studied the interaction between the peptide ADTC5 (Ac-CDTPPVC-NH_2) with the EC1-EC2 domain on E-cadherin [12]. Of the two structures of this peptide that distinguish is the amino acid structure, where on ADTC1 peptide there is substitution of alanine amino acid. The aim of this research is to determine the effect of alanine substitution to modulate of EC1-EC2 domain on E-cadherin.

2. Materials and Methods

The structure of OC ADTC1 peptide was built from PyMol program [13]. Meanwhile, the structure of CC ADTC1 peptide was performed by forming disulfide bond between S_{14} and S_{105} atoms on cysteine amino acid and simple molecular optimization using Avogadro program. The EC1-EC2 domain of E-cadherin protein obtained from Protein Databank with ID 2O72. The EC1-EC2 domain structure of E-cadherin, OC ADTC1 peptide, and CC ADTC1 peptide are depicted in Figure 1.

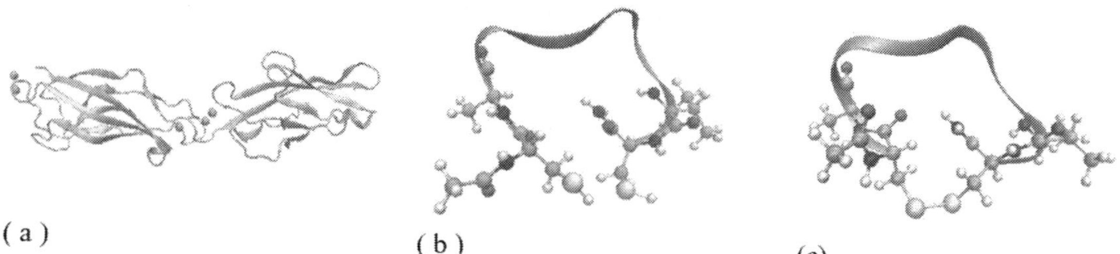

(a) (b) (c)

Figure 1. (a) The EC1-EC2 domain structure of E-cadherin, (b) CC ADTC1 peptide structure, and (c) CC ADTC1 peptide structure.

This study carried out in two steps. The first step was the optimization using GROMACS v.4.6.5 program. The optimization was done for two types of ADTC1 peptide: open-cyclic (OC) and close-cyclic (CC) on 20,000 and 120,000 ps, respectively. The OC and CC affected by its distance constraints and force restraints (Table 1). The most stable peptide was resulted by optimatization on 20,000 ps and the force restraint 12,000 kJ/mol^1.nm^2 applied to the optimatization on 120,000 ps, shown in Table 2. The second step was to determine the interaction properties using AutoDock 4.2 program [14]. The interaction was done for A-J box, and the best position of the binding site and binding energy on the OC and CC ADTC1 peptides against the EC1-EC2 domain of E-cadherin was selected.

Table 2. Variation of MD simulation the CC ADTC1 peptide for 120 ns

Code	Distance constraint of $S_{14}...S_{105}$ (nm)	force restraint (kJ/mol^1.nm^2)
E1	All bond freely rotatable	None
E2	All bond freely rotatable	4,000
E3	All bond freely rotatable	12,000

3. Result and Discussion

Molecular Dynamic (MD) Simulation

In the MD simulation of CC ADTC1 peptide performed preparation system, trajectory generation, and trajectory analysis. In the preparation system stage was done by adding solvents, ions, minimization energy, molecular dynamics on restraint position, and the unconstrained final equilibrium. The generate trajectory stage was done by MD simulation for 120 ns. In the trajectory analysis stage generated RMSD and total energy. Trajectory analysis performed RMSD analysis on Cα and total energy analysis, as depicted in Figure 2. Based on RMSD analysis of Cα on CC ADTC1 peptide, the movement of peptide molecules tends to be stable and does not undergo much change due to the presence of force restraint (Table 3). Force restraint of MD simulation of the peptide structure $C\alpha_1$ will undergo a non-significant change into $C\alpha_2$. Moreover, it can be proven by the change of distance between $S_{14}...S_{105}$ in the amino acid cysteine as shown in Table 4. Changes in peptide structure of CC ADTC1 affected the structure folding/unfolding, the peptide with folding structure is more stable due to it has the lowest energy [15].

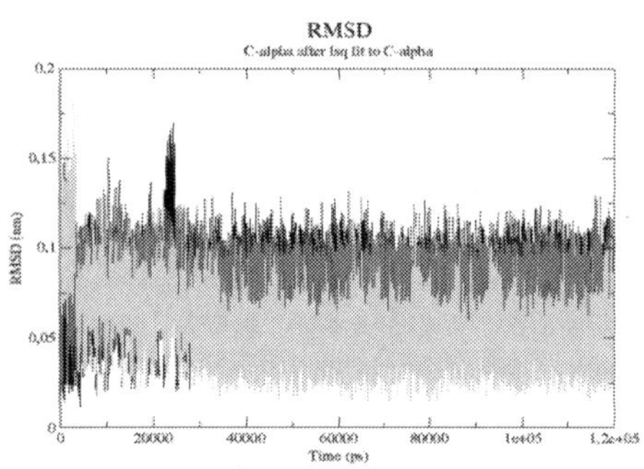

(a)

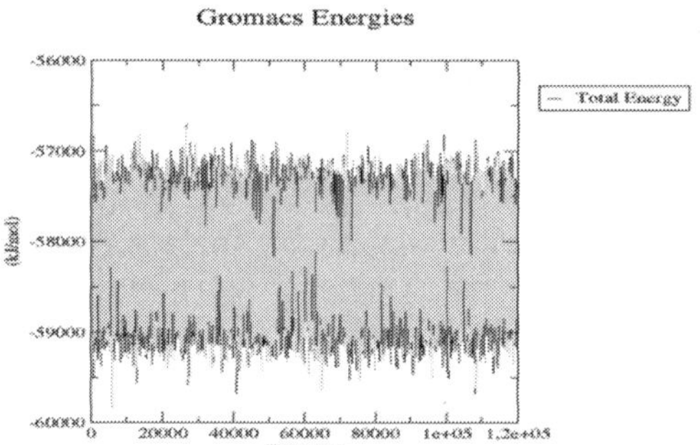

(b)

Figure 2. (a) The RMSD profile, (b) The total energy graph. Where variation force restraint of MD simulation the CC ADTC1 peptide to the black color: code E1 (None), red color: code E2 (4,000 kJ/mol^1.nm^2), green color: code E3 (12,000 kJ/mol^1.nm^2).

Table 3. Generated CC forms of the ADTC1 peptide by molecular dynamics on 120 ns

Code	Total Energy	RMSD

	(kJ/mol)	(nm)
E1	-59,675.75	0.0111 - 0.1702
E2	-59,605.48	0.0177 - 0.1473
E3	-59,851.93	0.0136 - 0.1886

Table 4. The movement of CC Peptide ADTC1 during MD simulation on 120 ns

	E1		E2		E3	
The movement of peptides per ns	Total Energy (kJ/mol)	$R_{S14...S105}$ (Å)	Total Energy (kJ/mol)	$R_{S14...S105}$ (Å)	Total Energy (kJ/mol)	$R_{S14...S105}$ (Å)
0	-57,875.71	2.02974	-57,697.34	2.02891	-57,730.98	2.02907
20	-58,144.59	2.02890	-58,157.43	2.02862	-58,097.78	2.02843
40	-58,422.90	2.02855	-58,450.05	2.02935	-58,522.11	2.02909
60	-57,663.82	2.02936	-57,914.05	2.03290	-57,831.62	2.02962
80	-57,532.91	2.02875	-57,826.46	2.02939	-57,596.36	2.02896
100	-58,574.47	2.02905	-58,335.01	2.02850	-58,494.67	2.02864
120	-58,185.25	2.02829	-58,212.71	2.02917	-58,172.73	2.02940

Furthermore, a total energy analysis was performed on CC ADTC1 peptide. Based on Table 3, optimization with the addition of force restraint 12,000 $kJ/mol^1.nm^2$ has generated the most stable energy of -59,851.93 kJ/mol. This is due to the effect of providing force restraint that causes rigid peptides structure, degrees of freedom of torque decrease and the probability of interaction with the electrostatic forces of the ions and solvents becomes smaller so that the conformation tends to be more stable. The stable total energy of the structure in code E3 was further investigate its interaction with E-cadherin domain using molecular docking.

Molecular Docking

After having obtained stable structure on optimization ADTC1 peptide using GROMACS, molecular docking was performed by using Autodock 4.2 program [14]. The preparation of molecular docking involved two steps: 1. Autogrid and 2. Autodock. In the Autogrid step was performed an evaluation on the EC1 domain of E-cadherin with a gridbox size of 50x50x50 on A-J boxes and grid spacing of 0.375 Å. Then, Autodock with the number of evaluation process of 5,000,000 was done as second step. Based on docking result that CC ADTC1 peptide from box F has the best interaction with binding energy of -24.89 kJ/mol (Table 5). This is also reinforced by the formation of hydrogen bonds between Lys25(O)...Thr4(HG1) and Asn27(ND2)...Cys8(NH) with a distance of 2.033 Å and 2.097 Å respectively. The interaction of CC ADTC1 peptide with EC1-EC2 domain from E-cadherin on Asp1, Trp2, Val3, Gln23, Ile24, Lys25, Ser26, Asn27, and Glu89 residues.

Table 5. Molecular docking of close-cyclic ADTC1 Peptide with EC1-EC2 domain of E-cadherin

Box type	Pose	ΔG (kJ/mol)	Ki (μM)

IOP Conf. Series: Materials Science and Engineering **349** (2018) 012050 doi:10.1088/1757-899X/349/1/012050

A	93	-16.15	1. 49 x 10^3
B	31	-14.14	3.31 x 10^3
C	57	-18.70	528.47
D	54	-16.40	1.34 x 10^3
E	131	-13.05	5.2 x 10^3
F	139	-24.89	43.2
G	109	-14.23	3.22 x 10^3
H	99	154.93	-
I	52	17.70	-
J	43	-22.43	118.44

The F box was validated using RMSD value ≤ 2 Å and the obtained binding energy of -26.36 kJ/mol and Ki of 24.09 μM was resulted. Validation results in F box is valid due to the value of RMSD ≤ 2 Å is 0.49 Å. The interaction of CC ADTC1 peptide with EC1-EC2 domain from E-cadherin involved Asp1, Trp2, Val3, Ile4, Ile24, Lys25, Ser26, Asn27, and Met92 residues (Figure 3). In the interaction of CC ADTC1 peptide with EC1-EC2 domain of E-cadherin formed two hydrogen bonds between Lys25(O)...Thr4(HG1) and Asn27(ND2)...Cys8(NH) with a distance of 1.983 Å and 2.177 Å respectively. In the previous study also obtained similar approach in investigating the interaction between ADTC5 peptide with EC1-EC2 domain of E-cadherin [12]. The difference between ADTC1 and ADTC5 peptides is the presence of an alanine amino acids on ADTC1 peptide sequence and the docking results of the interaction between CC ADTC5 peptide and EC1-EC2 domain of E-cadherin can be seen in Table 6.

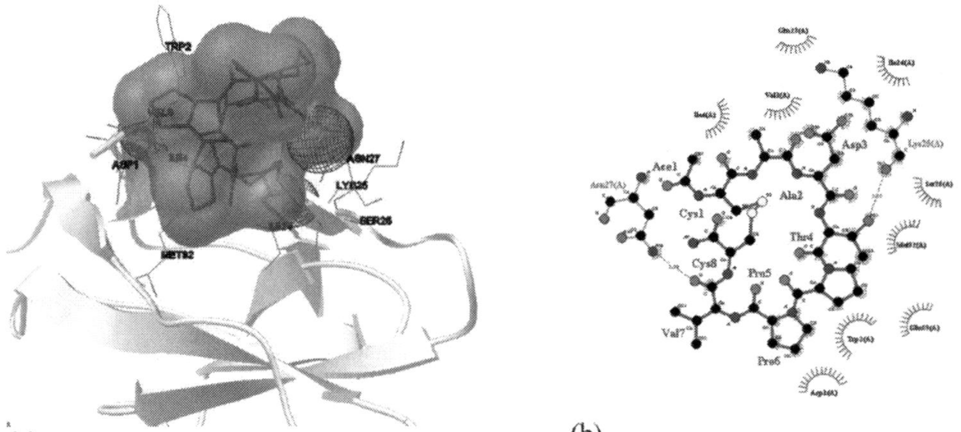

(a) (b)

Figure 3. The interaction between cyclic ADTC1 peptide and the EC1-EC2 domain of E-cadherin with used (a) Autodock program, and (b) Ligplus program.

Table 6. Molecular docking of close-cyclic ADTC5 Peptide with EC1-EC2 domain of E-cadherin [12].

ADTC5	Pose	ΔG (kJ/mol)	Ki (mm)

B1	26	-16.95	1.08
	57	-16.36	1.36
	73	-12.93	5.41
	147	-13.56	4.22
B2	23	-11.13	11.14
	71	-11.59	9.25
	80	-14.85	2.48
	100	-16.28	1.42
	124	-14.60	2.78
B3	23	-11.72	8.93
	38	-10.46	14.78
	124	-17.45	0.88

Based on Table 5 and Table 6 we can see the difference of binding energy of the CC ADTC1 and ADTC5 peptides. On the CC ADTC1 peptide has lower energy than ADTC5 which was -24.89 kJ/mol for ADTC1 peptide and after validation the binding energy became -26.36 kJ/mol. Meanwhile, the ADTC5 peptide has the lowest binding energy -17.45 kJ/mol. It assumes that alanine affected this interaction. The effect of alanine addition was estimated by the ability of its biological activity to modulate cadherin-cadherin interactions which has greater activity than CC ADTC5 peptide.

4. Conclusion

From the results of this interaction study it is shown that the CC ADTC1 peptide of the F box has the best interaction with binding energy of -26.36 kJ/mol and its energy is lower than the CC ADTC5 peptide. It is estimated that the effect of addition of Alanine amino acid on peptide ADTC1. The interaction of ADTC1 with EC1-EC2 domain from E-cadherin on Asp1, Trp2, Val3, Ile4, Ile24, Lys25, Ser26, Asn27, and Met92 residues.

Acknowledgment

Authors would thank Prof. Teruna J. Siahaan, Ph.D. (Phar-macy Department, University of Kansas, US), for the valuable direction and discussion of the cells and drug delivery systems. Thanks to Prof. Krzysztof Kuczera, Ph.D. (Chemistry Department, University of Kansas, US) for in depth discussion of a computational modeling. Last but not least, thanks to the Indonesian Directorate General of Higher Education, which has funded this research through the Research and Technology Funding scheme 2013.

Reference

1. Gao, J., et al., *Frictional forces and Amontons' law: from the molecular to the macroscopic scale.* Journal of physical Chemistry B, 2004.
2. Alaofi, A., et al., *Probing the interaction between cHAVc3 peptide and the EC1 domain of E-cadherin using NMR and molecular dynamics simulations.* Journal of Biomolecular Structure and Dynamics, 2017. **35**(1): p. 92-104.
3. Makagiansar, I.T., et al., *Improving the selectivity of HAV-peptides in modulating E-cadherin-E-cadherin interactions in the intercellular junction of MDCK cell monolayers.* Pharmaceutical research, 2001. **18**(4): p. 446-453.
4. Prasasty, V.D., U.S.F. Tambunan, and T.J. Siahaan, *Homology modeling and molecular dynamics studies of EC1 domain of VE-cadherin to elucidate docking interaction with*

cadherin-derived peptide. OnLine Journal of Biological Sciences, 2014. **14**(2): p. 155.

5. Zheng, K., et al., *Characterization of Multiple Stable Conformers of the EC5 Domain of E‐cadherin and the Interaction of EC5 with E‐cadherin Peptides.* Chemical biology & drug design, 2009. **73**(6): p. 584-598.
6. Sinaga, E., et al., *Increasing paracellular porosity by E-cadherin peptides: discovery of bulge and groove regions in the EC1-domain of E-cadherin.* Pharmaceutical research, 2002. **19**(8): p. 1170-1179.
7. Kiptoo, P., et al., *Peptide delivery*, in *Handbook of biologically active peptides*. 2013, Academic Press, San Diego, California. p. 1702-1710.
8. Sinaga, E., et al., *Modulasi Junction Antar Sel Menggunakan Peptida Kadherin Upaya Meningkatkan Penghantaran Obat.* Makara Journal of Science, 2004.
9. Zhao, H., et al., *11. Modulation of cell-cell adhesion by cadherin peptides.* Research Signpost, 2010.
10. Laksitorini, M.D., et al., *Modulation of Intercellular Junctions by Cyclic‐ADT Peptides as a Method to Reversibly Increase Blood–Brain Barrier Permeability.* Journal of pharmaceutical sciences, 2015. **104**(3): p. 1065-1075.
11. Laksitorini, M.D., *Design of Cyclic-ADT Peptides to Improve Drug Delivery to the Brain via Inhibition of E-Cadherin Interactions at the Adherens Junction.* 2012, University of Kansas.
12. Siahaan, P., et al., *Structural Stability of ADTC5 Peptide: Conformational Insights into Dynamics and Its Binding Mode.* Journal of Tropical Life Science, 2017. **7**(2): p. 138-145.
13. Makarewicz, T. and R. Kaźmierkiewicz, *Molecular dynamics simulation by GROMACS using GUI plugin for PyMOL.* Journal of Chemical Information and Modeling, 2013.
14. Morris, G.M., et al., *AutoDock4 and AutoDockTools4: Automated docking with selective receptor flexibility.* Journal of computational chemistry, 2009. **30**(16): p. 2785-2791.
15. Dill, K.A., *Dominant forces in protein folding.* Biochemistry, 1990. **29**(31): p. 7133-7155.

IOP Conf. Series: Materials Science and Engineering **349** (2018) 012051 doi:10.1088/1757-899X/349/1/012051

Extraction of gelatin from catfish bone using NaOH and its utilization as a template on mesoporous silica alumina

R Nuryanto[1,2], W Trisunaryanti[1], I I Falah[1], Triyono[1]

1Department of Chemistry, Faculty of Mathematics and Natural Sciences Gadjah Mada University Jl Sekip Utara Yogyakarta, Center of Java Indonesia 55281.
[2]Department of Chemistry, Faculty of Sciences and Mathematics, Diponegoro University, Tembalang Semarang

E-mail: nuryantorahmad@yahoo.com

Abstract Gelatin extraction from catfish bone using NaOH and its utilization as a template on a synthesis of mesoporous silica-alumina had been investigated. The extraction was prepared by immersing 25 g catfish bone in 125 mL of NaOH in concentration of 0.0; 0.05; 0.10; 0.15 and 0.20 M for 24 h, then washing with demineralized water until pH 7, followed by immersed the bone into 125 mL of 1 M HCl for 1 h, then washed using demineralized water into pH 5. To produce gelatin the bone was refluxed with 100 mL demineralized water at 70°C for 5 h then evaporated at 50°C. The dry gelatin was characterized using FTIR and electrophoresis (SDS-PAGE). The best performance of gelatin was produced by NaOH 0.10 M. The gelatin consists of amide A, B, I, II, III and molecular weight of 25-200kDa. Silica and Alumina material prepared from Lapindo mud extraction. Dry Lapindo mud crushed and filtered until pass 100 mesh, then reflux using 6 M HCl (1:4 w/V) at 90°C for 5h then filtered. The filtrate was consisting alumina solution adding with 6 M NaOH (2/3 V/V) them filtered. The filtrate then injected by CO_2 gas for 30 minutes and filtered, the residue was calcined at 500°C for 5h. The residual of Lapindo mud dried and refluxed with 6 M NaOH (1:4 w/v) at 90 °C. After 5h filtered and the filtrate added by HCl to pH 8 and filtered, the residual then dried. The Si and Al were then analyzed by XRF and consist of silica and alumina for 99.1 and 87.73%, respectively. Silica-alumina was prepared using silica and alumina extracted from Lapindo mud. 6 g of SiO_2 and 2 g of NaOH was immersed in 62 mL of demineralized water then added with alumina solution (0.204 g alumina in 30 mL demineralized water). The gelatin solution (5 g gelatin in 70 mL demineralized water) was dropped into the silica-alumina while stirring at 50°C for 4 h and aging for 24 h. The synthesized silica alumina was analysed using FTIR and surface area analyser. The FT-IR spectra indicated the TO_4 (T=Si, Al) vibration at wave number of 1049.28 and 1103.23 cm^{-1}. The synthesized silica-alumina showed mesoporous characters with a pore diameter of 41.18 nm and surface area of 32.76 m^2/g

1 Introduction

As the global demand for gelatin is continuously on the rise, many potential sources are being sought for combating this growing need. Majorly derived from pig skin, bovine hides, bones and others sources contributing 46%, 29.4%, 23.1% and 1.5%, respectively [1]. Gelatin is a mixture a mixture of peptides and protein produced by partial hydrolysis of collagen. The quality of gelatin depends on its physicochemical properties, rheological properties, and severity of the manufacturing method [2].

Extraction of fish gelatin from several fish such as cod [3], have skin [4], red and black tilapia [5] red snapper and grouper [2] African catfish [6] has been reported.

Technically, the term gelatin applies for a series of proteins obtained from collagen after partial hydrolysis, obtained from bones, skin, hides, ligaments, and cartilages, etc. In the conversion process of collagen to gelatin, acid or alkali pretreatment hydrolyze the crosslinking bonds between polypeptides and irreversible results in gelatin [7]. The gelatin is water soluble and forms thermo-reversible gels with the melting temperature near to the body temperature [8].

Mesoporous silica–alumina (MSA) belong to the frequently studied systems because their acidity allows various applications as catalysts and carriers. In contrast to zeolites, their characteristic properties are the amorphous character and larger pore diameters. These materials have often been synthesized by the sol–gel methods from organometallic precursors like tetraethyl orthosilicate (Dynasil-A) and aluminum alkoxides in the presence of pore-regulating agents, mostly tetrapropylammonium hydroxide (TPA-OH) and amines [9]. Gelatin from bovine as a template on silica mesoporous has investigated. The porous size of silica mesoporous is 6.08 nm [10].

Gelatin is a potential polymer to be used as a template for synthesis of mesoporous silica because it contains N-H functional groups tend to strongly interact with xylanol groups (Si-OH) on the silicate species via multiple hydrogen bonding. The pH of the solution can affect the concentration of ammonium ions inside gelatins molecules. Moreover, gelatin has good biocompatibility, surf activity biodegradability and non-toxic [10].

2 Material and methods

Catfish bone from home industry on kampung lele, Boyolali, Central Java. Lapindo mud from Sidoarjo, East Java. NaOH, HCl and acetic acid analytical grade were obtained from E-Merck (Germany)

2.1 Extraction of Gelatin

The extraction of gelatin was carried out according to Zelechowska *et al.* [11] and Zou and Regensteins [12] with slight modification. The bones were cleaned and washed with aquadest. Before gelatin extraction, 25 g of catfish bones was immersed with 0; 0.05; 0.10; 0.15 and 0.20 M NaOH solution (1:5, w/v) for 24h. After that, the mixture separated, and the solid portion was washed with aquadest into pH=7; then immersed with 1 M HCl (1:2.5, w/v) for 0,5h were repeated twice. The pretreated catfish bones were then refluxed in demineralized water (1:4, w/v) for 5 h at 70°C The mixture was filtered and filtrate dried at 50°C to produced gelatin. The gelatin was then characterized by FTIR and SDS-PAGE.

2.2 Extraction of Si and Al from Lapindo Mud.

Lapindo mud crushed and sieved with 100 mesh sieve to pass 100 mesh material. Then 75 g material reflux by 6 M HCl (1:4 w/v) at 90 °C for 5 h, then filtered. The Filtrate mixed with 6 M NaOH (2:3 v/v) then filtered. The liquid portion injected with CO_2 for 30 minutes and filtered, the residue was calcined at 500°C for 5h. The residual of Lapindo mud dried and refluxed with 6 M NaOH (1:4 w/v) at 90 °C. After 5h filtered and the filtrate added by HCl to pH 8 and filtered, the residual then dried. The Si and Al were then analyzed by XRF.

2.3 Synthesis of mesoporous Silica-alumina.

The synthesis began by mixed 6 g SiO_2 and 2 g NaOH dissolved in 62 mL demineralized water and stirred. 0.204 g Al_2O_3 dissolved in 30 mL demineralized water. Alumina solution was added to the silica solution. The gelatin solution was made from 5 g gelatin (immersed by NaOH 0.1 M) dissolved in 70 mL demineralized water at 50°C then added in the silica-alumina solution drops by drops while stirred for 30 minutes and aged for 24 h. After that mixture of Silica Alumina added with HCl drops by drops into pH 10.5 while stirred for 30 minutes, then filtered and solid dried in the oven 120°C for 24 h. Finally, the material calcined in the furnace at 550°C for 5 h, then analyzed by FTIR, GSA and XRD.

3 Result and Discussion

3.1 Characterization of Catfish bones Gelatin

Figure 1. All gelatins was showed five characteristic FTIR polypeptide band namely amide A, B, and I-III. Amide I band 1651cm^{-1} is mainly due to C=O stretching vibration coupled to contribution from the CN stretch, CCN deformation and in-plane bending modes [13]. The amide II band exhibited at a wave number of 1527 cm^{-1} is a resulted from an out of phase combination of a CN stretch and in-plane NH deformation modes of peptide group [13] [14]. The amide III were observed at wave number 1234 cm^{-1}, indicated the disorder in the gelatin molecules and were more likely associated with the loss of triple helix state [15][16]. The amide III bands represent the combination peaks between CN stretching vibrations and NH deformation from the amide linkages as well as the absorptions arising from wagging vibrations of CH_2 groups in the glycine backbone and proline side-chines [17].

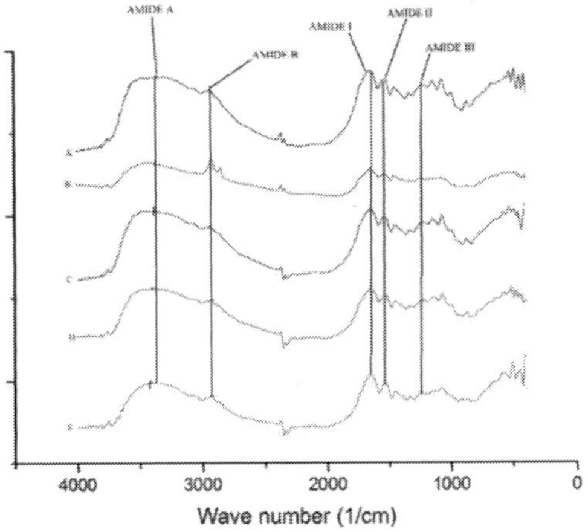

Figure 1. FTIR Spectra of catfish bone Gelatin with immersed in NaOH (A) 0.0 M (B) 0.05 M (C) 0.10 M (D) 0.15 M (E) 0.20 M.

The amide A band was found at 3387 assisted with the stretching vibrations of NH group coupled with hydrogen bonding. Normally, a free NH stretching vibrations are found in the range of 3400-3440 cm^{-1}, The position of this band shifted to more allow frequency because the NH group of a peptide is involved in hydrogen bonding [18]. Amide B was observed at 2931cm^{-1}corresponding to the asymmetric stretching vibrations of C=H as well as NH_3 [16].

Table 1: Molecular weight distribution of gelatin

NaOH concentration (M)	Molecular weight range (kDa)
0.00	Not recorded
0.05	15-200
0.10	25-200
0.15	< 10
0.20	<10

Table 1 showed the molecular weight distribution of extracted gelatin. The weight distribution was calculated with standard curve of SDS-PAGE data. Extracted gelatin with immersed in NaOH 0,05 M and 0,10 has molecular range 15-200 kDa and 25-200 kDa, respectively. The other NaOH concentration has a molecular weight less than 10 kDa. Gelatin extracted with immersed NaOH 0,1 M chosen with template because It has molecular weight higher than gelatin from immersed NaOH 0.05 M. The material is termed gelatin when having a minimum molecular weight of 30 kDa [10].

3.2 Extraction of Si and Al from Lapindo mud.

Extraction of Lapindo mud to obtain Si and Al, because it contains 46.7 % Si and 13 % Al [19]. The main purpose of acid extraction is to destroy the crystalline aluminosilicate phase, such as kaolinite, and boehmite, to release free aluminum ions. According to the thermodynamic calculation at room temperature [20], but the reaction rate is so slow that the aluminum minerals can be thought of as inert to acids at room temperature. Thus, the high-temperature acid leaching method was adopted, and hydrochloric acid was chosen as the leaching agent.

Another metal such Fe contains in the solution. Metal impurities could remove using precipitation method with added 6 M NaOH. Some impurities will be precipitated into M(OH)n except Al, the reaction of precipitation are as follows:

$Al_2Si_2O_5(OH)_4 + 6HCl \rightarrow 2\ AlCl_3 + 2SiO_2 + 5H_2O$

$2AlO(OH) + HCl \rightarrow 2\ AlCl_3 + 4\ H_2O$

$Fe_2O_3 + 6HCl \rightarrow 2FeCl_3 + 3H_2O$

Table 2. Yield and purity of Silica and Alumina from XRF analysis

	Silica (%)	Alumina (%)
Yield	24	7
Purity	99.1	87.73

Alumina can precipitate by injected using CO_2 gas with reaction as:

$M^{n+} + nOH \rightarrow M(OH)_n\downarrow$

$Al^{3+} + 4OH^- \rightarrow AlO_2^-$

$AlO_2^- + CO_2 + 2H_2O \rightarrow Al(OH)_3\downarrow + HCO_3^-$

The yield of alumina the extraction is 7 %, and the result of XRF has a purity of 87.73%. Another material included yield are Si 5.94%, Fe 2.35 %, and another material.

Silica was extracted from solid residue alumina extraction, by reflux using NaOH. Solid silica could obtain with added HCl in the filtrate. The yield of Silica from extraction is 24 % with purity of 99.1%, then other material impurities are Fe 0.51% and the other materials. The reactions of extraction are as follow:

$SiO_2 + 2NaOH \rightarrow Na_2SiO_3 + H_2O$

$Na_2SiO3 + HCl \rightarrow SiO_2 + 2\ NaCl + H_2O$

3.3 Synthesis of mesoporous Silica-alumina

Figure 2. illustrates the FTIR spectra of silica alumina. FT-IR spectra are collected in 3433 cm^{-1} is correspondent to overlapping of the O-H stretching bands of hydrogen-bonded water molecules (H–O–H···H) with SiO–H stretching of surface xylanols hydrogen-bonded to molecular water (SiO–H···H_2O). A band at 2924 cm^{-1} is symmetric and asymmetric fundamental stretching vibrations of CH_2 and CH_3 groups belonging to gelatin as a template. The band at 1635 cm^{-1} belongs to the scissor vibration arising from the proton vibration in the water molecule The bands at 1049 and 686.65 cm^{-1} represent the asymmetric and symmetric stretching vibrations corresponding to the inner TO_4 structure (T = Si, Al), respectively, whereas the bands at 1103 and 794 cm^{-1} represent the asymmetric and symmetric stretching vibrations corresponding to the external TO_4 structure (T = Si, Al), respectively [21].

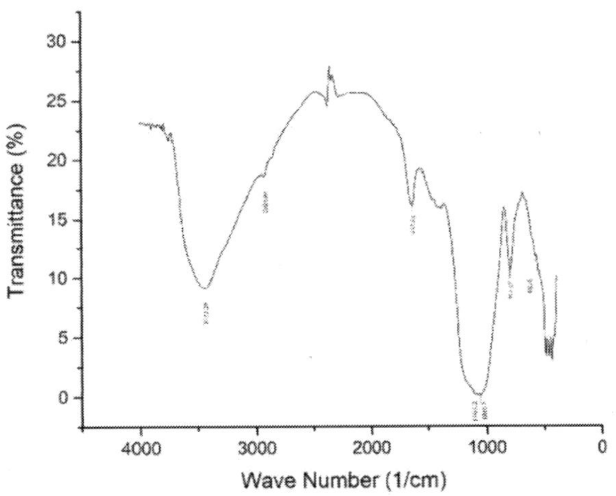

Figure 2. FTIR Spectra of Mesoporous Silica-Alumina

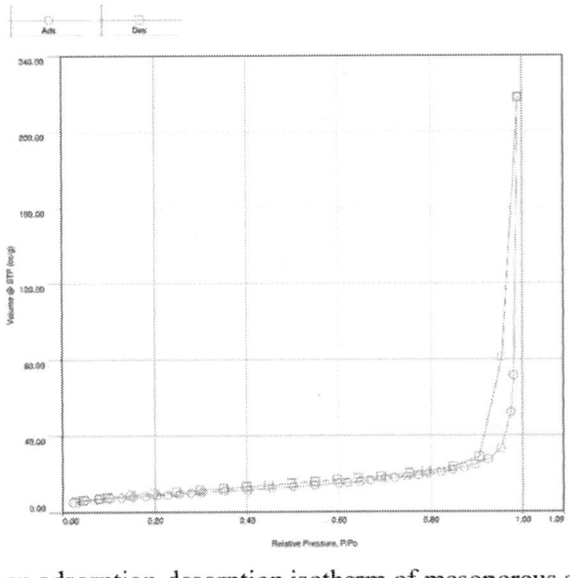

Figure 3. Nitrogen adsorption-desorption isotherm of mesoporous silica-alumina

Mesoporous indicated by exhibit type IV isotherm with hysteresis loops, indicating textural mesoporous (figure 3). The isotherm of Silica-alumina mesoporous exhibit increases at relative pressure p/p_0 0.9-1.0. The synthesized of mesoporous silica-alumina pore calculated by the BJH Method has pore diameter 41.18 nm. Specific surface area of the silica alumina material was calculated by BET method, the result of a specific surface area was 32.76 m^2/g. Pore volume determines by nitrogen adsorption at relative pressure (p/p0) 0.990, of mesoporous silica alumina was 0.337 $cm^3\ g^{-1}$. The crystallinity of silica alumina investigated with XRD was presented in figure 4. XRD pattern without peak around line 2θ so that the silica-alumina was indicating an amorphous material.

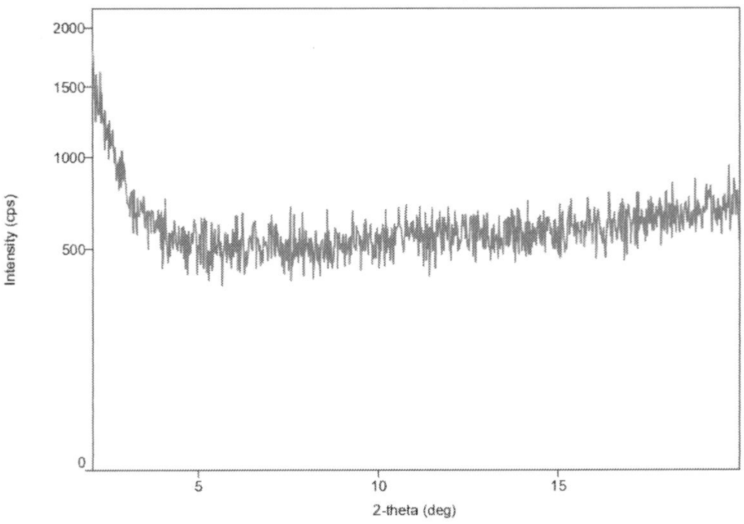

Figure 4. XRD pattern of Mesoporous Silica Alumina

4 Conclusion

Gelatin extracted from catfish bone by immersion in 0.1 used as mesoporous silica alumina template. An amorphous Mesoporous silica alumina mesoporous has been successfully preparing using catfish gelatin as a template. Based on adsorption-desorption, the Mesoporous Silica Alumina have pore diameter, specific surface area and pore volume of 41.18 nm, 32.76 m^2/g, and 0.337 cm^3 g^{-1}, respectively.

Acknowledgement
This research was conducted under the research grant of PUPT 2017 Universitas Gadjah Mada. The authors therefore thank to the Indonesian Ministry of Research, Technology, and Higher Education for the financial support (Contract number: 2454/UN1.P.III/DIT-LIT/LT/2017).

References

[1] Karim, A. A. and Bhat, R. (2009) 'Fish gelatin: properties, challenges, and prospects as an alternative to mammalian gelatins', *Food Hydrocolloids*. Elsevier Ltd, 23(3), pp. 563–576. doi: 10.1016/j.foodhyd.2008.07.002.

[2] Jeya Shakila, R. *et al.* (2012) 'Functional characterization of gelatin extracted from bones of red snapper and grouper in comparison with mammalian gelatin', *LWT - Food Science and Technology*. Elsevier, 48(1), pp. 30–36. doi: 10.1016/j.lwt.2012.03.007.

[3] Gudmundsson, M., & Hafsteinsson, H. (1997). Gelatin from cod skins as affected by chemical treatment. Journal of Food Science, 62(1), 37-47.

[4] Gomez-Guillen, M. C., Turnay, J., Fernandez-Diaz, M. D., Ulmo, N., Lizarbe, M. A., & Montero, P. (2002). Structural and physical properties of gelatin extracted from different marine species: a comparative study. Food Hydrocolloids, 16, 25-34

[5] Jamilah, B., & Harvinder, K. G. (2002). Properties of gelatins from skins of fish black tilapia (Oreochromis mossambicus) and red tilapia (Oreochromis nilotica). Food Chemistry, 77,81-84.

[6] Alfaro, A. T. *et al.* (2014) 'African catfish (Clarias gariepinus) skin gelatin: Extraction optimization and physical-chemical properties', *Food Research International*. Elsevier Ltd, 65(PC), pp. 416–422. doi: 10.1016/j.foodres.2014.05.070

[7] Yang, H., Wang, Y., Zhou, P., & Regenstein, J. M. (2008). Effects of alkaline and acid

pretreatment on the physical properties and nanostructures of the gelatin from channel catfish skins. Food Hydrocolloids, 22(8), 1541–1550.

[8] Norziah, M. H. *et al.* (2009) 'Characterization of fish gelatin from surimi processing wastes: Thermal analysis and effect of transglutaminase on gel properties', *Food Hydrocolloids*. Elsevier Ltd, 23(6), pp. 1610–1616. doi: 10.1016/j.foodhyd.2008.12.004.

[9] Vít, Z. and Šolcová, O. (2006) 'Synthesis and properties of mesoporous silica-alumina with narrow pore size distribution obtained without use of pore-regulating agents', *Microporous and Mesoporous Materials*, 96(1–3), pp. 197–204. doi: 10.1016/j.micromeso.2006.06.020.

[10] Trisunaryanti, W. *et al.* (2016) 'Extraction of gelatin from bovine bone and its use as template in synthesis of mesoporous silica', *Asian Journal of Chemistry*, 28(5), pp. 996–1000. doi: 10.14233/ajchem.2016.19561.

[11] E. Zelechowska, M. Sadowska and M. Turk, Isolation and some properties of collagen from the backbone of Baltic cod(Gadus morhua), Food Hydrocoll., 24, 325 (2010)

[12] P. Zhou and J.M. Regenstein, J. Food Sci., 70, C392 (2005)

[13] Bandekar, J. (1992). Amide modes and protein conformation. Biochimica et Biophysica Acta (BBA) – Protein Structure and Molecular Enzymology, 1120, 123–143.

[14] Lavialle, F., Adams, R. G., & Levin, I. W. (1982). Infrared spectroscopic study of the secondary structure of melittin in water, 2-chloroethanol, and phospholipid bilayer dispersions. Biochemistry, 21, 2305–2312.

[15] Friess, W., & Lee, G. (1996). Basic thermoanalytical studies of insoluble collagen matrices. Biomaterials, 17, 2289–2294.

[16] Sinthusamran, S., Benjakul, S. and Kishimura, H. (2014) 'Characteristics and gel properties of gelatin from skin of seabass (Lates calcarifer) as influenced by extraction conditions', Food Chemistry, 152, pp. 276–284. doi: 10.1016/j.foodchem.2013.11.109.

[17] Jackson, M., Choo, L. P., Watson, P. H., Halliday, W. C., & Mantsch, H. H. (1995). Beware of connective tissue proteins: Assignment and implications of collagen absorptions in infrared spectra of human tissues. Biochimica et Biophysica Acta Molecular Basis of Disease, 1270, 1–6.

[18] Doyle, B. B., Blout, E. R., & Bendit, E. G. (1975). Infrared spectroscopy of collagen and collagen like polypeptides. Biopolymers, 14(5), 937-957.

[19] Fadli., A.F., Tjahjanto.,R.T., and Darjito, 2013, Ekstraksi silika dalam lumpur lapindo menggunakan metode kontinyu, Kimia. Student Journal, Vol. 1, No. 2, pp. 182-187 Universitas Brawijaya Malang

[20] Dai, H., 1997. Thermodynamics discussion of alumina production by acid method. Nonferrous Min.Metall. 4, 40–44 .

[21] Zhao, J. et al. (2016) 'Synthesis and characterization of mesoporous zeolite Y by using block copolymers as templates', Chemical Engineering Journal. Elsevier B.V., 284, pp. 405–411. doi: 0.1016/j.cej.2015.08.143.

IOP Conf. Series: Materials Science and Engineering **349** (2018) 012052 doi:10.1088/1757-899X/349/1/012052

Molecular Docking Simulation of Neuraminidase Influenza A Subtype H1N1 with Potential Inhibitor of Disulfide Cyclic Peptide (DNY, NNY, LRL)

R P Putra[1], R Imaniastuti[1], M A F Nasution[1], Djati Kerami[2], U S F Tambunan[1,a)]

[1]Bioinformatics Research Group, Departement Of Chemistry, Faculty of Mathematics and Natural Science, Universitas Indonesia, Universitas Indonesia, Depok 16424 Indonesia

[2]Mathematics Computational Group, Department of Mathematics, Faculty of Mathematics and Natural Science, Universitas Indonesia, Depok 16424 Indonesia

Email: usman@ui.ac.id

Abstract: Oseltamivir resistance as an inhibitor of neuraminidase influenza A virus subtype H1N1 has been reported lately. Therefore, to solve this problem, several kinds of research has been conducted to design and discover disulfide cyclic peptide ligands through molecular docking method, to find the potential inhibitors for neuraminidase H1N1 which then can disturb the virus replication. This research was studied and evaluated the interaction of ligands toward enzyme using molecular docking simulation, which was performed on three disulfide cyclic peptide inhibitors (DNY, LRL, and NNT), along with oseltamivir and zanamivir as the standard ligands using MOE 2008.10 software. The docking simulation shows that all disulfide cyclic peptide ligands have lower Gibbs free binding energies ($\Delta G_{binding}$) than the standard ligands, with DNY ligand has the lowest $\Delta G_{binding}$ at -7.8544 kcal/mol. Furthermore, these ligands were also had better molecular interactions with neuraminidase than the standards, owing by the hydrogen bonds that were formed during the docking simulation. In the end, we concluded that DNY, LRL and NNT ligands have the potential to be developed as the inhibitor of neuraminidase H1N1.

1 Introduction

Influenza is a severe infectious disease caused by a virus and may infect humans and animals. In general, the common symptoms of this disease are fever, sore throat, headache, and cough. In more severe cases, it may lead to death for someone who infected by several dangerous types of influenza virus [1]. In 2009, influenza A subtype H1N1 virus became a pandemic, with a new viral strain identified in April 2009 and widely known as swine flu. This pandemic caused a fatal infection with more than 80 deaths in 40 countries, first detected in Mexico and spreading to various countries in South and North America, Europe and Asia [2]. One method for the prevention of H1N1 virus infection that has been suggested earlier is by vaccination. However, the current vaccine is thought to still have weakness and imperfections in protecting the body when an influenza virus occurs. This is caused by the mutation of the virus. This mutation may be due to antigenic drift or antigenic shift [3]. Based on its antigenic properties, influenza viruses have two functional surface glycoproteins namely hemagglutinin and neuraminidase (NA). The H1N1 virus is a virus subtype belonging to the type of influenza A virus. This kind of virus consists of sixteen hemagglutinins (H1-H16) and nine neuraminidases (N1-N9).

Theoretically, there are 144 subtypes, and until now 105 subtypes of influenza A virus have been found, and all are endemic to aquatic birds. However, some subtypes have been found in chickens and mammals [4]. At this time, various types of antivirals have been developed as a substitute for the vaccine method. Oseltamivir and zanamivir, which can block the development of H1N1 virus by inhibiting the NA, have been proposed and recommended the Centers for Disease Control and Prevention (CDC) and World Health Organization (WHO) to treat and cure the influenza flu patient[5]. Moreover, Oseltamivir is one of the antiviral drugs capable of inhibiting NA that functions in the viral replication process at the last stage of viral life cycle. The interaction between oseltamivir and NA is found in its interaction with the catalytic residue of the NA active site when the residual framework stabilizes the enzyme structure. However, it has also been reported that the oseltamivir resistance to influenza virus has increased [6]. Hence, the development of new antiviral to substitute oseltamivir as the anti-influenza agents is urgently needed.

In this study, the computational study of influenza A subtype H1N1 was conducted using the molecular docking methods [7]. The purpose of this study was to examine the interaction and conformation of three potential disulfide cyclic peptide inhibitors (DNY, NNY, LRL) with NA of influenza virus subtype H1N1 through molecular docking simulation methods. Also, this study also evaluates and determines potential ligands to be developed as a drug.

2 Materials and Methods

2.1 Sequence searching and preparation of the Influenza A virus (H1N1) neuraminidase

The NA sequence of influenza A virus (H1N1) used in this study can be downloaded from an existing database at NCBI via the site http://www.ncbi.nlm.nih.gov/genomes/flu/ by using a computer device connected to the internet. The NA sequence used has a GenBank code ACR08499.1. The homology modeling process for the three-dimensional (3D) structure of H1N1 NA can be done by using SWISS-Model which can be accessed via http://swissmodel.expasy.org/SWISS-MODEL.html [8–10]. After which, the 3D structure of H1N1 NA then downloaded with the PDB file format.

Visualization of the catalytic side of H1N1 NA was performed with MOE 2008.10 software through 'Site Finder' and 'Map' features. Furthermore, the process of geometry optimization and energy minimization of three-dimensional NA structures were carried out using the same software as well. The first step is protonation with 'Protonate 3D' feature to add the hydrogen atoms into the enzyme structure. Then, the MMFF94x force field was applied as the partial charge and optimization. When optimizing, the enzyme was solvated in 'Gas Phase' and RMS gradient of 0.05 kcal/mol.Å was applied. Other parameters were defined using the default protocols, and finally, the output files were saved in .moe file format.

2.2 Designing the 2D structure of the cyclic peptide ligands

The amino acid sequences of cyclic peptides that have been designed are then modeled into three-dimensional structures. This modeling step was done using ACD-Labs software, which then the results were stored in MDL Molfile (.mol) format. After which, the files were converted and optimized using VegaZZ software. Finally, the ligand database was made in MOE 2008.10 software by including all cyclic peptide design into MOE database viewer in .mdb file format. The optimization process begins by performing 'Wash' protocol for all cyclic peptides. Then, the partial charge of the ligands was loaded using 'Partial Charge' feature with the MMFF94 (modified) force field was applied to all ligands. Then, the 'Gas Phase' method was applied to the solvation mode of all ligands with RMS gradient 0.001 kcal/Å.mol. Finally, the output file was later saved in .mdb format.

2.3 Molecular Docking Simulation

The molecular docking simulation process was performed with the MOE 2008.10 through MOE → Simulation → Dock. The placement method of 'Triangle Matcher' was used with a repetition of energy readings per 1,000,000 positions and other parameters present in MOE. Moreover, 'London dG' was used as the Rescoring 1 function with the repetition population of 1000. The first repetition counted 30 times and the second set shows only one best result of 100 repetitions. After which, the Gibbs free binding energy ($\Delta G_{binding}$) value, hydrogen bond interactions and ligand's residues contact were observed and analyzed to determine the best ligand among all.

3 Results and Discussions

3.1 Sequence searching and preparation of the Influenza A virus (H1N1) neuraminidase

The sequence of NA data used as an enzyme target was determined using a database that presented on the NCBI website. This step was performed using the same parameters as we are done in previous work [11]. The NA sequence used was the Auckland NA sequence in 2009, NA [A/Auckland/1/2009 (H1N1)] with the genogram code of ACR08499.1. In this study, the resulting 3D NA structure has a 99.479% identical identity percentage of the template used (PDB ID: 3NSS). This significant percentage of identity value indicates that this 3D NA structure is good enough for use in later stages of research.

Determination of the catalytic site of NA was performed based on a study of viral sequences, followed by the literature study from NCBI sequence search results. From these results, it is known that NA has a catalytic side consisting of seven functional amino acids, namely Arg118, Asp151, Glu278, Arg293, Arg368, Tyr402, and Glu425. From the visualization results (**Figure 1**), it appears that the catalytic site residue is located on the surface of the enzyme and forms a large cavity.

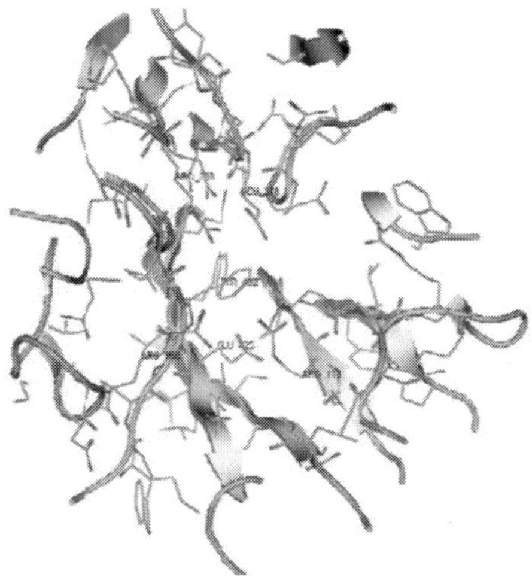

Figure 1. The visualization of catalytic site of Influenza A (H1N1) neuraminidase

NA optimization process needs to be done before molecular docking simulation with enzyme objective being in optimum condition when the simulation is done. In this study, the first step of the optimization process was done with 'Protonate 3D' program in MOE 2008.10 software, it is aimed to enzyme turn into the protonated state, followed by the addition of partial charge. Finally, the energy minimization of the enzyme was performed to avoid unrelated van der Waals contacts and to minimize high-energy steric effects of the structure [12]. The parameter setting of energy minimization was done in the optimal settings, with MMFF94x, RMS gradient of 0.05 kcal/mol.Å and gas phase solvation were used as the main parameters [13,14]. Later, the output file was saved in .moe format.

3.2 Designing the 2D structure of the cyclic peptide ligands

The design of amino acid peptide sequence was done by using ACD/ChemSketch software. The peptides, then called ligands, are the cyclic disulfide peptide generated from the previous studies, namely DNY, NNY, and LRL ligands [11]. In addition to these three ligands, zanamivir and oseltamivir which act as the standard ligands were designed as well. The ligands DNY, NNY and LRL are cyclic disulfide pentapeptide form comprising two amino cysteine acids with disulfide bonds, and three other amino acids, as it displayed in **Figure 2**. The cyclization form of amino acid itself was aimed to improve the stability of peptide molecules when introduced into the body, so proteases do not easily truncate them.

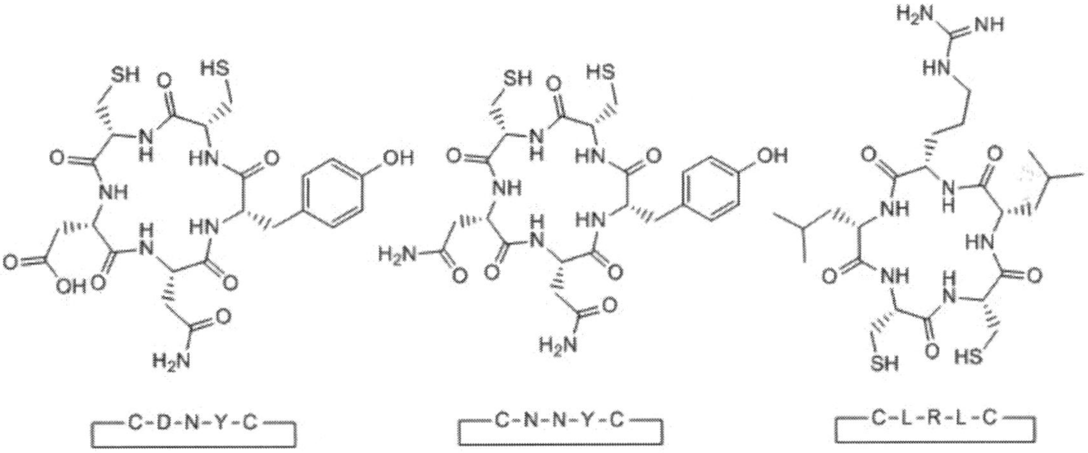

Figure 2. The scheme of the structured cyclic peptide designed in this research

The peptide and standard (oseltamivir & zanamivir) ligands already in the .mol format were imported into the MOE 2008.10 database viewer for optimization and minimization. The first step was washing for the entire ligand in order to improve the position of the hydrogen atoms found in the ligand to improve the ligand structure. Then, the partial charge was performed, followed by the energy minimization process with RMS gradient 0.001 kcal/mol Å [15].

3.3 Molecular docking simulation result and analysis

Molecular docking is a computational method to detect the exact bonding bond between the ligand and the receptor [16]. In this research, we used MOE 2008.10 as the primary tool for docking simulation purpose [17]. Docking process was performed on three potential peptide ligands (DNY, NNY, LRL) as well as the standard ligands (zanamivir and oseltamivir) against the catalytic side of NA [11]. In the process of docking, the enzyme is in rigid condition while the ligand is in a flexible condition so that it can be free to move and rotate. Based on the docking simulation results, it was found that the DNY ligand has the most negative $\Delta G_{binding}$ value and shows that DNY ligand has the strongest bond to the enzyme compared to the other two peptide enzymes (NNY and LRL) as well as the standard. The following results are displayed in **Table 1**.

Table 1. The result of molecular docking simulation and molecular interaction between the NA binding site and the ligands, red fonts indicate that this residue is the catalytic site of the NA

No.	Ligand Name	$\Delta G_{binding}$ (kcal/mol)	pKi	Interacted Amino Acid Residues
1	DNY	-7.8544	5.7243	Glu119, Ile149, Asp151, Asp151, Arg118, Arg152, Arg152, Ser247, Arg293, Asn344, Arg368, Arg368, Tyr402
2	NNY	-6.8994	5.0283	Glu278, Asn344, Gly345, Arg118, Arg156, Arg368, Tyr402
3	LRL	-6.8056	4.9599	Ile149, Asp151, Trp179, Trp179, Glu228, Tyr402, Ser247, Arg368, Arg368
S1	Zanamivir	-6.5858	4.7997	Glu119, Glu119, Ser247, Glu277, Glu278, Ser247, Arg293, Arg368, Arg368
S2	Oseltamivir	-4.1582	3.0305	Glu119, Glu278, Tyr402, Arg156

Docking process analysis is not only seen from $\Delta G_{binding}$ value, but also from the interaction between ligand with enzyme through residue contact especially hydrogen bond. The hydrogen bonds that occur in the ligand-enzyme complex can be identified and analyzed in the ligand interaction program using the MOE 2008 software as well. From the **Table 1**, DNY, LRL, and NNY ligands have an interaction with thirteen, seven, and nine amino acid residues in the catalytic site of NA, respectively. With the DNY, LRL, and NNY ligands forming six, four, and three hydrogen bonds. Additionally, the zanamivir and oseltamivir itself interact with nine and four amino acid residues, respectively, followed by 4 and two hydrogen bonds that created with the former and latter standard ligands. From these results, it can be analyzed that peptide ligands have better interaction and inhibition activity of NA enzyme compared to the standard ligands.

4 Conclusions

In this study, the DNY, LRL, and NNY ligands were deployed to know their inhibition activity and interaction in the catalytic site of the NA enzyme. Docking studies show that all three peptide ligands have better inhibition activities than both oseltamivir and zanamivir. With DNY ligand has the lowest $\Delta G_{binding}$ value at -7.8544 kcal/mol and pKi of 5.7243. Moreover, DNY has the highest number of interacted amino acid residues (thirteen interactions) and hydrogen bonds formed (six bonds). Therefore, further studies are needed on ligand affinity changes in the enzyme catalytic side sites and ADMET processes of ligands to metabolism in the body and taking into account the possibility of mutations in enzymes due to ligand binding.

Acknowledgments

The authors are very grateful to Direktorat Penelitian dan Pengabdian kepada Masyarakat (DIKTI), Kementerian Riset, Teknologi dan Pendidikan Tinggi (Kemenristek) and Directorate of Research and Community Engagement of Universitas Indonesia (DRPM UI) through Penelitian Unggulan Perguruan Tinggi (PUPT) No: 1121/UN2.R12/HKP.05.00/2016 for funding support this research. This study was designed and supervised by Usman Sumo Friend Tambunan and Djati Kerami. Riski Imaniastuti, Rendy Pramuda Putra, and Mochammad Arfin Fardiansyah Nasution have performed the experimental details and wrote the initial experimental report. All authors are responsible for drafting the final paper. Moreover, none conflict of interest in this research nor future publication were declared.

References

[1] WHO 2012 Situation updates - Avian influenza 6
[2] Centers for Disease Control and Prevention 2009 CDC H1N1 Flu | Origin of 2009 H1N1 Flu (Swine Flu)
[3] Robertson J S and Engelhardt O G 2010 Developing vaccines to combat pandemic influenza *Viruses* **2** 532–46
[4] Michaelis M, Doerr H W and Cinatl J 2009 An influenza A H1N1 virus revival - Pandemic H1N1/09 virus *Infection* **37** 381–9
[5] Rungrotmongkol T, Intharathep P, Malaisree M, Nunthaboot N, Kaiyawet N, Sompornpisut P, Payungporn S, Poovorawan Y and Hannongbua S 2009 Susceptibility of antiviral drugs against 2009 influenza A (H1N1) virus. *Biochem. Biophys. Res. Commun.* **385** 390–4
[6] Cheng P K C, Leung T W C, Ho E C M, Leung P C K, Ng A Y Y, Lai M Y Y and Lim W W L 2009 Oseltamivir-and amantadine-resistant influenza viruses A (H1N1) *Emerg. Infect. Dis.* **15** 966–8
[7] Leach A R 2001 Molecular modelling : principles and applications *Computers* **21** 784
[8] Biasini M, Bienert S, Waterhouse A, Arnold K, Studer G, Schmidt T, Kiefer F, Cassarino T G, Bertoni M, Bordoli L and Schwede T 2014 SWISS-MODEL: modelling protein tertiary and quaternary structure using evolutionary information. *Nucleic Acids Res.* **42** W252--8
[9] Arnold K, Bordoli L, Kopp J and Schwede T 2006 The SWISS-MODEL workspace: A web-based environment for protein structure homology modelling *Bioinformatics* **22** 195–201
[10] Bordoli L, Kiefer F, Arnold K, Benkert P, Battey J and Schwede T 2008 Protein structure homology modeling using SWISS-MODEL workspace *Nat. Protoc.* **4** 1–13
[11] Tambunan U S F, Amri N and Parikesit A A 2012 In silico design of cyclic peptides as influenza virus, a subtype H1N1 neuraminidase inhibitor *African J. Biotechnol.* **11** 11474–91
[12] Nurbaiti S, Hertadi R, Martoprawiro M A and Akhmaloka 2009 The Role of Interface Domain Interactions on Thermal Stability of DNA polymerase I ITB-1 *Open Struct. Biol. J.* **3** 16–25
[13] Singh H, Maria S S and Agarwal M 2006 Docking studies of tau protein *Lecture Notes in Engineering and Computer Science* pp 154–9
[14] Tambunan U S F, Nasution M A F, Azhima F, Parikesit A A, Toepak E P, Idrus S and Kerami D 2017 Modification of S-Adenosyl-L-Homocysteine as Inhibitor of Nonstructural Protein 5 Methyltransferase Dengue Virus Through Molecular Docking and Molecular Dynamics Simulation *Drug Target Insights* **11** 1–14
[15] Tambunan U S F, Parikesit A A, Nasution M A F, Hapsari A and Kerami D 2017 Exposing the Molecular Screening Method of Indonesian Natural Products Derivate as Drug Candidates for Cervical Cancer *Iran. J. Pharm. Res.* **16** 1115–29
[16] Teodoro M L, Phillips J and Kavraki L E 2001 Molecular docking: A problem with thousands of degrees of freedom *Proceedings - IEEE International Conference on Robotics and Automation* vol 1pp 960–5
[17] Vilar S, Cozza G and Moro S 2008 Medicinal chemistry and the molecular operating environment (MOE): application of QSAR and molecular docking to drug discovery. *Curr. Top. Med. Chem.* **8** 1555–72

IOP Conf. Series: Materials Science and Engineering **349** (2018) 012053 doi:10.1088/1757-899X/349/1/012053

Electrochemical disinfection of coliform and *Escherichia coli* for drinking water treatment by electrolysis method using carbon as an electrode

Riyanto[1,3] , W A Agustiningsih[2]

[1]Department of Chemistry, Faculty of Mathematics and Natural Science, Universitas Islam Indonesia, Jl. Kaliurang KM 14,5 Sleman Yogyakarta

[3]email: riyanto@uii.ac.id

Abstract. Disinfection of coliform and E. Coli in water has been performed by electrolysis using carbon electrodes. Carbon electrodes were used as an anode and cathode with a purity of 98.31% based on SEM-EDS analysis. This study was conducted using electrolysis powered by electric field using carbon electrode as the anode and cathode. Electrolysis method was carried out using variations of time (30, 60, 90, 120 minutes at a voltage of 5 V) and voltage (5, 10, 15, 20 V for 30 minutes) to determine the effect of the disinfection of the bacteria. The results showed the number of coliform and E. coli in water before and after electrolysis was 190 and 22 MPN/100 mL, respectively. The standards quality of drinking water No. 492/Menkes/Per/IV/2010 requires the zero content of coliform and E. Coli. Electrolysis with the variation of time and potential can reduce the number of coliforms and E. Coli but was not in accordance with the standards. The effect of hydrogen peroxide (H_2O_2) to the electrochemical disinfection was determined using UV-Vis spectrophotometer. The levels of H_2O_2 formed increased as soon after the duration of electrolysis voltage but was not a significant influence to the mortality of coliform and E.coli.

1. Introduction

The microorganisms that affect water quality are *Escherichia coli* (E. coli). These bacteria are classified as coliform and live normally in human and animal feces. Therefore, this bacterium is also called coliform fecal. According to the WHO report, at least 1.1 billion people do not have access to safe water [1]. According to the World Health Organization (WHO), microbiologically, the maximum permissible level of *Escherichia Coli* and Total Coliform Bacteria per 100 mL sample is 0 (cannot contain E. coli and coliform every 100 mL samples) as a quality measure of water. The maximum limit of *Escherichia coli* (E. coli) bacteria is < 2 APM/100 mL and should not contain pathogenic bacteria such as *Salmonella* and *Pseudomonas aeruginosa*. The requirement of drinking water that is to be safe to drink means free of pathogenic microbes and harmful substances and received in terms of color, taste, odor, and turbidity. In its management, water is susceptible to contamination from various microorganisms' especially coliform bacteria and E. coli bacteria [2].

Escherichia coli (E. coli) is one of the bacteria class of Coli used as water quality parameter. E. coli can cause water absorption in the intestinal wall is reduced so that diarrhea occurs. Some studies suggested that E. coli was the highest cause of death in infants caused by diarrheal disease [3]. Disinfection is one of the efforts to eliminate pathogenic bacteria in water. The most widely used disinfection is chemically by using chlorine compounds. Although this method was effective in reducing

IOP Conf. Series: Materials Science and Engineering **349** (2018) 012053 doi:10.1088/1757-899X/349/1/012053

bacteria and chlorine is cheap, chlorine can undergo chlorination resulting in toxic compounds causing carcinogenic, teratogenic effects, and damage to specific organs to humans . At present, inactivation of microorganisms in water generally uses chlorine or ozone, whereas chlorine produces residues that are toxic to aquatic biota. The use of excessive chlorine will cause side effects such as trihalomethanes that are carcinogenic chemicals that are suspected to cause cancer. Some disinfection methods are less efficient, hence an effective method is needed where the use of oxidative disinfection compounds is completely removed, especially with regard to chlorine, thereby eliminating all adverse consequences to human health. Many methods have been used for inactivation of microbial pollutants, including chlorination, ozone, ultra violet, and photocatalytic processes [4, 5].

An effective electrochemical disinfection system was developed as an alternative to conventional water treatment processes [6-9]. This is because electrochemical methods are more environmentally friendly, easy to operate and are known to disable various microorganisms from viral and algae bacteria [10]. Among electrochemical disinfection technologies, there is electro generation of disinfecting agents [11], bacterial electro sorption on the electrode surface using electricity [12], and electrophoresis [13]. Eliminating Escherichia Coli as an indicator of microbial contamination from water using electro photothermal method is an efficient method for water disinfection [14]. Prior research has been done by Jeong et al. [15], on the disinfection of E. coli electrochemical method using Pt anode by using inert media (free chloride phosphate buffer solution). The removal of microorganisms by electrolysis process using silver chloride electrode can kill coliform bacteria for 4 hours and E. coli bacteria for 2 hours [16].

In this paper will be delivered disinfection coliform and escherichia coli with the electrochemical method for drinking water with the carbon electrode. The use of carbon electrodes because carbon has a better effectiveness in an effort to reduce the content of metal ions, stable, does not react with electrolysis solution, otherwise, it is easy to obtain and cheap. This method uses carbon electrodes of dry battery rocks. Dry batteries commonly used by people in everyday life are a means that can store chemical energy and turn it into electric.

2. Experimental Section

2.1 Material and Instrumentation

The instrumentation used in this research is a set of electrolysis equipment, magnetic stirrer, and UV Vis Spectrophotometer Hitachi U2010. The materials used in this study were water samples, carbon electrode from dry battery, alcohol (Merck), H_2O_2 (Merck), $Na_2S_2O_3$ (Merck), HCl (Merck), KI (Merck), Ammonium (Merck), H_2SO_4 (Merck), starch *indicator* solution (Merck) and $K_2Cr_2O_7$ (Merck). The surface morphological structures were observed with the Scanning Electron Microscopy. Surface morphological analysis was performed on carbon electrode before and after electrolysis process. Characterization is done by the image results using SEM. Then from the figure is observed with EDS to know the composition of the elements contained in the carbon electrode.

2.2. Sampling Method

Water sampling by allowing water to go out for 5 minutes before measurement and sampling are carried out. The faucet is closed and heated the surface thoroughly, and then open again the faucet approximately one minute thoroughly, opened the faucet approximately one minute. Taken microbiological samples carefully, avoided from contamination, then filled with sterilized sample bottles without touching the surface. The glass storage vial previously sterilized by wrapping a glass bottle with a tight brown paper is then inserted in the oven for 1 hour at 121 °C. Plastered bottles with clear labels, then samples that are not analyzed immediately should be put in a cool or refrigerator (6-10 °C) and dark.

2.3. Water Electrolysis Process

The research has been done water electrolysis process with batch and continuous system. The research was conducted by connecting the DC potential source into the electrode which in this study used carbon electrode as cathode and anode. The carbon electrodes used in this process are obtained from carbon electrodes of the dry battery. The sample of water to be used was prepared by 350 mL into a closed beaker for electrolysis process. Then the electrode is paired with a DC device then the sample is electrolyzed. All tools used for bacterial analysis in this water should be sterilized first by wrapping brown paper and oven for 1 hour temperature 121°C. In the electrolysis of water the variation of electrolysis time, are 30, 60, 90, and 120 minutes with potential variations of 5, 10, 15 and 20 volts for 30 minutes. During the electrolysis process, Bunsen burners are used around the analysis site to avoid contamination from the outside and sprayed with 70% alcohol to avoid contamination from the outside. After completion of electrolysis, immediately moved into a sterile glass bottle, tightly closed, and labeled. If no direct analysis, then the bottle containing the sample should be immediately inserted into the refrigerator at a temperature of 4± 2°C. The MPN analysis is based on SNI 01-2332.1 (2006). Glass bottles containing water samples before and after electrolysis were immediately taken to the Microbiology Laboratory. Bacteriological testing includes the parameters of coliform bacteria and fecal coli. Scheme of electrochemical disinfection and its equipment can be seen in Figure 1.

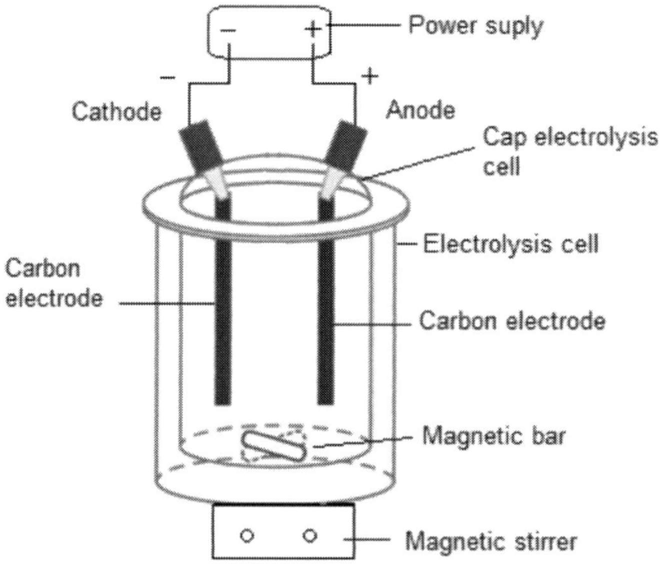

Figure 1. Scheme of the equipment used in the electrochemical disinfection.

2.4. Peroxide Content Analysis

Taken 1 mL of sample solution then diluted in 10 mL volumetric flask. Each flask containing the sample was added with 2 mL of HCl, 0.2 mL KI of 1.66 grams KI dissolved in a 10 mL volumetric flask, 0.2 mL ammonium molybdate in H_2SO_4, then sterilized for 20 min and added 0.2 mL of 1% starch of 0.05 grams of starch dissolved in 10 mL of distilled water. The measured the absorbance of the sample solution to calculate the hydrogen peroxide content in water with a maximum wavelength of 622 nm.

3. Result and Discussion

3.1 Determination of Total bacteria with Total Plate Count (TPC)

This electrolysis process is carried out with potential 5V for 1 hour. Obtain total plate count water sample that is before electrolysis of 670 colonies/gram and after electrolysis of 280 colony/gram. From these results managed to reduce the number of bacteria in the water though not perfect. Total bacteria test result with TPC method at 20V potential is shown in Figure 2.

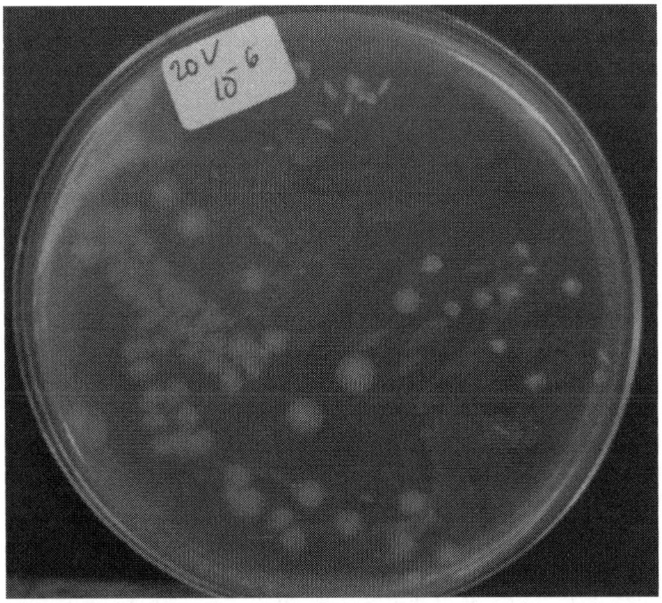

Figure 2. The result of the total bacteria test with TPC method at a potential of 20 V.

3.2. Purity of Carbon Electrode

The carbon electrodes used in this study were obtained from the dry battery. Dry batteries commonly used by people in everyday life because it is a means that can store chemical energy and convert it into electrical energy. The use of carbon electrodes has a better effectiveness in an effort to reduce the content of metal ions, stable, does not react with the solution in electrolysis, otherwise, it is easy to come by and cheap. The level of carbon purity in the battery is tested using SEM (Scanning Electron Microscope) as shown in Figure 3.

Figure 3. The image of SEM-EDS of the surface carbon material

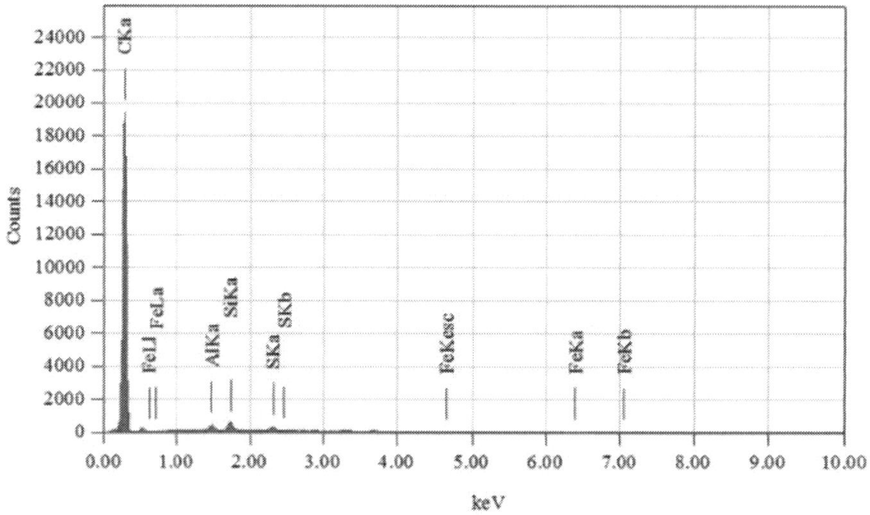

Figure 4. The EDS spectra carbon electrode.

Table 1. Element contains in carbon electrode using EDS

Element	Energy (keV)	Mass (%)	Atom (%)
C	0.277	98.31	99.33
Al	1.486	0.40	0.18
Si	1.739	0.66	0.28
S	2.307	0.43	0.16
Fe	6.398	0.21	0.05
Total		100	100

The morphology of the carbon surface with 5000x magnification in Figure 3 shows that carbon has an irregular shape, is not smooth, and has pores. Carbon in the battery contains various elemental compositions as in the SEM-EDS results in Figure 4 which are carbon (C), aluminum (Al), silicon (Si), sulfur (S), and iron (Fe). The purity of carbon element (C) in the battery is obtained from Table 1 of 98.31% carbon element mass (C).

3.3. Effect of Electrolysis Time on the Disinfection of Coliform and E.Coli Bacteria

Disinfection was done using electrolysis method using carbon electrode at anode and cathode with the variation of time 30, 60, 90, and 120 minutes at potential 5 V. Electrolysis process done in the closed condition and must be sterile to avoid contamination from outside. Based on the results of the research, the results obtained from the test of coliform and E. coli bacteria in Figure 5. Figure 5 shows the increase in the number of bacteria after electrolysis within 30, 60, 90, and 120 minutes at a potential of 5 V does not meet drinking water quality standard.

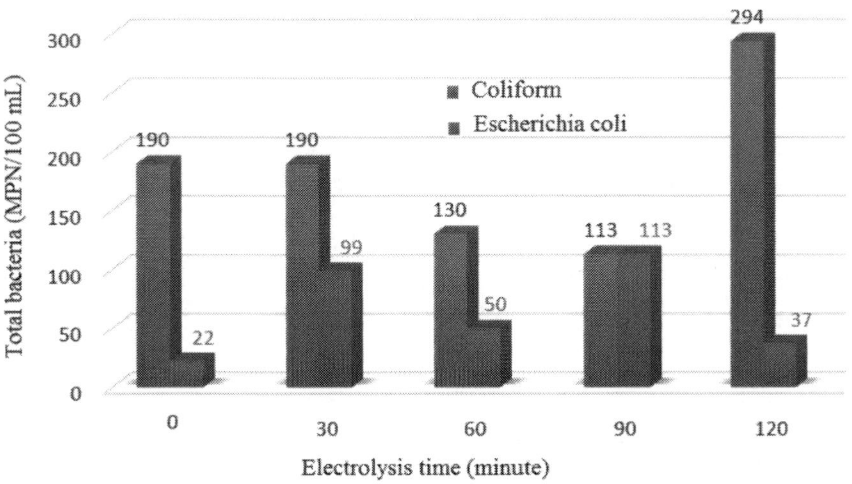

Figure 5. Effect of the electrolysis time to total bacteria coliform (Most Probable Number/MPN) and escherichia in water using potential 5 V

Figure 5 shows the longer electrolysis time the number of bacteria increases. This is due to the contamination from the outside which causes the number of bacteria to increase. In addition, due to the potential used is too low so it is not able to damage the cell walls of coliform and E. coli bacteria. While the levels of H_2O_2 formed at the time of electrolysis increases with the longer electrolysis time but does not affect the process of bacterial disinfection. Carbon electrodes used to experience saturation so it needs to be heated using the oven. Carbon has porous and can absorb on its surface, therefore the possibility of coliform and E. coli bacteria is absorbed into the pores of carbon in the living conditions and when reused for electrolysis the next possibility of bacteria from previously participated in the process of electrolysis so the number of bacteria increases with the longer the electrolysis process.

3.4. Potential Effects on Disinfection of Coliform and E. coli Bacteria

Disinfection of bacteria in water was carried out using electrolysis method with carbon electrode based on potential variations 5, 10, 15, and 20 V for 30 minutes. Based on the results of research conducted, obtained the results of bacterial coliform and stool coli as shown in Figure 6.

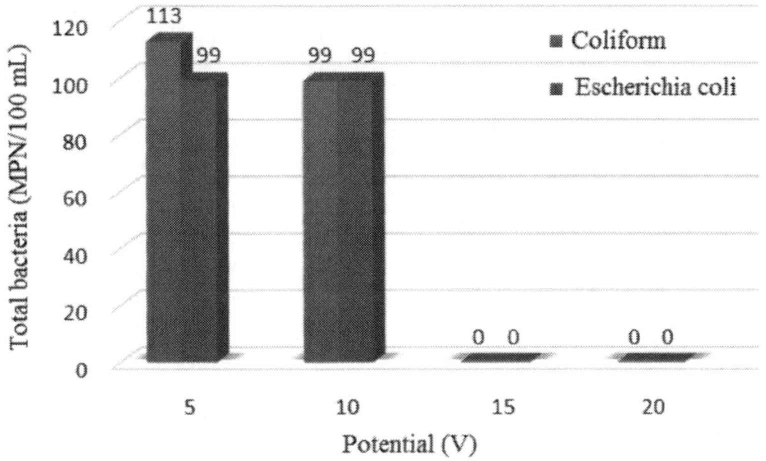

Figure 6. Changes in Number of Coliform and E.Coli Bacteria after Disinfection Process with Potential Variation for 30 minutes

Disinfection of coliform and E. coli bacteria uses electrolysis with potential variation as in Figure 6. Figure 6 shows that with potential variations can kill coliform and E. coli bacteria to meet the quality standard that is zero MPN/100 mL. Time and potential are two important variables in bacterial disinfection process by the electrolysis process. Gram-negative bacteria such as coliform and E. coli are more sensitive to electrical treatment than gram-negative bacteria [17]. The mechanism is potential bacterial death due to tissue damage. Provision of electricity with high potential causes the formation of pores in the bacterial cell membrane. In heat treatment (± 100 °C) results in severe damage to the cell organelle but does not break the breakdown of the cell wall as in the potential treatment. The greater the potential used the faster the time to kill the bacteria. According to Jeong [15], bacteria are killed due to the formation of hydrogen peroxide in electrolysis in anode oxidation reactions and cathode reduction as follow:

$2H_2O \rightarrow 4H^+ + 4e^- + O_2$

$4H_2O + 4e^- \rightarrow 2H_2 + 4OH^-$

The formation of a hydrogen atom (H) is one of the transitional steps in the formation of a hydrogen molecule in the cathodic reduction reaction with the reaction as follows:

$H^{\cdot} + O_2 \rightarrow HO_2$

$HO_2 + H^{\cdot} \rightarrow H_2O_2$

In the case, a metal electrode (M) can cause hydrogen peroxide to play a role in producing a hydroxyl radical reaction:

$H_2O_2 + M^{2+} \rightarrow M^{3+} + {}^{\cdot}OH + OH^-$

Hydrogen peroxide levels formed in the electrolysis process are not too much but with high potential can damage the cell wall of coliform and E. coli bacteria then hydrogen peroxide formed into the bacteria and reacts with the group negatively charged in the protein and disables the enzyme system.

3.5. Effect of Hydrogen Peroxide to Disinfection of Coliform and E.Coli Bacteria

Hydrogen peroxide (H_2O_2) is easily decomposed to form water (H_2O) and oxygen (O_2). The presence of metal ions in the cell cytoplasm of the microorganism can lead to the formation of superoxide radicals that will react with the negatively charged group in the protein and activate the enzyme system. Hydrogen peroxide can quickly kill bacteria. When hydrogen peroxide is in contact with bacteria, it quickly oxidizes the external components of the bacteria. Hydrogen peroxide is not only toxic to bacteria but can also hollow the outer membrane that protects the bacteria so that the bacteria die instantly.

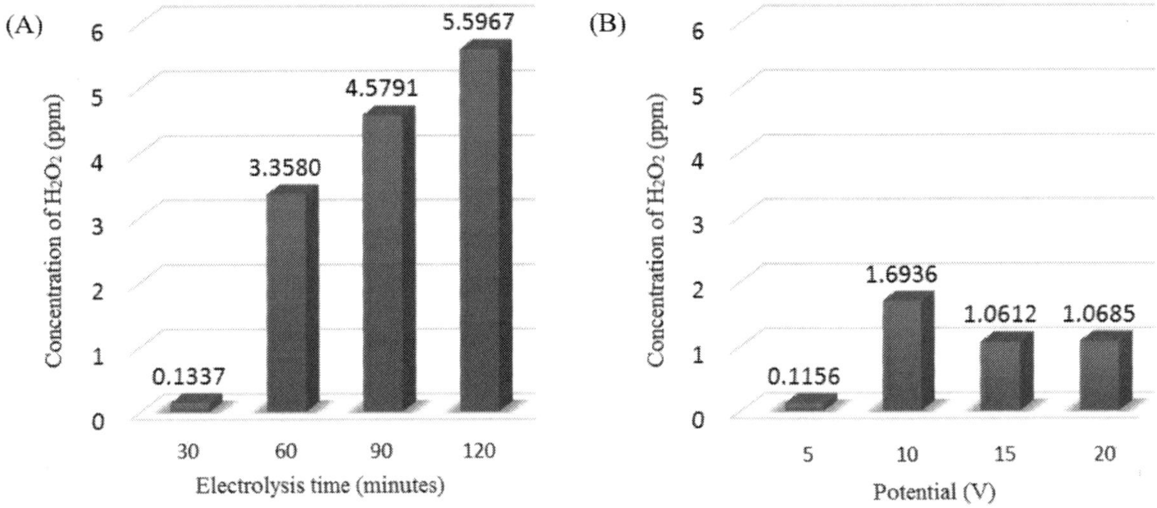

Figure 7. The effect of electrolysis time to concentration of H_2O_2 (A) and potential variation in H_2O_2 concentration (B)

The effects of time and potential on H_2O_2 concentrations are shown in Figure 7. The concentrations of H_2O_2 formed during electrolysis with time variations of 30, 60, 90, and 120 minutes were higher than in potential variations of 5, 10, 15, and 20 V for 30 minutes. The formation of OH radicals is the electrolysis of the solution at the anode and the cathode that the longer the electrolysis process, the more H_2O_2 formed.

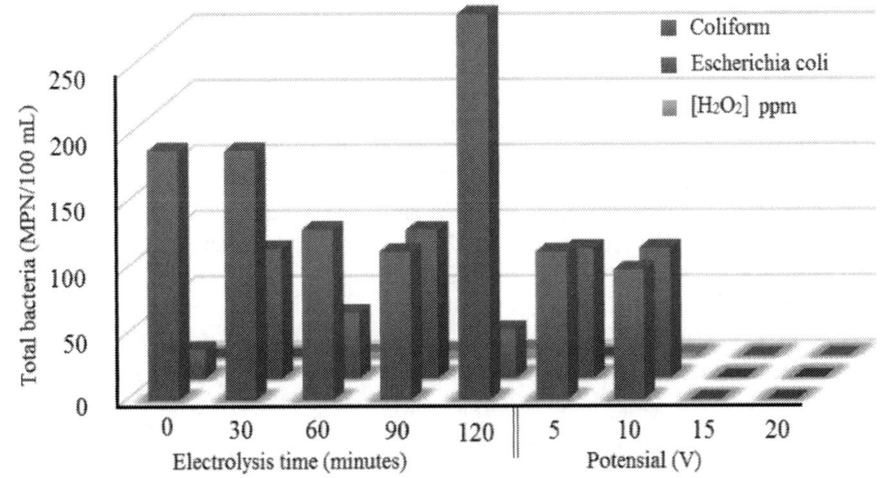

Figure 8. Disinfection Relationship of Coliform and E. Coli Bacteria to Hydrogen Peroxide Concentration (H_2O_2).

Disinfection of coliform and E. coli bacteria as shown in Figure 8 shows that time and potential variations can affect the formation of hydrogen peroxide in electrolysis. Disinfection at potential 15 and 20 V has little H_2O_2 but can kill bacteria until it meets the quality standard that is zero (0) MPN/100 mL. While at the time variation, has a greater H_2O_2 but can not kill coliform bacteria and E. coli so that hydrogen peroxide formed does not give a significant effect on the mortality of coliform and E. coli bacteria. In the process of disinfection of bacteria using electrolysis more influential on the potential used than the amount of hydrogen peroxide formed. The bigger the potential the shorter the time it takes to kill the bacteria.

Conclusion

Disinfection of bacteria electrolys effectively kills coliform and E. coli bacteria in clean water using carbon electrodes with potential above 15 V ie 0 (zero) MPN/100 mL. The levels of H_2O_2 formed increased with the length of electrolysis with a potential of 5 V but did not have a significant effect on coliform and E. coli mortality.

References

[1] Li Q, Mahendra S, Lyon D Y, Brunet L, Liga M V, Li D J and Alvarez J V 2008 *Water Res.* **42** 4591
[2] Levinson W 2008 *Review of Medical Microbiology* (Amerika: The McGraw-Hill Companies)
[3] Marquard D, Schneider-Barthold C, Dusterloh S, Scheper T and Lindner P 2017 *J. Biotechnol.* **259** 83
[4] Chong M N, Jin B, Chow C W K and Saint C 2010 *A Review. J. Wat. Res.* 4 **4** 2997
[5] Mezule 2009 *Desalination.* **248** 152
[6] Drees K P, Abbaszadegan M and Maier R M 2003 *Water Res.* **37** 2291
[7] Feng C, Suzuki K, Zhao S, Sugiura N, Shimada S and Maekawa T 2004 *Technol.* **94** 21
[8] Bergmann M E H and Koparal A S 2005 *J. App. Electrochem.* **35** 1321
[9] Mollah M Y A, Morkovsky P, Gomes J A G, Kesmez M, Parga J and Cocke D L 2004 *J. Hazard. Mater.* B **114** 199
[10] Diao H F, Li X Y, Gu J D, Shi H C and Xie Z M 2004 *Process Biochem.* **39** 1421
[11] Drogui P, Elmaleh S, Rumeau M, Bernard C and Rambaud A 2001 *Water Res.* **35** 3235
[12] Matsunaga T, Okochi M, Takahashi M, Nakayama T, Wake H and Nakamura N 2000 *Water Res.* **34** 3117
[13] Rowan N J, Macgregor S J, Anderson J G, Cameron D and Farish O 2001 *Appl. Environ. Microbiol.* **67** 2833
[14] Rezaee A, Kashi J J, Khataee A R and Niliahmadabadi A 2011 *J. Appl. Sci. Environ. Manage.* 15 **3** 439
[15] Jeong 2007 *J. Chemosphere.* **67** 652
[16] Miomir G, Pavlovic, Miroslav M, Pavlovic, Marija M, Pavlovic, Nebojsa D and Nikolic 2014 *Int. J. Electrochem. Sci.* **9** 8249
[17] Mazurek B, Lubicki P, Staroniewicz Z 1995 *IEEE Transactions on Dielectrics and Electrical Insulation.* 2 **3** 418

IOP Conf. Series: Materials Science and Engineering **349** (2018) 012054 doi:10.1088/1757-899X/349/1/012054

Treatment of Waste Lubricating Oil by Chemical and Adsorption Process Using Butanol and Kaolin

Riyanto[1,4], B Ramadhan[2], D Wiyanti[3]

[1-3]Department of Chemistry, Faculty of Mathematics and Natural Science, Universitas Islam Indonesia, Jl. Kaliurang KM 14,5 Sleman Yogyakarta

[4]email: riyanto@uii.ac.id

Abstract. Treatment of waste lubricating oil by chemical and adsorption process using butanol and kaolin has been done. Quality of lubricating oil after treatment was analysis using Atomic Absorption Spectrophotometer (AAS) and Gas Chromatography-Mass Spectrometry (GC-MS). The effects of the treatment of butanol, KOH, and kaolin to metals contain in waste lubricating oil treatment have been evaluated. Treatment of waste lubricating oil has been done using various kaolin weight, butanol, and KOH solution. The result of this research show metal content of Ca, Mg, Pb, Fe and Cr in waste lubricating oil before treatment are 1020.49, 367.02, 16.40, 36.76 and 1,80 ppm, respectively. The metal content of Ca, Mg, Pb, Fe and Cr in the waste lubricating oil after treatment are 0.17, 9.85, 34.07, 78.22 and 1.20 ppm, respectively. The optimum condition for treatment of waste lubricating oil using butanol, KOH, and kaolin is 30 mL, 3.0 g and 1.5 g, respectively. Chemical and adsorption method using butanol and kaolin can be used for decrease of metals contain in waste lubricating oil.

1. Introduction

It is estimated that more than 30.3 billion liters of used oil are produced every year. The waste of lubricant oil is generated mostly from vehicles and engines. This lubricant waste has a very high hazardous content from ash, carbon residue, asphaltenes, materials, metals, water, and other dirty materials produced during lubricants inside the machine. Used oils as hazardous wastes if disposed directly into the environment, especially rivers, seas and lakes create problems, such as the problem of disposing of used oil into water bodies not only contaminate water but also harmful to freshwater and marine life [1,2].

The waste of lubricant oil can be produced from various human activities such as industry, mining, and workshop. Lubricating oil wastes is including B3 waste is flammable so that if not handled the management and disposal will endanger human health and the environment. Currently, this liquid waste has been dumped into the environment and created many problems. Lubricating oil wastes are dumped on the ground, under a ditch, or sent to a landfill, which is eventually absorbed into the ground or floats on the water surface [3]. The management of this lubricating oil wastes to make used oil not pollute the environment and the nature of used oil becomes more harmless or even can be used again as new oil. Lubricating oil wastes management also aims to create a healthy environment for people's lives. Moreover, the handling of used oil well done, it will be able to provide benefits for industries that reuse used oil as a lubricant of various equipment, because used oil can still be used for lubricant again by way of different use before.

Components in lubricating oil wastes are hydrocarbons, polymer additives, carbon particles, water and metals [3]. The content of metals in lubricating oil wastes is not derived from lubricant base material but derives from additives added, metal, contaminants from petrol fuel containing lead, dust, dirt from the air, refrigerant and cooling water used to cool the machine. The process of lubricating oil wastes aims to make waste oil used to be a new oil that does not contain metal because the metal content in used oil is not useful.

Lubricating oil wastes processing is widely used, among others, by using a combination of solvent extraction and adsorption methods [3] using the addition of sulfuric acid, phosphoric acid, acetic acid and acids [4], using treatment of solvent/clay extraction [5]. Some of the processing that has been done, among others, used oil processing using various absorbent [6], and used oil recycling processes with new washing agents such as acetic acid belonging to Hamawand et al. [7]. According to Udonne, et al. [1] on the comparison of four recycled lubricant methods used: clay-processing, clay-distillation, acid treatment and charcoal processing methods or active clay. Recycling of engine oil waste is done using acetic acid. The recycling process takes place at room temperature. The advantage of using acetic acid is that it does not react or only reacts slightly with base oil [7]. Research from Abro et al. [8] on the ratio of engine lubricants enhanced by the extraction of composite solvents, solvents, and acid treatment methods.

A mineral may be referred to as kaolin clay if the kaolin mineral composition is more than 50% of the mineral composition [9]. Kaolin minerals can be found in nature in the form of pure kaolin and other kaolin minerals such as halloysite, nakrit and dicrit and other clay minerals such as smectite, ilit and mica as main components and feldspar and quartz as impurities [10].

Kaolin is a clay mineral with a 1:1 layer structure with a base unit composed of a tetrahedral sheet of SiO_4 and an octahedral sheet with Al^{3+} as an octahedral cation. In addition, kaolin widely used in the field of drugs, cosmetics, wine purifiers, and carrier materials for insecticides, pesticides, fungicides, crayon fillers, pencils, oil adsorbents, fertilizers, and also catalyst [10-12].

In this paper present of the treatment of waste lubricating oil by chemical and adsorption process using butanol and kaolin. Treatment of waste lubricating oil is using various kaolin weight, butanol, and KOH solution. After treatment of waste lubricating oil has been evaluated is the metal content of Ca, Mg, Pb, Fe and Cr. Analysis of the metal contains in waste lubricating oil before and after treatment has been done using Atomic Absorption Spectrophotometer (AAS).

2. Experimental Section

2.1 Materials and Instrumentation

The material has been used in this research is the waste of lubricant oil, butanol (Merck), sodium hydroxide (Merck), Kaolin (JP-100), nitric acid (Merck), hydrochloride acid (Merck), n-hexane (Merck) and acetonitrile (Merck). All chemicals used are analytical grade (pro analysis) and use distilled water. Instrumentation has been used in this research is furnace (Vulcan A-550), rotary evaporator (Buchi) and Atomic Absorption Spectrophotometer (AAS) (Perkin Elmer PinAcle 900T).

2.2 Treatment of Waste Lubricating Oil

Lubricating oil waste of 15 mL was put into a 250 mL beaker glass. Thereafter, 45 mL of n-butanol solution was added to dissolve the used oil and 2 g of KOH. After that, take 45 mL of n-butanol solution and mix it with oil and stir it evenly. KOH is incorporated into a mixture of used oil with n-butanol slowly gradually so as not to form the salt. After all mixed then stirred evenly using a magnetic stirrer for 1 hour with a temperature of 60 °C and filtered. Then the kaolin is put in the used oil. After stirring with a magnetic stirrer for 1 hour then stirring for 24 hours then filtered with Buchner funnel and using the vacuum pump. This filtering process is done for many times until the kaolin is completely lost. After that, an evaporation process is done to take n-butanol in the mixture. Block diagram of treatment of lubricating oil waste to new lubricating oil can be seen in Figure 1.

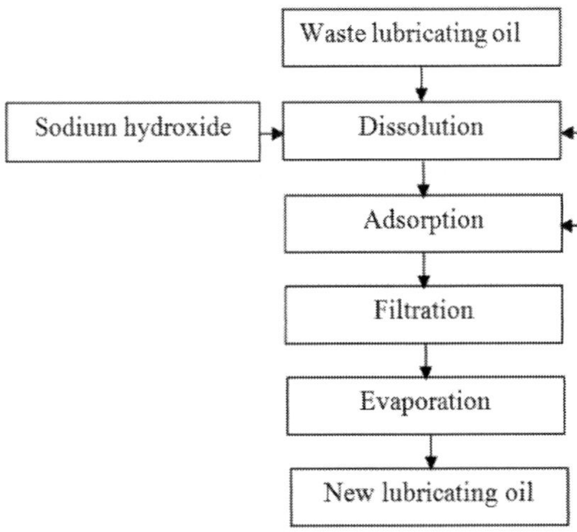

Figure 1. Block diagram of treatment of lubricating oil waste to new lubricating oil

2.3 Preparation of lubricating oil waste for metals analysis

The waste of lubricant oil sample weighed 5 g and put into porcelain cup and heated with oven with temperature 105 °C for 3 hours. After the oven and then put into the furnace with a temperature of 600 °C to ash. Samples added 5 mL of concentrated HNO_3 to dissolve the ash and add 5 mL of concentrated HCl solution. Then heat until the smoke in the solution was lost. The solution is put in a 100 mL volumetric flask and added distilled water until the mark and the sample is ready to be tested with AAS.

3. Result and Discussion

3.1 Effect of kaolin on metals in waste lubricating oil after process

The metals that are the main focus of this research are Ca, Mg, Pb, Fe, and Cr. The metals were studied using the ratio of butanol to used oil which is 3:1 where butanol is 45 mL and 15 mL of used oil waste sample. Initially, used waste lubricating oil was first tested before the research to find of metal content in waste oil samples used. In the sample of waste oil was first obtained Ca, Mg, Pb, Fe, and Cr is 1020.49; 367.02; 167.40; 36.76 and 1.80 mg/kg, respectively. The purpose of adding butanol solution is to break up the water content with the metal in waste oils so that the water content and metal content are separated and can be absorbed by kaolin, while the addition of KOH aims as the coagulant or it can also precipitate metals so it can be absorbed easily by kaolin.

Table 1. Effect of treatment using kaolin to waste lubricating oil

No	Metals	Before treatment in mg/kg	After treatment with kaolin (g) in mg/kg			Standard
			0.5	1.0	1.5	
1.	Ca	1020.49	0.72	0.17	2.38	1103.00
2.	Mg	367.02	58.90	9.85	8.34	870.80
3.	Pb	167.40	58.94	34.07	47.06	16.00
4.	Fe	36.76	90.21	78.22	4.15	2.76
5.	Cr	1.80	1.23	1.20	1.23	0.12

Figure 1 shown is the effect of treatment using kaolin to waste lubricating oil. In the first variation of using kaolin of 0.5 g obtained metal content on Ca metal of 0.72 mg/kg. While on metal Mg and Pb is 58.90 and 58.94 mg/kg, respectively. For the content of Fe and Cr are 90.21 and 1.23 mg/kg, respectively. In this first variation when compared with the metal content of the waste oil sample there was a decrease in Ca, Mg, Pb, and Cr metals. As for Fe metal, it experienced a significant increase from 36.76 mg/kg to 90.21 mg/kg. This may happen considering that in this process of using kaolin absorbent containing Fe metal also. It could be when the filtration process is less than the maximum so that the Fe metal in the sample of oil waste mixed with the absorbent kaolin. To minimize this it is advisable to use the softest filter paper with repetition of approximately three times, so it is expected that the kaolin absorbent can be filtered to the maximum.

3.2 Effect of treatment with acetic acid

This study used acetic acid solution to help break down or separate the water content with the compounds and metals contained in the waste oil. Table 2 shown of the effect of treatment using acetic acid to waste lubricating oil.

Table 2. Effect of treatment using acetic acid to waste lubricating oil

No	Metals	Treatment with acetic acid (mg/kg)	Standard
1.	Fe	70.84	2.76
2.	Ca	12.71	1103.00
3.	Mg	3.00	870.80
4.	Pb	5.56	16.00
5.	Cr	28.00	0.12

In Table 2 above, the effect of the addition of acetic acid, butanol, and KOH using kaolin absorbent on metal concentration in waste oil. This treatment can decrease the metal content obtained in the waste sample of waste oil maximally because it is expected with the addition of the acetic acid can help butanol in breaking or separating the water content contained in waste oil sample so that it can be absorbed maximally by kaolin absorbent. However, the results obtained are even less good compared with the results of the third variation where the results of the third variation can approach the standard metal content in the new oil. Thus it can be concluded that the addition of weak acids such as glacial acetic acid does not help to break down or separate the organic compound and the metal content contained in the waste oil sample.

Adsorption metal in kaolin is mechanism chemical and physical adsorption. Kaolin contains SiO_2 46.54%, Al_2O_3 39.50%, and H_2O 13.96% at the time of metal adsorption accompanied by H^+ release, then H^+ ion is substituted by the metal ion (cation) [13]. Figure 2 shown the mechanism of chemical adsorption of the Ca metal in kaolin

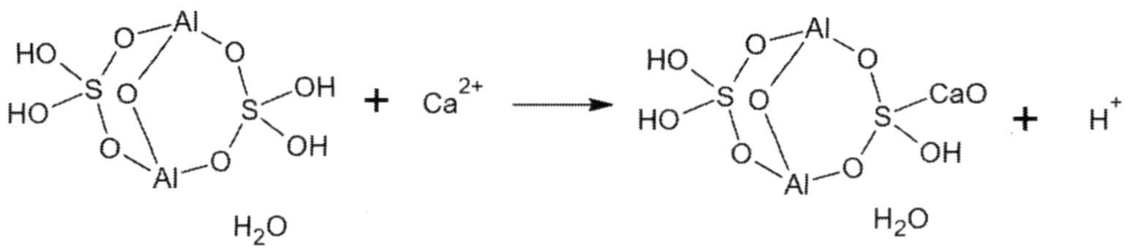

Figure 2. Mechanism of chemical adsorption of the Ca metal in kaolin

Kaolin has a skeletal structure, containing empty space occupied by cations and free water molecules allowing the exchange of ions and the absorption of chemical compounds. The evaporation of water content then the space occupied by the free water molecule becomes empty so that when adsorption is very possible the absorption of ions. Figure 3 shown is the mechanism of physical adsorption Ca metal in kaolin.

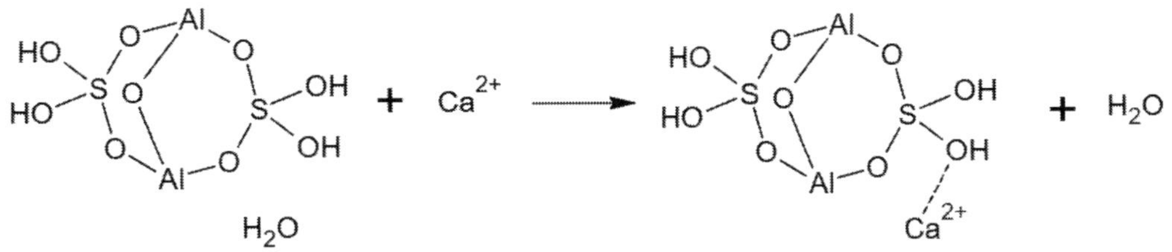

Figure 3. Mechanism of physical adsorption Ca metal in kaolin

3.3 Effect of butanol concentration

Metals content of used lubricating oil is not derived from lubricant base material but derives from additives added, metal, contaminants from petrol fuel containing lead, dust, dirt from the air, refrigerant and cooling water used to cool the machine. Table 3 shown the effect of treatment using butanol to metal content in waste lubricating oil

Table 3. Effect of treatment using butanol to waste lubricating oil

Metals	Lubricating oil waste (mg/kg)	Lubricating oil waste after treatment using butanol (mg/kg)			Standard (mg/Kg)
		2:1	3:1	4:1	
Fe	365.13	16.89	11.08	35.99	2.76
Ca	1019.00	6.76	2.51	nd*	1103.00
Mg	325.88	12.77	5.58	24.71	870.80
Pb	181.00	98.26	7.16	9.16	16.00
Cr	10.72	2.88	nd*	nd*	0.12

*nd is not detection

Variation of butanol with oil of butanol ratio: used oil (2:1) for Fe, Ca, Mg, Pb, Cr produced metal is 16,89; 6.76; 12,77; 98.26; 2.88 mg/kg, respectively. In the case of waste oil treatment using butanol solvent can significantly decrease Fe, Ca, Mg, Pb, and Cr content of metal in used oil and mean that butanol succeeds in reducing the metal in used oil which is characterized by the deposition of a layer of mud under a beaker. Maximum standards according to ASTM D5185 metal is Fe, Ca, Mg, Pb, Cr is 2.76, 1103.00, 870.80, 16.00 and 0.12 mg/kg, respectively. The Ca, Mg, and Cr metals do not exceed the standard but high concentrations of Fe and Pb metals are still high.

3.4 Effect of KOH concentration

Variations of butanol can be obtained that the best variation of butanol is 3:1 after that done KOH variation as much as 1, 2 and 3 g. The addition of KOH increases the ability of alcohol to remove sludge from waste oil. Table 4 shown is the effect of treatment using KOH concentration to waste lubricating oil.

KOH variation of 3 g for Fe, Ca, Mg, Pb and Cr is 74.32, 1.20, 8.83, 62.00, and 0.11 mg/kg, respectively. KOH variation of 3 g also decreased the concentration of each metal when compared with used oil, but when compared with variation KOH 1 and 2 g there is an increase in Fe and Pb high metal. The Pb metal does not exceed the ASTM D5185 standard, while the Fe metal exceeds the ASTM D5185 standard. The increase in the metal is probably due to KOH not reacting perfectly in the process so it is less effective at helping butanol increase sludge. The likelihood of the increase in metal concentration is also due to less optimal adsorption process so that metals that passes from kaolin adsorbent only slightly.

Table 4. Effect of treatment using KOH concentration to waste lubricating oil

Metals	Lubricating oil (mg/kg)	Lubricating oil waste after treatmentusing KOH concentration (mg/kg)			Standard (mg/kg)
		1 g	2 g	3 g	
Fe	365.13	39.30	11.08	74.32	2.76
Ca	1019.00	0.80	2.51	1.20	1103.00
Mg	325.88	5.43	5.57	8.83	870.80
Pb	181.00	28.51	7.16	62.00	16.00
Cr	10.72	nd	nd	0.11	0.12

nd is not detection

Based on Table 4 it can be concluded that KOH variations of 1, 2, and 3 g decrease for each metal, although there is a rising metal. The more KOH added in the waste oil treatment the better it is to lower the metal content of the used oil, but the most optimal addition of KOH is 2 g. The best KOH variation is 2 g. This is in accordance with the study Mohammed et al. which states that the addition of 2 g of KOH to butanol in the solvent has been shown to significantly increase the removal of sludge from waste oil [3].

Conclusions

The effect of butanol on the waste oil treatment is for the deposition of the mud layer under the beaker and acts as a solvent for the extraction-flocculation in reducing Fe, Ca, Mg, Pb, and Cr, while the KOH effect is as coagulant to increase the flocculation of the dirt, oil with the solvent, helping the butanol solvent in forming sludge. The best butanol variation used to reduce the Fe, Ca, Mg, Pb, and Cr on the used oil is 3:1, and for the best KOH variation is 2 g. The more butanol and KOH used, the better it is in reducing the metal content.

References

[1] Udonne J D 2011 *Journal of Petroleum and Gas Engineering.* **2** 12
[2] Udonne J D and Bakare O A 2013 *International Archive of Applied Sciences and Technology.* **4** 2 8
[3] Mohammed R R, Ibrahim A R, Taha H A and McKay G 2013 *Chem. Eng. J.* **220** 343
[4] Abu-Elella R, Ossman M E, Farouq R and Abd-Elfatah M 2015 *International Journal of Chemical and Biochemical Sciences.* **7** 57
[5] Emam A E and Shoaib M A 2012 *Refining of Used Lube Oilby Solvent/Clay and Acid/Clay-Percolation Processes* (Egypt: Department of Refining Engineering and Petrochemicals, Suez Canal University)
[6] Shakirullah M, Ahmad I and Saeed M 2006 *Journal of the Chinese Chemical Society.* **5 32** 335
[7] Hamawand I, Yusaf T and Rafat S 2013 *Energies.* **6** 1023
[8] Abro R, Chen X, Harijan K, Dhakan Z A and Ammar M 2013 *Chem. Eng.* **2013** 1
[9] Dombrowski T 2000 *The Origin of Kaolinite* (Westerville: American Ceramic Society)
[10] Ekosse G E 2000 *Applied Clay Science* **16 5-6** 301
[11] Belver C, Muoz M A B and Vicente M A 2002 *Chem.Mater.* **14** 2033
[12] Murray H H 2000 *Appl. Clay Sci.* **34** 39
[13] Bhattacharyaa G K and Gupta S S 2008 *Advances in Colloid and Interface Science.* **140** 114

IOP Conf. Series: Materials Science and Engineering **349** (2018) 012055 doi:10.1088/1757-899X/349/1/012055

Utilization of Android-base Smartphone to Support Handmade Spectrophotometer : A Preliminary Study

R Ujiningtyas[1], E Apriliani[1], I Yohana[1], L Afrillianti[1], N Hikmah[2], C Kurniawan[1]

[1]Chemistry Departement of Semarang State University
[2]Computer science Departement of Semarang State University

Email: restiujiningtyas@gmail.com

Abstract. Visible spectrophotometer is a powerful instrument in chemistry. We can identify the chemical species base on their specific color and then we can also determine the amount of the species using the spectrophotometer. However, the availability of visible spectrophotometer still limited, particularly for education. This affect the skill of student to have experience on handling the instrumentation. On the other hand, the communication technology creates an opportunity for student to explore their smart feature, mainly the camera. The objective of this research is to make an application that utilize the camera feature as a detector for handmade visible spectrophotometer. The software have been made based on android program, and we name it as Spectrophone®. The spectrophotometer consists of an acrylic body, sample compartment, and light sources (USB-LED lamp powered by 6600 mAh battery). Before reach the sample, the light source was filtered using colored-mica plastic. The spectrophone® apps utilize the camera to detect the color based on its RGB composition. A different colored solution will show a different RGB composition based on the concentration and specific absorbance wavelength. We then can choose one type of color composition, R or G or B only to be converted as an absorbance using -Log (Cs/Co), where Cs and Co are color composition of sample and blank, respectively. The calibration curve of metilen blue measured. In a red (R) composition, the regression is not linear (R^2=0.78) compare to the result of UV-Vis spectrophotomer model Spectroquant Pharo 300 (R^2=0.8053). This measurement result shows that The Spectrophone® still need to be evaluated and corrected. One problem than can we identify that the diameter of pick point of RGB composition is too wide and this will affect the reading color composition. Next, we will fix the problem and in advance we will apply this Spectrophone® in a wide scale.

1. Introduction

Regulation of the Minister of national education of the Republic of Indonesia Number 40 of the year 2008 contains about components on-site laboratories which include (1) building/space laboratory, (2) furniture, (3) equipment education, (4) tools and materials experiments, (5) educational media, (6) consumables, (7) other amenities [2]. Facilities and infrastructure is needed to facilitate in an activity. One of the facilities of the laboratory that can help researchers in conducting laboratory research is a instrument. Through these instruments, researchers be facilitated in analyzing the results

IOP Conf. Series: Materials Science and Engineering **349** (2018) 012055 doi:10.1088/1757-899X/349/1/012055

of the experiments have been performed. One of the instruments of the chemical is laboratory is visible spectrophotometer.

Visible spectrophotometer is a powerful instrument in chemistry. We can identify the chemical species base on their specific color and then we can also determine the amount of the species using the spectrophotometer. In qualitative analysis, organic compounds can be identified by use of spectrophotometer, if any recorded data is available, and quantitative spectrophotometric analysis is used to ascertain the quantity of molecular species absorbing the radiation. Spectrophotometric technique is simple, rapid, moderately specific and applicable to small quantities of compounds. The fundamental law that governs the quantitative spectrophotometric analysis is the Beer -Lambert law [1].

Beer-Lambert law: When beam of light is passed through a transparent cell containing a solution of anabsorbing substance, reduction of the intensity of light may occur. Mathematically, Beer- Lambert law is expressed as:

A = a b c, where

A = absorbance or optical density

a = absorptivity or extinction coefficient

b = path length of radiation through sample (cm)

c = concentration of solute in solution.

Both b and a are constant so a is directly proportional to the concentration c [1].

In absorption spectrometry, light is directed through a sample and the fraction of light that passes throught the sample is measured. The amount of light absorbed, or the absorbance A, is define as

$$A = -\log\left(\frac{I}{I0}\right)$$

where I is the intensity of light transmitted through the sample and I0 is the intensity of light transmitted through a blank. It is useful to use a color wheel (Figure 1) to estimate the color of light that is absorbed by a particular chemical species in solution.This approximation is done by noting the color on the wheel opposite the observed color of the compound. For example, if a particular species appears red in solution, it probably absorbs green light very well. The predictions made using the color wheel are not absolute. Its use is complicated by the fact that our eyes are not equally sensitive to all colors of light, in addition to other factors [3].

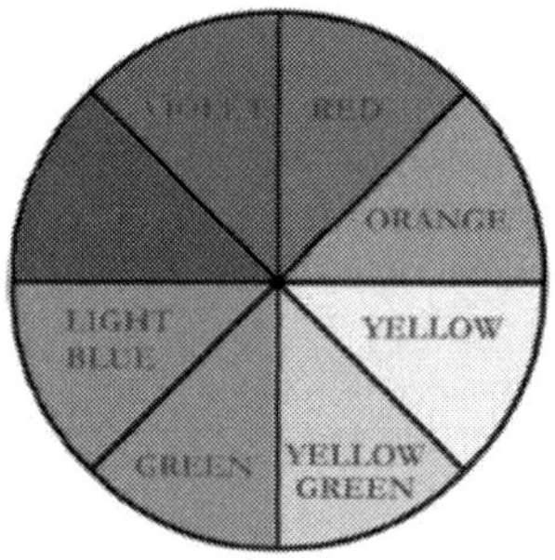

Figure 1. A color wheel used to estimate the color of light absorbed by a solution of a certain color.

However, the availability of visible spectrophotometer still limited, particularly for education. This affect the skill of student to have experience on handling the instrumentation. On the other hand, the communication technology creates an opportunity for student to explore their smart feature, mainly the camera. The objective of this research is to make an application that utilize the camera feature as a detector for handmade visible spectrophotometer. The software have been made based on android program, and we name it as Spectrophone®.

2. Method

2.1. Materials

The tools and materials needed for the manufacture of SpecPhone are: Smartphone, mirror, test tube, LED lamp, power bank, colored mica, black sticker and acrylic board.

2.2. Experimental

2.2.1 Spectrophotometer design

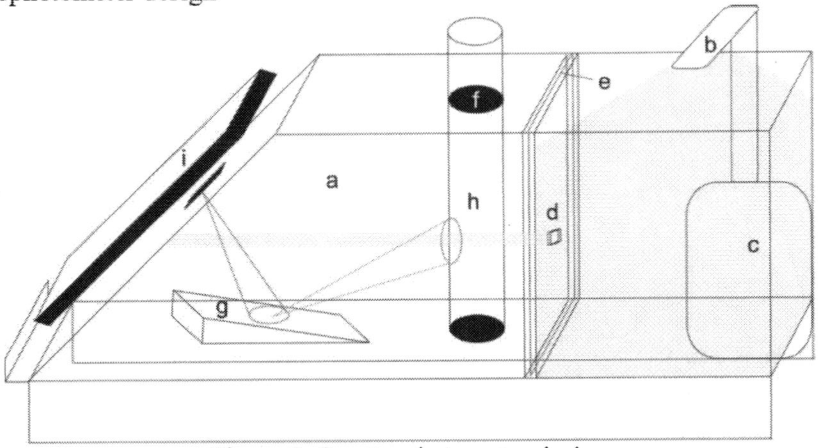

Figure 2. Spectrophotometer design

The Spectrophotometer design is as follows

a. The inside of the coated black stickers no color is reflected.
b. Light source coming from USB LED lamp is 5 watts.
c. Power bank is used as a source of electrical energy for lighting LED lamps with a capacity of 6600 mAh.
d. Regulate the width of the gap / slit which is a maximal hole 1 cm x 1 cm. Slit is made with two types that can and can be shifted so that the width of the hole can be varied.
e. Monochromator, this section is made of colored mica plastic that serves to select wavelengths that convert light coming from polychromatic light sources into monochromatic light. Xmax uses Spectroquant Pharo 300 spectrophotometer model.
f. The sample holder, as the sample venue is measured.
g. Mirror, serves to give the color shadow can be accepted by the camera lens.
h. Test tube size 15 cm to place the test solutions.
i. Smartphone as a catcher of sample color intensity.

2.2.2 Spectrophone interface

The spectrophotometer was completed with the software which can be easily installed. The user-software interface is shown in Figure 3.

IOP Conf. Series: Materials Science and Engineering **349** (2018) 012055 doi:10.1088/1757-899X/349/1/012055

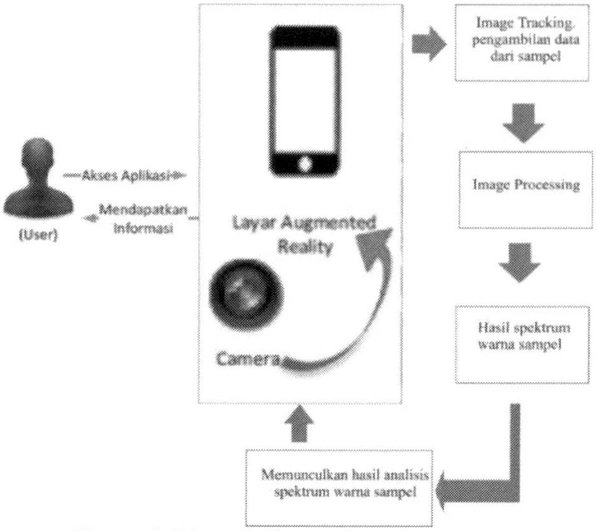

Figure 3. User-Software interface

3. Results and discussion

The spectrophotometer is show below in Figure 4. The Spectrophotometer consist of an acrylic body, sample compartment, and light sources (USB-LED lamp powered by 6600 mAh battery). Before reach the sample, the light source was filtered using colored-mica plastic. Colored-mica plastic used to revamp polikromatik rays of light into monochromatic. Acrylic body is coated with a black sticker so that the light from the outside can't get into there doesn't affect the camera RGB in reading.

Figure 5 show the appearance of Spectrophone® apps. This apps utilize the camera to detect the color based on its RGB composition. A different colored solution will show a different RGB composition based on the concentration and specific absorbance wavelength. We then can choose one type of color composition, R or G or B only to be converted as an absorbance using -Log (Cs/Co), where Cs and Co are color composition of sample and blank, respectively.

Figure 4. The Spectrophotometer

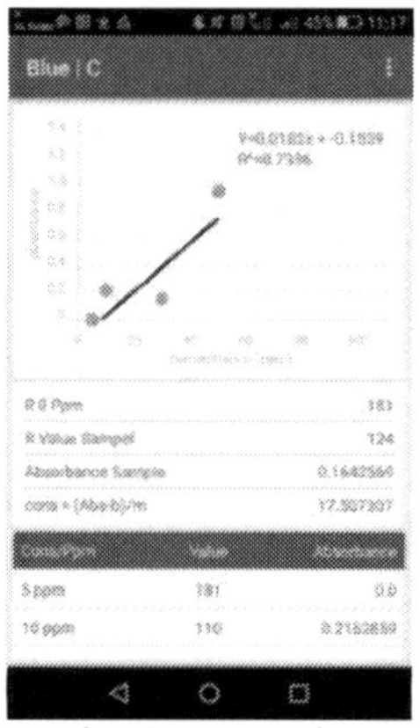
Figure 5. The Spectrophone® apps

4. Evaluation result

Once the application was developed, preliminary testing commenced. Testing is done by comparing the results of spectrophone with UV-Vis spectrophotomer Spectroquant Pharo model 300. Testing is done to determine the levels of Fe in water. Testing conducted in the laboratory of inorganic chemistry of engineering unnes. The method used is the method of calibration with variation concentration of metal blue by as much as 0 ppm 5 ppm, 10 ppm, 20 ppm, 30 ppm, and 50 ppm.

IOP Conf. Series: Materials Science and Engineering **349** (2018) 012055 doi:10.1088/1757-899X/349/1/012055

With the use of this procedure, the results of measurements of the levels of metilen blue using UV-Vis spectrophotometer show in figure 5 with the results of the calculations of the levels of metilen blue is 37.04 ppm. Meanwhile, the measurement results with the use of spectrophone can be seen in figure 6 with the result of levels metilen blue is 34.41 ppm in a Red composision.

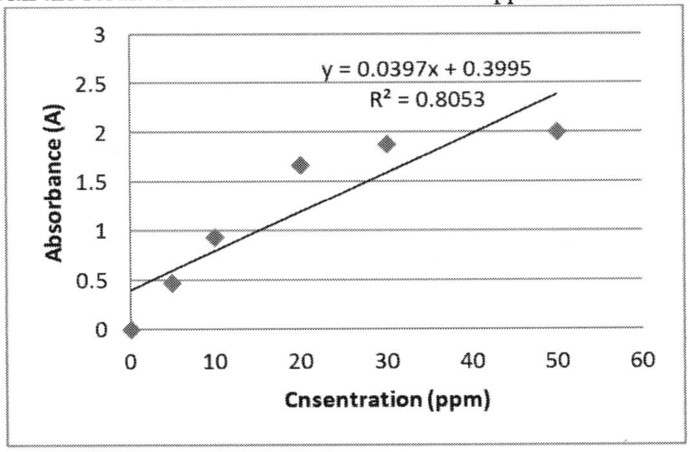

Figure 6. Result of Spectrophotometer UV Visible

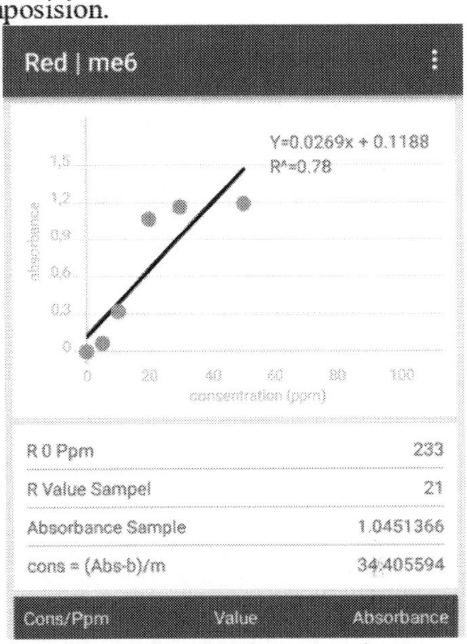

Figure 7. Result of Spectrophone

The calibration curve of metilen blue measured in a red (R) composition, the regression is not linear (R^2=0.78) compare to the result of UV-Vis spectrophotomer model Spectroquant Pharo 300 (R^2=0.8053). This measurement result shows that The Spectrophone® still need to be evaluated and corrected. One problem than can we identify that the diameter of pick point of RGB composition is too wide and this will affect the reading color composition. Next, we will fix the problem and in advance we will apply this Spectrophone® in a wide scale..

.

5. Conslusion

The experiment reported here provides a simple way to have explore absorption spectroscopy. Dari hasil pengujian yang telah dilakukan, konsentrasi metilen blue dalam air diukur dengan menggunakan spektrofotometer adalah 36.04 ppm sedangkan dengan spectrophone in a red composition sebanyak 34.51 ppm. This measurement result shows that The Spectrophone® still need to be evaluated and corrected. Next, we will fix the problem and in advance we will apply this Spectrophone® in a wide scale.

References

[1] Behera S, Ghanty S, Ahmad F, Santra S, and Banerjee S. 2012. UV-Visible Spectrophotometric Method Development and Validation of Assay of Paracetamol Tablet Formulation. *Journal Arial Bioanal Tecniques* **3** 1-6

[2] Permendiknas. 2008. Peraturan Mendiknas RI Nomor 40, Tahun 2008, tentang Standar Sarana dan Prasarana Sekolah Menengah Kejuruan/Madrasah Aliyah Kejuruan (SMK/MAK).

[3] Shakhashiri, B. Z. *Chemical Demonstrations: A Handbook for Teachers of Chemistry*. The University of Wisconsin Press: Madison, WI, 1983; Vol. 1, pp 262–266.

IOP Conf. Series: Materials Science and Engineering **349** (2018) 012056 doi:10.1088/1757-899X/349/1/012056

Synthesis and Characterization of Tetrakis(2-amino-3-methylpyridine)copper(II) Sulfate Tetrahydrate

S B Rahardjo[1], T E Saraswati[1], A Masykur[1], N N F Finantrena[1], H Syaima[2]
[1]Department of Chemistry, Faculty of Mathematics and Natural Sciences, Universitas Sebelas Maret, Surakarta, 57126, Indonesia
[2]Chemistry Graduate Program, Graduate School, Universitas Sebelas Maret, Surakarta, 57126, Indonesia

E-mail: sentotbr@yahoo.com

Abstract. The complex of Tetrakis(2-amino-3-methylpyridine)copper(II) sulfate tetrahydrate has been synthesized in a ratio of 1: 6 metal to ligand in methanol. The percentage of copper in the complex measured by Atomic Absorption Spectrometer (AAS) showed the complex formula was Cu(2-amino-3-metilpyridine)$_4$SO$_4$(H$_2$O)$_n$ (n = 3, 4, or 5). The analysis of TG/DTA showed that 1 mole of complex contains 4 moles of H_2O. The conductivity measurement indicated that the complex is in 1 to 1 electrolyte. The formula of the complex was estimated as [Cu(2-amino-3-metilpyridine)$_4$]SO$_4$·4H$_2$O. The complex was paramagnetic with μ_{eff} of 1.85 BM. The UV-Vis spectra showed a band peak at 730 nm with an electronic transition Eg⟶T$_2$g. IR spectral data indicated that the functional groups of N-pyridine 2-amino-3-metilpyridine coordinated to ion Cu(II). The geometry of the complex was probably square planar.

1. Introduction

Pyridine derivative which has antibacterial activity [1] is able to act as ligands. It is due to the availability of free electron pair to bond with a metal ion. Pyridine derivative ligands can form complexes with different types of donor atoms coordinated to the metal ion. In case the complex of [Ag(2-amino-3-methylpyridine)$_2$]NO$_3$, the ligand has an electron donor atom N in the primary amine and N-pyridine. Coordinated donor atoms are N in pyridine [2]. After complexing, the effectivity of ligands as antibacterial increase significantly. The complex of Cu(II)-2,4-dinitro-6(pyridine-2-ylmethylamino)methyl-phenolate has better bactericidal activity than complex of Fe(II), Zn(II), Ni(II) and 2,4-dinitro-6(pyridine-2-ylmethylamino)-methylphenolate as ligands [3]. Therefore, complexing copper ion with pyridine derivative ligands seems promising due to the improvement of both antibacterial activities.

Copper is one of transition metal elements with 3d^9 electron configuration that can form complexes because they have d orbitals are not fully charged electrons. Copper has the oxidation number of +1 and +2, copper in the form of Cu(II) is more stable than Cu(I). Cu(II) can form a wide variety of geometries in the complex include square planar [4], square pyramidal [5], and octahedral [6] but generally Cu(II) has a coordination number of four bonded to ligands forming complexes with square planar structure.

2. Experimental

2.1. Materials

All chemicals were purchased from E.Merck.

2.2. Physical measurements

The copper content was determined by Atomic Absorption Spectrometer (AAS) Shimadzu AA-6650. Infrared spectra were recorded on Prestige-21 Shimadzu spectrophotometers using KBr disc. The thermogravimetric analysis of the metal complex was recorded on Thermogravimetric/Differential Thermal Analysis (TG/DTA) Diamond Perkin Elmer analyzer. Spectra UV-VIS was performed on Shimadzu UV-3601 spectrophotometer. Molar conductivity (Λ*m) of 1 mM solution in methanol was measured on Jenway CE 4071 conductivity meter at 25 °C. The magnetic moment was measured with Auto Sherwood Scientific 10169 Magnetic Susceptibility Balance.

2.3. Synthesis of Cu(II) complex

Methanolic solution (10 mL) of $CuSO_4{\cdot}5H_2O$ (0.748 g; 3 mmol) was added dropwise to methanolic solution (20 mL) of 2-amino-3-methylpyridine (1.944 g; 18 mmol) with constant stirring for 3 hours. $CuSO_4{\cdot}5H_2O$ solution initially blue and then after the addition of ligand 2-amino-3-methylpyridine, the solution becae green. The solution was then concentrated 1/3 of the previous volume. Furthermore, the solution was diffused using diethyl ether and left to stand for 72 hours. The precipitate was filtered, then dried in a desiccator for 48 hours.

3. Result and Discussion

3.1. Electronic spectra

Figure 1 shows the shift of $CuSO_4{\cdot}5H_2O$ maximum absorption wavelength (818 nm) towards smaller wavelengths in the complex Cu(II)-2-amino-3-methylpyridine (730 nm). Shifting λ_{max} indicates the formation of the complex due to the substitutions of H_2O molecules with 2-amino-3-methylpyridine. It was estimated that 2-amino-3-methylpyridine have greater powers than the H_2O. Electronic spectra of the complex [Cu(2-amino-3-methylpyridine)$_4$]$SO_4.4H_2O$ shows a wide absorption at λ_{max} 730 nm (13700 cm^{-1}). A low-intensity absorption band of a copper complex at 740 nm is assigned to the Eg⟶T_2g transition which attributed to square planar geometry [7], so is 720 nm [8]. Therefore, it indicated the square planar geometry around metal ion in the formed complex.

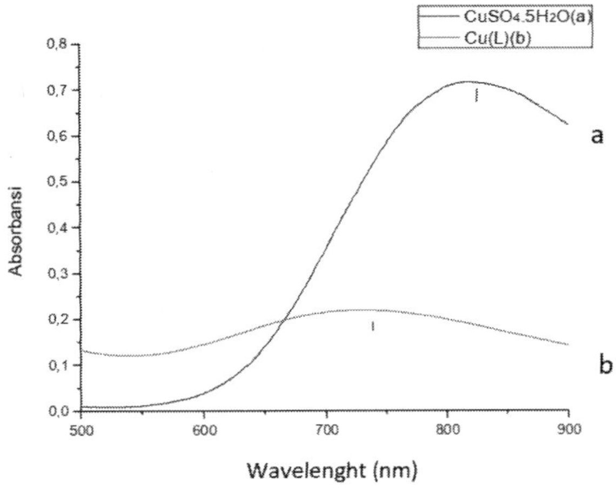

Figure 1. Electronic spectra of $CuSO_4{\cdot}5H_2O$ solution (a) and the complex solution (b)

3.2. Analysis of the amount of copper with Atomic Absorption Spectroscopy (AAS)

AAS was used to determine the total copper content of the complex shown in table 1. The result of copper content in the complex was 9.53±0.06%. This result of the measurements was compared to the calculation results of the proposed complex formula. It can be estimated that the complex formula is Cu(2-amino-3-methylpyridine)$_4$SO$_4$(H$_2$O)$_n$ (n=3, 4 or 5).

Table 1. Percentage of copper in the complex theoretically.

Empirical Formula	Molecular weight (g/mol)	% Cu
Cu(2-amino-3-metylpyridine)$_4$SO$_4$(H$_2$O)$_3$	645.5	9.84
Cu(2-amino-3-metylpyridine)$_4$SO$_4$(H$_2$O)$_4$	663.5	9.57
Cu(2-amino-3-metylpyridine)$_4$SO$_4$(H$_2$O)$_5$	681.5	9.32

3.3. Thermal analysis by TG/DTA

Thermogram of complex in figure 2 shows the endothermic reaction in the area of 129 °C followed by 10.89% mass reduction in the temperature range 135-174 °C, which corresponds to the loss of four molecules H$_2$O. Something similar happened at the complex Cu(II)-sepamet that showed a mass reduction in 79.9 to 146.4 °C 9.50 % indicating loss of four molecules of lattice water [9]. The complex [Cu(en)(phen)$_2$]Br$_2$·2phen·8H$_2$O, en=ethylenediamine, phen=[bis (1,10 Phenanthroline) also experienced a reduction in mass at 170 °C indicating the release of H$_2$O molecules as hydrates [10]. Based on the analysis TG/DTA can be estimated that the complex formula is Cu(2-amino-3-methylpyridine)$_4$SO$_4$.4H$_2$O.

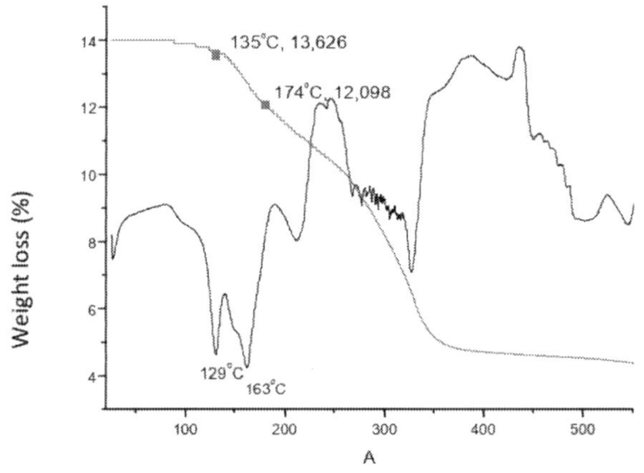

Figure 2. TG/DTA analysis of the complex

3.4. Analysis of electrical conductivity

The molar conductivity of the complex and standard solution were determined in methanol (10^{-3} M) at room temperature. The value of molar conductivity is shown in table 2. The molar conductance values of the complex lie in 12.7 ± 0.1 S.cm^2.mol^{-1}, thus indicating that is an electrolyte. By comparing the conductivity values of the complex sample with a molar conductivity of the standard solution, it can be seen that the charge ratio of cationic: anions= 1:1. This shows that the ion SO_4^{2-} is not coordinated to the central metal ion Cu(II) and only acts as an anion [11-12]. Thus, the formula is probably [Cu(2-amino-3-methylpyridine)$_4$]SO$_4$·4H$_2$O.

Table 2. Molar conductivity of metal salts and Cu(II)-ranitidine in water

Solution	Λ_m (S cm^2 mol^{-1})	Total ions
Metanol	-	-
$CuSO_4{\cdot}5H_2O$	**7.63 ± 0.2**	**2**
$NiSO_4{\cdot}6H_2O$	**4.40 ± 0.1**	**2**
$FeSO_4{\cdot}7H_2O$	**16,50 ± 0.1**	**2**
$CuCl_2{\cdot}2H_2O$	73.73 ± 0.2	3
$MgCl_2$	83.76 ± 0.15	3
$AlCl_3{\cdot}6H_2O$	192.26 ± 0.15	4
$FeCl_3{\cdot}6H_2O$	241.63 ± 0.15	4
The complex.	**12.7 ± 0.1**	**2**

3.5. Infrared (IR) analysis

In order to find binding modes of the ligand with transition metal ions, infrared spectra of compounds were recorded. Table 3 shows the infrared absorption of functional groups in the ligand and the complex. On comparison of infrared spectra of 2-amino-3-methylpyridine and the complex, the C=N band at 1583 cm^{-1} in ligand was shifted by 66 cm^{-1} to a higher frequency on complexation indicated coordination of N-pyridine in 2-amino-3-methylpyridine to the metal ion. It was evidenced by an absorption at 475 cm^{-1} which indicated the presence of Cu-N bond. A similar case also occurred in the complex [Cu_4(Salicilydene-Cefotaxime)$(OH)_6$] which has an absorption of N-pyridine at 1600 cm^{-1} and shift towards higher at 1640 cm^{-1} in the complex and there was a new absorption at 490 cm^{-1} indicated the presence of Cu-N bond [13]. In the complex [$Cu(L_5)_2Cl_2$], L_5=(E)-1-(2-amino hydroxy benzylidene) quinoline-2(1H)-one) also indicate the presence of Cu-N bond in the absorption at 471 cm^{-1} [14].

Table 3. Infrared absorption of ranitidine hydrochloride and the complex

Compounds	IR absorption (cm^{-1})			
	(C=N)	(C=C)	(N-H)	(Cu-N)
2-amino-3-methylpyridine	1583	1472	3193	-
		1453	3192	
[Cu(2-amino-3-methyl-	1649	1479	3187	475
pyridine)$_4$] $SO_4{\cdot}4H_2O$		1460	3157	

3.6. Magnetic properties (μ_{eff})

The μ_{eff} value measured for the present Cu(II) complex is 1.85 B.M., this value shows that the complex is paramagnetic which has one unpaired electron BM [15]. The square planar complex of Cu(II) showed magnetic moment BM 1.81 to 1.86 BM [16-17], indicated that [Cu(2-amino-3-methyl-pyridine)$_4$]$SO_4{\cdot}4H_2O$ was present in square planar geometry. Based on the above results, the structure in figure 3 is suggested for the complex.

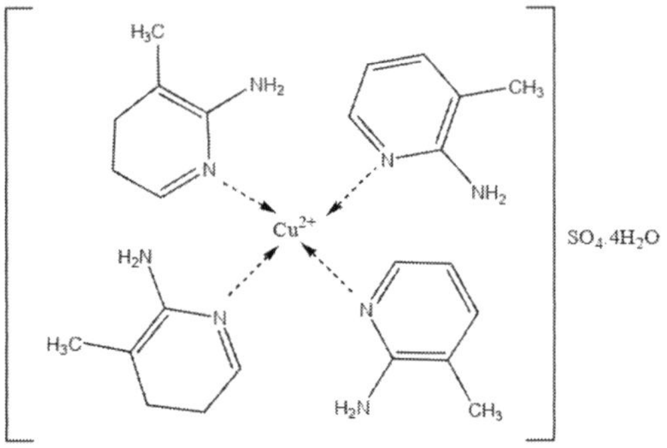

Figure 3. Suggested structure of $[Cu(2\text{-amino-3-methylpyridine})_4]SO_4{\cdot}4H_2O$

4. Conclusion

The complex $[Cu(2\text{-amino-3-methylpyridine})_4]SO_4{\cdot}4H_2O$ was able to be synthesized by mixing a methanolic solution of Cu(II) and a methanolic solution of 2-amino-3-methylpyridine solution with a mole ratio of 1:6 and stirring for 3 hours. The functional group coordinated to the Cu(II) was possibly N-pyridine, forming square planar geometry. The complex is paramagnetic has one peak at maximum UV-Vis absorption of 730 nm.

Acknowledgement
Authors gratefully express acknowledgments to Ministry of Research, Technology and Higher Education (RISTEKDIKTI) of the Republic of Indonesia for financial support.

References

[1] Islam I, Hossain A, Shah N, Barua H T, Kabir A, Khan M J, Mullick R 2015 Journal of Chemistry **2015** 1-8
[2] Abu-Youssef M A M, Solimana S M, Langerb V, Goharc Y M, Hasanena A A, Makhyouna M A, Zakyd A H, and Öhrström L R 2010 *Inorganic Chemistry* **49** 9788-9797
[3] Sharma N, Chaturvedi K 2014 *International Journal of Current Microbiology and Applied Science* **3(4)** 65-74.
[4] *Singh K, Thakur R, Kumar V* Journal of Basic and Applied Sciences **5** 21–30
[5] Saleemh F A, Musameh S, Sawafta A, Brandao P, Tavares C J, Ferdov S, Barakat A, Al Ali, A, Al-Noaimi M, Warad I 2016 Arabian Journal of Chemistry, Article in Press
[6] Chimaine F T, Yufanyi D M, Yuoh A C B, Eni D B, Agwara M O 2016 Cogent Chemistry 2 1253905
[7] Rawat S P, Choudhary M 2014 *International Journal of Inorganic Chemistry*
[8] Todar K 2011 *Staphylococcus aureus and Staphylococcal Disease* (USA: Science Magazine)
[9] Fitriyana Nur 2012 *Sintesis dan Karakterisasi Kompleks Cu(II) dengan Sepamet* (Surakarta: Jurusan Kimia Fakultas Matematika dan Ilmu Pengetahuan Alam UNS)
[10] Onawumi O O Esther, I O Adeoye and F A Oluwafunmilayo Adekunle 2013 *Open Journal of Inorganic Chemistry*, vol. **3**, hlm 26-33.
[11] Ababei L V, A Kriza, C Andronescu, A M Musuc 2011 *Journal of the Serbian Chemical Society* **2** 1-24
[12] Suresh M S and V Prakash 2010 *International Journal of the Physical Sciences* **5(14)** 2203-2211
[13] Anacona J R, Johan C, Ovidio A A 2013 Synthesis, Spectroscopic, and Magnetic Studies of Mono- and Polynuclear Schiff Base Metal Complexes Containing Salicylidene-Cefotaxime Ligand *International Journal of Inorganic Chemistry* Article ID 108740
[14] Al-Bayati R I H, Mahdi F R, Al-Amiery 2011 *British Journal of Pharmacology and*

Toxicology **2(1)** 5-11

[15] Huheey E, James A, Keiter E and Keiter L R 1993 *Inorganic Chemistry Fourth edition* (New York: Harper Collins College Publisher)

[16] Raman N, Jeyamurugana R B, Rajkapoorb B, Mituc L 2010 *J Iran Chem Soc* **7(4)** 917-933.

[17] Al-Shaalan N H 2011 *Molecules* **16** 8629-8645

IOP Conf. Series: Materials Science and Engineering **349** (2018) 012057 doi:10.1088/1757-899X/349/1/012057

Pinostrobin Derivatives from PrenylationReaction and their Antibacterial Activity against Clinical Bacteria

S D Marliyana[1], D Mujahidin[2], and Y M Syah[2]

[1]Department of Chemistry, Faculty of Mathematics and Natural Sciences, SebelasMaret University, Jl. Ir. Sutami 36 A Surakarta, Indonesia
[2]Department of Chemistry, Faculty of Mathematics and Natural Sciences, and Laboratory of Natural Products, Institut Teknologi Bandung, Jl. Ganesha 10 Bandung, Indonesia.

Abstract. *Kaempferia pandurata* (syn. *Boesenbergia rotunda*, *B. pandurata* (Roxb.)Schltr), locally known as "TemuKunci"in Indonesia, is one of the medicinal plants of the family Zingiberaceae. Phytochemical studies on the rhizome of *K. pandurata* showed the presence of flavonoid derivative, namely flavanones, which constitute as the main components of this plant. Bioactivity studies on this species exhibited various biological activities, such as antibacteria, anti-inflammatory, antitumor, antidiarrhea, antidisentri, anti-HIV, antioxidant, antipyretic, analgesic and insecticides. Among the biological activities, the antibacterial activity results are important as an attempt to answer the emergence of resistance of some bacteria against existing drugs, as well as the emergence of a number of outbreaks of disease caused by bacteria. Therefore, a search to find new compounds that are potential as an antibacterial is an urgent matter. The present study was aimed at the chemical transformation of pinostrobin (**1**) from *K. pandurata* rhizome and an antibacterial activity.The chemical transformation was performed through a prenylation reaction of pinostrobin (**1**) which is the main component of *K. pandurata* rhizome. The prenylation reaction was carried out by reacting pinostrobin (**1**) with prenyl bromide and potassium carbonat (K_2CO_3). The purification of product was done using the radial chromatography with mix solvent n-hexane and ethyl acetate (97.5:2.5; 9.5:0.5; 9.0:1.0.; 8.0:2.0). The purity test of isolated compound was analysedby TLC using different types of eluent. The identification of compounds was determined based on NMR data and mass spectra analysis. Five compounds were obtained from the prenylation reaction, i.e. monooxyprenylated pinostrobin (**2**), monooxyprenylated chalcone (**3**), diprenylated chalcone (**4**), triprenylated chalcone (**5**), and triprenylated cyclohexene chalcone (**6**). These compounds were tested for antibacterial activities against four clinical bacteria, namely *Bacillus subtilis, Staphylococcus aureus*, *Escherichia coli,* and *Pseudomonas aeruginosa*. The antibacterial test was conducted using microdilution method. The five compounds showed moderate antibacterial activity against clinical bacteria with MIC values in the range of 25 to 50 μg/mL.

1. Introduction

Flavanones, the important subgroup of flavonoid are polyphenolic compounds mainly present in plants as secondary metabolites and possess health benefits. These compounds are reported to exhibit a broad spectrum of biological activities such as antimicrobial, antiviral, antihypertensive, antioxidant, anti-inflammatory, and cytotoxic [1]. One of a group of flavanones is pinostrobin (**1**), the major compound that found in the rhizome of fingerroot (*Kaempferia pandurata*) and the herbal basis of Southeast Asia cooking [2]. Many pharmacological activities are attributed to pinostrobin (**1**), viz. antiulcer activity[3], antiinflammatory activity [4,5], antimicrobial activity [6,7], antimutagenic [8], anticancer [9], antioxidant activity [10].Among the biological activities, the antibacterial activity is important because of resistance of some bacteria against existing drugs.

The chemical modification of pinostrobin (**1**) opens new possibilities for creating new compounds with specific biological activity. Modified compound **1** is of interest for the creation of new antibacterial compounds. For this reason, we have done for a transformation of **1** with prenylated reaction. Our attention can be obtained the possibility antimicrobial activity from pinostrobin derivatives. Here in we describe a short synthesis of a series of simple pinostrobin derivatives and describe some preliminary antibacterial activity.

2. Experimental

2.1General procedures

Solvents for extraction and chromatography were distilled before use. All chemicals were purchased from E. Merck and Sigma Aldrich and used directly without any purification. Analytical thin-layer chromatography was performed using E. Merck silica gel 60 F254 aluminum plates. Purifications of compounds were carried out by liquid vacuum chromatography (silica gel 60G, E. Merck) and radial chromatography (silica gel 60 PF254, E. Merck). Prenyl bromide was purchased from Sigma Aldrich. Pinostrobin was isolated from *K. pandurate* rhizome. Mass spectra were recorded as LCT Premier XE Micromass MS Technologies (HRESIMS). ^{1}H and ^{13}C NMR spectra were taken in $CDCl_3$ at room temperature, on Agilent instrument 500MHz. Chemical shifts were reported in δ (ppm) using TMS as an internal standard, and coupling constants were expressed in Hertz. Melting points were recorded in Fischer apparatus and uncorrected. The following materials were synthesized and purified by the described procedures.

2.1.1 Synthesis of prenylated pinostrobin (**1**)

A mixture of pinostrobin (**1**) (300 mg, 1.11 mmol), prenyl bromide (0.26 mL, 2.22 mmol), and anhydrous K_2CO_3 (613.65 mg, 4.44 mmol) in anhydrous acetone (70 mL) was refluxed at 65 °C for 24hours. After cooling, the resulting product was filtered and the solvent removed under reduced pressure to afford the crude product (0.394 g). The crude productwas purified by radial chromatography with mix solvent n-hexane and ethyl acetate (97.5:2.5; 9.5:0.5; 9.0:1.0.; 8.0:2.0). Purity of the compounds was routinely checked by thin layer chromatography (TLC). TLC was performed on silica gel-coated plates, and the components were visualized under UV light or by spraying with cerium sulfuric acid (1.5 % b/v) followed by heating the plates at 100 °C in an oven.

2.1.2 Antibacterial activity

The minimum inhibitory concentrations (MIC) of the prenylation compounds were determined against four clinical bacterial strains, two representative gram-positive bacteria *Bacillus subtilis, Staphylococcus aureus*, and two Gram-negative bacteria *Escherichia coli, Pseudomonas aeruginosa*, by microdilution method recommended by CLSI Standard. The microorganisms were obtained from the STIKES Bandung, Indonesia. Cultures of test organism were maintained on Mueller-Hinton Agar (Oxoid).Protocol in liquid medium (Muller Hilton Broth) distributed in 96-well plates. Serial dilutions of the tested compounds were performed (concentrations from 150 to 0.97 µg/mL) in a 200 µL culture medium; afterward, each well was seeded with a 50 µL microbial suspension of 0.5 Mac-Farland density. In each test, a microbial culture control and a sterility control (negative) were performed. The plates were incubated for 24 h at 37°C. The lowest concentration which inhibited the visible microbial growth was considered as the MIC (µg/mL) value for the tested compound. The spectrophotometer was used for quantitative measurement of microbial growths. Chloramphenicol was used as standard drugs. The experiments were carried out in triplicate.

3. Results and discussion

3.1 Prenylation of pinostrobin (**1**)

The synthesis route of prenylated pinostrobin (**1**) outlined in Figure 1. As shown in Figure 1, the synthesis of pinostrobin (**1**) derivatives resulted in five prenylated compounds. The synthesis of pinostrobin (**1**) with prenyl side chains was carried out by the reaction of the correspondent starting material **1** with prenyl bromide, in the presence of anhydrous potassium carbonate (K_2CO_3). The method of prenylation obtained monooxprenylated chalcone (**3**) in 42.5% yield as the primary product. The diprenylated (**4**) and triprenylated chalcone (**5**) were a by-product of 5.4% and 0.8% yield, respectively. Furthermore, this method has also given monooxyprenylated pinostrobin (**2**) and compound **6** in 5.1% and 0.6% yield, respectively. In the case of compound **6**, the prenylation method has destroyed the aromatic ring of pinostrobin (**1**).

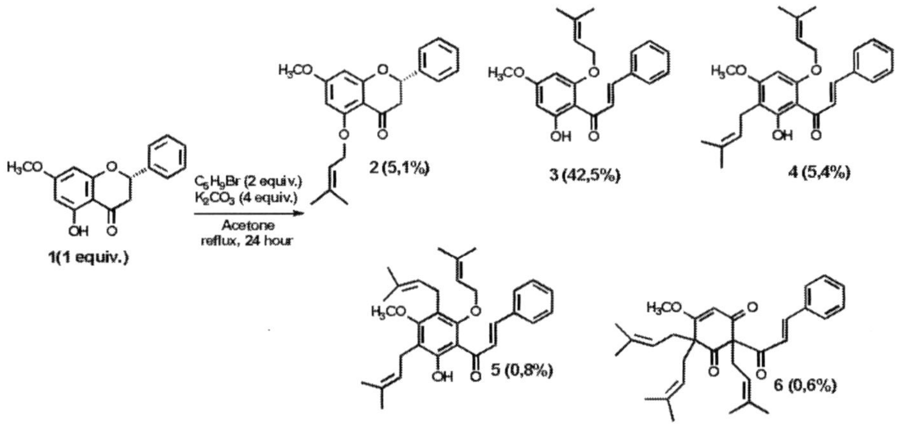

Figure 1. Scheme of the synthesis of prenylated pinostrobin (**1**).

*3.1.2 Structure elucidation of prenylated derivatives compounds **2-6***

The structure elucidation of compounds 2-6 was established on the basis of NMR and HRMS techniques. Five compounds were isolated: monooxyprenylated pinostrobin (**2**), monooxyprenylated chalcone (**3**), diprenylated chalcone (**4**), triprenylated chalcone (**5**), and triprenylated cyclohexene chalcone (**6**). The NMR and HRMS data of these compounds were:

Monooxyprenylated pinostrobin(2). Yield 5.1% as white solid; mp 115-116 °C; **^{1}H NMR** Spectrum ($CDCl_3$), δ (ppm): 5,36 (1H, *dd*, *J*= 2,8; 13,2 Hz, H-2); 2,96 (1H, *dd*, *J*= 13,2; 16,5 Hz, H-3ax); 2,74 (1H, dd, *J*= 2,9; 16,5 Hz, H-3eq); 3,76 (3H, *s*, 7-OCH_3); 6,11 (1H, *d*, H-6); 6,06 (1H, *d*, H-8); 7,32-7,43 (5H, *m*, H-2', H-3', H-4', H-5', H-6'); 4,58 (2H, *d*, H-1"); 5,52 (1H, *t*, H-2"); 1,72 (3H, *s*, H-4"); 1,76 (3H, *s*, H-5"). **^{13}C NMR** Spectrum ($CDCl_3$), δ (ppm): 79,1 (C-2); 45,7 (C-3); 106,2 (C-4a); 188,9 (C-4); 161,5 (C-5); 93,5 (C-6); 164,9 (C-7); 94,3 (C-8); 165,8 (C-8a); 137,5 (C-1'); 126,1 (C-2', C-6'); 128,7 (C-3', C-5'); 128,6 (C-4'); 55,5 (OCH_3-7); 66,1 (C-1"); 119,4 (C-2"); 138,8 (C-3"); 25,8 (C-4"); 18,3 (C-5"). ESI-HRMS (+) m/z: Anal. calc. for $C_{21}H_{22}O_4$ $(M+H)^+$:339.1596; found: 339.1588.

Monooxyprenylated chalcone(3) Yield42.5% yellow solid; mp 135-136 °C; **^{1}H NMR** Spectrum ($CDCl_3$), δ (ppm): 14,52 (1H, *s*, OH-6'); 5,96 (1H, *d*, *J*= 2,25 Hz, H-5'); 6,09 (1H, d, *J*= 2,25 Hz, H-3'); 3,82 (1H, *s*, 4'-OCH_3); 7,74 (1H, *d*, *J* = 15,6 Hz, H-α); 8,08 (1H, *d*, *J* = 15,6 Hz, H-β); 7,37-7,57 (5H, *m*, H-2, H-3, H-4, H-5, H-6); 4,50 (2H, *d*, H-1"); 5,63 (1H, *t*, H-2"); 1,86 (3H, *s*, H-4"); 1,76(3H, *s*, H-5"). **^{13}C NMR** Spectrum ($CDCl_3$), δ (ppm): 106,3 (C-1'); 162,0 (C-2'); 92,0 (C-3'); 168,9 (C-4'); 93,7 (C-5'); 166,3 (C-6'); 192,8 (C=O); 129,9 (C-α); 141,9 (C-β); 55,5 (OCH_3, C-4'); 135,8 (C-1); 128,2 (C-2, C-6); 128,8 (C-3, C-5); 128,0 (C-4); 65,6 (C-1"); 118,7 (C-2"); 139,9 (C-3"); 25,7 (C-4"); 18,3 (C-5"). ESI-HRMS (+) m/z: Anal. calc. for $C_{21}H_{22}O_4$ $(M+H)^+$:339.1596; found: 339.1604.

Diprenylated chalcone (4). Yield 5.4%yellow crystals; mp 105-106 °C; **^{1}H NMR** Spectrum ($CDCl_3$), δ (ppm): 14,26 (1H, *s*, OH-6'); 6,01 (1H, *s*, H-3'); 3,89 (1H, *s*, 4'-OCH_3); 7,74 (1H, *d*, *J* = 15,6 Hz, H-α); 8,06 (1H, *d*, *J* = 15,6 Hz, H-β); 7,37-7,58 (5H, *m*, H-2, H-3, H-4, H-5, H-6); 4,60 (2H, *d*, H-1"); 5,60

(1H, *t*, H-2"); 1,85 (3H, *s*, H-4"); 1,78 (3H, *s*, H-5"); 3,32 (2H, d, H-1'"); 5,24 (1H, *t*, H-2'"); 1,80 (3H, *s*, H-4'"); 1,70 (3H, *s*, H-5'"). ^{13}**C NMR** Spectrum ($CDCl_3$), δ (ppm): 106,4 (C-1'); 160,8 (C-2'); 87,3 (C-3'); 164,3 (C-4'); 109,9 (C-5'); 163,3 (C-6'); 193,1 (C=O); 129,7 (C-α); 141,4 (C-β); 55,4 (OCH_3, C-4'); 135,7 (C-1); 128,2 (C-2, C-6); 128,7 (C-3, C-5); 128,3 (C-4); 65,5 (C-1"); 118,7 (C-2"); 139,6 (C-3"); 25,7 (C-4"); 18,2 (C-5"); 21,3 (C-1'"); 122,7 (C-2'"); 131,2 (C-3'"); 25,7 (C-4'"); 17,7 (C-5'"). ESI-HRMS (+) m/z: Anal. calc. for $C_{26}H_{30}O_4$ $(M+H)^+$:407.2222; found: 407.2224.

Triprenylated chalcone(5). Yield 0.8% yellow crystals; mp115-116 °C;^{1}**H NMR** Spectrum ($CDCl_3$), δ (ppm): 13,05 (1H, *s*, OH-6');3,75 (1H, *s*, 4'-OCH_3); 7,85 (1H, *d*, *J* = 15,7 Hz, H-α); 8,01 (1H, *d*, *J* = 15,7 Hz, H-β); 7,39-7,64 (5H, *m*, H-2, H-3, H-4, H-5, H-6); 4,24 (2H, *d*, H-1"); 5,41 (1H, *t*, H-2"); 1,79 (3H, *s*, H-4"); 1,70 (3H, *s*, H-5"); 3,36 (2H, d, H-1'"); 5,25 (1H, *t*, H-2'"); 1,79 (3H, *s*, H-4'"); 1,57 (3H, *s*, H-5'"); 3,36 (2H, d, H-1""); 5,20 (1H, t, H-2""); 1,70 (3H, s, H-4""); 1,38 (3H, s, H-5""). ^{13}**C NMR** Spectrum ($CDCl_3$), δ (ppm): 119,1 (C-1'); 161,3 (C-2'); 112,3 (C-3'); 163,1 (C-4'); 120,1 (C-5'); 157,3 (C-6'); 193,5 (C=O); 129,8 (C-α); 142,7 (C-β); 61,0 (OCH_3, C-4'); 134,8 (C-1); 128,1 (C-2, C-6); 128,4 (C-3, C-5); 126,4 (C-4); 73,2 (C-1"); 119,5 (C-2"); 137,6 (C-3"); 25,3 (C-4"); 17,5 (C-5"); 22,8 (C-1'"); 123,7 (C-2'"); 131,3 (C-3'"); 25,2 (C-4'"); 17,5 (C-5'"); 22,6 (C-1""); 122,3 (C-2""); 130,7 (C-3""); 25,2 (C4""); 17,4 (C-5"").ESI-HRMS (+) m/z: Anal. calc. for $C_{31}H_{38}O_4$ $(M+H)^+$:475.2848; found: 475.2843.

Triprenylated cyclohexene chalcone(6). Yield0.6% yellow solid; mp 125-126 °C; ^{1}**H NMR** Spectrum ($CDCl_3$), δ (ppm):3,78 (1H, *s*, 4'-OCH_3); 3,10 (1H, m, H-5'); 6,94 (1H, *d*, *J* = 15,6 Hz, H-α); 7,56 (1H, *d*, *J* = 15,6 Hz, H-β); 7,35-7,52 (5H, *m*, H-2, H-3, H-4, H-5, H-6); 2,78; 2,43 (2H, *dd*, H-1"); 4,74 (1H, *t*, H-2"); 1,67 (3H, *s*, H-4"); 1,43 (3H, *s*, H-5"); 2,86 (2H, d, H-1'"); 4,81 (1H, *t*, H-2'"); 1,61 (3H, *s*, H-4'"); 1,57 (3H, *s*, H-5'"); 2,63; 2,36 (2H, dd, H-1""); 4,86 (1H, t, H-2""); 1,60 (3H, s, H-4""); 1,57 (3H, s, H-5""). ^{13}**C NMR** Spectrum ($CDCl_3$), δ (ppm): 79,4 (C-1'); 193,9 (C-2'); 106,4 (C-3'); 176,9 (C-4'); 58,5 (C-5'); 206,7 (C-6'); 190,4 (C=O); 131,2 (C-α); 144,6 (C-β); 56,7 (OCH_3, C-4'); 134,8 (C-1); 129,1 (C-2, C-6); 129,4 (C-3, C-5); 120,9 (C-4); 37,4 (C-1"); 118,5 (C-2"); 136,2 (C-3"); 26,6 (C-4"); 18,6 (C-5"); 36,6 (C-1'"); 119,0 (C-2'"); 135,5 (C-3'"); 26,4 (C-4'"); 18,3 (C-5'"); 32,9 (C-1""); 119,1 (C-2""); 135,6 (C-3""); 26,3 (C4""); 18,2 (C-5""). ESI-HRMS (+) m/z: Anal. calc. for $C_{31}H_{38}O_4$ $(M+H)^+$:475.2848; found: 475.2834.

3.2 Antibacterial activity

The five compounds of pinostrobin derivatives were subjected to antibacterial screening against four clinical bacterial strains, mainly *B. subtilis, S. aureus, E. coli, P. aeruginosa*, by employing broth dilution method [11] with chloramphenicol as standards. The antimicrobial activity was expressed regarding MIC values, which are the lowest concentrations able to inhibit the bacterial growth. The results presented in Table 1.

Table 1. Antibacterial activity of pinostrobin (**1**) derivatives

	MIC μg/mL			
Compound	*B. subtilis*	*S.aureus*	*E. coli*	*P.aeruginosa*
1	37.5	150	37.5	75
2	50	25	25	25
3	50	25	25	25
4	50	25	25	25
5	50	25	25	25
6	50	25	25	25
Chloramphenicol	1.17	1.95	3.125	12.5

[a]he concentration of compound **1** and compounds 2-**6**was300 μg/mL and 200μg/mL, respectively.
[b]The concentration of chloramphenicol was 200 μg/mL.

The MICs values of compound **1** were 37.5–150 μg/mL, meanwhile the compound derivatives (**2-6**) were 25-50 μg/mL. The results indicate that all of the compound derivatives were moderate antibacterial activity [12]. Substitution of the flavanone ring system with prenyl groups would increase their lipophilicity and consequently enhance their interaction with cellular membranes [13]. According to this analysis, it is evident that the prenyl substitution patterns can increase the antibacterial activity. All the tested compound derivatives, in general, showed significant antibacterial activity.

4. Conclusion

Synthesis of prenylated pinostrobin (**1**) which is the main component of *K. pandurata* rhizome was obtained five compound derivatives i.e monooxyprenylated pinostrobin (**2**), monooxyprenylated chalcone (**3**), diprenylated chalcone (**4**), triprenylated chalcone (**5**), and triprenylated cyclohexene chalcone (**6**). The *in vitro* antibacterial activity of these compounds was evaluated against four clinical bacteria, namely *Bacilus subtilis, Staphylococcus aureus, Escherichia coli,* and *Pseudomonas aeruginosa*. They showed moderate antibacterial activity.

5. Acknowledgments

The financial support from the Scholarship of Doctoral Degree (BPDN) from the Higher Education Directorate, Ministry of Education, Republic of Indonesia, is gratefully acknowledged.

6. References

[1] Patel, NK, Jaiswal, G and Bhutani, KK 2015 *Nat. Prod. Res.* 1478-6427
[2] Kicuntod, J, Khuntawee, W, Wolschann, P, Pongsawasdi, P, Kuawan, N, Rungrotmongkol, T 2015 *J. Mol. Graph. Model.***15**.30077-2
[3] Jain, P. 2015 *Nat Prod Res.* doi:10.1080/14786419.2015.1036269.
[4] Lee, C, Lee, JW, Jin, Q, Jang, DS, Lee ,S-J, Lee, D, Hong, JT, Kim, Y, Lee, MK, Hwang, BY. 2013.*Bioorg. Med. Chem.Lett.* **23** 4263–4266
[5] Sudsai, T, Prabpai, S, Kongsaeree, P, Wattanapiromsakul, C, Tewtrakul, S. 2014 *J. Ethnopharmacol***154** 453–461
[6] Massaro, CF, Katouli, M, Grkovic, T, Vu, H, Quinn, RJ, Heard, TA, Carvalho, C, Manley-Harris, M, Wallace, HM, Brooks, P. 2014 *Fitoterapia* **95** 247–257
[7] Wangkangwan, W, Boonkerd, S, Chavasiri, W, Sukapirom, K, Pattanapanyasat, K, Kongkathip, N, Miyakawa, T, Yompakdee, C. 2014. *Biosci.Biotechnol.Biochem.* **73** 1679–1682
[8] Atun, S, Arianingrum, R, Sulistyowati, E, Aznam, N. 2013 *Int. J. Chem. Anal. Sci.* **4** 3–8
[9] Kuete, V, Nkuete, AHL, Mbaveng, AT, Wiench, B, Wabo, HK, Tane, P, Efferth, T 2014 *Phytomedicine* **21** 1651–1657
[10] Eng-Chong, T, Yean-Kee, L, Chin-Fei, C et al. 2012 *Evidence-Based Complementary and Alternative Medicine* vol. 2012 Article ID 473637 25 pages
[11] Clinical and Laboratory Standards Institute (CLSI) 2012 Methods for dilution antimicrobial susceptibility tests for bacteria that grow aerobically; Approved Standard-9th edition, *CLSI Document M07-09***32** (2) Wayne PA USA
[12] Olajuyigbe, OO, and Afolayan, AJ 2012 *Molecules*, **17** 2103-2118
[13] Qing-hui, X,Ji-zhen, L, Jiang-hua, H, Xin, Z and Qi-sheng, H 2013 *Chem. Res. Chin. Univ.* **29**(4) 695—698

IOP Conf. Series: Materials Science and Engineering **349** (2018) 012058 doi:10.1088/1757-899X/349/1/012058

Curcuminoid content of *Curcuma longa* L. and *Curcuma xanthorrhiza* rhizome based on drying method with NMR and HPLC-UVD

S Hadi[1], A N Artanti[1], Y Rinanto[2], D S C Wahyuni[1*]

[1]Pharmacy Department, Faculty of Mathematics and Natural Sciences, Universitas Sebelas Maret, Indonesia, [2]Biology Education Program, Faculty of Teacher Training and Education, Universitas Sebelas Maret, Indonesia

*E-mail: dinarsari_cw@staff.uns.ac.id

Abstract. Curcuminoid, consisting of curcumin, demethoxycurcumin and bis demethoxycurcumin, is the major compound in *Curcuma longa* L. and *Curcuma xanthorrhiza* rhizome. It has been known to have a potent antioxidants, anticancer, antibacteria activity. Those rhizomes needs to be dried beforehand which influenced the active compounds concentration. The present work was conducted to assess the curcuminoid content of *C. longa* L. and *C. xanthorrhiza* based on drying method with Nuclear Magnetic Resonance (NMR) and High Pressure Liquid Chromatography (HPLC)-UVD. Samples were collected and dried using freeze-drying and oven method. The latter is the common method applied in most drying method at herbal medicine preparation procedure. All samples were extracted using 96% ethanol and analyzed using NMR and HPLC-UVD. Curcuminoid as a bioactive compound in the sample exhibited no significant difference and weak significant difference in *C. xanthorrhiza* and *C. longa* L., respectively. HLPC-UVD as a reliable analytical method for the quantification is subsequently used to confirm of the data obtained by NMR. It resulted that curcuminoid content showed no significant difference in both samples. This replied that curcuminoids content in both samples were stable into heating process. These results are useful information for simplicia standardization method in pharmaceutical products regarding to preparation procedure.

1. Introduction

Curcuma longa L. and *Curcuma xanthorrhiza Roxb* are representative of Indonesian traditional medicine as carminative, laxative, anthelmintic and a cure for liver ailment. [1]. They are also widely used in cosmetics and as a color dye. The main active components include curcuminoids, monoterpenoids and sesquiterpenoids [2]. Curcumin, demethoxycurcumin and bisdemethoxycurcumin belong to curcuminoids as the pharmacologically significant compounds and their chemical structures are shown in Fig. 1. These three curcuminoids have been proved to show strong anti-oxidant [3], anti-inflammatory [4], anti-bacterial [5], and anti-carcinogenic activities [6,7].

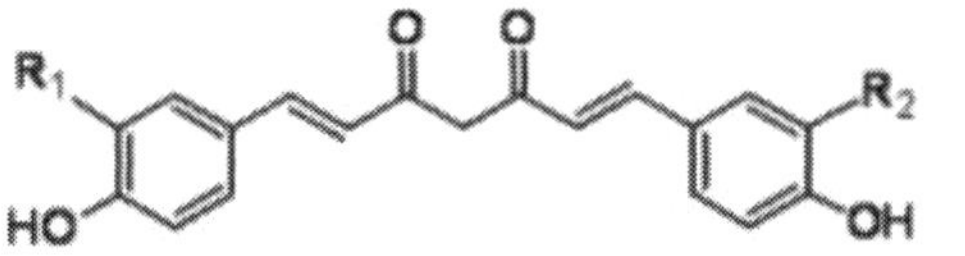

Compound	R_1	R_2
Curcumin	OCH_3	OCH_3
Demethoxycurcumin	OCH_3	H
Bisdemetoxycurcumin	H	H

Fig. 1: Structures of the three curcuminoids.

The post harvest processing of *Curcuma longa L.* and *Curcuma xanthorrhiza Roxb* rhizomes involves many steps, such as washing, cleaning, and drying. Drying is one of the most important methods of preservation and production of rhizomes, with major aim to prolong its storage life. Unfortunately, changes in the physical and biochemical structure are inevitable because the rhizomes are treated with thermal treatments.

Different types of dryers and drying methods have been adopted. Traditional drying method by direct sun drying could result in the degradation of the herb quality, for example loss of the of volatile oil (up to 25 per cent) by evaporation [8]. The traditional drying are also risky and result in mold growth, destruction of some heat sensitive pungent properties, and also influences on levels of bioactive compounds of sample. Previous study on turmeric rhizomes collected from 30 locations showed that conventional processing could maintain the intrinsic quality of sample up to a certain level, but extrinsic quality of product could not be achieved [9]. By considering the above point, it is necessary to investigate of different drying method as an effective alternative to traditional open sun drying. In this case, there is an underlying assumption that freeze-drying properly preserves the medicinal qualities of plants, and is superior to other drying methods. However, little systematic research has been done to verify these assumptions.

This study thus aimed to compare the freeze drying and oven drying method of fresh rhizomes of *Curcuma longa L.* and *Curcuma xanthorrhiza Roxb* with the specific objectives comparative evaluation of curcuminoids level of the sample. For this purpose, Nuclear Magnetic Resonance (NMR) and High Performance Liquid Chromatography-Ultra Violet (HPLC-UVD) Detector analytical method are applied for determination of curcuminoids dried by oven and freeze-drying method. NMR has been known as a simple preparation and produced good reproducibility; on the other hand, it has low sensitivity. In contrast to that, HPLC shows high sensitivity and resolution [10].

2. Methods

2.1 Chemicals and reagents
Standards of curcuminoids were purchased from CV Chemonic Pratama as commercial standard. All the other chemicals were of analytical grade and were purchased from E. Merck (Germany). The sample of *Curcuma longa L.* and *Curcuma xanthorrhiza Roxb.* was collected from local farmer in Tanjungrejo village, Nguter, Sukoharjo.

2.2 Drying experiment
The sample preparation protocol consisted of washing the rhizomes in running tap water in order to remove all the soil particles adhering to the rhizomes water. the rhizomes cut into piecesand finely powdered using a blender. Fresh powder sample was dried by two different drying methods, i.e. oven and freeze drying. For oven drying, temperature was set at 50°C and run for 3×24 hours. For frezee drying, a VirTis BenchTop Pro Freeze Dryer (SP Scientific), was set at temperature -109°C and run for 3×24 hours.

2.3 Extraction of plant materials for NMR analysis
Three replicates of each species were used for NMR profiling. The standard protocol of sample preparation and 1H-NMR profiling described by Kim et al. [9] was applied. Samples of 30 mg freeze-

dried plant material were weighed into a 2 ml microtube and extracted with 1.5 ml of a mixture of phosphate buffer (pH 6.0) in deuterium oxide containing 0.05% trimethylsilylproprionic acid sodium salt-d4 (TMSP) and methanol-d4 (1:1). Samples were vortexed at room temperature for 1 min, ultrasonicated for 20 min and centrifuged at 13000 rpm for 10 min. An aliquot of 0.8 ml of the supernatant was transferred to 5 mm NMR tubes for 1H-NMR measurement.

2.4 NMR analysis
^{1}H-NMR spectra was recorded at 21^{0}C on 400 MHz Agilent spectrometer. Deurated water was used as the internal lock. Each ^{1}H-NMR spectrum consisted of 128 scans requiring 10 min and 26 s acquisition time with the following parameters: 0.16 Hz/point, pulse width (PW) of 60° (11.3 µs), and relaxation delay (RD) of 2s. Two-dimensional J-resolved NMR spectra were acquired using 8 scans per 128 increments for F_1 (chemical shift axis) and 8 k for F_2 (spin-spin coupling constant axis) using spectral widths of 66 Hz and 5000 Hz respectively. Both dimensions were multiplied by sine-bell functions (SSB = 0) prior to double complex Fourier transformation. J-resolved spectra were tilted by 45°, symmetrized about F_1, and then calibrated to TMSP. ^{1}H-^{1}H correlated COSY spectra were acquired with a 1.0 s relaxation delay and 6361 Hz spectral width in both dimensions. The window function for the COSY spectra was Qsine (SSB = 0). MestRenova version 11.0.0 applied in order to identify metabolites in samples.

2.5 HPLC analysis
The HPLC system was consisted of a binary pump sytem, 20 µL Rheodyne sample injector, and UVD detector (LabAlliance). A Waters BioSuite™ pC_{18} column (4.6 mm×150 mm, 7 µm) is used for separation. The mobile phase was composed of methanol and run in isocratic mode. The flow rate was 1.0 mL/min and the injection volume was 20 µL. The mobile phase was degassed and filtered through a 0.22 µm membrane filter before use. The column temperature was maintained at 30 °C. The UVD wavelength was set at 420 nm.

Prior HPLC analysis, 0.1 g sample was quantitatively weighed and extracted with 10 mL methanol in an ultrasonic apparatus for 20 min at room temperature. After sonication, the supernatant liquid was filtered through a 0.45 µm membrane filter and used for subsequent chromatography analysis. External calibration curve was used as quanification method.

3. Results and discussion

3.1 NMR analysis
Curcuminoid was identified using 1D and 2D NMR spectra. Signals at δ 7.58 (d, J = 15.47 Hz), δ 7.15 (dd, J = 8.58, 2.83 Hz), δ 7.09 (d, J = 2.4 Hz), δ 6.90 (d, J = 8.31 Hz) and δ 6.70 (d, J = 15.47 Hz). A correlation between signals of proton δ 7.58 and δ 6.70 was observed in the COSY spectrum.
The results of effect of drying methods on curcuminoid contents of *Curcuma longa L.* and *Curcuma xanthorrhiza Roxb.* are tabulated in Figure 2. The curcuminoid between drying method did not show significant difference in C. xanthorrhiza (F=0.302, df=1, P=0.612). The curcuminoid exhibited weak significant difference in C. longa L between drying method (F=0.9.153, df=1, P=0.039). It has half level higher in freeze-drying method than in oven-drying method.

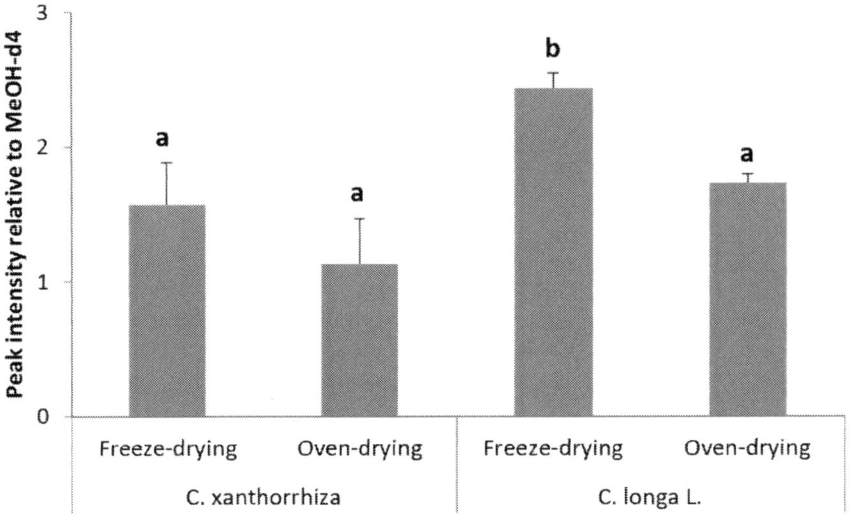

Figure 2: Variation of curcuminoid level of turmeric and temulawak samples dried with two different drying methods. Different letters indicate significant differences between treatment at $P \leq 0.05$

3.2 HPLC analysis

HPLC-UVD is a simple method and low cost for quantitative determination of curcuminoids with high sensitivity. In this study, curcuminoids was extracted from 2 different plant samples. The *C. longa L.* and *C. xanthorrhiza Roxb.* was extracted by sonication method. Peak of total curcuminoids under the optimum conditions described above identified in the chromatogram at t_R of 2.27±0.01 minutes. Isocratic run with methanol produces curcuminoids peak with a symmetry factor of 1,2 and could be well separated with other intererence peaks from matrice and detected within 10 min. Good selectivity and sensitivity were achived by the UVD detection. This method also showed wide linear range, good linearity as shown in **Table 1**.

Table 1: External calibration curve of HPLC-UVD

Compound	Conc. range [ng/µL]	Equation of regression	R^2
Curcuminoid	0.05-0.60	y = 99149x + 34433	0.999

3.3 Effect of drying method

HPLC-UVD analysis for the effects of drying methods on curcuminoids levels of *C. longa L.* and *C. xanthorrhiza Roxb.* are given in **Table 2**. The curcuminoid did not differ between drying method in *C. xanthorrhiza Roxb* and *C. longa* L. (F=0.098, df=1, P=0.770 and F=0.029, df=1, P=0.874).

These results are in accordance to previous study that curcuminoids did not change due to heat treatment [10, 11]. Tangkanakul et al. (2009) studied that curcuminoid content are remain stable due to normal heat treatment in *C. longa* L. (up to 100 °C) [12]. Freeze-drying is an excellent drying method to preserve heat-sensitive compounds. Thus, it is an ideal method to dry herbal plants with less compound degradation. Fortunately, curcuminoid is a stable compound due to heat treatment, therefore oven-drying method of *C. longa* L. and *C. xanthorrhiza* Roxb. as a raw material for herbal medicine can be applied regarding to preparation procedure.

Table 2. Curcuminoids content

Sample	Concentration [% w/w]*	
	Oven-drying	**Freezee drying**
C. longa L.	0,76 ± 0,06	0,76 ± 0,04
C. xanthorrhiza Roxb.	0,74 ± 0,02	0,73 ± 0,03

*based on dry weight

Conclusion

Curcuminoids as bioactive compounds of *C. longa* L. and *C. xanthorrhiza* Roxb. are remain stable due to heat treatment at preparation procedure. The high content of curcuminoids are important to be maintained during production process. Thus, these results are useful information for simplicia standardization method in pharmaceutical products. Further experiments may focus on other metabolites changing in *C. longa* L. and *C. xanthorrhiza* Roxb. due to heat treatment.

Acknowledgments

The authors are grateful to LPPM Universitas Sebelas Maret for providing financial assistance through the MRG grant in this research.

References

[1] R.C. Srimal, Turmeric: a brief review of medicinal properties, *Fitoterapia*. **1997**, 68, 483–493.
[2] R. Li, C. Xiang, X. Zhang, et al., Chemical analysis of the Chinese herbal medicine turmeric (Curcuma longa L.), *Curr. Pharm. Anal.* **2010**, 6, 256–268.
[3] H.Y. Choi, Antioxidant activity of Curcuma longa L., novel foodstuff, *Mol. Cell. Toxicol.* 5 (2009) 237–242.
[4] R.C. Lantz, G.J. Chen, A.M. Solyom, et al., The effect of turmeric extracts on inflammatory mediator production, Phytomedicine. **2005**, 12, 445–452.
[5] B.S. Park, J.G. Kim, M.R. Kim, et al., Curcuma longa L. constituents inhibit sortase A and staphylococcus aureus cell adhesion to fibronectin, *J. Agr. Food Chem.* 53 (2005) 9005–9009.
[6] M. Lopez-Lazaro, Anticancer and carcinogenic properties of curcumin: considerations for its clinical development as a cancer chemopreventive and chemotherapeutic agent, *Mol. Nutr. Food Res*. **2008**, 52, S103–S127.
[7] L. Das, M. Vinayak, Anti-carcinogenic action of curcumin by activation of antioxidant defence system and inhibition of NF-kappa B signalling in lymphoma-bearing mice, *Biosci. Rep.* **2012**, 32, 161–170.
[8] Jose, K.P. and Joy, C.M., Solar tunnel dryer assisted post harvest curing of ginger. In Vibhava Keralam *Proceedings of 2nd Swadeshi Vijnana Mela*, Centre for Bharatiya Marketing Development CBMD, Hyderabad, India. Kachru, **2005**, 100-103,
[9] Kim, H. K.; Choi, Y. H.; Verpoorte, R., NMR-based metabolomic analysis of plants. *Nature Protocols* **2010,** 5, (3), 536-549.
[10] Verpoorte, R., Choi, Y.H., Mustafa, N.R., and Kim, H.K., Metabolomics: Back to basics, *Phytochem. Rev.*, **2008**, 7 (3), 525–537
[11] Prathapan, A., Lukhman, M., Arumughan, C., Sundaresan, A., and Raghu, K.G., Effect of heat treatment on curcuminoid, colour value and total polyphenols of fresh turmeric rhizome, *Int. J. Food Sci. Technol.*, **2009**, 44 (7), 1438–1444
[12] Tangkanakul, P., Auttaviboonkul, P., Niyomwit, B., Lowvitoon, N., Charoenthamawat, P., and Trakoontivakorn, G., Antioxidant capacity, total phenolic content and nutritional composition of Asian foods after thermal processing, *Int. Food Res. J.*, **2009**, 16 (4), 571–580

IOP Conf. Series: Materials Science and Engineering **349** (2018) 012059 doi:10.1088/1757-899X/349/1/012059

Productive Liquid Fertilizer from Liquid Waste Tempe Industry as Revealed by Various EM4 Concentration

S Hartini[1], F Letsoin[2], A I Kristijanto[3]
[1,2,3]Satya Wacana Christian University, Salatiga, Indonesia

Email: sri.hartini@staff.uksw.edu

Abstract. Recently, using of productive liquid fertilizer assumed as a proper and practical fertilizer for plant productivity purposes. Various ways of enrichment of liquid fertilizer were done to achieve certain quality. The purpose of this research was to determine the proper additional formulation in the process of making productive liquid fertilizer based on the various concentration of EM4 as well as comparated the result with SNI. Liquid tempe waste were collected from some tempe industries at Sidorejo Kidul village, Tingkir district, Salatiga. The concentration of EM4 which were added to the tempe wastewater are 0%; 0.20%; 0.40%; 0.60%; 0.80%; 1.00% respectively. The pH, temperature, C total, N total, C/N ratio, and PO_4^{3-} were measured. Data was analyzed by using Randomize Completely Block Design (RCBD) with 6 treatments and 4 replications. Comparison between the average, the Honestly Significance Deference (HSD) 5% was used. The results showed that the addition of EM_4 indicated there were a significant progress. Moreover, the most effective formula to increase the quality of productive liquid fertilizer from liquid waste tempe was found in addition of 1.00% EM4 with the gained analysis value for the C total, N total, C/N ratio, and degree of PO_4^{3-} as follows : 4.395 ± 1.034%; 1.470 ± 0.081%; 3.01 ± 0.756; 685.28 ± 70.44 ppm . Associated with the need fulfillment of SNI hence can be concluded that result of Productive Liquid Fertilizer (PLF) from liquid waste tempe successfully fulfill SNI of liquid fertilizer for pH parameter and total N, only.

1. Introduction

Tempe is one of traditional fermented foods of Indonesia that has been knew globally [1]. Tempe is consumed practically all over Indonesia, although the consumption level still varies from one province to another. As a result of the many tempe industries, the waste production process bring an impact on the environment. According to [2], waste obtained from byproducts of tempe making, if not properly managed and only directly discharged into the waters will greatly disrupt the surrounding environment because it can damage the quality of groundwater, causing unpleasant odor, and trigger the growth of various bacteria Pathogens.

One way of processing tempe wastewater is to use it as a Productive Liquid Fertilizer (PLF). PLF is one of the many liquid fertilizers circulating in the market derived from organic materials, so often known as liquid organic fertilizer. Liquid organic fertilizers are mostly applied to leaves or also called foliar liquid fertilizers containing essential macro and micro nutrients such as N, P, K, S, Ca, Mg, B, Mo, Cu, Fe, Mn and organic materials [3]. Liquid organic fertilizer besides improving the physical, chemical and biological properties of the soil, can also help increase crop production and reduce the use of inorganic fertilizers and useful as an alternative to manure [4].

Productive Liquid Fertilizer (PLF).

Productive liquid fertilizer is a fertilizer derived from animals or plants that have undergone fermentation [5]. In the process of fermentation, organic compounds break down into simpler compounds such as sugar, glycerol, fatty acids and amino acids. Decomposition of organic compounds or decomposition can be done with the addition of a starter. The specialty of this organic fertilizer is that it can quickly overcome nutrient deficiencies, has no problem in nutrient leaching and is able to provide nutrients quickly compared to inorganic fertilizers. Liquid fertilizers generally do not damage the soil and plants, although used as often as possible. In addition, this fertilizer also has a binder, so that the solution of fertilizer given to the soil surface can be directly used by plants [6].

***Effective Microorganism* 4 (EM4).**

EM4 (Effective Microorganism 4) technology is a fermentation technology that uses a mixed culture of several microorganisms that the fermentation product is beneficial for plant growth. The natural fermentative microorganisms in EM4 consist of five groups of microorganisms, namely photosynthetic bacteria (*Rhodopseudomonas* sp.), Fermented fungi (*Saccharomyces* sp.), Lactic acid bacteria (Lactobacillus sp.), and Actinomycetes [7].

EM4 is useful to help speed up the process of making organic fertilizers and improve its quality. In addition, EM4 is also useful to improve soil structure and texture to be better and supply the nutrients needed by plants. Thus, the use of EM4 can make plants more fertile, healthy and relatively resistant to pests and diseases [6]. The organic matter is a food and energy source [7].

The purpose of this research was to determine the proper additional formulation in the process of making productive liquid fertilizer based on the various concentration of EM_4 as well as comparated the result with SNI.

2.Material and method

2.1. Material

Samples are wastewater from soaking water soybean, which were taken from four tempe craftsmen in Sidorejo Kidul Subdistrict, Salatiga. Chemicals used aquadest, Potassium Dichromate ($K_2Cr_2O_7$) 1 N, 0.5% aqueous Barium Chloride ($BaCl_2$) solution, concentrated sulfuric acid (H_2SO_4), Hydrochloric acid (HCl) 0.05 N, raw sacharose, PhosVer 3 Phosphate reagents, as well as Potassium 1, 2, and 3 reagents.

2.2. Method

2.2.1. Measuring on pH [9] and Temperature [10].

The pH value measurements were done using pH-meter on the liquid waste sample (as control) and the PLF product of various concentrations, while for temperature measured using a thermometer. Incubation time was performed during 1 week.

2.2.2. Total C Analysis [11]

Total C analysis is done by Walkley & Black method. It had performed on Laboratory of Agriculture Faculty-Satya Wacana Christian University.

2.2.3. Total N Analysis [12]

Total N analysis using Kjeldahl method.

2.2.4. PO_4^{3-} Analysis [13]

25 ml of sample and aquadest, each put into 2 pieces of cuvette. The first cuvette is blank while the second quvet is the sample. The spectrophotometer is then set to program 490 with a wavelength (λ) of 890 nm. The second cuvette plus 1 PhosVer 3 Phosphate reagent. After the reagent is added to the sample, the timer on the spectrophotometer is then activated. After the time on the timer is exhausted (2 minutes), the cuvette contains the blanks inserted in the spectrophotometer to zero the value on the LCD spectrophotometer screen, then the sample measured its phosphate content using a spectrophotometer.

2.2.5. Productive Liquid Fertilizer (PLF) Making.

Samples of 4 united tempe craftsmen (750 mL from each craftsman) were boiled first until boiling, after a cold sample, divided into 6 glass beaker 500 mL. Each beaker added EM4 with 0% concentration; 0.2%; 0.4%; 0.6%; 0.8%; 1%. After that closed for 1 week. Measurements were made for the total C and N parameters, C / N, P and K ratio after being PLF.

2.2.6. Data Analysis [8].

The research data were analyzed using Randomized Block Design (RBD) with 6 treatments and 4 replications. As the treatment is a variation of the addition of EM4 concentration into PLF, and the group is the time of analysis. Differences between treatment intervals were tested with Honesty Significant Different (HSD) with 5% significance level.

3. Result And Discussion

3.1. Temperature and pH between different concentrations of EM4 addition

The average temperature between the various concentrations of EM4 addition ranged from 25°C to 25.5 ± 0.29°C, while the pH ranged from 4.05 ± 0.0289 to 4.375 ± 0.0946 (**Table 1**).

Table 1. Temperature value and pH between different concentrations of em4 addition

	EM4 addition (%)					
	0	0.20	0.40	0.60	0.80	1
Temperature ($\overline{X}$ ± SE)	25.5 ± 0.29	25.25 ± 0.25	25.5 ± 0.29	25.25 ± 0.250	25 ± 0.000	25.25 ± 0.250
pH ($\overline{X}$ ± SE)	4.075 ± 0.025	4.05 ± 0.0289	4.2 ± 0.0707	4.25 ± 0.0866	4.175 ± 0.0250	4.375 ± 0.0946

Table 1 shows that the addition of various EM4 concentrations has no effect on temperature or pH. Standard pH of liquid fertilizer according to SNI between 4 - 9, so that the pH value obtained in this study has fulfilled the requirements of liquid fertilizer.

3.2. Total C (%) between different concentrations of EM4 addition

The total average C between the various concentrations of EM4 addition ranged from 1.393 ± 0.427% to 4.395 ± 1.034% (**Table 2**).

Table 2. Total average C between different concentrations of addition EM4

Total C Average	EM4 Addition (%)					
	0	0.20	0.40	0.60	0.80	1
$\overline{X}$ ± SE	1.393 ± 0.427	1.973 ± 0.587	2.350± 0.569	2.565± 0.574	3.15 ± 0.579	4.395± 1.034
W = 2.02	a	a	a	ab	ab	b

Note : - W = HSD 5%
- The numbers followed by the same letters show that the treatments are not significantly different, while the numbers followed by the unequal letters show between the different treatments significantly. This description also applies to Tables 3, 4 and 5.

Table 2 shows that total C did not increase until addition of EM4 concentration 0.4% and tends to start increasing in addition of EM4 concentration 0.6% and 0.8%. The highest total C was obtained at the addition of EM4 concentration of 1%, i.e. 4.395± 1.034% (**Fig.** 1).

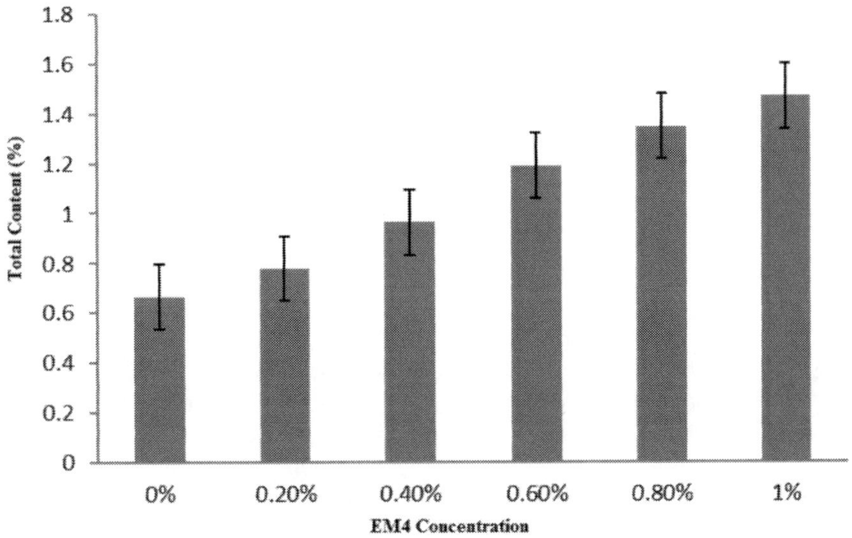

Figure 1. Bar diagram of total C amount of different concentrations of EM4 addition

Figure 1 shows that the addition of EM4 concentration increases the total C content, this is related to the EM4 fluid that has been found as a source of carbon in the form of molasses which is useful as a reservoir of C container contained in the EM4. Despite the increase, the total amount of C obtained is still far below the standard of liquid fertilizer quality (SNI total C of liquid fertilizer ranges between 9.8 - 32%).

3.3. Total N between different concentrations of EM4 addition

Average total N content between different concentrations of addition EM4 ranging from 0% to 1% (**Table 3**).

Table 3. Average total N content between different concentrations of addition EM4

	EM4 addition (%)					
	0	0.20	0.40	0.60	0.80	1
$\overline{X}$ ± SE	0.665±0.158	0.788± 0.163	0.963± 0.195	1.190± 0.103	1.348± 0.096	1.470±0.081
W = 0,31	a	a	a	b	b	b

From **Table 3** it can be seen that the addition of EM4 concentration of 0.2% and 0.4% did not increase the total mean of N. While the addition of EM4 from 0.6% to 1% concentration increased the total N and the same is between 1.190 ± 0.096 % to 1.470 ± 0.081% (**Figure** 2).

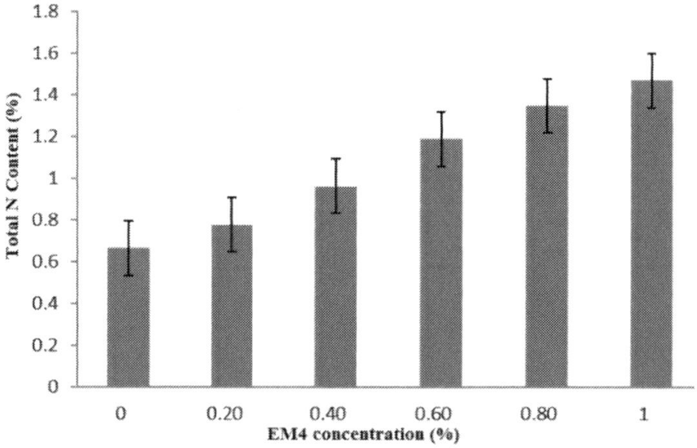

Figure 2. Bar diagram of total N between different concentrations of EM4 addition

The same increase in total N in the addition of EM4 concentration of 0.6% -1%, (**Fig. 2**) is associated with the increasing amount of EM4 volume added, the amount of microbial composition of the organic material increases as well, resulting in total inorganic N ($NH4^+$ and $NO3^-$) As the result of decomposition of organic material (protein) will increase also [14]. The total N total that produced have fulfilled SNI quality standard. SNI quality standard for liquid fertilizer is > 0.2%.

3.4. C / N Ratio Between Addition of EM4 Different Concentrations

The average C / N (X ± SE) ratio between the different concentrations of EM4 addition is the same and ranges from 2.123 ± 1.004 up to 3.010 ± 0.756 (**Table** 4).

Table 4. C / N Ratio (X ± SE) between different concentrations of EM4 addition (%)

	EM4 Addition(%)					
	0.60	0.80	0.20	0.40	0	1
$\overline{X}$ ± SE	2.123 ± 1.004	2.33 ± 0.718	2.39 ± 0.570	2.644 ± 0.335	2.65 ± 0.362	3.01 ± 0.756
W = 2.841	a	a	a	a	a	a

From **Table 4**, it is seen that the C / N ratio is the same among the various concentrations of EM4 addition (**Figure 3**).

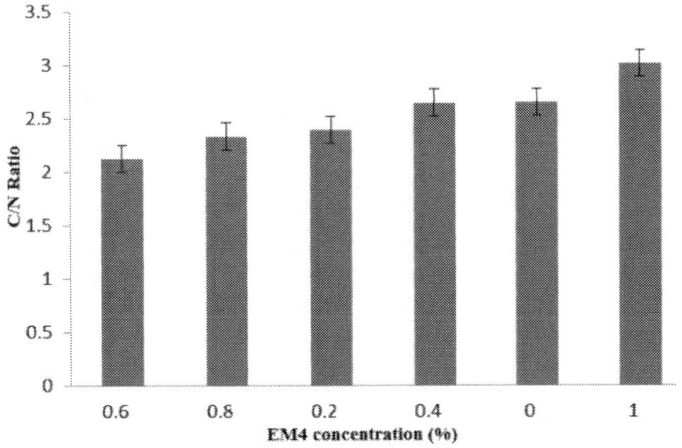

Figure 3. Bar diagram of C / N cycle between different concentrations of EM4 addition

Based on the research data, it is found that the C/N ratio of PLF of tempe liquid waste tested is still far below SNI 19-7030-2004 standard. SNI standard for C/N ratio ranges from 11 to 20. Research conducted in a C/N ratio of 1.71-3.73 with the addition of EM4 of 1%. The low C / N ratio due to the fermentation of tempe liquid waste has a low C organic content resulting in a low C / N ratio as well. This occurs because no additional carbon source is added to the fermentation of liquid tempe fertilizer, so the growth of microorganisms is hampered because it does not have enough energy source [5].

3.5. Phosphate level between different concentrations of EM4 addition

Average of PO_4^{3-} levels between various concentrations of EM4 addition ranged from 340 ± 0.0183 ppm to 685 ± 0.044 ppm (**Table 5**).

Table 5. Average of Phosphate Content between different concentrations of addition EM4

Average of Phospate Content	EM4 addition (%)					
	0	0,20	0,40	0,60	0,80	1
$\overline{X}$ ± SE	340 ± 0.0183	402.5 ± 0.0085	447.5 ± 0.0103	497.5 ± 0.023	557.5 ± 0.006	685 ± 0.044
W = 0.11	a	ab	abc	bc	c	d

From **Table 5** it appears that PO_4^{3-} increases in line with the increase in EM4 concentration and the highest at 1% EM4 addition concentration, that is 985.28 ± 70.44 ppm (**Figure** 4).

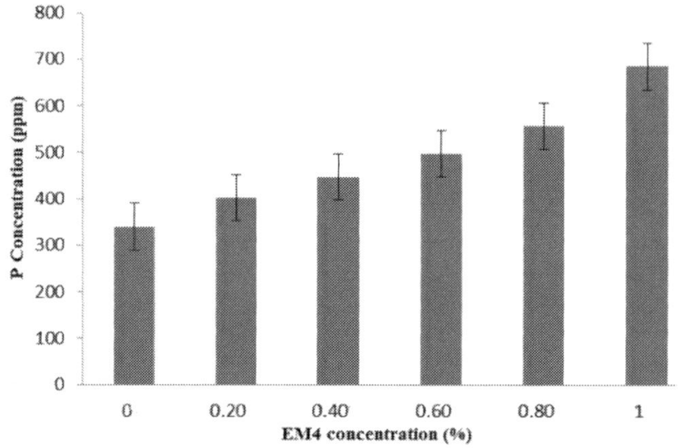

Figure 4. Bar chart of phosphate content between PLF with various concentrations of EM4 addition

The SNI content of PO_4^{3-} for organic liquid fertilizer is > 1000 ppm. The highest PO_4^{3-} level obtained from this study was 685.28 ± 70.44 ppm on the addition of EM4 1%, so PO_4^{3-} levels have not met the SNI requirements. The low phosphate content obtained is related to its presence in tempe waste in the form of phytic acid which is difficult to dissolve in water so it can not be overhauled by microbes present in EM4.

One way to break down phytate acid is by the addition of *Bacillus subtilis* bacteria. Supporting on the assertion that the activity of the phytase enzyme in *Bacillus subtilis* and Aspergillus sp. can hydrolyze phytic acid into inositol and orthophosphate [15]. The content of phosphorus in the compost is thought to be related to the N content in the compost material [5]. The larger the nitrogen contained, the multiplication of microorganisms that overhaul the phosphorus will increase, so that the phosphorus content in the compost material can increase.

4. Conclusions

The result of Liquid Productive Fertilizer (LPF) from tempe liquid waste successfully fulfill SNI of liquid fertilizer for pH parameter and total N.

References

[1] Nurdini, A. L., Nuraida L., Suwanto A. and Suliantari. 2015. Microbial growth dynamics during tempe fermentation in two different home industries. International Food Research Journal 22(4): 1668-1674. Journal homepage: http://www.ifrj.upm.edu.my
[2] Wiryani, E. 2009. Analisis Kandungan Limbah Cair Pabrik Tempe. eprints.undip.ac.id/2121/
[3] Lestari, B. I. (2005). Studi Pembuatan Briket Bioarang dari Sekam Padi dengan Proses Karbonisasi Menggunakan Tungku Sederhana. Laporan Tugas Akhir Jurusan Teknik Lingkungan FTSP–ITS. Surabaya.
[4] Parman, S., 2007. Pengaruh Pemberian Pupuk Organik Cair Terhadap Pertumbuhan dan Produksi Kentang (*Solanum tuberosum).Buletin Anatomi dan Fisiologi*, Vol XV: 21-31.
[5] Caseria, R. Y., Ruslan W. and Bambang S., 2014. Pengaruh penggunaan Starter Terhadap kualitas Fermentasi limbah cair tapioka sebagai alternatif pupuk Cair. Jurnal Sumberdaya Alam dan Lingkungan. *jsal.ub.ac.id/index.php/jsal/article/viewFile/122/104.*
[6] Hadisuwito, S., 2007. Membuat Pupuk Kompos Cair. Jakarta: Agro Media.
[7] Fitria, Y., 2008. Pembuatan Pupuk Organik Cair dari Limbah Cair Industri Perikanan
Menggunakan Asam Asetat dan EM4 (Effective Microorganism 4). Skripsi. Bogor : Institut Pertanian Bogor.

[8] Steel R G D, Torrie J H. Principles and Procedures of Statistic Biometrical Approach. 2nd. Japan. Mc Graw Hill International Book co, 1981, p.633.

[9] SNI, 2004. Air dan air limbah-Bagian 11 : Cara uji derajad keasaman (pH) dengan menggunakan alat pH meter. SNI 06-6989.11-2004.

[10] SNI, 2005. Air dan air limbah-Bagian 23 : Cara uji suhu dengan termometer. SNI 06-6989.23-2005

[11] Wang, X., Jiaping W. and Juan Z., 2012. Comparisons of Three Methods for Organic and Inorganic Carbon in Calcareous Soils of Northwestern China. PLoS One. 2012; 7(8): e44334.

[12] Wang H., Pampati N., McCormick W. M. and Bhattacharyya L., 2016. Protein Nitrogen Determination by Kjeldahl Digestion and Ion Chromatography. J Pharm Sci. 2016 Jun;105(6):1851-1857. doi: 10.1016/j.xphs.2016.03.039.

[13] HACH DR 2000. Analytical Procedures for HACH DR/2000 and DR/3000 Instrument.

[14] Megawati and Aji, K. W. 2014. Pengaruh Penambahan EM4 (Effective Microorganism-4) Pada Pembuatan Biogas dari Eceng Gondok dan Rumen Sapi. Jurnal Bahan Alam Terbarukan. ISSN 2303-0623.

[15] Savita, P. D., Yallappa M. and Suvarna V. C., 2017. Phytate Solubilizing Microorganisms and Enzyme Phytase to Combat Nutritional Problems in Cereal-Based Foods. Journal of Bacteriology & Mycology: Open Access. Volume 4 Issue 3.

IOP Conf. Series: Materials Science and Engineering **349** (2018) 012060 doi:10.1088/1757-899X/349/1/012060

Kinetic and mechanism formation reaction of complex compound Cu with di-n-buthildithiocarbamate (dbdtc) ligand

S Haryani, C Kurniawan, Kasmui

Chemistry department Faculty of mathematics and Natural Sciences, Universitas Negeri Semarang

haryanimail@gmail.com

Abstract. Synthesis of complex compound is one field of research which intensively studied. Metal-dithiocarbamate complexes find wide-ranging applications in nanomaterial and metal separation science, and have potential use as chemotherapeutic, pesticides, and as additives to lubricants. However, the information about is reaction kinetic and mechanism are very much lacking. The research and analyzes results show that reaction synthesis ligand DBDTC and complex compounds Cu-DBDTC. Optimum reaction condition of formation of complex compounds Cu with DBDTC at pH=3, [DBDTC] = 4.10^{-3} M, and the time of reaction 5 minutes. Based the analysis varian reaction of complex compounds at pH 3 and 4, diffrence significance at the other pH: 5; 5,5; 6; 6,5 ; 7; and 8. The various of mole with reactants comosition difference sigbificance, those the time reaction for 5 and 6 minutes diffrence by significance with the other time, it is 3,4,8, and 10 minutes. The great product to at condition pH 6, the time optimum at 5 minutes and molar ratio of logam: ligand = 1:2. The reaction kinetic equation of complex compound Cu with chelathing ligand DBDTC is $V=0.917106\,[Cu^{2+}]^{0.87921}\,[DBDTC]^{2.03021}$. Based on the kinetic data, and formed complex compounds estimation, the mechanism explaining by 2 stages. In the first stage formation of [Cu(DBDTC)], and then $[Cu(DBDTC)_2]$ with the last structure geomethry planar rectangle. The result of this research will be more useful if an effort is being done in reaction mechanism by chemical computation method for obtain intermediate, and for constant “k” in same stage, k1.k2. and compound complex constanta (β).

Introduction

The development of more reliable methods of analysis to determine the levels of metal ions at low concentrations including copper (Cu) is still continuing. The selection of low-concentration analysis method is also adjusted to the condition of the equipment in the laboratory in the hope of obtaining representative resultan [1]. An analytical method always involves the preparation stage of the sample, which includes dissolving the solid sample, pre-separation filtering, and concentration content. The process of pre-separation and concentration concentration is often done by extraction. Before the process of extraction takes place generally the metals with low concentrations are converted first into organic complex compounds that have polarity according to the polarity of the organic solvent extractor. Solvent extraction with the formation of chelate complex compounds proved to have remarkable success in the process of separation of metal ions [2-3].

The di-n-butyldietilyitiokarbamat (DDTC) is capable of forming a neutral and powerful complex compound with a number of metal ions such as Hg, Pb, Cd, Co, Ni, and Cu [4-7]. Generally the metal ions with dialkilditiokarbamat form a colored compound, and slightly soluble in water and its solubility will increase in a relatively less polar organic solvent if the chain is longer [8]. The complex can be determined directly by spectrophotometry After going through the extraction into an organic solvent. This DDTC ligand has long been used in analytic applications ie the extraction and determination of metal ions such as those of Wytenbach and Bajo [9]; Sanzolone and co-worker [10], Lo and co-worker [11]. Various things have been done on the use of ligand dialkilditiokarbamat, but still limited to DDTC, whereas other homologues such as di-n-propilditiokarbamate, And di-n-butilditiocarbamate can be used and also can be synthesized The latter two ligands have a good prophylactic because they have longer chains so their solubility is expected to increase, and the efficiency of the extraction is also increased, thus the complex compound formed by this extraction also has Better prospects [12-14]

The synthesis of complex compounds is always related to the development of analytical procedures that always attract the attention of researchers. In developed countries research in the field of synthesis of complex compounds has been done intensively and can now be synthesized various complex compounds that are useful in various industries, however information on kinetics and mechanisms of reaction is still very less. The kinetics information and the mechanism of complex forming reactions are very useful chemical data, for example in chromatographic separation, kinetic factors and complex stability are essential in regulating the interactions between metal complexes and components in chromatographic systems (stationary phase and mobile phase).

Based on the description of the above background then the problem to be solved in this research is to know how the kinetics equation of the reaction of the formation of complexes Cu with di-n-buthyldithiocarbamate.

Methods

The first stage, at this stage will be synthesized ligand di-n-buthyldithiocarbamate according to procedures Taufan and co-worker [15] that is from the reaction between di-n-butyl amin with karbondisulfida. Into the erlenmeyer flask placed on a large glass beaker filled with 5 mL of concentrated carbondisulfide was then added 6.25 mL concentrated NH_4OH, shaked gently. While continuously shaken, it was added dropwise in thick n-butylamin as much as 13.75 mL. Once the pasta is formed, the paste continues to stir slowly until it becomes a crystal. The formed crystals are dried in a wind-blown way. The synthesis results were tested for their efficacy by thin layer chromatography (TLC), and characterized by melting point determination, UV-VIS spectroscopy method and Infrared spectroscopy method. (2) The second stage, at this stage will be determined the factors that can affect the results of complex forming reactions.

The formation of complex Cu compounds with di-n-butilditiocarbamate was carried out by extraction, according to Taufan and co-worker (15) procedures. It is therefore necessary to specify the optimization of extraction which includes: pH of solution, reaction concentration, and reaction time. Analysis of reaction product was done by spectroscopic method of uv-vis and atomic absorption spectroscopy (SSA). The synthesis results were tested for their efficacy by thin layer chromatography (TLC) and characterized by uv-vis spectroscopy method and infrared spectroscopy and from the determination of coordination number using Job method.

Results and Discussion

Characterization of Complex Synthesis Result

Cu-DBDTC chelate complex is black after evaporation of the extractants. The extraction was carried out under optimum conditions of pH 6.5; Concentration of DBDTC 4 x 10^{-3} M; And 5 minute rection time. The purity test with TLC obtained only one spot which indicates there is only one component, this means the result of the reaction is relatively pure. Melting point of Cu-DBDTC compound is 74-76 ° C.

IOP Conf. Series: Materials Science and Engineering **349** (2018) 012060 doi:10.1088/1757-899X/349/1/012060

The formation of DBDTC complex compounds with Cu^{2+} metal ions is based on the interaction of Cu^{2+} ions with in-n-buthyldithiocarbamate ($DBDTC^-$) ions. The $DBDTC^-$ ion is obtained in the dissolution of ammonium dibutildithiocarbamate salt. DBDTC ions act as ligands forming neutral chelate complexes with Cu^{2+} ions in a chloroform solvent.

Kinetics Study of Reaction Formation Compound Cu-DBDTC Complex

The general aim of kinetic experiments is to determine the form of the reaction rate equations. The rate reaction equation in this study is determining use differentiation method. The rate equation of Cu-DBDTC complex can be written mathematically as follows

$$v = k[A]^m[B]^n \quad (1)$$

where k is the reaction rate constant, [A] is the concentration of Cu^{2+}, [B] is the concentration of DBDTC, m and n is the order of the reaction. The logarithm form of equation 1 is

$$\log v = \log k + m \log [A] + n \log [B] \quad (2)$$

The determination of the reaction order is determined by finding the initial rate one of reactant and the concentrations of other reactants are fixed. The graph of concentration versus time is obtained from the experimental data to determine the initial rate. A graph of initial rate versus concentration (Figure 1) is created to determine the value of intercept and slope, as well as the order of the reaction. Based on the graph, the value of reaction order in Table 1 is obtained.

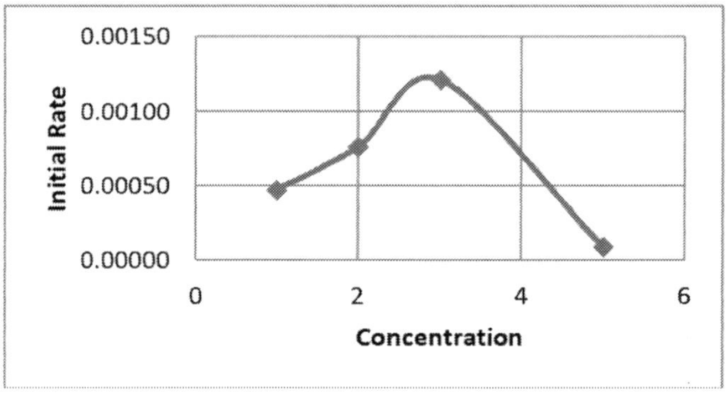

Figure 1. Graph intial rate vs concentration

Table 1. Reaction order at various pH

pH	Orde Reaksi	
	Cu^{2+}	DBDTC
6.0	1.03365	2.24132
6.5	0.87921	2.03021
7.0	0.88919	2.20839

The rate reaction constant can be obtained by substituting the initial concentration of the order of each order in equation 1 into equation 3. The reaction order price is selected at pH 6.5 because it is the optimum condition of Cu-DBDTC complex formation. The price obtained is k = 0.917106.

$$v = k\,[Cu^{2+}]^{0.87921}[DBDTC]^{2.03021} \quad (3)$$

by substituting the value of k in equation 3, is obtained

$$v = 0.917106\,[Cu^{2+}]^{0.87921}[DBDTC]^{2.03021} \quad (4)$$

The Mechanism of Cu-DBDTC Complex Formation Reaction

The theoretical study on the mechanism of reaction of Cu-DBDTC complex formation is based on 3 things: 1) data of reaction kinetics, 2) from value of n (coordination number) obtained by job method, and 3) from value of k. The mole ratio of Cu^{2+} and DBDTC is 1: 2 (based on equation 4) and the total order is 3. Based on this comparison then the complex molecular formula formed is Cu $(DBDTC)_2$ with a squere planar geometry structure. The mechanism of formation of complex compounds Cu $(DBDTC)_2$ is

$Cu^{2+} + DBDTC^- \rightleftharpoons [Cu(DBDTC)]^+$

$[Cu(DBDTC)]^+ + DBDTC^- \rightleftharpoons Cu(DBDTC)_2$

Then, the overall reaction is

$Cu^{2+} + 2DBDTC^- \rightleftharpoons Cu(DBDTC)_2$

The determination of n (coordination number) in this research by continuous variation method (job method). Graph obtained by entering data of mole fraction of DBDTC ligand and absorbance of Cu-DBDTC complex. Based on the graph obtained by curve intersection at 1.86 ≈ 2 approaching model of MLn stoichiometric structure with n = 2. Based on the result it can be concluded that the complex molecular formula is $Cu(DBDTC)_2$ and this is in accordance with the kinetic data. Another fact that supports the above reaction mechanism is the value of k at different pHs are contant, same as the order of reaction.

Cu $(DBDTC)_2$ complex compounds are theoretically expected to have 4 coordination numbers based on the valence bond theory. The Cu^{2+} ion has an electron configuration of $1s^2\ 2s^2\ 2p^6\ 3s^2\ 3p^6\ 3d^9$, thus the coordination bond between the bidentate ligand and the central atom is possible to form the sp^3 tetrahedron orbitals. However, this contradicts the results of the Huheey experiment (1983) which states there has never been any compounded Cu-shaped complex of tetrahedron.

The experiment results of Van Lenthe [16] showed that diethylditiocarbamate is strong ligand, so DBDTC which it homologous ligand in this study was also strong ligand. Theoretically it can be expected that the Cu complex species will have Cu molecular formula $(DBDTC)_2$ with square planar dsp2 hybridization.

Conclusion

The conclusions obtained from this study are: (1) the synthesis results of Cu-DBDTC complex compounds according to the character of the compound. (2) the optimum conditions of complex shielding between Cu^{2+} and Di-n-butyldithiocarbamate (DBDTC) were achieved at pH 6.5; [DBDTC] = 4 x 10-3 M; reaction time 5 minutes; and the mole ratio of the metal: mol ligand is 1: 2. (3) he kinetic equation obtained is $v = 0.917106\ [Cu^{2+}]^{0.87921}[DBDTC]^{2.03021}$. (4) Mechanism of reaction of complex formation of Cu $(DBDTC)_2$ covering 2 stages. The first stage is formed $[Cu\ (DBDTC)]^+$, then formed $[Cu\ (DBDTC)_2]$ with possible geometry structure is square planar.

However, it is still need to clarify the reaction mechanism by chemical computation method to prove the existence of intermediates, determination value of k at each stage (kl and k2), and the value of constituent of complex formation.

References

[1] S. Cotton, Lanthanide and actinide chemistry, John Wiley & Sons, 2007.
[2] M. Regel-Rosocka, F. Alguacil 2013.
[3] G. F. Nordberg, B. A. Fowler, M. Nordberg, L. Friberg, Handbook on the Toxicology of Metals, Academic Press, 2011.

IOP Conf. Series: Materials Science and Engineering **349** (2018) 012060 doi:10.1088/1757-899X/349/1/012060

[4] H. Seiler, A. Sigel, H. Sigel, Handbook on metals in clinical and analytical chemistry, CRC Press, 1994.
[5] G. G. Briand, N. Burford Chem. Rev. 1999, 99, 2601-2658.
[5] P. S Bagus, Y. S. Lee, K. S. Pitzer Chemical Physics Letters 1975, 33, 408-411.
[6] K. Dehnicke, A. Greiner Angew. Chem. Int. Ed. 2003, 42. 1340-1354.
[7] T. Moeller, D. F. Martin, L. C. Thompson, R. Ferrús, G. R. Feistel, W. J. Randall Chem. Rev. 1965, 65, 1-50.
[8] V. Spitsyn, L. Martynenko Russian Chemical Bulletin 1974, 23, 940-957.
[9] S. B. Castor, J. B. Hedrick Industrial Minerals volume, 7th edition: Society for Mining, Metallurgy, and Exploration, Littleton, Colorado 2006, 769-792.
[10] M. D. Regulacio, N. Tomson, S. L. Stoll Chem. Mater. 2005, 17, 3114-3121.
[11] I. Baba, I. Raya, B. M. Yamin Sains Malaysiana 2009, 38, 185-190.
[12] A. Ivanov, I. Egorova, M. Ivanov, O. N. Antzutkin, R. Tsvykh, Springer, 2014, pp. 16-20. [17] J. Mason Solid State Nucl. Magn. Reson. 1993, 2. 285-288.
[13] C. Fu, K. Ho Physical Review B 1983, 28, 5480.
[14] P. A. M. Dirac Proceedings of the Royal Society of London.Series A 1928, 118, 351-361.
[15] M Taufan, I Hastiawan, Y Mulyasih, *Proceeding of the Int. Sem on Chem. 2008. 175-179*
[16] E. Van Lenthe, E. J. Baerends J. Comput. Chem. 2003, 24, 1142-1156.

IOP Conf. Series: Materials Science and Engineering **349** (2018) 012061 doi:10.1088/1757-899X/349/1/012061

Increasing character value and conservation behavior through integrated ethnoscience chemistry in chemistry learning: A Case Study in The Department of Science Universitas Negeri Semarang.

Sudarmin*, Woro Sumarni

Chemistry Department, Mathematics and Natural Science Faculty Universitas Negeri Semarang, D6 Building 2nd Floor , Sekaran Campus, Gunungpati, Semarang, Indonesia 50229

*) Corresponding author . sudarmin@mail.unnes.ac.id

Abstract. The purpose of this study was to obtain a factual picture of the improvement of students' conservation character and conservation behavior through the application integrated ethnoscience chemistry learning. This research was a case study on students majoring in the Department Of Science Mathematics and Natural Science Faculty Universitas Negeri Semarang. The subjects of the study were 30 students attending ethnoscience course in one of the teacher education institutions in the even semester of the academic year of 2016/2017. The subjects were given chemistry learning integrated into ethnoscience for eight weeks. The technique of data collection was done by using attitude scale arranged based on Likert scale. The data were analyzed by using qualitative descriptive. The results showed that the integrated ethnoscience chemistry learning contributed positively to the improvement of the character value and conservation behavior

I. Introduction

In the 21st century there have been revolutionary changes in the field of science and technology that spread throughout the world. The changes that occur, in addition to giving positive impacts, also increasingly impact on the occurrence of new problems related to global issues that actually threaten human survival. Therefore, in order to live in the midst of modern society, besides requirement of adequate thinking skills, it is also required strong conservation character and behavior. Strong character will enable citizens to change all obstacles into challenges and opportunities and be able to leverage their knowledge to solve daily life's problems and also problems in the broader spectrum of life [**1**]. Similarly, for prospective teachers, by having strong conservation character and behavior, they will be the individuals who always try to protect and preserve cultural values and act in a real way related to the use of natural resources in a sustainable manner so that natural resources could be enjoyed by present and future generations.

Referring to Universitas Negeri Semarang (UNNES) ' Vision as Conservation-Oriented University that has been declared since March 2010, it has a strategic meaning in the context of character building. This is related to the meaning of conservation that is not just a physical connotation, but more broadly is value and culture [**2**]. Conservation, the choice of name attached to Semarang State University, has a very deep value content. Conservation is not only concerned with activities that are purely physical, related to the relationship between man and nature, but also includes a broad and universal value system. There are three activities in it, namely *protection, preservation, and sustainable use*. In such a context, rights and obligations become the main buffer of human attitudes and behavior, namely that what we gain must be in balance with what we give. It certainly is in the widest sense. The balance between rights and obligations is not only on the things that are economic, but in the relationship between humans and the natural environment. Breathing in the fresh air, enjoying the coolness of the trees, enjoying the charming birds singing; are the rights that we gain from the universe. Therefore, by taking the principle of protection, preservation, and wise use of the life order, whether physical, social, or cultural, this conservation is expected to establish and develop the students' character to become good citizens.

As a teacher education institution, one of the areas that can contribute significantly to improve the nation's competitiveness is through innovation in education and teaching. The emergence of Conservation Education course on curricular structure, it is appropriate to conduct collective consciousness on conservation values, on formal lecture activities. UNNES has formulated 11 conservation-based character values. The 11 conservation values are: religious, honest, caring, tolerant, democratic, polite, intelligent, strong, nationalist, loving homeland and responsible. Conservation values are very important for science learning, because in the next 50 years there will be 10 major problems faced by humankind that are 1) energy, 2) water, 3) food, 4) environment, 5) poverty, 6) terrorist and war, 7) disease, 8) education, 9) democracy, and 10) population. There are 5 of 10 problems namely energy, water, food, environment and disease that are closely related to science and can only be solved with new science concepts or green science [**3**].

There are four types of conservation character developed during the education process, namely (1) cultural values-based character education (moral conservation); (2) cultural-based character education (cultural conservation); (3) environmental-based character education (environmental conservation), and (4) character education based on self-potential (humanist conservation). When the four types of conservation characters developed are implemented in students' daily life through relationships with the universe; trees, birds, water, air, and fellow human beings; it will internalize those values to students as a moral knowing, moral feeling, and moral action. Based on moral analysis from [**4**], there are three development approaches namely Cognitive Moral Development, Affective Moral Development, and Behavior Moral Development.

The problem is not all academic community implements the values of character and conservation behavior either in their learning or life even though UNNES has been declared as a conservation university for seven years. The internalization of character values and conservation behavior has not been achieved optimally. The result of in-depth questionnaire and interviews of graduates in the last three years (2014-2016) has shown that only 34% of graduates implementing conservation character values (honest, intelligent, caring, and strong) and only 36% of graduates who have conservation behaviors (awareness of environmental protection, save energy and water) with good and very good category. Most students have enough conservation character and conservation values that are adequate and less. Students tend to be less honest (such as cheating on the exam, making a report not based on the data obtained, not mention the source clearly if the statement or opinion used refers to the source); less optimize ability to think, be easy to despair if they face difficulties, less care for people and the environment around them. Most students have low conservation behaviors which are indicated by less environmental-conscious behaviors (such as littering, much use of plastic bags, much use of air-conditioned rooms, and much use of motor vehicles) and energy waste behavior.

Based on the explanation mentioned before, it takes a way that can be utilized as a means of internalization of character value and conservation behavior for all academicians, especially lecturers and students, so that after going through the process of conservation values internalization, it is expected that the values will grow in every breath of spirituality. They are not only loving and responsible for the universe, but they are also doing it to their creator; creator of the universe. The intensive interactions between lecturers and students occur in classroom learning. The result of field observation has shown that the learning done so far only focuses on the concept mastery. Lecturers are less likely to give students opportunities to learn meaningfully and less emphasize on attaining character values and conservation behavior. Meaningful learning will occur if students are given the opportunity to reconstruct knowledge of a phenomenon and relate it to the concepts of science **[5]**.

The integrated ethnoscience learning is a strategy for creating learning environment and designing learning experiences that integrate culture as part of the learning process **[6, 7, 8]**. The education of science that concerns the wisdom of local culture, character and customs is a matter that needs to be addressed in the Curriculum in Higher Education and High School. Cultural integration into learning is a very important effort to do, because school learning that is appropriate for the 21st century is a multicultural approach to science **[9]**. Local culture approach learning that transforms (reconstructs) indigenous knowledge of society into scientific knowledge is definitely important, because there are not many Sains lecturers that have used it as one of learning resources **[10]**.

This research is conducted to cultivate students' awareness on the environment, support students to get good learning outcomes after learning, and get the character of moral conservation, culture, environment and humanism. The students as prospective teachers should be familiarized to follow the learning on campus with a learning approach that can internalize the value and spirit of integrated local wisdom Scientific Method.

Some researchers who examine the importance of local cultural aspects of science learning conclude that students 'cultural backgrounds have influences on students' learning processes and outcomes in schools **[11-14]**, In addition, it is also showed that ethnoscience that lives and develops in society is still in the form of concrete knowledge as a result of interaction between environment and culture The example is research from The Reconstruction of Society Indigenous Science into Scientific Knowledge in the Production Process of Palm Sugar **[15]**.

Considering the importance of efforts to improve the character and conservation behavior of prospective teachers, this study is implemented by utilizing local culture and wisdom (integrated ethnoscience learning) as a means to increase the value of character and conservation behavior of prospective teachers. The purpose of this research is to analyze the increase of character values and conservation behavior of prospective teachers through the application of integrated ethnoscience learning while providing experiences to students in learning process integrated into culture.

2. Method

This study is a case study based on observations on the character value and conservation behavior of prospective teachers before and after being given integrated ethnoscience learning. The subjects of the study were students of the ethnoscience course of Integrated Science Department in one of the teacher education institutions in Central Java with a number of 30 students. The study was conducted on lectures of even semester of academic year 2016/2017. To measure the character values and conservation behavior of students, it was used attitude assessment sheet related to character and conservation behavior of students in the form of check list that was declared "valid" by the expert. Assessment of attitude was done by 3 assessors by using rating scale with category ranging from not good (1) to very good (4). Reliability of the rating sheet was 0.718 as split-half method. Other instruments used to collect evidence during the study were diaries, lesson profiles, and the performance of study subjects **[16]**. The data were analyzed by using qualitative analysis technique.

To know the achievement level of character value and conservation behavior, it is classically used criteria as presented in Table 1.

IOP Conf. Series: Materials Science and Engineering **349** (2018) 012061 doi:10.1088/1757-899X/349/1/012061

Table 1. Criteria of Classical Achievement Level

Average Criteria Values	Average Criteria Values
$x \leq 35$	Low
$35 < x \leq 50$	Medium
$50 < x \leq 65$	High
$65 < x \leq 80$	Very High

To determine the achievement level of each indicator of character value and conservation behavior, it is used criteria as presented in Table 2.

Table 2. Criteria Level Achievement per aspect

Rata-Rata Nilai	Kriteria
$5 < x \leq 8$	Low
$8 < x \leq 12$	Medium
$12 < x \leq 16$	High
$16 < x \leq 20$	Very High

3. Research Results and Discussion

The learning applied in this study was chemistry learning integrated into ethnoscience which aimed to increase students' character value and conservation behavior. The cultural content and context were used as media in implementing chemistry learning integrated into ethnoscience that was relevant to the topic discussed in this study, namely colloid chemistry, solution and carbon compounds presented in Table 3.

Table 3. Content and context in relation to chemical matter

No	Topics related to chemical content	Context of local culture	aspects of chemical content developed
1.	Colloid Chemistry	making dhawet ireng making traditional herbal medicine making camcao making porridge from rice flour / sago / other tubers	Colloid making
2	Acid-base solution	with the use of lime betel to exhale fruits	Base properties
3	Carbon compounds	making palm sugar	Carbohydrate
		making coconut oil	Lipid
		utilization of fibrous vegetables for various food products	Usage of cellulose / fiber
		Producing tapioca flour, cornstarch, etc. as basic ingredients of traditional food	Extraction

There were five character values and conservation behavior developed in this study, they were 1) environmental care, 2) love environment, 3) creative, 4) hard-working, and 5) responsible. The observation of the five aspects of character values and conservation behaviors was conducted before and after the end of lecture.

The result of observation before learning allowed researchers to create profile of character and conservation behavior classically, to create profile of each indicator of character value and conservation behavior, and to identify issues related to mastery of student's character and conservation behavior. Some of the important notes that could be identified before the implementation of chemistry learning integrated into ethnoscience were 1) the students' conservation behavior concerning the aspects of environmental care and loving environment was still low, and 2) the students did not show attitudes related to conservation character value ie creative, responsible and hard-working. In addition, the work ethic of the student group was still dominated by some students so that not all students were fully responsible for their work. Therefore, in the implementation of learning, lecturers gave more emphasis on the five aspects of character value and conservation behavior. The five aspects of character values and conservation behavior with their indicators are presented in Table 4.

Table 4. The aspects of character values and conservation behaviors and their indicators

Rated aspect	Indicator
Conservation behavior	
Environmental care	Protect natural environment
	Genuine knowledge of society
	Save energy and water
	Clean and healthy life
Love environment	Care for the environment, the value of conservation and local wisdom
	Like to plant crops
	Glad to see the green environment
	Use products that do not require a lot of energy
Conservation Character Value	
Creative	Use creative ideas in finding solution
	Create something new
	Ask various questions
	Like to do experiments
Responsible	Perform individual tasks well
	Accept risk
	Not accuse others
	Use energy with responsibility
Hard-working	Serious in doing task / test / exam
	Complete the task as well as possible
	Not easily give up
	Study hard

The assessment result of students' character and conservation behavior classically before and after learning is presented in Table 5.

Table 5. The analysis result of characteristic value and conservation behavior of prospective teacher classically (n = 30, max value = 80, minimum value = 5)

No	Rated aspect	Before learning		After learning	
		Value	Category	Value	Category
1	Conservation character	33,7	low	60,3	high
2	Conservation behavior	37,8	medium	64,5	high

A good character consisted of moral knowing, moral feeling, and moral action, which were explained as habit in way of thinking, habit in heart, and habit in action (Lickona, 2013). In Table 5, it can be seen that the character value and conservation behavior of students before learning on the five aspects classically have a mean value in the medium category. After the implementation of chemistry learning integrated into ethnoscience, the five aspects reached a very high category.

These results showed that there was an increase in students' character value and conservation behavior after the introduction of integrated ethnoscience learning. The increase in student values showed that the learning implemented was suitable for improving the student's conservation character and behavior. This is in line with **[17]** and **[18]** that creative soft skills, hard-working and responsible in group can be created if learning is based on continuous improvement. Soft creative skills, hard-working, and responsible as parts of individual characteristics are developed through ongoing processes, until personal growth is achieved. The role of lecturers in learning is to help improving the mastery of soft skills. Lecturers should be sensitive to changes in soft skills of learning subjects. This result is also according to **[13]** who conveys that science learning integrated into ethnoscience can increase student interest and achievement.

The students' values of conservation character and behavior are also analyzed in every aspect. The achievement analysis result in each aspect of values of conservation character and behavior is presented in Table 6.

Table 6. The result analysis of aspects of students' conservation character and behavior (n = 30, maximum score = 20, minimum score = 5)

Rated aspect	**Before learning**		**After learning**	
	Value	**Category**	**Value**	**Category**
Environmental care	5,7	low	12,8	high
Love environment	8,2	medium	14,6	high
Creative	9,6	medium	13,2	high
Responsible	9,6	medium	15,8	high
Hard-working	6,6	low	14,4	high

Environmental awareness expresses general attitudes towards the quality of the environment embodied in the willingness of self to state actions that can improve and maintain the quality of the environment in any behavior related to the environment. In Table 6, it appears that the value of environmental care aspect as the conservation behavior aspect is in the low category before being implemented integrated ethnoscience learning, while the aspect of loving environment is in the medium category. This indicates that the internalization of conservation behavior cannot be formed by itself, but it must be given an advance understanding (explanation), made as a continuous obligation, until finally become a habituation **[14, 19, 21]**. After implementing integrated ethnoscience learning, both aspects of conservation behaviors are increased to be categorized high. This proves that the application of integrated ethnoscience learning makes the majority of students (84%) feel concerned about the surrounding environment, as many of the examples taught show how the relation of chemical concepts studied to local cultures is still widely maintained in society **[12, 21]**.

In Table 6 it also appears that the three student character values (creative, responsible, hard-working) have improved from the medium and low categories at the time before learning becomes high after learning. This shows that when lecturers elaborate knowledge in class, it is at once given the character value of being creative, hard-working, and responsible. The creative indicator is shown by the way students solve the given problem, whether students complete by memorizing the given examples or they use different techniques but getting the appropriate results. The hard-working indicator is shown by the tireless attitude in completing tasks to obtain results that can be accounted for, willing to surf the internet even though the place is less comfortable and takes a long time. The responsible indicator is demonstrated by acknowledging the mistakes made and not repeating the same mistakes, doing the tasks as instructed and reporting when they are done.

The conservation behavior of environmental care is a caring attitude toward the environment that is shown in the willingness to express actions that can improve and maintain the quality of environment in any behavior related to the environment. The results obtained in this study provide a picture that the character value and conservation behavior can be improved by applying integrated ethnoscience learning. But to note, the success of formation in students' conservation character and behavior cannot be evaluated immediately, it can only be seen in a long span of time. This is due to the character and behavior associated with the internalization of values and habituation that require a long process.

4. Conclution

The results of study and data analysis about the effort to increase the character value and conservation behavior of prospective teachers in attending integrated ethnoscience course, provide some information and findings formulated in the following conclusions:

First, the implementation of integrated ethnoscience learning is able to increase the students' value of conservation character. This learning makes students' willingness to ask increases. The courage of students in delivering papers seems more confident. Likewise, there is an increased willingness to respond to the issues being discussed either by lecturers or by students presenting the discussion material. In responding to questions from other friends, it is done clearly, objectively, scientifically in accordance with scientific procedures that have been understood.

Second, the implementation of integrated ethnoscience learning is able to improve students' conservation behavior, In the quality of providing answer, it is increased and can be shown in the use of relevance of reference book resources into reference material. The positive attitude of students' awareness of the environment has improved after the application of integrated ethnoscience learning model. Based on the research findings, the recommendation offered is that this model still needs to be developed further to test its effectiveness, it also needs to be socialized and disseminated among lecturers and students to improve the quality of learning.

Acknowledgement

Our thanks go to 1) Mathematics and Natural Science Faculty Universitas Negeri Semarang , which has provided funding for this research and 2) students who contribute to the completion of this research.

References

[1]. Rahayu, S. 2016. Mengembangkan Literasi Sains Anak Indonesia melalui Pembelajaran Berorientasi Nature of Science (NOS). Pidato Pengukuhan Jabatan Guru Besar Bidang Pendidikan IPA pada FMIPA disampaikan pada Sidang Terbuka Senat Universitas Negeri Malang tanggal 17 Maret 2016.

[2] Masrukhi. 2015. Mengembangkan Karakter Mahasiswa Berbasis Nilai Konservasi (Kasus Unnes Semarang). Tersedia: https://masrukhiunnes. wordpress. com/2015/ 01/26/ mengembangkan- karakter-mahasiswa-berbasis-nilai- konservasi-kasus-unnes-semarang/ diakses 20 Juli 2017

[3] Hjeresen, D.L., Schutt, D.L.;, Boese, J.M., 2000, Green chemistry and education. *J. Chem. Educ* 2000, 77, 1543–1544.

[4] Kurtines, W. M. 2002. *Morality, Moral Behavior, and Moral Development.* New York: John Wiley & Sons.

[5] Kipnis, M. & Hofstein, A. 2007. The Inquiry Laboratory As A Source for Development of Metacognitive Skills. International Journal of Science andMathematics Education, 6(3), 601-607.

[6] Aikenhead, G; Jegede, O.J. 1999. Cross-Cultural Science Education: A Cognitive Explanation of a Cultural Phenomenon. *Joumal of Research in Science Teaching, 36*, 269-287.
[7] Cobern,W.W & Loving,C.C. 2001. Defining "Science" in a Multicultural World: Implication for Science Education. *Science Education*. 85.
[8] Sudarmin. (2015). *Pendidikan Karakter, Etnosains dan Kearifan Lokal (Konsep dan Penerapannya dalam Penelitian dan Pembelajaran Sains)*. Semarang: FMIPA UNNES.
[9] Gunstone, A. (ed.). (2014). *Developing Sustainable Education in Regional Australia*. Melbourne: Monash University Publishing. 216 + viii pp
[10] Sudarmin, Zaenuri Mastur, Parmin. (2017). *Pengetahuan Ilmiah Berbasis Budaya dan Kearifan Lokal di Karimunjawa untuk Menumbuhkan Soft Skills Konservasi*. Jurnal Pendidikan Sains Pascaarjana Unesa. 6 (2). 1362-1368.
[11] Suastra, I. W. (2006). Perspektif Kultural Pendidikan Sains: Belajar Sebagai Proses Inkulturasi, *Jurnal Pendidikan dan Pengajaran IKIP Negeri Singaraja,34(3) Juli 2006.*
[12] Suastra, I. W.; Tika, K.; Kariasa, N. (2011). Efektivitas Model Pembelajaran Sains Berbasis Budaya Lokal untuk Mengembangkan Kompetensi Dasar Sains dan Nilai Kearifan Lokal Di SMP. *Jurnal Penelitian dan Pengembangan. 5(3)*, 258-273.
[13] Shidiq, A. S., 2016. *Pembelajaran Sains Kimia Berbasis Etnosains untuk meningkatkan minat dan prestasi belajar siswa*. Prosiding Seminar Nasional Kimia dan Pendidikan Kimia VIII. FKIP UNS. 14 Mei 2016
[14] Abidinsyah. (2013). Internalisasi Nilai Peduli Lingkungan melalui Pembelajaran Berbasis Kearifan Lokal (Studi di SD Antasari Besar 7 Banjarmasin). Bandung. *Dissertation:* Universitas Pendidikan Indonesia.
[15] Woro Sumarni, Sudarmin, Wiyanto, Supartono. (2016). The Reconstruction of Society Indigenous Science into Scientific Knowledge in the Production Process of Palm Sugar. *Jurnal of Turkish Science Education*. 13(4): 281-292, Tersedia http://www.tused.org
[16] Elliot, J, (1993), *Action Research for education change,* Philadelphia
[17] Kreitner, R. & Kinicki, A. 2008. *Organizational Behavior*, 8th. Boston: McGraw-Hill Irwin.
[18] Sudarmin, Ria Febu K, M. Nuswowat, W.Sumarni, (2017). Development of Ethnoscience Approach in The Module Theme Substance Additives to Improve the Cognitive Learning Outcome and Student's entrepreneurship. *Journal of Physics: Conf. Series **824** (2017) 012024: tersedia:* The 3rd International Conference on Mathematics, Science and Education 2016
[19] George, J. (2001). *Culture and Science Education*: Developing Word. http://www. id21. org/education/e3jg1g2.html. Diakses tanggal 24 April 2017.
[20] Rahmawati, Sri Utami (2014) Meningkatkan Karakter Peduli Lingkungan Melalui Sekolah Berbudaya Lingkungan Hidup :Studi Deskriptif di SMP N 7 Bandung. *thesis,* Universitas Pendidikan Indonesia.
[21] Lickona,T. 2013. *Educating for Character, Mendidik Untuk Membentuk Karakter*, terjemahan Juma Abdu Wamaungo, Jakarta: Bumi Aksara.

IOP Conf. Series: Materials Science and Engineering **349** (2018) 012062 doi:10.1088/1757-899X/349/1/012062

Formulation of Antibacterial Liquid Soap from Nyamplung Seed Oil (*Calophyllum inophyllum* L) with Addition of *Curcuma heyneana* and its Activity Test on *Staphylococcus aureus*

S Widyaningsih[1], M Chasani[1], H Diastuti[1], Novayanti[1]

[1]Chemistry Department, University of Jenderal Soedirman, Purwokerto, Indonesia

Email : sennysetiadi@yahoo.com

Abstract. The increasing demand of soap encourages new innovations of it's quantity, quality, benefits and raw materials. One of the innovations that has been developed is the use of nyamplung (*C. inophyllum* L) seed oil as basic ingredients of soap, with addition of active substances such as antibacterial and fragrance. This research aims to determine best formulation of liquid soap according to SNI 06-4085-1996 and its antibacterial activity after addition of n-hexane fraction of *C. heyneana* rhizome and fragrance from lavender flower oil at various concentrations, then a hedonic test was conducted to evaluate the panelist's preference towards the soap with the best formulation. This research has been done through several stages of soap making with saponification method, soap characterization according to SNI 06-4085-1996, determination of best formulation of liquid soap based on the effectiveness index method and antibacterial activity test of liquid soap to *S.aureus* with disc diffusion method and hedonic test using some parameters such as colour, odor, foam, clean power and effects after being used. The results showed that the best characteristic soap according to SNI 06-4085-1996 was soap with 0.2% antibacterial content and 2.5% perfume. The best soap characteristic showed total fatty acid 67.49%, free fatty acid 1.01%, neutral fat 7.24%, weight of type 1.05 g/mL, pH of 9.85 and foam stability 76.69%. The results of the antibacterial activity test of the best formula soap showed the inhibitory zone of 14.92 mm.

1. Introduction

Consumer demand for body cleanser that is increase, promotes new innovations on soap products, in terms of quantity, quality and function. Liquid soap is a preferred product than solid soap, because it is more hygienic in storage and more practical in use.

Researchers tried to look for alternative raw material for soap production from non-edible oils with abundant availability. One of them that have the potential to produce soap is oil derived from nyamplung seeds (*Calophyllum inophyllum* L.).

The research on produce sodium soap based on nyamplung seed oil (*C. inophyllum* L.) was reported that it have antibacterial activity [1]. In order to improve the quality and gain the public interest, we need to add antibacterial and fragrances. This research was done to develop the form of soap that was made from nyamplung seed oil by adding antibacterial additives and perfumes.

The antibacterial additive that was used for soap formulation is the n-hexane fraction of the *Curcuma heyneana*. It was known that *Curcuma heyneana* containing sesquiterpen, a terpenoid compounds with germakran framework that are potentially used as an antibacterial agent [2]. The addition of perfumes to antibacterial soap formulations is intended to attract public interest. People love the fragrance of the flowers. The essential oil of the flowers gives a certain flavor. One of them is lavender oil *(Lavandula agustifolia)* [3].

The research started with saponification of nyamplung oil using KOH base, then added antibacterial substance from n-hexane fraction of *Curcuma heyneana* and fragrance from lavender flower oil. The composition of the antibacterial was varied at 0.2, 0.4, 0.6, 0.8 and 1% while the perfume was varied at 0.5, 1, 1.5, 2and 2.5%. The results were characterized and compared with SNI 06-4085-1996. In order to test the antibacterial properties, the best characterized soap was tested with disc diffusion method against *Staphylococcus aureus* bacteria.

2. Experimental

2.1 Materials

The research used nyamplung seeds obtained from Gunung Selok Kroya, temu giring *(C. heyneana),* lavender flower oil, *S. aureus* bacteria, NA media (Nutrient Agar), NB media (Nutrient Broth) and tetracycline. Organic solvents n-hexane, acetone, methanol, sodium sulphate anhydrous, potassium hydroxide, carboxyl methyl cellulose (CMC), sodium lauryl sulfate (SLS), aquades, ethanol absolute, 70% alcohol, methyl red indicator, phenolphthalene (pp), hydrochloric acid and sulfuric acid.

2.2 Preparation of the n-hexane fraction of temu giring rhizomes [2]

Temu giring is cut into small pieces, dried, and smoothed with blender. The temu giring rhizome powder is macerated with acetone for 3x24 hours. The maseration results are filtered using a buchner funnel. The resulting filtrate is concentrated under low pressure using a rotary evaporator. The concentrated acetone extract was extracted by partitioning with n-hexane: methanol (1: 1). Partition extract result obtained two fractions of n-hexane fraction and methanol fraction. The obtained n-hexane fraction is concentrated at low pressure using a rotary evaporator. The n-hexane fraction of concentrated acetone extract was added anhydrous Na2SO 4 to bind water and continued by filtration.

2.3 Preparation of soap

The nyamplung seed oil was stirred and heated until 70-80°C then was added KOH 30% (w/v). The reaction mixture is continuously heated while stirring at 500 rpm for 1 hour. The temperature is lowered to 60 and added aquadest (1:1 ratio of the reaction product) while stirring until homogeneous soap is formed. Furthermorecarboxyl methyl cellulose (CMC) and sodium lauryl sulfate (SLS) were added to the product soap with various concentrations. Variations of n-hexane fraction of tilapia rhizome and natural perfume were added and remained distilled at 500 rpm for 10-15 min. Stirring is stopped and the magnetic stirer is immediately taken from the soap solution. Soap products poured into containers and closed tightly. Variations of antibacterial and perfume concentrations can be seen in Table 1.

Table 1. Variations of antibacterial and perfume concentrations

n-hexane fraction of temu giring rhizome (w/w)	Lavender perfume (w/w)				
	P1 (0.5%)	P2 (1.0%)	P3 (1.5%)	P4 (2.0%)	P5 (2.5%)
A1 (0.2 %)	A1P1	A1P2	A1P3	A1P4	A1P5
A2 (0.4 %)	A2P1	A2P2	A2P3	A2P4	A2P5
A3 (0.6 %)	A3P1	A3P2	A3P3	A3P4	A3P5
A4 (0.8 %)	A4P1	A4P2	A4P3	A4P4	A4P5
A5 (1.0 %)	A5P1	A5P2	A5P3	A5P4	A5P5

2.4 Characterization of soap [4]

Characterization of soap includes parameters: total fatty acid, free fatty acids, unsabsorbed fats or neutral fats, pH test, weight type, foam stability

2.5 Data analysis result of soap characterization

Data analysis used was factorial completely randomized design. Factors tested were n-hexane fraction content of acetone extract of temu giring (C. heyneana) and perfume oil of lavender. The data obtained were analyzed using Statistical Product Services Solution (SPSS) 16.0 for windows. The results that showed a real difference will be analized by DMRT test (Duncan's Multiple Range Test) with 95% confidence level ($\alpha = 0{,}05$).

2.6 Determination of the best soap formulation

The best antibacterial soap formulations were determined by the effectiveness index method [6]. This method is done by sorting the variables by priority and contribution to the results. Each variable has a weight according to its contribution with relative numbers 0-1. The value weights are obtained from each variable by dividing the weight of the variable by the sum of all the variable weights. The best formulated soap is a soap that has the highest product value. The value of the product is calculated by multiplying the value of effectiveness (EV) by the weight of the value (WV).

2.7 Antibacterial activity test

The antibacterial activity test was performed by pouring 15 mL of Nutrient Agar (NA) medium at ± 40 °C into sterile petri dish then left at room temperature until the medium to solidify. A number of bacterial cultures in Nutrient Broth (NB) liquid medium are taken and dispersed in Nutrient Agar (NA) medium. The volume of bacteria taken based on the absorbance results at 600 nm. If the absorbance value is less or equal to 0.5 then is taken 100 µL bacterial culture, whereas if 0,6-1,0 is taken 50 µL. The suspension of the test bacteria on Nutrient Agar (NA) medium is distreaked on the spread plate by using drugalsky, then allowed to dry for 15 minutes at room temperature. After drying, a paper disc with a diameter of ± 6 mm is placed over the NA medium. Samples and controls were taken as much as 10 µL and dripped onto paper disc then incubated for 24 hours at 37 ° C. Then it was measured the inhibitory diameter formed around the disc paper using calipers.

3. Results and Discussion

3.1 Antibacterial liquid soap

The process of producing antibacterial liquid soap is done using semi-hot method (semi-boiled). The soap produced from the saponification process that is diluted and added with the additive. The additives are carboxyl methyl cellulose (CMC), sodium lauryl sulfate (SLS), antibacterial substances from the n-hexane fraction of the temu giring rhizome (*C. heyneana*) and fragrance from lavender flower essential oil . The purpose of addition of temu giring rhizome (*C. heyneana*) and the perfume of the lavender flower oil is to give antibacterial and aromatic effects to the soap.

The characterization of liquid soaps are total fatty acids, free fatty acids, neutral fat, pH, type weight and foam stability. Then the result were compared with SNI 06-4085-1996 [4].

The highest of total fatty acid was shown by control soap without antibacterial and perfume addition of 70.52%. While total amount of fatty acid in formulation soap showed the range 66,30% - 69,70% (Fig.1). The lower amount of total fatty acids in the formulation soap is due to the higher n-hexane fractional content of temu giring rhizome (*C. heyneana*) and the fragrance of the lavender flower essential oil. The characterization results show that the total fatty acid amount in antibacterial liquid soap in accordance with SNI 06-4085-1996 [4].

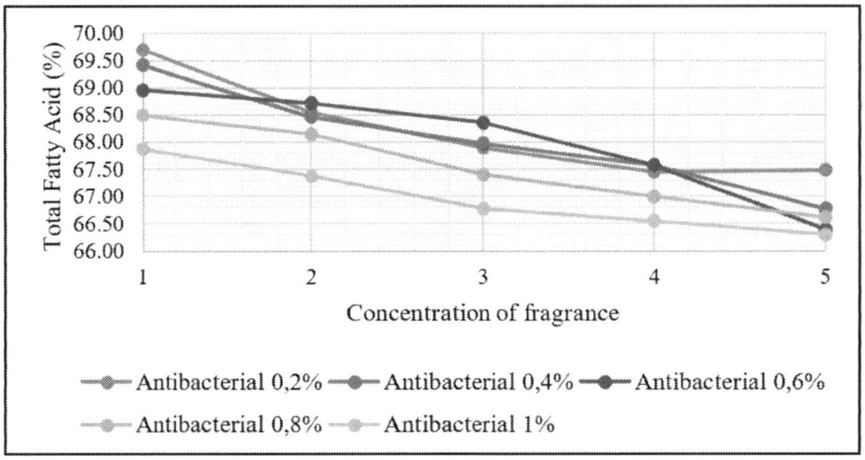

Figure 1. Total fatty acid at variation of fragrance concentration

The result of data analysis showed that there were significant differences on soap with 1% antibacterial addition and 2.5% perfume which showed total fatty acid amounting to 66.30%.

Formulated soaps have a range of free fatty acids 0.40% - 1.20% (Fig.2). The high amount of free fatty acids are caused by acidic terpenoid compounds [6,7] that were contained in additives. The terpenoids will react with the KOH solution, resulting a decrease of the alkaline content which will soap the free fatty acids. Result of free fatty acid amount on antibacterial liquid soap in accordance with SNI 06-4085-1996 [4].

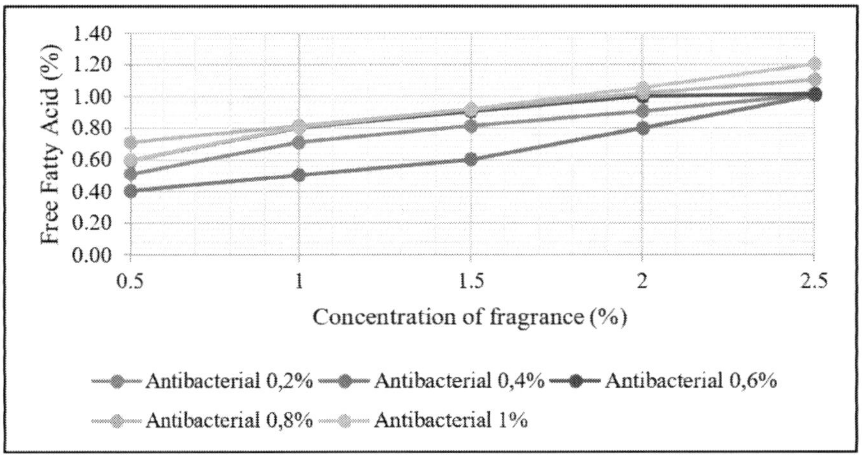

Figure 2. Free fatty acid at variation of fragrance concentration

The result of data analysis of free fatty acid showed a significant effect on the amount of free fatty acid in the formulation soap. The significant differences occurred in soap with 1% antibacterial addition and 2.5% perfume.

Soap formulations show the neutral fat in a range of 5.36% - 7.25% (Fig.3). The high amount of neutral fat in soap formulation is caused by the content of neutral fat components in the soap. The neutral fat components are generally sterols, dyestuffs, hydrocarbons and complex lipids [8]. The additives that is used contain terpenoid compounds. Terpenoid compounds are generally regarded as lipids and include either neutral fat [6]. In addition, the n-hexane fraction of the temu giring rhizome (*C. heyneana*) contains a dye of curcumin which can affect the amount of neutral fat in the formulation soap. The result of the amount of neutral fat in antibacterial liquid soap is not in accordance with SNI 06-4085-1996 [4].

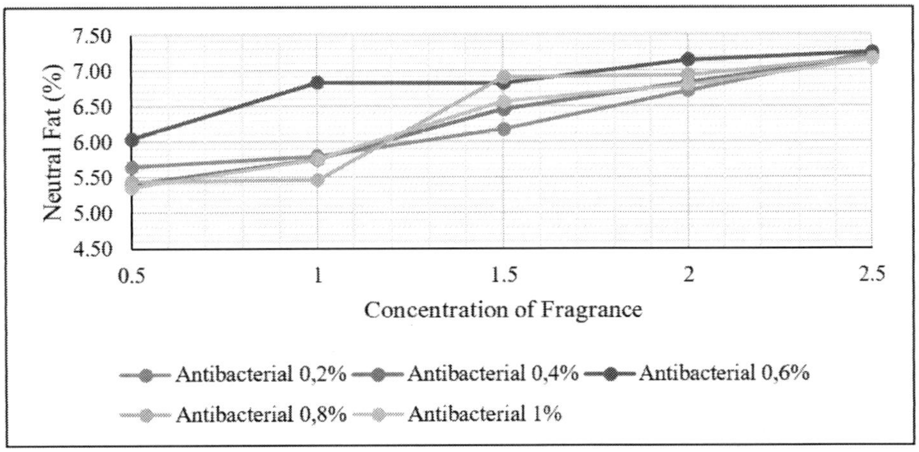

Figure 3. Natural fat at variation of fragrance concentration

Result of data analysis of neutral fat quantity showed a significant effect to neutral fat amount on soap formulation. The significant differences occurred in soap with 0.2% antibacterial addition and 2.5% perfume.

The pH measurement results show that the control soap has a pH range of 9.65 - 9.85 (Fig.4). While the formulation soap showed a pH range of 9.55 - 9.85. The higher pH value of the control soap is caused by the soap that is formed from the saponification process of triglycerides with alkaline base (KOH), where KOH is one of the strongest bases [9]. The lower pH value of the formulation soap are caused by the presence of terpenoid compounds in additive which can react with KOH. Terpenoid compounds are one of the acidic compounds because they contain hydroxyl and carbonyl groups [7,8]The result of measurement of pH value in soap in accordance with SNI 06-4085-1996. pH value of soap that is safe for the skin is 8-11 [10].

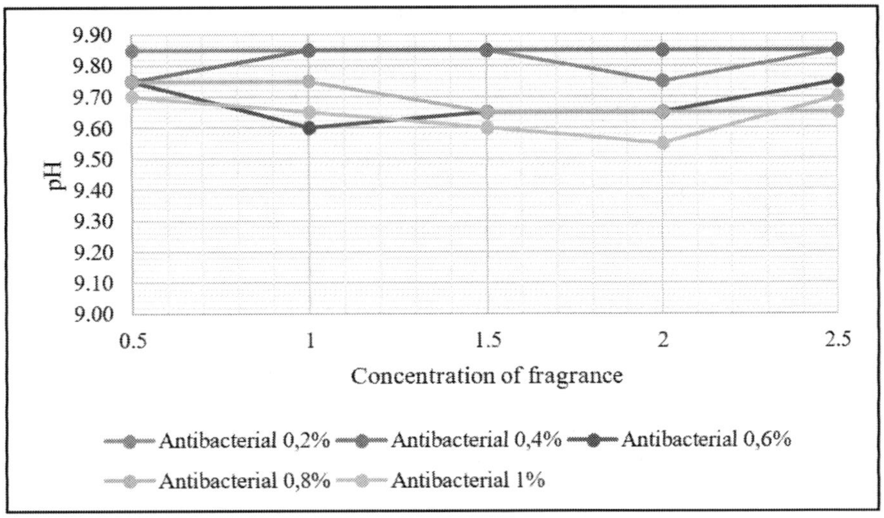

Figure 4. pH at variation of fragrance concentration

Result of data analysis showed a significant effect to pH on soap formulation. The significant differences occurred in soap with 0.8% antibacterial addition and 2.5% perfume.

The density of antibacterial liquid soap is influenced by the type and concentration of raw materials that was added. The higher the value of molecular weight of the raw material added, the higher the density [11]. The result shows that the density of soap in accordance with SNI 06-4085-1996. Result of data analysis showed a significant effect to density of soap formulation (Fig 5). The significant differences occurred in soap with 1% antibacterial addition and 2.5% perfume.

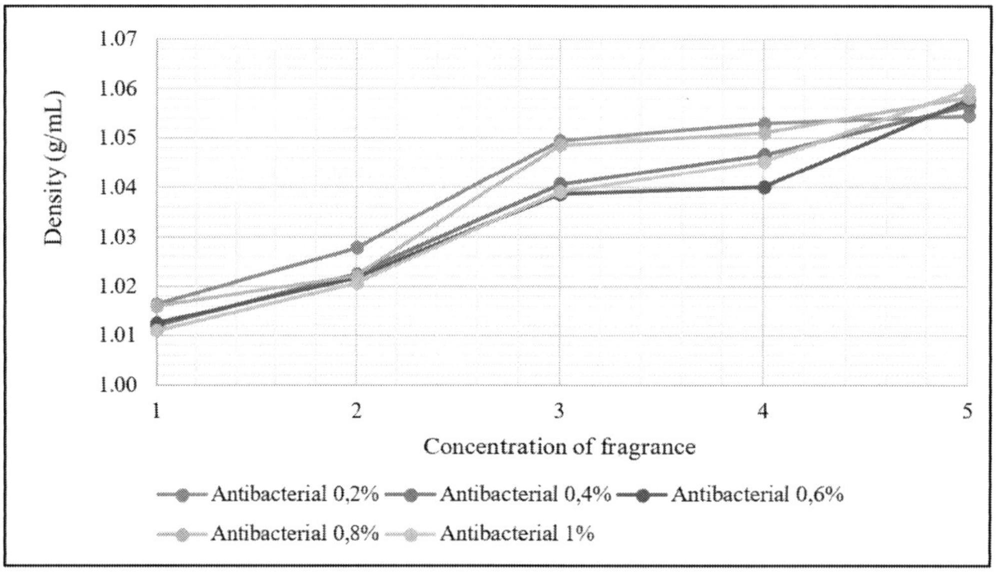

Figure 5. Density at variation of fragrance concentration

Soap formulation results show the value of foam stability with a range of 72.92% - 80.52% (Fig 6.). Addition of additive causes the foam stability value relatively higher. This is caused by the presence of saponin content in the additive that can produce foam [12]. One factor that affecting foam stability is the type of fatty acid in soap formulations. The nyamplung seed oil contains oleic acid which can produce a stable and soft foam, palmiic acid and stearic acid which has the stabilizing properties of the foam. Result of data analysis showed a non effect to foam stability.

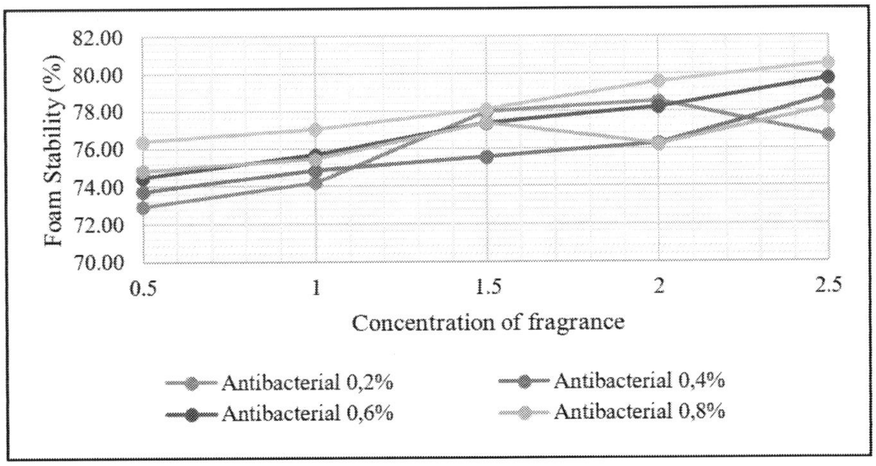

Figure 6. Foam stability at variation of fragrance concentration

3.2 The Best Soap Formulation

The best antibacterial liquid soap formulations were determined by the effectiveness index method [6]. This effectiveness index method can be done by sorting the variables based on the priority and contribution to the results, whereby the best antibacterial liquid soap formula has the highest product value. The order of priority and contribution to the results in this study were total fatty acids (%), free fatty acids (%), neutral fats (%), pH, type-weights and foam stability (%). Calculation result of soap product value of formula can be seen in Fig 7.

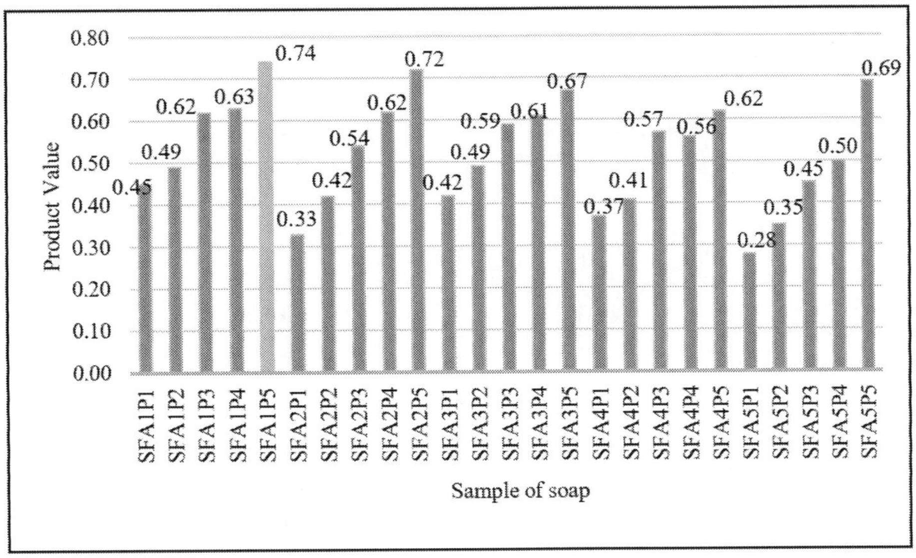

Figure 7. Product value of soap

Based on the Fig.7 shows that the best characterized soap formulation according to SNI 06-4085-1996 is soap with 0.2% antibacterial addition and 2.5% deodorizer with product value of 0.74.

The antibacterial activity test

The antibacterial activity test on the best soap formula according to SNI 06-4085-1996 was done on gram-positive bacteria i.e *S. aureus* bacteria. The n-hexane fraction of the temu giring rhizome (*C. heyneana*) showed an antibacterial activity with a inhibitory diameter of 5.34 - 15.23 mm. The highest antibacterial activity was shown by the n-hexane fraction of the temu giring rhizome (*C. heyneana*) at a concentration of 1% with a resistor diameter of 15.23 mm. Antibacterial activity on essential oil fragrance of lavender flower showed 13.19 mm inhibition diameter. This suggests that the higher concentrations of the added substances can increase antibacterial activity [13]. Some things that can affect antibacterial work include the concentration or intensity of antibacterial substances, the number of microorganisms, temperature, species of microorganisms and the presence of organic materials [14]. The terpenoids will accumulate in the lipid of bacterial cell membrane and cause disruption of the structure and function of the cell membrane that was caused by swelling and changes in the permeability of bacterial cell membranes [15,16].

The results of the antibacterial activity test on soap formulation showed that the addition of n-hexane fraction of temu giring rhizome (*C. heyneana*) and fragrance of lavender flower oil decreased antibacterial activity when compared in its single state. The effect of component interactions is due to the n-hexane fraction of the temu giring rhizoma (*C. heyneana*) and the fragrance of lavender flower oil that they become antagonistic because the combination of compounds has a smaller therapeutic effect or excludes the activity of other compounds [17,18].

Table 2. Result of antibacterial activity test on soap formulation

Sample	Inhibitory diameter (mm)
Aquades (-)	-
Tetracycline (+)	27.69
The fraction of n-hexane 0.2%	5.34
The fraction of n-hexane 0.4%	7.07
The fraction of n-hexane 0.6%	10.77
The fraction of n-hexane 0.8%	14.42
The fraction of n-hexane1%	15.23
Essential oil of lavender 2.5%	13.19
SK	14.89
SKA1	14.43
SKA2	14.51
SKA3	14.80
SKA4	15.09
SKA5	15.54
SKP5	15.70
SFA1P5	14.92
SFA2P5	14.58
SFA3P5	14.25
SFA4P5	13.43
SFA5P5	13.04

4. Conclusion

Antibacterial liquid soap with the best characterization according to SNI 06-4085-1996 is antibacterial liquid soap with n-hexane fraction of temu giring rhizoma content of 0.2% and fragrant content of 2.5% lavender flower essential oil and has antibacterial activity against *S. aureus* of 14.92 mm.

References

[1] Chasani M, Widyaningsih S and Mubarok A 2015 *Jurnal Molekul*. 10 (1): 66 – 73.
[2] Diastuti H, Syah Y M, Juliawaty L D, and Singgih M 2014 *Indonesian Journal of Chemistry* 14 (1): 32–36.
[3] Morgan T J, Morden W E, Al-Muhareb E, Herod A A, and Kandiyoti R 2006 *Energy and fuels*. 20 (2): 734-737.
[4] Standar Nasional Indonesia 1996 *Sabun Mandi Cair* (Jakarta: Badan Standarisasi Nasional)
[5] De Garmo E D, Sulivan, and Canada J R 1984 *Engineering Economics* (New York: Mc. Millan Publishing Company)
[6] Fessenden R J dan Fessenden J S 1982 *Kimia Organik jilid 2* (Jakarta: Erlangga)
[7] Rowe C R, Paul J S, dan Marian E 2009 *Handbook Of Pharmaceutical Excipiens Sixth Edition*. (London: Pharmaceutical Press)
[8] Hernani T K, Bunasor, dan Fitriati 2010 *Bulletin Litro*. 21 (2): 197-200.
[9] Hambali E, Suryani A, dan Rifai M 2004 *Jurnal Teknik Industri Pertanian*. 14(2): 74-79.
[10] Endoga 2009 *Jurnal Engineering Applied Scince*. 4 (2): 110-113.
[11] Musy R, Enawati K, dan Sutoyo 2003 *Jurnal Ilmu Kedokteran*. 35 (4).
[12] Putra A M P, Rustifah, and Arsyad M 2015 *Jurnal Ilmiah Manuntung*. 1 (1) 68-74.
[13] Miksusanti, Fitriya, dan Nike M 2011 *Jurnal Penelitian Sains*. 14: 40-47.
[14] Pelczar M J dan Chan E C S 2007 *Dasar-Dasar Mikrobiologi* (Jakarta: Universitas Indonesia Press)
[15] Sikkema J, De Bont J A M, and Poolman B 1994 *Journal of Biological chemistry* 269 (11): 8022-8028.
[16] Ardani M, Pratiwi S U T, Hertiani T 2010 *Majalah Farmasi Indonesia*. 21(3): 191 – 201.
[17] Abreu A C, Mc Bain A J, Simoes M *Natural Product Report*. 29: 1007-1021.
[18] Faleiro M L 2011 *Formatex* 1143-1156.

IOP Conf. Series: Materials Science and Engineering **349** (2018) 012063 doi:10.1088/1757-899X/349/1/012063

The Effect of Acetone Amount Ratio as Co-Solvent to Methanol in Transesterification Reaction of Waste Cooking Oil

T S Julianto, R Nurlestari

Department of Chemistry, Faculty of Mathematics and Natural Sciences, Universitas Islam Indonesia, Yogyakarta, Indonesia

tatang_shabur@uii.ac.id

Abstract. The production of biodiesel from waste cooking oil by transesterification reaction using acetone as co-solvent has been carried out. This research studied the optimal amount ratio of acetone as co-solvent to methanol in the transesterification process using homogeneous alkaline catalyst KOH 1% (w/w) of waste cooking oil at room temperature for 15 minutes of reaction time. Mole ratio of waste cooking oil to methanol is 1:12. Acetone was added as co-solvent in varied amount ratio to methanol are 1:4, 1:2, and 1:1, respectively. The results of fatty acid methyl esters (FAME) were analysed using GC-MS instrument. The results showed that the optimal ratio is 1:4 with 99.93% of FAME yield.

1. Introduction

Biodiesel is an alternative energy that can be used as fuel like fossil fuel. Biodiesel is obtained from vegetable oil or animal oil so it can be renewed. Since biodiesel is a non-fossil oil it is certainly free from sulfur, which will have a positive impact on the environment. Biodiesel has advantages compared to petroleum diesel fuel derived from petroleum, among others, has a better lubrication properties that can extend engine life, is a safe fuel, easy to handle, and non-toxic, has a relatively low exhaust gas.

Biodiesel is produced by a transesterification reaction between triglycerides and alcohols such as methanol (methanolysis) using a strong base catalyst such as NaOH and KOH. One of the problems in the transesterification reaction of biodiesel is the non-homogeneity of the reactants involved. Methanol is not soluble with triglycerides. Conventionally to increase the solubility of reactants was carried out with temperature 65°C for 2 hours [1] or with using microwave-assisted reactor [2]. The use of high energy in reaction for a long time leads to inefficiency in biodiesel production. The use of co-solvent becomes an appropriate alternative to solve the problem of solubility. A one-phase reaction can be formed by adding a solvent that can increase the solubility of the oil, the solvent hereinafter referred to as co-solvent [3]. Co-solvent is very soluble with alcohol, fatty acids and triglycerides. The used co-solvent should not contain water and the more co-solvent added is better because it will increase the solubility of the oil. The selected co-solvent has a boiling point close to methanol which can facilitate the separation process at the end of the reaction.

Several co-solvents has been used for transesterification reaction including n-hexane, diethyl ether, acetone, 2-propanol, tetrahydrofuran, or ethyl acetate. Acetone was found to be the best co-solvent in transesterification reaction [4].

Our objective in this research is to understand the effect of acetone amount ratio as co-solvent to methanol in waste cooking oil transesterification reaction.

2. Materials and methods

2.1. *Material*

Acetone, methanol, and Potassium hydroxide (KOH) is purchased by Merck Millipore (Germany). Waste cooking oil was obtained from several restaurants around campus of Universitas Islam Indonesia Yogyakarta that have already washed with alcoholic washing method [5].

2.2. *Methods*

2.2.1. Transesterification reaction. The transesterification stage is carried out by reacting washed waste cooking oil with methanol containing 1% homogeneous potassium hydroxide (KOH) catalyst into a three-neck flask. Mole ratio methanol/oil is 1:12. Furthermore, acetone is added as co-solvent. The transesterification process is carried out by varying the co-solvent acetone amount ratio to methanol. The mole ratio of acetone to methanol is done by 3 variations: 1: 4, 1: 2 and 1: 1 mol. The entire mixture is stirred using a magnetic stirrer at a constant speed of ± 600 rpm at room temperature (27°C) for 15 minutes.

2.2.2. Separation and identification of Fatty Acid Methyl Esters (FAME). The entire mixture is inserted into the separating funnel and silenced so that the mixture will form two layers (two phases). The lower layer is glycerol which is a by-product of the transesterification reaction process, while the top layer is the methyl esters obtained. The top layer is washed using aquadest and evaporated to separate the excess methanol. The yield of total methyl esters was analysed using GC-MS (Shimadzu QP 2100 SP with RTX-5 column).

3. Results and Discussion

In this research the co-solvent used is acetone. Acetone is an aprotic solvent having intermediate polarity so as to dissolve high-polarity methanol and non-polar use of crude oil triglyceride into a homogeneous reaction system. This phase change increases the rate of transeterification reaction between methanol and triglyceride of cooking oil with the appearance of a homogeneous sodium hydroxide as base catalyst. Another important property of acetone is as aprotic solvent so that acetone has ability to stabilize the methoxide ion formed. This methoxide ion is an active intermediate compound in the transeterification reaction that attacks triglycerides to form methyl esters through the mechanism of S_N2.

To determine the optimum condition of transesterification reaction by using co-solvent acetone, the mole ratio of acetone to methanol was evaluated.

3.1.1. Effect of amount of acetone to methanol to yield of transesterification product

Molar ratio of acetone to methanol is one of the important aspects affects yield of transesterification product. This study aims to obtain the optimum amount ratio required to obtain a homogeneous reaction conditions resulting in the most transesterification product. Amount ratio of acetone to methanol was varied over 1:4, 1:2, and 1:1.

Table 1. Yield of transesterification product of waste cooking oil with varied amount ratio of acetone to methanol

No	Weight of WCO (g)	Amount ratio of acetone to methanol	Weight of transesterification products (g)	Yield of transesterification products (%)	Density (g/mL)
1	10	1:4	6.183	61.83	0.87
2	10	1:2	7.270	72.70	0.87
3	10	1:1	4.823	48.23	0.88

Table 1 shows that the addition of acetone as a co-solvent in the transesterification reaction causes a decrease in the yield of transesterification products. Most of products is obtained at a ratio of 1:2 (acetone: methanol). The addition of acetone with a ratio above 1: 2 causes a decrease in the yield of transesterification products.

3.1.2. Effect of acetone amount ratio to methanol to FAME yield

The yield of FAME was determined via GC-MS analysis. The GC-MS obtained data of FAME composition in biodiesel.

Table 2. Biodiesel composition and FAME yield in varied acetone amount ratio to methanol via GC-MS analysis

No.	**Retention time (minutes)**	**Chemical Constituents**	**Area Percentages (%)**		
			1:4	**1:2**	**1:1**
1	11.40	Methyl Laurate	0.06	0.05	0.05
2	13.86	Methyl Miristate	0.76	0.80	0.85
3	15.91	Methyl Palmitoleate	0.31	0.33	0.35
4	16.15	Methyl Palmitate	38.30	38.21	38.32
5	17.14	Methyl Heptadecanoate	0.07	0.08	0.08
6	17.97	Methyl Oleate	54.92	54.6	54.47
7	18.14	Methyl Stearate	4.64	4.79	4.83
8	19.62	Methyl Hexadecadienoate	0.08	0.08	0.09
9	19.79	Methyl-9,10-Methylene Hexadecanoate	0.31	0.31	0.30
10	19.99	Methyl Aracate	0.38	0.41	0.40
11	21.69	Methyl Docosanoate	0.05	0.06	0.07
12	23.38	Methyl Lignoserate	0.05	0.05	0.06
FAME Yield			**99.93**	**99.77**	**99.87**

Table 2 shows that the lowest ratio (1: 4) gives FAME with highest yield of 99.93% followed by the ratio of 1: 2 of 99.77% and the ratio of 1: 1 of 99.87%. This suggests that the addition of small amount of acetone as co-solvent can increase the amount of methyl esters so that the transesterification reaction process is more efficient. Based on the composition of the transesterification result, the highest peak value of chromatogram is methyl palmitate, methyl oleate and methyl stearate compounds.

The greater amount of acetone in the mixture does not give a significant percentage difference in yield of methyl esters, but causes a decrease in the yield of methyl ester on the weight of the waste cooking oil. Increasing the amount of acetone causes the separation of methyl ester products to glycerol and the residual triglycerides becomes more difficult to be marked by the length of time required for the emergence of two layers of organic layer containing methyl esters and glycerol layer.

4. Conclusion

The amount of acetone ratio as a co-solvent to methanol in the transesterification reaction gives effect to the yield of biodiesel and FAME yield. The highest FAME yield was produced at room temperature conditions for 15 minutes with a ratio of 1: 4

References

[1] Wang Y, Ou S, Liu P, Xue F, Tang S. Comparison of two different processes to synthesize biodiesel by waste cooking oil. J Mol Catal A: Chem 252 (2006):107–12.

[2] Zhang S, Zu Y-G, Fu Y-J, Luo M, Zhang D-Y, Efferth T. Rapid microwave-assited transeterification of yellow horn oil to biodiesel using a heterogeneuous solid catalyst. Bioresour Technol 101 (2010) 931–6.

[3] Mahajan, S., Konar, S.Kdan Boocock, D.G.B., 2006, StandardBiodiesel from Soybean Oil by a Single Chemical Reaction,J.Am.OilChem.Soc.83:641– 645.

[4] Thanh, L.T., Okitsu, K., Sadanaga, Y., Takenaka, N., Maeda, Y., and Bandow, H., A New Co-Solvent Method for The Green Production of Biodiesel Fuel –Optimization And Practical Application, Fuel 103 (2013) 742–748

[5] Julianto, T.S. and Suratmi, The Effect Of Concentrations And Volumes of Methanol in Reducing Free Fatty Acid Content of Used Cooking Oil as Biodiesel Feedstock, International Conference on Chemistry, Chemical Process and Engineering (IC3PE) 2017, AIP Conf. Proc. 1823, 020130-1–020130-5

IOP Conf. Series: Materials Science and Engineering **349** (2018) 012064 doi:10.1088/1757-899X/349/1/012064

Synthesis, characterization, and application of novel Zn(II)-ionic imprinted polymer for preconcentration of Zn(II) ions from aqueous solution

T Wirawan[1,2*], G Supriyanto[2,3], A Soegianto[4]

[1]Department of Chemistry, Faculty of Mathematics and Natural Sciences, University of Mulawarman, Samarinda, Indonesia
[2]Department of Chemistry, Faculty of Science and Technology, University of Airlangga, Surabaya, Indonesia
[3]Laboratory of Sensor and Biosensor, Institute of Tropical Deseases, University of Airlangga, Surabaya, Indonesia
[4]Department of Biologyy, Faculty of Science and Technology, University of Airlangga, Surabaya, Indonesia

[*]Corresponding author: twirawan@yahoo.co.id

Abstract. Preparation of a new Zn^{2+} ion-imprinted polymer (Zn-IIP) is presented in this report. The Zn-IIP are prepared by precipitation polymerization using 8-hydroxyquinoline (8HQ) as a ligand, methacrylic acid (MAA) as functional monomer, and ethylene glycol dimethacrylate (EGDMA) as a cross-linker has been prepared. The benzoyl peroxide and ethanol/acetonitrile (2:1) mixture were used as initiator and porogen, respectively. Precipitation polymerization was carried out by heating in a water bath at 60°C for 8 hours. After polymerization, cavities in the polymer particles corresponding to the Zn^{2+} ions were created by leaching the polymer with 2 mol L^{-1} HNO_3. The polymer was washed with aquabidest and dried in an oven at the temperature of 60°C for 24 hours. The Zn-IIP was characterized by Fourier Transform Infrared Spectrophotometry (FT-IR) and Scanning Electron Microscopy (SEM). The synthesized Zn-IIP was used as a new adsorbent for solid phase extraction (SPE) of Zn(II) prior to Flame Atomic Absorption Spectrometry (FAAS) determination. The experimental parameters for SPE, such as pH of the sample, loading rate, and elution volume, have been optimized. The effect of pH of the sample on the extraction of analyte was studied in batch mode. The effects of loading rate and elution volume on the extraction of analyte were studied in dynamic mode by loading of the sample through IIP-SPE cartridge containing 100 mg of the synthesized Zn-IIP. The imprinted polymer (Zn-IIP) have bands at 3433.06 cm^{-1} (O-H), 1508.23 cm^{-1} (C=N aromatics), 1284.5 cm^{-1} (C-N aromatics), 1056.9 cm^{-1} (C-O phenol), 1724.24 cm^{-1} (C=O), and 1639.38 cm^{-1} (conjugated C=O with C=C). The Scanning Electron Microscopy (SEM) images of IIP and IIP show that the IIP is seen to have more cavities than NIP. The optimum pH for quantitative Zn(II) retention was 5.5, and the elution was completed with 2 mL of 1.0 mol L^{-1} nitric acid. The optimum loading rate was 0.5 mL min^{-1}. The recovery of Zn(II) from solution samples after its SPE extraction on IIP with 50-fold theoretical preconcentration was 94.60-104.50%. The LOD and LOQ with 50-fold theoretical preconcentration obtained were 0.0073 mg L^{-1} and 0.0244 mg L^{-1}, respectively.

1. Introduction

Zinc is one of the most important essential trace elements in human nutrition, the second most abundant transition metal in the body, and highest concentrations occur in the brain [1]. Zinc deficiency in the human body causes several disorders including growth retardation, diarrhea, eye and skin lesions, immunity depression, and malfunction of wound healing. On the other hand, excessive zinc intake also causes various acute and chronic adverse effects [2,3]. Zinc is used extensively in several industries, such as mining, metal coating, and battery production. Zinc is also found in industrial paints, ceramics, wood, fabrics, drugs, sun blocks, and deodorants [4].

Several analytical techniques including spectrophotometry [5], spectrofluorometry [6], voltammetry [7], optical sensor [8], electrochemical sensors [9], flame atomic absorption spectrometry [10], electrothermal atomic absorption spectrometry [11], and inductively coupled plasma optical emission spectrometry [12] have been used for the direct determination of zinc in different matrixes. Most of these methods suffer from chemical interferences, poor limits of detection, competition effects, disturbance of solution equilibrium, long equilibrium time, or poor preconcentration and poor selectivity [13]. In an effort to improve sensitivity, several methods such as preconcentration techniques have been developed. Several procedures, such as liquid–liquid extraction [14], cloud point extraction [15,16], co-precipitation [17,18], and solid phase extraction, have been proposed and applied for the enrichment and separation of zinc.

In recent years, SPE has been successfully used for the separation and sensitive determination of metal ions, mainly in water samples. Solid phase extraction has proven to be a fast and inexpensive technique for the determination of trace metals with a use of low sample volumes. High preconcentration factor and low sample volume requirements were major advantages of the technique [19]. Solid phase extraction (SPE) is a sensitive, inexpensive, and fast sample preparation process that involves partitioning between a liquid phase. The analytes were separated from the solution by sorption on the solid sorbent according to their physical and chemical properties [20].

The solid phase extraction method is an alternative to replace solvent extraction. Solid phase extraction is a chromatographic sample preparation technique that is used for the separation, purification, and preconcentration of chemical compounds from solution [21]. In SPE, the sample is detained or adsorbed and later eluted adsorbent with elution in one stage or gradient. There are several advantages compared to solvent extraction, including reducing the use of hazardous solvents, requires a shorter time, do automation [22], facilitate storage and transport [23], the component desired can be taken out by using a selective solvent [24], and can be used for the separation and purification [25]. One of the most promising groups of new sorbents is an ionic-imprinted polymer (IIP). The ion-imprinting idea is based on the formation of a complex of given metal ion (serving as a template) with appropriate ligand and its subsequent immobilization in a cross-linked polymeric matrix. Upon metal ion extraction from the polymetric body spatially pre arranged binding sites are left in the sorbent structure. Therefore, due to their predetermined selectivity thus obtained synthetic materials are able to recognize and preferentially bind the metal ion which was used as a template in the polymerization process which offers new possibilities in separation science [26].

Recently, IIP has been growing rapidly as an adsorbent for SPE because it has the advantage on the enrichment factor and high selectivity. Applications of IIP as adsorbent in the SPE has been widely reported in the extraction of arsenic [27], iron [28,29], cadmium [30,31], mercury [32], copper [33], lead [34,35], and zinc [36-38].

In the preparation of IIP, suitable ligand groups were brought into contact with metal ions to form a complex that was then covered with a crosslinked polymer. After removal of the template (metal ion), the uptake of metal ions is strongly enhanced in many cases and also more selective. The selectivity of IIP mainly because of the following factors [39]: i) the specificity of interaction of ligand with the metal ion, ii) the coordination geometry and the coordination number of metal ion, iii) the charge of the metal ion, and iv) to a large extent on the size of the metal ion.

In this paper, we report the synthesis, characterization, and application of Zn-IIP for zinc preconcentration by SPE. Zinc concentration was analyzed using Flame Atomic Absorption Spectrophotometer (FAAS). The IIP was synthesized using 8-hydroxyquinoline (8HQ) as complexing ligands, methacrylic acid (MAA) as the functional monomer and ethylene glycol dimethacrylate (EGDMA) as agents of cross-linking. Characterization of the IIP was investigated using Fourier Transform Infrared Spectrophotometry (FTIR) and Scanning Electron Microscopy (SEM). The experiment parameter such as pH of the solution, flow rate, volume of HNO_3 for elution was investigated. This method was applied for the enrichment of trace amount of zinc(II) ion from aqueous solutions..

2. Experiment section

2.1. Materials

The materials used were: $ZnSO_4.7H_2O$ (Merck, Germany), HNO_3 (Merck, Germany), 8-hydroxyquinoline (8HQ) (Sigma Aldrich, US), methacrylic acid (MAA) (Sigma Aldrich, US), ethylene glycol dimethacrylate (EGDMA) (Sigma Aldrich, US), benzoyl peroxide (Merck, Germany), ethanol (Merck, Germany), and acetonitrile (Merck, Germany).

2.2. Instrumentation

The characteristic of functional groups among adsorbent was confirmed by Fourier Transform Infrared Spectrophotometer (Perkin Elmer Frountier-89485), The surface morphological image of adsorbent was obtained by Scanning Electron Microscopy (FEL, Inspect-S50). The concentration of zinc was analyzed using an atomic absorption spectrophotometer (Perkin Elmer, AA 700). A pH-meter (Mettler Toledo) was used for the pH measurements.

2.3. Procedure

2.3.1. Preparation of IIP

The ionic imprinted polymer was synthesized following the method proposed by Barciela-Alonso *et al.* [40]. Therefore, 0.0287 g (0.1 mmol) $ZnSO_4.7H_2O$ and 0.0363 g (0.25 mmol) 8-hydroxyquinoline were mixed in 60 mL of porogen (2:1 ethanol-acetonitrile). This solution was stirred for 30 min with a magnetic stirrer. Finally, 0.34 mL (4 mmol) of methacrylic acid (monomer), 0.0485 g (0.2 mmol) of benzoyl peroxide (initiator), and 3.77 mL (20 mmol) of ethylene glycol dimethacrylate (cross-linker) were added, the glass tube was purged with nitrogen for 5 min and immediately sealed. Precipitation polymerization was carried out by heating in a water bath at 60°C for 8 hours while stirring with a magnetic stirrer. The polymer obtained was filtered, washed with ethanol, and oven-dried for 2 hours at 60°C. The imprinted zinc ions were leached from the polymer by using 100 mL of 1 mol L^{-1} HNO_3 solution in magnetic stirrer for 30 min. The adsorbent was washed with distilled water, filtered, and dried at 100°C for 24 hours. The same procedure, except the presence of $ZnSO_4.7H_2O$ and leaching, was applied for preparing non-imprinted polymers (NIP). The polymer was prepared by the same procedure, except the presence of $ZnSO_4.7H_2O$, 8-hydroxyquinoline, and leaching. Ionic imprinted polymer and non-ionic imprinted polymer were characterized by FTIR and SEM.

2.3.2. Effect of pH

The effect of pH solution was tested by shaking 0.01 g IIP and 20 mL of the Zn(II) solution (4.5 mg L^{-1}) at various pH for 10 min. The pH of the sample solutions was varied from pH 4 to 8. The IIP was separated by filtration and concentration of Zn(II) in the filtrate was determined by FAAS.

2.3.3. Effect of eluent volume

A cartridge with an i.d. of 5 mm was filled with 0.1 g of dry IIP. 50 mL of 2.5 mg L^{-1} Zn(II) solution, adjusted to pH 5.5, was passed through the cartridge. Then the solution of 1 mol L^{-1} HNO_3 was passed to desorb Zn(II) from the cartridge. The concentration of Zn(II) was determined by FAAS. The volume of HNO_3 was varied from 0.5 to 5.0 mL.

2.3.4. Effect of flow rate

A cartridge with an i.d. of 5 mm was filled with 0.1 g of IIP. 50 mL of 2.5 mg L^{-1} Zn(II) solution, adjusted to pH 5.5, was passed through the column at various of flow rate. The flow rate was varied from 0.2 to 2.2 mL min^{-1}. Then, Zn(II) was eluted from the cartridge with 2 mL of 1 mol L^{-1} HNO_3 and measured by FAAS.

2.3.5. Solid phase extraction

A cartridge with an i.d. of 5 mm was filled with 0.1 g of dry ionic-imprinted polymers. The sorbents were conditioned by passing 3 mL of buffer pH 5.5. For the loading of the analyte, Zn(II) standard solution was adjusted to pH 5.5 and passed through the cartridge at a flow rate of 0.5 mL min^{-1}. Then, Zn(II) was eluted from the cartridge with 2 mL of 1 mol L^{-1} HNO_3 measured by FAAS.

3. Result and discusion

The new ionic imprinting material was prepared by using 8-hydroxyquinoline (8HQ) as complexing ligands, methacrylic acid (MAA) as the monomer, and ethylene glycol dimethacrylate (EGDMA) as agents of cross-linking. The polymer was synthesized by precipitation polymerization technique. The synthesis of IIP was carried out in three step (Figure 1): (1) The complexation Zn(II) with 8HQ, (2) the polymerization of monomer in the presence of the template, and (3) the removal of the template after copolymerization from the polymer.

3.1. Characterization of IIP

The FTIR spectra of complexing ligand (8HQ), monomer (MAA), crosslinker (EGDMA), polymer, and IIP are presented in Figure 2. The spectra of 8HQ (Figure 2e) has a characteristic peaks at 3433.06 cm^{-1} (O-H), 1508.23 cm^{-1} (C=N aromatics), 1284.5 cm^{-1} (C-N aromatics) and 1056.9 cm^{-1} (C-O phenol). The spectra of MAA (Figure 2d) has a characteristic peaks of 1527.52 cm^{-1} (C=C), 1473.51 cm^{-1} (C=C aromatic), 1203.5 cm^{-1} (C-O), and 1693.38 cm^{-1} (conjugated C=O with C=C). The spectra of EGDMA (Figure 2c) has characteristic peaks at 1245.93 cm^{-1} (C-O), 1724.24 cm^{-1} (C=O), and 1639.38 cm^{-1} (conjugated C=O with C=C). The polymer formed by the addition of the C=C double bond into a single C-C bond, shown in the FTIR spectra of polymers (Figure 2b) that are absence peak at 1527.52 cm^{-1} (C=C) when compared to spectra of MAA. The IIP formed indicated by the appearance of a peak at 1492.8 cm^{-1} (C=C aromatic) of IIP spectra (Figure 2a) compared with the spectra of polymers.

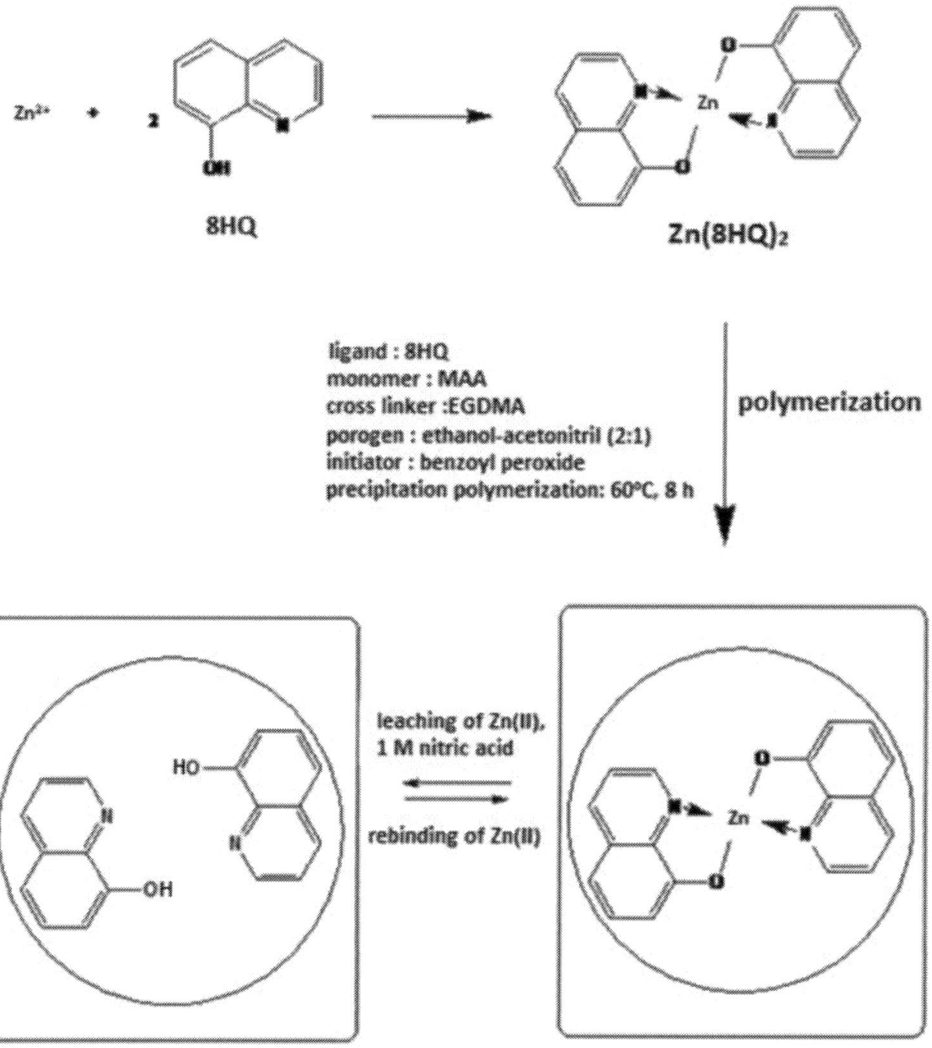

Figure 1. Illustration of IIP formation

Scanning electron microscopy (SEM) images are presented in Figure 3. There are obvious differences between the scanning micrograph of NIP and IIP. The textural characteristics were closely examined at 20k x magnification. The particles of both NIP and IIP were present a spherical shape. It should be spherical shape when IIP was synthesized by precipitation polymerization. The IIP is seen to have more cavities than NIP, this is because Zn in IIP has been eluted.

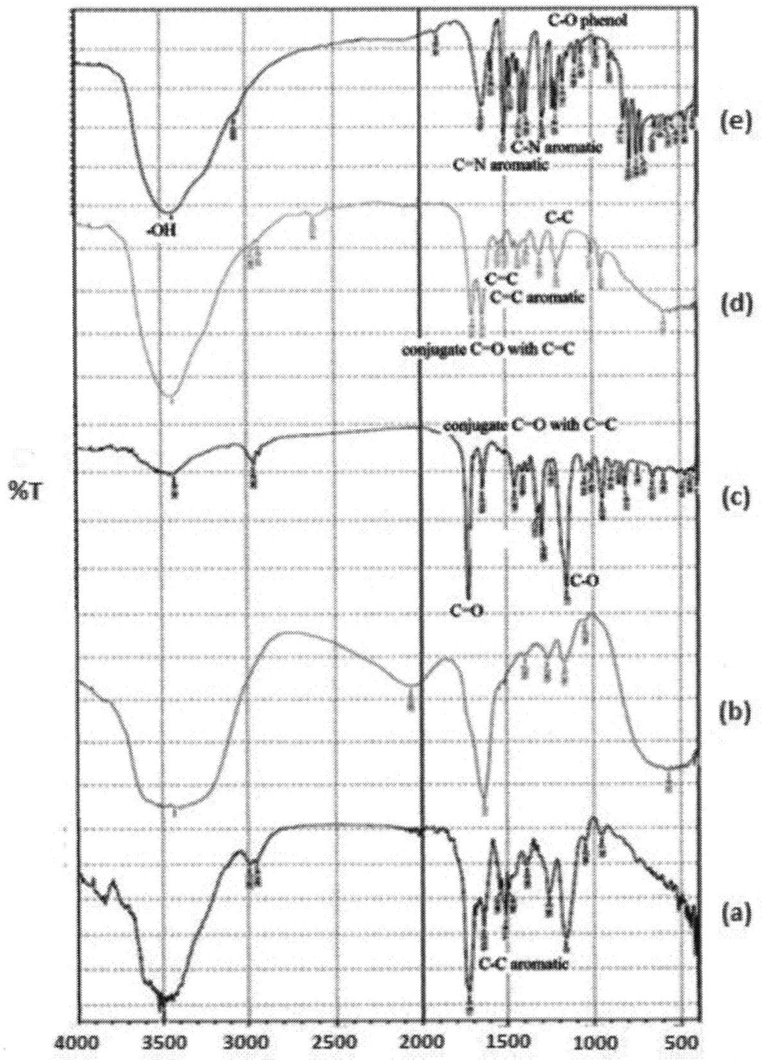

Figure 2. FTIR spectra of (a) IIP, (b) polymer, (c) EGDMA, (d) MAA, and (e) 8HQ

a b

Figure 3. SEM images of (a) NIP and (b) IIP

3.2. Influence of initial pH on Zn(II) adsorption

Among all other parameters, pH of solution has been found to be the most important one. It not only influences the speciation of metal ions but also the ionization of surface functional groups of adsorbent [41]. The influence of the initial pH on adsorption of Zn(II) on the IIP was studied to determine the optimum pH. The adsorption of Zn(II) as a function of the pH is shown in Figure 4. The adsorption of Zn(II) increased greatly between pH 4.5 to 5.5 and then gradually decreased. In a highly acidic environment, several functional groups become protonated and act as positively charged species, resulting in reduced attraction between the metals and the minerals [4,42]. With increasing pH, electrostatic repulsion decreases due to a reduction of positive charge density on the sorption sites thus resulting in an enhancement of metal adsorption [43]. At higher pH values OH^- ions compete for Zn(II) with the active sites on the surface of the adsorbents [44]. However, at an alkaline environment, the solubility of metals decreases allowing precipitation, which may complicate the sorption process. Zn(II) ions react with OH^- ligand to precipitate to $Zn(OH)_2$.

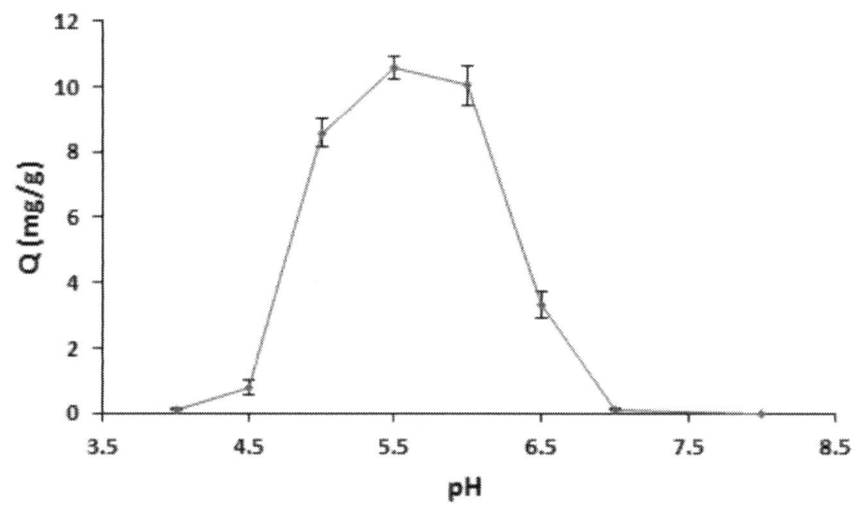

Figure 4. Influence of pH on the adsorption of Zn(II) on IIP

3.3. Influence of eluent volume

The influence of the eluent volume on adsorption of Zn(II) on the IIP was studied. After sample loading, the adsorbed Zn(II) were eluted by passing 1 mol L^{-1} HNO_3 at different volume (from 0.5 to 5 mL) and the result was presented in Figure 5. The eluted of Zn(II) was increased greatly from 0.5 to 1.5 mL and then increased slightly from 1.5 to 2 mL. The volume above of 2 mL was constant and volume 2 mL was chosen as optimum volume. This result was similar to the result obtained by Ebrahimzadeh *et al.* [45], increasing in volume and concentration of eluent volume caused higher recovery. The eluting solution must be correctly chosen to ensure stronger affinity of the trace element for the solvent, to ensure disruption of its interaction with the sorbent [19].

3.4. Influence flow rate

In set sorption time applying high sample flow rate may seem favorable in terms of obtaining high enrichment factor due to higher sample volume loaded onto the column. However, sorption effectiveness is limited by the sorption kinetics, may be deteriorated [26]. The sample flow-rate through the sorbent should be low enough to enable efficient retention of the analytes, and high enough to avoid excessive duration [19]. Different experiments were performed in order to find the optimum flow rates for the Zn(II) standard solution loading. The solution of Zn(II) standard was passed at a different flow rate (from 0.2 to 2.25 mL min^{-1}). As shown in Figure 6, concentration of Zn(II) decrease for flow rates larger than 0.5 mL min^{-1}. In order to decrease the time needed for analysis, the optimum flow rate was chosen as 0.5 mL min^{-1}.

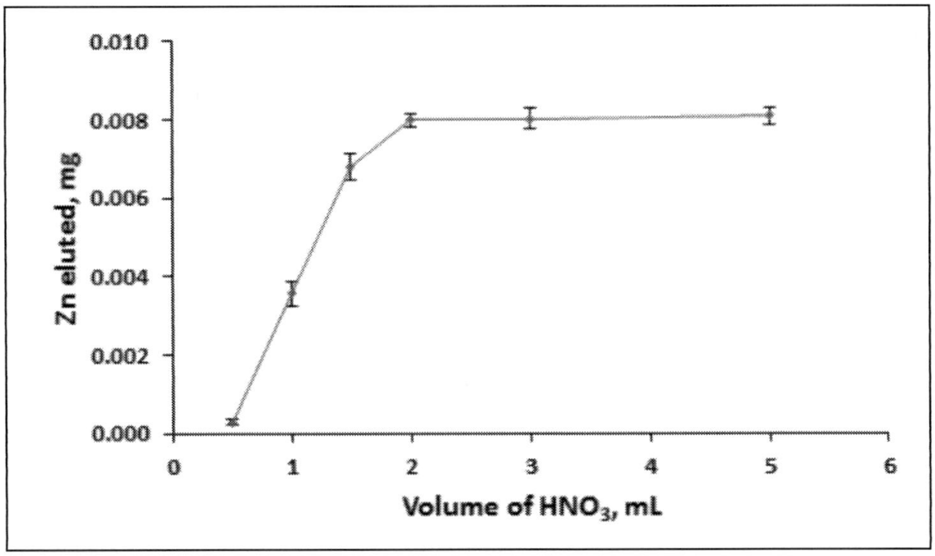

Figure 5. Influence of eluent volume on the elution of Zn(II) on IIP

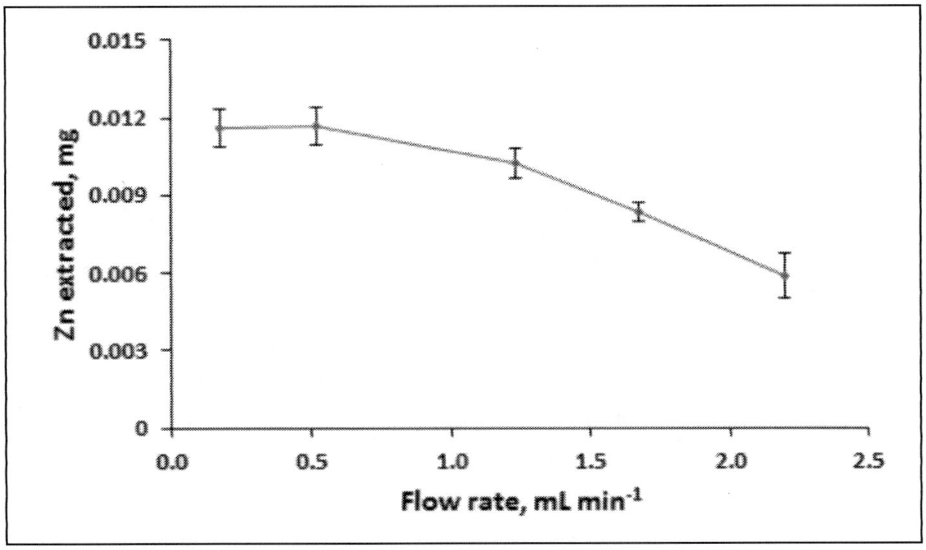

Figure 6. Influence of eluent flow rate on the extraction of Zn(II) on IIP.

3.5. Analytical characterization

The limit of detection (LOD) and limit of quantification (LOQ) were calculated according to LOD=(3SD)/m and LOQ=(10SD)/m, where SD is the standard deviation and m is the slope of the calibration graph (Figure 7 and Figure 8). The LOD and LOQ obtained were 0.3157 mg L^{-1} and 1.0523 mg L^{-1} for without preconcentration, and 0.0073 mg L^{-1} and 0.024 mg L^{-1} for with 50-fold theoretical preconcentration.

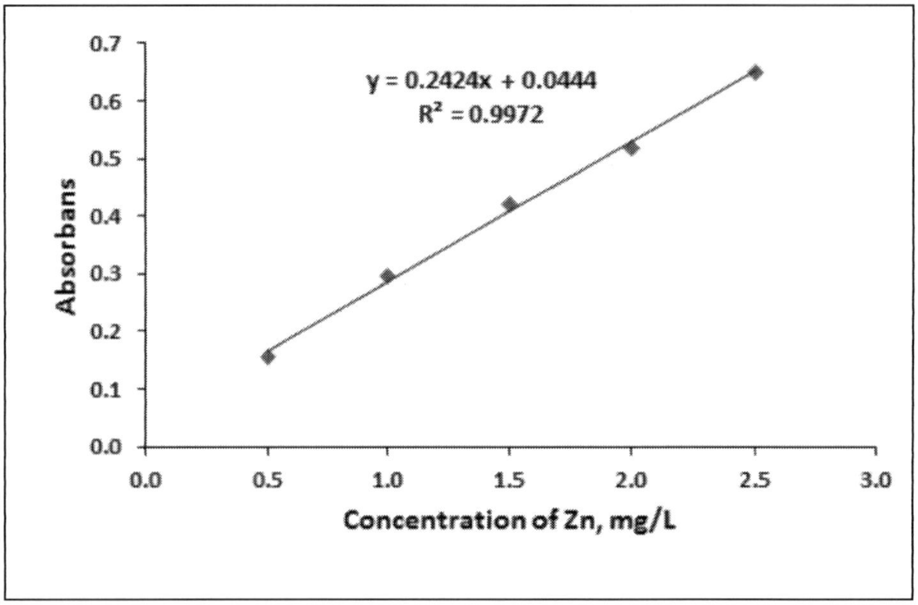

Figure 7. Calibration graph of Zn solution without preconcentration

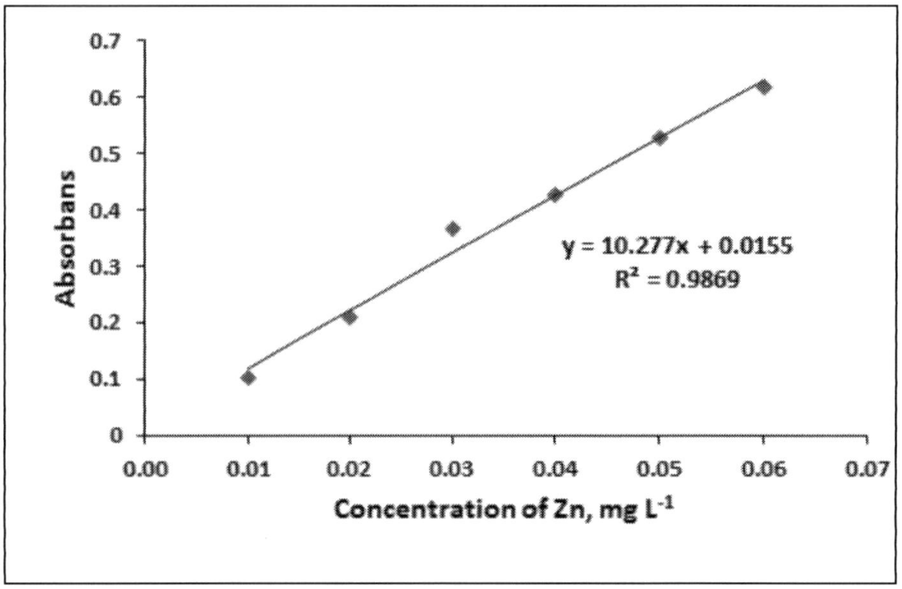

Figure 8. Calibration graph of Zn solution with 50-fold theoretical preconcentration.

The synthesized IIP was then applied as an adsorbent for the analysis of zinc(II) by SPE-AAS (Table 1). Analytical recovery was assessed by analyzing five of zinc standard solutions 0.01-0.05 mg L^{-1} and preconcentration by SPE. The analytical recovery obtained was 94.60-104.50%.

Tabel 1. Recovery of Zn(II) from solution samples after its SPE extraction on IIP (sample: pH: 5.5; flow rate: 0.5 mL min^{-1}; elution: 2 mL of 1 mol L^{-1} HNO_3)

No.	Added Zn (mg L^{-1})	Found Zn (mg L^{-1})	Recovery (%)
1	0.01	0.00950	95.00
2	0.02	0.01964	98.20
3	0.03	0.02838	94.60
4	0.04	0.04180	104.50
5	0.05	0.05012	100.24
Average			98.51
SD			4.0798

The comparison of pH, load flow rate, and detection limit in the present work with those reported in the literature are given in Table 2. Generally, the detection limit obtained by this work is comparable to those of reported or better than them.

Table 2. Comparison of the proposed method with some methods reported in literature.

Sample	Adsorben	Detection	pH load flow rate eluen	LOD	References
Cr, Cu, Ni, Pb, Zn	Silica gel-bound amines phase modified with p-dimethylamino-benzaldehyde (p-DMABD)	ICP-OES	4 1.5 mL min^{-1} 2.0 mL of 0.5 M HCl	6.5 µg L^{-1}	[46]
Cu, Ni, Zn	Multiwalled carbon nanotubes (MWCNTs) impregnated with di-(2-ethyl hexyl phosphoric acid) (D_2EHPA) and tri-noctyl phosphine oxide (TOPO)	FAAS	5 2 mL min^{-1} 10 mL of 2 1 mol L^{-1} HNO_3	50 µg L^{-1}	[47]
Fe, Zn	Multiwalled carbon nanotubes (MWCNTs)	HR-CS GFAAS	9 1 mL.min^{-1} 1 mL of 1 mol L^{-1} HNO_3	0.5 µg L^{-1}	[48]
Ag, Cd, Cu, Zn	magnetic nanoparticles (MNPs) coated by 3-(trimethoxysilyl)-1-propantiol and modified with 2-amino-5-mercapto-1,3,4-thiadiazole	ICP-OES	6 - 1 mL of 1 mol L^{-1} HCl and 4% (w/v) thiourea	0.11 µg L^{-1}	[49]
Zn	IIP was sinthesized by using morin, 4-vinyl-pyridine (VP), ethylene-glycol dimethacrylate (EDMA) and 2,20-	ICP-OES	4 - 10 mL of 1:1 (v/v) mixture of CH_2Cl_2–DMSO	2.9 µg.L^{-1}	[50]

Sample	Adsorben	Detection	pH load flow rate eluen	LOD	References
	azobisisobutyronitrile (AIBN).				
Zn	IIP was prepared via precipitation polymerization using 8-hydroxyquinoline (8HQ) as a ligand, methacrylic acid (MAA) as funtional monomer, and ethyleneglycol dimethacrylate (EGDMA) as a cross-linker	FAAS	5.5 0.5 mL min^{-1} 2 mL of 1 mol L^{-1} HNO_3	7.3 μg.L^{-1}	Present method

4. Conclusion

Adsorbent IIP-ionic imprinted polymer (Zn-IIP) can be synthesized through a precipitation polymerization process using 8-hydroxyquinoline as a complexing ligand, methacrylic acid as a monomer and ethylene glycol dimethacrylate as a cross-linker. Ethanol-acetonitrile (2:1) and benzoyl peroxide were used as a porogen and an initiator, respectively. IIP adsorbent can be used as an adsorbent for the analysis of zinc(II) by SPE-FAAS. The optimum conditions for the SPE process were at pH 5.5 and the volume of 1 mol L^{-1} HNO_3 solution required to elute was 2 mL. The accuracy of the method of SPE-FAAS was 94.60-104.50%. The LOD and LOQ with 50-fold theoretical preconcentration obtained were 0.0073 mg L^{-1} and 0.0244 mg L^{-1}, respectively.

References

[1] Que EL, Domaille DW and Chang CJ 2008 Metals in Neurobiology: Probing Their Chemistry and Biology with Molecular Imaging *Chem. Rev.* **108** 1517–49

[2] Scherz H and Kirchhoff E 2006 Trace elements in foods: Zinc contents of raw foods: comparison of data originating from different geographical regions of the world *Journal of Food Composition and Analysis* **19** 420–33

[3] Zhao J, Hana B, Zhanga Y and Wang D 2007 Synthesis of Zn(II) ion-imprinted solid-phase extraction material and its analytical application *Anal. Chim. Acta.* **603** 87–92

[4] Malamis S and Katsou E 2013 A review on zinc and nickel adsorption on natural and modified zeolite, bentonite and vermiculite: Examination of process parameters kinetics and isotherms *J. Hazard. Mater.* **252**, 428–61

[5] Sabel CE, Neureuther JM and Siemann S 2010 A spectrophotometric method for the determination of zinc, copper, and cobalt ions in metalloproteins using Zincon *Analytical Biochemistry* **397** 218–26

[6] Prat MD, Guiteras J, Campano R and Beltran JE 1991 Spectroflourimetry method for simultaneous determination of zinc and cadmium, *Journal of Flourescence* **1(4)** 267-72

[7] Pinto L and Lemos SG 2013 Multivariate optimization of the voltammetric determination of Cd, Cu, Pb and Zn at bismuth film. Application to analysis of biodiesel, *Microchemical Journal* **110** 417–24

[8] Abbasitabar F, Zare-Shahabadi V, Shamsipur M and Akhon M 2011 Development of an optical sensor for determination of zinc by application of PC-ANN *Sensors and Actuators B* **156** 181–6

[9] Singh P, Singh AK and Jain AK 2011 Electrochemical sensors for the determination of Zn^{2+} ions

based on pendant armed macrocyclic ligand *Electrochimica Acta* **56** 5386–95
[10] Ramakrisnha TV, Robinson JW and West PW 1967 Determination of copper, cadmium, and zinc by atomic absorption spectroscopy *Anal. Chim. Acta* **37**, 20-6
[11] Török P and Žemberyová M 2011 A study of the direct determination of Cd, Cr, Cu, Pb and Zn in certified reference, materials of soils by solid sampling electrothermal atomic absorption spectrometry *Spectrochimica Acta Part B* **66** 93–7
[12] Souza AL, Lemos SG and Oliveira PV 2011 A method for Ca, Fe, Ga, Na, Si and Zn determination in alumina by inductively coupled plasma optical emission spectrometry after aluminum precipitation *Spectrochimica Acta Part B* **66** 383–8
[13] Fan HT, Li J, Li ZC and Sun T 2012 An ion-imprinted amoni-functionalized silika gel sorbent prepared by hydrothermal assisted surface imprinting technique for selective removal of cadmium(II) from aqueous solution *Applied Surface Science* **258** 3815-22
[14] Silvestre CIC, Santos JLM, Lima JLFC and Zagatto EAG 2009 Liquid–liquid extraction in flow analysis: A critical review *Anal. Chim. Acta* **652** 54–65
[15] Tabrizi AB 2007 Cloud point extraction and spectrofluorimetric determination of aluminium and zinc in foodstuffs and water samples *Food Chemistry* **100** 1698–1703
[16] Silva EL, Roldan PS and Giné MF 2009 Simultaneous preconcentration of copper, zinc, cadmium, and nickel in water samples by cloud point extraction using 4-(2-pyridylazo)-resorcinol and their determination by inductively coupled plasma optic emission spectrometry *J. Hazard. Mat.* **171** 1133–8
[17] Uluozlu OD, Tuzen M, Mendil D and Soylak M 2010 Coprecipitation of trace elements with Ni^{2+}/2-Nitroso-1-naphthol-4-sulfonic acid and their determination by flame atomic absorption spectrometry *J. Hazard. Mat.* **176** 1032–7
[18] Feist B and Mikula B 2014 Preconcentration of some metal ions with lanthanum-8-hydroxyquinoline co-precipitation system *Food Chemistry* **147** 225–9
[19] Camel V 2003 Solid phase extraction of trace elements *Spectrochimica Acta Part B* **58** 1177–233
[20] Ca´mara-Martos F and Moreno-Rojas R 2016 Zinc: Properties and Determination *Encyclopedia of Food and Health* Elsevier
[21] Wells MJM, Rener DD and Wells-Knecty C W 1994 Development and optimization of a solid phase extraction scheme for determination of the pesticide metribuzin, antrazine, metalochlor, and esfenvalerate in agricultural run off water *J. Chromatogr, A* **659** 337-48
[22] Johnson WE, Fendinger NJ and Plimmer JR 1991 Solid phase extraction of pesticide from water: posible interferences from dissolved organic material *Anal. Chem* **63** 1510-3
[23] Hennion MC and Coquart V 1993 Comparation of reversed-phase extraction sorbents for on-line trace enrichment of polar organic compounds in environmental aqueous samples *J. Chromatogr. A.* **642** 211-24
[24] Manes G F, Molto JC and Pico Y 1993 Solid phase extraction in multi residu pesticide analysis of water *J. Chromatogr. A.* **642** 135-61
[25] Snyder LR, Kirkland JJ and Glajch J L 1997 *Practical HPLC Method Development* Second edition John Wiley & Sons, Inc. New York
[26] Gawin M, Konefał J, Trzewik B, Walas S, Tobiasz A, Mrowiec H and Witek E 2010 Preparation of a new Cd(II)-imprinted polymer and its application to determination of cadmium(II) via flow-injection-flame atomic absorption spectrometry *Talanta* **80** 1305-10
[27] Tsoi YK, Ho YM and Leung KSY 2012 Selective recognition of arsenic by tailoring ion-imprinted polymer for ICP-MS quantification *Talanta* **89** 162-8
[28] Özkütük EB and Karabörk M 2007 Fe^{3+}-Imprinted polymeric systems *Hacettepe J. Biol. & Chem.* **35** 195-202
[29] Utku S, Yilmaz E, Türkmen D, Uzun L, Garipcan B, Say R and Denizli A 2008, Ion-imprinted thermosensitive polymers for Fe^{3+} removal from human plasma *Hacettepe J. Biol. & Chem.* **36** 291-304
[30] Buhani, Narsito, Nuryono and Kunarti ES 2010 Production of metal ion imprinted polymer from

mercapto–silica through sol–gel process as selective adsorbent of cadmium *Desalination* **251** 83-9
[31] Singh DK and Mishra S 2009 Synthesis, characterization and removal of Cd(II) using Cd(II)-ion imprinted polymer *J. Hazard. Mater.* **164** 1547–51
[32] Singh DK and Mishra S 2010, Synthesis and characterization of Hg(II)-Ion imprinted polymer; Kinetic and isotherm studies *Desalination* **257** 177-83
[33] Dakova I, Karadjova I, Ivanov I, Georgieva V, Evtimora B and Georgiev G 2007 Solid phase selective separation and preconcentration of Cu(II) by Cu(II)-imprinted polymethacrylic microbeads *Anal. Chim. Acta* **584** 196-203
[34] Zhang M, Zhang Z, Liu Y, Yang X, Luo L, Chen J and Yao S 2011 Preparation of core-shell magnetic ion-imprinted polymer for selective extraction of Pb(II) from environmental samples *Chem. Eng. J.* **178** 443-50
[35] Liu Y, Liu Z, Gao J, Dai J, Han J, Wang Y, Xie J and Yan Y 2011 Selective adsorption behavior of Pb(II) by mesoporous silica SBA-15-supported Pb(II)-imprinted polymer based on surface molecularly imprinting technique *J. Hazard. Mater.* **186** 197–205
[36] Shamsipur M and Besharati-Seidani A 2011 Synthesis of a novel nanostructured ion-imprinted polymer for very fastand highly selective recognition of copper(II) ions in aqueous media *React. Funct. Polym.* **71** 131–9
[37] Kim M, Jiang Y and Kim D 2013 Zn^{2+}-Imprinted porous polymer beads; synthesis, structure, and selective adsorption behavior for template ion *React. Funct. Polym.* **73** 821-7
[38] Behbahani M, Salarian M, Bagheri A, Tabani H, Omidi F and Fakhar A 2014 Synthesis, characterization and analytical application of Zn(II)-imprinted polymer as an efficient solid-phase extraction technique for trace determination of zinc ions in food samples *Journal of Food Composition and Analysis* **34** 81-89
[39] Rao TP, Daniel S and Gladis JM 2004 Tailored materials for preconcentration or separation of metals by ion-imprinted polymers for solid-phase extraction (IIP-SPE) *Trends in Analytical Chemistry* **23(1)** 28-35
[40] Barciela-Alonso MC, Plata-García V, Rouco-López A, Moreda-Piñeiro A and Bermejo-Barrera P 2014 Ionic imprinted polymer based solid phase extraction for cadmium and lead pre-concentration/determination in seafood *Microcim. J.* **114** 106-10
[41] Lee SH, Jung CH, Chung H, Lee MY and Yang TW 1998 Removal of heavy metals from aqueous solution by apple residue *Process Biochemistry* **33** 205-11
[42] Farooq U, Kozinski JA, Khan MA and Athar M 2010 Biosorption of heavy metal ions using wheat based biosorbents – A review of the recent literature, *Bioresource Technology* **101** 5043–53
[43] Jordao CP, Braganc R, Fernandes A, Ribeiro KL, Nascimento BS and Barros PW 2009 Zn(II) adsorption from synthetic solution and kaolin wastewater onto vermicompost *J. Hazard. Mat.* **162** 804–11
[44] Bhattacharya AK, Mandal SN and Das SK 2006 Adsorption of Zn(II) from aqueous solution by using different adsorbents *Chem. Eng. J.* **123** 43–51
[45] Ebrahimzadeh H, Behbahani M, Yamini Y, Adlnasab L and Asgharinezhad AK 2013 Optimization of Cu(II)-ion imprinted nanoparticles for trace monitoring of copper in water and fish samples using a Box–Behnken design *Reactive & Functional Polymers* **73** 23–9
[46] Cui Y, Chang X, Zhu X, Luo H, Hu Z, Zou X and He Q 2007 Chemically modified silica gel with p-dimethylaminobenzaldehyde for selective solid-phase extraction and preconcentration of Cr(III), Cu(II), Ni(II), Pb(II) and Zn(II) by ICP-OES *Microchemical Journal* **87** 20–6
[47] Vellaichamy S and Palanivelu K 2011 Preconcentration and separation of copper, nickel and zinc in aqueous samples by flame atomic absorption spectrometry after column solid-phase extraction onto MWCNTs impregnated with D2EHPA-TOPO mixture *J. Hazard. Mater.* **185** 1131–9
[48] Krawczyk M, Jeszka-Skowron M and Matusiewicz H 2014 Sequential multi-element determination of iron and zinc in water samples by high-resolution continuum source graphite

furnace atomic absorption spectrometry after column solid-phase extraction onto multiwalled carbon nanotubes *Microchemical Journal* **117** 138–43

[49] Mashhadizadeh MH and Karami Z 2011 Solid phase extraction of trace amounts of Ag, Cd, Cu, and Zn in environmental samples using magnetic nanoparticles coated by 3-(trimethoxysilyl)-1-propantiol and modified with 2-amino-5-mercapto-1,3,4-thiadiazole and their determination by ICP-OES *J. Hazard. Mater.* **190** 1023–29

[50] Khajeh M, Kaykhaii M, Mirmoghaddam M and Hashemi H 2009 Separation of zinc from aqueous samples using a molecular imprinting technique *Intern. J. Environ. Anal. Chem.* **89(13)** 981–992

Blend membrane of succinic acid-crosslinked chitosan grafted with heparin/PVA-PEG (polyvinyl alcohol-polyethylene glycol) and its characterization

V D A Sangkota[a], R A Lusiana[b*], Y Astuti[b*]

aPost graduate Department of Chemistry Faculty of Mathematics and Natural Sciences, Universitas Diponegoro, Semarang 50275, Indonesia
[b*]Department of Chemistry, Faculty of Mathematics and Natural Sciences, Universitas Diponegoro, Semarang 50275, Indonesia

Corresponding author: retno.lusiana@live.undip.ac.id

Abstract. Crosslinking and grafting reactions are required to modify the functional groups on chitosan to increase the number of its active groups. In this study, crosslinking reaction of succinic acid and grafting reaction of heparin on chitosan were conducted to produce a membrane as a candidate of a hemodialysis membrane. The mole ratio between chitosan and succinate acids was varied to obtain the best composition of modified materials. By blending all the material composition with PVA-PEG, the blend was transformed into a membrane. The resulted membrane was then characterized by various test methods such as tests of thickness, weight, water uptake, pH resistance, tensile strength and membrane hydrophilicity. The results showed that the best composition of the membrane reached in the addition of 0.011 gram of succinic acid proved by its highest mechanical strength compared to the other membranes.

Keywords: chitosan; membrane; cross-linking; grafting.

1. Introduction

Hemodialysis is an important process in medical treatment aiming to remove toxic substances from the blood of a patient with acute renal failure. The main component of hemodialysis equipment is semi-permeable membrane which can selectively transport permeates having low molecular weight, such as urea and creatinine from the blood [1]

The hemodialysis membrane should be compatible, porous, hydrophilic and has functional groups that are capable of capturing permeates through the formation of hydrogen bonds. Chitosan, a natural polymer, is developed into a raw material of membrane because of its ability to be dissolved in a dilute acid to form a thin-film membrane [2]. Furthermore, Hoven, et al., [3] stated that chitosan is biodegradable, inert, non-toxic and biocompatible. The content of –OH and -NH_2 groups of chitosan makes this material becomes reactive, which means it can be easily modified with the other groups to obtain a material suitable for the application needs. Lusiana, et al., [2] used three methods to modify chitosan, i.e. grafting, cross-linking and blending. These three modifications improved the strength of the chitosan membrane by increasing its number of functional groups and porosity, and reducing the negative charge on the membrane surface. Mitra et al., [4] also modified chitosan with succinic acid/collagen by cross-linking. The result showed that the –COOH group at both ends of the succinic acid increased the active side of the chitosan membrane.

To improve the mechanical strength of the membrane, it is blended with synthetic polymer like PVA (polyvinyl alcohol). PVA is hydrophilic, elastic, non-toxic, non-carcinogenic and biocompatible with human body [5] Polyethylene glycol (PEG) is also excellent for chitosan blend since it is a water-soluble polymer and less toxic. Zen, et al.[6] examined the modification of chitosan with PEG. The investigation showed that modification using PEG increased the porosity and the membrane mechanical properties. The modification of membrane with PVA/PEG blends is needed to increase the elasticity and porosity of the membrane so that it can be used in the hemodialysis process which takes approximately 4-6 hours.

To reduce red blood clots due to the incompatible membrane to the blood, heparin is added as a grafting agent. Heparin is anticoagulant in blood and it is non-toxic. Chitosan modification using heparin was performed by Badr, et al., [7]. They showed that the resulted membrane was selective against sodium and an increase in its compatibility to blood was observed. The purpose of this study is to investigate the results of the synthesized chitosan membrane in terms of its physical properties through various physical characterization tests.

2. Experimental method

2.1. Materials

Materials used in this study were chitosan (Surindo Bioteck, Cirebon, DD 87%), succinic acid, polyvinyl alcohol (PVA), polyethylene glycol (PEG), heparin, disodium phosphate (Na_2HPO_4), phosphoric acid (H_3PO_4), acetic acid (CH_3COOH) and sodium hydroxide (NaOH) 2 M. All materials were pro analysis and purchased from Merck.

2.2. Tools

The physical characterization tests of the resulted membrane included OHAUS analytical balance to measure the membrane weight, thickness meter (Mitutoyo) to measure the thickness of the membrane, and tensile strength analyzer (zwick 2.0,5) to measure the tensile strength of the membrane. Moreover, pH resistance was measured using pH meter (pH-009 type), while water uptake and membrane hydrophilicity were determined using contact angle method.

2.3. Procedures

2.3.1. The synthesis of succinic acid crosslinked chitosan membranes (CS-Succ)

CS-Succs with eight different compositions were prepared by dissolving 1.5 gram of chitosan in 100 ml of acetic acid 1%, and then added with succinic acid (Table 1) with a constant stirring for 4 hours at 60 °C to achieve an optimal ratio and then added with 5 ml of acetic acid 10% during the reaction process. The next step was pouring 10 ml of the mixed solution into a petri dish and dry it in an oven at 50-70 °C for 24 hours. After the printed membrane was dry, NaOH 2 M was added to remove it from the dish. Subsequently, it was washed using distilled water until pH 7 or neutral was achieved. The synthesized membrane was characterized to identify the quality of the membrane.

Table 1. Ratio of chitosan monomer and succinic acid.

Material type	Chitosan (g)	Succinic acid (g)	Chitosan mol ($3.75x10^{-5}$ mmol)	Succinic acid mol ($3.75x10^{-5}$ mmol)
CS-Succ1	1.500	0.037	1.000	8.333
CS-Succ2	1.500	0.028	1.000	6.250
CS-Succ3	1.500	0.022	1.000	5.000
CS-Succ4	1.500	0.018	1.000	4.167
CS-Succ5	1.500	0.016	1.000	3.571
CS-Succ6	1.500	0.014	1.000	3.125
CS-Succ7	1.500	0.012	1.000	2.778
CS-Succ8	1.500	0.011	1.000	2.500

2.3.2. Blending CS-Succ membrane with PVA/PEG

The CS-Succ compound fabricated in the previous reaction was then added with 32 mL of PVA and 18 mL of PEG and stirred with a constant stirring for 2 hours. After that, the membrane was produced by pouring 10 ml of the mixture into the petri dish and dried in an oven at 50-70 °C. The dried membrane was removed from the petri dish using NaOH 2 M and washed using distilled water until pH 7 or neutral was achieved. The resulted membrane was then physically tested.

2.3.3. Grafting of CS-Succ/PVA-PEG membrane with heparin

The CS-Succ/PVA-PEG solution was added 1 ml of Heparin 100 iu and stirred for 2 hours. After that, the membrane was produced by molding the 10 mL of the mixture into the petri dish. The mold was then dried in an oven at 50-70 °C. To remove the mold from the dish, the dried membrane was added by NaOH 2 M. The resulted membrane was then washed using distilled water until pH 7 or neutral was achieved.

2.4. Membrane characterisation

2.4.1. Thickness and Weight Tests

The thickness of all membranes synthesized from different compositions was measured using a thickness meter, while the weight of the membranes was determined using an analytical balance. Every membrane was measured before and after the use for the next process.

2.4.2. pH Resistance Test

Before testing, the membrane was weighed (w_1) and immersed in various solutions of pH 3, 5, 7, 9 and 11. They were kept for 6 hours, after that being weighed using the analytical balance (w_2).

2.4.3. Water Uptake Test

At the beginning of the process, the membrane was weighed using an analytical balance and immersed in water for 6 hours. After immersing, the membrane was weighed again. The weight difference taking from the weight before and after the immersion was used to determine the water uptake adjusted according to the following formula:

$$\% \text{ water uptake} = \frac{Wt}{Wo}$$

W_o = the weight of the initial membrane

W_t = the weight after the immersion

Tensile Strength Test

The tensile strength test was performed using the membrane with a size of 15 x 1.5 cm. The both ends of membrane were clamped in a mechanical test equipment, then pulled them up to break the membran and its tensile strength was observed.

3. Results and discussion

3.1. The synthesis of CS-Succ compounds

Crosslinking reaction was undertaken using an acid catalyst to facilitate a proton transfer to the succinic acid molecules. Heating of the mixture at 50°C aimed to accelerate the crosslinking process. The mechanism of the crosslinking reaction using an acid catalyst usually begins with transferring protons from or into the substrate molecules. In an acidic condition, $-NH_2$ groups of chitosan underwent protonation resulting in $-NH_3^+$, which was then easily attacked by free carboxylic groups, whereas the hydroxyl groups initiate an esterification with succinic acid [8]. The hydroxyl groups would subsequently react with the carboxylate group of succinic acid. The mechanism of the crosslinking reaction of chitosan with succinic acid can be explained by two-stage reactions: 1) chitosan underwent protonation due to the acidic condition coming from acetic acid (Scheme 1); 2) the –OH group of succinic acid molecules was bound to the protonated chitosan by hydrogen bonding with the amine group of chitosan (Scheme 2). This two-stage reaction mechanism is shown in Figure 1.

The success of the crosslinking reaction between chitosan and succinic acid was proved by the FTIR spectra as are shown in Figure 2. The FTIR spectra indicated the difference between pure chitosan and succinic acid cross-linked chitosan. The FTIR spectrum of chitosan membrane showed the vibration modes at 3425 and 2877 cm^{-1} attributed to -OH and -NH groups, respectively. Twin peaks at 1651 and 1597 cm^{-1} indicated the presence of $-NH_2$ group. In the mean time peaks at 1427, 1381 and 1327 cm^{-1} are represented as C-O group.

FTIR spectrum of succinic acid cross-linked chitosan membrane showed the appearance of a peak at 3448 cm^{-1} assigned to -OH group. However, the twin peaks attributed to $-NH_2$ group as present in pure chitosan disappeared. Nevertheless, a single peak at 1720 cm^{-1} attributed to C=O group of carboxylates appeared. This shift indicated that there was a change of $-NH_2$ group from primary to secondary or tertiary due to the inclusion of the -COOH group from succinic acid. It can be concluded that the crosslinking reaction of succinic acid to the chitosan was carried out properly.

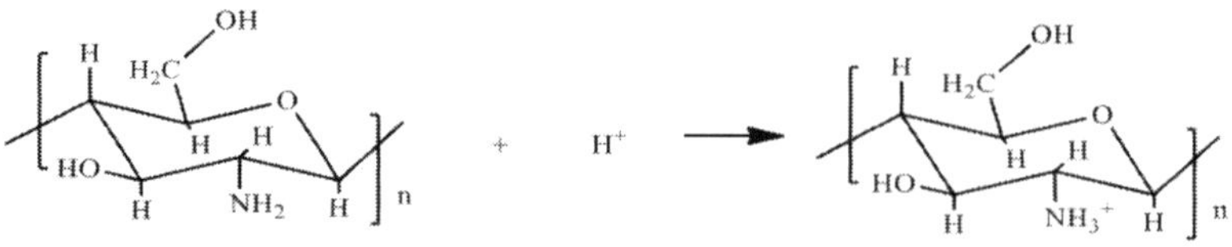

Scheme 1

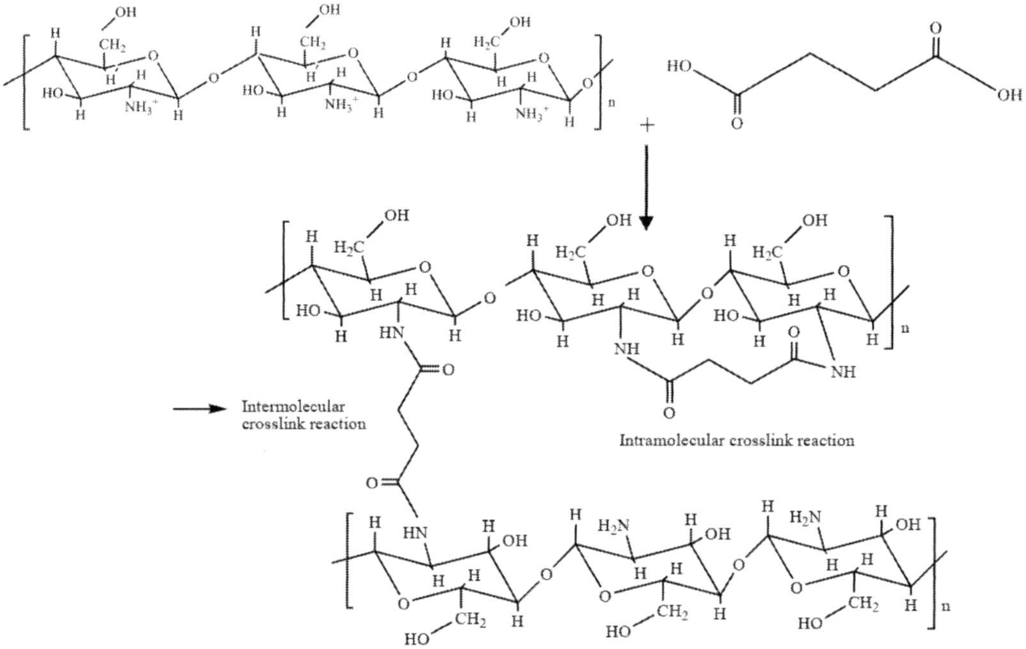

Scheme 2

Figure 1. The cross-linking reaction mechanism between chitosan and succinic acid.

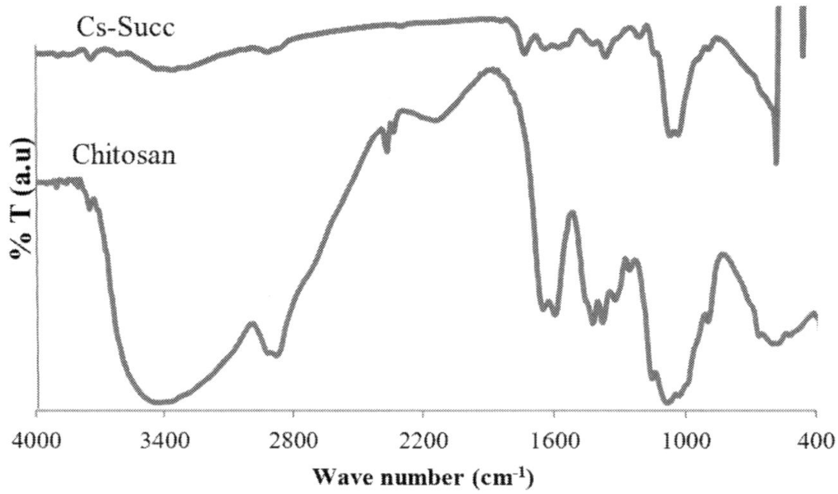

Figure 2. The FTIR spectra of chitosan and CS-Succ.

3.2. Characterization

3.2.1. Tests of Membrane Thickness and Weight

Membranes have different weights and thicknesses for each type of membranes indicating the mass content of the substances of the membranes. Membranes having high mass content would increase their weight and thickness as is shown in Table 2.

Table 2. The results of measurement of thickness and weight on various membranes.

Material Type	Weight (gram)	Thickness (mm)			
		I	II	III	Average
CS-Succ1	0.147	0.550	0.570	0.620	0.580
CS-Succ2	0.145	0.380	0.330	0.340	0.350
CS-Succ3	0.139	0.330	0.290	0.260	0.290
CS-Succ4	0.136	0.360	0.420	0.410	0.400
CS-Succ5	0.134	0.280	0.200	0.220	0.230
CS-Succ6	0.134	0.310	0.290	0.440	0.210
CS-Succ7	0.132	0.180	0.170	0.220	0.190
CS-Succ8	0.132	0.190	0.140	0.180	0.170

3.2.2. pH Resistance Test

The membrane resistance to pH is shown in Table 3 and Table 4. The results showed that the membrane was not resistant to acid pH indicated by significant decrease in the membrane mass at pH 3 and 5, while at basic pH membrane, there was no significant change in membrane mass at pH 7, 9, 11 and 13.

Table 3. pH resistance test of CS-Succ1 membrane.

pH	CS-Succ1		CS-Succ1/PVA-PEG		CS-Succ1/ PVA-PEG/Hep	
	W_0	W_t	W_0	W_t	W_0	W_t
3	0.0236	0.0003	0.0424	0.0035	0.0360	0.0003
5	0.0236	0.0031	0.0424	0.0165	0.0360	0.0005
7	0.0236	0.0259	0.0424	0.0428	0.0360	0.0366
9	0.0236	0.0265	0.0424	0.0419	0.0360	0.0368
11	0.0236	0.0240	0.0424	0.0442	0.0360	0.0363
13	0.0236	0.0252	0.0424	0.0424	0.0360	0.0375

Table 4. pH resistance test of CS-Succ8 membrane.

pH	CS-Succ8		CS-Succ8/PVA-PEG		CS-Succ8/PVA-PEG/Hep	
	W_0	W_t	W_0	W_t	W_0	W_t
3	0.0248	0.0067	0.0384	0.0146	0.0386	0.0064
5	0.0248	0.0024	0.0384	0.0132	0.0386	0.0162
7	0.0248	0.0261	0.0384	0.0379	0.0386	0.0391
9	0.0248	0.0248	0.0384	0.0395	0.0386	0.0396
11	0.0248	0.0281	0.0384	0.0398	0.0386	0.0393
13	0.0248	0.0230	0.0384	0.0384	0.0386	0.0396

3.2.3. Water Uptake Test

The water uptake test aimed to determine not only the ability of the membranes to adsorb water but also the membrane hydrophilicity. The results of the water uptake test of all membranes can be seen in Figure 3.

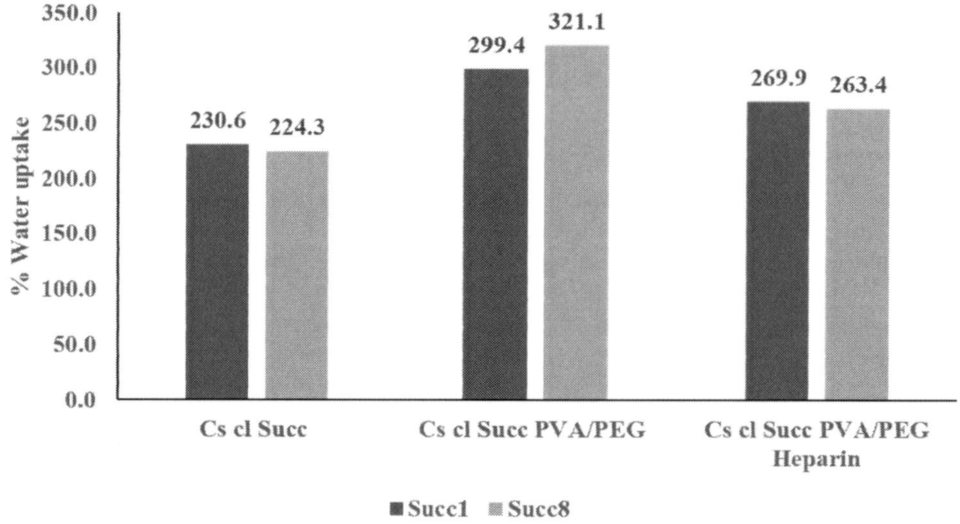

Figure 3. Water uptake of various membranes

Membrane blended with PVA and PEG showed the highest percentage of water uptake. This is because the membrane is more hydrophilic than others which may be due to the presence of PVA which is also hydrophilic [8].

3.2.4. Tensile Strength Test

The measurement of the tensile strength of the membranes with different composition showed the strength and elasticity of the membranes as reflected by the measurement of tensile strength presented in Figure 4.

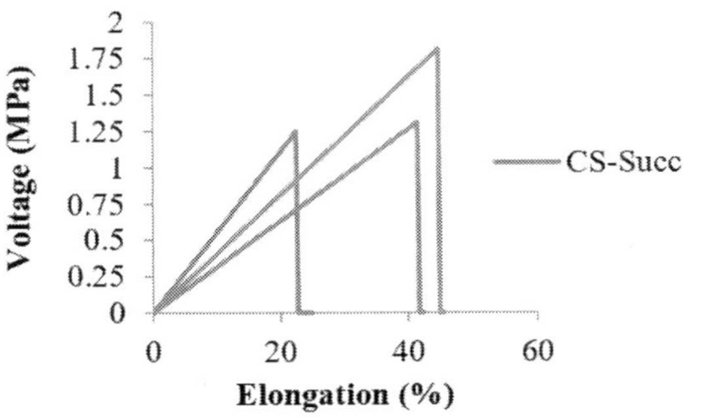

(a)

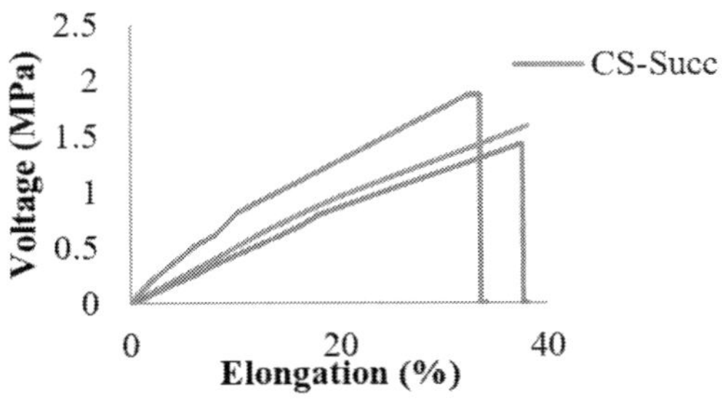

(b)

Figure 4. Tensile strength of (a) CS-Succ1 and (b) CS-Succ8.

The CS-Succ/PVA-PEG membrane had the highest elasticity when compared to other membranes. The CS-Succ blended with PVA/PEG increased the level of membrane elasticity. This is due to the presence of PVA/PEG which is more elastic and mechanically robust when compared to the chitosan polymer itself. By increasing the mechanical properties, it is expected that the resulting membranes can withstand the pressure when being used for transport processes. Further, it is open a way to apply this membrane for hemodialysis.

4. Conclusion

The crosslinking reaction of chitosan with succinic acid was carried out to increase the number of active groups of chitosan, while blending the membrane with PVA/PEG was conducted to improve membrane elasticity. The results showed that the best composition of the membrane occurred in the addition of 0.011 gram succinic acids with the highest mechanical strength compared to other membranes.

References

[1] Lin, W. C., Liu, T. Y., & Yang, M. C., 2004, Hemocompatibility of polyacrylonitrile dialysis membrane immobilized with chitosan and heparin conjugate, Biomaterials., 25, 1947-1957

[2] Lusiana, R. A., Siswanta, D., Mudasir, & Hayashita, T., 2013, The Influence of PVA.cl.Citric Acid/Chitosan Membrane Hydrophicility on the Transport of Creatinine and Urea. Indo. J. Chem, 13(3)., 262-270

[3] Hoveen, V. P., Tangpasuthadol, V., Angkitpaiboon, Y., Vallapa, N., & Kiatkamjornwong, S., 2007, Surface-charged chitosan: Preparation and protein adsorption, Carbohydrate Polymers 68 ,44-53

[4] Mitra, T., Sailakshmi, G., Gnanamani, A., & Mandal, A. B., 2013, Studies on Cross-linking of Succinic Acid with Chitosan/Collagen, Materials Research., 16(4), 755-765

[5] Salehi, E., & Madaeni, S. S., 2013, Influence of poly(ethylene glycol) as pore-generator on morphology and performance of chitosan/poly(vinyl alcohol) membrane adsorbents, Applied Surface Science., 1-5

[6] Zeng, M., Fang, Z dan Xu, Chengwei., 2004, Novel method of preparing microporous membrane by selective dissolution of chitosan/polyethylene glycol blend membrane, Journal of Applied Polymer Science, 91, 2840–2847

[7] Badr, I.H.A., M. Gouda, R. Abdel-Sattar dan Hossam E.M.S., 2014, Reduction of thrombogenicity of PVC-based sodium selective membrane electrodes using heparin-modified chitosan, Carbohydrate Polymers, 99,783-790

[8] Mohsen, A. M., Aly, A. S., Hrdina, R., Montaser, A. S., & Hebeish, A., 2011, Eco Synthesis of PVA/Chitosan Hydrogels for Biomedical Application, Journal of Polymers and the Environment., 19(4), 1005-1012

IOP Conf. Series: Materials Science and Engineering **349** (2018) 012066 doi:10.1088/1757-899X/349/1/012066

Development of assessment instruments to measure critical thinking skills

W Sumarni, K I Supardi, N Widiarti

Department of Chemistry, Semarang State University, Indonesia

[1)] Email : woro@mail.unnes.ac.id

Abstract. Assessment instruments that is commonly used in the school generally have not been orientated on critical thinking skills. The purpose of this research is to develop assessment instruments to measure critical thinking skills, to test validity, reliability, and practicality. This type of research is Research and Development. There are two stages on the preface step, which are field study and literacy study. On the development steps, there some parts, which are 1) instrument construction, 2) expert validity, 3) limited scale tryout and 4) narrow scale try-out. The developed assessment instrument are analysis essay and problem solving. Instruments were declared valid, reliable and practical.

1. Introduction

According to commission of 2013 Curriculum, learning that must be implemented is the learning that can develop higher order thinking skills (HOTS) that included skills of problem solving, creativity, critical, and decision making. The achievement of high order thinking skills surely can't be separated from the assessment that must be implemented as integral parts of a learning process, to know the developing and students' learning result, and also to fix the learning. [1] [2]. It is needed test instrument, measurement and assessment that valid and reliable to assess critical thinking skills as an HOTS indicators that are expected as continue effect in Chemistry learning.

One of the motivations for the development Critical Thingking Skills assesment as a way to prepare students to be more successful in college and in the workforce, the underlying assumption being that students are not prepared currently to do so upon graduation from high school [3]. The transition from teacher-centered learning to student-centered learning should include complex goals such as fostering higher order thinking skills. In the past, the most common way of assessing students was the traditional form of a summative test. This sort of test usually examined content knowledge and did not assess higher order thinking skills [4]. In recent years, many researchers have reported about reforms that emphasized teaching for improving higher order thinking skills, such as asking complex questions, generating argumentation, constructing graph, and analyzing case-based articles through various innovative 'real world' activities [5] [6] [7].

Thinking critically is skill activity that can be done well or otherwise, and a good critical thinking will fulfill various intellectual standards, like clarity, relevance, adequacy, coherence etc [8]. Critical thinking skills is one of the thinking skills that defined by [9] as self-guided, self- disciplined thinking which attempts to reason at the highest level of quality in a fair-minded way. Ennis (in [10] gave a definition – critical thinking is thinking by argument and reflection by pressing the decision making about what to believe and to do. The purpose of critical thinking is to make people away from wrong and rushed decision so it can be guaranteed. According to Costa (in [11] in thinking learning there are 3 ways, they are teaching for thinking, teaching of thinking and teaching about thinking. The same thing was stated by [12] in Foundation of Critical Thinking, critical thinking skills is an intellectual process about conceptualization, implementation, analysis, synthesis and evaluation actively and expert towards information that gained from observation, experience, reflection, thinking or communication as a guidance to believe and to do. This skills is marked by intellectual values that is universal, which are clarity, accuracy, consistency, carefulness, suitability, right evidence, a good thinking, deepness, width, and justice.

Evaluation to assess students' result was integrated done, which was not only by assessing what they had learnt, but also by assessing how the influence and students' skills in doing every steps in problem solving. Thus, the assessment began when the problem was laid until the final result assessment. The teacher assessed the students' skills of thinking, comprehension, and the successful in doing each steps of problem solving, how the students guide themselves to work and students' influence in team work.

Evaluation of critical thinking skills need measuring instrument that can draw perfectly the true condition. From the result of literacy study, it was gained some instruments to measure critical thinking skills on students, among others Academic Profile, California Critical Thinking Dispositions Inventory (CCTDI), Collegiate Assessment of Academic Proficiency (CAAP), Cornell Critical Thinking Test (CCTT), Critical Thinking Assessment Battery (CTAB) [13]. California Critical Thinking Skill Test (CCTST) is one of the instruments that developed by Facione since 1900 that was accurately believed to measure critical thinking skills of nursing students. Question items on CCTST reflected moderate critical thinking skills of nurses, Evidence- Centered Concept Map (ECCM) constructed and studied by [14] as instrument of computer based thinking skills assessment.

During this time, evaluation that was done by teachers/lecturers is generally the skills of reminding fact, concept or procedure. The teachers/lecturers often forget that the good evaluation is the evaluation that complete and can reveal every skills of assessed subjects [15]. Evaluation towards skills of presentation, performance evaluation and the skills of knowledge application on a product making, skills of synthesizing information and giving reason based on known information, often being ignored. From the learning result evaluation if that is implemented well it will form the critical thinking skill and creativity. These two skills underlying problem solving skill that was expected on students.

The developed assessment instrument were two test questions which were analysis essay test and problem solving test. Essay test was used because this type of test have characteristic to explore students opinion according to the skills they have, so the essay test supported the existence of critical thinking skills test subject. The used text to answer questions paid attention to problematic text criteria (text than can be exploited to think critically). Thus, those texts have criteria among other, (a) the content explain the things that is not suitable to real life/illogical, (b) the story content explanation is not followed by clear reason, (c) there is no continuity in describing a thing, (d) there is no appropriate between the content and the title, (e) given information in the text have particular purposes, and (f) lack of completeness in content explanation.

Developed cognitive instruments have provided the content and construct validity, reliability, practicality and good legibility. It can be seen from the test results of the test group, namely the expert assessment instruments, reading experts, practitioners, and students in the questionnaire assessment guidelines that have been given. Terms of construct validity of assessment instruments that have been developed include the portion of outline parts accuracy against the critical thinking concept. That aspect was obtaining the average score of 83.7%, so it can meet good construct validity. The developed assessment instrument includes aspects of critical thinking competencies which include, (1) developed the text interpretations, and (2) reflected and evaluated the text.

This is consistent with the theory expressed by Facione in [16] which said that critical thinking become objectives and assessment of self-regulation that results in interpretation, analysis, evaluation, and conclusions. Ennis in [8] added that critical thinking is thinking that makes sense and reflective focusing to decide what should be trusted or do. Therefore, aspects of critical thinking skills include, aspects of interpret, reflect and evaluate the text in the developed assessment instrument.

The developed assessment instrument must meet the requirements of reliability. Aspects of reliability in the developed assessment instrument include instructions and rubric assessment. The level of reliability of the developed assessment instrument was good qualified. It can be seen from the test results to the group test on the reliability aspects of the assessment instrument. Instruments of thinking skills, not only emphasizes on understanding concepts but rather on aspects of the synthesis, analysis and evaluation, so that it has a relatively low constancy [17] [1] [18] Reliability with an alpha value above 0.6 then declared reliable. By referring to the reality on the field and the opinion of experts about the very need for the availability of raw instrument of critical thinking skills, especially in chemistry learning before it is implemented, then the problem solved in this research is to develop a valid and reliable standard instrument of critical thinking skills in chemistry learning.

2. Methods

The study was conducted in State Senior High Schools in Semarang Regency, Indonesia. This type of research includes into Research and Development (R & D), referring to the [19] which is adapted to the needs of research. In this study, assessment levels differed according to the revised version of Bloom's [20] is at an intermediate, apply, and the three higher levels of Bloom's (analyze, evaluate, and create) require higher-order cognitive skills [21][22]. The taxonomy was originally designed to be a cumulative hierarchy where mastery of the higher levels of the taxonomy were prerequisite to performance on higher levels[20].

Aspects of critical thinking skills is composed of indicators according to [18] which includes: focus the questions, ask and answer questions, identify the assumptions, analyzing arguments, defining the term and consider a definition, consider whether the source is trustworthy or not, make and determining the consideration, observe and consider the report of observation, determine a course of action, inducing and consider the results of induction, and interact with others. Likewise for aspects of problem-solving ability is composed of indicators namely the students are able to identify the problem, stating a causal connection, able to apply the concept in accordance with the problem, curious, able to create charts or images to solve a problem, explain some possibilities as a solution, open-minded, make decisions, able to work carefully, unwilling to speculate and able to reflect on the effectiveness of the problem-solving process.

In the first step of the development phase is an expert validation, validation is conducted in the form of content and construct validity of assessment instruments both test and non-test. According to [23]; [1], stated that the thinking skills instrument should have the good construct and content validity before used. Therefore, the expert validity becomes an important element to start the development. Validation results declared as valid after conducted revision in writing and legibility, compatibility between indicators of critical thinking skills with test items, compatibility of Bloom revision cognitive taxonomy used on each test item, the accuracy of the concept, the accuracy of images used in test items, the accuracy of test items presentation in problem solving items, simplifying the use of scientific terms.

In the next step, the test items reliability testing in school was conducted at the pre-trial phase. The instrument was declared valid by expert and have a high reliability coefficient been tested on a limited scale of 13 students, the sample was taken by purposive sampling. The data was collected using research instruments in the form of expert validity sheet, questionnaires sheet, essay tests analysis and problem solving tests. Qualitative data processed using a simple presentation technique then conducted a categorization of test items reliability by using alpha Cronbach. To see the instruments practicality can be seen from the student and teachers questionnaire responses percentage which was higher than 80%.

3. Results And Discussion

The products developed are CTS based assessment in acid-base solution material. This assessment instrument was in the form of essay analysis and problem solving test items consist of 15 questions with a rubric assessment for teachers, where the item tests were designed or developed in order to measure students' critical thinking skills while simultaneously measuring the indicators of learning that has been made in accordance with Competence Standard-Basic Competence (CS-BC). Essay questions have been proposed [24]; [25] for effectively capturing students' meaningful understanding of science concepts, and problem solving test used to develop students' scientific literacy and higher order thinking skills through context-based chemistry[26]. The research has been compiled test instruments to measure critical thinking skills is test description. Inside it contains a description of the situation, followed by a question that leads to certain indicators of critical thinking skills. This is in line with the opinion of [27], which is very appropriate description shaped test to measure students' critical thinking skills.

To find out the characteristics of a CTS-based assessment that has been developed, it was conducted a test towards the conformance aspects of the material content with CS-BC, the material content conformity with CTS indicators, legibility and attractiveness of assessment instrument, student response, and the response of teachers to the assessment instrument developed.

The initial stage in the CTS based development of assessment on acid-base solution material was the needs analysis or preliminary studies which included a literature study and field study. Literature study generated a learning device in the form of concept analysis, syllabus, and CS-BC mapping about acid-base material.

At the second stage the preparation of the product assessment based on material CTS acid-alkaline solution called draft 1. At this stage the first thing to do is prepare questions outline in accordance with CS-BC. Thus produced an outline in which there were basic competencies, critical thinking skills activities, indicators of student competence, questions and answers. Outline contained indicators of critical thinking skills and SBC curriculum indicators. In this case there were two question models namely the essay analysis test (EAT) and problem solving test (PST) models and so there were two different outlines. Beside the questions, there was a student activity observation sheet regard to the critical thinking skills for learning.

Based on the outline, conducted a preparation of essay analysis and problem solving test items as many as 15 questions. Each item has been designed to measure students' critical thinking skills to any learning indicators are formulated. Expert or experts who validate the instrument is a chemist, an expert in critical thinking skills and educational research experts and field practitioners to validate the contents namely the chemistry teacher of high schools in Semarang regency.

The validity result of essay instruments analyzes the acid-base materials was stated very valid for all test items by all the experts that have been validated. In preparing this evaluation sheets, each stage conducted repairs and improvements based on guidance/input from material experts and expert of evaluation.

Table 1. Results of Validity of Essay Analysis Test (EAT)

No.	Description	Validator				Percentage (%)	Validation Criteria	Information
		1	2	3	4			
1	Instruction Aspects: Instructions of students' critical thinking skills instrument is clearly stated	3	3	4	3	87.5	Very Valid	Slightly Revised
2	Scope of Problem Aspects:							
	The formulation of a logical test items (realistic)	3	3	4	4	100	Very Valid	Slightly Revised
	Conformity of assessment instrument content with critical thinking indicator	4	4	4	4	100	very valid	No revisions
	Conformity of assessment instruments with KD	4	4	4	4	100	Very Valid	No Revisions
	Items according to the level of critical thinking skills	3	4	3	3	87.5	Very Valid	No Revisions
3	Aspects of Language:							
	The formulation of the sentence of communicative test items.	3	3	3	3	75	Valid Enough	Revision

The validation results of an acid-base materials problem solving instrument was stated very valid for all items by all the experts that have been validated. Problem solving test instruments can be used for research. Validation of a questionnaire was conducted by critical thinking skills expert and educational research experts. The results of the questionnaire validation declared valid seen by the scores given by experts.

Table 2. Results of Validity of the Problem Solving test

No.	Description	Validator				Percentage (%)	Validation Criteria	Information
		1	2	3	4			
1	Instruction Aspects: Instructions of students' critical thinking skills instrument was clearly stated	3	4	4	3	87.5	Very Valid	No Revisions
2	Scope of Problem Aspects:							
	The formulation of a logical test items (realistic)	4	4	4	4	100	very valid	No revisions
	Conformity of assessment instrument content with critical thinking indicator	4	4	4	4	100	Very Valid	No Revisions
	Conformity of assessment instruments with KD	4	4	4	4	100	Very Valid	No Revisions
	Items according to the critical thinking skills level	3	4	3	3	87.4	Very Valid	No Revisions
3	Aspects of Language:							
	The formulation of the sentence of communicative test items.	3	3	3	3	75	Valid Enough	Revision

Table 3. Results of Validity of responses questionnaire

Validator	Total Points Earned	Category
Critical thinking skills Expert	18	Very Valid
Educational research Expert	18	Very Valid

The revised assessment form subsequently tested at State Senior High Schools in Semarang Regency, Central of Java. The trial is addressed to the chemistry teacher and 30 students of class X. For teachers a questionnaire given on various aspects of conformity assessment based on CTS with CS-BC and CTS indicators as well as interviewed the teacher's response to the developed CTS based assessment.

Measurement of the essay analysis test and problem solving tests reliability performed by using Alpha Cronbach. The results of the overall reliability of the essays test instrument of acid-base analysis declared as reliable with a coefficient of 0.71. The reliability of the test instrument problem solving declared as reliable with a coefficient of 0.79, thus concluded otherwise reliable instrument with very high reliability criteria for the essay analysis and problem solving tests item. Instruments or products which declared as reliable in the trial will be continued on a limited test phase. The results of the reliability test limited scale of the test instrument essays acid-base analysis revealed reliable with a coefficient of 0.74 with high reliability criteria. Instrument problem solving test reliability declared as reliable with a coefficient of 0.82, thus concluded instrument declared as reliable with very high reliability criteria.

All the results related to the items under study indicate that properly designed objective questions are valid and reliable tools for assessing students' meaningful understanding of acid-base concepts. This finding supports the suggestions [28]; [29] about the effectiveness of the conceptual questions on assessing aspects of their understanding.

The assessment instruments practicability related to the ease of assessment instruments, this aspect includes the appearance and layout of the assessment instrument, part of the contents of the assessment instrument, and applicability of assessment instrument. The look and layout of the assessment instrument lies in the content, presentation, and graphic. Student responses viewed using a questionnaire distributed after the end of the study. Data can be shown in Table 4.

Table 4. Students' Response toward Critical Thinking Skills Assessment Instrument

Category	many Students	Proportion
Very high	2	$^{2}/_{40}$
High	27	$^{27}/_{40}$
Enough	9	$^{9}/_{40}$
Low	2	$^{2}/_{40}$

The results of student responses on the implementation class showed that the student response was high, evidenced by from 40 students 27 students gave high response, 2 students responded very high, and the rest was sufficient and low. It showed a positive response from the students about the use of the instrument of students' critical thinking skills. This may imply that the developed critical thinking skills instruments assessed as practical. The use of adequate assessment, such as performance tasks, even during early stages, such as pre-service programs rather than just in professional development in-service programs.

The teacher's response is given to the development of students' critical thinking skills instrument. Before responding, teachers were asked to assess the developed instruments and then given a respondents' questionnaire after it. The result of the teacher's response to students' critical thinking skills instruments was discovered that the average response of teachers to use the instrument of critical thinking skills was 85%, thus concluded as practical instrument. The result of teachers' response to use the critical thinking skills instrument obtained a high response. This means that according to the high schools chemistry teachers in Ungaran city, critical thinking skills instruments was practical to be used due to in accordance with the school curriculum and graduation standards. Some teachers also want to learn to prepare the instrument and apply it in the learning process.

At the third stage, the data of product practicality. From the trial against the practicality aspect of the assessment instrument with a pilot group overall the average feasibility achieved the percentage of 83.33%, so that the instrument can be categorized as met the good assessment practicality aspect. And the fourth stage, the legibility data of products. From the trial of aspects of the assessment instrument legibility with the pilot group overall the average feasibility achieved the percentage of 80.75%, so that the assessment instrument can be categorized as met the good legibility.

From the test results, also discovered some advantages of assessment instrument. In the aspect of the text selection can broaden the student knowledge as assessment instrument users, because the text used considering the aspect of reading situations and variations in the type of text. Those items presented were considering the aspect of critical thinking skills to be measured and the type of tests used to uncover these capabilities was vary. Assessment instrument was equipped with an attractive cover, introduction, assessment instrument components overview, questions introductory, and instructions. Nevertheless, the assessment instrument also has some drawbacks. In the aspect of text selection has not noticed the difficulty level of the text in the presentation, so that need for improvements in the designated text. At the critical thinking skills in the assessment instrument, evaluative questions were still lacking. In the aspect of language use was still lack of proper spelling, still contained the term on items that have an impact on student difficulties (as users of the product) to be difficult to understand the purpose questions on items. At the writing of the source of the text was incompletely presented. Fourth, readability assessment instrument associated with the use of language in the assessment instrument.

Below is a table 4 which showed the conducted limited testing results. The criteria tested in the sheet covering the aspects of conformity with the contents of CS-BC, and aspects of the appropriateness of the contents with CTS indicators, construction aspects, and aspects of CTS based assessment legibility on the subject material of acid-base solution. For the fourth aspects the average ratings was $\geq 80\%$ with a very high category. So it can be said that the developed assessment sheet has been eligible to be tested. The feasibility assessment sheet showed that it has met the criteria of a good assessment sheet.

In Table 2, the results table limited testing has been done on 30 students and 1 teacher of chemistry. Average results obtained -rata is $\geq 77\%$ with a high category for all four aspects: the suitability of the contents of the CS-BC, and aspects of the appropriateness of the contents with CTS indicators, construction aspects, and aspects of CTS based assessment legibility on the subject material of acid-base solution.

Furthermore, from the limited test results were also obtained students' response to the learning instruments developed. The results of student responses indicate that students are happy to development results instrument. It can be seen from the students' answers which showed a positive response to the instruments that have been developed. Students said that this instrument did not like the previous questions which only ask students to answer the questions based on concept or material. While the questions developed were not only asked students to answer based on the concept, but also to challenge the students to be able to give a simple explanation, build basic skills, concluded, elaborating and implementing strategies and tactics.

For the level of questions' difficulty, most students said that the problems provided were neither too difficult nor easy. Suggestions for improvement, they said that it needed to be better clarified for example the picture and the words used should not be too long. The teacher response to developed CTS-based assessment taken through the teachers responses at questionnaire on development of CTS based assessment on acid-base solution material can received the instruments that have been developed, according to the teachers, this instrument was good and can added knowledge and increased the existed test item types. This instrument also used critical thinking indicators thus be able to help the students to train students' critical thinking skills.

In these discussions presented a study of the CTS-based assessment characteristics that has been developed and constraints faced in the development of the developed CTS-based assessment. CTS based assessment characteristics on acid-base solution material was seen under the aspect of conformity with the CS-BC and can measured CTS students indicators. Based on the attractiveness aspects, CTS-based assessment has the attractiveness of the high level. The images used in the matter. Based on the legibility aspects, CTS-based assessment has a very high level of legibility. These assessments have high validity, both of the validity of conformity between the content of CTS-based assessment and the CS-BC and also the validity of conformity between the SBC-based assessment content validation and SBC indicator. In the legibility validation test, this assessment also has a high validity. Developed CTS-based assessments aim to train students to be able to have good critical thinking skills.

Constraints faced when application of the assessment results namely this assessment for teachers in schools has not given questions to train students' CTS, so that students' answers were shorter and focus on the points in question, without giving a reason or an explanation of the answer, whether in the CTS-based assessment was demanding descriptive answers. This assessment should be made on going and can be completed in groups. Another obstacle is this assessment entirely in the form of essay questions that require the student's ability that includes conceptual and procedural knowledge.

Thinking skills is not an instant learning outcome that can be directly measured by two to three times of learning then declared as good or not. Based on the results of the research (Afcariono, 2008; Richmond, 2007; Woolf et al., 2005), said that it takes a process and not a short exercise to be able to change one's thinking skills.

Conclusion

Based on the results, findings and discussion are obtained from the activities of preliminary studies, limited testing and test-wide scale, it has composed an analysis essay test (AET) instrument as well as problem solving test (PST), which was developed in the form of a description about its semi-open - ended is valid and reliable. Both types of instruments the test has been declared valid for all test items according to the expert assessment, and reliable with the reliability coefficient of analysis essay test instrument of acid-base of 0.74 and the problem solving test of 0.82. The developed Analysis Essay Test (AET) and Problem Solving Test (PST) also have to meet the criteria of practicality and good legibility and received a positive response from both students and teachers. It can be concluded that the assessment instrument developed has met the quality criteria of a good test, so it can be used as a form of alternative assessment instrument for teachers to uncover students' critical thinking skills. The two assessment schemes was developed based on a different approach, therefore, it can provide different information regarding students' critical thinking skills. A combination of the two assessment schemes offers the potential for multifarious evaluation of students' conceptual understanding and critical thinking skills.

However, since the developed assessment instrument yet include all the indicators and all the learning topics, it still needed further research on the development of critical thinking skills assessment instruments based on the other indicators whether the indicators propounded by Ennis or indicators developed by other experts, as well as another topics in chemistry lessons. Continuous support of the teachers was found as a critical. The importance of supporting the teachers is aligned, the use of adequate assessment, such as performance tasks, even during early stages, such as pre-service programs rather than just in professional development in-service programs.

Acknowledgement
Here I say a regards to all those who have helped in the implementation of this study, especially the students participating in the learning related to critical thinking skills.

References

[1] J. Docktor and K. Heller, "Robust Assessment Instrument For Student Problem Solving.," in The NARST 2009 Annual Meeting., 2009.
[2] S. Suwandi, Model-Model Penilaian dalam Pembelajaran., Surakarta: Yuma Pustaka, 2011.
[3] A. Porter, J. McMaken, J. Hwang and R. Yang, " Common core standards: the new U.S. intended curriculum," Educational Researcher, 40(3), p. 103–116, 2011.
[4] J. Osborne and R. Millar, "Science education for the future: which way now?," Prim Sci Rev 52, p. 21–23, 1998.
[5] C. P. Dwyer, M. J. Hogan and I. Stewart, "An evaluation of argument mapping as a method of enhancing critical thinking performance in e-learning environments," Metacognition Learning 7, p. 219–244, 2012.

[6] Y. Dori and I. Sasson, "Chemical understanding and graphing skills in an honors case-based computerized chemistry laboratory environment: the value of bidirectional visual and textual representations.," J Res Sci Teach 45, pp. 219–250 DOI 10.1007/s11409-012-9092-1, 2008.
[7] K. I. Supardi, W. Sumarni and N. Widiarti, "Pengembangan Model Pembelajaran Kimia dengan Aktivitas Aesop's Berorientasi Chemoentrepreneurship (CEP) untuk Meningkatkan Keterampilan Berpikir Siswa dan Kinerja Guru Kimia.," LP2M, Semarang State University, Semarang. Unpublished, 2013.
[8] A. Fisher, Berpikir Kritis: Sebuah Pengantar., Jakarta: Erlangga, 2009.
[9] L. Elder and R. Paul, "Critical Thinking Development: A Stage Theory with Implications for Instruction. [Online]," 2008. [Online].
[10] I. Z. Hassoubah, Developing Creatif and Critical Thinking Skill (Cara Berpikir Kreatif dan Kritis)., Bandung: Nuansa, 2004.
[11] Liliasari., " Membangun keterampilan berpikir manusia Indonesia melalui pendidikan sains, Pidato pengukuhan Jabatan Guru Besar Tetap dalam Ilmu Pendidikan IPA pada FPMIPA UPI," UPI, Bandung, 2005.
[12] M. Scriven and R. Paul, "Critical Thinking.," in The 8th Annual International Conference on Critical Thinking and Education Reform, C.A., 1987.
[13] B. S. Stein, A. F. Hynes and J. Unterstein, " Assessing Critical Thinking Skills. Paper presented at," in SACS/COC Annual Meeting, Tennessee Technology University, December 6, 2003. Retrieved July 23, 2015, from http://iwebtech.edu/cti/SACS%20presentat, 2003.
[14] Y. Rosen and M. Tager, "Computer -based Assessment of Collaborative Problem Solving Skills: Human -to-Agent versus Human-to-Human Approach. Retrieved August 18,2015," 2013. [Online].
[15] Harjito, Nurhayati S. and S. Hadisaputro, "Implementasi Authentic Assesment Pada Pembelajaran Berbasis Proyek (Project Based Learning) Untuk Peningkatan Softskill Mahasiswa. Research Report Hibah Bersaing," LP2M Semarang State University, Semarang, 2013.
[16] W. Kuswana, Taksonomi Berpikir., Bandung: : Remaja Rosdakarya, 2011.
[17] J. Carson, "A Problem with A Problem Solving : Teaching Thinking Without Teaching Knowledge.," The Mathematics Educator. Vol 17 (2), pp. 7-14., 2007.
[18] Ennis, R. H., "Critical Thinking Assessment.," College of education, The Ohio State University, 32 (3), pp. 179-186., 1993.
[19] M. Gall, J. Gall and W. Borg, Education Research : An introduction Seventh ed., Boston: Allyn and Bacon, 2003.
[20] D. R. Krathwohl, "A revision of Bloom's Taxonomy: An overview," Theory Into Practice, 41(4), p. 212–218, 2002.
[21] A. Crowe, C. Dirks and M. P. Wenderoth, "Biology in bloom: Implementing Bloom's Taxonomy to enhance student learning in biology," CBE - Life Sciences Education, 7, p. 368–381, 2008.
[22] U. Zoller, "Are lecture and learning compatible? Maybe for LOCS: unlikely for HOCS (SYM).," Journal of Chemical Education 70, p. 195–197, 1993.
[23] R. H. Ennis and E. Weir, "The Ennis Weir Critical Thinking Essay Test.," 1985.
[24] A. M. L. Cavallo and L. E. Schafer, " Relationships between students' meaningful learning orientation and their understanding of genetics topics," Journal of Research in Science Teaching,31, p. 393–418., 1994.
[25] M. Nieswandt and K. Bellomo, "Written extended-response questions as classroom assessment tools for meaningful understanding of evolutionary theory," Journal of Research in Science Teaching, 46, p. 333–356, 2009.
[26] S. Avargil, O. Herscovitz and Y. J. Dori, "Teaching Thinking Skills in Context-Based Learning: Teachers' Challenges and Assessment Knowledge," J Sci Educ Technol 21, p. 207–225, 2012.
[27] Mundilarto, Penilaian Hasil Belajar Fisika, Yogyakarta: Pusat Pengembangan Instruksional Sains., 2010.

[28] K. Salta and C. Tzougraki, "Conceptual versus algorithmic problem-solving: Focusing on problems dealing with conservation of matter in chemistry," Research in Science Education, 41(4), p. 587–609., 2011.
[29] O. L. Liu, H.-S. Lee, C. Hofstetter and M. C. Linn, " Assessing knowledge integration in science: Construct, measures, and evidence," Educational Assessment, 13(1), p. 33–55, 2008.
[30] C. Kothari, Research Methodologi, Methods and Techniques., India: New Age International Limited (P) Publisher., 2004.
[31] B. Woolf, T. Murray, D. Marshall, M. Bruno, T. Dragon, M. Mattingly, K. Kohler, M. and J. Sammons, "Critical Thinking Environments for Science Education.In C. K. Looi & G. McCalla & B. Bredeweg & J. Breuker (Eds.),," in 12th International Conference on Artificial Intelligence in Education, AIED 2005, Amsterdam, The Netherlands, July 2005, 2005.

IOP Conf. Series: Materials Science and Engineering **349** (2018) 012067 doi:10.1088/1757-899X/349/1/012067

Fractionation of Java Citronella Oil and Citronellal Purification by Batch Vacuum Fractional Distillation

W T Eden[1,2,*], D Alighiri[1,2], E Cahyono[1,2], K I Supardi[1,2,3], and N Wijayati[1,2]

[1] Essential Oil Study Center, Faculty of Mathematics and Natural Sciences, Universitas Negeri Semarang, Indonesia
[2] Chemistry Department, Faculty of Mathematics and Natural Sciences, Universitas Negeri Semarang, Indonesia
[3] Department of Postgraduate Program, Universitas Negeri Semarang, Indonesia

*Corresponding author: willytirzaeden@mail.unnes.ac.id

Abstract. The aim of this work was to assess the performance of a vacuum fractionating column for the fractionation of Java Citronella Oil (*Cymbopogon winterianus*) and citronellal purification during batch mode operation at vacuum -76 cmHg and reflux ratios 5:1. Based on GC-MS analysis of Java Citronella Oil is known that citronellal, citronellol, and geraniol has yielded 21,59%; 7,43%; and 34,27%, respectively. Fractional distillation under reduced pressure and continued redistilled are needed to isolate the component of Java Citronella Oil. Redistilled can improve the purity, then distillate collected while the temperature changed. In the first distillate yielded citronellal with a purity of 75.67%. The first distillate obtained residue rhodinol product will then be carried back to separation into citronellol and geraniol. The purity of citronellol reached 80,65% purity, whereas geraniol reached 76.63% purity. Citronellal Purification resulting citronellal to 95.10% purity and *p*-menthane-3,8-diol reached 75.95% purity.

1 Introduction

Java citronella oil is the essential oil from citronella grasses (*Cymbopogon winterianus*) from Java Island, Indonesia. One part of Indonesia which is abundant of raw essential oil source especially citronella oil in Central Java. Central Java has the potential production of essential oils are quite large. Some essential oils produced from refining industry owned by the government, private and community sectors [1]. Java citronella oil is popularly known by its characteristic insect and mosquito repellent and used in perfumery, soaps, detergents, and industrial polishes [2,3]. The oil is used extensively as a source of perfumery chemicals such as citronellal, citronellol, and geraniol. These compounds are used extensively in soap, candles and incense, perfumery, cosmetic, and flavoring industries throughout the world. The oil of java citronella is composed mostly by monoterpenes (±80%) (major components being aldehydes and alcohols), although sesquiterpenes are also found. Citronellal, citronellol, and geraniol are the major terpenoids found in the Java Citronella Oil [2,4,5]. Citronella oil is also a plant-based insect repellent. These can be used as conventional mosquito repellents with little or no harmful effects [6]. Citronellal or rhodinal is the major component of the monoterpene fraction of citronella oil and gives the essential oil of citronella its characteristic lemon odor [7], is also used in many chemical syntheses [8]. Geraniol (3,7-dimethyl octa-trans-2, 6-dien-1-ol) oil, the main ingredient in the essential

IOP Conf. Series: Materials Science and Engineering **349** (2018) 012067 doi:10.1088/1757-899X/349/1/012067

oil of ginger, lemon, lavender, orange, rose, palmarosa and other plants, is acyclic monoterpene alcohol [9]. Geraniol is a colorless, odor pleasant compound (aroma of roses) [2]. Citronellol (3,7-dimethyl oct-6-en-1-ol), an essential oil component of lemongrass, citronella grass, and also presents an odor of roses. It is used as a flavoring agent in food and beverages, fragrances, and an insect repellent [2, 10].

Most terpenes such as citronellal, citronellol, and geraniol are thermally unstable, decomposing or oxidizing at high temperatures or the presence of light or oxygen [11]. Therefore, separation of component active from java citronella oil is needed a mode of vacuum condition to decrease temperature operation. The fractional distillation is one of unit operation that aims the separation of two or more substances using vacuum state by the volatility difference between them. This process depends on the pressure and temperature of the system, as well the physical and chemical characteristics of the components to be separated [12]. Batch vacuum distillation works at mild temperatures and offers flexibility to operate with different oils in the same industrial plant. The main advantages of batch processes are the flexibility and versatility and also the possibility of working with small volumes, thus allowing performing raw material tests before large-scale processing [13]. The objective of this work is to evaluate the use of the batch vacuum fractional distillation to separate terpenoid compounds of the Java Citronella Oil. The Java Citronella Oil was isolated by using steam distillation from grasses of *Cymbopogon winterianus* plants collected in Boyolali District, Central Java, Indonesia.

2 Methods

2.1 Raw Material

Grasses of *Cymbopogon winterianus* was collected from Boyolali District, Central Java, Indonesia. This herb was selected, dried, and prepared for one day. Other chemicals were purchased from Merck.

2.2 Steam Distillation Method

The essential oils were collected by the steam distillation method. The grasses of *Cymbopogon winterianus* were packed into the kettle sitting on a perforated plate above the boiling water. The essential oils were volatilized with boiling water at temperature 100°C for 6 hours or more. After the steam distillation process, the oil will be collected and separated used the separatory funnel which can be used to separate the immiscible liquids of two layers such as oil and water. The oils which have lighter density than water will be visible at the top of the separatory funnel. Wait for the water and oil separated completely and formed two layers and oil was collected. Solid of Na_2SO_4 was used to adsorb the residual of water that fused with the oil. The quantity of the oil determined according to the yield. The yields of essential oils were calculated using the formula:

$$\%\text{Yield} = \frac{\text{Weight of Oil (g)}}{\text{Weight of } Cymbopogon\ winterianus\ Grasses\ Taken\ \text{(g)}} \times 100\%$$

2.3 Batch Vacuum Fractional Distillation Method

The rhodinol, citronellol, geraniol, and *p*-menthane-3,4-diol oils (Figure 1) obtained by fractional vacuum distillation of java citronella oil (*Cymbopogon winterianus*) (distillate fractions DF-33 P = -76 cmHg R = 5:1).

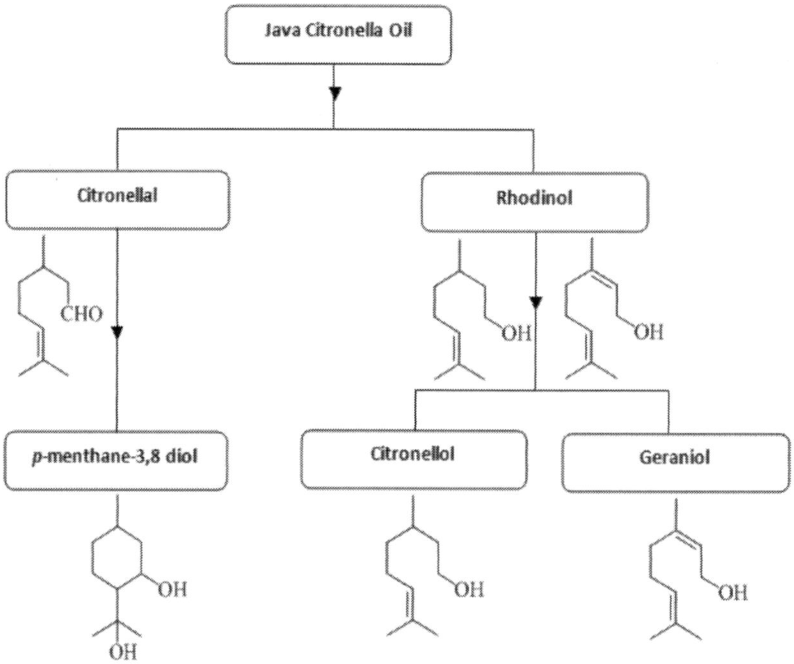

Figure 1.Fractional Distillation Mechanisms of Java Citronella Oil to produce Citronellal, Rhodinol, Citronellol, Geraniol, and *p*-menthane-3,4-diol Oils

The batch vacuum fractional reactor built in stainless steel SUS 316 L. The capacity reactor of 25 kg consists of reflux column, mixer, condenser, and tank product. All part connected to a vacuum pump (Figure 2).

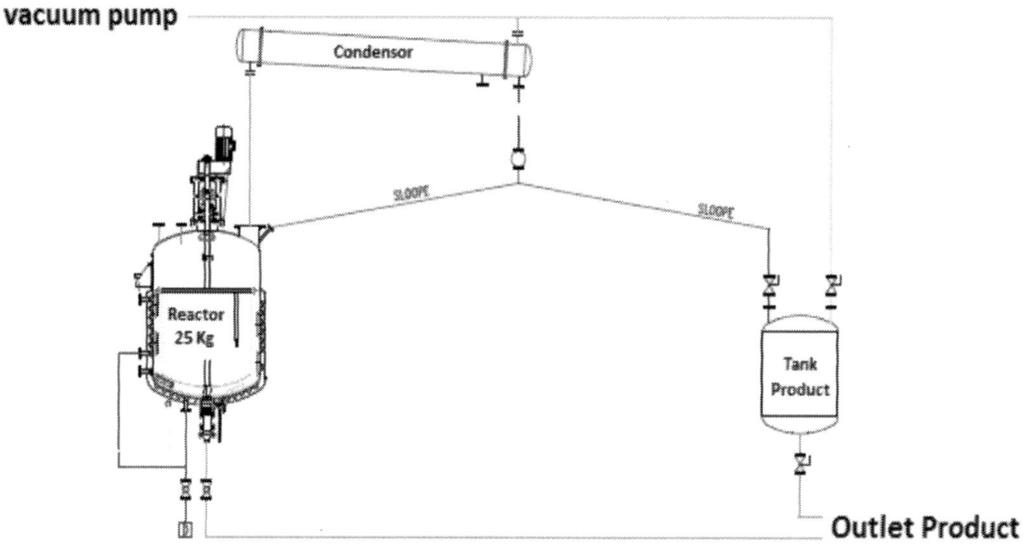

Figure 2.Batch Vacuum Fractional Distillation Reactor

2.4 *Physical properties*

The specific gravity and refractive index of the Java citronella oil extracted from the *Cymbopogon winterianus* grasses, citronellal, rhodinol, citronellol, geraniol, and *p*-menthane-3,4-diol oils were measured. The specific gravity measured at 25°C and refractive index measured at 20°C.

2.5 *GC/MS Analysis*

Gas chromatography-mass spectrometry (GC–MS) analysis was carried out on an Agilent Technologies GC–MS instrument equipped with a GC 7890A gas chromatograph, an MS 5975C VL MSD mass spectrometer detector and provided with an HP-5MS capillary column. The data acquisition and data processing were performed using the MSD Chemstation E.01.01.335 (Agilent) software.

3 Result and Discussion

The term essential oil is used to refer to highly volatile substance isolated by steam distillation from an odoriferous plant of a single botanical species. This method is the favorite method for essential oil extraction process was steam distillation [14]. In this study, the java citronella oil from the grasses of *Cymbopogon winterianus* was extracted by the hot steam that applied in this method. The final of steam distillation process will be condensed and separated into the liquid form that was the mixture of oil and steam. The oils were less dense than water and could be separated using proper method and instruments.

3.1 *Physical Properties of Java Citronella Oil*

The java citronella oil was obtained from grasses of *Cymbopogon winterianus* by using steam distillation methods are pale yellow to yellow when freshly distilled. This essential oil yielded varied from 0.956% to 1.06% and shown in Table 1. The physical properties of java citronella oil shown in Table 2. They are only slightly soluble in water and dissolve fairly well in ethanol and mixed very well with vegetable oils, glycerol, and propylene glycol.

Table 1. The Yield of Java Citronella Oil (*Cymbopogon winterianus*)

Parameter	Java Citronella Oil				
	I	II	III	IV	V
Sample Weight (kg)	5.00	5.10	5.10	5.00	5.05
Content of Oils (g)	47.80	50.45	52.90	51.90	53.03
Yield (%)	0.956	0.989	1.04	1.04	1.06

Table 2. The Physical Properties of Java Citronella Oil (*Cymbopogon winterianus*)

Parameter	Java Citronella Oil				
	I	II	III	IV	V
Appearance	Oily liquid				
Color	Pale Yellow			Yellow	
Odor	sweet, warm floral, woody, citrusy				
Refractive Index at 20 °C	1.467	1.467	1.469	1.468	1.469
Specific Gravity at 25 °C	0.879	0.877	0.880	0.880	0.881

3.2 Chemical Composition of Java Citronella Oil

The combination of gas chromatography and mass spectrometry (GC-MS) allows rapid identification of essential oil components, provided that these compounds are already known and their mass spectra available in a library. According to the data of GC-MS analysis, we investigated essential oil in *Cymbopogon winterianus* grasses consist mostly of terpenoid compounds. The chemical composition of java citronella oil shown in Table 3. The main component of Java citronella oil are citronellal, citronellol, and geraniol has yielded 21,59%; 7,43%; and 34,27%, respectively. The Chromatogram of java citronella oil shown in Figure 3.

Table 3. Chemical Compositions of Java Citronella Oil (*Cymbopogon winterianus*)

Composition	RT	Percent (%)
β-myrcene	4.274	0.10
α-pinene	5.007	6.739
β-cis-ocimene	5.216	1.903
Linalool	6.311	1.07
Neo-allo-ocimene	7.321	0.18
β-citronellal	7.991	21.59
(-)-L-isopulegol	8.272	0.36
Capraldehyde	10.011	0.26
β-citronellol	11.421	7.43
cis-3,7-Dimethyl-2,6-octadienal	11.518	0.44
cis-geraniol	12.852	34.27
α-citral	12.944	0.61
Geranyl formate	14.403	0.10
Citronellyl acetate	16.418	1.63
Nerylacetate	16.699	0.04
Geranyl acetate	17.395	4.89
(-)-*β*-elemene	18.103	0.72
β-caryophyllene	18.978	0.88
α-bergamotene	19.545	0.72
α-humulene	19.978	0.13
(E)-β-farnesene	20.064	0.13
Methyl iso-eugenol	20.577	3.00
(-)-germacrene D	20.766	1.18
α-farnesene	21.155	2.90
α-muurolene	21.317	0.16
α-amorphene	21.668	0.27
δ-cadinene	21.917	0.75
Elemol	22.452	1.07
Geranyl butyrate	22.673	1.24

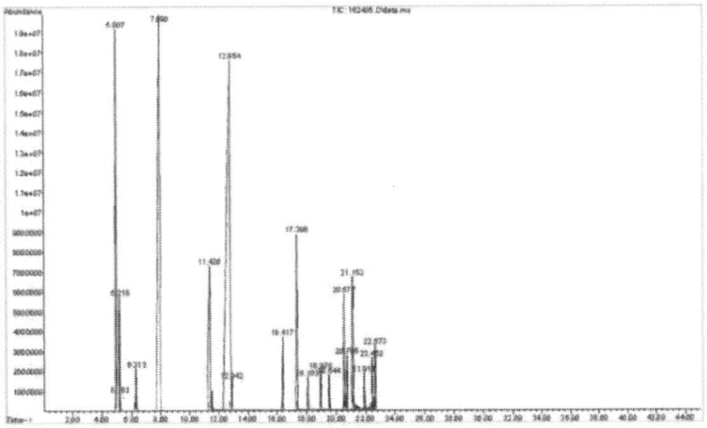

(a)

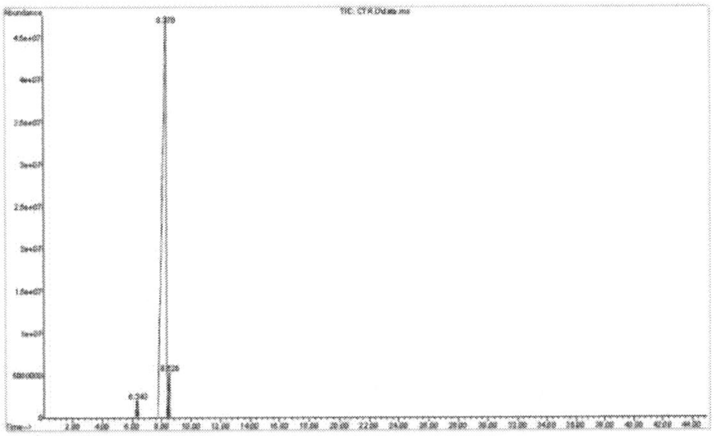

(b)

Figure 3. (a) Chromatogram of Java Citronella Oil from Steam Distillation of *Cymbopogon winterianus* Grasses and (b) Chromatogram of Citronellal Oil

3.3 Physical Properties of Active Component From Java Citronella Oil

The terpenoids are, by far, the most important group of java citronella oil. They are defined as substances composed of isoprenes (2-methyl butadiene) units such as citronellal, citronellol, and geraniol. Fractional distillation under reduced pressure and continued redistilled are needed to isolate this terpenoids compounds that are active component in java citronella oil. According to this method, we can be isolated the main active component in java citronella oil. They are citronellal, citronellol, geraniol, and also *p*-menthane-3,4 diol. Citronellal is responsible for the characteristic of odor in java citronella oil. It shows an intense citrus-like odor but seems to be less sweet and fruity than citral. Geraniol and citronellol are known as the rose alcohols because of their occurrence in rose oils and also because they are the key materials responsible for the rose odor character in java citronella oil. *p*-menthane-3,8-diol, also known as *para*-menthane-3,8-diol or PMD is a mixture of the cis and trans isomers of *p*-menthane-3,8-diol. This compound is obtained by purification of citronellal through fractional distillation under pressure. PMD is structurally similar to menthol so that the odor like menthol that gave minty and eucalyptus-like aroma. Each of active component has the characteristic of physical properties that shown in Table 4.

Table 4. The Physical Properties of Active Component from Java Citronella Oil (*Cymbopogon winterianus*)

Parameter	Active Component of Java Citronella Oil*				
	C-al	R-ol	C-ol	G-ol	PMD
Appearance	Oily Liquid	Oily Liquid	Oily Liquid	Oily Liquid	Oily Liquid Slightly Viscous
Color	Colorless	Light Yellow	Pale Yellow	Pale Yellow	Pale Yellow
Odor	Citrus-like aroma, floral rosy waxy, slightly sweet, green, and aldehyde	Sweet, rosy, floral, waxy, citronellal like with fresh citrus	Herbaceous rose with fruity citrus nuances, floral, and slightly sweet	Sweet rose, floral, fruity and fresh note	Minty, herbaceous, eucalyptus-like aroma
Refractive Index at 20 °C	1.448	1.479	1.462	1.476	1.470
Specific Gravity at 25 °C	0.851	0.905	0.857	0.885	0.965

*Note: * C-al: Citronellal; C-ol: Citronellol; PMD: p-menthane-3,8-diol; R-ol: Rhodinol; G-ol: Geraniol*

3.4 *Chemical Composition of Active Component From Java Citronella Oil*

The active component of Java citronella oil obtained through fractional distillation method. Redistilled can improve the purity, then distillate collected while the temperature changed. The identification of the active component fraction from java citronella oil was carried out by gas chromatography-mass spectrometry (GC-MS). In the first distillate yielded citronellal with a purity of 95.10%. The chromatogram of citronellal oil shown in Figure 4. The first distillate obtained residue rhodinol product will then be carried back to separation into citronellol and geraniol with the purity of 19.33% and 34.31%, respectively. The purity of citronellol reached 80,65% purity, whereas geraniol reached 76.63% purity. The *p*-menthane-3,8-diol obtained from refining citronellal resulting citronellal and *p*-menthane-3,8-diol reached 75.95% purity.

4 Conclusion

Java citronella oil is rich in citronellal, geraniol, and citronellol with purity of 21,59%; 7,43%; and 34,27%, respectively. Batch vacuum fractional distillation under reduced pressure yielded citronellal with a purity of 75.67%, and the residue rhodinol consist of citronellol and geraniol component with the purity of 19.33% and 34.31%, respectively. Purification of rhodinol yielded citronellol and geraniol with the purity of 80,65% and 76.63%, respectively. Purification resulting citronellal in 95.10% purity and *p*-menthane-3,8-diol reached 75.95% purity.

Acknowledgment
The authors acknowledge Penelitian Produk Terapan Fund from Kemenristekdikti for financial support that makes possible the development of our research.

Reference

[1] Alighiri, D., Eden, W.T., Supardi, K.I., Masturi, and Purwinarko, A., 2017, Potential Development Essential Oil Production of Central Java, Indonesia, *Journal of Physics: Conference Series*, **824** (1), 012021.

[2] Guenther, E., Althausen, D., and Sterrett, F.S., 1975, *The Essential Oils* (Florida: Krieger Publishing Company) p 87–90.

[3] Rocha, S.F.R., Ming, L.C., and Marques, M.O.M., 2000, The influence of drying temperature on the yield composition of citronella (*Cymbopogon winterianus* Jowitt) essential oil. *Revista Brasileira Plantas Medicinais*, **3**, 73–78.

[4] Patra, N.K., Singh, H.P., and Kalra, A., 1997, Isolation and development of a geraniol-rich cultivar of citronella (*Cymbopogon winterianus*). *J. Med. Aromat. Plant Sci*, **19**, 672–676.

[5] Silva, R.M., Ximenes, R.M., Martins da Costa, J.G., Kalyne, L., Leal, A.M., de Lopes, A., and deBarros Viana, G.S., 2010, Comparative anticonvulsant activities of the essential oils (EOs) from *Cymbopogon winterianus* Jowitt and *Cymbopogon citratus* (DC) Stapf. in mice. *Naunyn-Schmiedeberg's Arch. Pharmacol*, **381** (5), 415–426.

[6] Mahidol, C., 2004, Malaria: integrated approaches for prevention and treatment. *Acta Tropica* **89**, 265–269.

[7] Bauer, K., Garbe, D., and Surburg, H., 2001, *Common Fragrance and Flavor Materials: Preparation, Properties and Uses, Fourth Edition* (Germany: Wiley-VCH Verlag Gmbh).

[8] Lenardão, E.J., Botteselle, G.V., Azambuja, F., Perin, G., and Jacob, R.G., 2007, Citronellal as key compound in organic synthesis. *Tetrahedron*, **63** (29), 6671–6712.

[9] Liu, J., Zhang, W., Du, G., Chen, J., and Zhou, J., 2013, Overproduction of geraniol by enhanced precursor supply in *Saccharomyces cerevisiae*, *Journal of Biotechnology*, **168** (4), 446–451.

[10] Murakami, A., Furukawa, J., Kawasaki, Y., and Ota, R., 2013, *Flavoring agent with natural fruit-like hop aroma and manufacturing method of beverages*. (Japan: Kirin Brewery Company, Ltd.), pp 19.

[11] Silvestre, W.P., Agostini, F., Muniz, L.A.R., and Pauletti, G.F., 2016, Fractionating of green mandarin (*Citrus deliciosa* Tenore) essential oil by vacuum fractional distillation, *Journal of Food Engineering*, **178**, 90–94.

[12] Vives, D., 1981, Principles of Unit Operations, 2nd Edition (Foust, Alan S.; Wenzel Leonard A.; Clump, Curtis W.; Maus, Lois; Anderson, L. Bryce), *Journal of Chemical Education*, **58** (4), A154.

[13] Adari, P.V.R.K. and Jana, A.K., 2008, Comparative control study of a high-purity ternary batch distillation. *Chemical Product and Process Modeling*, **3** (1), 26.

[14] Masango, P, 2005, Cleaner Production of Essential Oils by Steam Distillation, *Journal of Cleaner Production*, **13** (8), 833–839.

IOP Conf. Series: Materials Science and Engineering **349** (2018) 012068 doi:10.1088/1757-899X/349/1/012068

Transformation of Indonesian Natural Zeolite into Analcime Phase under Hydrothermal Condition

W W Lestari,[1*] D N Hasanah,[1] R Putra,[1,2] R R Mukti,[3] K D Nugrahaningtyas[1]

[1]Research Group of Porous Materials for Sustainability, Chemistry Department, Faculty of Mathematics and Natural Sciences, Sebelas Maret University, Jl. Ir. Sutami No. 36A, Kentingan, Jebres Surakarta, 57126, Indonesia.
[2]Master of Chemistry Program, Graduate School, Sebelas Maret University, Jl. Ir. Sutami No. 36A, Kentingan, Jebres Surakarta, 57126, Indonesia.
[3]The Division of Inorganic Physical Chemistry, Faculty of Mathematics and Natural Sciences, Institute Technology Bandung, Jl. Ganesha No.10 Bandung, 40132, Indonesia.

[*]Corresponding Author. Email: witri@mipa.uns.ac.id

Abstract. Natural zeolite is abundantly available in Indonesia and well distributed especially in the volcano area like Java, Sumatera, and Sulawesi. So far, natural zeolite from Klaten, Central Java is one of the most interesting zeolites has been widely studied. This research aims to know the effect of seed-assisted synthesis under a hydrothermal condition at 120 °C for 24 hours of Klaten's zeolite toward the structural change and phase transformation of the original structure. According to XRD and XRF analysis, seed-assisted synthesis through the addition of aluminosilicate mother solution has transformed Klaten's zeolite which contains (mordenite and clinoptilolite) into analcime type with decreasing Si/Al ratio from 4.51 into 1.38. Morphological analysis using SEM showed the shape changes from irregular into spherical looks like takraw ball in the range of 0.3 to 0.7 micrometer. Based on FTIR data, structure of TO_4 site (T = Si or Al) was observed in the range of 300-1300 cm^{-1} and the occupancy of Brønsted acid site as OH stretching band from silanol groups was detected at 3440-3650 cm^{-1}. Nitrogen adsorption-desorption analysis confirmed that transformation Klaten's zeolite into analcime type has decreased the surface area from 55.41 to 22.89 m^2/g and showed inhomogeneous pore distribution which can be classified as micro-mesoporous aluminosilicate materials.

1. Introduction

Natural zeolites in Indonesia are spread out over several large islands, especially those with volcanic areas, such as Java, Sumatera and Sulawesi with the main phase content of mordenite (ca. 60-70%) and the rest are clinoptilolites and amorphous quartz [1–4]. Zeolite is divided into hundreds of topology types [5], one type of zeolite topology, mordenite, that is often used as a catalyst in the industrial world [6]. It can be used as a catalyst in cracking reaction, isomerization [6] and combinations of hydrogenation cracking reaction [2,3]. Natural zeolite from Klaten, Central Java contain mordenite and clinoptilolite with Si/Al ratio of 4.96 [7]. However, the existence of clinoptilolite and other impurities obtained in Klaten's natural zeolite need to be purified such as through dealumination or acid activation.

Another way to an obtained pure phase of zeolite might be carried out by phase transformation or through seed-assisted synthesis.

The activation process could be conducted via physically and chemically treatments [8]. The physical activation process is carried out by heating (calcination) to evaporate water trapped in zeolite pores so that the amount of pore and its specific surface area increases. Chemical activation can be proceed by employing hydrochloric acid, sulfuric acid, hydro-fluoride acid, and nitric acid solution to purify the pore surface, remove the impurities and re-arrange the exchangeable atoms [9]. Additionally, activation can also be performed by ammonium chloride or ammonium nitrate addition as an advanced process after catalytic treatments aimed for cation exchange [10]. These activation processes can improve purity and crystallinity of mordenite. However, based on Material Safety Data Sheet (MSDS) [11] strong acids like HF and HCl are corrosive, so better when it is avoided.

Another method possibly be used to improve the purity of mordenite is the seed-assisted synthesis. This method is more effective and environmentally friendly. The process of transformation of natural materials through hydrothermal synthesis by employing natural minerals as seed in the aluminosilicate parent solution was conducted by Mukti *et al.* [12]. This hydrothermal seed-assisted synthesis method has proven to transform natural minerals instead of zeolite into mordenite-type zeolite with high crystallinity [12]. This method was also successfully performed by Mignoni *et al.* [13] which transforms kaolin material into mordenite.

Transformation of natural mineral, like kaoline zeolite into a specific phase of zeolite currently become an interest. This transformation has spurred many researchers towards improving the quality of natural zeolites, synthesizing novel zeolites, and employing them in new applications. Analcime is a type of zeolite which abundantly available in nature. In a recent report, Sandoval *et al.* [14] synthesized of zeotype ANA framework by a hydrothermal reaction using natural clinker as starting material. Other type of ANA was also synthesized by Atta *et al.* [15] using a hydrothermal technique using rice husk ash and meta-kaolin as sources of silica and alumina with optimal condition sodium hydroxide (Na_2O/Al_2O_3 = 2.6) after aging for 3 days and reaction time of 24 h at temperature of 180 °C. A successful synthesis of analcime was also reported by Hsiao *et al.* [16] under microwave-assisted hydrothermal route with commercial mica powders, composed of sericite and pyrophyllite, were treated by Na_2SiO_3 as the mineralizer. Herein, we report for the first time the transformation of natural zeolite (NZ) from Klaten into analcime phase via seed-assisted synthesis under the hydrothermal condition.

2. Experimental Section

2.1. Materials

NZ was obtained from Katen, Central Java (particle size 170/240 mesh), Tetraethylorthosilicate (TEOS, 99%), Aluminium Hydroxide [$Al(OH)_3$] (98%), Natrium Hydroxide (NaOH, 99%) were provided from Merck, Germany and aquadest (Bratachem, Indonesia). All reagents were used as purchased without any further purification.

2.2. Preparation of Mother Solution and Hydrothermal Reaction

The mother solution was prepared according to modified literature procedure published by Kim *et al.* [17]. NaOH (1.52 g, 38 mmol) was dissolved in 3.2 mL aquadest. Into this solution $Al(OH)_3$ (0.57 g, 7.3 mmol), aquadest (25.78 mL) and TEOS (3.93 g, 18.89 mmol) were carefully added. The mixture was then stirred for 30 minutes. The obtained mother solution was called as Z-1 (without seed-assisted). As variation, all reagent was measured in the same composition, but the addition of TEOS was modified (5 g, 24 mmol) then the sample was called as Z-2 and Z-3. In each Z-2 and Z-3 were added 1.62 g Klaten's natural zeolite (NZ) as seed.

Sample Z-1 and Z-2 were placed into an autoclave (50 mL) and continued with reaction under a hydrothermal condition at temperature 170 °C for 24 h in an Oven Memmert UN30. The obtained gels were filtered and washed with aquadest until a white precipitate was detected, then dried at room temperature for 24 h and continued with drying at 120 °C for 6 h. The variation of hydrothermal reaction was also conducted at 120 °C for 24 h according to the procedure reported by Azizi *et al.* [18] dried at room temperature for 24 h and continued heated at 80 °C for 10 h. The obtained sample was notified as Z-3.

2.3. Materials Characterization

XRD diffractograms were obtained from Rigaku miniflex 600 dan X-ray Diffraction Philip type X'Pert MPD using Cu–Kα radiation (λ = 1.5406 Å) to confirm the expected structure, phase purity and the crystallinity of the obtained materials. Elemental composition and Si/Al ratio of the materials were recorded using X-ray Fluorescence Spectrometry BRUKER S2 Ranger. Morphological analysis was observed using Scanning Electron Microscopy (SEM) FEI type Inspect S50. To determine the surface area, porosity dan pore size distribution analysis were conducted in liquid N_2 at 77.3 K using an Autosorb-iQ (Quantachrome Co., USA). FTIR analysis using Shimadzu IR Prestige-21 using the KBr technique in the range 4000–400 cm^{-1} were performed to assign the functional groups found in the materials.

3. Results and Discussion

3.1. Materials characterization

X-Ray diffractogram of Z-1, Z-2 and Z-3 are presented in Figure 1. Seed-assisted synthesis of Z-2 and Z-3 resulted dominant diffraction peaks in high suitability with characteristic for analcime phase (ANA) at 2θ = 15.79° (211), 25.93° (400) and 30.51° (322) with parameter structure of symmetry of space group P-1 ANA, crystal system: triclinic, lattice parameter: a = 13.1605, b = 13.0149, and c = 13.0335 Å, angle between axis: α = 90.178 β = 86.849 γ = 88.601 according to ICSD #156598. Affording to the database of zeolite structures (IZA) [19] the framework of ANA having a 19.2 T/1000 $Å^3$ density greater than the MOR with a density of 17 T/1000 $Å^3$, so that the zeolite structure changes from the MOR topology to ANA is acceptable. Moreover, the excessive presence of aluminum causes the distance adjacent so as to increase the density of the framework [3]. In contrast, hydrothermal synthesis (with the same parameter condition) but without the addition of natural zeolite as a seed (Z-1) showed unidentified phase as shown in XRD analysis (Figure 1).

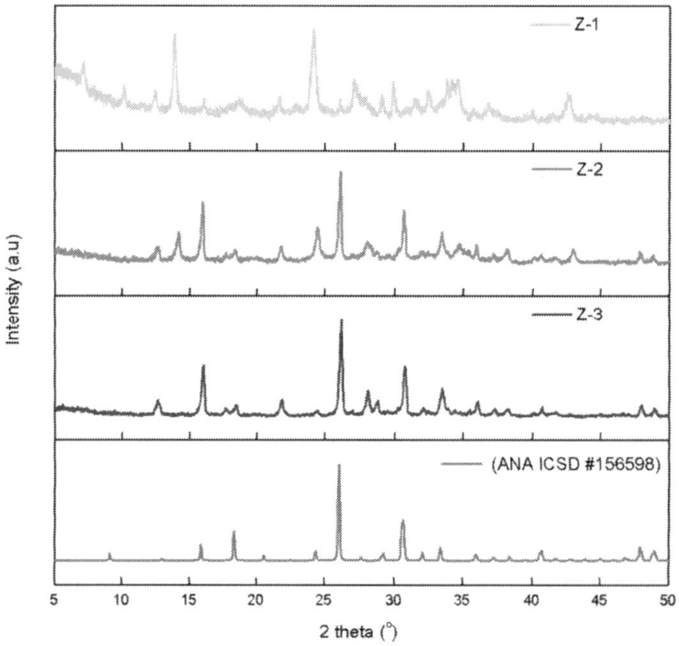

Figure 1. Diffractogram of Z-1, Z-2 and Z-3 in comparison with analcime standard pattern

Metal oxide contents and Si/Al ratio of NZ, Z-2 and Z-3 were observed using XRF spectroscopy and listed in Table 1. NZ itself has Si/Al ratio of about 4.51, it was transformed into Z-3 (in analcime phase) with Si/Al ratio 1.38, while Z-2 has Si/Al ratio 1.11.

Tabel 1. Metal oxides content and Si/Al ratio of NZ transformed into Z-2 and Z-3

Element	Concentration		
	NZ	Z-2	Z-3
SiO_2	**71.43%**	**38.01%**	**45.80%**
Al_2O_3	**12.84%**	**30.20%**	**22.88%**
CaO	5.53%	2.45%	4.57%
Fe_2O_3	4.34%	1.30%	2.28%
MgO	1.82%	-	-
K_2O	1.64%	0.36%	0.81%
MnO	0.07%	0.03%	0.04%
	Si/Al = 4.51	**Si/Al = 1.11**	**Si/Al = 1.38**

The surface morphological analysis using SEM shows that NZ has a totally different image from Z-1 and Z-3 after hydrothermal treatment (Figure 2). NZ has bigger size and irregular shape, while Z-1 appear as cleaner plaque or looks like a rectangular thin sheet, Z-3 present as a spherical-looks like a takraw ball in the range of 0.3 to 0.7 micrometer with a unflat surface.

Figure 2. SEM image of NZ (a), Z-1 (b) and Z-3 (c) at magnification of 20,000x

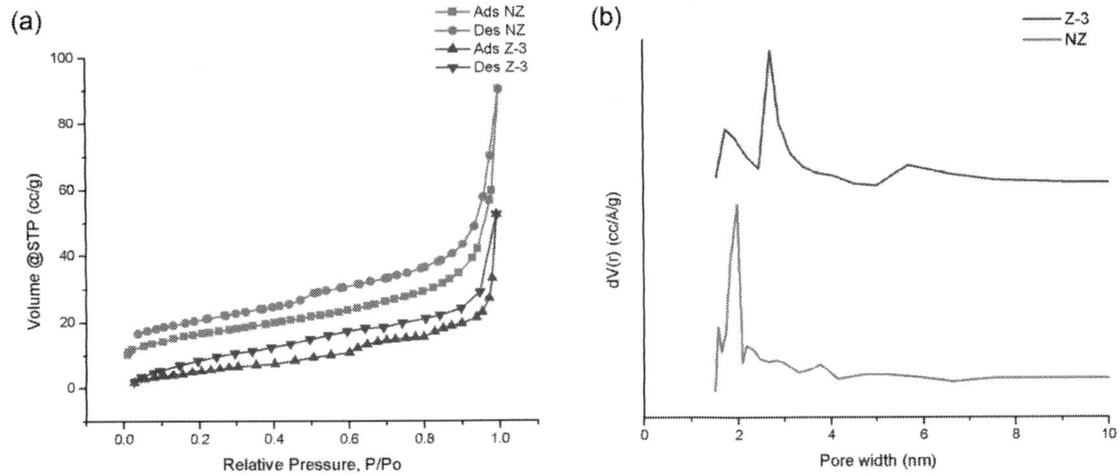

Figure 3. Nitrogen adsorption-desorption isotherm (a) and pore size distributions (b)

NZ exhibited type I adsorption isotherm typical of microporous materials with H4 type hysteresis loop and synthesized ANA phase (Z-3) exhibited type IV adsorption isotherm typical of mesoporous materials (figure 3(a)). This data is confirmed from BJH desorption branch result as shown in figure 3 (b) and proved that the materials are classified as crystalline micro-mesoporous (2-50 nm) aluminosilicate materials. The transformation of NZ into ANA (Z-3) lead to decrease BET surface area and pore volume total up to 22.888 m^2/g and 0.08134 cc/g as shown in Table 2.

Tabel 2. Surface area and porosity of NZ and Z-3

Materials	BET surface area (m^2/g)	Volume total pore (cc/g)	Average pore size (Å)
NZ	55.413	0.09249	33.3815
Z-3	22.888	0.08134	71.0794

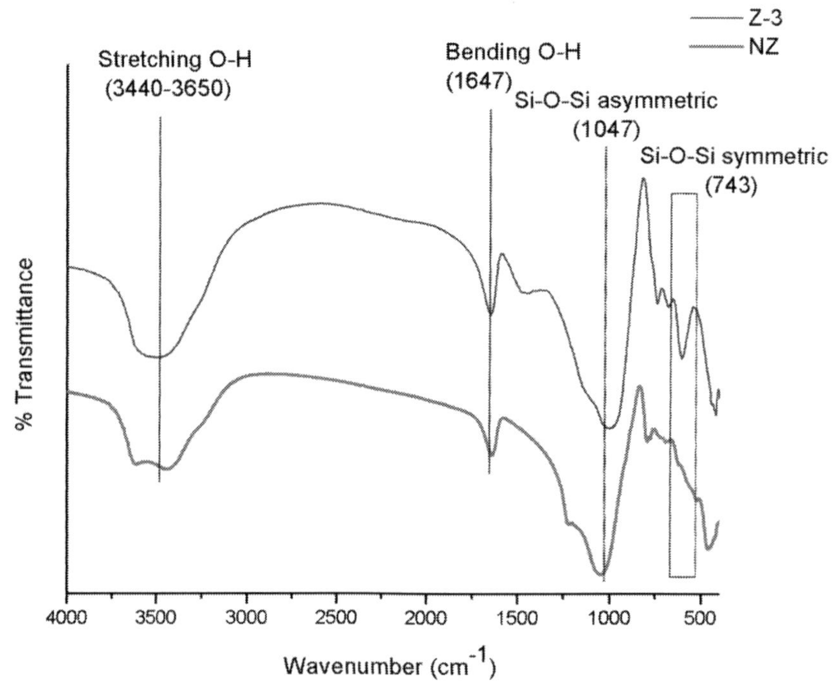

Figure 4. FTIR spectra of NZ and Z-3

There is no significant shift observed in FTIR spectra (Figure 4) of both materials, NZ and Z-3. The absorption spectra corresponding to the synthesized analcime obtained in this study are similar to previous results reported by Sandoval *et al.* and Atta *et al.* [14,15]. However, the present of absorption band at 743 cm^{-1} which is characteristic of symmetric stretching of Si-O-Si for the Z-3 sample has appeared indicated that synthesizing analcime is in a good agreement with the aluminosilicate materials. Furthermore, these spectra indicated that transformation of NZ into ANA phase (Z-3) under hydrothermal condition does not affect the change of functional group in zeolitic structure, although the frameworks are changed as confirmed previously by XRD data (Figure 1). Peaks in the range of 300-1300 cm^{-1} corresponded the frame vibration (T-O-T unit; where T is SiO_4 and AlO_4 tetrahedron) which is characteristic of aluminosilicate compounds such as zeolites. OH-groups as bridging unit (such as ≡Al(OH)Si≡ groups with Brønsted acidic character) give bands in the fundamental stretching region at about 3440-3650 cm^{-1} (free bridging OH-groups) which are responsible for important catalytic properties of the hydrogen forms of zeolites. Furthermore, the fundamental vibrations of bending OH-groups are approximately observed at 1600 cm^{-1}. Detail absorption peak and its assignment are presented in Table 3.

Tabel 3. Interpretation of IR Spectra of NZ and Z-3

Interval Wavenumber (cm^{-1})	NZ (cm^{-1})	Z-3 (cm^{-1})	Interpretation of vibration types
3400-3700	3444.05	3498.06	Stretching O-H
1600-1850	1639.56	1647.28	Bending O-H
950-1250	1047.39	1002.06	Asymmetric Si-O-Si
750-820	796.64	743.59	External symmetric O-Si-O/Al-O-Al
650-720	692.47	683.79	Internal symmetric O-Si-O/Al-O-Al
420-500	461.01	441.72	Bending Si-O/Al-O
300-420	446.54	420.50	Pore opening

4. Conclusions and Outlook

Seed-assisted synthesis of NZ under a hydrothermal condition at 120 °C for 24 h was successfully transformed the NZ which contain MOR and HEU phase into ANA (Z-3). This phenomenon totally changes the Si/Al ratio of the obtained material up to 1.38 and the morphology turns from irregular shape into a spherical-looks like a takraw ball. Purification of NZ into MOR or transformation into other type of zeolite such as ZSM-5 would be an interesting further investigation by reducing the $Al(OH)_3$ composition in mother solution or increasing the NaOH concentration. Moreover, further study about the application of the obtained ANA phase in this work for environmental or catalytic application is currently on progress in our group.

Acknowledgments

The authors thank to L'Oréal-UNESCO For Women in Science (FWIS national fellowship 2014 awarded to W.W.L) and Hibah MRG PNBP UNS 2017 project number 623/UN.27.21/PP/2017 for the financial support to perform the present study.

References

[1] Trisunaryanti W, Shiba R, Miura M, Nomura M, Nishiyama N and Matsuka M 1996 *J. Japan Pet. Inst.* **39** 20–25
[2] Trisunaryanti W, Purwono S and Putranto A 2008 *Indones. J. Chem.* **8** 342–347
[3] Trisunaryanti W, Syoufian A and Purwono S 2013 *J. Chem. Chem. Eng.* **7** 175–180
[4] Trisunaryanti W, Triwahyuni E and Sudiono E 2005 *Teknoin* **10** 269–282
[5] Treacy M M J and Higgins J B 2007 *Collection of Simulated XRD Powder Patterns for Zeolites* (Amsterdam: Elsevier) pp 133-134
[6] Fernandes L, Monteiro J, Sousa-Aguiar E, Martinez A and Corma A 1998 *J. Catal.* **177** 363–377
[7] Lestari D Y 2010 *Proc. Semin. Nas. Kim. Dan Pendidik. Kim* vol 2010 (Yogyakarta: Yogyakarta State University) p 6
[8] Priatna K, Suharto S and Syariffudin A 1985 Prospek Pemakaian Zeolit Bayah Sebagai Penyerap NH_4^+ dalam Air Limbah *Laporan Teknik Pengembangan* **69** PPTM Bandung
[9] Suyartono and Husaini 1991 *Bull. PPTM* **13** p 15
[10] Trisunaryanti W, Triwahyuni E and Sudiono S 2005 *Indo. J. Chem.* **5** 48–53

IOP Conf. Series: Materials Science and Engineering **349** (2018) 012068 doi:10.1088/1757-899X/349/1/012068

[11] Material Safety Data Sheet: Hydrofluoric Acid, 48% - 52% 2017 Science Lab
[12] Mukti R R, Wustoni S, Wahyudi A and Ismunandar 2013 *Indones. J. Chem.* **13** 278–282
[13] Mignoni M L, Petkowicz D I, Fernandes N R C and Pergher S B C 2008 *Appl. Clay Sci.* **41** 99–104
[14] Sandoval M V, Henao J A, Ríos C A, Williams C D and Apperley D C 2008 *Fuel* **88** 272–281
[15] Atta A Y, Jibril B Y, Aderemi B O and Adefila S S 2012 *Appl. Clay Sci.* **61** 8–13
[16] Hsiao Y H, Ho T Y, Shen Y H and Ray D 2017 *Appl. Clay Sci.* **143** 378–386
[17] Kim G J and Ahn W S 1991 *Zeolites* **11** 745–750
[18] Azizi S N, Daghigh A A and Abrishamkar M 2013 *J. Spectrosc.* **1** 5
[19] Ferraris G, Jones D W and Yerkess J, 1972 *Zeitschrift Für Krist* **135** 240–252.

IOP Conf. Series: Materials Science and Engineering **349** (2018) 012069 doi:10.1088/1757-899X/349/1/012069

Comparison of Conventional and Microwave-assisted Synthesis of Benzimidazole Derivative from Citronellal in Kaffir lime oil (*Citrus hystrix* DC.)

W Warsito[1,2,*], Noorhamdani A.S[3], Suratmo[1,2], R Dwi Sapri[1,2], D Alkaroma[1], A Z Azhar[1]

[1] Department of Chemistry, Faculty of Science, Brawijaya University, Jl Veteran Malang, 65145, Indonesia
[2] Essential Oil's Institute, Brawijaya University, Jl Veteran Malang 65145, Indonesia
[3] Department of Medical Faculty, Brawijaya University, Jl Veteran Malang 65145, Indonesia

*Corresponding author: warsitoub88@yahoo.com

Abstract. Simple method has been used for the synthesis of benzimidazole derivative from citronellal in kaffir lime oil under microwave irradiation. These compounds were synthesized also by conventional heating for comparison. In addtion, microwave-assited synthesis was also compared between using to dichloromethane and methanol solvents with variation of reaction time for 30 to 70 minutes and 4 to 12 h for conventional heating. The 2-citronellyl benzimidazole compound synthesized were characterised by FT-IR, GC-MS, ^{1}H and ^{13}C NMR spectroscopy. Comparison between conventional and microwave-assisted synthesis was done by comparing between correlation of reaction time and percentage yield. The time optimum of microwave-assisted and conventional synthesis using dichloromethane solvent respectively at 60 minutes (yield 19.23%) and 8 hours (yield 11.54%). In addition, microwave-assited synthesis increasing 157.81 times compared by conventional heating. While using methanol solvent tends to increase linearly however the percentage of yield only 0.77 times of synthesis using dichloromethane solvent.

1. Introduction

Benzimidazole, a fused heterocycle bearing benzene and imidazole has gained considerable attention in the field of contemporary medicinal chemistry. The moiety is of substantial importance because of its wide array of pharmacological activities [1]. At the core of these molecules have basic and acidic characters that can form salts with different acids. Optimization of the structure of benzimidazole has been widely practiced, such as changes in substituents in aryl groups, benzyl groups and or indole groups that can provide various bioactivity [2].

Benzimidazole and its derivatives are reported to be physiologically and pharmacologically active and some applications are found in the treatment of several diseases including epilepsy, diabetes and infertility [3]. These compounds show a wide range of biological activities like anti-inflammatory as well as analgesic activity [4,5]. antibacterial and antifungal [6,7], antitubercular [6], antiviral [8], anthelmintic and antiproliferative [9,10], antimalarial [11], antiulcerative, antioxidant [12,13], antiproliferative [14], antihypertensive [15, 16], antitumor [17,18]. Some benzimidazoles have also been evaluated as cholinesterase inhibitors [19,20], drugs against parasites [21,22] while some derivatives have been synthesized and evaluated for inhibition of HIV-1 [23].

The efficient and economical methods of synthesizing benzimidazole by condensation reaction between ortho phenylene diamine and various compounds in the presence of various reaction condition [24]. Nowadays, microwave-assisted organic synthesis is gaining widespread acceptance in drug discovery laboratories. Microwave technology, by accelerating chemical reactions from hours or days to minutes, provides quick results [25]. In this study, benzimidazole derivative of citronellal in kaffir lime oil were synthesized utilizing both microwave and conventional methods. The performance of two methods was compared and studying the reaction kinetics.

2. Material and Methods

2.1 Materials and Instrumentation

The kaffir lime oil was obtained from Essential Oil's Institute, Brawijaya University. The products were prepared by steam distillation.

All chemicals and solvents were supplied by Sigma Aldrich and were used without further purification. Microwave reaction was carried out in domestic microwave oven (LG 1200W 245 MHz). GC/MS 5977 Agilent Tech. with HP-5MS capillary column used to make TIC and mass spectra. Nuclear magnetic resonance (NMR) spectra were recorded on JEOL ECS 400 MHz NMR spectrometer using methanol d4 as solvent. Chemical shifts were reported in parts per million relative to internal standard tetramethylsilane. IR spectra of the compounds were recorded on Perkin Elmer FT-IR spectrometer with KBr pellets.

2.2 Methods

The Essential Oil Extraction

The kaffir lime oils were obtain by steam distillation. 375 kg of leaf and twig mix (w/w) 1: 1 distilled with vapor pressure 2.5-4 bar for 4h. The essential oil was characterized by refractometer, picnometer and GC type HP 5895 with HP-5MS capillary column.

Conventional Method

A mixture of 6.0 mL of kaffir lime oils (5.1 g), 1.4 g o-phenylenediamine and 10 mL dichloromethane were fed into a round bottom flask. The reflux condensation was carried out for 4 hours. Then, the mixture was cooled to room temperature and the nex cooled in the refrigerator for 24 hours. The crystals are filtered and washed with cold dichloromethane, dried in air and stored. Variations of reflux condensation have been done for 4, 6, 8, 10, 12 and 14 hours.

Microwave Method

A mixture of 6.0 mL (5.1 g) of kaffir lime oils, 1.4 g o-phenylenediamine and 10 mL dichloromethane were fed into a round bottom flask and microwaved at 1200W for 60 min. The reaction was monitored by TLC silica with ethylacetate: n-hexane: methanol (3: 2 : 1) as eluent. Then, the mixture was cooled to room temperature and then kept in the refrigerator for 24 hours. The crystals are filtered and washed with cold dichloromethane, dried in air and stored. Variations of reaction time in microwave have been done for 30, 40, 50, 60 and 70 minutes. Another variation in synthesis used methanol solvent.

3. Result and Discussion

Kaffir lime oil as a citronellal source was obtained from steam distillation. The characteristic of kaffir lime oil that is liquid yellow with unique aromatic scent and refractive index, specific gravity, percentage of rendemen, percentage of ccitronellal are 1,439; 0.85 g/ml, 0.5% and 60.40% respectively.

2-Citronellyl benzimidazole were synthesised by both conventional synthesis and microwave-assisted synthesis by synthetic schemes (Fig. 1). The products of benzimidazole compounds synthesized using dichloromethane solvent appear purer in the form of white crystals, while using methanol obtained yellow crystals. The structures of the synthesized compounds were confirmed using IR, GC-MS and NMR. Character of benzimidazole compound synthesized on the IR spectra contained a distinctive band at 1661 cm^{-1} and 1271 cm^{-1} as C = N str and C-N str of imidazole group, C = C str of benzene group at 1622 cm^{-1} and 1456 cm^{-1} and (C = C str) of alkene group at 1379 cm^{-1}. GC-MS analysis showed pure (tR 15.339 minute) and m/z 242 as molecular ion (M^+) and m/z 145 as base peak of mass-spectra. The NMR spectroscopic character exhibited ^{1}H-NMR and ^{13}C-NMR were characterized by 11 proton spectra types and 16 carbon-13 spectra as shown in Fig.2.

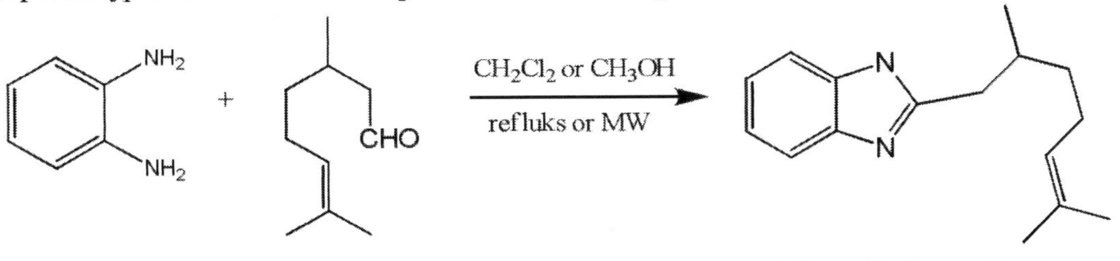

o-phenylenediamine citronellal 2-Citronellyl benzimidazole

Fig. 1 Synthesis benzimidazole derivative compound

The comparative data of the synthesized compounds are provided graph in Fig.3. The reaction time for the synthesis of all compounds by conventional heating methods was 4 to 10 h, in comparison with the microwave heating (30-60 min), an obvious many-fold time reduction. Overall approximately, the rate of synthesis of benzimidzole by microwave method is greater than the conventional heating method. The time optimum of microwave-assisted syntheses using dichloromethane solvent at 60 minute with yield 19.23% and increase157.81 times compared by conventional hating.

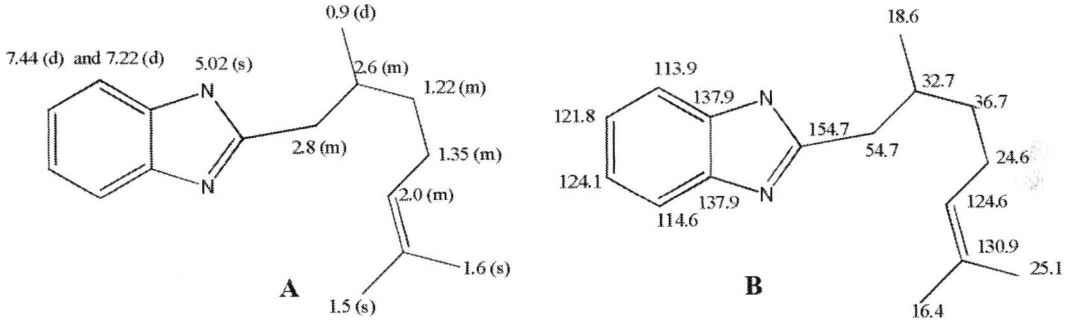

Fig 2. NMR spectra pattern of 2-citronellyl benzimidzole
A. δ (ppm) of ^{1}H-NMR spectra B. δ (ppm) of ^{13}C-NMR

In addition, the microwave-assisted synthesis using methanol tends to increase linearly with increasing reaction time with percent yield only 0.77 times of synthesis using dichloromethane solvent.

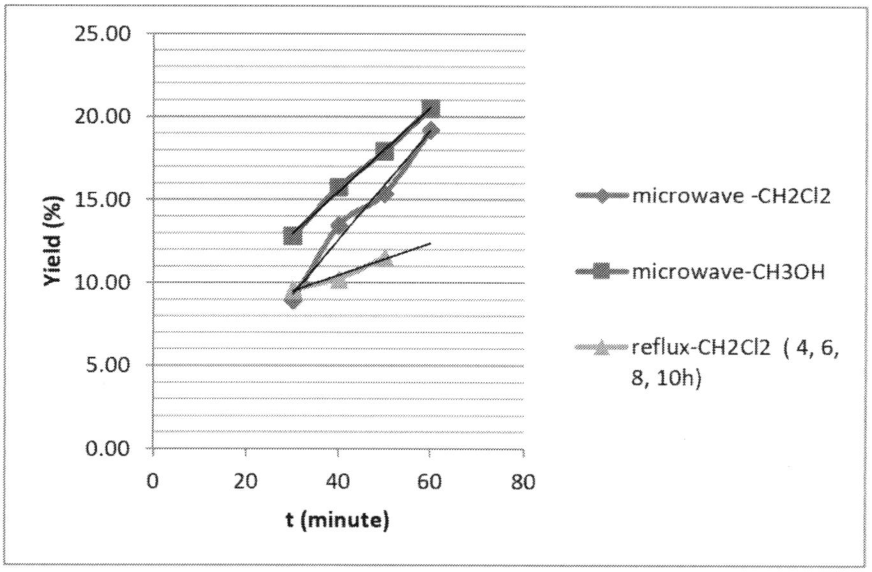

Figure 3. Correlation of time reaction and % yield of 2-Citronellyl benzimidazole synthesis

4. Conclusion

The time optimum of microwave-assisted and conventional synthesis using dichloromethane solvent respectively at 60 minutes (yield 19.23%) and 8 hours (yield 11.54%). In addition, the microwave-assisted synthesis increasing 157.81 times compared by conventional heating. While using methanol solvent tends to increase linearly however the percentage of yield only 0.77 times of synthesis using dichloromethane solvent.

Acknowledgments

On this occasion, thank you to: (1) Directorate General of Higher Education that has funded this research; (2) Essential Oil's Institute that gave a facilities for the research activities in essential oils.

References

[1] Wasim Akhtara, Mohemmed Faraz Khan, Garima Verma, M. Shaquiquzzaman, M.A. Rizvib, Syed Hassan Mehdi, Mymoona Akhter, M. Mumtaz Alam, Therapeutic evolution of benzimidazole derivatives in the last quinquennial period, European Journal of Medicinal Chemistry 126 (2017) 705-753.

[2] Oztekin Algul, Andre Kaessler, Yagmur Apcin, Akin Yilmaz and Joachim Jose , Comparative Studies on Conventional and Microwave Synthesis of Some Benzimidazole, Benzothiazole and Indole Derivatives and Testing on Inhibition of Hyaluronidase, Molecules 2008, 13, 736-748.

[3] Pavel Kopel, Dorota Wawrzak , Vratislav Langer, Kristyna Cihalova, Dagmar Chudobova, Radek Vesely, Vojtech Adam and Rene Kizek, iological Activity and Molecular Structures of Bis(benzimidazole) and Trithiocyanurate Complexes, Molecules 2015, 20, 10360-10376; doi:10.3390/molecules200610360.

[4] Mohammed A.F., Abdel-Moty S.G., Hussein M.A., Abdel-Alim A.A., Design, synthesis and molecular docking of some new 1,2,4-triazolobenzimidazol-3-yl acetohydrazide derivatives with anti-inflammatory-analgesic activities, Arch. Pharm. Res. 36 (2013) 1465-1479.

[5] Eman S. Nossier, Hoda H. Fahmy, Nagy M. Khalifa, Wafaa I. El-Eraky and Marawan A. Baset, Design and Synthesis of Novel Pyrazole-Substituted Different Nitrogenous Heterocyclic Ring Systems as Potential Anti-Inflammatory Agents, Molecules 2017, 22, 512; doi:10.3390/molecules22040512.

[6] Ranjith, P.K.; Rajeesh, P.; Haridas, K.R.; Susanta,N.K.; Row, T.N.G.; Rishikesan, R.; Kumari, N.S., Design and synthesis of positional isomers of 5 and 6-bromo-1-(phenyl)sulfonyl-2-(4-nitrophenoxy)methyl-1H-benzimidazoles as possible antimicrobial and antitubercular agents. Bioorg. Med. Chem. Lett. 2013, 23, 5228–5234.
[7] Kankatea R.S., Gide P.S., Belsare D.P., Design, synthesis and antifungal evaluation of novel benzimidazole tertiary amine type offluconazole analogues, Arab. J. Chem. (2015). http://dx.doi.org/10.1016/j.arabjc.2015.02.002.
[8] Evans C.W., Atkins C., Pathak A., Gilbert B.E., Noah J.W., Benzimidazole analogs inhibit respiratory syncytial virus G protein function, Antivir. Res. 121 (2015) 31-38.
[9] Mavrova A.T., Wesselinova D., Vassilev N., Tsenov J.A, Design, synthesis and antiproliferative properties of some new 5-substituted- 2-iminobenzimidazole derivatives, Eur. J. Med. Chem. 63 (2013) 696e701.
[10] Hranjec M., Pavlovic G., Zamola G.K., Synthesis, crystal structure determination and antiproliferative activity of novel 2-amino-4-aryl-4,10-dihydro[1,3,5]triazino[1,2-a] benzimidazoles, J. Mol. Struct. 1007 (2012) 242-251.
[11] Toro, P.; Klahn, A.H.; Pradines, B.; Lahoz, F.; Pascual, A.; Biot, C.; Arancibia, R. Organometallic benzimidazoles: Synthesis, characterization and antimalarial activity. Inorg. Chem. Commun. 2013, 35, 126–129.
[12] Haque, R.A.; Iqbal, M.A.; Asekunowo, P.; Majid, A.; Ahamed, M.B.K.; Umar, M.I.; Al-Rawi, S.S.; Al-Suede, F.S.R. Synthesis, structure, anticancer,and antioxidant activityof para-xylyl linked bis-benzimidazolium salts and respective dinuclear Ag(I) N-heterocyclic carbene complexes (Part-II). Med. Chem. Res. 2013, 22, 4663–4676.
[13] Nile, S.H.; Kumar, B.; Park, S.W. In VitroEvaluation of Selected Benzimidazole Derivatives as an Antioxidant and XanthineOxidase Inhibitors. Chem. Biol. Drug Des. 2013, 82, 290–295.
[14] Kralova, V.; Hanusova, V.; Stankova, P.; Knoppova, K.; Canova, K.; Skalova, L. Antiproliferative effect of benzimidazole anthelmintics albendazole, ricobendazole, and flubendazole in intestinal cancer cell lines. Anti-Cancer Drugs 2013, 24, 911–919.
[15] Jain, A.; Sharma, R.; Chaturvedi, S.C. A rational design, synthesis,characterization, and antihypertensive activities of somenew substituted benzimidazoles. Med. Chem. Res. 2013, 22, 4622–4632.
[16] Sharma M.C., Sharma S., Sahu N.K., Kohli D.V., 3D QSAR kNN-MFA studies on 6-substituted benzimidazoles derivatives as nonpeptide angiotensin II receptor antagonists: a rational approach to anti-hypertensive agents, J. Saudi. Chem. Soc. 17 (2013) 167-176.
[17].Kahveci, B.; Mentes, E.; Ozil, M.; Ulker, S.; Erturk, M. An efficient synthesis of benzimidazoles via a microwave technique and evaluation of their biological activities. Mon. Chem. 2013, 144, 993–1001.
[18].Purushottamachar, P.; Godbole, A.M.; Gediya, L.K.; Martin, M.S.; Vasaitis, T.S.; Kwegyir-Afful, A.K.; Ramalingam, S.; Ates-Alagoz, Z.; Njar, V.C.O. Systematic Structure Modifications of Multitarget Prostate Cancer DrugCandidate Galeterone ToProduce Novel Androgen Receptor Down-Regulating Agents as an Approach to Treatment of Advanced Prostate Cancer. J. Med. Chem. 2013, 56, 4880–4898.
[19] Yoon, Y.K.; Ali, M.A.; Wei, A.C.; Choon, T.S.; Khaw, K.Y.; Murugaiyah, V.; Masand, V.H.; Osman, H. Synthesis, characterization, and molecular docking analysis of novel benzimidazole derivatives as cholinesterase inhibitors. Bioorg. Chem. 2013, 49, 33–39.
[20] Zhu, J.M.; Wu, C.F.; Li, X.B.; Wu, G.S.; Xie, S.; Hu, Q.N.; Deng, Z.X.; Zhu, M.X.; Hong, X.C.; Luo, H.R. Synthesis, biological evaluation and molecular modeling of substituted 2-aminobenzimidazoles as novel inhibitors of acetylcholinesterase and butyrylcholinesterase. Bioorg. Med. Chem. 2013, 21, 4218–4224.
[21] Ceballos, L.; Virkel, G.; Elissondo, C.; Canton, C.; Canevari, J.; Murno, G.; Denegri, G.; Lanusse, C.; Alvarez, L. A pharmacology-based comparison of the activity ofalbendazole and flubendazole against Echinococcus granulosus metacestode in sheep. Acta Trop. 2013, 127, 216–225.

[22] Perez-Villanueva, J.; Hernandez-Campos, A.; Yepez-Mulia, L.; Mendez-Cuesta, C.; Mendez-Lucio, O.; Hernandez-Luis, F.; Castillo, R. Synthesis and antiprotozoal activity of novel 2-[2-(1H-imidazol-1-yl)ethyl sulfanyl]-1H-benzimidazole derivatives. Bioorg. Med. Chem. Lett. 2013, 23, 4221–4224.

[23] Roth T, Morningstar ML, Boyer PL, Hughes SH, Buckheit RW, Michejda CJ,. Synthesis and Biological Activity of Novel Non nucleoside Inhibitors of HIV1 Reverse Transcriptase 2-Aryl-Substituted Benzimidazoles. J Med Chem 1997; 40: 4199 4207.http://dx.doi.org/10.1021/jm970096g.

[24] Shatha Ibrahim Alaqeel, Synthetic approaches to benzimidazoles from o-phenylenediamine : A literature review, Journal of Saudi Chemical Society (2017)21, 229–237.

[25] Shah, J.J and Krishnapriya Mohanraj, Comparison of Conventional and Microwave-assisted Synthesis of Benzotriazole Derivatives, Indian J Pharm Sci 2014; 76(1): 46-53.

IOP Conf. Series: Materials Science and Engineering **349** (2018) 012070 doi:10.1088/1757-899X/349/1/012070

Study of Catalyst Variation Effect in Glycerol Conversion Process to Hydrogen Gas by Steam Reforming

Widayat[1,2], R Hartono[1], E Elizabeth[1], A N Annisa[1]

[1]Department of Chemical Engineering, Faculty of Engineering, Diponegoro University, Jalan Prof Soedarto, SH, Tembalang, Semarang, Central Java, Indonesia, 50725.

[2]Laboratory of Material Technology, Central of Research and Service Diponegoro University, Jalan Prof Soedarto, SH, Tembalang, Semarang, Central Java, Indonesia, 50725.

E-mail: widayat@live.undip.ac.id

Abstract. Along with the economic development, needs of energy being increase too. Hydrogen as alternative energy has many usages. Besides that, hydrogen is one source of energy that is a clean fuel, but process production of hydrogen from natural gas as a raw material has been used for a long time. Therefore, there is need new invention to produce hydrogen from the others raw material. Glycerol, a byproduct of biodiesel production, is a compound which can be used as a raw material for hydrogen production. By using glycerol as a raw material of hydrogen production, we can get added value of glycerol as well as an energy source solution. The process production of hydrogen by steam reforming is a thermochemical process with efficiency 70%. This process needs contribution of catalyst to improve its efficiency and selectivity of the process. In this study will be examined the effect variation of catalyst for glycerol conversion process to hydrogen by steam reforming. The method for catalyst preparation was variation of catalyst impregnation composition, catalyst calcined with difference concentration of hydrochloric acid and calcined with difference hydrochloric acid ratio. After that, all of catalyst which have been prepared, used for steam reforming process for hydrogen production from glycerol as a raw material. From the study, the highest yield of hydrogen gas showed in the process production by natural zeolite catalyst with 1:15 Hydrochloric acid ratio was 42.28%. Hydrogen yield for 2M calcined natural zeolite catalyst was 38.37%, for ZSM-5 catalyst was 15.83%, for 0.5M calcined natural zeolite was 13.09% and for ultrasonic natural zeolite was 11.43%. The lowest yield of hydrogen gas showed in catalyst 2Zn/ZSM-5 with 11.22%. This result showed that hydrogen yield product was affected by catalyst variation because of the catalyst has difference characteristic and difference catalytic activity after the catalyst preparation process.

1 Introduction

Nowadays, energy production is always linked with clean emissions by an environmentally friendly raw material, like hydrogen. The increasing of energy needs doesn't support by a suffice energy source which is from the fossil materials, so availability of energy is insufficient to satisfy human needs [1]. Hydrogen as alternative energy has many usages. Besides that, hydrogen is one source of energy that is a clean fuel, but process production of hydrogen from natural gas as a raw material has been used for a long time. Therefore, there is need new invention to produce hydrogen from the others raw material [2]. Glycerol, a byproduct of biodiesel production, is a compound which can be used as a raw material for hydrogen production. Glycerol can converted to hydrogen (H_2), carbon monoxide (CO), carbon dioxyde (CO_2), methane (CH_4) and the others trace of hydrocarbon. The products are used as an alternative fuel [3]. Considering of the increasing need for renewable fuels throughout the world, there is increasing production of biodiesel too. So, an excess of glycerol will be available in the world [4]. By using glycerol as a raw material of hydrogen production, we can get added value of glycerol as well as an energy source solution.

The process production of hydrogen by steam reforming is a thermochemical process that can produce up to 70% efficiency [6], but this process need contribution of catalyst to improve its efficiency. Catalyst for steam reforming process must have good catalytic activity, high selectivity to hydrogen, and resistance to coke formation [7]. The catalyst for steam reforming process has been investigated by [8]. An alumina supported Ni catalyst was used in catalytic steam reforming and get the hydrogen yield 30%. It was prepared by wet impregnation and result good structure for catalytic process [8]. In this research will be examined the effect of catalyst variation such as natural zeolite catalyst 1:15 HCL ratio [12], 2-Zn/ZSM-5 [9], 2M calcined natural zeolite [10], 0,5M calcined natural zeolite [10], ZSM5 and ultrasonic natural zeolite [11].

2 Methods

2.1 Material

Glycerol used as reactant in this research is in technical grade from Indrasari Chemical Store. Aquadest as reactant was bought from Membrane Research Center (MeR-C) Diponegoro University. Natural zeolite was bought from Wonosari Gunung Kidul and synthetic catalyst ZSM-5 was made by Latourette (1990) method used for catalytic processes. $ZnSO_4.7H_2O$ in pro analysis grade used to add zinc metal component by impregnation ZSM-5. Hydrochloric acid in technical grade used for chemical treating the natural zeolite. Pure nitrogen gas used for a gas carrier of the process.

2.2 Catalyst preparation and characterization

The catalysts used in this research are based from natural zeolite and also synthetic catalyst ZSM-5. Firstly, the synthetic catalyst was made by Latourette (1990) method. Then 2Zn-ZSM5 was made by adding zinc component 2% weight/weight catalyst to ZSM-5 through wet impregnation method. The natural zeolite was being reduced into 100 mesh. Then hydrochloric acid solution with concentration 4M used as the solvent. Then, silver nitrate solution was also added for washing indicator. After that, by used three-neck flask then it was put in ultrasonic cleaner type JP-060S using 40 kHz frequency for the ultrasonic process [11]. Other variances of the natural zeolite are treated using the ratio of hydrochloric of 1:15, 0.5 M and also 2 M. After being chemically treated, the catalyst product was being filtered and washed using distillate water. At last, the catalysts were dried in 110°C for 12 hours and calcined using furnace on the isothermal temperature at 500°C for 5 hours in inert condition by adding nitrogen gas. Characterization of the catalyst has been done by [9] for 2Zn/ZSM-5 catalyst and by [11] for ultrasonic zeolite catalyst. The 2M calcined zeolite catalyst and 0.5M calcined zeolite catalyst [10] and zeolite 1:15 Hydrochloric acid catalyst was done by [12]

2.3 Catalytic testing for steam reforming process

The catalyst performances are being tested through steam reforming process. The catalyst was placed in catalytic fix bed reactor. At first, the feed was made by mixing glycerol and aquadest with the volume ratio of 1:9. Then, the feed was being vaporized in 200°C before injected into the reactor. The nitrogen also injected to the reactor as the gas carrier. The reactor was previously heated to 500°C as the operation condition of the reactor for ten minutes of each process. Then, the product was passed through the cooler and the gas product being tested using gas analyzer and also hydrogen analyzer to get the gas composition.

Yield and the selectivity of each product being calculated by equation 2 and equation 3:

$$\text{Yield} = \frac{\text{hydrogen mass}}{\text{feed mass}} \times 100\% \quad (1)$$

$$\text{Selectivity} = \frac{\text{hydrogen mass}}{\text{gas product mass}} \times 100\% \quad (2)$$

3 Results and Discussions

3.1 Characterization of the catalyst

Characterization of the catalyst has been done by [9] for 2Zn/ZSM5 catalyst and by [11] for ultrasonic zeolite catalyst. The 2M calcined zeolite catalyst and 0.5M calcined zeolite catalyst by [10] and zeolite 1:15 Hydrochloric acid catalyst was done by [12].

Table 1. BET Result Surface Area of the Catalyst

Catalyst	Surface Area (m^2/g)	Reference
ZSM-5	20	[13]
2-Zn/ZSM-5	0.402	[9]
0.5M Calcined Natural Zeolite	101.943	[10]
2M Calcined Natural Zeolite	130.657	[10]
Ultrasonic Natural Zeolite	79.09	[11]
Zeolite 1:15 HCl	199.760	[12]

3.2 Hydrogen yield for various catalyst types

Figure 1 shows the highest yield could be achieved by zeolite catalyst which had been calcined by HCl 1:15 molar ratio, with hydrogen yield 42.48%. This high yield could be achieved because the natural zeolite already being chemically treated. From the studies before, the zeolite framework could be changed along with the treatment. Acid treatment could cause dealumination of the zeolite sample, which means it leached Al framework from zeolite, and it could increase the catalyst Si/Al ratio [14]. Along with the increasing of the Si/Al ratio, a catalyst could be more stable at high temperature. Then, for another catalyst, the results are 38.75% for 2M Calcined Natural Zeolite, 15.83% for ZSM-5, 13.09% for 0.5M Calcined Natural Zeolite, 11.43% for Ultrasonic Natural Zeolite and 11.22% for 2-Zn/ZSM5.The performance of catalyst 2-Zn/ZSM5 is affected by its surface area. From BET data it shows that the surface area of 2-Zn/ZSM5 is only 0.402 m^2/g, where it is much smaller than the surface area of natural zeolite that around 20 m^2/g. From the study, yield of the process is increase during the increasing of the surface area of the catalyst.

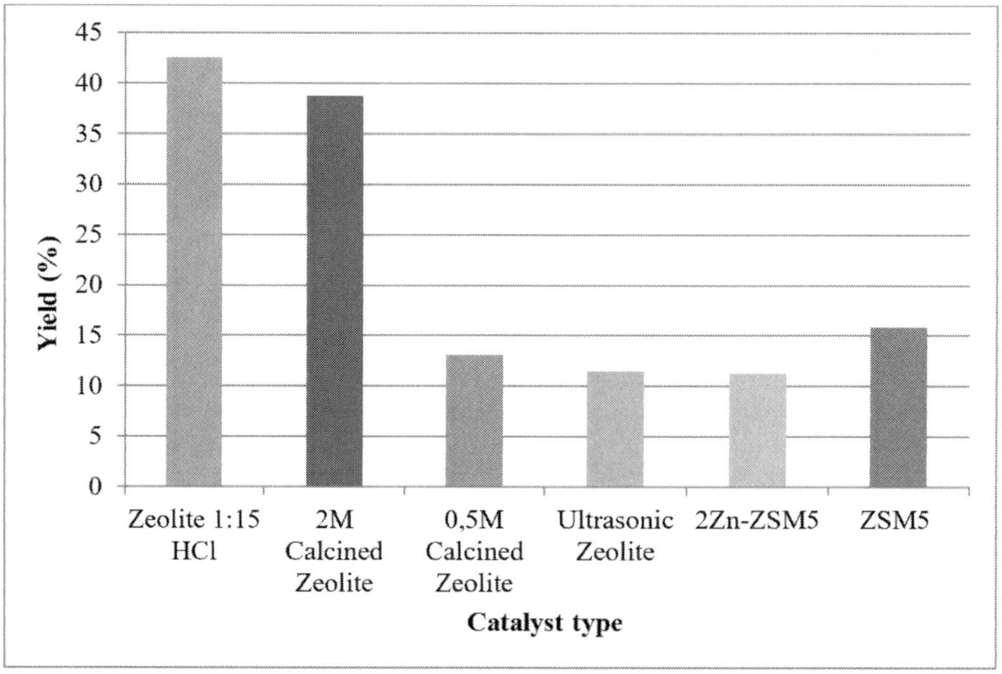

Figure 1. Yield of hydrogen for various catalyst types

3.3 Hydrogen selectivity for various catalyst types

Catalysts for steam reforming processes are currently a research subject, mainly focused on the improvement of the catalytic activity, the selectivity to hydrogen, and the resistance to coke formation [7].

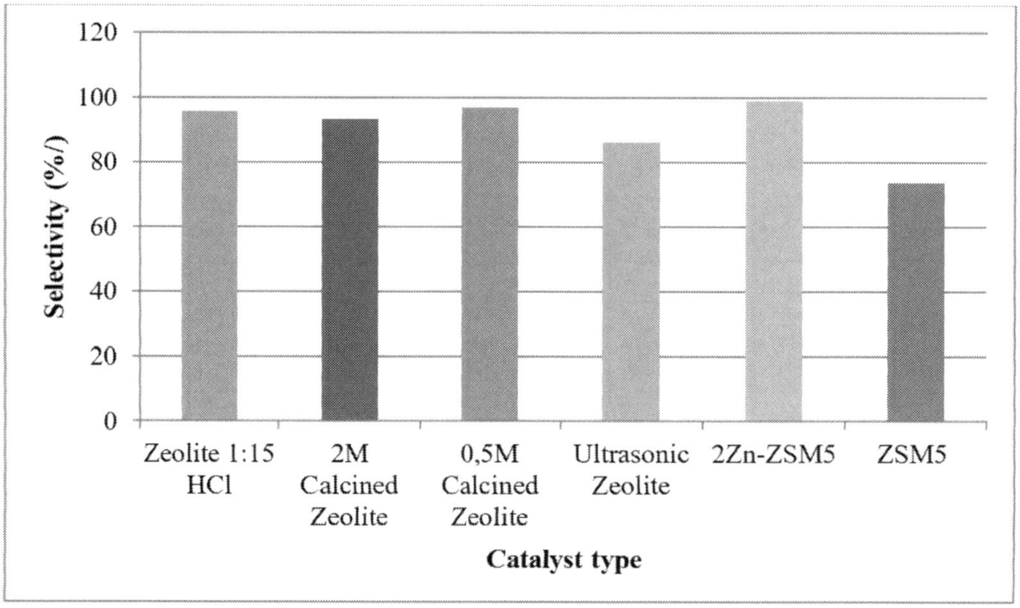

Figure 2. Selectivity of hydrogen for various catalyst types

From the water and glycerol compound reaction, can be produced H_2, CO, CO_2, CH_4, and hydrocarbon component and condensate product as this following reaction:

Steam reforming of glycerin:

$$C_3H_8O_3(g) + 3H_2O(g) \leftrightarrow 3CO_2(g) + 7H_2(g) \quad (3)$$

Reverse water gas shift reaction:

$$CO(g) + H_2O\ (g) \rightarrow CO_2(g) + H_2(g) \quad (4)$$

With H_2, CO_2, CO, and CH_4 which can be obtained from steam reforming process, so this can influence the product selectivity, where the selectivity is variant along with the catalyst variances. The highest selectivity of the steam reforming process was by 2Zn-ZSM5 catalyst, with selectivity percentage 98.77%. The selectivity for 0.5M Calcined Natural Zeolite 96.75% and 95.51% for Natural Zeolite catalyst 1:15 HCl ratio. The selectivity for 2M Calcined Natural Zeolite 93% and 85.9% for Ultrasonic Natural Zeolite. The lowest selectivity of the catalyst was 73.81% for ZSM-5.

The highest selectivity is influenced by catalyst mechanism and properties. The stability state in high temperature for each catalyst is influenced by the Si/Al ratio, where commonly natural zeolite classified as intermediate Si /Al zeolites (~2-5), while the ZSM5 classified as high Si/Al zeolites (~10-100) [15]. Hence, 3Zn-ZSM5 catalyst shows the highest selectivity because of its ability to keep stable at high temperature. While the higher Si/Al ratio of a catalyst, the more stable it will be. Adding metal component into zeolite catalyst can change its structure and characteristic [16].

4 Conclusions

Glycerol steam reforming process to produce hydrogen was influenced by many factors, one of them is catalyst variances. According to the yield percentage, the result showed that Natural Zeolite catalyst 1:15 HCl ratio gave the highest yield because of its large active surface. But, according to the selectivity, 2Zn-ZSM5 catalyst gave the highest selectivity due to its small Si/Al ratio. Furthermore, metal impregnation can make that catalyst more stable because of the structure and characteristic change.

5 References

[1]. Trevor L L V, A R Richard and M Fan. The Progress in Water Gas Shift and Steam Reforming Hydrogen Production Technologies. *International Journal of Hydrogen Energy*. **39**: 16983-17000 (2014)

[2]. Niels R U. Hydrogen Production By Steam Reforming Of Hydrocarbons. *Prepr.Pap.-Am.Chem. Soc.Div.Fuel Chem*. **49(2)**:906 (2004)

[3]. Adhikari S, Fernando S and Haryanto A. A comparative thermodynamic and experimental analysis on hydrogen production by steam reforming of glycerin. *Energy Fuel*. **21(4)**:2306 (2007)

[4]. Dasari M A, Kiatsimkul P, Sutterlin W R and Suppes G J. Low-pressure hydrogenolysis of glycerol to propylene glycol. *Appl Catalysis*. **281**:225-231 (2005)

[5]. Narasimha R R and S S Patel Production of Hydrogen from Glycerol via Steam Reforming Process Using Ni/ZSM-5 Catalyst. *International Conference on Multidisciplinary Research & Practice*. **1(8)**:438

[6]. Xing L Y and R Hino. *Nuclear Hydrogen Production Handbook* . United States: Taylor & Francis Inc; 2011

[7]. Alicia C, J A Calles, L García-Moreno and A J Vizcaíno. Production of Renewable Hydrogen from Glycerol Steam Reforming over Bimetallic Ni-(Cu,Co,Cr) Catalysts Supported on SBA-15 Silica. *Catalysts*. **7**:55 (2017)

[8]. Diana LM, Maria M, Monica D. Catalytic Glycerol Steam Reforming for Hydrogen Production. *AIP Conference Proceedings*. **1700**:060001-1-060001-5 (2015)

[9]. Widayat, dkk. Biofuel production by catalytic cracking method using Zn/HZMS-5 catalyst.

IOP Conf. Series: Materials Science and Engineering **349** (2018) 012070 doi:10.1088/1757-899X/349/1/012070

ARPN Journal of Engineering and Applied Sciences. **12**: 6347-6351 (2017)

[10]. Widayat, H Susanto, H Satriadi. Optimization process oh H-Zeolite catalyst preparation with surface response method. *Green Chemistry Section 1:Material Chemistry.* pp. 124-129 (2015)

[11]. Widayat, H Susanto and H Satriadi. The Effect of ultrasonik Irradiation on Preparation Zeolite Catalyst from Natural Mineral. **ISSN:** 2354-5917 (2014)

[12]. Widayat, Adit R W, Lukman H F, Ndaru O. Synthesis H-Zeolite catalyst by impregnation KI/KIO_3 and performance test catalyst for biodiesel production. *IOP Conf. Series:Materials Science and Engineering.* **107** (2016)

[13]. Widayat, A Roesyadi and M Rachimoellah, (2009), The Effect of Time dealumination and Solvent Concentration in Synthesis of Zeolite Catalyst and Catalytic Test for DiEthyl Ether Production Process, AIP Conference Proceeding Internation Workshop on Advanced Material for New and Renewable Energy editor L.T handoko and Masbah R T Siregar pp.106-112

[14]. Hernawan, Satriyo KW, Roni M, and Diah. Modification of GunungKidul Natural Zeolite as Bioethanol Dehydrating Agents *Energy Procedia.* **65**: 116-120 (2015)

[15]. E M Flanigen. *Pure Appl Chem.* **52**:2191–2211 (1980)

[16]. Pramatha P and Prabir K D. *Handbook of Zeolite Science and Technology Zeolites: A Premier.* Ohio: Marcel Dekker, Inc; 2003

IOP Conf. Series: Materials Science and Engineering **349** (2018) 012071 doi:10.1088/1757-899X/349/1/012071

Analysis Study of Stevioside and Rebaudioside A from *Stevia rebaudiana* Bertoni by Normal Phase SPE and RP-HPLC

Y Martono [1], A Rohman[2], S Riyanto[2], S Martono [2]

[1] Chemistry Department, Faculty of Science and Mathematics, Universitas Kristen Satya Wacana Salatiga, Jl Diponegoro 52-60 Salatiga 50711, Central Java, Indonesia

[2] Faculty of Pharmacy, Universitas Gadjah Mada, Yogyakarta 55221, Indonesia

Email: yohanes.martono@staff.uksw.edu

Abstract: Solid Phase Extraction (SPE) method using silica as sorbent for stevioside and rebaudiosida A analysis in *Stevia rebaudiana* Bertoni leaf have not been performed. The aim of this study is to develop SPE method using silica as sorbent for Reverse Phase-High Performance Liquid Chromatography (RP-HPLC) analysis of stevioside and rebaudiosida A in *S. rebaudiana* leaf. The results of this study indicate that the optimal conditions for normal phase SPE (silica) are conditioned with 3.0 mL of hexane. The sample loading volume is 0.1 mL. Cartridge is eluted with 1.0 mL acetonitrile: water (80: 20, v/v) to separate both analytes. The cartridge is washed with chloroform and water of 0.3 mL respectively. The developed SPE sample preparation method meets the accuracy and precision test and can be used for the analysis of stevioside and rebaudioside A by RP-HPLC.

1 Introduction

Stevia rebaudiana is an herb plant originating from South America (Paraguay and Brazil). *S. rebaudiana* contains steviol glycoside compounds. Major steviol glycoside compounds in *S. rebaudiana* are stevioside (6-12%) and rebaudioside A (1-4%) [1]. Stevioside and rebaudioside A are widely used as a natural sweetener in some countries [2]. On the other hand, various regulations regulate the permissible levels of stevioside and rebaudioside A and its limits [3]. The economic value of the purified extracts of stevioside and rebaudioside A is viewed from the quantitative aspect of both analytical compounds [4] . Therefore, the quantitative aspect is very important in the development of studies related to the active compounds stevioside and rebaudiosida A.

Various methods of stevioside and rebaudioside A analysis have been developed. Most of the quantitative methods of stevioside and rebaudiosida A are based on the High Performance Liquid Chromatography (HPLC) method. The normal phase HPLC (NP-HPLC) method developed can completely separate the chromatograms of stevioside and rebaudioside A [5,6]. However, column conditioning takes a long time; in addition, method reproducibility is less due to repeating the retention time of the analysis so variably [7]. Reversed phase HPLC (RP-HPLC) method using gradient elution system can also separate both the stevioside and rebaudioside A chromatograms properly [3,4,8–10] The RP-HPLC method using isocratic elution system developed is more practical in operation, efficient in acetonitrile consumption and rapid in running system [1,7,11].

Sample preparation is crucial step in the analysis of active compounds. Sample preparation methods employed in HPLC analysis include solid phase extraction (SPE) phase. The SPE method has advantage of removing impurities that can reduce complexity of sample matrix and preventing emulsion occurred in extract solution. Disadvantages of SPE method are more expensive and may occur biased by the solvent dilution factor [7]. The SPE method developed is all using reverse phase (C-18) (RP-SPE) [4,7,12]. Normal phase SPE (silica) method (NP-SPE) has not been developed for the analysis of stevioside and rebaudioside A compounds using RP-HPLC. In addition, every SPE method requires optimization of operational conditions because each method and sorbent will provide different characteristics and adsorption capacity. This research was aimed to develop sample preparation method using normal phase SPE for stevioside and rebaudiosida analysis. The method developed was then applied for RP-HPLC analysis of both analytes in *S. rebaudiana* leaf.

2 Methods

2.1 Plant Materials

The leaves of *S. rebaudiana* (Bert.) are obtained from the Bandungan, Poloboga, Tajuk, Tawangmangu, and Pakis, Central Java, Indonesia. The identification of plant species of *S. rebaudiana* was done by the Faculty of Pharmacy UGM Pharmaceutical Biology Division under the supervision of taxonomist Drs. Djoko Santosa, M.Si.

2.2 Standard and Chemicals

The standard of stevioside and rebaudioside A used was purchased from WAKO, Japan that have purity grade of 99.0% based on Certificates of Analysis (CoA). Solvents for mobile phase of HPLC were acetonitrile (Liquid Chromatography grade), methanol (Liquid Chromatography grade). The solvents for extraction were ethanol (pro analysis grade). All these solvents were purchased from E. Merck, Germany. SPE cartridge used was Lichrolut Si (40-63 μm) purchased from E.Merck, Germany.

2.3 Conditioning and Optimization of sample loading

Silica sorbent of SPE was activated and conditioned using 3.0 mL of hexane. Standard solutions were used as sample loaded. Optimization of sample loading was performed based on volumes of standard. Standard solutions loaded were varying of 0.5; 0.2; and 0.1 mL. Each eluent volume of the standard solution adsorbed was analyzed using RP-HPLC method. The optimal volume loaded is a volume that gives a result of more than 99% of the adsorbed analyte compound in the stationary phase.

2.4 Optimization of Washing

A number of solvents involving chloroform, water, acetonitrile and mixed water: acetonitrile (8: 2, v/v and 9: 1, v/v) are used to wash impurities. A volume of solvent wash (0.3-5.0 mL) was eluted at stationary phase. The optimal solvent is a solvent which can remove the impurity compound without eluting the analytes. Each eluent optimization of the washing solution was analyzed by RP-HPLC method.

2.5 Optimization of Eluting

A number of solvent including pure methanol, mixtures of water: acetonitrile (2: 8, v/v), methanol: acetonitrile (1: 1, v/v) and mobile phase for eluting the analytical compound. Solvent volume of 1.0-5.0 mL is eluted on the stationary phase. The optimum solvent is a solvent which can remove the whole analytes without carrying the impurity compound. Each eluent optimization of the elution solution was analyzed by RP-HPLC method.

IOP Conf. Series: Materials Science and Engineering **349** (2018) 012071 doi:10.1088/1757-899X/349/1/012071

2.6 RP-HPLC Condition [1].

HPLC used was Knauer, GmBH, Germany. Stationary phase used was Eourosphere C-18 (250 × 4.6 mm., 5 μm) with guard column. Mobile phase was mixture of methanol 10% in water : acetonitrile (65 : 35, v/v). TFA added into a mixture was 0.01% (v/v). Mobile phase was homogenized and degassed using ultrasonic. Flow rate of mobile phase applied was 0.6 mL/min. Separation was detected using UV detector at 210 nm. Sample volume injected was 20 μl using Rheodyne 7726i injector.

3 Result and discussion

Normal phase SPE using silica sorbent for sample preparation of stevioside and rebaudioside A analysis by HPLC has not been developed. The polar silanol group in silica interacts with the analytes through the primary interaction of hydrogen bonds of hydroxy groups of glycosides [13]. The interaction of the hydroxy group with a strong silanol group on the sorbent surface can retain the analytes. Elution strength and / or eluent polarity can be adjusted to break the interaction of the analyte compound with sorbent.

3.1 Optimization of Sample Loading

Silica as a stationary phase of normal phase SPE is conditioned prior to use. The solvent used for this conditioning is hexane. The sample loading volume of 0.1 mL is the optimal volume in which all compounds of the analyte are adsorbed in the stationary phase. In a 0.5 mL sample application, non-retained rebaudioside A and stevioside were 99.23% and 77.52% respectively. In the application of 0.2 mL of sample volume, rebaudioside A and stevioside which were not retained in silica were 67.25% and 63.40%, respectively. The results of this study indicate that each SPE material has a specific maximum adsorption capacity to the type of analytical compound and operational conditions of SPE (Sigma-Aldrich, 1998). The chromatogram profile of sample volume optimization applied to the normal phase SPE (silica) is shown in Figure 1.

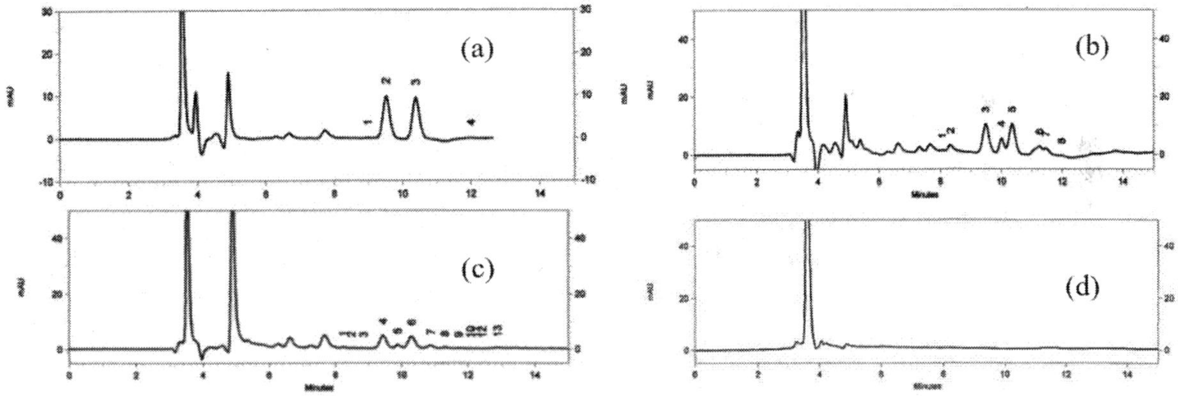

Figure 1. Chromatogram profile of sample loading optimization of normal phase SPE (silica). Eluent analysis injected to RP-HPLC of (a) standards mixture (2. rebaudioside A, t_R = 9,533 min and 3. stevioside, t_R= 10,400 min); (b) loaded sample volume of 0,5 mL (3. rebaudioside A, t_R = 9,483 min and 5. stevioside, t_R= 10,350 min); (c)) loaded sample volume of 0,2 mL (4. rebaudioside A, t_R = 9,433 min and 6. stevioside, t_R= 10,300 min) (d)) loaded sample volume of 0,1 mL (analytes is not detected in eluent).

3.2 Optimization of Eluting

In contrast to RP-SPE, the normal phase (SPE) phase separation mechanisms developed in this study are selective elution, which elutes the analytical compound and retains impurities [13]. After the sample is applied to the stationary phase, then the elution of the analytes is performed. Several solvents optimized for the elution of the analytes are methanol, acetonitrile : water (8 : 2, v/v), methanol :

acetonitrile (1 : 1, v/v) and mobile phase. The optimum solvent for the elution of a normal-phase SPE silica is acetonitrile: water (8 : 2, v/v) with 1.0 mL volume. The organic solvent-water mixture with a larger organic solvent ratio is the appropriate eluting solvent for the selective elution mechanism [13].

An acetonitrile: water (8: 2, v/v) mixture has the most appropriate polarity and elution strength among other solvents to break the strong hydrogen bond interaction between the analytes and the sorbent without breaking the interaction of the impurity compound with the sorbent. These results indicate that a selective elution separation mechanism can be achieved for NP-SPE (silica). Important factor concerned is the suitability of the polarity and strength of the eluting solvent to be sufficiently strong to break the interaction of the analytes with sorbent but weak to break the interaction of impurity compounds with the sorbent. The elution optimization profile of the analyte compound with several solvents is presented in Figure 2.

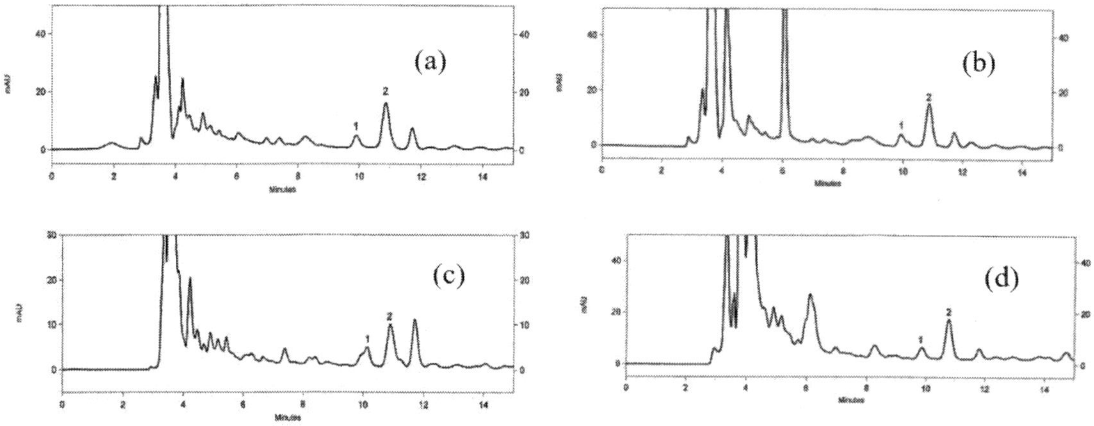

Figure 2. Chromatogram profile of analytes eluting optimization of normal phase SPE (silica). Eluting solvent optimized of (a) acetonitrile : water (8 : 2, v/v) (1. rebaudioside A, t_R = 9,900 min and 2. stevioside, t_R= 10,850 min); (b) methanol : acetonitrile (1 : 1, v/v) (1. rebaudioside A, t_R = 9,933 min and 2. stevioside, t_R= 10,850 min); (c) methanol (1. rebaudioside A, t_R = 10,133 min and 2. stevioside, t_R= 10,833 min); (d) mobile phase of RP-HPLC (1. rebaudioside A, t_R = 9,850 min dan 2. stevioside, t_R = 10,767 min).

3.3 *Optimization of Washing*

The washing step after elution of the analytes is proposed to find out whether there are still compounds of analyte left in the stationary phase. In the initial optimization, the washing step is performed prior to the elution of the analyte compound. However, the analytes is still eluted in the water. In addition, the analytes interaction to the sorbent becomes stronger when the impurity compound is washed using chloroform.

Washing is done with a non-polar solvent chloroform, followed by acetonitrile : water (2: 8, v/v) and followed by water. The result shows that no analytes are detected in the washing step. The chromatogram profile of washing step optimization in normal phase SPE (silica) is shown in Figure 3.

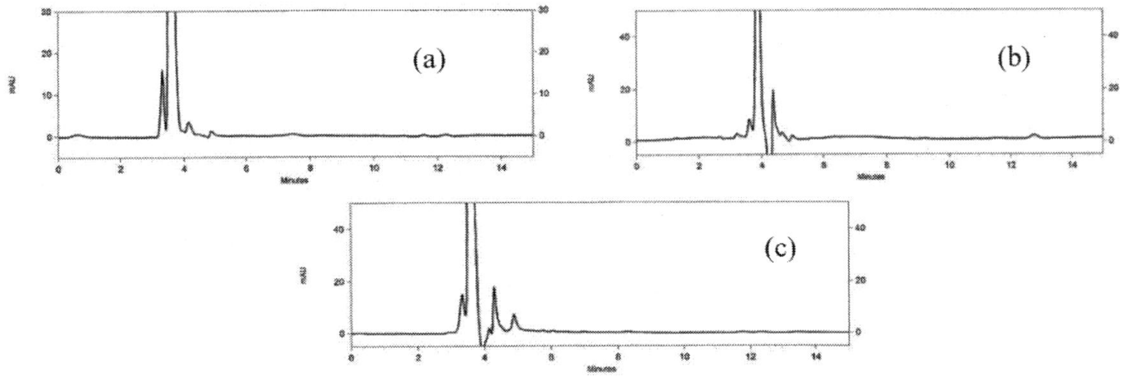

Figure 3. Chromatogram profile of washing optimization of normal phase SPE (silica). Washing solvent optimized of (a) 2,0 mL chloroform; (b) 1,0 mL acetonitrile : water (2 : 8, v/v); and (c) 0,6 mL water. There is no analytes detected in washing eluent.

3.4 Accuracy and Precision Test

Accuracy test is performed using the recovery test by standard addition method. Precision is determined on the basis of the relative standard deviation (RSD) of three levels replicated samples which have been spiked with the standard. Accuracy and Precision of developed NP-SPE method are presented in Table 1 and 2. The developed method is meet accuracy and precision and indicate that the SPE method can be used for RP-HPLC analysis. The NP-SPE (silica) method developed is more efficient in its implementation than RP-SPE (C-18) because the elution of the analyte compound is performed after loading stage. The developed NP-SPE method is become shorter in the process.

3.5 Method Application

The developed NP-SPE method is then applied to analyze ethanolic extract of *S. rebaudiana* leaf using RP-HPLC. The optimum normal phase SPE step is a sample application volume of 0.1 mL followed by eluting of the analytes compound using 1.0 mL acetonitrile: water (8: 2, v/v). Chromatogram of sample analysis is shown in Figure 4.

The results of this study indicate that sample preparation in normal phase (silica) SPE was proven to clean the matrix complex of samples of ethanolic extract of *S. rebaudiana* leaf without loss of analytes (Figure 4). SPE normal phase (silica) developed method has a selective elution mechanism which is eluting the analytes first using particular solvent and holding the impurity compound in the stationary phase.

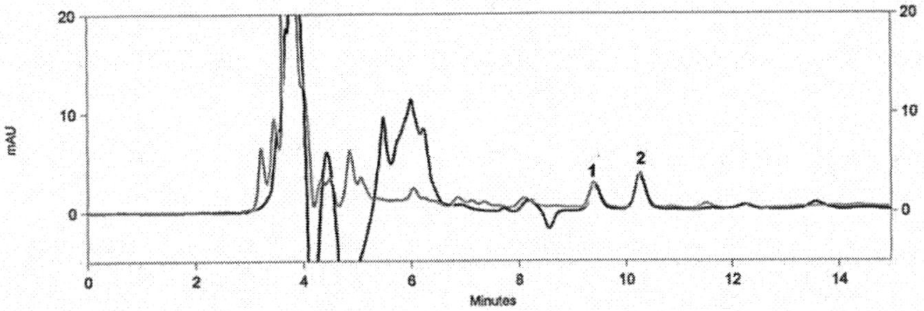

Figure 4. Profile of overlay chromatogram between non SPE (black line) and silica SPE (green line) RP-HPLC analysis of *S. rebaudiana* leaf.

Table 1. Accuracy and precision test of rebaudioside A determination using normal phase SPE (silica) method developed for RP-HPLC analysis

average of standard spiked (mg)	Reb A (mg) founded in 500 mg of leaf extract	theoretical of Reb A concentration (mg)	Reb A founded in SPE sample	*recovery* (%)	RSD (n = 3)
0,0215	0,0052	0,0267	0,0268	100,47	3,34
0,0493	0,0112	0,0605	0,0618	102,62	0,97
0,0766	0,0066	0,0832	0,0838	100,76	1,17

Reb A = rebaudioside A
Stv = stevioside

Tabel 2. Accuracy and precision test of stevioside determination using normal phase SPE (silica) method developed for RP-HPLC analysis

average of standard spiked (mg)	Stv (mg) founded in 500 mg of leaf extract	theoretical of Stv concentration (mg)	Stv founded in SPE sample	*recovery* (%)	RSD (n = 3)
0,0294	0,0142	0,0436	0,0442	101,32	1,61
0,0643	0,0100	0,0743	0,0737	99,21	0,91
0,1104	0,0158	0,1262	0,1321	104,69	1,54

Acknowledgments

We would like to thank Ministry of Research, Technology and Higher Education and Universitas Kristen Satya Wacana who funded this research through Doctoral Dissertation Research Program 2016.

References

1. Martono Y, Riyanto S, Rohman A, and Martono S. 2016 *AIP Publishing* p. 080001. Available from: http://scitation.aip.org/content/aip/proceeding/aipcp/10.1063/1.4958509
2. Liu J, Li J, and Tang J. 2010 *Food Bioprod Process.* **88** 215.
3. Well C, Frank O, and Hofmann T 2013 *J Agric Food Chem.* **61** 11312.
4. Gardana C, Scaglianti M, and Simonetti P 2010 *J Chromatogr A.* 1217 1463.
5. Kolb N, Herrera JL, Ferreyra DJ, and Uliana RF. 2001 *J Agric Food Chem.* **49** 4538.
6. Dacome AS, da Silva CC, da Costa CEM, Fontana JD, Adelmann J, and da Costa SC. 2005 *Process Biochem.* **40** 3587.
7. Bergs D, Burghoff B, Joehnck M, Martin G, and Schembecker G. 2012 *J Für Verbraucherschutz Leb.* **7** 147.
8. Cacciola F, Delmonte P, Jaworska K, Dugo P, Mondello L, and Rader JI. 2011 *J Chromatogr A.* **1218** 2012.

9. Jaworska K, Krynitsky AJ, and Rader JI. *J AOAC Int.* **95** 1588.
10. Zhao Y-G, Cai M-Q, Chen X-H, Pan S-D, Yao S-S, and Jin M-C. 2013 *Food Res Int.* **52** 350.
11. Martono Y, Riyanto S, Rohman A, Martono S. 2015 *Int J Pharm Clin Res.* **8** (5) suppl **397**
12. Woelwer-Rieck U, Lankes C, Wawrzun A, Wüst M. 2010 *Eur Food Res Technol.* **231** 581–8.
13. Supelco 1998 *Guide to Solid Phase Extraction* (Bulletin 910: SIgma-Aldrich).

2, 4, 6-Trithiol-1, 3, 5-Triazine-Modified Gold Nanoparticles and Its Potential as Formalin Detector

Y Yulizar[1*], H A Ariyanta[2], L Rakhmania[1], and M A E Hafizah[3]

[1] Department of Chemistry, Faculty of Mathematics and Natural Sciences, Universitas Indonesia, Kampus UI Depok 16424, Indonesia
2Department of Pharmacy, Faculty of Pharmacy, Universitas Gunadarma, Kampus UG Depok 16424, Indonesia
[3]Postgraduate Program of Material Science, Faculty of Mathematics and Natural Sciences, Universitas Indonesia, Jakarta 10430, Indonesia

[*]E-mail: yokiy@sci.ui.ac.id

Abstract. Stabilized gold nanoparticles (AuNP) have been successfully prepared by a modification of ligand 2, 4, 6- trithiol- 1, 3, 5- triazine (TT). TT has three thiol groups and nitrogen atoms on the aromatic ring that can **interact** and stabilize AuNP. TT modified AuNP (AuNP/TT) was characterized using UV-Vis spectrophotometer, particle size analyzer (PSA) and transmission electron microscopy (TEM). The characterization showed that AuNP/TT stable at a maximum wavelength (λ_{maks}) of 537 nm with the particle diameter of 9.41 nm. The increased acidity (pH) causes the protonated thiol groups of TT marked with a visual change of colloidal AuNP/TT from purple to blue, causing AuNP and TT bonds weakened. In this study, the AuNP/TT was reacted with formalin. This interaction shows that AuNP/TT has a potential as an efficient detector of formalin, marked by changes in the diameter of the particle, colloidal color, and maximum wavelength shift.

1 Introduction

The application of nanotechnology in food safety (detection system, biosensor, etc.) has now been widely used as the focus of research because it is closely related to human survival [1-5]. Gold nanoparticles (AuNP) have absorption at a visible wavelength, so it is often used as an ideal method to detect the presence of hazardous chemicals in foods such as borax, formalin, and melamine, one of them through colorimetric sensors. In previous studies, we have studied the interaction between AuNP with melamine as an ideal colorimetric sensor [6]. AuNP detects the presence of melamine through colloidal color changes that occur due to hydrogen bonding between the melamine and acid groups in the modified AuNP. The method offers several advantages over other detection techniques that require complicated sample preparation, expensive cost, and limited detection capabilities. Firstly, this method does not require expensive and complicated instruments, which simplify operations and reduce costs. Secondly, this method allows the detection concentration as low as 2.5 ppb in short enough time [7]. Common methods used for qualitative and quantitative analysis of formalin include titrimetry, photometry, chromatography, NMR ^{13}C, HPLC/spectrophotometry [8,9].

On the other hand, it has been well known that nanoparticles are unstable because they have high surface energy [10,11] and cause aggregation [12] and thus need to be stabilized with an appropriate modifier.

Some functional groups, such as cyano (CN), mercapto (SH), and amino (NH_2) are known to have a high affinity for AuNP [13]. A 2, 4, 6-Trithiol-1, 3, 5-triazine (TT) is a ligand with three active thiols (-SH) sites [14] potentially binding to gold nanoparticles. The existence of TT on the AuNP surface is expected to inhibit the aggregation of nanoparticles electrostatically between thiol groups to improve stability. In this study, the preparation of TT ligand modified AuNP ((AuNP/TT) was interacted with formalin. The reducing agent used in the optimization of AuNP synthesis is $NaBH_4$. AuNP/TT is interacted with formalin to investigate its potential as a fast and efficient foodstuff detector.

2 Experimental

2.1 Materials

$HAuCl_4$ used in this research was synthesized by dissolving 99.9% Au metal from PT. Antam in aqua regia solvent with HNO_3 and HCl ratio of 1:3. All chemicals were of analytic grade and were used without further purification. Formalin was obtained from Merck and 2, 4, 6-trithiol-1, 3, 5-triazines from Sigma Aldrich. MilliQ water (18.2 Ω cm^{-1}) was used to make aqueous solutions.

2.2 Synthesis of Gold Nanoparticles (AuNP)

Five mL of $HAuCl_4$ $1.0x10^{-4}$ M solution was added into one mL of $NaBH_4$ 0.004 M solution then stirred using a magnetic stirrer for 2 min. The same experiment was repeated for $NaBH_4$ concentrations of 0.007; 0.023; 0.030; and 0.070 M. The characterization was performed using UV-Vis spectrophotometer to find λ_{max} value, and PSA to determine the particle size.

2.3 Modification of AuNP with TT

Optimum AuNP colloid (optimization 2.2) were added to 1.0 mL of TT in various concentrations of $1.33x10^{-6}$; $6.67x10^{-6}$; $1.33x10^{-5}$; and $6.67x10^{-5}$M, while stirring with a magnetic stirrer for 2 min. The formed mixture was characterized using UV-Vis spectrophotometers, PSA, and TEM. Furthermore, the optimum AuNP/TT was tested for stability in pH variations of 1.0-8.0. HCl and NaOH solutions with concentrations of 0.1 M were used by varying the pH of reaction.

2.4 Interaction of AuNP/TT with Formalin

About 0.2 mL of formalin 1.0×10^{-3} M was added into the AuNP-TT colloid under optimum conditions. The mixture was stirred using magnetic stirrer for 3 min. The absorbance of the mixture was observed using UV-Vis spectrophotometer.

3 Results and discussion

3.1 Synthesis of AuNP with various concentration of $NaBH_4$

AuNP synthesis optimizations were performed by varying the concentration of reducing agent, $NaBH_4$. The variation of reducing agent concentration aims to control the size of the AuNP formation. The nanoparticle size control will improve the functionalization and selectivity of AuNP [15-17]. Based on previous research, it is known that the parameters affecting the growth, shape, and structure of nanoparticles are the reactants concentration (reductants), type of capping agent or stabilizer, pH of the solution, and reaction time [17]. The formation of AuNP is visually marked by a color change from a colorless ($HAuCl_4$ solution) to a ruby red colloid [18-21]. Colloidal color changes indicate that the Au^{3+} ion source has been reduced to Au^0 [22].

Spectroscopic characterization was shown in Figure 1. AuNP formation was characterized by the appearance of λ_{max} spectral peaks in the range 500-530 nm [23-26] from the peak of $HAuCl_4$ solution spectrum at 200 nm [27].

IOP Conf. Series: Materials Science and Engineering **349** (2018) 012072 doi:10.1088/1757-899X/349/1/012072

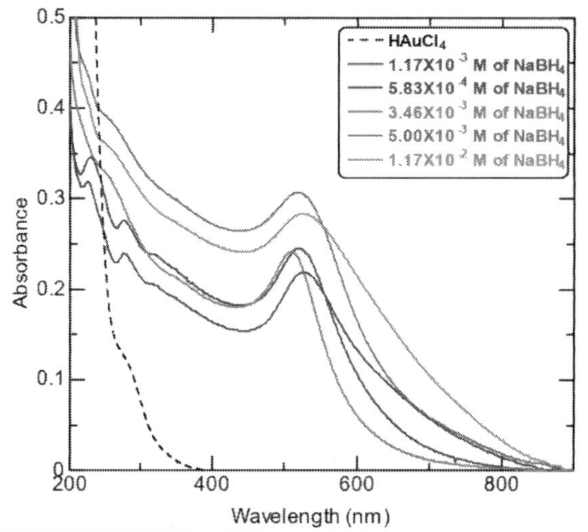

Figure 1. UV-Vis absorption spectra of AuNP with $NaBH_4$ concentration variation; $[HAuCl_4] = 1.0 \times 10^{-4}$ M

The graph of $NaBH_4$ concentration variation in Figure 1 shows that $NaBH_4$ 3.46×10^{-3} M has the smallest λ_{maks}, 509 nm. The λ_{max} value indicates that the size of AuNP has the smallest size. If the particle size decreases, the Bohr radius distance also decreases, so that electrons will be trapped and led to the band gap energy increases, and λ_{maks} shifted to smaller (higher energy) [31]. In addition to the smallest λ_{max} value, the use of 3.46×10^{-3} M $NaBH_4$ also yields a considerable absorbance value, that is 0.242. The magnitude of absorbance shows the large number of the formed AuNP. The greater absorbance value, the more amount of light absorbed by AuNP, indicates many AuNP was formed.

The observation using PSA aims to determine the average size distribution of AuNP. Table 1 show Au^{3+} reduction using $NaBH_4$ 3.46×10^{-3} M produced AuNP with the smallest size of 8.64 nm.

Tabel 1. PSA Data: AuNP size with $NaBH_4$ concentration variation

$NaBH_4$ Concentration (M)	Size (nm)	% Volume	Peak width (nm)
5.83×10^{-4}	10.89	100.00	6,88
1.17×10^{-3}	11.54	99.90	7,89
3.46×10^{-3}	8.64	100.00	8,78
5.00×10^{-3}	11.68	99.20	4,39
1.17×10^{-2}	18.02	99.00	4,74

Based on observations using UV-Vis spectrophotometer and PSA were then used 3.46×10^{-3} M $NaBH_4$ for AuNP synthesis in further treatment.

3.2 Modification of AuNP with TT

The TT ligand has three thiols that can stabilize AuNP (AuNP/TT) through Au-S interaction. The optimization of AuNP/TT modification was performed by varying the ligand concentrations at 1.33×10^{-6}; 6.67×10^{-6}; 1.33×10^{-5}; and 6.67×10^{-5} M. Figure 2 shows the difference in absorbance and λ_{max} of AuNP before and after modified TT ligand. The λ_{maks} shift occurred from 509 to 537 nm, while the absorbance decreased from 0.255 to 0.190.

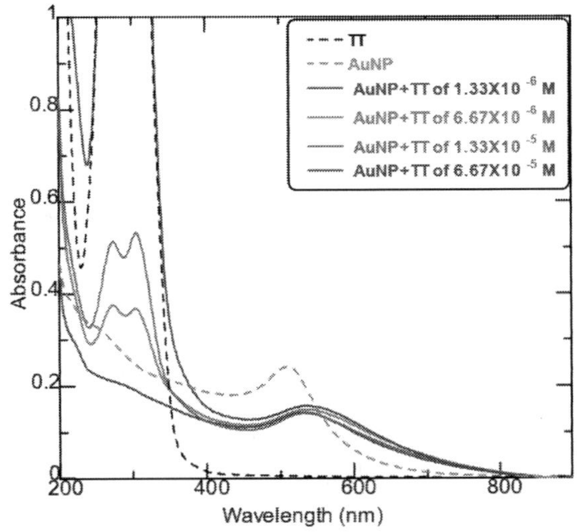

Figure 2. UV-Vis absorption spectra of AuNP/TT with various TT concentrations.

To facilitate the observation, λ_{max} and absorbance curves against-various TT concentrations were shown in Figure 3.

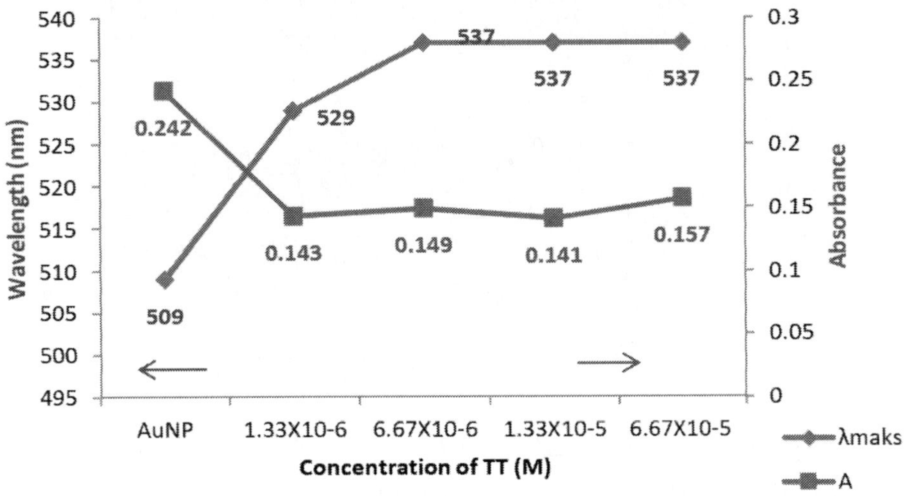

Figure 3. A curve of λ_{max} and absorbance changes of AuNP/TT against TT concentration variation.

Figure 3 shows that TT 6.67x10^{-6} M was the optimum TT concentration, the λ_{max} of AuNP/TT peak constantly at 537 nm. The various TT concentrations did not give significant effect to λ_{max} shift. This suggests that the excess of TT as a modifier does not affect AuNP stability. TT was bound to AuNP surface through interactions between S and N atoms in heterocyclic of TT on AuNP surface. Due to the symmetry of thiol groups and the active site of N atom, the interaction between TT and AuNP involves a tridentate coordination link [28], as illustrated in Figure 4.

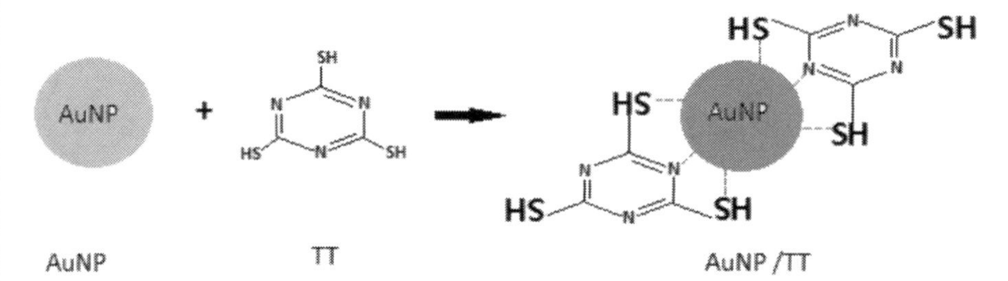

Figure 4. Illustration of the TT ligand binding on the AuNP surface

PSA measurements show that the optimum AuNP modified by TT 6.67x10^{-6} M was 9.41 nm in size. The resulted AuNP/TT morphology can be seen using TEM in Figure 5 show nanosphere with 5-10 nm in size .

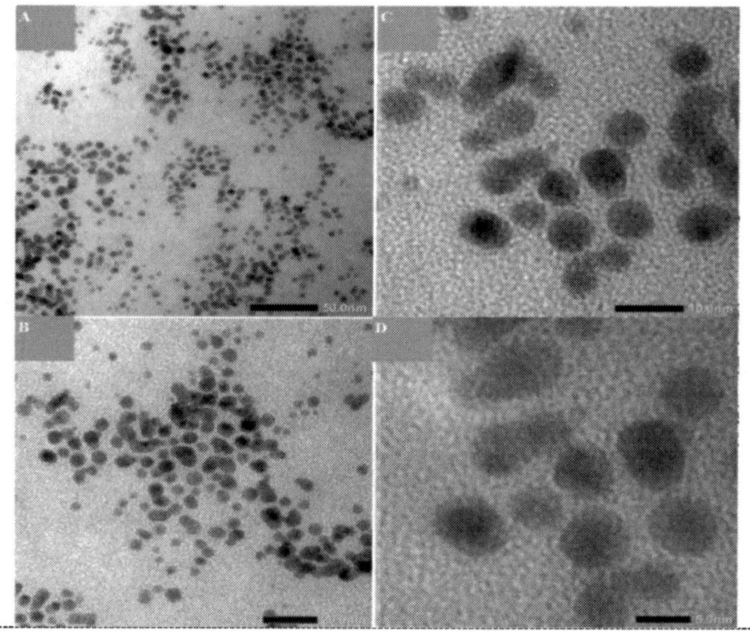

Figure 5. AuNP/TT characterization at TT 6.67x10^{-6} M using TEM at scale bar: a. 50 nm, b. 20 nm, c. 10 nm, and d. 5 nm.

Furthermore, pH variation of AuNP/TT colloid was started from pH 8.0 then decreased at pH 7.0; 6.0; 5.0; 4.0; 3.0; 2.0 and 1.0. Observations were conducted through λ_{max} shifts, absorbance changes, and

AuNP/TT size distributions. Increased acidity causes the electrons in thiol group attached stronger to H atom so that electrons are difficult to donate to AuNP. It causes the release of TT from AuNP, as evidenced from the UV-Vis absorption at pH 2.0 and 1.0 (Figure 6) and particle size growth due to agglomeration (Figure 7). The optimum condition of AuNP/TT modification was observed at pH 8.0.

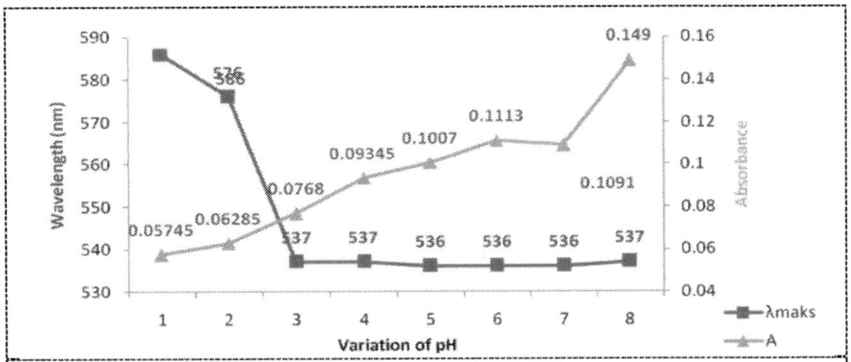

Figure 6. λ_{max} shifts and absorbance changes of AuNP/TT against various pH

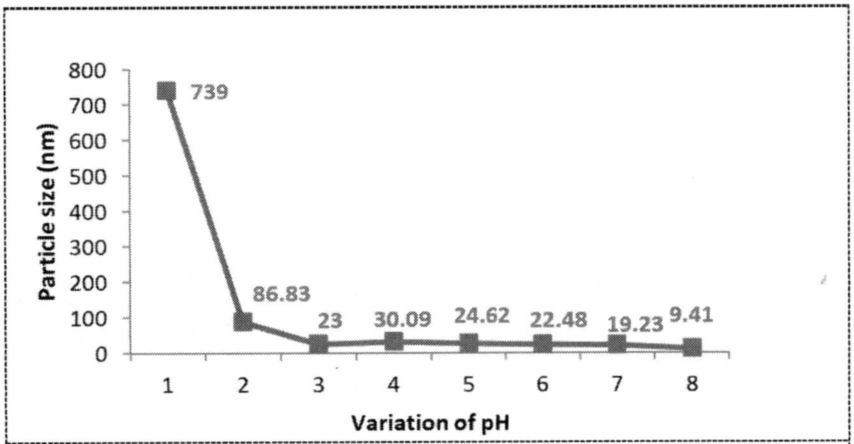

Figure 7. Particle size changes of AuNP/TT against various pH

TT was a ligand with three thiol groups and nitrogen atoms in the aromatic ring. The addition of thiol compound to a metal nanoparticle will form a stable compound [29].Thiol will form a single layer on the surface of the metal nanoparticle (self-assembly monolayer) which means the ligand has modified the surface of nanosphere or three-dimensional space [30]. In this modification, there is an interaction between SH-group with stable metal nanoparticles. The observed results with UV-Vis spectrophotometer were shown in Figure 8.

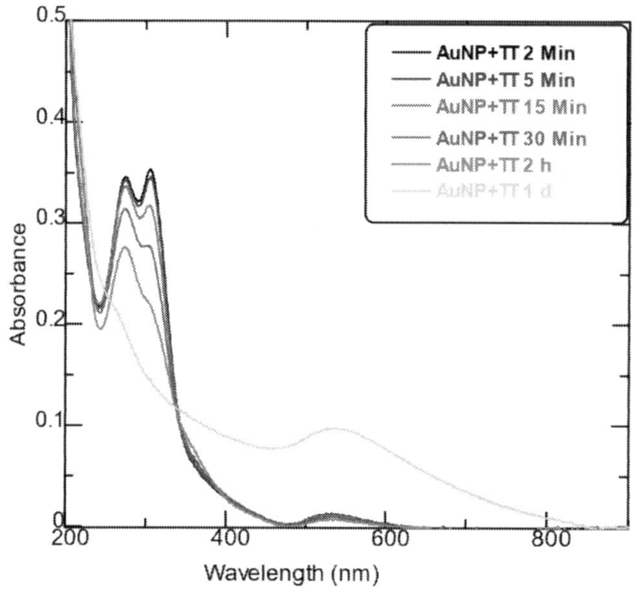

Figure 8. The UV-Vis absorption spectra of AuNP/TT against reaction time

Figure 8 shows the AuNP/TT formation reaction time of 2 min to 2 hours, and no λ_{max} of AuNP/TT changes at 537 nm, indicating that the modified nanoparticles were stable. The peaks at 280 nm and 320 nm appearing in Figure 8 are peaks of $HAuCl_4$ and TT, respectively. Increasing reaction time causes both peaks to decrease as they are used in AuNP/TT formation shown in absorption spectra increase at 537 nm. However, from the first day there was a significant increase in absorbance of AuNP/TT to 0.089. The stability of AuNP/TT occurs as a result of electrostatic stabilization of AuNP protected by the self-assembly monolayer of -SH functional group of TT [30]. Electrostatic stabilization can increase colloidal stability so that the size and its distribution of AuNP can be maintained.

3.3 Interaction of AuNP/TT with Formalin

Interactions of AuNP/TT with formalin can be visually observed through colloidal color changes characterized using UV-Vis spectrophotometers. Observations were conducted from the changes of λ_{max}, absorbance value, and colloidal size.

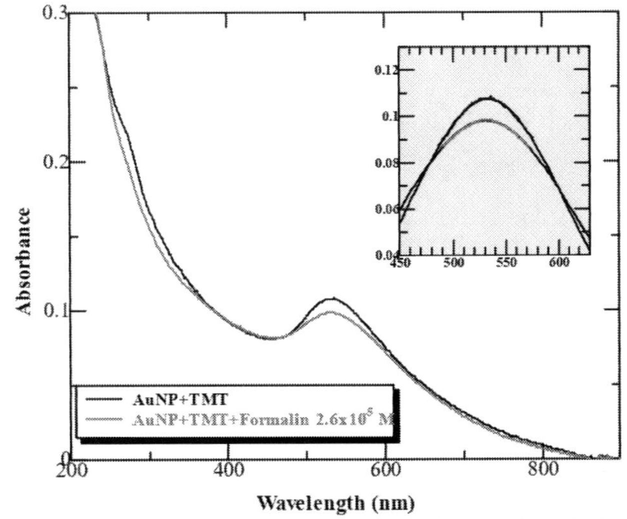

Figure 9. UV-Vis absorption spectra of AuNP/TT interaction with formalin

The UV-Vis absorption spectra (Figure 9) showed a change in λ_{max} ($\Delta\ \lambda_{max}$ = -1.90 nm) and a change in absorbance value (Δ Abs = - 0.01) when AuNP/TT was reacted with formalin. Similarly, showed a change in the particle size (Δ particle size = 2.29 nm) of AuNP/TT before and after reaction with formalin measured using PSA as shown in Table 2. This confirms an interaction between AuNP/TT and formalin in the form of AuNP/TT-formalin complex through hydrogen bond. The thiol group as the active site of TT, in the presence of hydrogen atom in formalin can perform a intermolecular hydrogen bond. Therefore, AuNP/TT has a good enough potential to detect the presence of formalin by a simple method based on AuNP/TT size changes, λ_{max} shifts, and absorbance value change as a colorimetric detector.

Table 2. Changes in λ_{max}, absorbance and size of AuNP/TT after interaction with formalin

	λ_{max}	Abs	Particle Size (nm)	$\Delta\ \lambda_{max}$	Δ Abs	Δ Particle Size (nm)
AuNP/TT	533.03	0.11	9.41			
AuNP/TT-Formalin	531.13	0.10	11.70	-1.90	-0.01	2.29

Abs = Absorbance, Δ Abs = Absorbance change

(Negative $\Delta\ \lambda_{max}$ indicate a shift toward the hypochromic)

4 Conclusions

AuNP has been successfully modified with a TT ligand acted as a stabilizer agent to improve AuNP stability. The optimum condition of AuNP/TT modification was conducted at TT concentration of 6.67×10^{-6} M, $NaBH_4$ concentration of 3.46×10^{-3} M, pH 8.0 with a particle size of 9.41 nm. The release of TT from AuNP occurs at pH 1.0 and 2.0. The formed AuNP/TT exhibits a good interaction with formalin, characterized through changes in particle size, maximum wavelength, and absorbance value. Therefore, AuNP/TT has a potential as formalin detector using colorimetric sensor method.

Acknowledgement
This work was funded by hibah penelitian unggulan perguruan tinggi (PUPT) 2017, No. 2720/UN2.R3.1/HKP.05.00/2017 from the Directorate general of research and development strengthening, Ministry of research, technology, and higher education, Indonesia through the Directorate of research and community services, Universitas Indonesia.

References
[1] Barbara S R, Javier G S, Javier H B and Miguel A R D 2017 *TrAC Trends in Analytical Chemistry* **96** 172
[2] Yan Z, Zhihong Z, Dan D and Yuehe L 2016 *Journal of Electroanalytical Chemistry* **781** 147
[3] Keith W, Subrayal M R, Azadeh N and Suresh N 2014 *Trends in Food Science & Technology* **40** 183
[4] Mian M S, Jabar Z K K, Shahzad Y and Muhammad S R 2017 *Photodiagnosis and Photodynamic Therapy* **19** 67
[5] Xiaojia H and Huey-Min H 2016 *Journal of Food and Drug Analysis* **24** 671
[6] Yulizar Y, Ariyanta H A, Abdurrachman L 2017 *Bulletin of Chemical Reaction Engineering & Catalysis* **12** 212
[7] Kelong A, Yanlan L and Lehui L 2009 *Journal of The American Chemical Society* **131** 9496
[8] Detlef E, Willem A and Klaus-Peter W 2007 *Analytical Chemistry* **79** 2096
[9] Wenhong L, Hui L, Yuan Z and Catharina Y W A 2001 *Journal of Chromatography B: Biomedical Sciences and Applications* **753** 253
[10] Fahed M K and Ahmed A 2017 *Applied Surface Science* **401** 65
[11] Nanda K K, Maisels A, Kruis F E, Fissan H and Stappert S 2003 *Physical Review Letters* **91** 106102
[12] Ernest M H, Tanapon P and Gregory V L 2010 *Journal of Environmental Quality* **39** 1909
[13] Muriel K C, Neil S C, Mark S, Simon G J M, Laurence B L, Adrian R and Bruce L 2001 *Journal of The American Chemical Society* **123** 10411
[14] Jin Michael J P, Natalie K P, Matthias Z, Allen D H, Sam L and Stephen D P 2007 *Inorganic Chemistry Communications* **10** 631
[15] Andrea R T, Susan H, Peidong Y 2008 *Small* **4** 310
[16] Ariyanta H A and Yulizar Y 2016 *IOP Conf. Series: Materials Science and Engineering* **107** 012002
[17] Yulizar Y, Foliatini and Hafizah M A E 2016 *Arabian Journal of Chemistry* **doi:** 10.1016/j.arabjc.2016.05.013
[18] Turkevich J, Garton G and Stevenson P C 1954 *Journal of Colloid Science* **9** 26
[19] Ashok B, Bhagyashree J, Ameeta R K and Smita Z 2010 *Colloids and Surfaces B: Biointerfaces* **80** 45
[20] Kumar V G, Gokavarapu S D, Rajeswari A, Dhas T S, Karthick V, Kapadia Z, Shrestha T, Barathy I A, Roy A, Sinha S 2011 *Colloids and Surfaces B: Biointerfaces* **87** 159
[21] Ali D M, Thajuddin N, Jeganathan K, Gunasekaran M 2011 *Colloids and Surfaces B: Biointerfaces* **85** 360
[22] Santhoskumar J, Rajeshkumar S and Kumar S V 2017 *Biochemistry and Biophysics Reports* **11** 46
[23] Philip D and Unni C 2011*Physica E* **43** 1318
[24] Yallapa S, Manjanna J, Dhananjaya B L 2015 *Spectrochimica Acta Part A: Molecular and Biomolecular* **137** 236
[25] Devi P R, Kumar C S, Selvamani, Subramanian N and Ruckmani K 2015 *Materials Letters* **139** 241
[26] Yulizar Y, Kadja G T M, Safaat M 2016 *Reaction Kinetics, Mechanisms and Catalysis* **117** 353
[27] Xi L, Zhou-jun W and Chang-jun L 2010 *Nanoscale Research Letters* **5** 124
[28] Vitor M Z, Ronaldo A T, Koiti A and Henrique E T 2008 *Inorganic Chemistry* **47** 2934

[29] Naoki T, Yukihide S, Toshiharu T, Mikio M, Toshihiro T, Hiroshi W, Werner B, Helmut B and Günter S 2001 *Applied Organometallic Chemistry* **15** 178.
[30] Love J C, Estroff L A, Kriebel J K, Nuzzo R G and Whitesides G M 2005 *Chemical Reviews* **105** 1103.
[31] Koole R, Groeneveld E, Vanmaekelbergh, Meijerink A, Donega C D M 2014 Size Effects on Semiconductor Nanoparticles *Nanoparticles (Chemistry and Materials Science)* ed Donega C D M (Heidelberg: Springer) chapter 2 pp 13–51

IOP Conf. Series: Materials Science and Engineering **349** (2018) 012073 doi:10.1088/1757-899X/349/1/012073

Determination of Urease Biochemical Properties of Asparagus Bean (*Vigna unguiculata ssp sesquipedalis* L.)

Zusfahair[1], D R Ningsih[1], A Fatoni[1], and D S Pertiwi[1]

[1] Department of Chemistry, Faculty of Mathematics and Natural Sciences Jenderal Soedirman University, Purwokerto 53123, Indonesia.

Email: zusfahair@gmail.com

Abstract. Urease is enzyme that plays a role in nitrogen metabolism during plant germination. Plants that produce a lot of urease are grains. This study used asparagus bean as source of urease. The purpose of this research is to learn the effect of germination time on the activity of urease enzyme from asparagus bean and its biochemical properties. The research was started by germination of asparagus bean on day 2, 4, 6, 8, 10 and 12. Asparagus bean sprouts were extracted using acetone and separated by centrifugation to obtain the crude extract of urease. The biochemical properties of the crude extract of urease was further determined including: the effect of temperature, pH, substrate concentration, and metal addition to urease activity. The urease activity is determined by the Nessler method. The germination time of asparagus bean in yielding urease enzyme reached the optimum activity on the 8th day with activity value of 593.7 U/mL. The biochemical properties of urease from asparagus bean have optimum activity at 35 °C, pH 7.0 and substrate concentration 0.125% with activity value of 600 U/mL. Addition of $CaCl_2$, $SnCl_2$ and $ZnCl_2$ metals decrease the activity of urease.

Introduction

Urease has been isolated and characterized from several types of bacteria, fungi and plants [1]. Urease is commonly found in seeds such as chickpeas (*Cicer arietinum* L.) [2], *Pisum sativum* L. [3], *Cajanus cajan* [4] and cotton *(Gossypium hirsutum)* seeds [5]. Urease is one form of enzyme that plays a role in nitrogen metabolism at plant germination [3]. The urease enzyme can catalyze the pathogenic reaction of urea in plant cells. In the germination stage it will produce growth hormones followed by activation of hydrolytic enzymes such as α-amylase lipase, peptidase and urease until optimum growth is reached when the beans have perfectly formed the roots, stems and leaves.

Urease is also important in human bodies due to the fact that many urinary tract and gastroduodenal diseases [6], including in cancer [7]; are related in some ways to this enzyme. Along with the increased use of urease enzymes, it is necessary to find the source of urease enzyme from raw material that is easily obtained. One ingredient that has the potential to be explored as a source of urease enzymes is the asparagus bean. Asparagus bean are abundant and easy to cultivate in Indonesia. So far, the use and utilization of asparagus bean is minimal and mostly as vegetables. In this study asparagus bean will be used as a source of urease. The use of asparagus bean as a source of enzymes is expected to increase the economic value of asparagus bean.

In this research urease enzyme isolation process was conducted from asparagus bean seeds which has been germinated by urease then extracted using acetone solvent and separated by centrifugation method so that the crude extract of urease is produced. The crude extract of the urease is determined by its biochemical properties including: the influence of various temperature variations, pH, substrate concentration, and metal addition to urease enzyme activity.

Methods

Tools and materials

The tools that used in this study are glassware, micropipette (Soccorex), tip (Biologix), digital scales (Ohause), pH meters (Hanna Instrument), cuvet, UV-Vis spectrophotometer (Shimadzu UV-1800) Incubator (Memmert), centrifuge (Quantum), bath, refrigerator, mortar and pestle, plastic wrapping, stirrer, cotton, and muslin fabric, stirrer, funnel and filter paper. The materials that used in this study were asparagus bean (*Vigna unguiculata ssp sesquipedalis* L.) seeds obtained from Purwokerto Wage Market, $NaH_2PO_4.2H_2O$, $Na_2HPO_4.2H_2O$, $C_6H_8O_7.2H_2O$, $Na_3C_6H_5O_7.2H_2O$, $C_4H_{11}.O_3N$, HCl, $CuSO_4$, $CaCl_2$, $SnCl_2$, $ZnCl_2$, $CO(NH_2)_2$, aceton, H_2SO_4, Na-Wolframat, Nessler reagents and aquadest.

Procedures

Isolation of urease from asparagus bean [3]

Isolation was done on the seeds of asparagus bean, where as much as 200 grams of seeds were soaked in water for 6 hours, then drained and put into a plastic container filled with wet cotton, then covered with wrapping and left in the dark room on room temperature and germination was done with variation of time 0, 2, 4, 6, 8, 10 and 12 days. The next step was urease extraction.

The urease extraction was carried out as follows: As much as 20 grams of spouts was taken and mashed with mortar and pestle then soaked in 80 mL of acetone 20% in cold temperature (4 °C) and stirred with stirrer for 3 hours to produce 2 layers of filtrate and suspension. The filtrate was separated by means of a muslin cloth. The filtrate obtained was centrifuged at 7,000 rpm for 15 minutes at 4 °C. The resulting supernatant is used as a crude urease extraction.

Urease activity test of crude extract [8]

The urease activity test was performed by 1 mL urea concentration of 0.1% plus 1,95 mL of buffer phosphate solution of pH 7 inserted into sample tube. A total of 0.05 mL of a crude extract solution of the urease enzyme was added. The tube was incubated at 35 °C for 15 minutes. The tube was cooled with ice. As much as 1 mL 2/3N of H_2SO_4 solution was added to the sample tube to stop urease enzyme activity and 1mL Na-Wolframat was added to perfect H_2SO_4 work. The blank tube was filled with 3 mL of distilled water. Both tubes were centrifuged for 15 minutes and supernatant was taken by filtration process. Deuteronomy was done 3 times. A total of 1.5 mL of solution of each sample tube and blank was taken. Each sample solution and blanks were then added with 250 μL of Nessler reagent. The solution uptake was then measured by using UV-Vis spectrophotometry at λ 500 nm.

Determintation of urease activity was done by using the following formula:

$$\text{Activity unit} = \frac{[S]-[B] \text{ x } 5 \text{ mL}}{0{,}05 \text{ mL x } 1{,}5 \text{ mL x } 15}$$

Explanation:

[S] = concentration of sample
[K] = concentration of control
5 = volume of solution (mL)
0.05 = volume of enzyme (mL)
1.5 = test volume (mL)
15 = incubation time (menit)

The urease estimation was performed using the standard ammonium sulfate curve. One unit of activity is defined as "the amount of ammonia formed μmol per mL of hydrolysis of urea by urease enzyme within 15 minutes".

Determination of urease biochemical properties of asparagus bean seeds

Determination of the biochemical properties of urease was begun by examining the effect of incubation temperature variations on urease activity. Determination of urease activity was performed in similar way as in activity test but conducted at variation of incubation temperature of 25, 30, 35, 40 dan 45 °C. The enzyme reaction mixture was adjusted to 0.2 M phosphate buffer solution pH 7, with 15 minutes of reaction time. At the optimum temperature, the effect of substrate pH variation on urease activity was further tested. Variations of urea pH substrates used were pH 5.0; 6.0; 7.0; 8.0 and 9.0 in 0.2 M buffer solution. The effect of substrate concentration on urease activity was tested on concentration variation; 0.05; 0.075; 0.1; 0.125 and 0.15%.

The effect of metal addition to urease activity was done by the addition of $CaCl_2$, $SnCl_2$, and $ZnCl_2$ with varying concentrations of 10^{-3}-10^{-8} M. The enzyme activity was tested by adding 1 mL of urea 0.1% and enzyme solution (0.05 mL Enzyme + 0.1 mL metal + 1.85 mL 0.2 M bufer). The enzyme reaction mixture is adjusted under optimum conditions using the addition of the specified metal.

Results and Discussion

Effect of Asparagus Bean Growing Period on Urease Activity

Isolation of urease in this study was carried out through the stage of germination of asparagus bean seeds. Long dried seeds were soaked in water. The process of immersion makde the cells in the plant tissue become actively growing due to imbibisi process, which is the absorption of water into plant tissue cavities. Seeds were grown using wet cotton and in dark conditions for 0, 2, 4, 6, 8, 10 and 12 days. Asparagus bean sprouts were then extracted and centrifuged at cold temperatures to prevent the occurrence of denaturation due to hot temperatures. The supernatant obtained was a crude extract of urease enzyme which then tested for its activity. The effect of germination time on urease activity can be seen in **Figure 1.**

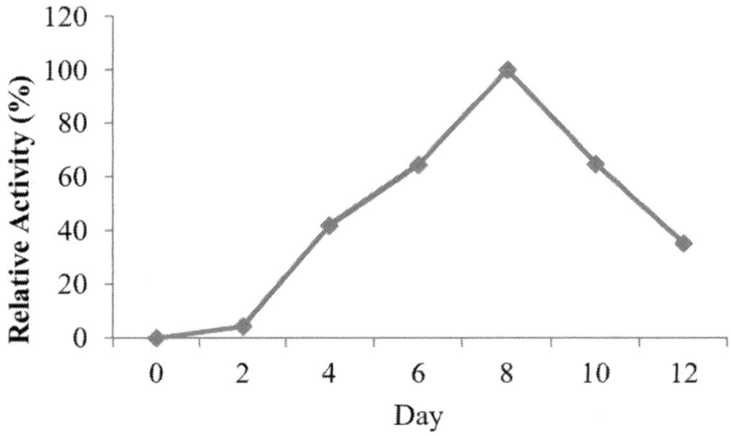

Figure 1. Curve of asparagus bean germination time on relative urease activity

Figure 1 shows the urease activity getting significantly increased after the 2nd day until the 8th day. In this situation, the germination process began, where the embryo in the seeds grew into new plants, thus triggering the growth of shoots from within the seed. The process of plant cell division was done with the help of growth hormone that also activated hydrolytic enzymes such as amylase, phosphatase, lipase, peptidase and urease. The optimum urease activity took place on the 8th day of germination with activity value of 593.7 U/mL. In these circumstances urease had high activity in catalyzing the chemical reaction and most urea was hydrolyzed by urease to produce the nitrogen used as plant growth [9]. The germination time for 8 days was used for further research.

Determination of Urease Biochemical Properties of Asparagus Bean Seeds

Effect of incubation temperature variation on urease activity of asparagus bean

Activity of enzyme is influenced by several factors one of them is temperature. In this study the variation of incubation temperature used was 25, 30, 35, 40 and 45 °C. The effect of temperature on urease activity can be seen in **Figure 2**.

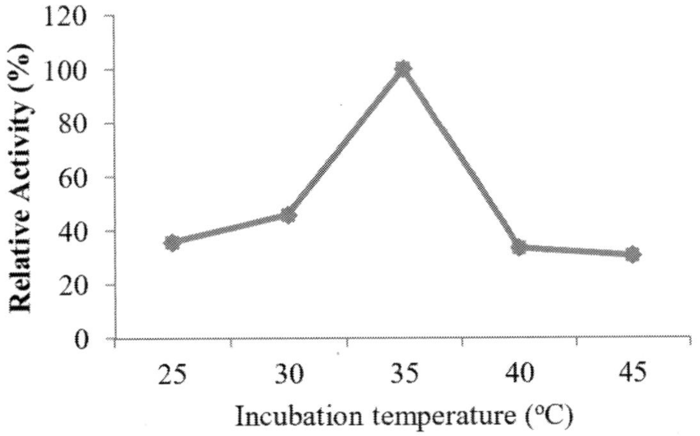

Figure 2. Curve of relation of the incubation temperature to the relative urease activity of the asparagus bean

Figure 2 shows the activity of urease enzyme at 25 °C to 35 °C has increased significantly. In these circumstances there is an increase in collisions between enzyme molecules and substrate. The resulting collision is small or rare due to the small kinetic energy resulting in low activity. The activity of urease reaches optimum condition at 35 °C with activity value equal to 612.4 U/mL. Activity can be increased due to increased kinetic energy which increased the chance of collision between enzyme molecules with substrate and forming complex of enzymes, so that the resulting product increases. The activity decreased significantly after reaching the optimum. According to [10], temperature which rises above optimum temperature results in decreased enzyme activity. Decrease in activity is caused by the enzyme denaturation process by high heat. This is because high temperatures will break secondary bonds such as hydrogen bonds that retain enzymes at their natural temperature, so that the secondary and tertiary structures of the enzyme will be partially damaged followed by a decrease in activity.

Research by [3] states that the optimum temperature of urease isolated from *Pisum sativum* L. is at 40 °C. Pervin et al., (2013) state that the optimum temperature of urease isolated from the chickpea seed (*Cicer arietinum* L.) is at 48 °C. Most enzymes have optimum activity in the 20 - 50 °C temperature range. Enzymes that actively work at that temperature range belong to the mesozyme group [11].

Effect of pH variation on urease activity of asparagus bean

In this study, the variation of pH being used was 5.0; 6.0; 7.0; 8.0 and 9.0. Data on the effect of pH on urease activity can be seen in **Figure 3**.

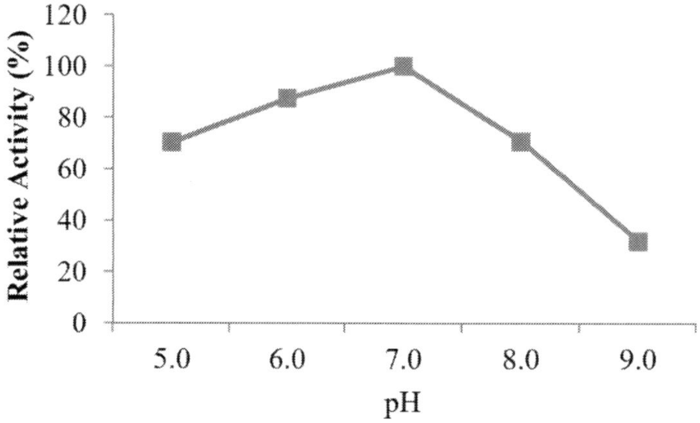

Figure 3. Curve of the relation of pH of the substrate to the relative activity of urease from asparagus bean

The data in **Figure 3** shows the urease activity rose sharply at pH 5.0 to 7.0. In this state, the interaction between the enzyme and the substrate was still low so that the resulting activity was small. This is because the conformation of the enzyme was not yet ideal. The pH value of the environment affects the effectiveness of the active side of the enzyme in forming the enzyme-substrate complex. Changes in pH lead to changes in ionization levels in enzymes or substrates that affect activity. Activity increased until it reaches the optimum condition produced at pH 7.0 with activity value of 613 U/mL, when the optimum pH of enzyme conformation was in ideal condition. This causes the interaction between enzyme and substrate to be maximal in urea degradation process and formed the product. The enzyme has an active site with certain groups acting as a catalyst in the formation of the enzyme-substrate complex. Enzyme activity is closely related to its structure, structural changes will cause changes in enzyme activity.

Enzyme activity decreased sharply after reaching optimum condition. This is due to the change in acidity which causes decreased activity. In an atmosphere that is too acidic or alkaline, the conformation of the enzyme will change and cause disruption of the activity of the enzyme. Conformation of altered enzymes can lead to decreased enzyme activity due to enzyme conformation different from substrate conformation [12].

The optimum pH value of urease activity isolated from chickpeas (*Cicer arietinum L.*) was at pH 7.2 [13], *Cajanus cajan* was pH 7.5 [4]. Based on the results of this study it can be seen that the optimum pH for some urease enzymes generally lies in neutral pH.

Effect of substrate concentration variation on urease activity of asparagus bean

In this research, the variation of substrate concentration used were 0,05; 0.075; 0.1; 0.125; and 0.15%. Data of effect of substrate concentration on urease activity can be seen in **Figure 4.**

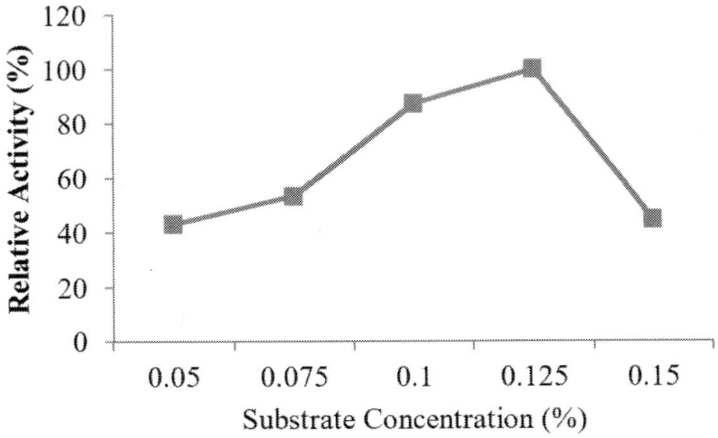

Figure 4. Curve of relation of substrate concentration to relative urease of asparagus bean

The data in **Figure 4** shows the urease activity rose sharply with increasing substrate concentration and reached optimum at 0.125% concentration. When the substrate concentration was enlarged, substrate binding by the enzyme will increase. Thus, the concentration of the enzyme-substrate complexes was greater and the activity of the enzyme will also increase. Relative activity decreased at 0.15% concentration. According to [14], the more substrate concentration, the more the active side of the enzyme in contact with the substrate also increases, resulting in more substrates being hydrolyzed into the product. But too much product causes inhibition because the product will stick to the enzyme's aloesteric side so that the enzyme's active side can no longer be occupied by the substrate. This resulted in decreased enzyme activity when substrate addition with greater concentration after reaching optimum condition.

Effect of metal addition to urease activity of asparagus bean

This study used a variety of metals such as metal $CaCl_2$, $SnCl_2$ and $ZnCl_2$ with variation of concentration 10^{-3}- 10^{-8} M. The data of effect of metal addition to urease enzyme activity can be seen in **Table 1**.

Table 1 The effect of metal concentration on relative activity (%) and inhibition (%)

Type and Metal Concentration (M)	Relative Activity (%)	Inhibition (%)
No metal	100	0
$CaCl_2$ 10^{-3}	37	63
$CaCl_2$ 10^{-4}	46	54
$CaCl_2$ 10^{-5}	56	44
$CaCl_2$ 10^{-6}	64	36
$CaCl_2$ 10^{-7}	77	23
$CaCl_2$ 10^{-8}	83	17
$SnCl_2$ 10^{-3}	48	52
$SnCl_2$ 10^{-4}	58	42
$SnCl_2$ 10^{-5}	67	33
$SnCl_2$ 10^{-6}	78	22
$SnCl_2$ 10^{-7}	85	15
$SnCl_2$ 10^{-8}	96	4
$ZnCl_2$ 10^{-3}	50	50
$ZnCl_2$ 10^{-4}	57	43
$ZnCl_2$ 10^{-5}	69	31
$ZnCl_2$ 10^{-6}	78	22
$ZnCl_2$ 10^{-7}	88	12
$ZnCl_2$ 10^{-8}	91	9

The data in **Table 1** shows that the higher metal concentrations added, the greater the inhibition to urease activity. The greatest inhibition occured in the addition of $CaCl_2$ metal at a concentration of 10^{-3}M. The results of this study were comparable to that of the [15] on the urease enzyme isolation from *Rhizopus oryzae* which states that the addition of $CaCl_2$ metal with increasing concentration (10^{-3}-10^{-5} M) decreases the activity of urease enzymes.

Addition of $ZnCl_2$ concentration increased caused the activity of seed urease *Cicer arietinum* L. also decreased. This is because the $ZnCl_2$ metal will inhibit the enzyme binding to the substrate so that the resulting product becomes less and less. Addition of some heavy metals one of which is Zn^{2+} produces an effect in decreasing the activity of urease. The inhibition that occured was resulted from the reaction between metal molecules and the active part of the enzyme ie thiol (-SH) [13]. The inhibitor binds the enzyme to cause the enzyme not to bind to the substrate on its active side (competitive inhibitor). The competition that occurs depends on the concentration of substrate and inhibitor. Inhibitors did not only work to bind the active side of the enzyme, but also to the enzyme section so as to alter the three-dimensional conformation of enzyme-building proteins. This condition does not produce substrate-enzyme complexes so that the urease activity decreased.

Conclusion

Based on the result of the research, it can be concluded that: time of germination of asparagus bean in yielding of enzyme urease reach optimum activity on day 8 with activity value equal to 593,7 U/mL. The biochemical properties of urease from asparagus bean seeds have optimum activity at 35 °C, pH 7.0 and 0.125% substrate concentration with activity value of 600 U/mL. The addition of $CaCl_2$, $SnCl_2$ and $ZnCl_2$ metals decrease the activity of urease.

Acknowledgements

The authors gratefully acknowledge the research funds provided by the LPPM Unsoed as Riset Unggulan Unsoed based on Kept 1246/UN23.14/PN.01.00/2017.

References

1. Follmer, C., *Insights into the role and structure of plant ureases.* Phytochemistry, 2008. **69**(1): p. 18-28.
2. Sana, N.K., et al., *Partial Purification and Characterization of Urease from Germinating Chickpea (Cicer arietinum L.) Seed.* Research & Reviews in BioSciences, 2009. **3**(1).
3. EL-Hefnawy, M.E., et al., *Extraction, purification, kinetic and thermodynamic properties of urease from germinating Pisum Sativum L. seeds.* BMC biochemistry, 2014. **15**(1): p. 15.
4. Banerjee, S. and A. Aggarwal, *Isolation, partial purification, characterization and inhibition of urease (EC 3.5. 1.5) enzyme from the Cajanus cajan seeds.* Asian Journal of Bio Science, 2012. **7**(2): p. 203-209.
5. Menegassi, A., et al., *Urease from cotton (Gossypium hirsutum) seeds: isolation, physicochemical characterization, and antifungal properties of the protein.* Journal of agricultural and food chemistry, 2008. **56**(12): p. 4399-4405.
6. Konieczna, I., et al., *Bacterial urease and its role in long-lasting human diseases.* Current Protein and Peptide Science, 2012. **13**(8): p. 789-806.
7. Sekichi, F., et al. *Correlation between the abundance of Helicobacter pylori-derived urease and expression levels of PAR1 or PAR2 in the gastric tissue isolated from gastric cancer patients.* in *journal of pharmacological sciences*. 2009.
8. Jayaraman, J. and J. Jayaraman, *Laboratory manual in biochemistry.* 2004: Kalyani Publishers.
9. Goldraij, A. and J.C. Polacco, *Arginase is inoperative in developing soybean embryos.* Plant physiology, 1999. **119**(1): p. 297-304.
10. Baehaki, A., et al., *Purifikasi dan karakterisasi protease dari bakteri patogen Pseudomonas aeruginosa.* Jurnal. Teknol. dan Industri Pangan, 2008. **19**(1).
11. Saropah, D.A., A. Jannah, and A. Maunatin, *Kinetika Reaksi Enzimatis Ekstrak Kasar Enzim Selulase Bakteri Selulolitik Hasil Isolasi Dari Bekatul.* ALCHEMY: Journal of Chemistry, 2013.
12. Sauro, H.M., *Enzyme kinetics for systems biology.* 2011: Future Skill Software.
13. Pervin, S., et al., *Research & Reviews in.* 2013.
14. Simanjuntak, M. and J. Silalahi, *Biokimia.* 2003.
15. Geweely, N.S., *Purification and characterization of intracellular urease enzyme isolated from Rhizopus oryzae.* Biotechnology, 2006. **5**(3): p. 358-364.

IOP Conf. Series: Materials Science and Engineering **349** (2018) 012074 doi:10.1088/1757-899X/349/1/012074

Effect of potentials and electric charges for copper and indium depositions to the photocurrent responses of $CuInS_2$ thin films fabricated by stack electrodeposition followed by sulfurization

Gunawan,[1] A Haris,[1] H Widiyandari,[2] D S Widodo[1]

[1]Chemistry Department, Faculty of Sciences and Mathematics, Diponegoro University, Semarang Indonesia
[2]Physics Department, Faculty of Sciences and Mathematics, Diponegoro University, Semarang Indonesia

E-mail: gunawan@live.undip.ac.id

Abstract. Effect of potentials and electriccharges of copperand indium depositions to the photocurrent responses of $CuInS_2$ thin film fabricated by electrodeposition followed by sulfurizationwere investigated. The characterization and elemental compositions of as-deposited Cu/In and $CuInS_2$ thin films used X-RD and EDAX. Photocurrent responses of the obtained $CuInS_2$ thin films were analyzed by linear sweep voltammetry (*LSV*) in europium solution under chopped irradiation. Photocurrent responses showed that fabricated $CuInS_2$ thin films had *p*-type photoresponses. Indium deposition on copper was influenced by copper morphology resulted by a certain potential deposition. Increasing potential of indium deposition on copper and electric charge of copper deposition reduced photocurrent for the former but did not change photocurrent of $CuInS_2$ for the latter.

1. Introduction

Chalcopyrite of copper indium sulfide ($CuInS_2$) has been used as a photocathode in the research of photoelectrochemical water splitting to produce H_2[1,2]. Since this ternary semiconductor compound has optimal band gap value for sunlight radiation (1.5 eV) [3], large absorption coefficient of $3x10^5$ cm^{-1} below 1000 nm [4] assures absorption of the entire incident photon flux in an absorber layer as thin as a few microns.Ideal absorber of $CuInS_2$ was prepared by stack deposition of Cu and In metallic layers on Mo glass substrate by vacuum methods such sputtering and evaporation [5,6], then the film was sulfurized using H_2S gas or elemental sulfur [7,8]. However, the methods need high cost for the equipment [9].

Electrodeposition is one of the promising technologies for fabrication of low–cost photovoltaic absorber filmsbecause of thelow–cost equipment, scalability of large scale, and good control over film composition and morphology. $CuInS_2$ film fabricated by stack electrodeposition of Cu and In bilayer then followed by sulfurization has been applied as the most usual process [10-13]. However, for deposition of Cu and In on Mo glass substrate is very challenging due to difficulty of indium to deposit on copper layer. In this study, therefore, we examined fabrication of a $CuInS_2$ film by stack electrodeposition from a separate solution of copperand indium containing electrolytes. Effects of applied potentials on film compositions and morphologies were studied in detail. In addition, we

IOP Conf. Series: Materials Science and Engineering **349** (2018) 012074 doi:10.1088/1757-899X/349/1/012074

attempted to fabricate ternary $CuInS_2$ films by sulfurization of the as-electrodeposited film; electronic property of the thus-obtained films was also examined by using photo electrochemical measurement.

2. Experiment Methods

Material synthesis

Electrodeposition was carried out potentiostatically in a vertical three–electrode setup consisting of an Ag/AgCl as reference electrode, a Pt–sheet as counter electrode, and a Mo–coated glass substrate (Mo/glass) as a working electrode. Before the deposition, the Mo/glass substrate was pre–cleaned by sonication in acetone. The working electrode area was 0.7 cm^2. The distance of the working electrode and counter electrode was kept constant at about 2 cm. The copper ion electrolyte used consisted of 50 mM $CuSO_4$, 150 mM trisodium citrate, and 242 mM citric acid. The pH was adjusted 2.4 using H_2SO_4. Indium electrolyte solution consisted of 30 mM $InCl_3$, 242 mM citric acid, and 36 mM trisodium citrate. Electrodeposition was performedby scan potential from 0 to-1500mV(*vs.* Ag/AgCl) for both copper and indium, separately. Then for copper, deposition was carried out at fixed potentials of -200 and -400 mV, while for indium the potentials were -760, -850 and -980 mV with electric charges of 730 mC and 840 mC for copper and indium deposition, respectively. The electric charge was changed for copper and indium depositions as stated in the discussion. The electrodeposition was done at room temperature without stirring using a Hokuto Denko HSV–110 potentiostat-galvanostat. The temperature of all deposition solutions was kept at 24±1 °C by immersion in a thermostatted water bath. Theas-deposited film was finally subjected to sulfurization at temperature of 560 °C under H_2S (5% in Ar) flow for 10 min. The film, then, was etched using KCN to remove Cu_xO [14-16].

Characterizations and photocurrent measurement

Elemental compositions of thus–obtained films were determined using a Hitachi TM3000 scanning electron microscope (SEM) equipped with a Swift ED3000 energy dispersive X-ray spectrometer (EDX). Crystal structureswere measured by X-ray diffraction (XRD)analysis using Rigaku Mini Flex X-ray diffractometer (CuKα, Ni filter). Surface and cross-sectional morphologies of the films after sulfurization were examined using a Hitachi S–5000 FEG field emission scanning electron microscope (FE–SEM). Photoelectrochemical measurements were carried out using a three–electrode cell in 0.2 M $Eu(NO_3)_3$ electrolyte. A Pt wire and an Ag/AgCl electrode in saturated KCl served as the counter electrode and reference electrode, respectively, and Hokuto Denko HB–151 potentio–galvanostat was employed to obtain linear sweep voltammograms.

3. Results and discussion

Preparation of semiconductor of $CuInS_2$ by stack electrodeposition of the precursor was done by scanning the voltage applied, and the voltammogram of copper and indium deposition can be seen in Figure 1. Figure 1a shows that the valley of reduction potential of copper deposition is around -117 mV, while from Figure 1b for indium is around -814 mV. Therefore for deposition, we used -200 and - 760 mV. The more negative potential we used caused the more hydrogen evolution, and this will reduce the electrodeposition efficiency.

Effect of the more negative potential (-400 mV) causes copper that deposited on molybdenum has dark color compared to the potential of -200 mV with a white red color. Besides that, the indium will be difficult to be deposited on the more negative copper deposited on molybdenum. Meanwhile, indium deposition on the copper layer is done at the potential of -200 mV of copper deposition.

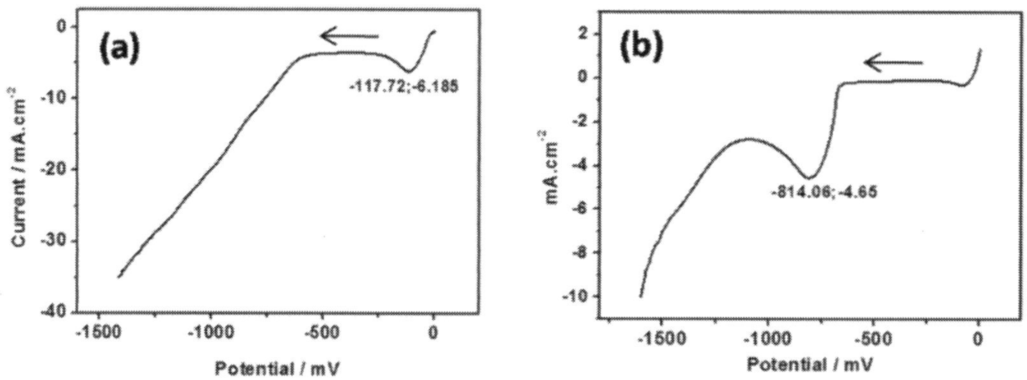

Figure 1. Voltammogram profiles of copper (a) and indium (b) depositions.

Copper deposition at different potentials is also characterized using SEM and XRD as shown in Figure 2 and 3, respectively. Surface SEM images show copper deposited at -200 mV produced bigger grains than deposited at -400 mV (Figure 2). While, from XRD spectra we can see different patterns of XRD of the copper deposited at -200 mV and -400 mV (*vs.* Ag/AgCl) as shown in Figure 3 that is the appearance of CuO pattern at 35° for copper deposited at -400 mV, this is also supported with the dark deposition of copper. It seems the crystal orientation of copper deposited -200 mV is favor for indium deposition.

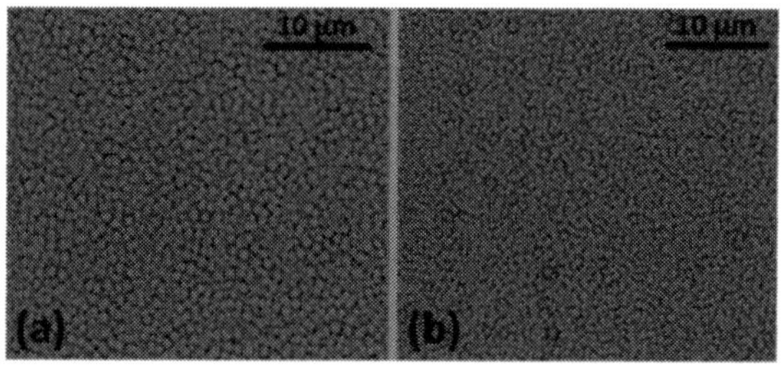

Figure 2. SEM images of Cu depositions at potentials of (a) -200 mV and (b) -400 mV.

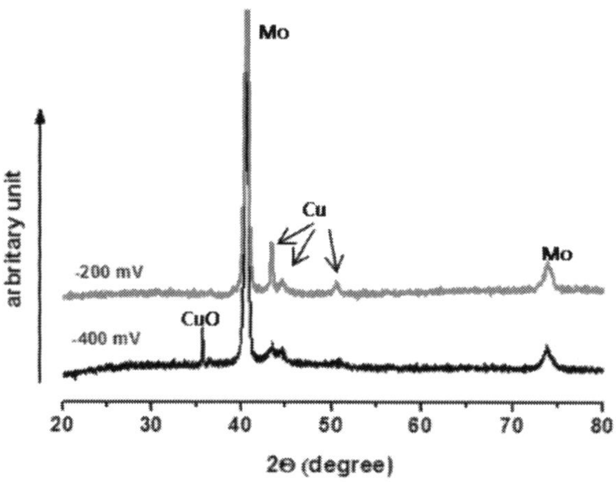

Figure 3. XRD-patterns of copperat different deposited potential.

Evaluation of as-deposited Cu/In (deposited at-200 mV and -760mV (*vs.* Ag/AgCl) for copper and indium, respectively) using SEM, and EDAX are shown in Figure 4 and Table 1. Surface SEM image of the film shows the porous deposition of indium on copper layer (meaning the indium did not cover completely the surface of copper). Besides that indium did not cover homogeneously on the thin copper layer (as shown from cross section of as-deposited Cu/In SEM image). Elemental composition of as-deposited Cu/In can be seen in Table 1.

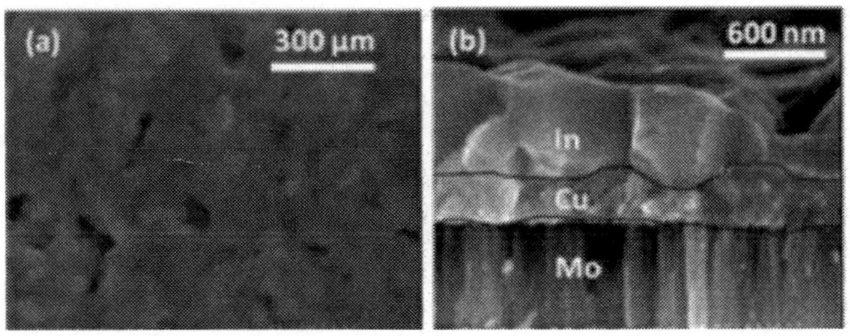

Figure 4. SEM images of Cu/In deposited at -200 mV and -760mV (Vs. Ag/AgCl) for copper and indium, respectively.

Table 1. Elemental composition of as-deposited Cu/In

Metal	Composition [%]
Copper	41.087
Indium	58.913

Sulfurization of as-deposited Cu/In (deposited at -200 mV and -760 mV (*vs.* Ag/AgCl) for copper and indium, respectively) produced $CuInS_2$ and the chalcopyrite properties appear after being etched using KCN as impurities remover in the semiconductor as confirmed at Table 2. However, the impurities were undetected from XRD pattern since the patterns are similar with $CuInS_2$ after etched with KCN (Figure 5).

SEM image shows the big grain of $CuInS_2$ with porous appearance. Cross section image of the film seen that the film was peeled off near the substrate in some part. And $CuInS_2$ has more than 1 µm thick as shown in Figure 6.

Table 2. Elemental composition of $CuInS_2$ before and after KCN etching

Metals	Percentage	
	Before KCN etching	After KCN etching
Copper	34.09	27.65
Indium	20.87	23.86
Sulfur	44.44	48.49

Analysis using XRD of as deposited Cu/In (Figure 3) appears the peaks of 43.4° and 50.5° as well as peaks of 34.5° and 43.1° correspond to copper and Cu/In, respectively. XRD patterns of $CuInS_2$ in Figure 6 affirm the chalcopyrite properties [17], and the sulfurization completely converted as-deposited Cu/In become $CuInS_2$ (written CIS in the figure) as shown the different pattern of the both.

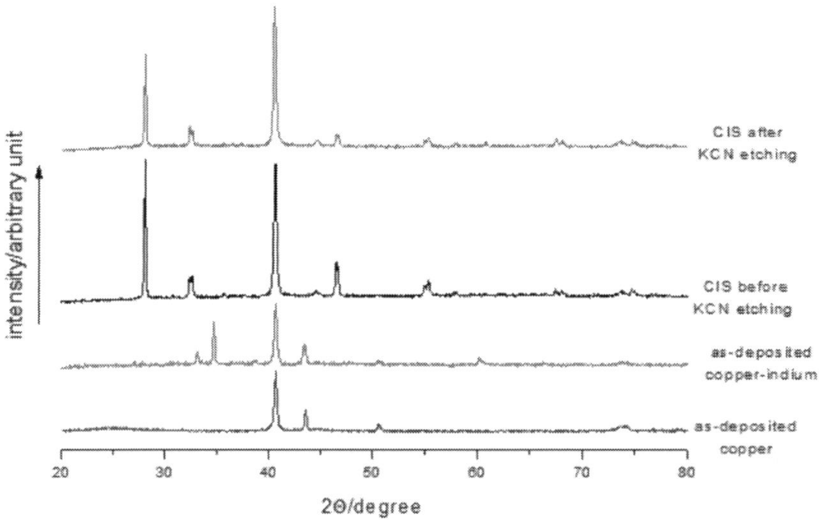

Figure 5. XRD patterns of as-deposited copper and Cu/In as well as $CuInS_2$ before and after etched using KCN.

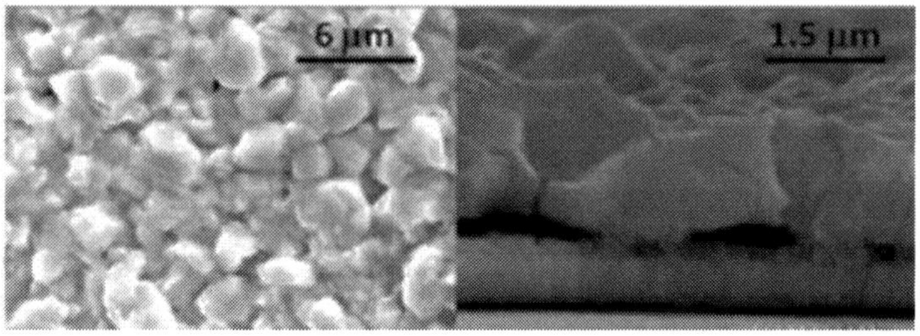

Figure 6. SEM images of $CuInS_2$ after etched using KCN.

The obtained $CuInS_2$film, then, was tested for its photocurrent properties using europium solution under chopped illumination as shown in Figures 7 and 8. Figure 7 shows that the increase of more negative potential for indium deposition reduced the photocurrent of $CuInS_2$ film, this is due to the lower indium deposited on copper as confirmed by hydrogen appearance on the working cathode. As our previous work [18], we obtained the higher photocurrent of $CuInS_2$ using close the plateau of indium deposition, namely 760 mV. Meanwhile by increasing the electric charge (as shown in Figure 8) of copper deposition from 730 mC to 1 C did not change significantly the photocurrent of $CuInS_2$ thin film (indium deposition was kept constant at 840 mC). This can be the elemental composition of indium was quite enough to react with copper to form $CuInS_2$ semiconductor. When the electric charge was made double for both copper and indium deposition, we got the thicker film with lower photocurrent.

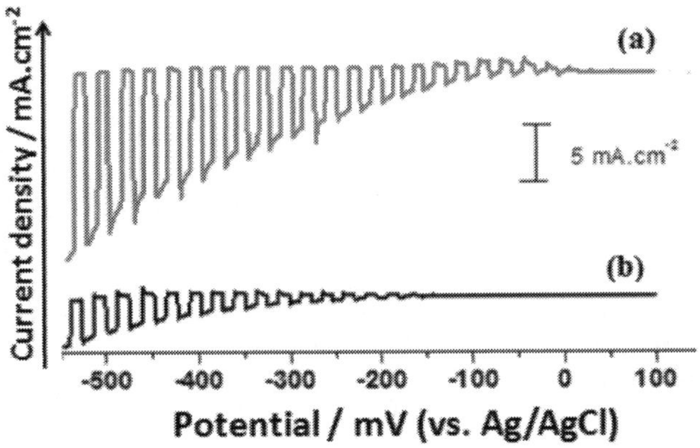

Figure 7. linear sweep voltammograms of $CuInS_2$films (when indium was deposited using potentials of (a) -850 and (b) -980 mV under chopped light illumination in a 0.2 M $Eu(NO_3)_3$ electrolyte.

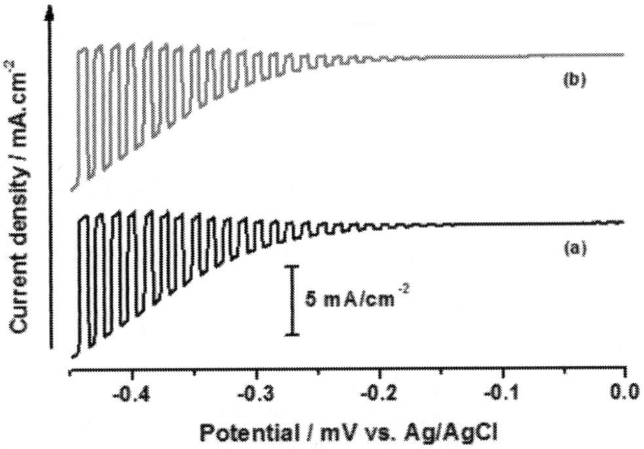

Figure 8. linear sweep voltammograms of $CuInS_2$films (when coppers were deposited using electric charges of (a)730 and (b) 1000 mC, respectively, under chopped light illumination in a 0.2 M $Eu(NO_3)_3$ electrolyte.

4. Conclusion

In the present study, Effect of potentialsandelectriccharges of copperand indium depositions to the photocurrent responses of $CuInS_2$ thin film fabricated by electrodeposition followed by sulfurization were examined. Deposition of the desired copper and indium films was achieved by fixing appropriate applied potentials and electric charges, so that $CuInS_2$ with high photocurrent was obtained. Further works, especially work focused on obtaining highly crystalline $CuInS_2$ films having relatively high onset potential and photocurrent, which are believed to be optimum for photoelectrochemical water splitting to produce H_2, are now in progress.

Acknowledgement
Financial support From Diponegoro University (No.: 276-33/UN7.5.1/2017), Ministry of Research and Higher Education Republic of Indonesia is acknowledged.

References

[1]Septina W, Gunawan, Ikeda S, Harada T, Higashi M, Abe R, Matsumura M., 2015, *J. Phys. Chem. C* **119** 8576–8583.

[2] Ikeda S, Nonogaki M, Septina W, Gunawan G, Harada T and Matsumura M 2013 *Catal. Sci. Tecnol.* **3** 18849-1854.

[3] Green MA, Emery K, King DI, Igari S, Warta W, 2002, *Prog. Photovolt. Res. Appl.* **10** 355-360.

[4] Akkari FC, Brini R, Kanzari M, Rezig B, 2005, *J. Mater. Sci.* **40** 5751-5755.

[5]Bandyopadhyaya S, Chaudhuri S, Pal AK, 2000, *Sol. Energy Mater. Sol. Cells* **60** 323-339.

[6] Klopmann CV, Djorjevic J, Rudigier E, Scheer R, 2006, *J. Cryst. Growth* **289** 121-133.

[7] Klaer J, Bruns J, Henninger R, Siemer K, Klenk R, Ellmer K, Braunig D, 1998, *Semicond. Sci.Technol.* **13** 1456-1458

[8] Siemer K, Klaer J, Luck I, Bruns J, Klenk R, Braunig D, 2001, *Sol. Energy Mater. Sol. Cells* **67** 159-166.

[9] Hibberd CJ, Chassaing E, Liu W, Mitzi DB, Lincot D, Tiwari AN, 2010, *Prog. Photovolt. Res. Appl.* **18** 432-452.

[10] Nakamura S, Yamamoto A, 2003, *Sol. Energy Mater. Sol. Cells* **75** 81-86.

[11] Lockhande CD, Hodes G, 1987, *Sol.Cells* **21** 215-224.

[12] Herrero J, Ortega J, 1990, *Sol. Energy Mater. Sol. Cells* **20** 53-65.

[13] Antony A, Asha AS, Yoosuf R, Manoj R, Jayaraj MK, 2004, *Sol. Energy Mater. Sol. Cells* **81** 407-417.

[14] OgawaY, JagerWaldauA, HuaTH, HashimotoY, Ito K, 1996, *Appl. Surf. Sci.* **92** 232–236.

[15] Bar M,. Schubert B.A, Marsen B, Krause S, Pookpanratana S, Unold T, Weinhardt L, Heske C, Schock H.W, 2011, *Appl. Phys. Lett.*, **99**152111.

[16]. Tanaka T, Sueishi T, Saito K, Guo QX, Nishio M, Yu KM, Walukiewicz W, 2012, *J. Appl. Phys.* **111** 053522.

[17] Ikeda S, Nakamura T, Lee SM, Yagi T, Harada T, Minegishi T, Matsumura M, 2011, *ChemSusChem*. **4**, 262-268.

[18] Gunawan, Septina W, Ikeda S, Harada T, Minegishi T, Domen K, Matsumura M, 2014, *Chem. Commun.* **50**, 8941.

IOP Conf. Series: Materials Science and Engineering **349** (2018) 012075 doi:10.1088/1757-899X/349/1/012075

Modification of Natural Zeolite with Fe(III) and Its Application as Adsorbent Chloride and Carbonate ions

Suhartana*, Emmanuella Sukmasari, Choiril Azmiyawati

Diponegoro University, Department of Chemistry, 50275 Semarang, Indonesia.

*E-mail: suhartana@live.undip.ac.id

Abstract. The aim of the research is to natural zeolite with Fe(III) using anion exchange process to improve the anion exchange capacity. Natural zeolite was activated using HNO_3 1 N and then mixed with $FeCl_3$ solution and refluxed followed by oven and calcination at a temperature of 550°C. The influence of Fe(III) to zeolite was characterized by FTIR while presence of Fe in zeolite characterized by AAS. Zeolite and Zeolite-Fe adsorption capacity of chloride and carbonate anions were determined through adsorption test by variation of pH and contact time. In advanced, and then to determining the Fe adsorbed concentration at Zeolite using UV-Vis spectrophotometer. FTIR analysis result showed that the addition of Fe does not affect the zeolite's structure but change the intensity of the zeolite spectra. The Fe concentration in Zeolite-Fe of 714 mg L^{-1}, indicate that Fe was present in the zeolite. Both Zeolite and Zeolite-Fe adsorbtion results showed that optimum pH of Chloride anion is 2, with adsorption capacity 2,33 x 10^{-3} gg^{-1} and optimum contact time is 8 minutes. While Zeolite and Zeolite-Fe adsorbtion results showed that optimum pH of Carbonate anion is 5, with adsorption capacity 5,31 x 10^{-3} gg^{-1} and optimum contact time is 8 minutes.

1. Introduction

Zeolite is a natural compound found in many parts of Indonesia. Zeolite consists of a cluster of alumina and a group of silica-oxide, each of which is tetrahedral and interconnected by the oxygen atom to form a three-dimensional framework. Zeolite is used as a carrier molecular because its crystal structure is porous and has a large surface area, composed by a silica-alumina skeleton, has high thermal stability, is cheap, and its presence is quite abundant [1].

The best method to look at the ability and quality of zeolites is from ion exchange capabilities. This capability is expressed as the value of cation exchange capacity. The value of cation exchange capacity depends on the degree of substitution of the number of Al^{3+} and Si^{4+} atoms which results in a negative charge on the zeolite framework. Although the capacity of zeolite adsorption capacity against large cations, natural zeolites do not show any affinity for anions [1]. The negative charge on the zeolite structure causes the zeolite to have a small ability to absorb the anion. One way of modifying zeolites to improve the anion exchange properties is by modification with transition metal cations, one of which is Fe(III). Ion Fe(III) can enter empty zeolite cavities and replace balancing cations in interchangeable zeolites [2]. The modified zeolite with Fe(III) is no longer negatively charged because it has binds to Fe so it can be used to absorb the anions [3].

Chloride and bicarbonate is one of the causes of water hardness, whether it is permanent hardness ($CaCl_2$, $MgCl_2$, and $CaSO_4$, $MgSO_4$), as well as the temporary hardness ($Ca(HCO_3)_2$, and $Mg(HCO_3)_2$). In industrial boilers (boyler, boil feed water), water is highly avoided because it will result in crust on the pipe / boiler, so its existence should be handled properly. Anion chloride and bicarbonate are present in ground water as soluble salts, both from cations derived from alkali metals, alkaline earths and transition metals [4, 5].

The objective of this study was to obtain Fe(III) modified natural zeolite and to determine the optimum adsorption capacity chloride and bicarbonate anion. Interaction Fe zeolite with chloride and bicarbonate anion based on pH variation and contact time.

2. Experimental Section

2.1. Materials

The materials used in this research are Bayat natural zeolite, HCl 6 N, $FeCl_3.6H_2O$, aquades, 0.1 M HCl, 0.1 M NaOH. The equipment used is a set of glass tools, erlenmeyer, mistype glass, a mesh size of 100 mesh, an analytical balance, oven, vacuum pump, shaker, reflux apparatus, pH meter, filter paper, micro biuret, furnace, Fourier Transform Infrared Spectroscopy (Shimadzu Prestige 21), Atomic Absorption Spectrophotometry (Perkin Elmer 1300).

2.2. Preparation of Zeolite

Natural zeolite from Bayat of Klaten Regency was crushed then sieved with 100 mesh strainer to homogenize the size then washed with distilled water. After washing with distilled water several times, the zeolite is filtered and then dried with oven at 110°C for 4 hours. Zeolite mashed by crushing until obtained powder back.

2.3. Activation of Zeolite

Zeolite which passes 100 mesh sieve added HCl 6 N solution with ratio 1:10 and soaked for 24 hours. The zeolite is then filtered and rinsed with distilled water to a neutral pH (equal to pH of aquadest). Furthermore, the zeolite was dried with oven at 110°C for 4 hours.

2.4. Modification of Zeolite

A total of 20 grams of zeolite plus 100 mL $FeCl_3$ 0.2 M then the solution was refluxed for 4 hours at 85°C, the result was then filtered and dried in an oven at 110oC for 4 hours then calcined 550°C for 2 hours. To know the effect of adding Fe(III) to zeolite spectra, characterization using Fourir Transform Infrared (FTIR) instrument was done by comparing Fe(III) and zeolite activated zeolite spectra. In addition, AAS instruments are also used to determine Fe content in zeolite-Fe.

2.5. Adsorption of Anion chloride and carbonate

For the pH variation, 0.25 grams of Fe(III) modified zeolite were mixed with 10 mL Chloride anion concentration of 150 ppm or with 10 mL Carbonate anion concentration of 150 ppm. Furthermore, stirring using shaker with variations of pH 2, 4, 6, 8, and 10 for 60 minutes. Then the mixture was filtered, the filtrate and the starting solution were analyzed anion chloride concentration using a Mohr methods by $AgNO_3$ solution as standard.

Created the same treatment against activated natural zeolites. for variation of time. Furthermore, stirring using shaker with 10 mL Chloride anion concentration of 150 ppm or with 10 mL Carbonate anion concentration of 150 ppm variation of contact time 5, 10, 30, and 90 minutes at optimum pH then filtered. The filtrate and the starting solution were analyzed contain anion carbonate concentration using a titrimetric methods by EDTA solution as standard. Created the same treatment with activated natural zeolite.

3. Results and Discussion

This research was modified to natural zeolite from Bayat sub district, Klaten Regency with the addition of Fe_2O_3 as an adsorbent of Chloride and Carbonate ions. Stages of discussion include activation of natural zeolite, modification of active zeolite with Ferric Chloride and NaOH Solution, characterization of zeolite with FTIR spectrophotometer and its adsorbent ability test.

3.1. Activation of Natural Zeolite.

Natural zeolite carried out destruction and sieving measuring 100 mesh, with the aim of enlarge the surface area of natural zeolite so that the adsorption ability can be more optimal [6]. Then the zeolite was immersed in 3 M HF solution for 15 minutes. This is done to dissolve the remaining impurities on the zeolite [6]. Zeolite already soaked with 3 M HF solution having an acid pH of about 3-4, then pH zeolite neutralized to pH 7 by washing using aquadest and dried at 120°C. Zeolite which is free from impurity oxide, ion exchange process with ammonium ion. This ion exchange aims to form NH_4^+ zeolite. The initial exchange occurs between the cations within the zeolite NH_4^+ with of ammonium chloride (NH_4Cl). The ion exchange between the cations in the zeolite with NH_4^+ aims to replace all the cations in the zeolite because in the zeolite there are still alkaline or alkaline earth cations such as K^+, Na^+, Ca^{2+} and Mg^{2+} which act as zeolite balancers [7]. Furthermore the alkali or alkaline earth may be interchangeable with the ammonium cation so that all of the cations will be converted to NH_4^+. According to selectivity NH_4^+ has a high selectivity value of NH_4> Ag> Ba> Na> Sr> Ca> Li> Cu [2, 8]. Then to form zeolite-H, after ion exchange process then heating at temperature 300°C because at that temperature NH_3 will loose and formed zeolite-H [2, 9].

3.2. Modified Zeolite with Fe_2O_3.

The activated zeolite formed into zeolite-H was dissolved into aquadest 70°C. This mixing process is intended to allow the zeolite and the material to be added to interact thoroughly and will speed up the reaction when $FeCl_3$ is added [10]. The next process is the addition of NaOH solution. Addition of NaOH in the solution is done dropwise and with stirring done. The water temperature is kept at 70°C intended. so $FeCl_3$ and NaOH immediately form $Fe(OH)_3$, because the Ksp price is very small [4]. The stirring process is done for 2 hours. This stirring is intended to allow the zeolite and Fe $(OH)_3$ to react in a homogeneous manner [11]. After stirring, to remove NaCl then added aquades. This addition process is intended to allow the NaCl present in the mixture to dissolve with water. After this stage screening and washing using water at a temperature of 70°C. Washing is done to remove the NaCl still contained in the solution. To determine the presence of Cl^- ion, it was tested with a silver nitrate solution ($AgNO_3$). $AgNO_3$ solution with Cl- will then form a white precipitate [5, 12]. The reaction is as follows:

$$NaCl\ (aq) + AgNO_3\ (aq) \rightarrow \underset{(white)}{AgCl\ (s)} + NaNO_3\ (aq)$$

The next stage of the Cl liberation process is drying. This drying process is done to remove water. The calcination process is carried out at 450°C. At that temperature is expected $Fe(OH)_3$ will be changed to Fe_2O_3. Fe_2O_3 will form at high temperatures which are then expected to bond with zeolites either physically or chemically. In its basic state, the zeolite has a balance cation on its skeleton. When the H-zeolite having a H-balancing cation has been modified with Fe_2O_3 to zeolite-Fe [13].

The comparison of physical appearance between natural zeolite, active natural zeolite and zeolite-Fe can be seen with different properties. Fe_2O_3 modified zeolite provides a more yellow- brown color than the unmodified zeolite. This confirms the formation of zeolite-Fe. Acquired natural zeolite, activated natural zeolite and Fe modified zeolite with differences in physical appearance are shown in Figure 1.

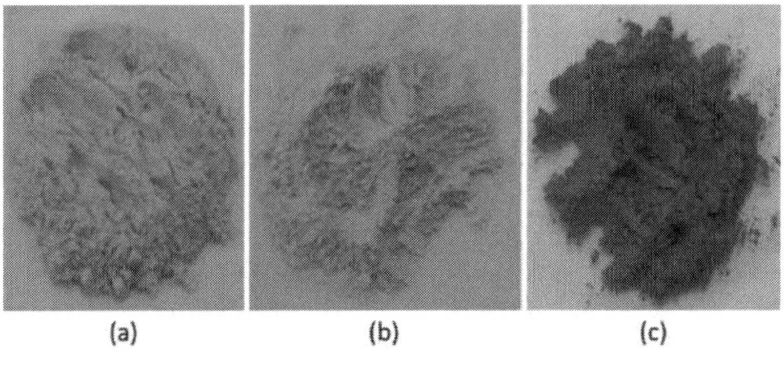

Figure 1. Color difference between (a) Natural zeolite, (b) Activated natural zeolite, and (c) Zeolite-Fe

3.3. FTIR Analysis Result

An important method to characterize zeolite frame structure and to know the effect of Fe_2O_3 is by infrared spectra analysis, especially Fourier Transformation-Infra Red (FTIR). In general, the range of 300-1300 cm^{-1} wave numbers are tetrahedral bonds, namely O-Si-O and O-Al-O. In the band 900-1250 cm^{-1} is an asymmetric range, the symmetric range is shown on the band 650-850 cm^{-1}, the internal Si-O / Al-O (Fe-O) bends appear in the 410-500 cm^{-1} region while External will appear at 700-780 cm^{-1} [3]. The infrared spectra analyzed by FTIR of Fe zeolite and activated natural zeolite are shown in Figure 2.

IOP Conf. Series: Materials Science and Engineering **349** (2018) 012075 doi:10.1088/1757-899X/349/1/012075

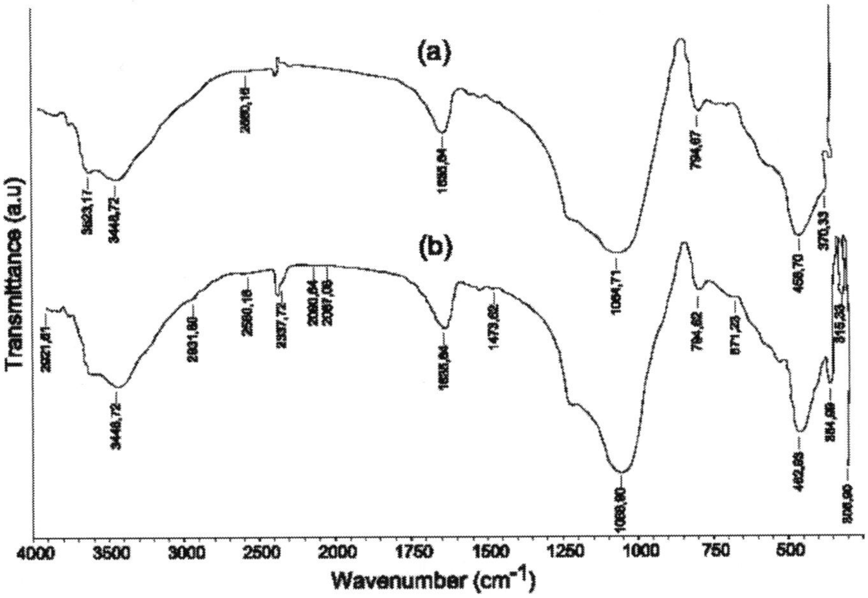

Figure 2. FTIR spectras of (a) zeolite-Fe and (b) activated natural zeolite

The spectra produced by zeolite-Fe have a similar pattern with the activated zeolite. This shows that Fe_2O_3 reacted to zeolites does not alter the original structure of the zeolite. In the determination of existing functional groups in the structure of zeolite is divided into two [14]. The first functional group is the internal vibration associated with Fe_2O_3 which is the main structural unit. The second functional group is the external vibration associated with the tetrahedral connection [6].

FTIR spectra interpretation can be known from several absorptions. The absorption band of the wave number region 420-300 cm^{-1}, in the absorbed activated zeolite is at 324.04 cm^{-1}, which in which the absorption is included in the opening of the zeolite pore. The opening of the pore includes the external vibrational vibration spectra of the zeolite [14]. At ZA-Fe at 317.1 cm^{-1} this shows the presence of a certain amount of Fe-O on the surface of the zeolite. Peak produced by FeO has a very large intensity and the resulting curve is very sharp. The decrease in absorption in wave numbers between 400-350 cm^{-1} in ZA-Fe confirms the presence of van der wals forces between zeolites and FeO. When Fe_2O_3 is combined with zeolite via Fe-O-Si or Fe-O-Al, the presence of a Fe-O-binding force bounded by a surface bond on the zeolite structure causes the energy absorbed in FTIR become weak [13]. Another opinion of the region of the wave number also indicates the opening of zeolite pores [14].

The decrease in absorption also occurs not only in wave numbers between 400-350 cm^{-1}. In the uptake between 500-420 cm^{-1} which is the Fe-O SiO / AlO bending Vibration also decreases between ZA and ZA-FeO. The size of the area produced in ZA-FeO (448.1 cm^{-1}) is smaller than that of ZA (455.20 cm^{-1}). This is due to the reduced bending between SiO / A1O. The amount of bending energy of SiO / A1O buckling is reduced, possibly due to Fe-binding of -O of SiO / AlO. Wavelength 790 cm^{-1} is the vibration of O-Si-O / O-Al-O symmetry. A stretching vibration involving a tetrahedral atom can be seen in the 650-820 cm^{-1} region. In ZA and ZA-Fe each have the same uptake at 794.67 cm^{-1} This resilient vibration shows the bonding composition of Si-Al zeolite framework [14].

3.4. AAS Analysis Result on Zeolite- Fe

To account Fe concentrations was bonding in zeolite should be analyzed by Atomic Absorption Spectroscopy (AAS). Result of the Fe in zeolites-Fe and activated natural zeolites are shown in Table 1.

Table 1: Fe concentrations in zeolite-Fe and activated natural zeolite

Adsorbent	Concentration (mg L^{-1})
Natural zeolite	2.164
Activated Zeolite-Fe	11.786

An increase in Fe concentration in Fe zeolite. The greater the concentration the amount of Fe in the zeolite is greater, it indicates that Fe(III) has entered into zeolite. In activated natural zeolites there is a high enough Fe concentration, probably because Fe is still present as an impurity in the zeolite.

3.5. Adsorption of Ion Chloride and Carbonate Using Zeolite-Fe.

Measurements taken to determine the results of these two variations using titrimetric methods. The principle of the titrimetric reaction is to react between the sample (which is not yet known) and the standard solution (which is already known). In the chloride ion analyzer, according to Vogel and Svehla [5], the prepared NaCl sample was analyzed with 3-5 drops of potassium chromate indicator (K_2CrO_4), then titrated with a silver nitrate solution ($AgNO_3$), in case of sediment color change from white precipitate to red brick , then the reaction has reached the equivalent point . The reaction is as follows.

$$NaCl\ (aq) + AgNO_3\ (aq) \rightarrow \underset{(white)}{AgCl\ (s)} + NaNO_3\ (aq)$$

$$K_2CrO_4\ (aq) + AgNO_3\ (aq) \rightarrow \underset{(red\ brownish)}{AgCrO_4\ (s)}$$

In the analysis of carbonate ions, according to Vogel and Svehla [5] the carbonate ion sample is ready to be analyzed ($CaCO_3$) given 3-5 drops of Eriochrome Black T (EBT) indicator, then titrated with Ethylen Diamine Tetra Acetate (EDTA) solution. Titration has reached the equivalent point if there is a change of analytical solution, from red wine to sky blue [15, 16]. The reaction is as follows.

$$Ca^{2+} (aq) + H_2Y^{2-} (aq) \rightarrow (CaY)^{2-} (aq) + 2H^+ (aq)$$
(red wine)

$$Ca^{2+} (aq) + EBT (aq) \rightarrow (Ca - EBT)^{2+} (aq)$$
(blue sky)

3.6. Adsorption Capacity of Zeolite-Fe on Anion Cl^-

Curves effect of pH on the adsorption capacity of the zeolite-Fe and activated natural zeolite for anion Cl- shown in Figure 3. To know and compare the adsorption capacity of zeolite activated Fe to chloride anion also measured the concentration of chloride ion adsorbed with various variations of pH. It is expected to obtain optimum pH conditions, whether in acidic, neutral or alkaline conditions.

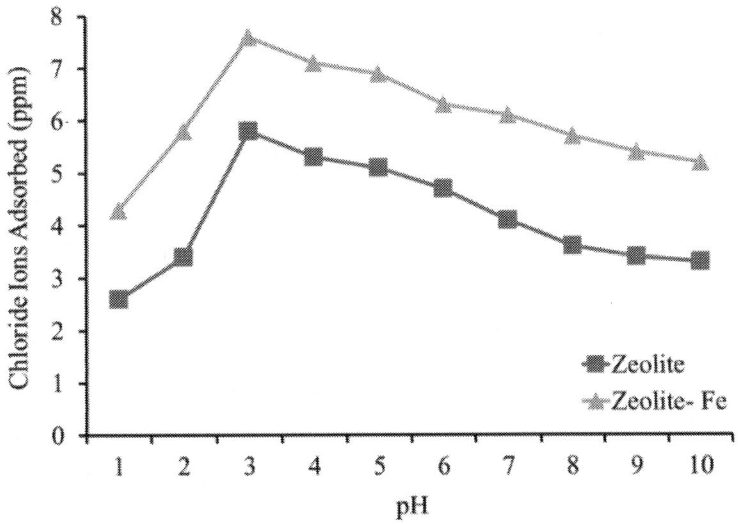

Figure 3. Activated zeolite adsorption capacity and zeolite-Fe to Chloride anion versus variation of pH

To investigate and compare the adsorption capacity of zeolite activated Fe to the carbonate anion, also carried out exactly the same as the chloride anion. While the results obtained is also a graph, which states the relationship between pH and carbonate anion was adsorb.

The effect curve of pH variation on zeolite-Fe adsorption capacity and activated zeolite nature for CO_3^{2-} anions is shown in Figure 4.

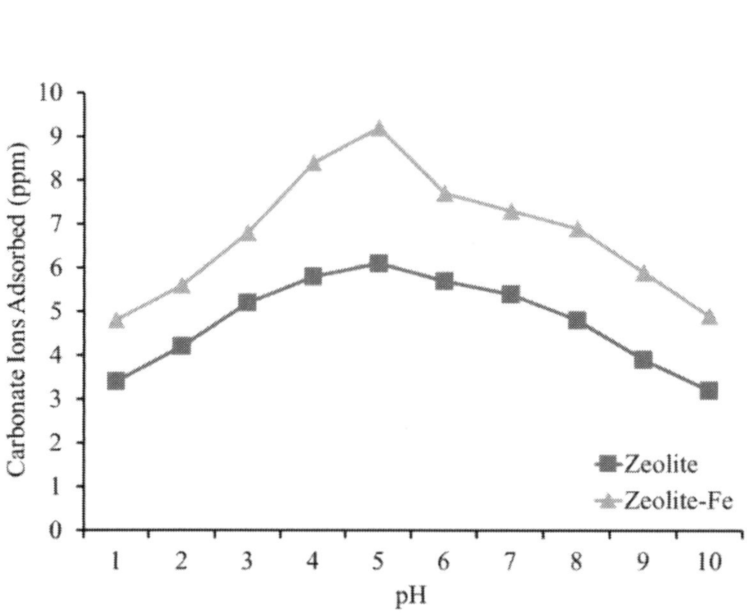

Figure 4. Activated zeolite adsorption capacity and zeolite-Fe to CO_3^{2-} anion versus variation of pH

It is seen that the adsorption of anion chloride and carbonate using Fe zeolite as well as activated zeolite reaches the highest adsorption capacity at different pH. At adsorption of chloride ions at pH 2-3, and adsorption of carbonate ions at pH 5. This difference is due to the different response, from both anions, in binding H-ions. after the optimum pH, the higher the pH of the solution, the adsorption capacity decreases. It shows that at pH 2-3 (for chloride) and pH 5 (for carbonate) is the optimum pH due to the highest adsorption capacity of anion chloride and carbonate both for activated zeolite and Fe-zeolites [10].

The optimal adsorption occurs at pH 2-3, according to Vogel and Svehla [5], at low pH (acid) chloride ions prefer to bind H ions in the form of HCl. While at higher pH (base) chloride ions will occur to compete H ions with iron ions in the form of divalent cations. At the high pH of the solution (base), the presence of hydroxyl ions (OH-) according to Crabtree and Hamilton [17], is likely to inhibit the interaction with Fe oxide on the zeolite surface. Since Ksp $Fe(OH)_2$ is very small ($2x10^{-23}$) there may also be a precipitation reaction of Ferrohydroxide. So also for carbonate ions. The optimal adsorption occurs at pH 5, according to [5], at low pH (acid) carbonate ions prefer to bind H ions in the form of bicarbonate ions (HCO_3^-). While at higher pH (base) carbonate ions will occur to compete H ions with iron ions in the form of divalent cations. At the high pH of the solution (base), according to Crabtree and Hamilton [17], the presence of hydroxyl ions (OH-) is likely to inhibit the interaction with Fe oxide on the zeolite surface. Because Ksp $FeCO_3$ is very small ($2,1x\ 10^{-8}$) it is possible to have a precipitation reaction between the carbonate ion and the ferrous cation, forming a ferrocarbonate precipitate ($FeCO_3$) [16].

The curve effect of contact time variation on the adsorption capacity of the anion chloride and carbonate using the adsorption zeolite and the activated zeolite is almost the same, shown in Figure 5 and Figure 6.

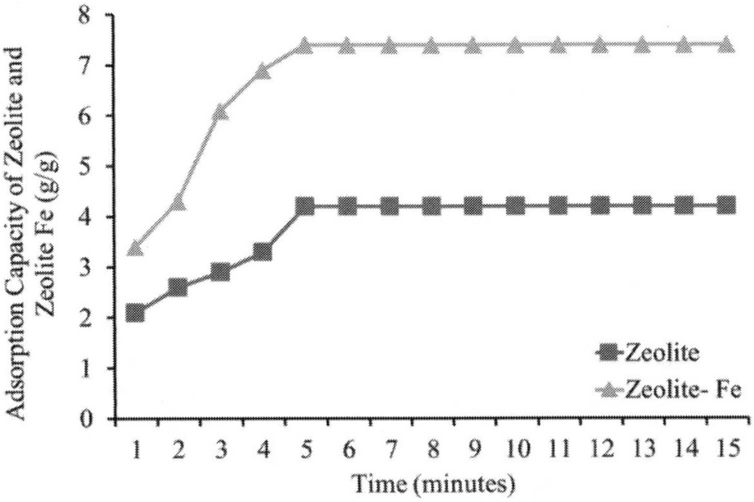

Figure 5. Activated zeolite adsorption capacity and zeolite-Fe to Cl^- anions versus time

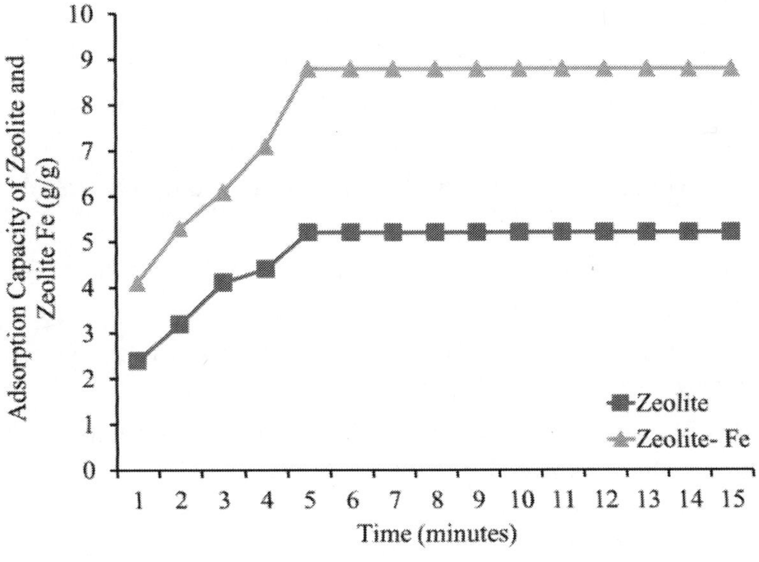

Figure 6. Activated zeolite adsorption capacity and zeolite-Fe to CO_3^{2-} anions versus time

It is seen that in general the adsorption of both anions (both chloride and carbonate) using zeolite-Fe and activated natural zeolite reached the highest adsorbed concentration at5 minute for Cl^-, after passing the 5 th minute of constant adsorbed concentration. The results show that the contact time between the ion and the adsorbent will affect the amount of ions bonded to the adsorbent, where the longer the contact time then the number of metal ions adsorbed also more and more [3]. Chloride anions and Carbonate anions is different result, because ionic radii from two anions is significant different [15, 18]. However, after exceeding the optimum time, the active side of the adsorbent has completely bonded with the ion and there is equilibrium. In both pH and contact time curves on the adsorbent adsorption capacity of both anions (both chloride and carbonate) it can be seen that the adsorption capacity between activated zeolite and zeolite-Fe has a relatively small difference. This suggests that the formation of iron oxide (Fe oxide) in zeolites has no significant effect on the adsorption capacity of the zeolites on the two anions [11].

4. Conclusions

Zeolite-Fe is obtained as evidenced by changes in spectral intensity of zeolite and Fe content in Fe zeolite is greater than activated natural zeolite. The zeolite-Fe adsorption capacity of chloride anion and carbonate is greater than activated natural zeolite, because 9.622 ppm of cations Ferrous was adsrobed at nature zeolite. For Chloride ions with an adsorption capacity of 5.7 ppm (Activated Zeolite) and 7.6 ppm (Zeolite- Fe) optimum adsorption at pH 3 contact time of reaction 5 minute and Carbonate ions with an adsorption capacity of 5.7 ppm (Activated Zeolite) and 7.6 ppm (Zeolite- Fe) optimum adsorption at pH 3 contact time of reaction 5 minute.

References

[1] S. Wang, Y. Peng, Natural zeolites as effective adsorbents in water and wastewater treatment, Chemical Engineering Journal, 156 (2010) 11-24.

[2] D. Danabas, T. Altun, Effects of zeolite (clinoptilolite) on some water and growth parameters of rainbow trout (Oncorhynchus mykiss Walbaum, 1792), aquaculture, 6 (2011) 10.

[3] H. Hamdan, Introduction to Zeolites: Synthesis, Characterization, and Modification, Universiti Teknologi Malaysia, Kualalumpur, (1992).

[4] J. Huheey, E. Keiter, R. Keiter, O. Medhi, Inorganic Chemistry, Pearson Education, New Delhi, 2011.

[5] A.I. Vogel, G. Svehla, Textbook of Macro and Semimicro Qualitative Inorganic Analysis, Longman Scientific & Technical, 1987.

[6] J.D. Sherman, Ion exchange separations with molecular sieve zeolites, in: Zeolites: Science and Technology, Springer, 1983, pp. 583-623.

[7] S.G. Poulopoulos, V.J. Inglezakis, Adsorption, ion exchange and catalysis: design of operations and environmental applications, Elsevier, 2006.

[8] P. Hemalatha, P.P. Rao, Adsorption of Cr (VI) ions from aqueous solutions using calcined brick powder (CBP), International journal of environmental sciences, 2 (2012) 2321.

[9] A. Joy, V. Ramamurthy, Chiral photochemistry within zeolites, Chemistry-A European Journal, 6 (2000) 1287-1293.

[10] M. Kragović, A. Daković, Ž. Sekulić, M. Trgo, M. Ugrina, J. Perić, G.D. Gatta, Removal of lead from aqueous solutions by using the natural and Fe(III)-modified zeolite, Applied Surface Science, 258 (2012) 3667-3673.

[11] G. Du, Z. Li, L. Liao, R. Hanson, S. Leick, N. Hoeppner, W.-T. Jiang, Cr (VI) retention and transport through Fe(III)-coated natural zeolite, Journal of hazardous materials, 221 (2012) 118-123.

[12] F.a.H. Treadwell, WT, , Analytical Chemistry, , , John Wiley & Sons,, New York., 1971,.

[13] T. Perraki, A. Orfanoudaki, Mineralogical study of zeolites from Pentalofos area, Thrace, Greece, Applied Clay Science, 25 (2004) 9-16.

[14] L. De Araujo, C. Cavalcante Jr, K. Farias, I. Guedes, J.M. Sasaki, P.d.T.C. Freire, F.E.d.A. Melo, J. Mendes-Filho, Synthesis of cubic Y zeolite using a pulsed microwave heating system, Materials Research, 2 (1999) 105-109.

[15] D. Skoog, D. West, F. Holler, Fundamentals of analytical chemistry Saunders College Pub, Fort Worth, (1992).

[16] A. Vogel, 4 edition, The English Language Book Society and Longman, New York., A Textbook of Quantitative Inorganic Analysis, , book, 4 (1978,).

[17] R.H. Crabtree, D.G. Hamilton, H-H, C-H, and Related Sigma-Bonded Groups as Ligands, in: F.G.A. Stone, R. West (Eds.) Advances in Organometallic Chemistry, Academic Press, 1988, pp. 299-338.

[18] F.a.H. Treadwell, WT,, Analytical Chemistry, John Wiley & Sons, New York., 1971.

Institute of Physics
Dirac House, Temple Back
Bristol BS1 6BE UK

ISSN: 1757-8981
ISBN 978-1-5108-6265-4

9 781510 862654

CURRAN ASSOCIATES INC.
proceedings
.com